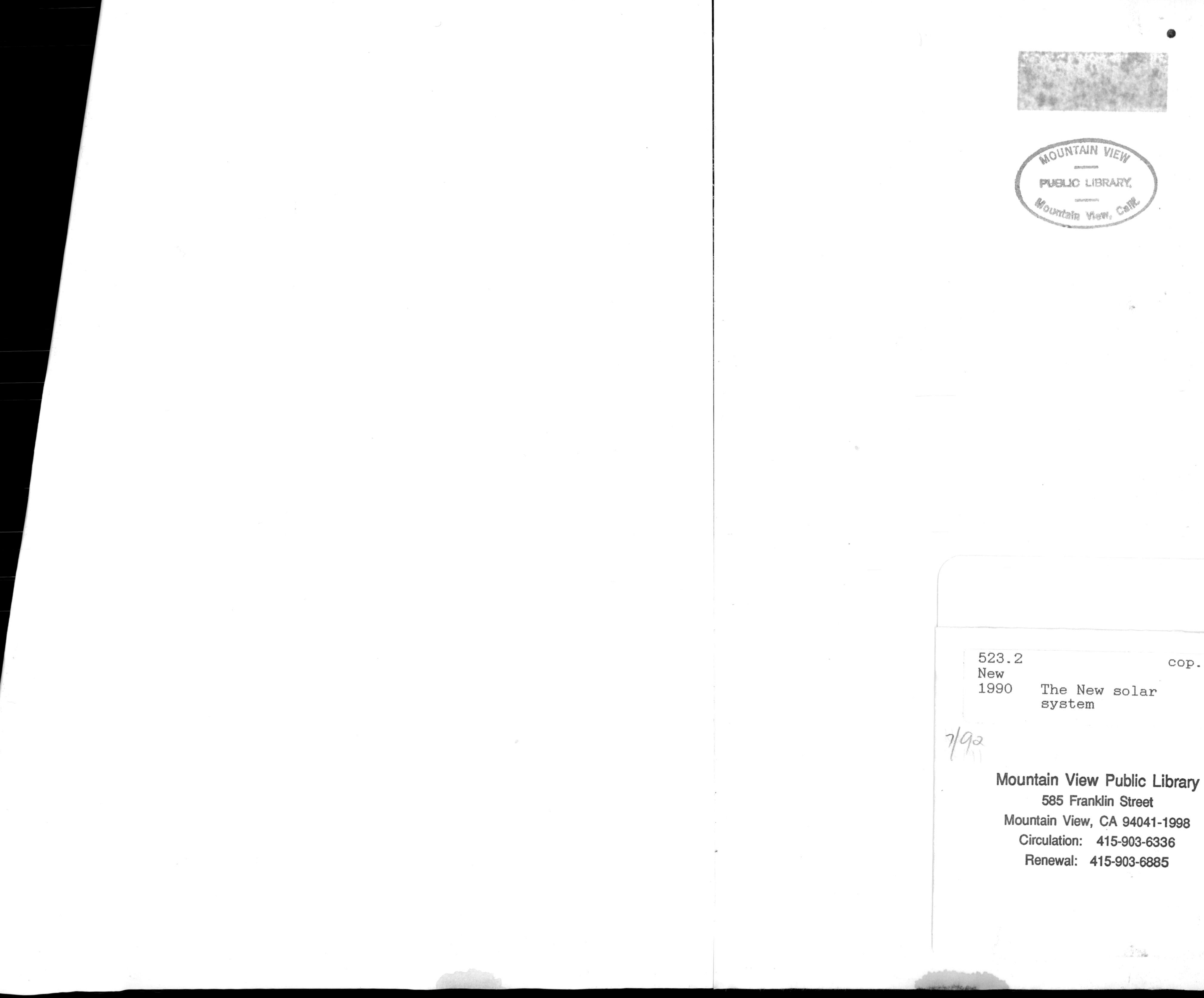

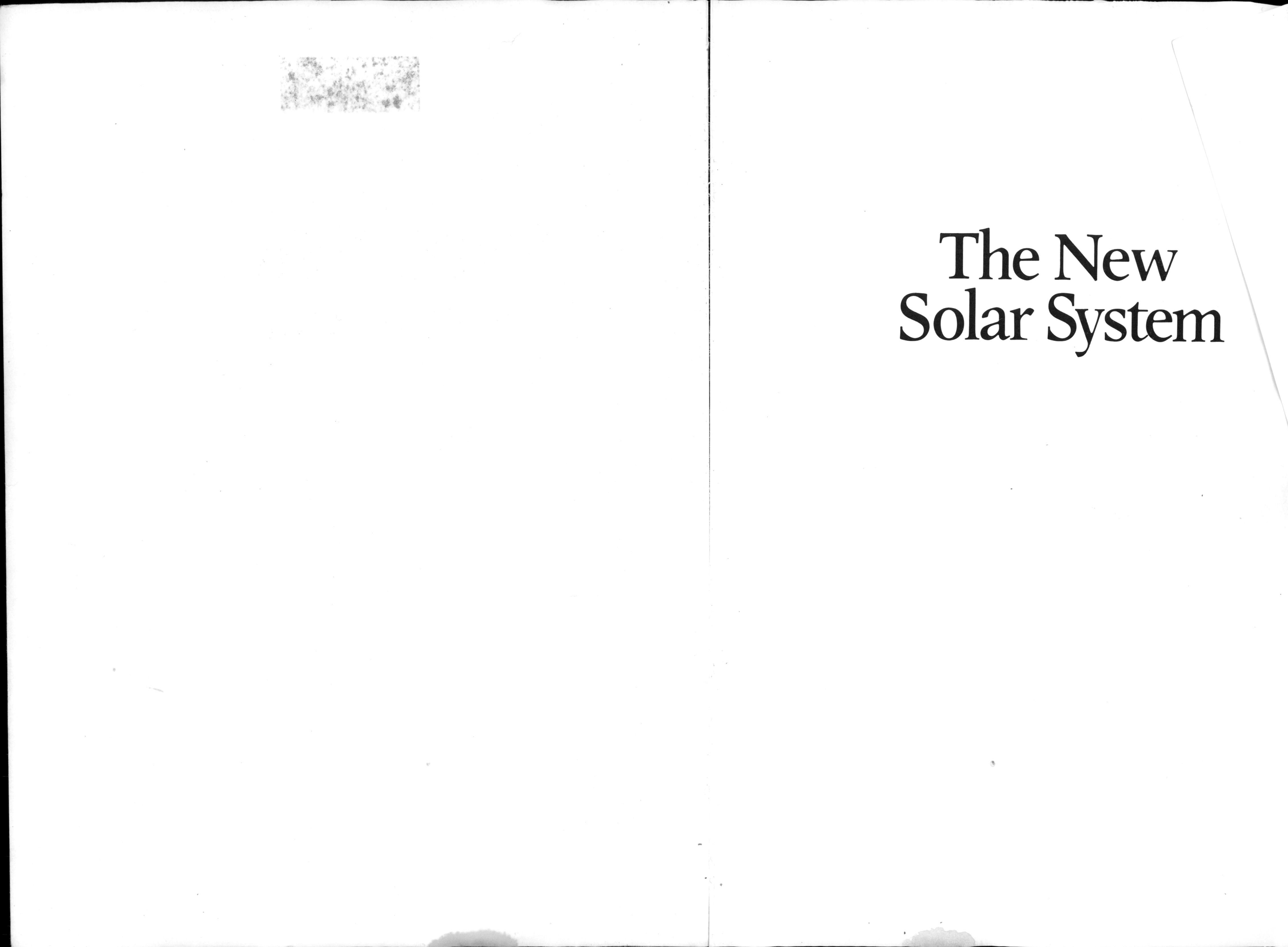

The New Solar System

The New Solar System

THIRD EDITION

edited by

J. Kelly Beatty
Andrew Chaikin

introduction by

Carl Sagan

CAMBRIDGE UNIVERSITY PRESS
Cambridge
New York Port Chester Melbourne Sydney
&
SKY PUBLISHING CORPORATION
Cambridge, Massachusetts

Published by Sky Publishing Corporation
49 Bay State Road, Cambridge, Massachusetts 02138
and by the Press Syndicate of the University of Cambridge
The Pitt Building, Trumpington Street, Cambridge CB2 1RP
40 West 20th Street, New York, NY 10011, USA
10 Stamford Road, Oakleigh, Melbourne 3166, Australia

First published 1981
Second edition 1982
Third edition 1990

Printed in the United States of America

Library of Congress cataloging in publication data

The New solar system / edited by J. Kelly Beatty and Andrew Chaikin. – 3rd ed.
p. cm.
ISBN 0-521-36162-1. – ISBN 0-521-36965-7 (paperback). – ISBN 0-933346-55-7 (Sky Pub. Corp.). – ISBN 0-933346-56-5 (Sky Pub. Corp. : paperback)
1. Solar system. I. Beatty, J. Kelly. II. Chaikin, Andrew, 1956– .
QB501.N47 1990
523.2–dc20 89-38040 CIP

British Library cataloguing in publication data

The New solar system. – 3rd ed.
1. Solar system
I. Beatty, J. Kelly II. Chaikin, Andrew, *1956–*
523.2

ISBN 0 521 36162 1 hard covers
ISBN 0 521 36965 7 paperback

SE

Table of Contents

Preface

FORTY YEARS AGO, nearly a decade before the first artificial satellite was launched, Gerard P. Kuiper edited and published what would become a standard reference in planetary astronomy: *Atmospheres of the Earth and Planets.* At that time, when world-class telescopes like the 5-meter reflector at Palomar Mountain were revealing the majesty of the distant universe, planets were far from center stage in astronomy. Observatories had unwritten rules that no more than 10 percent of the telescopes' observing time be devoted to studies of our planetary neighbors, including comets. But Kuiper, who had already distinguished himself with research on double stars and stellar evolution, had in the 1940s begun to devote his considerable talent to the planets. Working with the limited tools of the era, utilizing his skills as an observer and his rigorous scientific judgment, Kuiper amassed a body of knowledge that formed the basis for modern planetary science.

In *Atmospheres* he and several other researchers summarized what was then known about the planets and their satellites. It is startling how much of what was tentatively presented there has since been confirmed: that large impacts helped shape the surfaces of Mercury and the Galilean satellites, that the rings of Saturn are composed of frozen water, that clouds of ammonia crystals float atop the atmospheres of Jupiter and Saturn, and a host of other findings. Of course, the book also had its share of erroneous predictions. Kuiper insisted, for example, that the polar caps of Mars were water-ice rather than frozen carbon dioxide; we now know that both are present, and that the broad seasonal caps observable from Earth are in fact carbon dioxide frost.

By 1961, on the eve of the first interplanetary missions, so much had been learned about Earth's neighbor worlds that the task of summarizing it was almost hopelessly formidable. Undaunted, however, Kuiper and Barbara Middlehurst published another landmark work that year, *Planets and Satellites*, the third installment in a four-volume set intended to survey the state of knowledge about the entire solar system. In its preface Kuiper argued that, even in the era of planetary exploration, telescopic observations from Earth would remain important. Today, even though our robotic emissaries have surveyed at close range every one of the known planets except Pluto, it is clear that he was right. Were it not for the ongoing scrutiny of the solar system by patient observers here on Earth, much of what you will read in these pages would still lie waiting to be discovered. We would know nothing of the Sun's quivering sphere, Earth-crossing asteroids, and Pluto's satellite Charon, to cite but three examples.

When Kuiper died in 1973, he left a void not easily filled. No single person has matched his influence on and command of planetary science. But it is also true that the solar system has since grown on us. Today the study of "a planet" involves a host of complex disciplines: astronomy, of course, but also physics, chemistry, geology, mathematics, fluid dynamics, biology, and others. Few scientists possess a complete working knowledge of one planet, let alone the entire solar system; such a person would truly be a Renaissance scholar. Consequently, the task confronting us in preparing *The New Solar System* — to summarize the current state of understanding — could only be accomplished by bringing together specialists in a wide variety of fields.

Perhaps the most important ingredient was the realization the solar system is no longer a collection of individual bodies that can be addressed in isolation. It is instead an interrelated whole, whose parts must be studied comparatively. The organization herein is a direct consequence of this development. Although there are a few chapters on individual worlds such as the Moon and Mars, you will find that the four gas giants — Jupiter, Saturn, Uranus, and Neptune — are discussed collectively in chapters on their interiors and on their atmospheres. Similarly, there is one chapter on the surfaces of the terrestrial planets and another on their atmospheres. Smaller bodies require no fewer than five chapters to be covered adequately.

Each of the authors endeavored not only to provide the most up-to-date information available, but also to identify the gaps in our understanding that beg for further investigation. Their presentations are not intended to be entirely self-consistent. Many topics are the subject of disagreement or even outright feuding. For others there may be a consensus of theoretical opinion that lacks observational confirmation.

The final chapter of *Planets and Satellites*, written by

Kuiper, is entitled "Limits of Completeness." In it he took stock of the known worlds, and "completeness" was equated with knowing how *many* worlds orbited the Sun. Today we usually think of completeness in a broader sense, and we are reminded again and again that it is unattainable. Since the second edition of *The New Solar System* was published in 1982, there have been many new developments — so many, in fact, that most chapters had to be entirely rewritten for this third edition. Advances in solar astronomy have transformed our understanding of the Sun. New planetary magnetic environments have been charted. Scientists have made great progress in deciphering the riddles of Mars' climatic history. Robotic eyes have let us glimpse the black heart of Halley's Comet. New seismic techniques have probed the Earth's interior with unprecedented detail (our home planet receives its own new chapter in this edition). We have updated the collection of maps at the book's end to include the newly surveyed satellites of Uranus. And a special pictorial supplement following Chapter 9 offers highlights of the Voyager 2 encounter with Neptune. As much as we would have liked to include detailed results of that spectacular reconnaissance, doing so would have required substantial revisions to many chapters and forced a long delay in publication. It turns out that Neptune offers another lesson in the limits of completeness.

This third edition was two years in the making. It was begun at a time when space exploration was at a standstill following the *Challenger* disaster and concluded after Voyager 2's final planetary encounter with Neptune. In that time we drew upon the talents of many helpful individuals, only a few of whom can be singled out here. We thank Stuart Goldman, Richard Tresch Fienberg, and Leif Robinson of Sky Publishing, and Simon Mitton of Cambridge University Press for editorial support; Brian O'Leary, though not among the book's editors this time, played a key role in creating the earlier editions. Our thanks also to William Bonney, Susan Gilday, Charles Baker, Ron Arruda, Deborah McGonagle, Carla Thompson, Leslie Williams, and Karyn Bickford for their production help; and to Steven Simpson and David Gardner for their outstanding illustrations. The graphic teams at Color Prep and Shea Brothers provided excellent camera work when we needed it most. And finally we must thank the entire staff of *Sky & Telescope* and Cheryl Beatty, who persevered for many months while the editors were painstakingly creating this work.

In assembling this book, we have attempted to bring the fruits of recent planetary exploration to a wide audience. This is neither a textbook nor a "coffee-table" volume — it lies somewhere in between. By the same token, we have encouraged the authors to avoid both incomprehensible details and sweeping generalizations. Above all, we strove to make this enjoyable reading for those with either casual or professional interest.

September 1989

J. Kelly Beatty
Andrew Chaikin

Introduction

Carl Sagan

A LITTLE LESS than 5 billion years ago, something happened. A cloud of interstellar gas and dust, perhaps triggered by a nearby exploding star, collapsed and condensed to form the solar system. The central mass in the cloud contracted under its own gravity and heated, until temperatures became so high that thermonuclear reactions were initiated and the early Sun was born. Subsidiary, smaller lumps of matter did not achieve such high temperatures and pressures and did not become stars. Shining by reflected light, they evolved into the planets, satellites, asteroids, and comets that today comprise the Sun's entourage.

In one of those smaller lumps of matter, the heating of the interior and the infall of still-smaller bodies generated an atmosphere. One variety of gas condensed on the surface, forming protective lakes and oceans. Chemical reactions in the air and water produced complex organic compounds, which eventually – about 4 billion years ago – resulted in a well-organized system capable of making identical copies of itself from the surrounding molecular building blocks. The world on which these events happened, third from the Sun, is of course the Earth, and we are some of the descendants of that first self-replicating molecular system.

We humans are new to this solar system. We tend to think in short time scales. But the surfaces of many other objects in the solar system date back billions of years to a very different epoch, a time of titanic collisions and catastrophes. We have a tendency to think of other worlds as like our own and imagine other epochs to be as placid as this one. This is not the case. We inhabit a solar system rich in wonders: a stifling hot world, where heat is trapped in a thick atmosphere surmounted by clouds of sulfuric acid; an extinct volcano three times the height of Mount Everest; delicate circumplanetary rings twisted in astonishing, braided patterns; a world criss-crossed with elaborate straight and curved valleys; hurtling balls of ice, trailing tails of gas and dust as long as the distances between the planets; an underground ocean of molten sulfur. There exists a giant planet with an interior of liquid metallic hydrogen, into which a thousand Earths would fit; a planet-size moon surrounded by an atmosphere thicker than ours, an unbroken haze layer made up of complex organic molecules, and possibly covered by an ocean of liquid hydrocarbons;

Figure 1. **The subtle yet profound relationships between matter and life provide the theme for "DNA Embraces the Planets."**

and a beautiful, small, blue-and-white world on which organic molecules have evolved to produce creatures able to contemplate the diversity of worlds and the passage of ages, slowly becoming aware of the richness that the solar system offers, in space and time.

Voyager 2, in its exploration of Neptune, its system of rings, and its moons, completed a preliminary reconnaissance of all the known planets save Pluto. This is an historic moment in the evolution of the solar system: the first time, so far as we know, that one world in the Sun's family has achieved some fair understanding of the others. For human history, this moment is certainly more significant than the discovery of America by Christopher Columbus, and perhaps as significant as the colonization of the land by the first amphibians about 400 million years ago. We are leaving the world of our origins and have begun – first by machines, but later, almost certainly, with spacecraft piloted by human beings – exploring other worlds.

This is an effort where the benefits, while very real, are long range. A consistent commitment to planetary exploration has been difficult to organize. Now many nations are beginning to explore other worlds – not just the United States and the Soviet Union, but also Japan, the constituent members of the European Space Agency, and perhaps others. It seems likely that if we do not destroy ourselves – thereby ending this promising local experiment in the evolution of matter – we will continue our explorations, garnering more information on the atmospheres, surfaces, interiors, and possible biology within our solar system. We will discover what else is possible. We will better understand our own world by comparing it with others. We will send our little vehicles bravely into the turbulent atmosphere of Jupiter, across the exotic surface of Mars, through the tails of comets, past the asteroids, settling gently down onto the unimaginable landscape of Titan. We will return samples from some of these worlds. We will send people – perhaps Americans and Soviets traveling together – on behalf of the human species to make the first footfalls on Mars. These are all tasks well within our engineering capability. We need only a steadfast determination and a continuation of that ancient and honorable human exploratory tradition.

But even the complete exploration of our solar system is unlikely to settle the uncertainties surrounding those tumultuous events from which the Sun and planets emerged some 5 billion years ago. For that, we need to search for other examples of planetary systems. This is possible from the ground, but it will be easier from large telescopes in orbit around Earth. There is a real chance that, by the middle of the 21st century, we will have performed not only a deep reconnaissance of all of the worlds from Mercury to Pluto and beyond, but also a systematic survey of the planetary systems – if such exist – of hundreds of nearby stars. We will then be able to say, with a fair likelihood of being correct, something of the cosmic generality and origins of our solar system, our planet, and ourselves. This is an enterprise with deep significance for every inhabitant of the planet Earth.

1
The Golden Age of Solar-System Exploration

Noel. W. Hinners

THE EXPLORATION of the solar system by spacecraft has now spanned more than three decades, producing a wealth of basic discoveries and new data at a rate unparalleled in history. Robotic spacecraft have transformed our view of the planets from one largely of fuzzy, shimmering telescopic images to one dominated by crisp global perspectives of truly new worlds, simultaneously displaying raw, simple beauty and the awesome end-products of powerful, complex natural forces. Lunar exploration by spacecraft and astronauts, along with all-important laboratory analysis of returned samples, provided relatively detailed "ground truth" that allows us to glimpse the history of the Sun and a chronology of the processes at work in the first billion years of the solar system. It is a picture unlikely to be updated in the foreseeable future.

During this period of unprecedented exploration, Earth-based telescopic observations led to the discoveries of the rings of Uranus, the satellite of Pluto, a host of Earth-crossing asteroids, and invaluable compositional information on comets, asteroids, and planetary atmospheres from Mercury to Pluto. Finally, intricate terrestrial laboratory studies of meteorites, the only true "free lunch" in space science, have helped to develop the factual link between pre-solar-system stellar explosions and the birth of the Sun and the planets. Interestingly, some of those meteorites appear to have originated on the Moon and possibly Mars; the lack of both geologic context and proof of the site of origin, however, means they cannot fully substitute for samples obtained firsthand.

The totality of planetary exploration over the last three decades, summed up in this book, easily justifies, I believe, the appellation "The Golden Age of Solar System Exploration." No previous exploratory effort can match it in terms of rate of basic discovery, immediate impact on scientific thinking, eventual significance to society, and rapid communication of the results to Earth's inhabitants. This in no way denigrates previous eras of discovery, the most important of which, in aggregate, was the early exploration of our planet and the determination of its true place in the solar system and in turn of the solar system's place in the galaxy and universe. Consider, however, that this occurred over thousands of years in a combination of tedious, lengthy geographical and astronomical ventures including: Minoan

Figure 1. **Truly a space-age photograph, this view of the Earth and Moon was taken by Jupiter-bound Voyager 1 on September 12, 1977. Since the Moon is actually a rather dark object, computer processing was used to make its image appear three times brighter.**

voyages in the Mediterranean (2000–500 BC); Phoenician navigations around Africa and the west coast of Europe (1500–500 BC); Alexander's conquest of the Persian empire (about 330 BC); the religiously motivated travels of Hsuan-tsang to India (about 600 AD); river and ocean voyages of the Vikings (about 1000); Marco Polo's expeditions to China (about 1300); major ocean voyages by Portuguese, Spanish, French, and English explorers (1500s and 1600s); Cook's Australasian explorations (late 1700s); and the Lewis and Clark expedition (early 1800s). These explorations and others, one by one, were integrated into a partial global picture of the physical attributes of our home planet. The accomplishment was only partial because even the first "orbit" of the Earth by Magellan's expedition, a three-year venture begun in September 1519, did not reveal the whole planet. Only in the last 200 years has the extent of the polar regions become evident, and not until the past 40 years has the vast ocean bottom begun to yield its essential topographic and geologic secrets.

Progress in establishing our astronomical context was similarly slow: more than a century elapsed between Copernicus' development of the heliocentric nature of the solar system in the early 1500s and the confirming observations and analyses of Galileo, Tycho, and Kepler. Not until the early 20th century did our galactic "address" become known, and even today our yardstick of cosmic distance (and thus age) has a nasty habit of gaining or losing billions of light-years at a clip.

The pulse of planetary exploration parallels in many ways that which is also evident in other sciences, especially biology, astronomy, physics, and terrestrial geology – in fact, one might easily wish to reference all of these in aggregate as the Golden Age of Science. This spurt in many sciences is not simply coincidence; common denominators link their phenomenal growth: a solid foundation in mathematics and physics, new technologies, emphasis on mass education, realization of the long-term importance of basic research, popularization of science, potential near-term economic benefits, governmental "discretionary" budgets, and political competition. These factors interact in often subtle and complex ways to determine the health of the scientific process and the potential for future research.

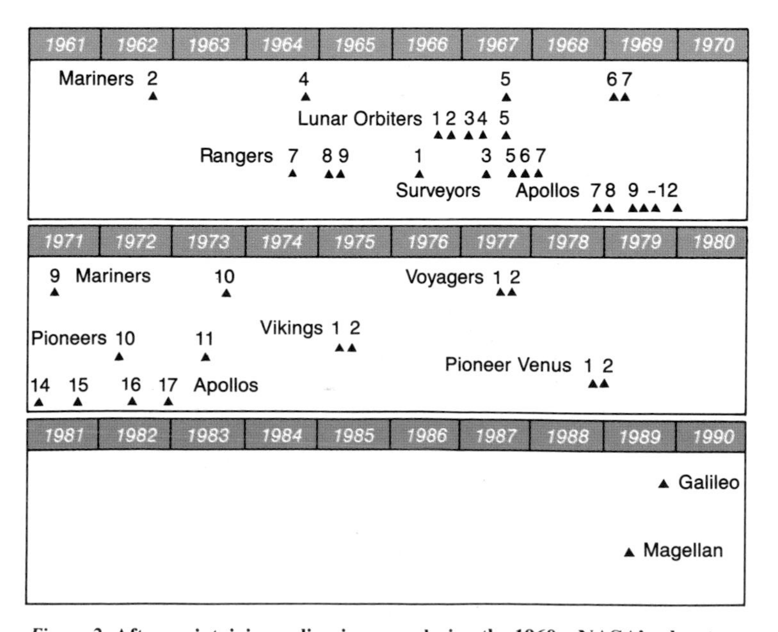

Figure 2. **After maintaining a dizzying pace during the 1960s, NASA's planetary exploration program began to wane by the mid-1970s. In fact, not a single U.S. spacecraft was boosted into interplanetary space from late 1978 until 1989 – a hiatus of 12 years.**

Unlike planetary exploration, however, other sciences are apt to continue a relatively steady pace of activity because of significant differences both in exploitation of the common denominators and in underlying motivations. It is these differences which have led to a precipitous decline in the rate of planetary exploration and which may well change its nature in the future. That would leave those missions accomplished to date, along with what few are under way or contemplated for the 1990s, to delineate the Golden Age of Solar System Exploration.

One can rationally inquire: with such an impressive and fruitful record of planetary exploration, why is it that at this writing the United States has just launched its first planet-bound spacecraft since 1978 (Figure 2), another that was approved long ago has sat on the ground for years (along with its graying scientists and engineers), and only one rather modest mission, the Mars Observer, is under construction? To understand this dilemma better, and to prepare viable plans for the future, we must examine and understand the origin and evolution of NASA's solar-system exploration program.

WHY EXPLORE THE SOLAR SYSTEM?

Solar-system exploration must be viewed with two minds, one asking why it exists, the other how it has been accomplished. To confuse the two, as frequently happens, makes it difficult to understand the problems of the future. In this section, we deal with the basic reasons behind the program; later, we examine how it grew and evolved.

The United States has undertaken a solar-system exploration program for four dominant reasons: national prestige, vision, knowledge, and applications. These are listed in order of the priority they held in the late 1950s, when the national space program was formally initiated; subsequently the emphasis has shifted, such that today national prestige is scarcely recognized as a significant factor by "decision makers."

National prestige, a product of how we view ourselves and how others view us as a nation, is an elusive commodity and difficult to measure. Surely, it is dependent upon perceptions by the citizenry and its elected representatives. How those perceptions come about is dictated in large part by how the news media perceive and report them and how the politician chooses to present them. Few would deny that the Soviet Union's launch of Sputnik 1 in 1957 caused immediate national concern and soul-searching. Since 1945 the U.S. had enjoyed the psychological luxury of being "first" in science and technology; we thought our democracy to be the best political system, one in which science and technology could flourish. In contrast, conventional wisdom in the early 1950s held that the Soviet political and industrial system, despite its earlier-than-anticipated development of nuclear weaponry, was somehow archaic, backward, and relatively undeveloped. The collective shock and wounded national ego, exacerbated by the realization of potential military applications of rocket and satellite technology, resulted in

Spacecraft	Launch	Encounter	Object	Event
Luna 2	12 Sep 1959	15 Sep 1959	Moon	impact with surface
Luna 3	4 Oct 1959	7 Oct 1959	Moon	photograph of farside
Mariner 2	27 Aug 1962	14 Dec 1962	Venus	flyby
Ranger 7	28 Jun 1964	31 Jun 1964	Moon	photographs at close range
Mariner 4	28 Nov 1964	14 July 1965	Mars	flyby
Venera 3	16 Nov 1965	1 Mar 1966	Venus	impact with surface
Luna 9	31 Jan 1966	3 Feb 1966	Moon	photographs from surface
Luna 10	31 Mar 1966	3 Apr 1966	Moon	orbiter
Surveyor 1	30 May 1966	2 June 1966	Moon	controlled soft landing
Lunar Orbiter 1	10 Aug 1966	14 Aug 1966	Moon	photographic orbiter
Zond 5	15 Sep 1968	18 Sep 1968	Moon	round trip with life forms
Apollo 8	21 Dec 1968	24 Dec 1968	Moon	human crew (no landing)
Apollo 11	16 Jul 1969	20 Jul 1969	Moon	humans explore surface; samples returned to Earth
Luna 16	12 Sep 1970	20 Sep 1970	Moon	automated sample return
Venera 7	17 Aug 970	15 Dec 1970	Venus	soft landing
Mariner 9	30 May 1971	13 Nov 1971	Mars	long-life orbiter
Mars 3	28 May 1971	2 Dec 1971	Mars	soft landing
Pioneer 10	3 Mar 1972	3 Dec 1972	Jupiter	flyby
Mariner 10	3 Nov 1973	29 Mar 1974	Mercury	flyby (also on 21 Sep 1974 and 16 Mar 1975)
Venera 9	8 Jun 1975	22 Oct 1975	Venus	photographs from surface
Viking 1	20 Aug 1975	20 Jul 1976	Mars	photographs from surface; search for life forms
Pioneer Venus 1	20 May 1978	4 Dec 1978	Venus	long-life orbiter
Voyager 1	5 Sep 1977	5 Mar 1979	Jupiter	flyby
Pioneer 11	5 Apr 1973	1 Sep 1979	Saturn	flyby
Voyager 1		13 Nov 1980	Saturn	flyby
Vega 1	15 Dec 1984	11 Jun 1985	Venus	atmospheric balloon
ICE (ISEE 3)	12 Aug 1978	11 Sep 1985	comet	pass through plasma tail of Comet Giacobini-Zinner
Voyager 2	20 Aug 1977	24 Jan 1986	Uranus	flyby
Vega 1		6 Mar 1986	comet	photographs of Comet Halley
Voyager 2		25 Aug 1989	Neptune	flyby

Table 1. **Over the past three decades, the Soviet Union (italic type) and United States have amassed an impressive list of milestones as their space explorations have extended ever-farther from the Earth.**

near-immediate decisions to beef up science education programs; establish a civil space program; and reinvigorate the military applications of space activity.

Our initial, politically driven thrust into space was reinforced and extended in 1961 when President Kennedy stated his goal of sending American astronauts to the surface of the Moon and safely returning them to Earth by the end of that decade. Kennedy's decision, and the consequent initiation of the Apollo program, were driven largely by the loss of face suffered from continuing Soviet space firsts (including Gagarin's orbital flight in 1961) and partly from the Bay of Pigs fiasco.

It is in the above political context that the planetary program was born and evolved and to which we will return later. For the moment, consider that the first widely recognized success in the American planetary program – Ranger 7's close-range photography of the Moon in July 1964 – was much more political than scientific. In showing that the U.S. could accomplish a major technological objective, coming as it did after a long series of failures, this achievement regained some of our lost prestige, eased the passage of the NASA budget, and led to a more favorable climate for the conduct of the Apollo program. For 12 years thereafter, the drive to put humans in space via the Gemini, Apollo, and Skylab programs overshadowed robotic explorations of the planets in public attention and international acclaim. Automated probes regained the spotlight with the Viking landings on Mars in 1976, by which time the entire political climate had changed to one in which the Soviets were no longer viewed as serious competition and the public had become somewhat numbed to space activity. Although news articles frequently highlighted the fact that the U.S. was the first to conduct experiments on the Martian surface, in contrast to several Soviet failed attempts, editorial comment focused more on the remarkable technological and scientific accomplishment per se, and on the exploration ethic and its place in American culture. That pattern continued with the Pioneer mission to Venus and the Voyager missions to Jupiter, Saturn, and beyond.

Political motivation is not, of course, limited to the United States space program. That of the Soviet Union is well known. Its early planetary program was, compared to NASA's, more preoccupied with achieving "firsts" (Table 1), and the Soviets made much ado of the planting of the Soviet flag and Lenin pendants on the Moon and Venus with the implied superiority of the Soviet political system. Aside from their exploration of Venus, a systematic, primarily science-based, long-term Soviet strategy and program plan for solar-system exploration did not exist much before the mid-1980s. In fact, lunar and Martian exploration attempts ceased relatively soon after the spectacular American successes. That changed dramatically in recent years, starting with the impressive Soviet missions to Venus during the early 1980s, flybys of Comet Halley in 1986, and a renewed, ambitious program Mars of exploration commencing with the 1988 Phobos mission. (Unfortunately, both Phobos craft failed to accomplish their planned objectives.) As significant as these are scientifically, the Soviets are beating us at our own game by openly inviting the world to view their missions firsthand and by incorporating experiments from other (mainly European) nations. Moreover, countries and organizations that used to depend upon cooperation with the U.S. to participate in planetary missions now have, or soon will, the capability to conduct their own flights. Both the European Space Agency and Japan mounted successful missions to Comet Halley, and apparently they feel the political attributes that accrue to a planetary program are significant and worth the financial investment.

The growing interest in planetary exploration worldwide stands in stark contrast to that of this country, which forwent the opportunity to conduct a flight to Comet Halley. Indeed, the United States effectively ended up with *no* viable planetary program during the 1980s. The competitive nature of science and the role of science as a driving force behind technological advances has apparently not been recognized either by the Congress or recent administrations. Rather, the prevailing view (encouraged by some in the scientific community) is that science is conducted for the benefit of all humankind and transcends political boundaries. To that end, American political leaders have encouraged NASA to seek significant foreign participation in its scientific missions. However, in view of the changing nature of the foreign capability and attitudes, and of the complexity of managing joint endeavors, this policy begs for reexamination.

After a hiatus of several years following the Apollo-Soyuz mission in 1975, formal cooperation in space exploration with the Soviet Union resumed in April 1987 with the signing of an agreement and the establishment of five working groups, one of which is focused on planetary exploration. Soviet delegations have openly solicited cooperation with the U.S. in joint ventures to Mars, starting with the automated return of surface samples and leading toward human exploration of the planet. But the U.S., fearing technology transfer and remaining uncertain about the long-term viability of Soviet "glasnost" (openness), has taken a go-slow approach. This strategy may be sensible and realistic, but it hardly matches the Soviet publicity coup.

Vision, a concept frequently invoked as a reason for conducting planetary exploration, carries specific justifications that run the gamut from such simplistic vagaries as the "exploration imperative" in human nature and "climb the mountain because it's there" to more erudite, philosophical discourses. But "vision" cannot easily be used as a selling point for new explorations and in fact can even backfire. It is easy, in the political world, for one who speaks of vision to be labeled as an idle dreamer or science-fiction nut. For our purposes, therefore, I prefer to define vision as the perception of a challenge to be met and knowledge to be gained, wrapped up in the belief that we possess the will and the means to pursue those objectives, and capped off with the conviction that, in so doing, humankind is well served.

This definition fits well Kennedy's 1961 commitment to Apollo: "If we are to win the battle that is going on around the world between freedom and tyranny, if we are to win the battle for men's minds, the dramatic achievements in space which occurred in recent weeks should have made clear to us all, as did the Sputnik in 1957, the impact of this adventure on the minds of men everywhere who are attempting to make a determination of which road they should take.... We go into space because whatever mankind must undertake, free men must fully share." It also fits well what must have driven Robert Goddard through decades of isolated, dedicated research on rockets. Although both cases involved vision, there is a spectrum of talent necessary to enable such endeavors as space exploration. Goddard and Kennedy represent the extremes: the early researcher who develops the scientific and technological basis, frequently with a particular goal in mind; and the political leader who, for whatever reasons, can galvanize the necessary resources, resolve, and patience to pull it off.

One might cogently argue that an earlier phase is required, one in which someone plants the basic concept in the public consciousness, a seed that stimulates future generations to make once-crazy dreams come true. John Wilkins was such a person, publishing *Discovery of a World in the Moone* in 1638, two centuries before another, Jules Verne, gave his great, albeit unpowered, boost to the concept of lunar exploration.

In the first 15 years of planetary exploration, it was primarily the scientific community who invoked vision as a selling point, though in the eyes of a wary public or body politic this approach was sometimes seen as more self-serving than genuinely altruistic. Journalist Tony Reichardt, in describing Voyager's pictures of Saturn, summed up rather eloquently what may constitute one aspect of vision: "A momentary elevation of our sight, a halt to routine and preoccupation, a spark of the old, nearly forgotten emotions of wonder all evoked by strange visions sent by a robot from a planet a billion miles away.

"From the great optical distance our own situation comes into focus for a second; we feel refreshed for having seen something 'new' yet enduring and bigger than our own lives. The illusion may be fleeting, but it's a good one – of a common humanity united in a single effort toward something grandly mysterious.

"In a better world this would be reason enough to have an active space program. Never mind the inventory of newly discovered rings and moons, or methane atmospheres or low-density ammonia ice particles. What is involved here is something decidedly more spiritual, something to do with a nation's character and its sense of purpose."

Knowledge. Planetary exploration's greatest contribution is certainly its addition to the great store of our collective knowledge. In terms of sheer bulk, its data is overwhelming; in terms of new information, its results are mind-boggling. Both the data and information are immediate returns whose true value cannot yet be adequately assessed. That must await their more complete conversion into the cache of human knowledge, whereby we have a sensible understanding of the natural processes involved in creating the objects or phenomena observed. However, even if our comprehension is not yet complete, we already realize much of the *significance* of what has been observed, be it for scientific or exploitive purposes. Indeed, the evidence summed up in this book shows convincingly that we are making demonstrable progress toward understanding the origin and evolution of the solar system, better understanding Earth through comparative studies, and deciphering the relationship of life and the chemical history of the Sun's family.

Applications. The belief that practical benefits can be

obtained from planetary exploration is largely a matter of faith, and in support of this belief one must inevitably resort to calling upon history. For example, most people are convinced (or can be) that the exploratory expeditions of Lewis and Clark eventually led to the development of the western United States. But such historical analogies do not completely satisfy the casual inquisitor, so proponents of planetary exploration must then speculate somewhat about its possible downstream benefits. Commonly used examples are using lunar materials to construct permanent bases there, capturing Earth-crossing iron-nickel asteroids to supplement Earth's dwindling resources, and colonizing Mars for future human expansion. Whether these or other scenarios come to pass is moot; the key point is that only by conducting the exploration and research in the first place will we ever have the opportunity to make intelligent choices.

Specifics aside, there exist several general practical benefits of planetary exploration. An increasingly popular thesis maintains that by learning something about the origin, evolution, and physical state of the planets, we will better understand the Earth itself. This approach to solar-system research, often termed *comparative planetology*, contains the implicit assumption that a more complete scientific understanding of the Earth is of more practical value than, say, an understanding of Venus per se.

The concept of comparative planetology has demonstrable intrinsic merit: there is no way to develop a decent comprehension of the origin and evolutionary history of a single, highly evolved, complex planet (Earth) by studying it in isolation from the class of objects of which it is but one member. Even though the other planets may also be complex, the fact that they differ significantly in size and composition and have evolved differently enables one to see directly the effects of the various initial conditions. Ultimately, we are able to construct more plausible models of the origin and evolution of planetary bodies both in general and in particular.

These same methods are successfully applied toward explaining present-day conditions. For example, general circulation models of the Earth's atmosphere (known as GCM's in the trade) are relatively crude but rapidly improving in their ability to reproduce observed atmospheric behavior. They are also being used to determine how the growing levels of carbon dioxide in Earth's atmosphere are enhancing a greenhouse effect that will alter our short- and long-term climate. A GCM that takes into account such factors as planetary rotation, solar input, atmospheric density, and chemistry actually can be better tested by applying it to other, simpler planets, then comparing the predicted results with terrestrial reality.

Such applications of planetary exploration must never be construed as the *raison d'être*, but rather as valuable and essential spin-offs that enhance the enterprise as a whole. The time needed to assimilate planetary data and information is too long to yield short-term benefits. But this is typical of many scientific fields; basic research and geologic exploration, in particular, frequently take even longer to yield so-called practical applications.

BEGINNINGS

Conceptually, planetary exploration began with the dreamers of yore. They, of course, did not worry much about the practicality of the venture, and it wasn't until the late 1800s and early 1900s that Konstantin Tsiolkovskiy, Hermann Oberth, and Robert Goddard established the theoretical and experimental basis for escaping Earth's gravity. Unfortunately, as with many good ideas and inventions, it took the technological stimulus of war to effect the transition from the mind and laboratory to the field. Thus did true planetary exploration have its origin in the same root mass as the rest of the space program: the development of rocket technology during World War II, as epitomized by von Braun's V-2 missile, and in the postwar development of intermediate and long-range ballistic missiles. It then took but a relatively small step in scale and efficiency to progress from lobbing atomic weapons one-quarter of the way around the world to putting a satellite into orbit or having it leave Earth altogether.

Planetary studies cannot, or at least ought not, be conducted without planetary scientists. As a breed, they came to the space program relatively late (post-Sputnik). Atmospheric scientists had gained a head start in the late 1940s and early 1950s through experiments with sounding rockets and captured V-2's, and not surprisingly it was these "sky" scientists (together with politicians and technologists) who laid the first plans for exploring outer space. In its proposal for a national space program, dated November 21, 1957, the Rocket and Satellite Research Panel made no mention of the planets; another proposal submitted one month later envisioned only crude robotic exploration of the Moon followed by lunar missions with human crews. Even a year later, the newly formed space-science division of NASA had no plans to study any solid bodies (despite their inclusion in the agency's charter), and not until late 1959 was a formal organization formed for lunar and planetary programs.

This gradual evolution finally spawned a five-flight Pioneer "lunar program" sponsored by the Defense Department's Advanced Research Projects Agency, and a pair each of Venus flybys and lunar orbiters sponsored by NASA. But the discovery of the Van Allen radiation belts by Explorers 1 and 3 soon altered these plans: Pioneer flights were recast to probe the radiation belts more fully, and the successful Soviet probe Luna 1 provided the political impetus to redirect NASA's program toward Moon-only objectives.

The first coherent plan for planetary exploration emerged in April 1959 from one of NASA's early acquisitions – the Jet Propulsion Laboratory (JPL) in Pasadena, California. This extremely ambitious program would have utilized a planetary spacecraft called Vega, to be launched by powerful Atlas-Vega and Saturn 1 boosters. However, the Atlas-Vega was soon abandoned for the smaller Atlas-Agena B, and the Vega concept itself became eclipsed by an updated NASA lunar program involving the "kamikaze" Ranger series and, later, the more sophisticated soft-landing Surveyors. The impetus for this new emphasis on lunar programs was both political and scientific: here was an opportunity to impress the world with quick technological success, while leading scientists had begun to impress upon NASA the importance of lunar science to the secrets of planetary origins.

The Apollo program, announced in 1961, changed the program's character dramatically. Because of the vast number of engineering and scientific unknowns involved in putting astronauts on the Moon, both Ranger and Surveyor became supporting players to Apollo. They were soon joined

slingshot effect of one planet's gravity to send spacecraft on to the next one in the sequence. But tight budgets forced NASA to scrap the Grand Tour concept in favor of other projects, even though the Office of Management and Budget had endorsed it. According to Robert S. Kraemer, who headed the agency's planetary program at the time, the OMB then made the rare offer of additional funds so that NASA could send two smaller, Mariner-class flights to Jupiter and Saturn – spacecraft that ultimately became known as Voyagers 1 and 2.

The third example, actually an addendum to the restructured Grand Tour, was the Mariner Jupiter-Uranus (MJU) mission, proposed within NASA for launch in 1979. But MJU would have cost $400 million and required a costly two-year extension of the Titan-Centaur program beyond the 1977 Voyager launches. These and other factors, plus the potential for sending one of the two Voyagers to Uranus, combined to bury the MJU mission. (Notably, Uranus and Neptune eventually became official Voyager mission objectives, but that did not happen until *after* the twin spacecraft left Earth.) Finally, there is Magellan and its precursor, the Venus Orbiting Imaging Radar (VOIR) mission. As originally conceived, VOIR would likely have cost $500 to $600 million. But in 1982 the Reagan administration – and science adviser George Keyworth in particular – balked in response to NASA's proposal, which led to its restructuring into the Magellan program and cost saving of about 50 percent.

It is within the OSSA that planetary missions compete for funds with those that deal with space physics, solar-terrestrial interactions, astrophysics, and life sciences. Some of these competitors are simpler, less-expensive spacecraft that orbit the Earth, carry fewer experiments, and have less-involved operations. Furthermore, most planetary missions are assigned to JPL, which is operated for NASA by the California Institute of Technology. JPL's personnel costs must be included in each mission's price tag, whereas other mission categories utilize civil-service personnel whose salaries are absorbed elsewhere in the NASA budget. Not surprisingly, planetary programs historically have consumed the lion's share of the agency's space-science funding.

Figure 3. **Faced with tight budgets during the early 1970s, NASA chose not to undertake a "Grand Tour" of the outer solar system. Had it done so, instrument-laden spacecraft would have completed visits to Jupiter, Saturn, Uranus, Neptune, and Pluto by late 1988. Fortunately, the long-lived Voyager spacecraft have managed to accomplish most of the Grand Tour's original objectives.**

But the situation is changing (Figure 4). In recent years billion-dollar projects like the Hubble Space Telescope (HST) and the Advanced X-ray Astronomical Facility (AXAF) have been chosen over high-priority planetary endeavors such as the Comet Rendezvous and Asteroid Flyby (CRAF) mission. On a dollar-for-dollar-how-much-science-do-you-buy basis, and on popular appeal, space astronomy continues to pose powerful competition to planetary missions. One might think intrinsic scientific merit is the sole discriminant in establishing OSSA's budget priorities, but in fact the distribution of funds for new starts is determined as much by whose turn it is as anything else. While this may seem arbitrary, the truth is that any proposal good enough to survive the weeding-out process within its particular scientific discipline would yield high-quality scientific results if successfully flown. However, there simply is not enough money to do all the good science desired, and as soon as a mission enters the budgetary process, someone in OSSA must choose from among "apples and oranges" (for example, planetary versus astronomical missions).

The trick, then, from the vantage point of both administrators and mission advocates, is to figure out why today an apple may taste better than an orange. "Well, if we can afford only one," they might say, "why not an apple today, an orange tomorrow?" This method of initiating major new space-science missions is a reasonable way to operate *if* it doesn't take too long to cycle through the rotation. But the current budgetary environment is such that each discipline must wait at least four to five years between its major new starts. At this low level of activity there is a real danger both of losing the engineering and operations talent so essential to the conduct of highly sophisticated planetary missions and of having the scientists drift off to more viable, if not as stimulating, careers.

For 20 years or so, many space scientists believed that as each of NASA's most expensive projects – Apollo, Skylab, and Space Shuttle – wound down, there would be room in the agency's budget to accommodate a burst of new science missions. That is, a "wedge" would appear between NASA's anticipated overall budget and the declining expenditures for existing programs. However, each participant in the budgetary process – from scientist to space-station buff to Office of Management and Budget examiner – eyes this windfall as his or her preserve. In reality, the long-sought funding wedge has never materialized for the space sciences, nor is one likely in the future. Space Shuttle operations have cost much more than was projected at the project's inception in 1972, and now the Space Station *Freedom*, authorized in 1987, assures that NASA will continue to have one very expensive project dominating its budget for the foreseeable future.

It is easy, and occasionally justifiable, to blame the space sciences' budget woes on costly human-exploration projects. However, I believe that the other factors discussed are of greater significance, and that, in the long run, all of NASA's robotic spaceflight programs are accommodated to some extent by the manned program's budgetary umbrella.

The role of launch vehicles cannot be underestimated. In the long cradle-to-grave life of a planetary mission, the hour or so of critical launch activity takes on a deserved eminence, and planetary exploration owes a debt of gratitude to the rocketeers. But planetary programs have likewise been held hostage to snags in launch-vehicle development. The most visible recent example – and certainly the most convoluted – involves the Galileo mission to Jupiter.

When approved in 1977, Galileo was to be the first planetary mission launched by the Space Shuttle. A solid-fuel booster eventually called the Inertial Upper Stage (IUS) was to be carried up with the spacecraft and, once clear of the Shuttle's cargo bay, propel it toward Jupiter. But the Shuttle and IUS programs fell so far behind schedule that planners were forced to postpone Galileo's launch from 1982 to 1984. The timetable continued to slip until, in 1981, NASA was forced to substitute an existing liquid-fuel booster, the Centaur, for the IUS. Even that plan was not without problems, as the Centaur would need considerable modification to work compatibly with the Shuttle. In particular, NASA's Johnson Space Center questioned the safety of carrying the Centaur and its highly volatile propellants from the start. The impasse was temporarily resolved by a Congressional mandate to use the Centaur.

The loss of *Challenger* in January 1986 ultimately settled the entire debate in a most unexpected and tragic way. The Centaur modification program was canceled soon thereafter and the IUS resurrected as the upper stage to be used on the Shuttle's planetary missions. With a tad of good luck, the Galileo will be launched in 1989, 12 long years after the program's initiation. But since the IUS is much less powerful than the Centaur, a special trajectory has been designed for Galileo that will take it first to Venus, back to Earth, and back to Earth again during its first three years in space (Figure 5). This series of gravity assists will give it enough velocity to reach Jupiter in late 1995 (Figure 6), almost *20 years* after the project's approval. Moreover, each flip-flop on the choice of booster necessitated costly design changes in the Galileo spacecraft; each postponement of its launch caused its already-amassed team of scientists and engineers to "march in place" for years longer than expected. As a consequence, Galileo's overall cost has risen from the initial expectation of about $420 million to nearly $1.4 billion – yet little of this three-fold increase can be blamed on the program itself. Meanwhile, the Shuttle tragedy has caused NASA, following the lead of the U.S. Air Force, to go back to using expendable launch vehicles in situations whenever astronauts are not required.

Political and popular support are certainly important and necessary ingredients at all steps in the planetary-mission approval process. Given a president lacking a positive stance on the space program, NASA probably would not receive major funding approval from the Office of Management and Budget. Once NASA's budget is submitted to Congress, it then must garner the support of four oversight subcommittees and their hierarchal parent committees. Despite some interesting tussles and an occasional floor show, the Congress has generally supported NASA budgets largely as proposed by the president. When serious congressional disagreement has arisen concerning planetary missions, it has usually been tied to peripheral problems of overall and specific science priorities in the budget, the inherent value of basic science and exploration, the relative balance within NASA between science and applications, and the availability of launch vehicles.

Most of the American public is favorably disposed toward planetary exploration, which has certainly been a positive influence on the overall political support of, and thus the progress of, the program to date. Members of the executive and legislative branches and their staffs are also aware of the widespread support given planetary missions in the news media. They could hardly have avoided seeing the extensive

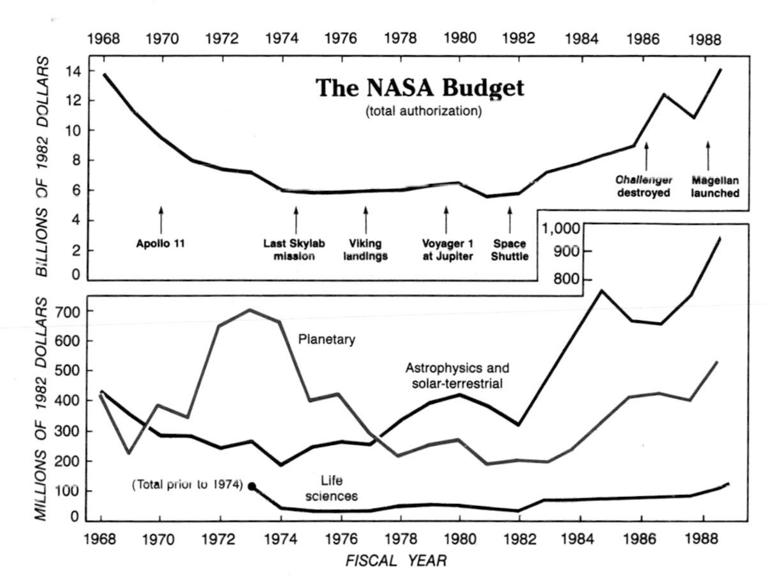

Figure 4. **When corrected for inflation, the buying power of NASA's overall budget has dropped dramatically since fiscal year 1968, which was in the midst of the Apollo program. The peak in funding for planetary exploration during the 1970s resulted primarily from the Viking and Voyager projects, but since then the program has not fared particularly well. (These figures do not include a "transition quarter" that was added to accommodate a fiscal-year revision between 1976 and 1977.)**

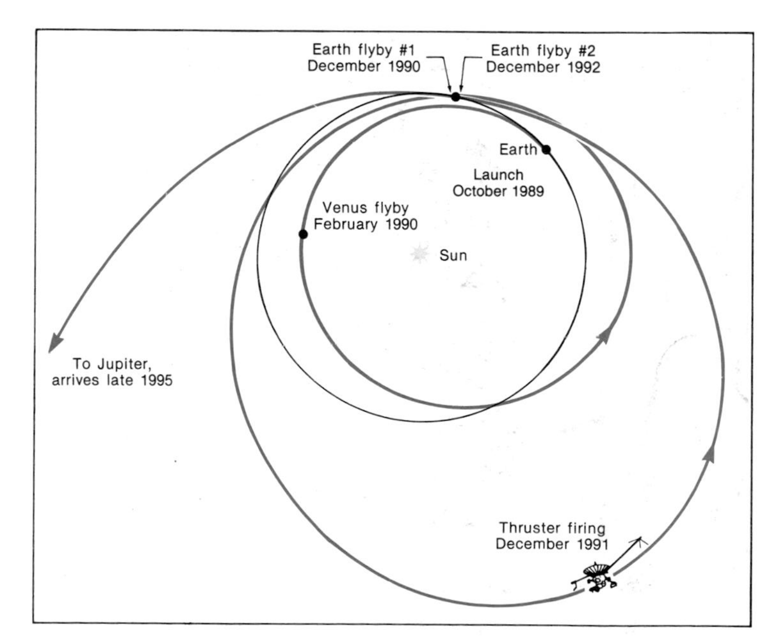

Figure 5. **Undaunted by the lack of a sufficiently powerful Shuttle-compatible rocket stage, Galileo mission planners devised this VEEGA (Venus-Earth-Earth gravity assist) trajectory to get the spacecraft to Jupiter. Over three years, one close flyby of Venus and two of Earth will accelerate the probe toward its distant target.**

coverage of the Viking and Voyager results, especially the spectacular photography. Beyond all this interest, which can be viewed as passive advocacy, there exist dozens of organizations that maintain strong or central interest in overall space exploration. The enthusiasm of such space-interest groups frequently swells to the level of active advocacy, and while their combined influence is difficult to evaluate, without it much of the civil space-exploration program might fade from existence.

FUTURE PLANETARY EXPLORATION

The results from U.S. planetary exploration to date have been undeniably spectacular, whether measured in terms of human achievement, basic scientific discovery, or technological success. However, one cannot help but recognize there has been an overall decline in the rate of planetary exploration during the 1980s, despite all the programmatic complexities and the changing circumstances. Indeed, not a single American spacecraft sped off toward interplanetary space in the decade following two 1978 launches of the Pioneer Venus mission. To reverse this disturbing trend – a desire of many of us – we must fully recognize and allow for increased mission sophistication and costs, tighter budget constraints, the transition of objectives from basic discovery to more detailed exploration, and increased competition from other fields of science.

The essential first ingredient of any revitalization of planetary exploration is a well-formulated implementation strategy and plan. To this end, NASA commissioned the Solar System Exploration Committee (SSEC) in 1980 to develop a viable approach within explicit budget constraints. The SSEC's response built upon the detailed science strategy already developed by the National Academy of Sciences via COMPLEX; specifically, it strongly endorsed those missions already in preparation (Galileo and Magellan). Using those as a starting point, the SSEC next proposed a "core program" that it felt could be achieved with an annual budget of about $300 million (in fiscal 1984 dollars, to be adjusted upward in later years based on inflation). This core program involves two basic elements: (1) a continuation of basic research at a vigorous pace, including planning and technology development for future missions; and (2) a

Figure 6. **If its mission proceeds as planned, on December 7, 1995, the Galileo orbiter will sail over the volcanic plains of Jupiter's satellite Io at a range of only 1,000 km. This is the only encounter that the spacecraft will make with Io; the innermost of the four Galilean satellites lies within a zone of intense magnetospheric radiation that would ultimately prove lethal to the spacecraft's delicate electronics.**

steady, well-paced sequence of planetary missions. Despite all this robust activity, the SSEC felt its core program could be achieved with less money than had been the norm for planetary exploration during the 1960s and 1970s (Figure 7).

By what magic would the SSEC accomplish its objectives with a constant, modest budget? The clue lay in two key assumptions. First, NASA must achieve *program stability*, which avoids the inefficiencies of start-stop cycles or slowdowns. Given adequate funding, the SSEC reasoned, planetary missions would come along frequently enough to permit significant inheritance in hardware design as well as more efficient use of highly trained personnel like those at JPL. A cornerstone of this cost-saving approach would be the development of a modular spacecraft, the Mariner Mark II, which can be easily reconfigured for different missions or to accept new technological developments (Figure 8). Indeed, this is the rationale behind NASA's fiscal 1990 proposal to build spacecraft simultaneously for the Cassini and CRAF (Comet Rendezvous and Asteroid Flyby missions).

The second key element is the creation of a new class of inner-solar-system missions termed Planetary Observers. These would be relatively low in cost (achieved by adapting the designs of existing spacecraft used in orbit around Earth), more frequent in rate, and less ambitious in their objectives (that is, used to address a specific, well-defined set of scientific questions). The concept is analogous to the Explorer program, which serves an equivalent role in space physics and astronomy. The Mars Observer is the first mission of this class.

The SSEC believed that the core program would, if implemented, enable the U.S. to maintain a position of leadership in planetary exploration. It also recognized that modest budgets do not provide for the "big" missions, such as those that would return samples of Mars or a comet to Earth or make major studies of the outer planets. In scope, and thus cost, such endeavors would be closer to the Vikings and Voyagers of the 1970s and therefore should be considered by NASA separately from the steady pacing of the core-program missions.

The SSEC has not been alone in advocating a revitalized U.S. planetary-exploration effort. In 1987, the well-respected physicist and astronaut Sally Ride led a task group that addressed how to establish the long-term direction for the U.S. civil space program – a course that would retain (or regain, depending on the emotional or political bent of the observer) its leadership in space exploration. As documented in its 1987 report to NASA's administrator, Ride's task group laid the foundation for the evaluation of four "leadership initiatives": (1) establishing an inhabited outpost on the Moon, (2) sending humans to Mars, (3) exploring the solar system, and (4) mounting an intensive study of Earth. The first two, while clearly endorsing a continued human presence in space, would also involve planetary exploration to some degree, as of course would the third. And the fourth embraces the multidisciplinary approach fostered by the planetary-exploration program.

The Ride task force's third leadership initiative is essentially an endorsement of the SSEC core program combined with a Mars rover and sample return (MRSR) mission from the augmented program. While recognizing the MRSR concept as a powerful scientific undertaking in its own right, the Ride panel argues that it also is "a necessary precursor to the human exploration of Mars." The real thrust of the Ride report concerning planetary exploration is its overall focus on Mars (it proposes *three* MRSR missions). The task group obviously believed that such an interplanetary assault would showcase our national

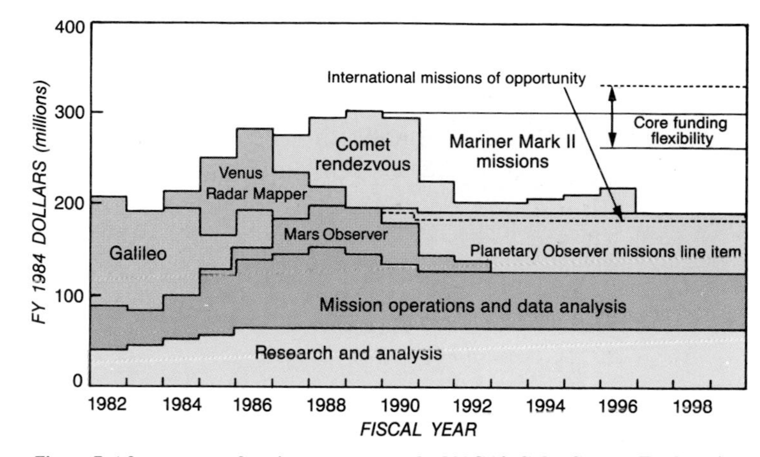

Figure 7. **After a comprehensive two-year study, NASA's Solar System Exploration Committee (SSEC) concluded in 1984 that a well-balanced exploration program could be achieved at moderate cost – about $300 million per year, as compared to more than double this amount spent during the mid-1970s. The key to its approach was establishing a "core program" of well-paced missions through the year 2000 that would be economical by virtue of their common hardware elements and shared organizational resources.**

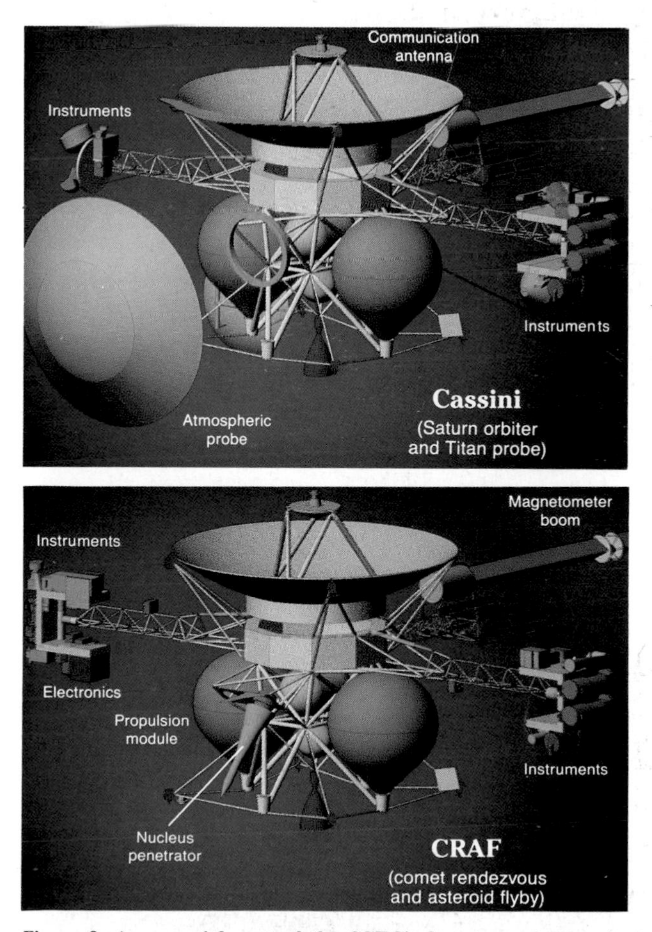

Figure 8. **An essential part of the SSEC's long-term strategy is development of the Mariner Mark II, a sophisticated spacecraft that can be adapted in modular fashion to accomplish a broad range of mission objectives.**

capability best, while simultaneously offering us abundant opportunities to exercise international cooperation from a position of strength.

A similar view is expressed in the 1988 report *Space Science in the Twenty-First Century* by the Academy's Space Science Board. It too calls for a Mars focus in parallel with a broad-based, balanced exploration program. If by now the reader wonders why all these studies, which are dominated by scientists, endorse both the intensive exploration of Mars *and* an underlying balanced program, here's why: they are deathly afraid that an expensive Martian sample-return mission, and the human exploration following it, will be done at the expense of the rest of the scientific program. This may seem a paranoid approach, but, as the saying goes, "Just because you're paranoid doesn't mean they're not out to get you."

One might question whether getting samples of Mars back to Earth really is an essential precursor to human exploration. Of course not. But asked a bit differently, the question becomes, "Would an automated sample-return mission to Mars significantly *enhance* subsequent human exploration?" In my view, the answer is yes. Although the following analogy is not perfect, the Apollo 11 lunar science tasks might have been quite different and more productive had a Soviet Luna sample been in hand five years before. But wouldn't such a successful Luna mission have taken away a major chunk of the underlying rationale for Apollo – and similarly for the future human exploration of Mars? Worry not. Since neither our missions to the Moon nor our comparable plans for Mars were (will be) driven by scientific merit, it seems all the more sensible to do everything possible to use humans most effectively when they finally do set foot on the red planet.

The Congress has also taken an active hand in shaping future space direction. In 1984 it charged the Reagan administration with appointing a National Commission on Space (NCOS) to formulate "a bold agenda to carry America's civilian space enterprise into the 21st century." Led by Thomas O. Paine, a past NASA administrator who in 1969 proposed sending astronauts to Mars as the successor to Apollo, the NCOS indeed met its charge. Its 1986 report, *Pioneering the Space Frontier*, advocated a vigorous, across-the-board civil space program. In the arena of planetary exploration the NCOS essentially echoed the SSEC's conclusions, but it also recommended accelerated human exploration of the Moon, Mars, and asteroids, with a concomitant emphasis on the utilizing these bodies' resources as a means of reducing the costs of exploration.

Lest planetary-exploration buffs think that they are alone in their frustration, they should look over a 1986 report by NASA's Space and Earth Science Advisory Committee (SESAC), *The Crisis in Space and Earth Science*. It details problems for all the other space sciences that sound exactly like those addressed for so long in the planetary program. Thus the playing field of demand has been leveled! One intriguing nuance of the SESAC study was the recognition that the overall budget situation may not improve and that, if it doesn't, an overall reduction in space-science efforts may be in store. Such a major reprioritization of program content, SESAC suggests, would be beyond purview of NASA and the scientific community alone. Rather, such a decision would require a "consensus of the American people, and thus of their representatives in the executive and legislative branches of government."

However, no obvious mechanism exists to develop such a consensus. It remains the job of NASA and the space-science community to make the best case possible for the benefits which can accrue from the programs and to set the priorities. And indeed that is the basis for the OSSA's "Strategic Plan 1988," which methodically lays out the overall approach by establishing programmatic themes, decision rules, and priorities within themes, followed by checks of program viability, technological readiness, personnel, launch-vehicle availability, and the all-important budget. OSSA's highest priorities for solar-system exploration are to get the Mariner Mark II program moving (via CRAF and Cassini), and to follow up the Mars Observer with a Lunar Observer (to survey the Moon's surface mineralogy and other properties). More visionary goals are a long way from realization. But even though something like global atmospheric experimentation on Mars seems fanciful now, we can still focus on missions to provide the scientific basis that will enable us to make utilization decisions farther downstream.

Whatever is concluded in the near-term regarding specific missions, a few obvious facts stand out. No amount of remote sensing will ever yield the kind of information attainable by analyzing planetary materials in terrestrial laboratories. The return of samples to Earth from Mars, Venus, comets, and asteroids is essential to making the desired leap in basic knowledge of the composition and physical state of those bodies. Let us then get on with developing the techniques of sample return, fully recognizing that this should be done prior to proposing piloted missions to these worlds.

We should not become overly depressed by the current state of affairs regarding planetary exploration. In contemplating the human footprint on the Moon, stream channels on Mars, volcanoes on Io, and braided rings at Saturn, recall how tortuous the route has been to get there. Let us use the current situation to stimulate innovation, focus our resources, plan the options, and be ready to respond to new exploration challenges. Let us not idly await a hoped-for opportunity or for the spirit of the "exploration imperative" to strike; opportunities are more often a result of creation than random occurrence. Most things happen because determined people want them to.

2
The Sun

Robert W. Noyes

WE LIVE right next to a star. Our 150-million-km distance from the Sun is only about 108 times the Sun's diameter. By astronomical standards, this is close indeed: the next nearest star, Proxima Centauri, is some 270,000 times more distant. (If the Sun and Earth were 1 foot apart, Proxima would lie 51 miles away.) Our relative closeness to the Sun is no accident – its radiation is necessary for our very existence. For that reason alone we would like to know all we can about our star.

There are other reasons to be interested in the Sun. One is that its outpouring of energy, magnetic fields, and particles varies in ways that have important consequences for Earth. Some effects of solar variability are spectacular, like the aurora borealis (northern lights). Others are of considerable practical concern, such as the interruptions of radio communications following great solar flares, or the probable effects solar activity has on weather and climate.

Astronomers have another reason to study the Sun. Many of the complex messages we read from observations of the distant universe can be deciphered through comparison with the much more detailed script provided by solar observations. For this reason the Sun has sometimes been described as "the Rosetta stone of astronomy."

THE SUN'S SOURCE OF ENERGY

The Sun is a run-of-the-mill star. Its mass, radius, surface temperature, and energy output (Table 1) lie in the middle of the ranges observed on other stars. Its chemical composition is also typical (Table 2). And there is nothing unusual about its age of around 4.5 billion years; stars this old abound in the Milky Way, as do much younger and much older stars. However, we know the Sun's vital statistics far better than those of any other star. Building on this detailed knowledge, and on our less certain knowledge for a host of other stars, astronomers some time ago developed a general understanding of stellar structure and evolution.

Figure 1 shows what astronomers think is going on inside the Sun. A very hot central core produces the Sun's energy by the fusion of hydrogen into helium. This energy is carried outward by radiation through the inner 70 percent of the Sun's radius and primarily by convection through the outer 30 percent. At the surface the energy is emitted as the sunlight that warms Earth.

The *luminosity* of a star – that is, the total amount of energy it produces every second – depends mainly on its mass and the requirement that forces in the interior (pressure and gravity) must balance out everywhere. The inward pressure at a star's core is simply the weight of the overlying mass. To keep the star from collapsing, the temperature in the core must be high enough to create an identical outward gas pressure. For stars like the Sun, gas pressure is proportional to temperature, and to achieve the required outward pressure, the temperature in the Sun's core must be about 15,000,000° K. In general, the hotter a star's core is, the greater the rate of nuclear fusion, which produces the star's luminosity. A star more massive than ours will have a higher

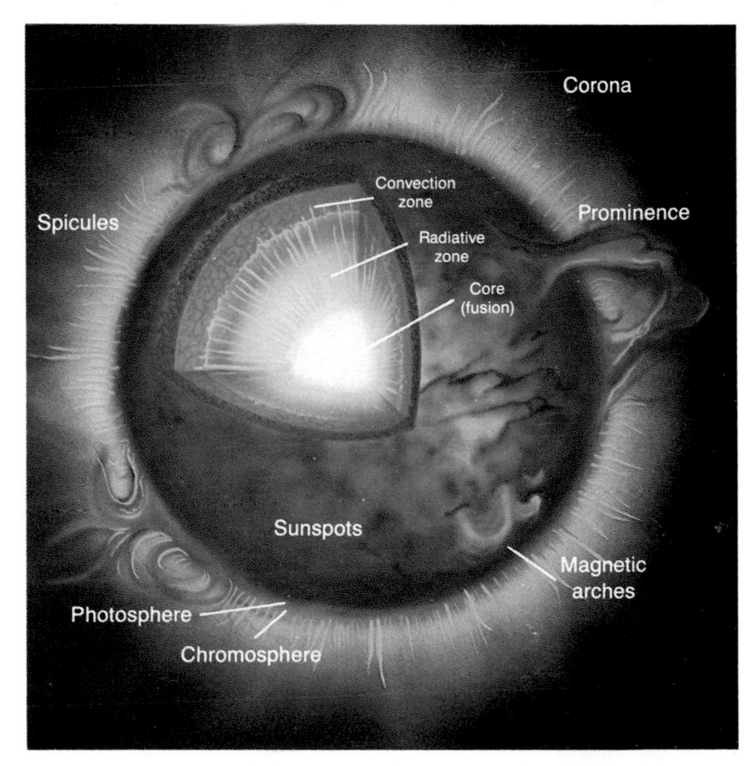

Figure 1. **A cross-section of the Sun's interior. Energy produced through fusion in the core is carried outward, first by a countless series of absorptions and emissions in the radiative zone and then by convection. Convective motions within the electrically conductive solar interior generate magnetic fields that express themselves at the surface as sunspots, prominences, and magnetic active regions.**

Mass	1.989×10^{33} g
Radius	695,000 km
Surface temperature	5,770° K
Luminosity	3.827×10^{33} ergs/sec
Age	4.5 billion years
Principal chemical constituents	
by number of atoms: Hydrogen	92.1 percent
Helium	7.8
Oxygen	0.061
Carbon	0.030
Nitrogen	0.0084
Neon	0.0076
Iron	0.0037
Silicon	0.0031
Magnesium	0.0024
Sulfur	0.0015
All others	0.0015

***Table 1.* Vital statistics of the Sun. Astronomers classify it as a *G*2 star, a rather average type found abundantly in the galaxy. By coincidence, the Sun's nearest stellar neighbor, the Alpha Centauri system, contains a brilliant *G*2 star as well.**

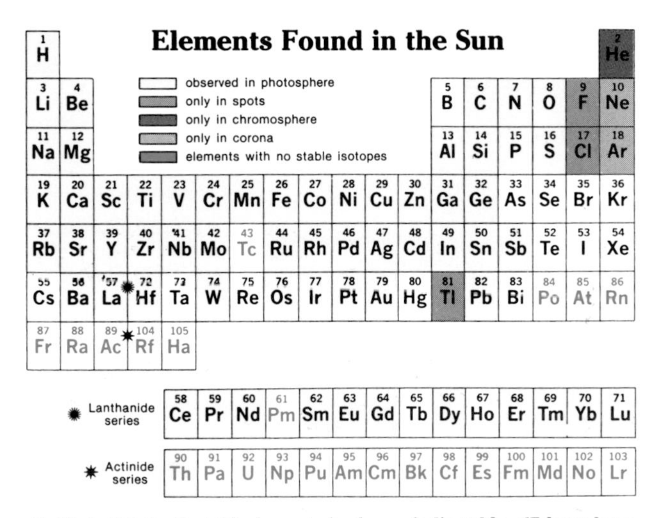

***Table 2.* Of the first 92 elements in the periodic table, 67 have been found in the Sun – and one of them (helium) was first discovered there.**

core pressure, temperature, nuclear-energy production rate, and luminosity. The most massive stars, with about 100 times the Sun's mass, are a million times more luminous than the Sun. Stars at the other extreme, with but a tenth the Sun's mass, are a thousand times less luminous.

The nuclear process that produces the Sun's energy is primarily the proton-proton reaction chain, in which four protons, or hydrogen nuclei, are successively fused into one alpha particle, or helium nucleus. However, one alpha particle is about 0.7 percent less massive than four protons, and this mass difference is converted into energy according to Einstein's famous equation $E = mc^2$. Every second the Sun's nuclear fire processes about 700 million tons of hydrogen into helium "ashes." In doing so 0.7 percent of this matter – 5 million tons – disappears as pure energy, and every second the Sun becomes that much lighter.

So far, fusion has consumed a significant fraction (some 50 percent) of the hydrogen originally in the Sun's core. Clearly, something dramatic can be expected to happen before another 4.5 billion years come to pass, and we shall return to this point at the end of the chapter. The most massive stars, which burn hydrogen a million times faster than the Sun does, exhaust their fuel in only a few million years. A solar system surrounding such a star would surely not have enough time to produce the gradual evolution of life that has occurred on Earth.

Because the Sun's core has by now converted a large fraction of its original protons into alpha particles (at a four-to-one exchange rate), the sum total of protons and alpha particles has decreased. But the gas pressure is proportional to the number density of particles, and the only way to keep the pressure high enough to balance the weight of overlying material has been for the core temperature to increase. This heating, in turn, has produced a noticeable increase in the Sun's luminosity, by about 40 percent since nuclear reactions started 4.5 billion years ago.

RADIATION AND CONVECTION

Another factor governs the rate of energy generation in a stellar core, and that is how quickly energy leaks to the surface through the cooler gases enveloping the core. To the extent that this gaseous envelope is a poorer or better insulator against leakage (just as in houses with varying amounts of insulation), then more or less energy needs to be produced in the central furnace to maintain the required internal temperature.

Deep in Sun's interior, heat generated by nuclear reactions is carried outward by countless emissions and reabsorptions of energy in a process known as *radiative diffusion*. The diffusion rate depends both on the temperature gradient and the *opacity* of matter to radiation. Opacity, or the resistance of matter to radiative energy flow, is rather like the insulating quality of a house's outer walls; doubling the insulation efficiency doubles the temperature difference (gradient) between inside and outside, for the same heat flow through the walls. To maintain a given heat loss to the outside, the house's interior must become hotter. In the same way, at depths within a star where the opacity is higher, there is a bigger drop in temperature outward across those layers. The temperature gradient adjusts itself so that the total amount of heat that is carried outward is the same at all depths (and is equal to the solar luminosity measured at the surface).

By about 70 percent of the way from the center to the surface, the temperature in the solar interior has declined from its 15,000,000° K value in the core to about 1,500,000°. At this relatively low temperature the opacity is very large indeed. As a result, the rate of temperature drop outward becomes very steep, so much so that a new and much more efficient energy-transport process, called *convection*, comes into play. In convection the heat is bodily carried outward by rising hot eddies. For these to develop the temperature gradient must be so steep that an eddy will remain hotter than its surroundings as it rises and cools by expansion. A region with such a steep temperature gradient is termed convectively unstable. In the outer 30 percent of the Sun's radius, known as the *convection zone*, convective instabilities produce a state of continuous agitation. The convective motions are so efficient that they carry essentially all the solar energy outward to the surface, where it radiates to space.

Convective motion has a very important effect that will be discussed later: the seemingly chaotic pattern of rising and falling gases somehow causes the ordered patterns of solar magnetism, which in turn produce sunspots, flares, and a host of other fascinating features on and above the solar surface.

PROBLEMS WITH NEUTRINOS

Armed with appropriate equations describing all of the above physical effects – pressure equilibrium throughout the interior, nuclear-energy generation, radiative diffusion, opacity, and convection – theorists have constructed numerical models of the solar interior. They have also examined how the interior must have evolved over the aeons, as the core gradually burned hydrogen into helium and produced changes in the interior's structure, radius, and luminosity. Figure 2 shows details of the solar interior, as we understand them today.

About 25 years ago a group of experimenters set out to check current theory by direct observation, using an extraordinary instrument that can directly "see" the results of nuclear interactions in the solar core. The detector, a huge tank filled with 100,000 gallons of ordinary dry-cleaning fluid (perchloroethylene or C_2Cl_4), is located more than 1 km beneath the surface of the Earth in a mine at Homestake, South Dakota (Figure 3). This strange detector is designed to capture solar neutrinos, a type of elementary particle created when particles such as protons or neutrons interact.

Neutrinos have surprising properties. For example, they appear to have no mass, just like the photons that constitute ordinary light. Being massless, neutrinos once created cannot sit still; they travel through space at the speed of light. Unlike photons, however, neutrinos undergo almost no interactions with other forms of matter. Thus a neutrino, born in the solar core during the proton-proton reaction chain, will zip out of the Sun with almost no likelihood of interacting with any of the huge number of overlying atoms. This property is what makes it possible to peer directly into the solar core, for any neutrinos detected here will have originated directly in the core a scant 8 minutes earlier (light's travel time from the Sun to Earth).

Of course, if the Sun is nearly transparent to neutrinos, so is any device on Earth. From the predicted neutrino-creation rate in the solar core, one can calculate that about 10^{21} solar neutrinos should pass through the Homestake detector every day – but that only six of them interact with the fluid inside the tank! Because the detection rate is so small, all sources of spurious signals, such as cosmic rays, must be eliminated. Thus, to obtain adequate shielding, the tank was placed far beneath the surface of the Earth. (The neutrinos, of course, penetrate solid ground virtually unimpeded; indeed, they reach the tank equally well whether they pass through 1 km of Earth at midday or 13,000 km of it at midnight.)

The perchloroethylene in the Homestake detector consists largely of chlorine atoms, and every day six of them interact with neutrinos and in doing so are converted to atoms of radioactive argon. Half of the argon so formed will decay back to ordinary chlorine in about 35 days. If the number of solar neutrinos reaching the tank per day remains constant, after about 100 days the combination of creation and decay should reach a balance with about 60 radioactive argon atoms in the tank. Finding and counting these stragglers out of a total of some 10^{31} atoms is a truly extraordinary task, involving extremely sensitive radiochemical analysis.

The results are surprising. The experimenters find only

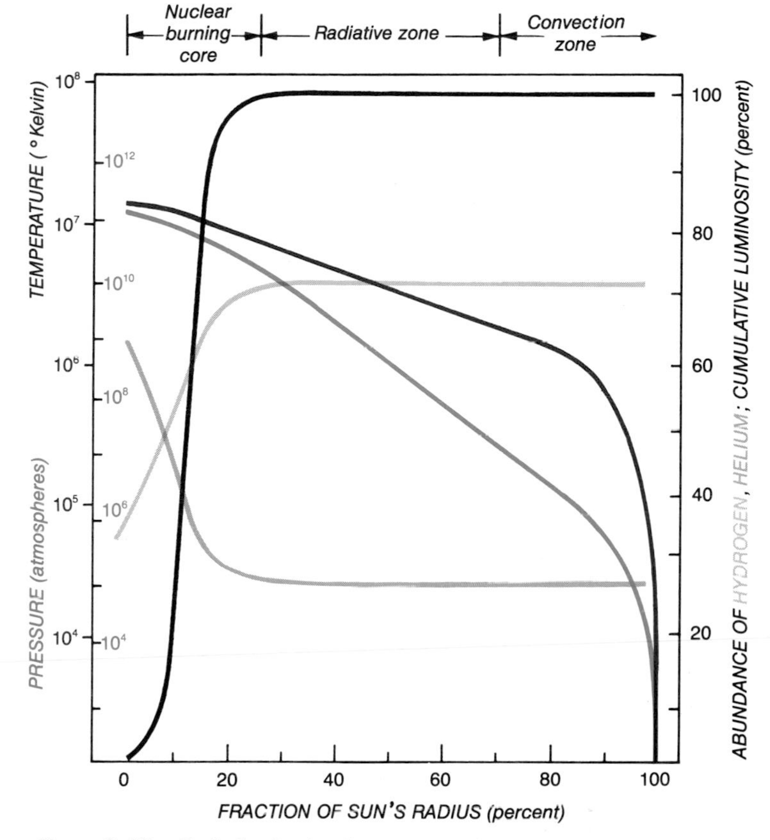

Figure 2. The Sun's luminosity (energy output), temperature, pressure, and hydrogen-to-helium ratio all vary with depth in its interior. Nearly all the energy that powers the Sun is produced in its core, which takes up only about 1.5 percent of our star's volume.

Figure 3. The underground neutrino detector in the Homestake mine at Lead, South Dakota, contains 100,000 gallons of perchloroethylene. Neutrinos from the solar core convert a few of the liquid's chlorine atoms into radioactive argon, which can be extracted and measured.

about one-third as much radioactive argon as expected from the Sun's predicted neutrino flux. This so-called "neutrino problem" has plagued theorists and experimenters for many years. There seems to be something basically wrong, either with our understanding of stellar structure or the physics that governs the creation and decay of neutrinos.

However, there *is* some room for error in stellar-structure theory. The Homestake detector can record only neutrinos with very high energy, the ones produced by a high-temperature version of the proton-proton nuclear-burning process. Thus, a slightly drop in the core's temperature would reduce the rate of neutrino formation. The core will, in fact, be a little cooler than theory predicts if we have underestimated the number density of hydrogen atoms by overestimating the initial abundance of helium. Alternatively, we could invoke a mixing process by which fresh, unburned hydrogen is swept into the core while the helium "ashes" are partially removed, thereby allowing the core to produce the observed luminosity but remain slightly cooler.

There are other, more exotic, ways to resolve the neutrino problem. For example, a theoretically predicted new class of *w*eakly *i*nteracting *m*assive *p*articles (WIMPs) possibly formed during the expansion of the early universe and may have been swept up by the Sun over time and fallen into its core. Such particles, able to travel large distances between interactions with other matter, would efficiently conduct heat out of the core and enable it to burn at a slightly lower temperature. Of course, the problem may lie in the physics of neutrinos themselves. It has been theorized that as the neutrinos travel outward through the massive body of the Sun they undergo a resonance with electrons. In doing so, a fraction of them are converted into a kind of neutrino to which the Homestake detector is insensitive.

Whatever the solution to the problem, the search for it has excited astrophysicists for many years. The answer may come from new detectors sensitive to the dominant, lower-energy solar neutrinos produced by the proton-proton reaction. It may come from the new technique of helioseismology, discussed later in this chapter. It may come from particle physics, as the properties of neutrinos and their interactions with matter are explored further. Or it may come from cosmology and the intercession of WIMPs.

CONVECTION AT THE SOLAR SURFACE: GRANULATION AND SUPERGRANULATION

Although we cannot see directly into the interior of the Sun to test our theories of solar structure, we can study our star's surface in great detail. A cursory view of it through a large telescope (Figure 4) shows a bright disk mottled, perhaps, by a few dark *sunspots* (Figure 5). A greatly magnified view of a particularly sharp image shows bright individual grains, called *granules*, surrounded by dark lanes (Figure 6). These granules are the tops of convective elements that bring heat to the surface from as deep as 30 percent of the way to the Sun's center.

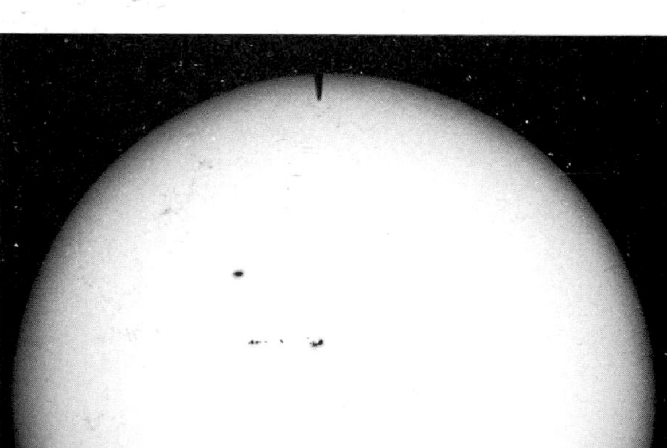

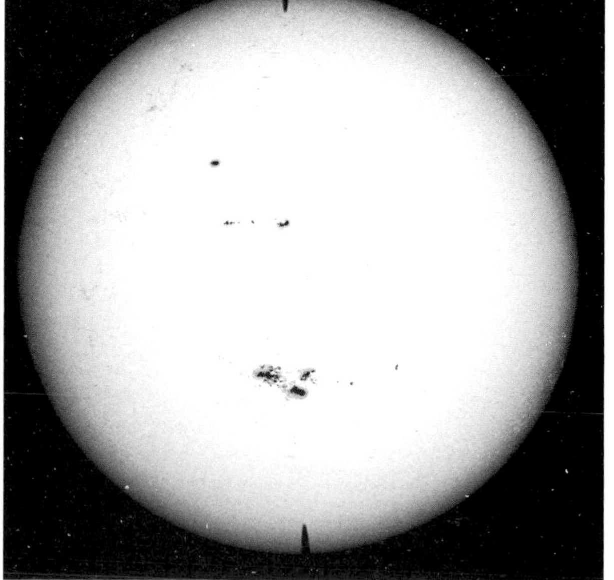

Figure 4. **On July 1, 1988, a huge naked-eye group of sunspots dominated the Sun's southern hemisphere. They appeared during the most rapid buildup of solar activity on record.**

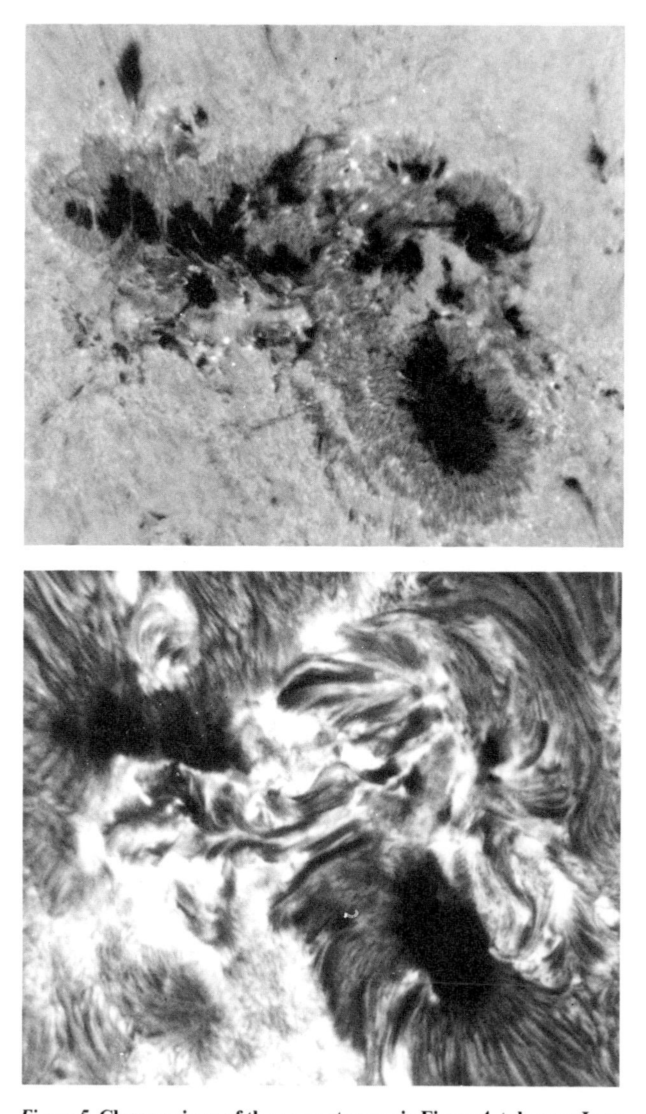

Figure 5. **Closeup views of the sunspot group in Figure 4, taken on June 28, 1988. The upper photograph approximates a white-light view. The corresponding one below it, taken in the hydrogen-alpha emission at 6563 angstroms, shows how material in the Sun's lower atmosphere traces the magnetic field lines emerging from this active region.**

Motion pictures show that granules change their appearance rapidly; they form and reform in a few minutes, even though they are typically as large as Texas. With a spectrograph we can measure their vertical velocities (as Doppler shifts), and we find that the changing pattern is accompanied by violent motions. The granules erupt upward with a speed of perhaps 500 meters per second (about 1,800 km per hour), after which their material overturns and flows downward at a comparable speed into the dark intergranular lanes. When the motion picture is sped up, the changing pattern looks very similar to a pot of bubbling oatmeal. Indeed, the analogy is not bad – in oatmeal, as in the solar convection zone, heat is supplied from below and bodily carried upward by the process of convection.

Images as sharp as those in Figure 6 are obtained only very rarely from the ground because each granule appears only about 1 arc second across, which is near the limit of photographic resolution due to blurring by Earth's atmosphere. This situation is frustrating to solar physicists, for the important information about how convection works lies just beyond what can be readily observed. In 1985 an experiment aboard the Spacelab 2 mission (Figure 7) on the Space Shuttle made a major advance by photographing the granulation pattern in the absence of atmospheric blurring, even if only for a few tens of minutes. Now there are ambitious plans to place a long-lived solar laboratory in space to study granulation and its interaction with magnetic fields.

The characteristic size of a granule, about 1,000 km across, is due to a region of extreme convective instability that extends about the same distance beneath the solar surface. There is also a larger-scale pattern of convection, termed *supergranulation*, that is probably due to a less extreme instability lying at a depth of several tens of thousands of kilometers. Roughly 30,000 km across, these supergranules are not at all easy to see because they carry very little thermal energy upward and thus do not appear noticeably brighter than their surroundings. However, they are quite visible in maps that show variations in radial velocity from point to point on the solar surface (Figure 8).

As described later, convection near the surface plays a key role in moving around the magnetic fields that penetrate the solar surface and, thereby, produce magnetic heating of the overlying atmosphere. Deeper down, the interaction of convective motions with rotation probably causes weak magnetic fields to become amplified, creating strong fields that rise to the surface. Although we cannot presently detect the deep-lying magnetic fields, we can easily see their manifestations at the surface; these are the dark sunspots and the magnetic *active regions* that accompany them.

SUNSPOTS, SOLAR ROTATION, AND THE ORIGINS OF SOLAR MAGNETISM

Sunspots are the best known solar features. On occasion they can be glimpsed with the naked eye, as ancient Chinese records attest. But they became much more intensively studied after Galileo turned a telescope on them in 1610. Individual spots live for only a few weeks, and their rate of appearance is approximately cyclical. The so-called *sunspot cycle* (Figure 9) is based on sunspot numbers maintained systematically since the mid-19th century and on historical reconstructions before that. While sunspot numbers rise and fall every 11 years or so, the periodicity is not regular nor are the amplitudes of different cycles consistent.

Of particular interest is the interval roughly from 1640 to 1710, when sunspots and their cycle apparently all but

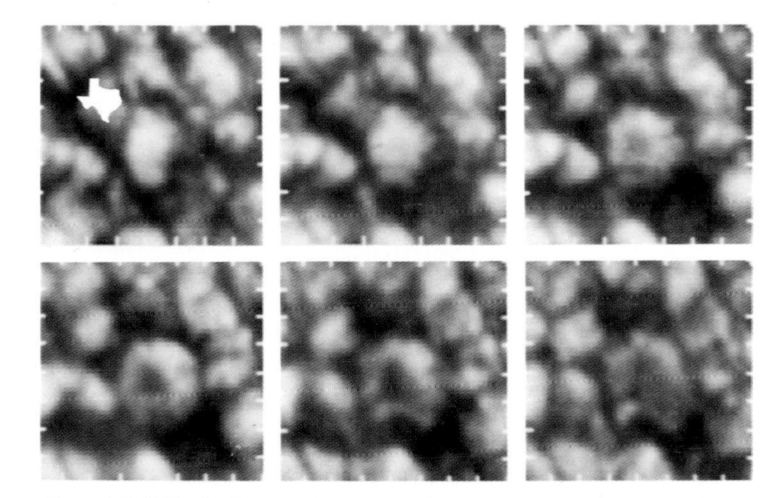

Figure 6. **Individual solar granules are about the size of Texas (inset). Many of them expand outward rapidly at the end of their lives, as is evident in this sequence of images taken 2 minutes apart by the SOUP instrument of Spacelab 2 (see Figure 7). Each field is 8 arc seconds across.**

Figure 7. **Spacelab 2, which flew aboard the Space Shuttle in the summer of 1985, carried an independently pointed platform onto which four instruments were mounted. They are (clockwise from upper right) a helium abundance experiment, an ultraviolet irradiance monitor, a high-resolution ultraviolet telescope, and a solar optical universal polarimeter (SOUP). The last of these acquired the images shown in Figure 6.**

Figure 8. **A "Doppler image" of the Sun, in which light-colored areas denote matter moving toward the Earth and dark ones matter that is moving away. The pock-marked pattern indicates that the surface is being pushed sideways wherever the tops of supergranulation cells meet it. Because this motion is primarily horizontal, it is most evident near the Sun's limb but does not show up as a line-of-sight velocity near the center of the disk.**

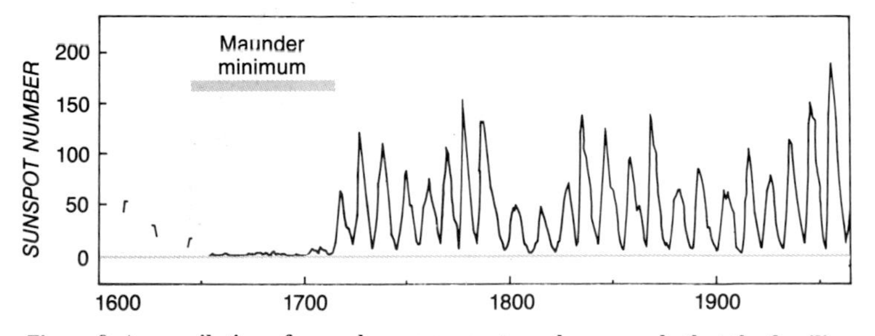

Figure 9. **A compilation of annual mean sunspot numbers reveals that the familiar 11-year sunspot cycle was absent throughout much of the 17th century. This dearth of sunspots has become known as the Maunder Minimum.**

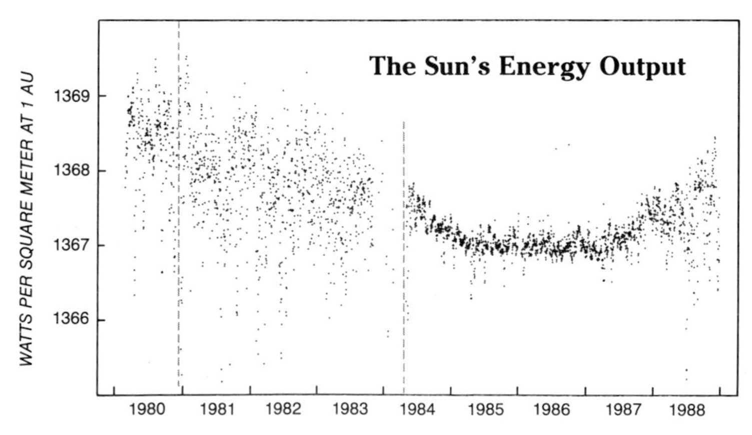

Figure 10. **Since early 1980, a radiometer on the Solar Maximum Mission satellite has provided a nearly continuous record of the Sun's brightness at all wavelengths. But this so-called "solar constant," plotted here, actually varies somewhat, depending on the number of sunspots and other active regions present on the Sun's disk. Between the vertical dashed lines, some of the data's scatter is due to instrumental trouble.**

disappeared. This interval, now known as the Maunder Minimum, has excited considerable interest among solar physicists, for it clearly poses a problem: how to explain both the regularity of the sunspot cycle and, at the same time, its large irregularities. The Maunder Minimum has also attracted the interest of climatologists, for it coincided rather closely with a period of colder-than-average temperatures in northern Europe, sometimes called the "little ice age." There is considerable debate whether the timing of the prolonged cold wave and the Maunder Minimum was a coincidence, or whether it represents the Sun's direct effect on Earth's climate.

One thing is certain, however: the Sun's output does change. Observations from NASA's Solar Maximum Mission satellite have revealed that slight, irregular variations occur on a time scale of days. Also, the spacecraft data show that between 1980 and 1986, during the decline of the last solar cycle, the Sun's energy output gradually fell about 0.04 percent; in 1986, when a new magnetic cycle began, the output began to rise again (Figure 10). Some scientists argue that such changes, if sustained over a long period of time, could drastically affect our climate by shifting normal weather patterns.

Sunspots exhibit other regular patterns. One of the clearest is revealed by the so-called "butterfly diagram" (Figure 11), which shows the solar latitude at which spots appear throughout successive cycles. It demonstrates that at the start of a new cycle, after a minimum in the number of spots, the new spots begin to form at high latitudes. Then, as the cycle progresses, spots gradually form closer and closer to the solar equator. This regular variation must strongly hint at the nature of sunspots, if we could only decipher it.

The best clues come from detailed studies of the spots themselves. We know that sunspots appear dark only because they are relatively cool compared to their surroundings – about 3,800° K, compared to the ambient photospheric temperature of about 5,800° K. Thus, in the visible spectrum, sunspots radiate only about 20 percent as much light per unit area as does the photosphere. Nevertheless, their radiation is significant: if somehow placed alone in the nighttime sky, a large spot would appear as a reddish star 10 times brighter than the full Moon.

The reason sunspots are dark is undoubtedly connected with their strong magnetic field, first measured by George Ellery Hale in 1908 through application of what is called the Zeeman effect, which causes a splitting of spectral lines. Solar gases have rather high electrical conductivity (about equivalent to that of copper) because they are very hot and thus ionized. This means that the Sun's magnetic field lines are "bonded" to the conducting gas: the field lines move with the gas, and the gas can flow along the field lines but not across them. In sunspots, the vertical magnetic fields are so strong (some 3,000 gauss) that they prevent any horizontal gas motions and thus inhibit the overturning motions of convection. This greatly reduces the energy that can be brought upward within the sunspot and makes the surface of the spot much cooler than a region free of magnetic fields.

The Zeeman effect allows solar astronomers to create maps or *magnetograms* of not only the strength of the magnetic field from place to place but also its polarity. Magnetic fields are sprinkled all over the Sun (Figure 12) but are concentrated in the belts where sunspots occur and are particularly strong in and near sunspots.

Sunspots tend to form in pairs. By measuring the polarity of each one, Hale discovered some important relations that provide a real key to understanding the nature and origin of solar magnetic fields. First, the spots of each pair tend to have opposite polarities (Figure 13). This is also the case for the associated magnetic fields (Figure 12) that accompany active regions. Second, during a given 11-year cycle, all the spot pairs in the northern hemisphere have the same polarity configuration: the leading spot, in the sense of solar rotation, is positive while the trailing spot is negative. In the southern hemisphere, the polarity of paired spots is exactly reversed. However, Hale found that during the subsequent cycle, all the polarities switch. This means that the 11-year sunspot cycle is actually part of a 22-year magnetic cycle.

With so many clues to the nature of sunspots and solar magnetism, it would seem a simple matter for a good detective to solve the mystery. However, the Sun is so complex that even these clues have not given us more than some helpful hints as to where the answer lies. It is thought that magnetic fields are created beneath the solar surface by the combined actions of convection and *differential rotation*. This expression refers to the fact that, unlike the Earth, the gaseous Sun does not rotate rigidly. Its equatorial regions rotate considerably faster than do higher latitudes: the rotation period at the equator is about 25.4 days, while near the poles it is as long as 36 days. This behavior has been extensively studied by measuring the motions of sunspots and the line-of-sight motion near the solar limb.

The differential rotation of the solar surface is large. If Earth had comparable shears, huge earthquakes would split its crust apart in a small fraction of one turn. Of course, the gaseous Sun can accommodate such shears, but nevertheless they have major effects on the magnetic fields that permeate the interior, causing them to stretch and in so doing to become stronger. This process appears fundamental to the functioning of the 11-year solar-activity cycle.

To understand the role differential rotation plays in generating magnetic fields, we need to know how it varies not just with latitude on the surface but at various depths in the interior, where the magnetic fields are generated. It has often been assumed that the Sun's interior rotates rather faster than its surface layers do. Such an assumption led to a qualitative picture, first suggested by Eugene Parker in 1950 and then amplified by Horace Babcock in 1961, of what may be happening. They envisioned a solar *dynamo*, whereby the combined actions of differential rotation and convection strengthen weak, deep-lying magnetic fields and bring them to the surface (Figure 14). There they appear as sunspot pairs and bipolar active regions, both of which obey Hale's polarity relation. These surface fields later subside, but the process leaves behind a weak remnant with a polarity opposite the original one – the "seed" field for a new 11-year cycle.

The dynamo theory explains both the Hale polarity laws and the butterfly diagram. However, it does not explain why the solar cycle departs from the clockwork regularity that is the norm rather than the exception in astronomical phenomena. Nor does it explain prolonged periods of nearly nonexistent solar activity, such as apparently occurred during the Maunder Minimum.

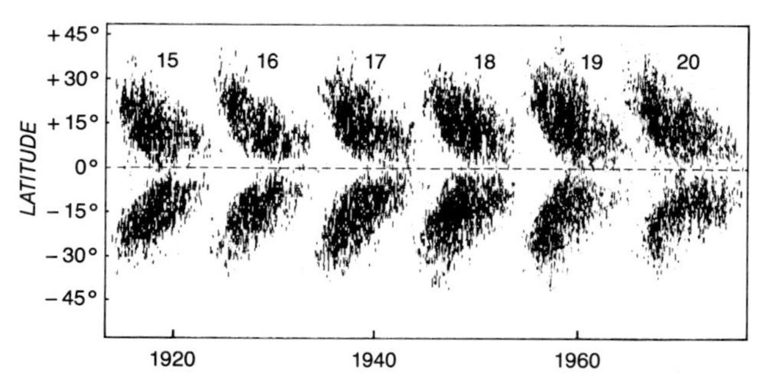

Figure 11. **The migration of sunspots from high latitudes at the beginning of a cycle to near the Sun's equator at its end is evident in what solar scientists call a "butterfly" diagram.**

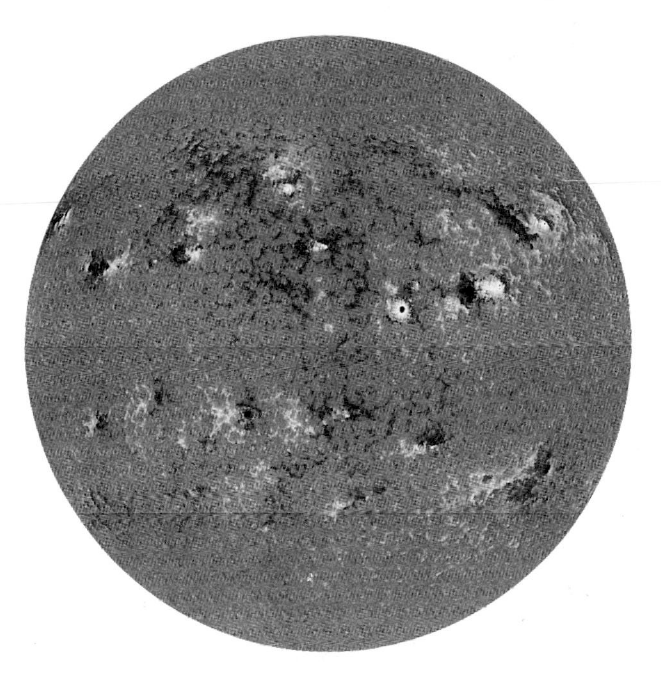

Figure 12. **A map of the strength and polarity of solar magnetic fields, acquired in May 1979 near the maximum of magnetic activity. Note how darker areas (locations of positive magnetic polarity) are frequently paired with lighter ones (negative polarity).**

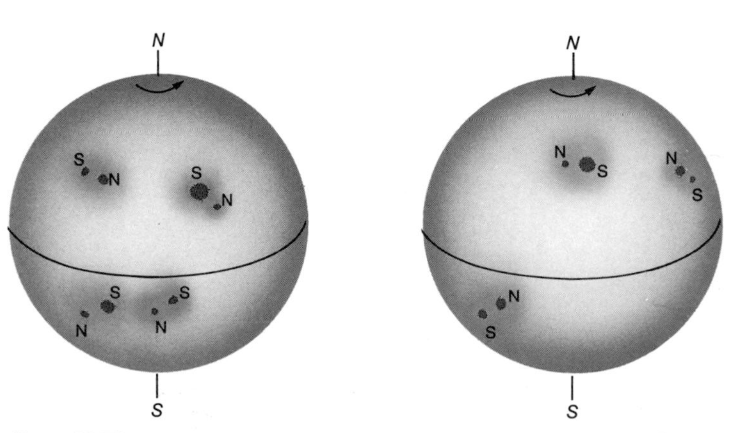

Figure 13. **The magnetic polarities of leading (right) and trailing (left) spots switch between the Sun's northern and southern hemispheres. Furthermore, all the spots' polarities assume the opposite sign between one 11-year cycle and the next, accompanied by a reversal of the Sun's general magnetic field.**

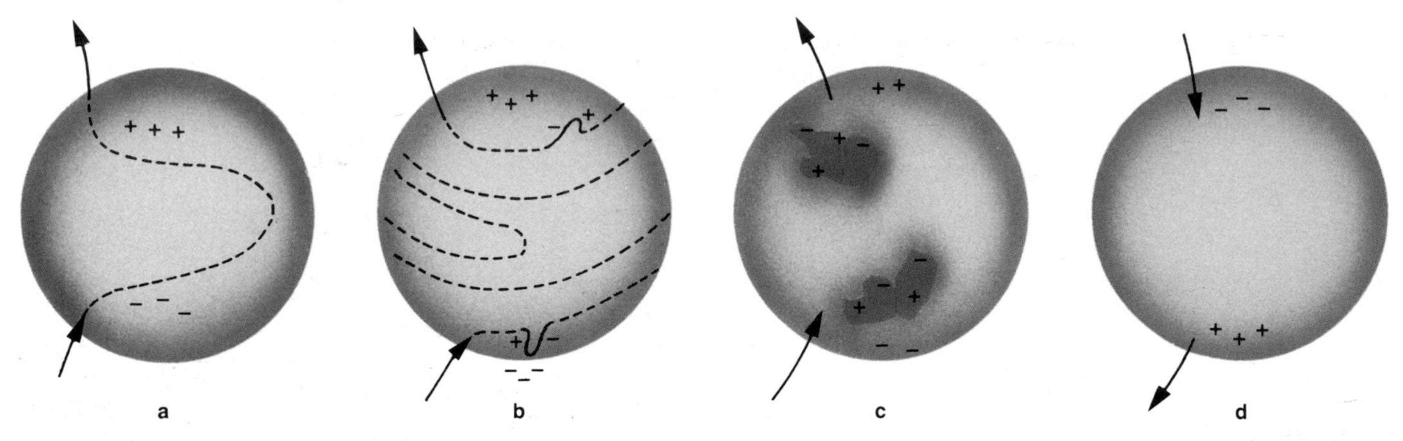

Figure 14. **Babcock's model of the sunspot cycle. As an 11-year cycle begins *(a)*, magnetic field lines run primarily from south to north, but differential rotation begins to stretch them beneath the faster-spinning equatorial region. In time the stretching wraps the lines several times around the Sun *(b)*, causing them to intertwine and intensify, and ultimately to rise locally as buoyant loops. Each rising loop spawns an active region once it breaks the surface *(c)*, creating bipolar active-region groups that obey Hale's polarity laws. As the cycle nears its end *(d)*, the leading regions drift toward the equator, where their opposite polarities mix and cancel; the trailing regions drift poleward, where their polarity cancels and replaces the existing fields of opposite polarity. The entire sequence then repeats, except that all polarities have been reversed. Thus, a complete magnetic cycle takes 22 years to complete.**

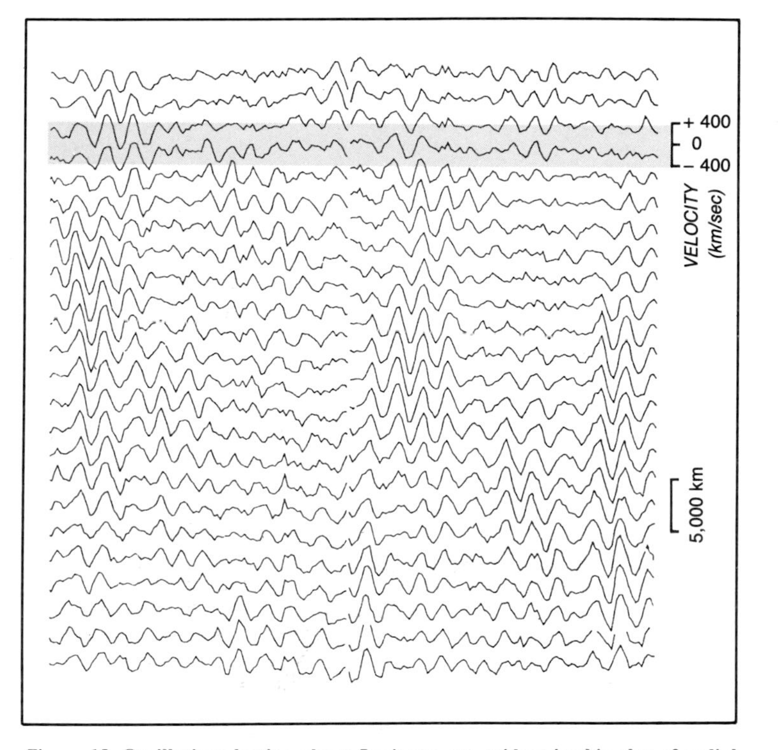

Figure 15. **Oscillations lasting about 5 minutes are evident in this plot of radial motion along 80,000 km of the solar surface. These motions reflect the superposition of the Sun's 10 million different acoustical resonances (see Figure 16). The modes go in and out of phase with one another at a given location, creating the wavelike swells; the in-and-out motion typically involves radial velocities of 400 to 500 m per second.**

There are other reasons for discomfort with the dynamo picture. One is our inability to pinpoint the exact depth in the Sun where the dynamo operates. If the magnetic fields are generated in the convection zone proper, they should rise quickly to the surface, because magnetic pressure forces inflate their surroundings like a balloon. Convection should carry magnetized regions to the surface too rapidly for the fields to become amplified to the strengths we observe. A possible way out of this predicament is that the fields may be generated just below the convection zone in the stable radiative zone (Figure 1). From there they may leak out only gradually before popping to the surface.

Another concern is that we do not really know the exact variation of differential rotation throughout the convection zone, even though this variation is fundamental to the dynamo picture. What is needed is a direct probe of the deep interior. Fortunately, a new technique has appeared that is now allowing us to measure the Sun's internal structure and motions more or less directly.

HELIOSEISMOLOGY

Helioseismology resembles terrestrial seismology in that a myriad of waves at the Sun's surface are analyzed to learn what is happening in the deep interior. Of course, solar "seismic" waves differ from their terrestrial analogues in that there are no "sunquakes" to set them off. The waves in the Sun appear to be generated continuously throughout the convection zone, probably by the never-ending turbulent motions that occur there. Put another way, the Sun rings like a bell, but not one struck by a clapper; rather it hums steadily as if being continually struck by tiny grains in a sandstorm. These acoustic waves are the same as ordinary sound waves, except for their long period. If sound could travel across the vacuum of interplanetary space, and if our ears were attuned to very low frequencies, we would hear a continuous rumbling note at a pitch some 16 octaves below the lowest note on a piano keyboard.

On the solar surface the waves appear as up-and-down oscillations of gases with a period near 5 minutes. Their maximum line-of-sight velocities are about 0.4 km per second, enough to be readily detectable as Doppler shifts by modern spectrographs (Figure 15). These waves, however, are actually made up of millions of superimposed oscillating patterns, each pulsing at only a fraction of a meter per second. Each pattern represents a different *mode* of oscillation of the Sun as a whole (Figure 16). The precise values of the oscillation periods vary from mode to mode, and it is from measurements of many different periods that solar physicists are able to extract information about the solar interior.

A mode's period depends on the physical characteristics within a corresponding "resonant cavity," a region of the solar interior within which the acoustic wave is confined by the laws of propagation. The top of the cavity is the solar surface itself, where the upward-moving waves are reflected

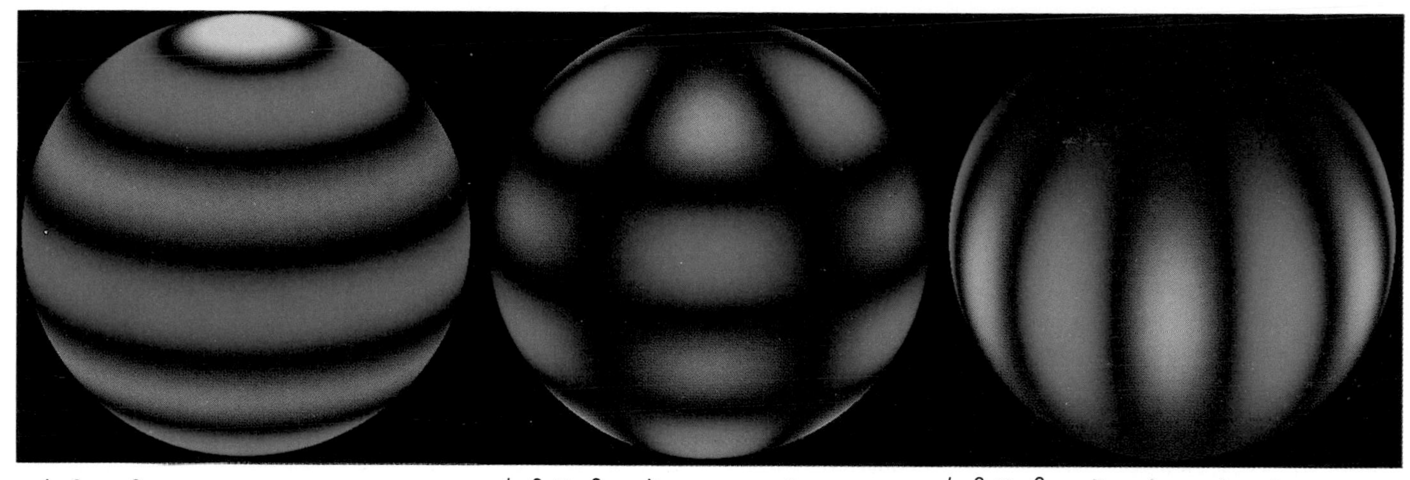

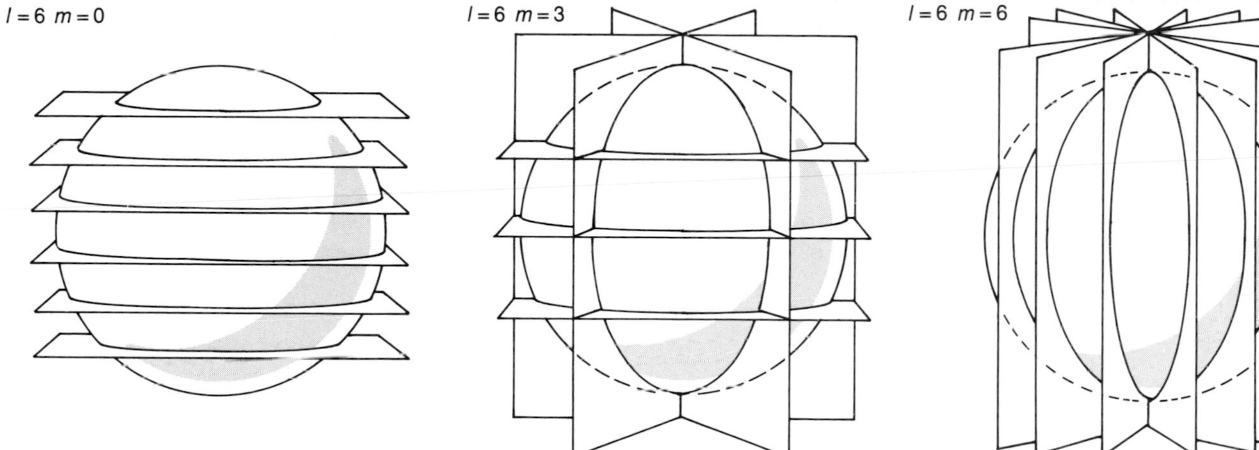

Figure 16. **Solar oscillations can be very complex but are more easily understood when classified by their *l* and *m* values, which represent the number of nodes (stationary points) belonging to each oscillation. A vibrating violin string, for example, has at least two nodes (one at each end) and usually others. The *l* parameter denotes the total number of planes slicing through the Sun, with *m* denoting how many of those pass through longitudinally. The computer-generated globes show how the solar surface would move if it were oscillating according to the accompanying diagrams. Dark regions correspond to the nodal boundaries; green denotes areas moving radially outward and yellow those moving inward.**

downward by the very steep drop-off in density there. The bottom of the cavity differs for various modes, lying deeper for those that propagate nearly straight into the interior and shallower for ones that move more nearly horizontally. We can tell the direction of propagation by looking at the oscillation pattern on the solar surface: nearly horizontal waves have a smaller distance between their wavefronts than those aligned more nearly vertically.

Oscillations have been seen with a horizontal wavelength as short as a few thousand kilometers, and such waves are confined to the very outer parts of Sun near the top of the convection zone. At the other extreme are modes with an effectively infinite horizontal wavelength – the whole Sun oscillates in and out like a pulsating balloon. Such waves propagate straight downward and penetrate to the very center of our star.

Many musical instruments, such as a pipe organ or clarinet, produce their characteristic tones through similar resonant cavities. As with musical instruments, the biggest solar cavities have the lowest pitches resonating within them. Also like musical instruments, each solar cavity can support a number of different overtones – waves with successively shorter wavelength (higher frequency) – such that two or more wavelengths can fit exactly within the cavity (Figure 17).

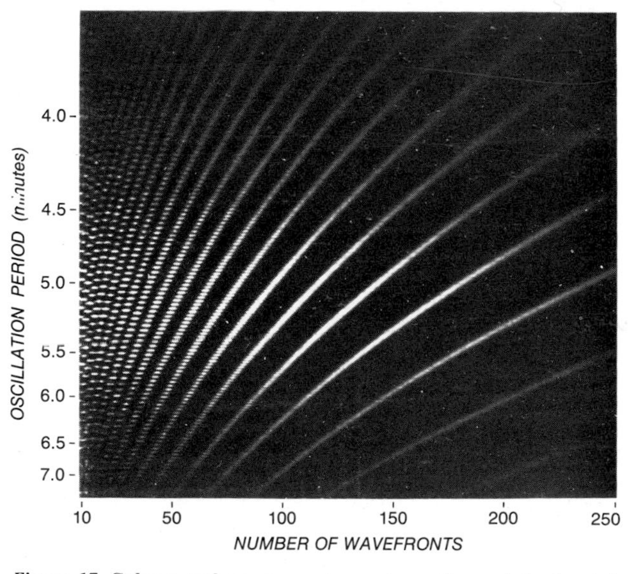

Figure 17. **Solar sound waves create overtones for each horizontal wavelength. In this diagram, the distance between wavefronts seen at the surface – their horizontal wavelength – is plotted on the horizontal axis. Bright spots occur where strong oscillations are observed, and each "ridge" marks a different overtone of the oscillation. The sloping of each ridge toward lower left illustrates the correspondingly lower frequencies associated with deeper cavities in the Sun.**

By studying the precise frequencies of many modes of solar oscillation, we can determine the average speed of sound within resonant cavities, ranging all the way from shallow ones confined near the surface to the entire volume of the Sun. Helioseismology uses that information to determine how the sound speed varies from point to point in the solar interior. And, because the sound speed depends on a number of fundamental physical quantities, we may infer how they too vary with depth in the Sun.

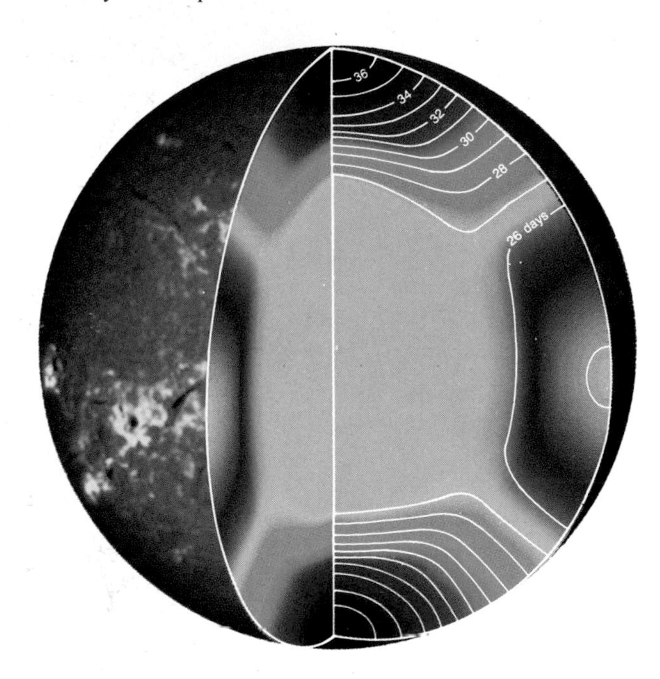

Figure 18. **Until recently, the rate of rotation of the Sun's interior could only be inferred from theoretical models. But by using helioseismology, astronomers have probed how our star spins inside. Most surprising is that the rotation of the surface, which varies from 25 days at the equator to 36 days at the poles, persists inward for at least 200,000 km. Deeper down, below the convective zone, everything appears to rotate with a period of 27 days – though the spin rate of the core itself remains undetermined.**

Figure 19. **The Sun's corona, or atmosphere, is usually kept from our view by the blinding light of the Sun itself. However, it is actually quite luminous and presents a stunning appearance during total solar eclipses. Kazuo Shiota recorded this photograph from Kenya during the unusually long eclipse of June 30, 1973. The polar brushes and long sweeping equatorial streamers are typical features when solar activity is low.**

The two most important quantities governing the sound speed are the temperature and the composition of the solar gases. Higher temperature translates into greater speed, and a higher helium-to-hydrogen ratio translates into lower speed. Applying these relations to observations, helioseismologists determined early on that the depth of the solar convection zone is rather greater than had been previously supposed. They also deduced that the Sun must have formed with about 25 percent helium by mass. This abundance, higher than some other estimates, is comforting because it is close to that expected from theories of element synthesis in the early universe. On the other hand, the increased proportion of helium causes the temperature in the solar core to be so high that neutrinos are produced at a rate some three times greater than observed. Therefore, the first results from this powerful analytical technique have not resolved the neutrino problem.

Another application of helioseismology relates to the internal rotation of the Sun. The effective speed of sound increases if a wave is propagating in the direction of rotation, simply because the velocity of the solar material is added to that of the wave. Just the reverse happens to waves propagating opposite to the direction of rotation. From the difference between the frequencies of the two waves, we can extract the average rotation rate over the resonant cavity.

Before the advent of helioseismology, it had been speculated that the Sun's deep interior may be rotating much faster than its surface, as a relic of the rapid spin our star was born with. (Stars much younger than our present Sun take only a few days to rotate.) An increase of rotation inward would also conform with the standard picture of the dynamo discussed earlier. Recently, however, helioseismologists have found that the convection zone appears to be rotating at a rate rather similar to that of the surface layers (Figure 18). This finding is important, both for calculations of how convection interacts with the mean rotation of the Sun to yield differential rotation, and also as a step toward the ultimate understanding of the solar dynamo. Solar physicists are optimistic that helioseismology will eventually lead to a detailed understanding of how rotation and the magnetic dynamo work.

THE SOLAR ATMOSPHERE: CHROMOSPHERE, CORONA, AND SOLAR WIND

At the moment the Sun's bright disk disappears behind the Moon during a total solar eclipse, there is a flash of ruby-colored light above the lunar limb. This is the *chromosphere*, so named because of its deep red color. Because it is much fainter than the Sun's disk, it cannot be seen until the Moon blocks out the direct sunlight. After a few more seconds the chromosphere also disappears, and the still-fainter pearly white structure of the *corona* is revealed, stretching millions of kilometers into space (Figure 19). These tenuous outer regions of the solar atmosphere have an importance for astronomy far exceeding their own contribution to the bulk of the Sun. After all, the mass of the solar chromosphere and corona make up only a few trillionths of our star's total mass. However, they have many interesting and important properties.

The chromosphere and corona are much hotter than the 6,000° K photosphere. They are heated from below, by the interaction of the turbulent motions and magnetic fields in the photosphere, in a way not yet fully understood. However, the effects of heating are very clear: the temperature rises from its minimum of about 4,500° K at the base of the chromosphere to about 8,500° K at its top.

The chromosphere's color is due to hydrogen's strong red-light emission at 6563 angstroms, the hydrogen-alpha wavelength. Above about 8,500° K hydrogen becomes ionized and no longer emits much energy. Not only does the red light disappear above this temperature, but the gas can no longer cool itself efficiently. Thus the temperature rises very abruptly to more than 1,000,000° K over a very short height interval, called the *transition zone*. Above it the hot corona stretches far into interplanetary space. Its eerie glow, seen during an eclipse, results from sunlight scattered by electrons that have been stripped from their parent atoms because of corona's high temperature.

Why is the corona so hot? This question has dogged solar researchers for decades. At one time it was thought the heating was mainly due to shock waves generated by the same sound waves that give rise to the solar oscillations. However, we now know that the corona is riddled with magnetic fields, which appear to play a dominant role in the heating process. The fields extend upward from the photosphere and expand with height. Because the corona is so rarified, its pressure is not great enough to withstand the magnetic fields, which force the gas to conform to their structure. There is a great deal of small-scale magnetic structure in the corona, shown not only in white-light photographs taken during solar eclipses but also in remarkable images at X-ray wavelengths (Figure 20).

When the photospheric magnetic fields are jiggled, the perturbations must propagate into the corona along the field lines, like waves on a taut string. The coronal gases, bonded to the field lines, are shaken by these motions. Although the details are not yet fully understood, it appears that such "magnetohydrodynamic waves" are a prime source of coronal heating. This idea explains why the hot coronal structures seen in Figure 20 outline magnetic loops.

Considerable free energy can be stored in magnetic fields. In this sense "free" means energy above what would exist in the absence of distortions of the field caused by electric currents or other forces acting on the magnetized atmosphere. This energy can, in principle, be released abruptly if the fields relax suddenly to a lower energy state. A simple analogy is a twisted rubber band, which if twisted too much will suddenly develop a kink and reduce its total twist. In the case of solar magnetic fields, such an energy release is thought to explain the violent explosions in the solar atmosphere known as *solar flares*.

In a few seconds a large flare releases tremendous amounts of energy into the solar atmosphere – equivalent to more than a billion one-megaton thermonuclear explosions. Some of this energy rains down on the photosphere, causing heating and a host of dynamical effects. It may eject substantial amounts of material (as much as 10 billion tons) outward into interplanetary space (Figure 21) where many hours later it may interact with Earth and its magnetosphere. Some of the energy may accelerate electrons, protons, and other atomic nuclei to very high (relativistic) velocities. The electrons, for example, are ejected at speeds up to one-third the speed of light, emitting radio bursts as they interact with other electrons in the corona.

A solar flare is truly an impressive event, one that is important for many aspects of life on Earth. When its high-energy particles impact Earth's magnetic field they can create a host of important terrestrial effects, including the aurora borealis, interruption of radio communications, induced currents in electrical power grids large enough to blow out transformers, and confusion in birds' navigation. It would be

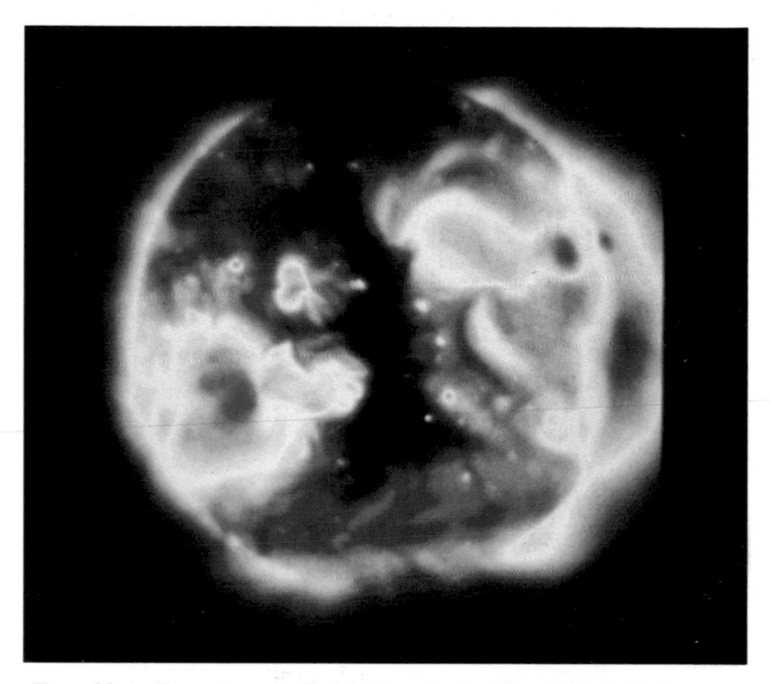

Figure 20. **An X-ray photograph of the Sun, obtained from NASA's Skylab space station on June 1, 1973. A huge dark coronal hole runs from the north pole down across the equator. The X-ray emission is essentially confined to upward-arching loops, large and small; two huge loops just left of center are pointing straight at the Earth.**

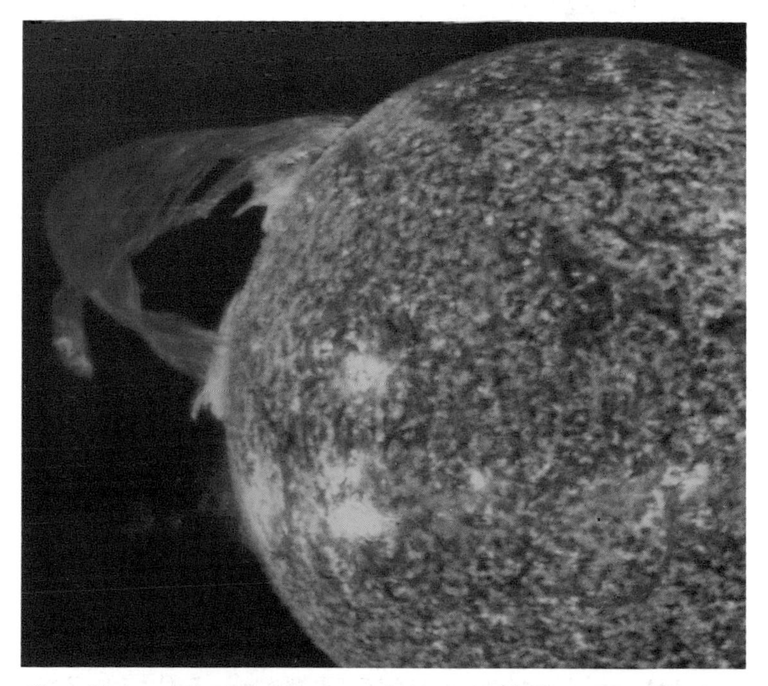

Figure 21. **An eruptive prominence, propelled by magnetic forces, lifts off from the Sun. Footpoints of the growing prominence stand in the photosphere and chromosphere; twisted lines of force connect them. This Skylab far-ultraviolet image was made at the 304-angstrom line of singly ionized helium.**

very nice to be able to predict the time and energy of such an outburst, but like Earth's weather, such forecasts are extremely difficult.

The magnetic-field lines that overlie active regions and produce loops, X-ray emission, and flares have a characteristic geometry: they arch up from a magnetic footpoint of one polarity in the photosphere, pass through the corona, and then arch down to a footpoint of the opposite polarity (Figure 12). However, when the magnetic field, weakening with height, arches very high in the corona, the pressure of the hot gases can force it far out into interplanetary space (Figure 22). These regions are called *coronal holes* – places where the corona is no longer "bottled up" by the magnetic fields and is free to expand outward. Energy is deposited in the coronal holes, quite possibly by the same mechanisms that heat the magnetically confined regions, and much of it goes into accelerating the gases outward, forming what is known as the *solar wind*. The coronal material left behind is both lower in density and cooler than material in the magnetically confined corona. Thus it emits much less high-temperature radiation; when photographed in X-rays, a coronal hole appears essentially black compared to magnetically closed regions (Figure 20).

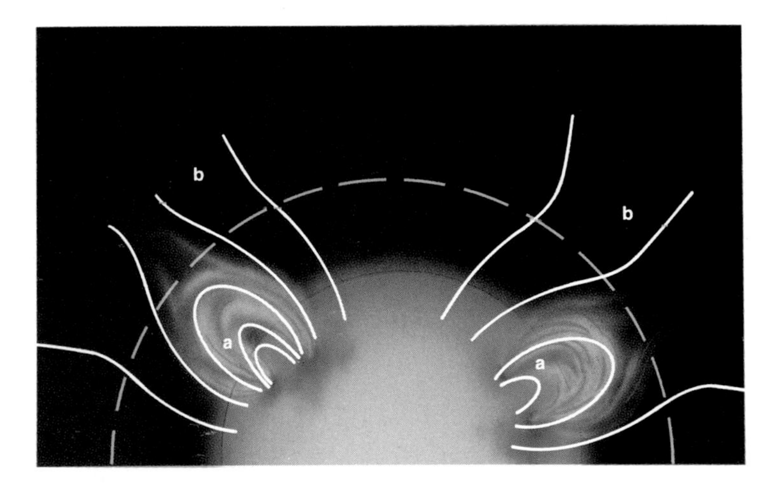

Figure 22. **In regions where the Sun's magnetic field is "closed" *(a)*, field lines loop through the corona but are connected at both ends to the surface. However, field lines that loop above an estimated height of 400,000 km (dashed line) are swept into interplanetary space by the expanding solar wind *(b)*. Such regions of "open" field lines are termed coronal holes.**

Solar magnetic fields do more than accelerate the solar wind though coronal heating; they also transfer angular momentum from the body of the Sun to the wind. Inside 30 solar radii the magnetic fields are so strong that the wind co-rotates with the Sun. Beyond 30 solar radii the magnetic field has weakened to the point that it can no longer direct the wind, which is then able to flow radially outward and drag the field lines away with it. As a result, the wind leaves the vicinity of the Sun in the shape of an Archimedean spiral (see Chapter 3).

About five days after a coronal hole rotates past the center of the disk, a high-speed stream of solar wind reaches the Earth, where it can be detected by orbiting spacecraft. The stream may have a velocity up to about 700 km per second and a density of 10 to 100 particles per cm^3. Spacecraft data show that the direction of the magnetic field carried by the solar wind invariably matches that of the field observed in the solar photosphere beneath the responsible coronal hole. Often a very large coronal hole develops at one of the Sun's poles (Figure 20). If that pole is tipped toward Earth, a very intense solar wind can be encountered here. It appears that intense solar winds blow out of the polar coronal holes and bend around toward the ecliptic plane. When the solar wind ejected from coronal holes strikes the magnetosphere of Earth, its particles may cause long-lived auroral displays. While these are not as spectacular as those following large solar flares, they are more common and may recur at 27-day intervals as solar rotation sweeps the same stream past our planet.

THE HISTORY OF THE SUN AND ITS MAGNETIC ACTIVITY

The rate at which the solar wind removes angular momentum from the Sun is very small today, but in the distant past it was much larger – enough to have slowed the rotation of at least the outer layers of the Sun. In fact, astronomers now think that the transfer of angular momentum to stellar winds is the major reason that stars like the Sun rotate more slowly as they age. This spin-down weakens the dynamo activity that generates the magnetic fields, which in turn lessens the distance out to which the coronal wind is forced to co-rotate and reduces the loss of angular momentum. That is why the *rate* of spin-down today is much less (by perhaps 1,000 times) than it was when the Sun was young and rapidly rotating.

This scenario results from the synthesis of data from many other stars that have convection zones, as the Sun does, but differ in age, mass, and rotation rate. As in the Sun, the combination of rotation and convection appears to generate magnetic fields, "starspots," and chromospheric and coronal heating. Of course, we cannot see the starspots directly, since all stars except the Sun appear as points of light with no resolvable surface detail. But as they rotate, their light varies as dark spots or bright emission patches are carried into view. Observations of the chromospheric emission tell us the mean level of magnetic activity, and observations of its periodic modulation tell us the star's rotation rate (Figure 23).

Taken in combination, these data suggest that the rotation period of stars with convection zones and magnetic activity declines approximately as the inverse square root of their age. It also appears that magnetic activity declines in step with the rotation rate. For example, when the Sun was one-fourth its present age, it rotated about twice as fast and had roughly twice its current level of activity. Finally, cycles very similar to the Sun's seem to be common in stars that are old and relatively slow rotators. Younger stars have much higher levels of activity and also much more chaotic long-term behavior; there may be regular cycles in these stars as well, but if so they are masked by large, irregular variations.

From such data and many theoretical studies, astronomers have gleaned a general picture of how the Sun's rotation and magnetic activity have evolved. When the Sun became a stable star some 4.5 billion years ago, it rotated rather rapidly, perhaps once in a few days. Its internal dynamo very effectively generated magnetic fields, which produced large sunspots, powerful solar flares, intense coronal ultraviolet and X-ray emissions, and a strong wind. This wind caused a rapid spin-down and weakening of magnetic activity, so that when the Sun was about 1 billion years old, its rotation rate

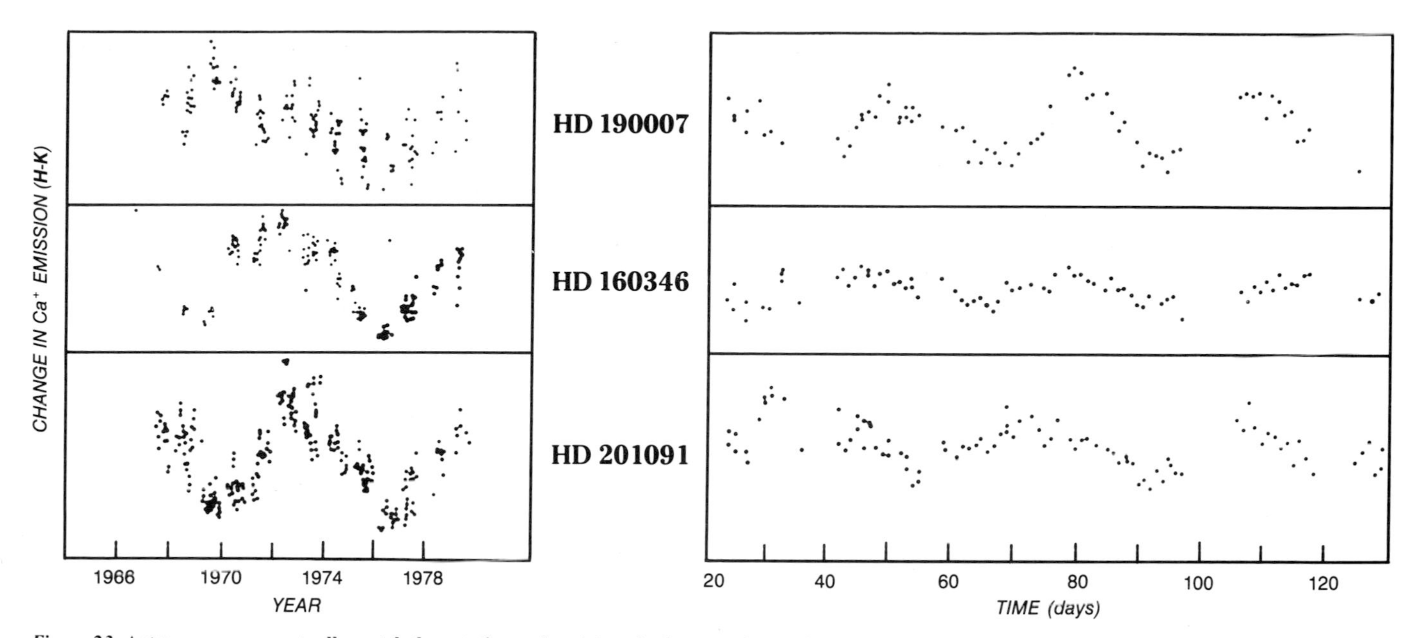

Figure 23. **Astronomers can actually watch the rotation and activity of other stars by monitoring two spectral absorptions due to ionized calcium, the H and K Fraunhofer lines. The lines appear more or less pronounced depending on how many magnetic active regions are present on the hemisphere facing Earth. At left are 1966–1980 records of H-K data for three stars, showing long-term cycles in activity, while those at right demonstrate the effects of rotation.**

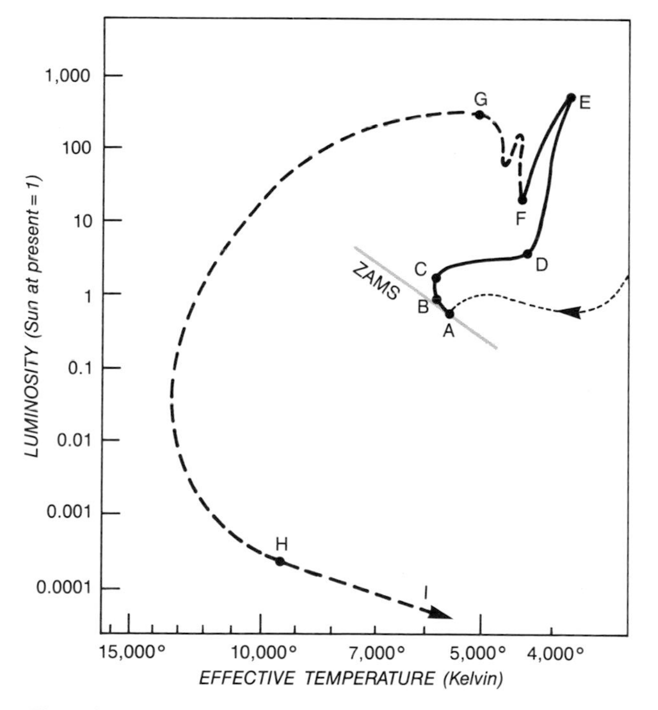

Figure 24. **The evolution of the Sun, as depicted on a Hertzsprung-Russell diagram (a plot of temperature versus luminosity). The newly born Sun *(A)* is on the zero-age main sequence line, or ZAMS, on which all stars lie when their nuclear fires first ignite. Presently the Sun is at *B*. The light-gray line indicates the Sun's history prior to full starhood, and other labels denote milestones in the Sun's evolution that are described in the accompanying text.**

had slowed until the period was about two weeks. Consequently, its spottedness, magnetic activity, and flares all fell to about twice their present level. The magnetic production was highly variable, and the Sun's activity cycle, even if quasi-periodic, would have been masked by shorter-term fluctuations. Only after the rotation had slowed still further, to about 20 days, did the activity cycle become as we see it today.

Whether there remains a rapidly spinning deep core, as theory suggests, may be learned when helioseismology returns data powerful enough to measure the rotation rate near the very center of the Sun.

THE FUTURE

Compared to its younger days, our parent star has now settled down to a rather placid existence. In Figure 24, point *A* on the curve indicates the luminosity and temperature of the newborn Sun, 4.5 billion years ago. Today, at *B*, the Sun is some 40 percent more luminous, for the reasons discussed earlier. This, in turn, has caused our star to become about 300° K hotter at its surface, and about 6 percent greater in radius than when it was born. Aside from this gradual brightening, we anticipate that the Sun will continue to be much as it is now for several billion years more, though its magnetic activity will probably decline a bit as the relatively feeble solar wind slows our star somewhat more. However, from both astrophysical theory and observations of other stars, we know that ultimately there are dramatic changes in store for the Sun.

The trend of increasing luminosity, radius, and surface temperature will continue for another 5 billion years, until the Sun reaches *C*, with a luminosity about twice its present value and a radius some 40 percent larger. By then most of the hydrogen in the core will be exhausted, forcing most of the energy production to take place farther out in a hydrogen-burning shell. The core itself, having an insufficient source of energy to maintain its gas pressure, will start to contract – even as the Sun's outer layers continue to expand.

Over the next 1.5 billion years the surface should enlarge to about 3.3 times its present size *(D)*, while the surface cools to about 4,300° K. As seen from Earth, the Sun will appear as a bloated orange-red disk. However, with the surface of Earth by then about 100° hotter than at present and all the water having long boiled away, it is extremely doubtful that any creatures will view this sight. Next, the Sun will rapidly swell and brighten to become a red giant star. Within about 250 million years – only about 2 percent of its then 12-billion-year age – it should expand to 100 times its present size and grow 500 times more luminous *(E)*, despite a temperature drop to about 3,500° K. From the lava seas covering the now-molten Earth, the Sun's looming dull-red disk will fill nearly the entire sky.

A dramatic event will occur at *E*, when the core temperature has reached about 100,000,000° K, enough to ignite the helium "ashes" left over from earlier hydrogen burning. As helium burns, it is transformed into carbon and oxygen, liberating the excess nuclear-binding energy as heat, just as the hydrogen once did. The helium ignition occurs very suddenly, and calculations indicate this "helium flash" should produce shock waves that blast as much as a third of the Sun's body off into space. During this brief but dramatic episode, the temperature of the core will soar to some 300,000,000° K, then cool to about 100,000,000°, after which the Sun should continue to burn helium steadily *(F)*.

Thereafter, the Sun will evolve in ways that are not easy to calculate, largely because the amount of mass ejected in the helium flash is uncertain. We do know it should gradually become more luminous once again, approaching but not quite reaching its maximum brightness as a red giant. And its surface should become hotter, appearing orange-red for a second time *(G)*.

Throughout its career as a giant star, the Sun is also destined to continue losing mass. This is due both to gravity's tenuous hold when the Sun expands to such a large size and to the pressure exerted by the flow of radiation from such a luminous star. The very strong solar wind will rob the Sun of mass even beyond what was lost in the helium flash. The outer layers will gradually dissipate until the massive and dense core itself is exposed: a dim, white-hot sphere no bigger than Earth *(H)* with a surface temperature of perhaps 10,000° K. At this point our Sun will have become a white dwarf, similar to the many that astronomers have already discovered and identified as endpoints of stellar evolution.

The white-dwarf Sun will be a bizarre object. Despite its small size, it should contain about half the Sun's present mass – equivalent to 100,000 Earths. Its density, about 2,000,000 g/cm^3, will equal that obtained by compressing several automobiles into a thimble. This extraordinary density is achieved because the very strong gravitational field inside the core of the once-giant Sun squeezes the material there into a so-called degenerate state, where the ordinary laws relating density to gas pressure and temperature do not apply.

Once the white dwarf state is reached, it appears the action is over. We think the Sun will cool over many billions of years, eventually radiating away the heat still stored inside it but creating no new energy by nuclear burning *(I)*. Its ultimate fate will probably be to circulate around the galaxy as a black dwarf, perhaps still accompanied by the cinderlike remnants of the planets it once warmed.

3

Magnetospheres, Cosmic Rays, and the Interplanetary Medium

James A. Van Allen

ONE OF THE MOST ENGAGING detective stories of modern science has been recognition of the existence and importance of solar "corpuscular radiation," that is, gaseous material shed by the Sun that is distinct from its light and other forms of electromagnetic radiation. This outpouring of ionized gases and their associated magnetic field constitutes the interplanetary medium and is now usually referred to by Eugene Parker's collective term *solar wind.*

In their great 1940 treatise, *Geomagnetism*, Sydney Chapman and Julius Bartels document the hypothesis that fluctuations in the Earth's magnetic field (magnetic storms) are caused by solar corpuscular streams. A decade later, the German astronomer Ludwig Biermann suggested that the streams were not simply intermittent bursts but were instead a continuous phenomenon (a conclusion he based on observations of comets' bluish gas tails). Over the past 30 years, instrumented spacecraft have confirmed some, but not all, of the early inferences concerning solar corpuscular streams and have provided a wealth of detailed knowledge about them.

The solar wind consists of a hot (about 100,000° K) *plasma* – an electrically neutral mixture of ions (principally protons with minor but significant proportions of heavier ions) and electrons. Its source is the Sun's atmosphere, or corona, and it is continuously present in interplanetary space. After escaping from the gravitational field of the Sun, this gas flows radially outward at a typical speed of 450 km per second to distances now known to be at least as great as 47 AU or 7.0 billion km – the current distance of Pioneer 10. The average speed of the flowing gas is remarkably independent of distance from the Sun, but marked fluctuations in this speed induce a variety of collisionless shock phenomena as fast streams overtake slow ones.

Even though the solar corona ejects gas radially outward into interplanetary space, the continuous stream takes on, by virtue of the Sun's rotation, the approximate form of an Archimedean spiral, as first suggested by Chapman (Figure 1). At the orbit of the Earth, the spiral makes an angle of about 45° with a radial line from the Sun but becomes nearly perpendicular at the orbit of Saturn (9.5 AU) and beyond.

At the Earth's orbit the number density of ions and electrons in interplanetary space is typically five particles per cm^3 under quiet conditions. This population density diminishes as the inverse square of the heliocentric distance, but sporadic order-of-magnitude fluctuations can occur in response to varying solar activity. Atomic collisions in this exceedingly dilute interplanetary medium are rare; however, complex electric and magnetic interactions cause energy exchanges among the constituents of the gas, and these give rise to a host of wave-particle phenomena. Some examples are Alfvén waves in the magnetic field (generated by the oscillations of the ions around their equilibrium positions), electrostatic oscillations, ion-acoustic waves, and bursts of radio "noise."

The ionized, electrically conducting gas carries with it an entrained magnetic field that also originates in the solar corona (see Chapter 2). Since the dynamics of the

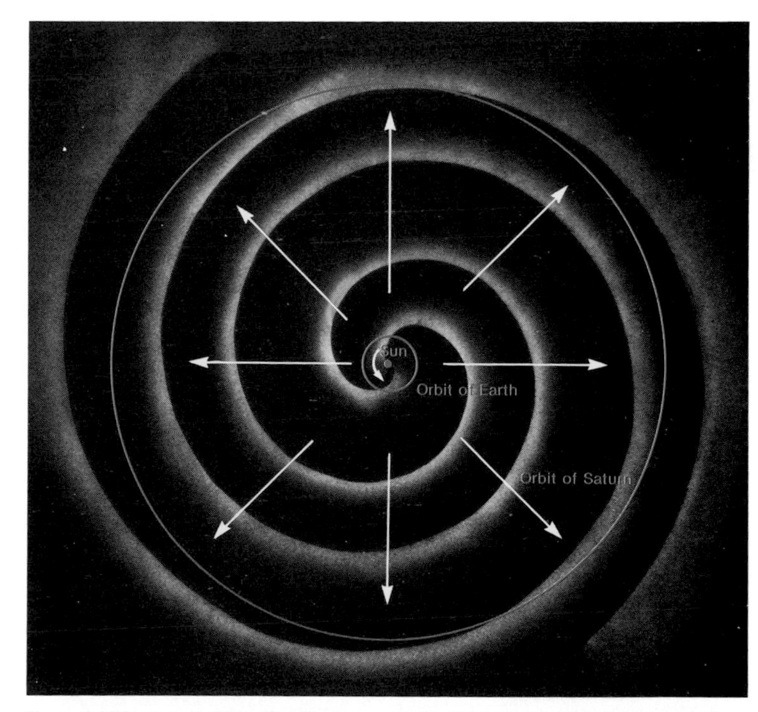

Figure 1. **This view, looking directly down onto the Sun's equatorial plane, shows the radially moving solar wind (arrows) and the Archimedean spiral formed by the continuous ejection of particles from a particular point on the rotating Sun. On average, the interplanetary magnetic field carried outward with the particles is aligned parallel to the spiral.**

interplanetary medium are dominated by the mass motion of the ionized gas it contains, the magnetic field lines become stretched out approximately parallel to the "wave fronts" of the corpuscular stream. The magnitude and direction of this field fluctuate from point to point, but it generally parallels the theoretical Archimedean spiral. At 1 AU the magnetic field strength is typically 5 gammas, or 0.00005 gauss (100,000 gammas equal 1 gauss), whereas it is many orders of magnitude greater at localized points in the solar chromosphere. Because the field lines assume a tightly wrapped spiral form far from the Sun, the magnetic field's strength does not decrease as quickly as the number density of the plasma does; rather, it falls off as the inverse of heliocentric distance.

Direct observations of the solar wind have been confined, thus far, to a thin pancake-shaped region near the plane of the Earth's orbit (the ecliptic plane), which is approximately the equatorial plane of the Sun. A space mission to high solar latitudes is ballistically difficult and has not yet been achieved, but this situation should change during the 1990s with the Ulysses spacecraft of the European Space Agency. The velocity necessary to leave the ecliptic plane will come from a flyby of Jupiter, during which the giant planet will act as a gravitational slingshot. In the meantime, the detailed properties of the solar wind at high solar latitudes remain largely conjectural – though we have gained some general insights by observing the ionized tails of comets and by looking in directions far from the ecliptic plane and noting how the interplanetary medium induces scintillation in radio signals from remote stellar sources.

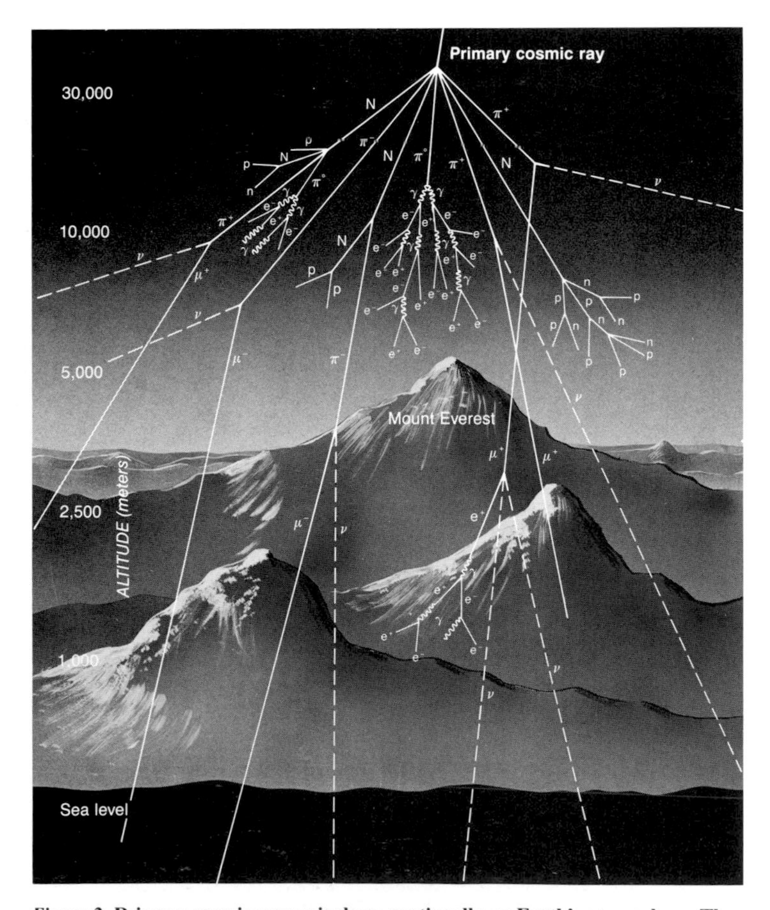

Figure 2. **Primary cosmic rays rain down continually on Earth's atmosphere. They collide with atomic nuclei *(N)* in the air and produce a shower of secondary particles, many of which subsequently decay or undergo further collisions: protons *(p)*; neutrons *(n)*; electrons *(e^-)*; positrons *(e^+)*; negative, positive, and neutral pions *(π)*; negative and positive muons *(μ)*; neutrinos *(ν's with dashed lines)*; and gamma rays *(γ's with wiggly lines)*. Yet despite the ever-widening cascade of particles and radiations, only a small fraction of these by-products ever reach the ground.**

COSMIC RAYS

In 1912 Victor Hess found that the intensity of a weak but mysterious penetrating radiation increased with altitude in the Earth's atmosphere. He surmised that it must come from outer space, a conjecture that was soon confirmed. Investigations of this phenomenon, called *cosmic radiation* after the suggestion of Robert Millikan, have occupied generations of physicists – despite the fact that only a few cosmic-ray particles per second pass through an area of 1 cm^2 in the Earth's vicinity. We now realize that cosmic radiation entering our atmosphere consists of atomic nuclei (principally protons) and a much smaller contribution of electrons. These are termed "primary" particles, and their energies range from tens of millions to many billions of electron volts. (By comparison, a photon of sunlight carries about 2 electron volts of energy.) Because primary cosmic rays and their progeny are so energetic, they have had an important role in modern high-energy particle physics.

The compositional distribution of cosmic-ray nuclei resembles the relative abundance of nuclei in "universal" matter, as estimated from astrophysical evidence. This is not surprising, since it is widely believed that cosmic-ray particles, sometimes called galactic cosmic rays, are accelerated in violent astrophysical events such as supernovae and star flares, and that they become accelerated further during many subsequent encounters with interstellar plasma clouds and shock waves. These theories are supported by our observations of energetic particles emitted sporadically by the Sun and their acceleration by interplanetary shock waves. However, the solar particles acquire only some tens of thousands or millions of electron volts and are thus considered low-energy cosmic radiation.

At ground level, instruments detect only cosmic-ray "secondaries," which result when primaries having energies of many billions of electron volts interact with molecules of gas in our atmosphere (Figure 2). As first shown by Scott Forbush in 1954, the intensity of secondaries varies in synchronism with the 11-year cycle of solar activity. The maximum cosmic-ray intensity occurs near the time of minimum sunspot (or solar-flare) activity and vice versa, and his measurements during the period 1937–1952 showed a peak-to-peak variation of about 4 percent. At balloon altitudes the cyclic variation becomes considerably more pronounced, especially at high latitudes. In a continuous set of measurements acquired in interplanetary space near the Earth from 1967 to 1972, made by one of the author's instruments, the maximum cosmic-ray intensity was 80 percent higher than the minimum level. Thus, for lower-energy particles, whose secondaries are unable to penetrate Earth's atmosphere, the magnitude of the variation is much greater.

The mystery of this cyclic variation in cosmic-ray intensity has intrigued many investigators. Undoubtedly, the effect is attributable to the Sun – even though such particles do not, to any significant degree, *originate* in the Sun. The prevailing view is that the entrained magnetic field in the outflowing solar wind acts as a modulating agent. Apparently, cosmic rays enter the outer solar system, then diffuse inward through the turbulent magnetic field of the interplanetary medium. The process may be likened to a school of fish swimming upstream in a turbulent river. The interplanetary field tends to deflect the particles and convect them outward. Thus, cosmic rays are less populous in the inner solar system than in the nearby interstellar medium. Low-energy particles are repelled most easily, so they are not well represented among the cosmic rays that manage to penetrate inward as far as Earth's orbit.

The challenge, therefore, is to determine how well the interplanetary magnetic field repels cosmic-ray particles of a given energy throughout the planetary system and to find the boundary at which the Sun's modulating influence ends. Instruments on Pioneers 10 and 11 have made the principal contributions to this subject (Figure 3). Moving outward from the Sun, the average increase in total cosmic-ray intensity is 1.5 to 2 percent per AU, and the outer boundary is now known to lie outside 47 AU. The measurements are continuing as all four Pioneers and Voyagers proceed out of the solar system. We now believe that the solar wind merges with the nearby interstellar medium and loses its identity at an estimated distance on the order of 100 AU. The boundary at which this merging begins is called the *heliopause* and the region inside it the *heliosphere* (Figure 4).

Locating the heliopause is an important objective of the Pioneer and Voyager missions. Their passage out into the interstellar medium will be revealed by a number of observable effects. First, the solar wind, flowing radially outward, will be replaced by an interstellar wind of different speed, ionic composition, temperature, and direction of flow. Second, the intensity and spectrum of the cosmic radiation will become constant, no longer modulated by the magnetic irregularities of the solar wind. This expectation is based on the presumption that the interstellar medium contains a smoothed average of contributions for many extremely remote sources. Third, the total cosmic-ray intensity will be greater than that at any point within the heliosphere.

Designing a plasma instrument sensitive enough to detect the dwindling wisps of solar wind at such great distances has proven a formidable challenge. On the other hand, since relatively low-energy cosmic rays become more abundant farther away from the Sun, observations of such cosmic rays may offer the best potential for revealing the location of the heliopause. Even so, the transition to the interstellar medium may occur not at an exact distance but over perhaps many AU and at a mean distance that fluctuates with the 11-year cycle of solar activity.

PLANETARY MAGNETISM

One of the most fundamental characteristics of a planetary body is its state of magnetization. It was shown many years ago that the magnetic field of the Earth must arise primarily from a system of electrical currents flowing in its deep interior. In other words, the Earth is a huge electromagnet (Figure 5), and this general concept has been extended to all other planetary bodies. The necessary electrical currents are maintained by what is called a self-excited *dynamo* (see Chapter 6). Theories of such a dynamo are complex, and their predictive value is quite limited. But all of the plausible theories require two basic features: a rotating body, and a hot, fluid, electrically conducting interior within which convective motion occurs.

Firm knowledge of the state of magnetization of planetary bodies is a key objective of missions to the planets. The Earth's external magnetic field has been extremely well characterized, as have its short- and long-term variability. At

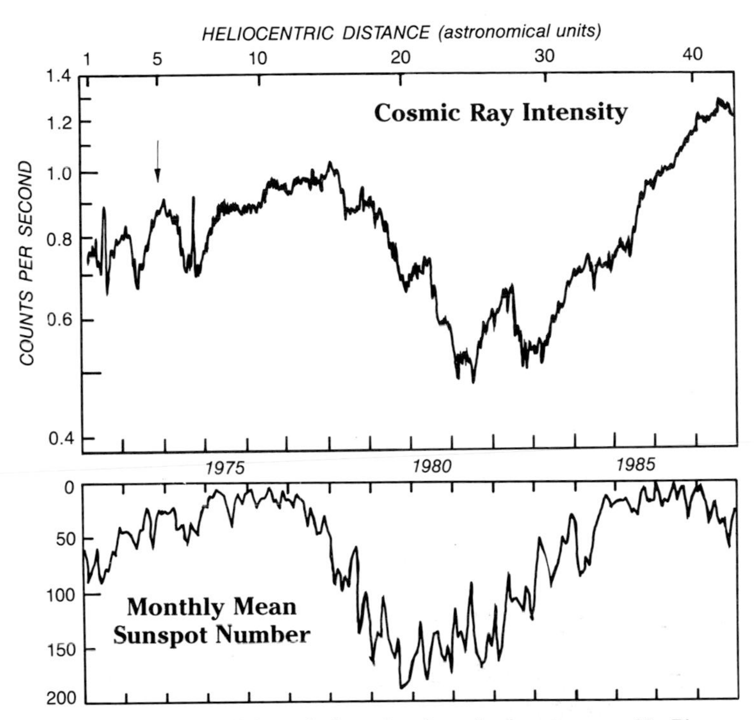

Figure 3. **The upper panel shows the intensity of cosmic rays as measured by Pioneer 10 during the first 16 years after its launch on March 3, 1972. The spacecraft passed close to Jupiter (arrow) on December 4, 1973. The lower panel shows the monthly mean sunspot number, plotted upside down to accentuate the inverse correlation between cosmic-ray intensity and sunspot number. Note the increasing time delay in this relationship as the spacecraft becomes progressively more distant from the Sun.**

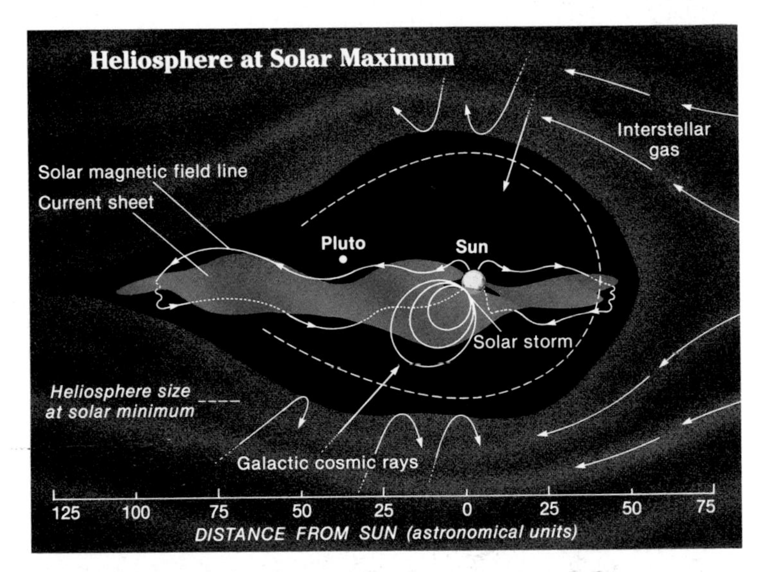

Figure 4. **This conjectural cross-section depicts the interplanetary environment out to more than twice the distance to Pluto. Within the heliosphere the solar wind flows radially outward until it encounters the irregular and fluctuating heliopause; the heliosheath is the transition region to the interstellar medium beyond. The current sheet defines the boundary between regions of opposite polarity in the interplanetary (solar) magnetic field. Weak galactic cosmic rays are deflected away by the heliosphere, though those with highest energy penetrate all the way to the inner solar system.**

a less detailed level, spacecraft have measured the magnetic properties of seven other planetary bodies: the Moon, Mercury, Venus, Mars, Jupiter, Saturn, and Uranus. In addition, some crude upper limits have been determined for the Galilean satellites of Jupiter and for Titan, the largest satellite of Saturn.

The simplest general characterization of a planet's magnetism is its *equivalent dipole magnetic moment*. This is a vector quantity that takes into account the intrinsic strength of the field, its tilt with respect to the planet's rotation axis, and its offset from the planet's geometric center. The dipole moment divided by the cube of the planet's radius yields the average strength of the field along the magnetic equator. For example, at the Earth's magnetic equator this value is 0.305 gauss or 30,500 gammas. To be able to generate a field this strong at its surface, thousands of kilometers from its core, the Earth must have an impressively large magnetic moment, 7.92×10^{25} gauss cm^3. By comparison, solenoid electromagnets in physics laboratories have typical magnetic moments only of the order of 100,000 gauss cm^3.

MAGNETOSPHERE OF THE EARTH

Using simple radiation detectors aboard the Explorer 1 and 3 satellites in 1958, I and my students discovered that a huge population of energetic charged particles surrounds the Earth and remains durably trapped there by our planet's magnetic field. No one had predicted such an effect; indeed, those early instruments were designed for a comprehensive survey of cosmic-ray intensities above the atmosphere. The discovery was soon confirmed and greatly extended by our research group and by many other American and Soviet investigators.

This early work identified two distinctively different "radiation belts." Each has the shape of a toroid and encircles the Earth in such a way that its central plane approximately coincides with the Earth's magnetic equatorial plane. Earth's belts are optically transparent and are not visible in the ordinary sense. The particle distribution itself does not shield the Earth in any significant way.

In 1907 Carl Størmer showed theoretically that an energetic, electrically charged particle can be permanently trapped, or confined, within the magnetic field of a dipole (Figure 6). In any static magnetic field, the force on a moving charged particle is directed at right angles to both the direction of motion and the magnetic vector. This is called the *Lorentz force;* its magnitude is proportional to the particle's velocity, electrical charge, and the magnetic-field strength. The Lorentz force bends the trajectory of the particle but does not change its energy, making the particles move in helical (spring-shaped) paths. A trapped particle travels quickly back and forth in magnetic latitude along this helical trajectory and also drifts slowly in longitude; eventually it sweeps out a toroidal volume encircling the dipole. In the Earth's field, the helices of electrons drift eastward and those of positively charged ions westward; a particle's specific trajectory depends on its energy and electrical charge. Størmer showed that if charged particles are somehow injected onto such orbits they remain trapped forever. However, he considered only an idealized case – the planets' electromagnetic environments are considerably more interesting.

(The phrase "radiation belts" is a generic term having historical roots. It does not imply radioactivity, as is sometimes stated mistakenly. It *is* theoretically possible to trap ions of the radioactive elements in a dipole field, but nearly all of the particles actually present in natural radiation belts derive from hydrogen and other common, stable atoms.)

In the years following 1958, scientists discovered an immense variety of physical phenomena within the volume of space dominated by Earth's magnetic field. Thomas Gold is the scientist who suggested the term *magnetosphere* to describe this entire region. The magnetosphere, and especially its outer reaches, is dynamic and constantly changing. "Low-energy" particles (with less than about 10,000 electron volts, which we collectively call plasma) have a much greater influence on gross physical phenomena than do their relatively rare "high-energy" counterparts. But the latter pose hazards to living things, and their distribution places practical limits on the regions of space around the Earth where human crews and animals are safe from excessive radiation exposure. The most readily accessible region of safe flight lies at altitudes below 400 km. By contrast, the radiation dosage within the equatorial region of the inner radiation belt, at an altitude of about 2,500 km, is especially severe – even electronic instrumentation has a limited useful lifetime there (Figure 7).

Understanding of the Earth's magnetosphere has been advanced to a fairly satisfactory level, with the help of dozens of instrumented spacecraft and intensive interpretative and

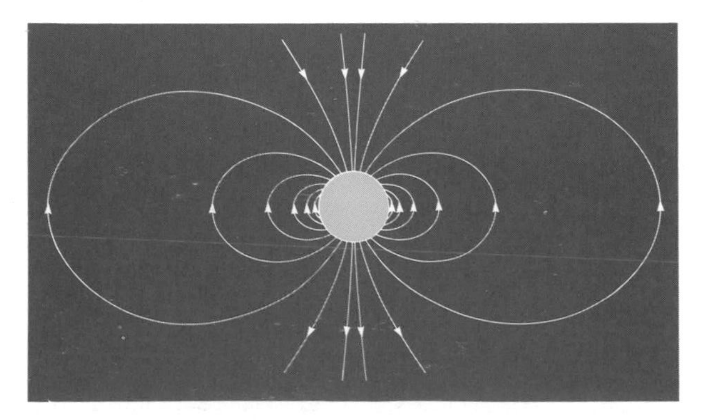

Figure 5. **An idealized representation of the external magnetic field of a planet caused by a system of electrical currents in its interior. The lines for a dipole field are shown, but in reality planets possess higher-order (quadrupole, octupole) fields that create a more complex pattern of field lines, especially near the planet itself.**

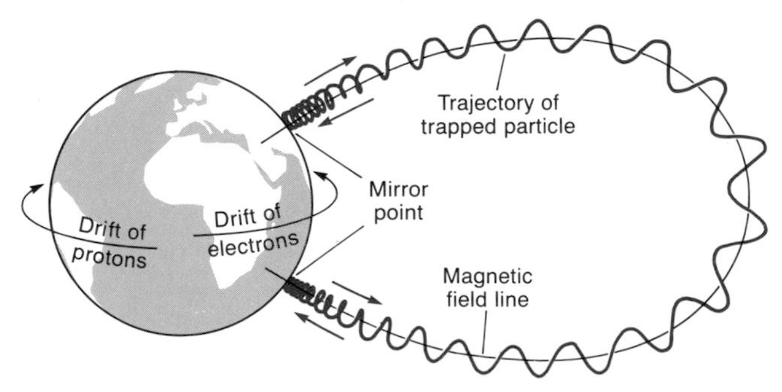

Figure 6. **When an electrically charged particle becomes trapped in the magnetic field of a planetary dipole, it experiences three types of motion: (1) circling around a magnetic field line ("cyclotron" motion), (2) bouncing back and forth between so-called mirror points in each hemisphere, and (3) a longitudinal drift around the planet.**

	Rotation period (days)	*Dipole moment (gauss cm³)*	*Field at equator (gauss)*	*Polarity same as Earth's?*	*Angle between axes*	*Typical magnetopause distance* (R_p)	*Plasma sources*
Mercury	58.65	3×10^{22}	0.002	yes	< 10°	1.1	W
Venus	243.01R	$< 10^{21}$	<0.0003	?	?	1.1	W,A
Earth	1.00	7.9×10^{25}	0.305	yes	11°.5	10	W,A
Mars	1.03	(1.4×10^{22})	(0.0004)	?	?	?	?
Jupiter	0.41	1.5×10^{30}	4.2	no	9°.5	60–100	W,A,S
Saturn	0.44	4.3×10^{28}	0.20	no	< 1°	17–25	W,A,S
Uranus	0.72R	3.8×10^{27}	0.23	no	58°.6	18–25	W,A
Neptune	0.67	2×10^{27}	0.06–1.2	no	46°.8	25–30	W,A,S

Table 1. **The solar system's planets exhibit a wide range of rotation rates (*R* means retrograde) and magnetic properties. Earth's magnetic polarity is that of a bar magnet with its south magnetic pole in the Northern Hemisphere. *Angle between axes* refers to the magnetic and rotational axes. *Typical magnetopause distance* is in the direction of the Sun and given in planetary radii. Under *plasma sources*, the letter *W* is for solar wind, *A* for atmosphere, and *S* for satellites or rings. Parenthetical values are very uncertain.**

theoretical work. Magnetospheric physics is employed in the analysis of such phenomena as auroras, geomagnetic storms, heating of the upper atmosphere by particle "precipitation," propagation of so-called whistler waves from lightning discharges, and the generation and propagation of a rich variety of electrostatic and electromagnetic waves originating in plasma instabilities. Taken together, these processes make the Earth a natural radio source that radiates on the order of 100 million watts into interplanetary space.

ORIGIN OF PLANETARY MAGNETOSPHERES

Our knowledge of the Earth's magnetosphere provides the basis for understanding the electromagnetic properties of other planets. However, as our spacecraft emissaries have found, each of these magnetospheres exhibits distinctive and unique features (Table 1). In general, a planetary magnetosphere is that region surrounding the planet within which its own magnetic field dominates the behavior of electrically charged particles. The term does not imply spherical shape but is used in a looser sense, as in the phrase "sphere of influence."

The solar wind has a negligible effect on the movements of large bodies such as planets. But it has profound effects in their immediate vicinity, creating an amazing assortment of physical phenomena. Because the solar wind is itself an ionized gas or plasma, it behaves like an electrically conducting fluid. It is also magnetized; that is, it contains systems of electrical currents that survive from their origin in the solar corona. These two properties of the solar wind, when combined with its bulk flow, are essential to most magnetospheric phenomena.

In his 1873 treatise, *Electricity and Magnetism*, James Clerk Maxwell described theoretically what was later recognized as the essence of the interaction of the solar wind with a magnetized planet. Maxwell envisioned a magnetic dipole whose axis is parallel to a flat conducting plate of infinite size. He then calculated the system of eddy currents that would be induced on the plate as it was moved toward the dipole. The magnetic field associated with these currents was found to be identical to that of a second (virtual) magnetic dipole, exactly like the real one, located at the same distance from the plate but on the opposite side. The effect of this "mirror dipole" is to limit the magnetic field of the real dipole to the half-space containing it; to produce two neutral

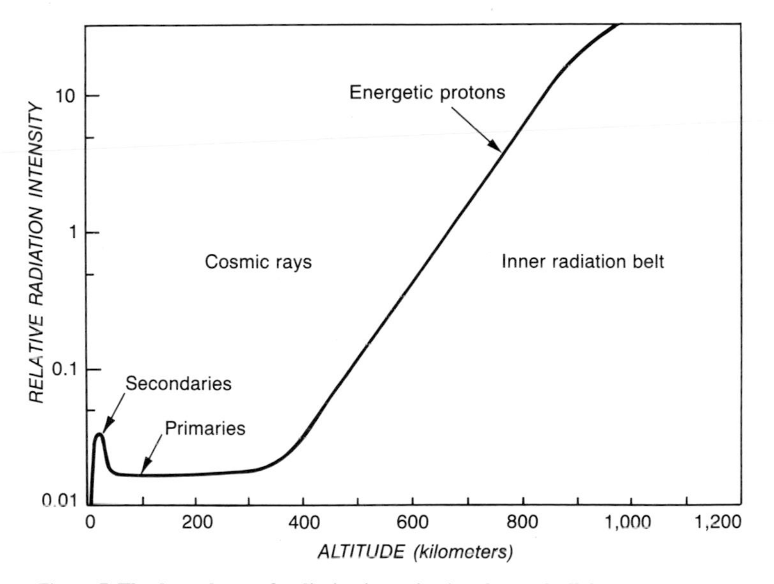

Figure 7. **The dependence of radiation intensity (as observed off the coast of Brazil) with altitude. At this location the Earth's inner belt of trapped radiation comes the closest to the surface.**

points in that field; and to otherwise modify the form of the field (Figure 8). In addition, the plate and the dipole repel each other by a calculable force.

The real situation is somewhat similar, but the solar wind, being a dilute fluid rather than a rigid plate, flows around the sides of a magnetic barrier of limited lateral dimensions. Like the plate, the solar wind is repelled by a planet's magnetic field; the point at which this repulsive force balances the solar wind's pressure is called the stagnation point, and this is where the solar wind comes closest to the center of the planet. A planet may possess a magnetosphere containing electrically charged particles in durably trapped orbits only if the stagnation point is located well above the sensible upper limit of the planet's atmosphere. Otherwise, a well-developed magnetosphere will not form, though certain plasma-physical and electromagnetic effects such as the acceleration of charged particles will still occur. The distance from the stagnation point to the center of the planet depends on the planet's magnetic moment (which does not vary) and the solar-wind pressure (which does). For the Earth, this standoff distance is usually about 64,000 km (10 times the

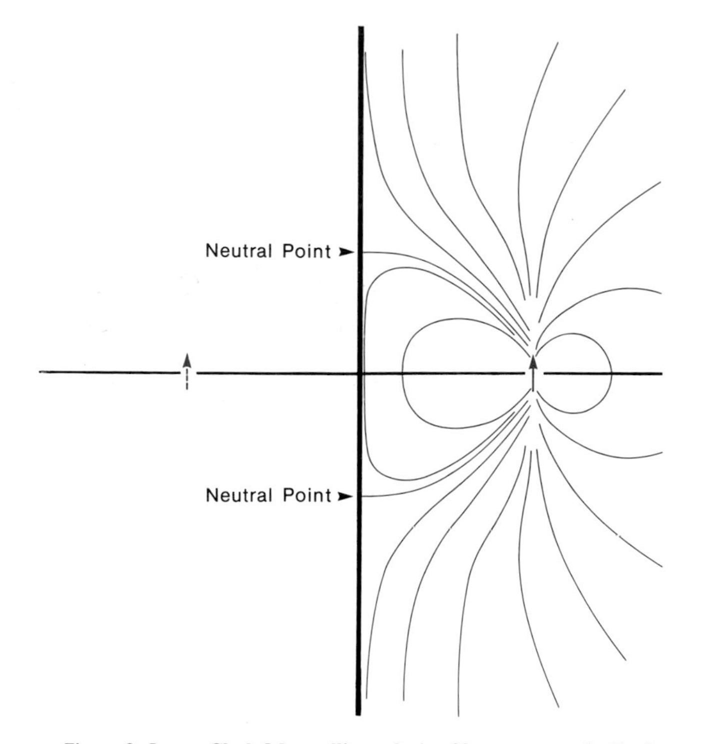

***Figure 8*. James Clerk Maxwell's analysis of how a magnetic dipole interacts with a metal plate of infinite dimension. His work has a direct bearing on how the solar wind interacts with a planet's magnetic field. The heavy black line represents a rigid conducting plate that is moving toward the magnetic dipole at the right (red arrow). Maxwell's "mirror dipole" is represented by the light arrow at the left. The resulting magnetic field to the right of the plate becomes distorted as shown, with two neutral points on the plate.**

Earth's radius) on its sunward or "upwind" side – far above the atmosphere.

The magnetic field accompanying the solar-wind plasma merges with that of a planet and stretches it out to produce a long, turbulent *magnetotail*, or wake, on the downwind side of the planet (Figure 9). Spacecraft have determined that the Earth's magnetotail is several million kilometers long. Moreover, the solar wind induces electric fields and turbulence that distort the planet's outer magnetosphere and permit a small fraction of the surrounding interplanetary plasma to be injected into it.

Fluctuations in the solar wind and its magnetic field also induce electric and magnetic fields that randomly accelerate and decelerate particles. The affected particles diffuse both inward and outward across the magnetospheric boundary (the *magnetopause*). Those that diffuse inward become trapped in the inner magnetosphere, where the field is stronger. Some particles ultimately reach the planet's atmosphere, occasionally precipitating into it with enough energy to trigger auroral emissions from gases at high altitudes (Figures 10,11); others work their way back out and are lost to space.

Thus, the entire process of producing and populating a magnetosphere with energetic particles is dynamic and quite complex. The residence time of a magnetically trapped particle varies widely – from hours to years. Additional low-energy ions and electrons come from the ionosphere of the planet and the ionospheres and atmospheres of the planet's satellites. In the Earth's case, such an atmosphere-derived plasma was discovered in the Earth's magnetic field in 1953, during Owen Storey's study of the radio whistlers that result from lightning discharges. These particles are also subject to the dynamical processes sketched above.

One "home-grown" source of magnetospheric particles seems to be the neutrons produced when galactic cosmic rays and solar energetic particles bombard the gases in Earth's atmosphere. As they fly off into space, a small fraction of these neutrons decay into energetic protons and electrons, thereby injecting charged particles directly into the magnetosphere. The residence times of such particles in trapped orbits depend in part on the ambient field strength; in the strong field present in Earth's inner magnetosphere, residence times for protons having energies of tens of millions of electron volts are of the order of a decade. So even though the decay of neutrons provides only a weak source of particles, they remain trapped long enough to accumulate in substantial numbers.

INNER-PLANET MAGNETOSPHERES

The Moon. The first extraterrestrial body to be investigated firsthand by spacecraft was the Moon, an object studied intensively by American and Soviet flybys, orbiters, and landers. Yet the Moon has no general magnetic field – its magnetic moment is at least 10 million times weaker than that of the Earth. The Moon must therefore lack an internal dynamo, because of its slow rotation and the almost certain absence of a hot fluid core. Curiously, experimenters have discovered *localized* regions on the Moon with surface magnetic fields of 5 to 300 gammas, but these appear to be geologic anomalies exhibiting residual magnetization that has survived from long ago. How the primordial Moon could have developed such magnetized patches remains a nagging problem.

In its orbit about the Earth, the Moon passes through the Earth's magnetotail for a few days each month near the time of full Moon; for the remainder of the month it is outside the magnetosphere and immersed in the solar wind. Because the Moon lacks both a general magnetic field and an atmosphere, the solar wind strikes the lunar surface and actually accumulates there. Extending from the antisolar side is a long plasma void (or plasma umbra) shaped like an ice-cream cone, with its apex downwind and with the Moon itself as the scoop of ice cream (Figure 12). Although this plasma void gradually fills in downstream because of the lateral motion of the particles in the solar wind, it still may be the most nearly perfect vacuum in the solar system. The mere existence of such a plasma void implies a system of weak electrical currents along its boundary; these currents were observed during the late 1960s by the lunar-orbiting spacecraft Explorer 35. In addition, the varying magnetic field entrained in the solar wind induces a system of transient electrical currents within the Moon itself, as was observed by the long-lived Apollo magnetometers placed on its surface.

Overall, the plasma-physical phenomena found at the Moon are a large-scale example of how the solar wind interacts with an inert, though slightly conductive, object.

Venus. American and Soviet spacecraft have observed the magnetic properties of Venus at increasing levels of detail. These investigations include Mariner 2 (1962) – the first-ever close approach to another planet – followed by Mariner 5 (1967), Mariner 10 (1973), the Venera series (1967–1983),

Vegas 1 and 2 (1985), and, perhaps most valuable of all, the Pioneer Venus mission, whose orbiter (Pioneer 12) has been circling the planet since 1978 and is still transmitting data.

The magnetic moment of Venus is at least 25,000 times weaker than the Earth's, despite our "sister" planet's comparable size and probable fluid core. But Venus rotates very slowly (sidereal period: 243 days), and it appears that its system of internal currents is correspondingly weak or nonexistent on a large scale. Nonetheless, the solar wind is prevented from reaching the surface by Venus' dense atmosphere and by magnetic eddy currents induced in its conducting ionosphere. Thus, the planet has a well-developed bow shock and causes many interesting plasma effects, but it possesses no population of durably trapped particles (Figure 13). Venus represents a situation intermediate between those of the Earth and the Moon, but more similar to the latter.

Mars. Sparse data from Mariner 4 (1965) and the Soviet orbiters Mars 2, 3, and 5 (1971–1974) and Phobos 2 (1989) represent our only observations to date of the magnetic properties of Mars. Its magnetic moment is at least 5,000 times weaker than that of the Earth. As at Venus, the solar wind interacts with the planet's conducting ionosphere, creating a weak bow shock and other plasma phenomena. Mars' sidereal rotation period is 24.6 hours, only slightly longer than ours, which should be fast enough to drive an internal dynamo. But the planet's radius is only 53 percent of the Earth's, and the magnetic evidence suggests that Mars lacks a hot, fluid core.

Mercury. During 1974 and 1975, the Mariner 10 spacecraft made three successive flybys of Mercury, providing the first and thus far only closeup observations of the innermost planet. The sidereal rotation period of Mercury is nearly 59 days, which might be considered too slow to sustain dynamo activity. Nonetheless, the planet has a well-determined though weak general magnetic field, with a magnetic moment similar in magnitude to the upper limit cited above for Venus. A bow shock has been observed (despite the absence of an appreciable atmosphere and ionosphere), as has the transient acceleration of charged particles and other plasma phenomena. But the Mercurian magnetic field is too weak to maintain a belt of trapped particles.

JUPITER

As early as 1955, Bernard Burke and Kenneth Franklin reported the first persuasive evidence that Jupiter is a source of sporadic bursts of radio noise at a frequency of 22.2 megahertz (decametric wavelengths). These are termed "nonthermal emissions" because they arise from physical processes other than those associated with heat. (Any object in the universe whose temperature is above absolute zero "glows" or radiates energy at radio and other wavelengths.) Soon thereafter, another type of nonthermal radiation from Jupiter was discovered in a quite different portion of the spectrum, from 300 to 3,000 megahertz (decimetric wavelengths). Jupiter's decimetric radiation has an intensity and spectral form that are nearly constant with time. It originates not near the planet's atmosphere, as does the decametric radiation, but within a toroidal region encircling the planet (Figure 14); the central plane of this torus tilts about 10° to the planet's equatorial plane. In 1959, Frank Drake and Hein Hvatum interpreted the decimetric emission

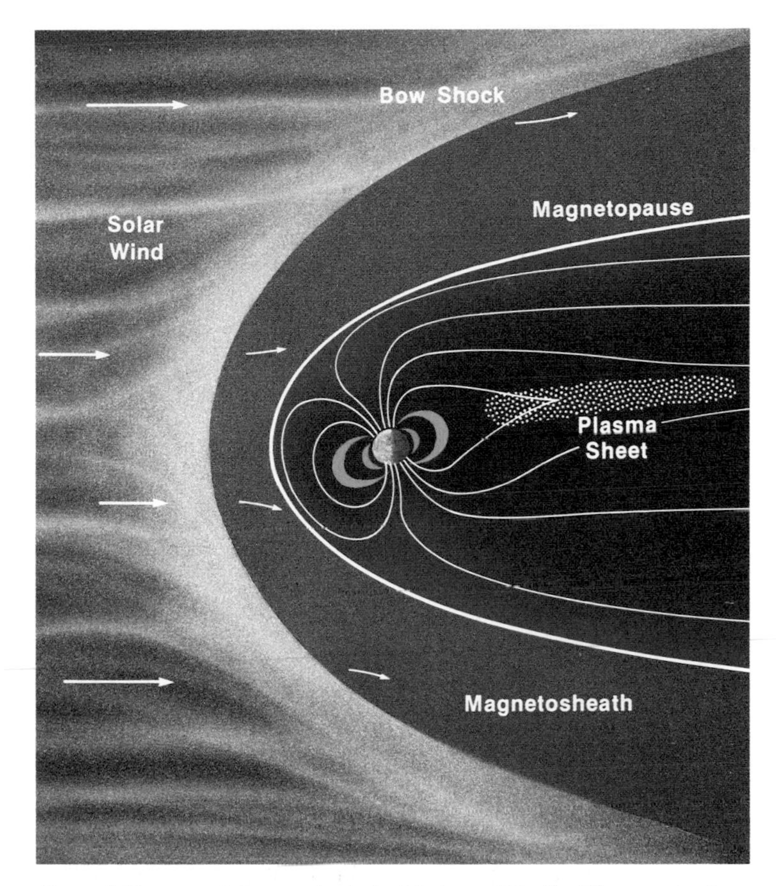

Figure 9. **The general shape and principal features of the Earth's magnetosphere. Note the inner and outer radiation belts (blue) near the Earth and the magnetotail, which extends to the right. Compare the magnetic field with that shown in Figure 8.**

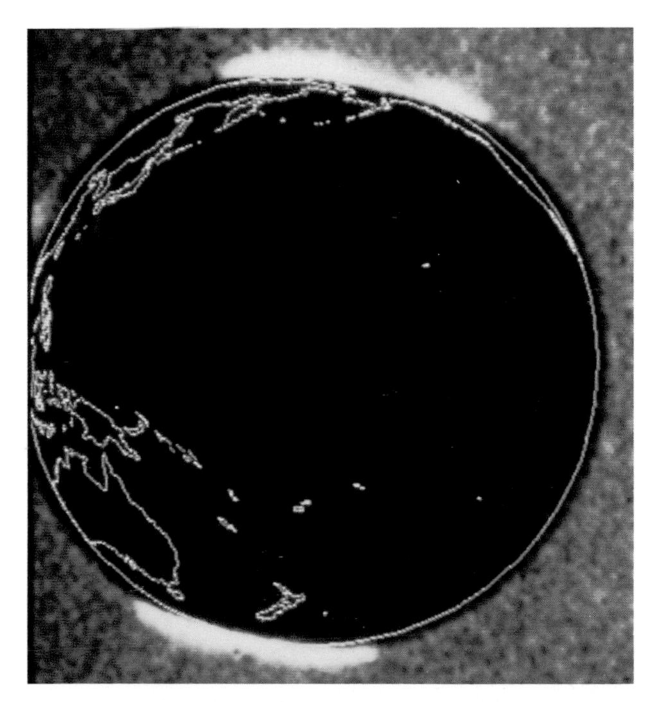

Figure 10. **This extraordinary image from NASA's Dynamics Explorer 1 satellite shows both the aurora borealis ("northern lights") and the aurora australis ("southern lights"). These glowing ovals or rings are about 500 km wide, 4,500 km in diameter, and centered roughly on the Earth's magnetic poles. Green lines show major land areas; Australia is at lower left and North America at upper right.**

Figure 11. **A spectacular aurora glows eerily over the Earth between Australia and Antarctica. Astronaut Robert Overmyer photographed this and 17 other displays during a Space Shuttle mission in May 1985. The colors of an aurora depend on its altitude. Its bottom may be no more than 100 km high, involving blue and red light given off by nitrogen molecules. However, most displays occur from 110 to 240 km up and are various shades of green (dominated by 5577–angstrom emission from oxygen atoms). Above that, up to 400 km high (rarely to 1,000 km), the aurora has a ruby-red glow from oxygen ions emitting at 6300 and 6364 angstroms. The faint reddish-brown band of light along the horizon is called airglow, which occurs when atoms in the upper atmosphere combine to form molecules.**

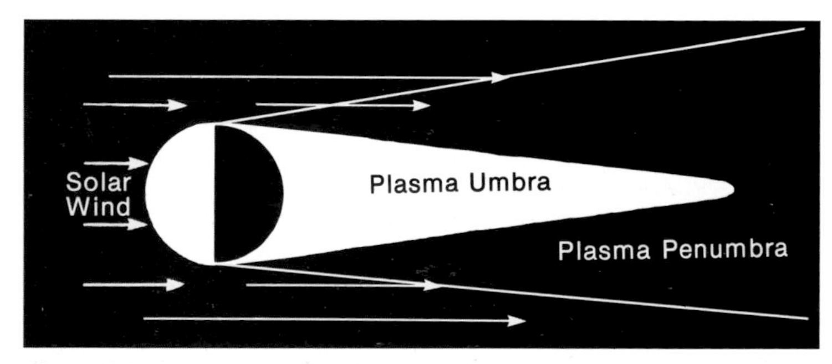

Figure 12. **The interaction of the solar wind with the Moon is much simpler than it is with planets having magnetic fields or atmospheres. The outward flowing solar-wind plasma strikes the lunar surface directly, and its flow past the Moon creates an umbra or shadow virtually devoid of any matter whatsoever.**

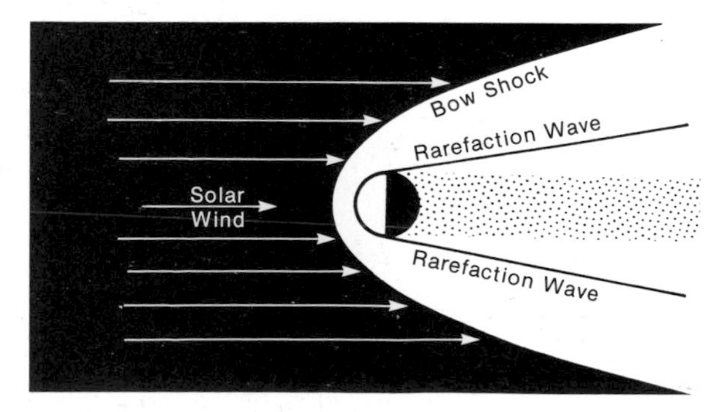

Figure 13. **Venus has no appreciable magnetic field, but the interaction of its electrically conductive ionosphere with the solar wind produces a well-developed bow shock nonetheless. Rarefaction waves are a common aerodynamic effect observed when a high-speed flow is interrupted by an inert, spherical body.**

as synchrotron radiation emitted by electrons trapped in the external magnetic field of the planet (as occurs in the radiation belts around Earth) and moving at relativistic velocities.

These ground-based observations set Jupiter apart from all other planets and provided an impetus for sending spacecraft to probe its radiation belts directly. Our first two probes of the Jovian magnetosphere were successful far beyond anyone's expectations – even of those of us who participated in the missions. Pioneer 10 encountered Jupiter in early December 1973, and it is still providing good interplanetary data as it continues to escape the solar system. Pioneer 11 arrived at Jupiter almost exactly one year later and continued onward to Saturn, where it arrived in the late summer of 1979. It also continues to operate well. Instruments on Pioneers 10 and 11 showed that Jupiter's decimetric radio emissions do indeed come from relativistic electrons trapped there (Figure 15). The probes also measured the configuration and structure of the bow shock and the energy spectra, angular distributions, and positional distributions of energetic electrons and protons within the planet's magnetosphere.

The general form of the Jovian magnetosphere resembles that of the Earth, but its dimensions are at least 1,200 times greater (Figure 16). In fact, if one could see the Jovian magnetosphere from the Earth, it would appear several times larger than the full Moon. Voyager observations later showed that the magnetotail of Jupiter extends behind it to the orbit of Saturn and beyond – at least 650 million km! These enormous dimensions result from a magnetic moment 19,000 times greater than the Earth's and from the fact that the solar-wind pressure there (5.2 AU from the Sun) is only about 4 percent of its value here at 1 AU. The magnetic moment vector of Jupiter is tilted about 9°.5 to its rotational axis, in agreement with the evidence from ground-based radio observations. It has the same sense as the rotational angular momentum vector; that is, both (magnetic and rotational) north poles lie in the same hemisphere, opposite to the situation for the Earth. But in the context of our incomplete understanding of planetary magnetism and in light of the paleomagnetic evidence for many reversals of the Earth's magnetic moment over geologic time, such relationships are not considered permanent or fundamentally significant.

Beyond the basic characteristics sketched above, the magnetosphere of Jupiter exhibits a rich variety of special and unique features. The outer magnetosphere occupies an enormous disk-shaped region, the result of two principal mechanisms. First, the great mass of low-energy plasma trapped within the magnetic field exerts a kinetic pressure on the magnetic field, inflating it in the same manner as an air-filled balloon. Because the field is weakest in its equatorial plane, the outward distension is most prominent there. Second, the magnetic field rotates as fast as the planet's interior (9 hours 55.5 minutes), and the plasma interacting with the field is forced to circle Jupiter with this period as well. Therefore, centrifugal force pushes the plasma outward. The first of these effects dominates out to about 20 planetary radii and the second at greater distances. The combination of these two effects produces a distinctive disk of plasma, or *magnetodisk*, lying roughly in the planet's magnetic equatorial plane.

Analyzers on the Voyager spacecraft showed that Jupiter's plasma derives principally from sulfur dioxide, hydrogen sulfide, and other gases vented by volcanic eruptions on the large, tidally heated satellite Io (see Chapter 13). Once ionized, these gases contribute a unique feature to Jupiter's magnetosphere: a doughnut-shaped region of low-energy particles called the Io torus.

Io, its gaseous torus, and the other inner satellites play an essential role in limiting the total population of trapped particles by absorbing or scattering those that diffuse into the inner magnetosphere. (The Moon does not have such an effect on the Earth's magnetosphere because it has no active volcanoes and is too far away.) Nonetheless, the characteristic energies of such particles at Jupiter are about an order of magnitude greater than those in the Earth's inner magnetosphere, and the population of the trapped radiation is several orders of magnitude denser. Even during the brief encounters of both Pioneers 10 and 11, trapped radiation caused the failure of several transistor circuits and the darkening of exposed optics. For each spacecraft, the cumulative exposure to electrons energized to at least 500,000 electron volts was about 400,000 rads. For comparison, a dosage of only 400 rads, if distributed throughout the human body, is sufficient to cause very severe radiation sickness or death.

The power for populating and maintaining the magnetosphere of Jupiter is derived principally from the rotational energy of the planet and the orbital energy of Io, whereas the power source of Earth's magnetosphere is

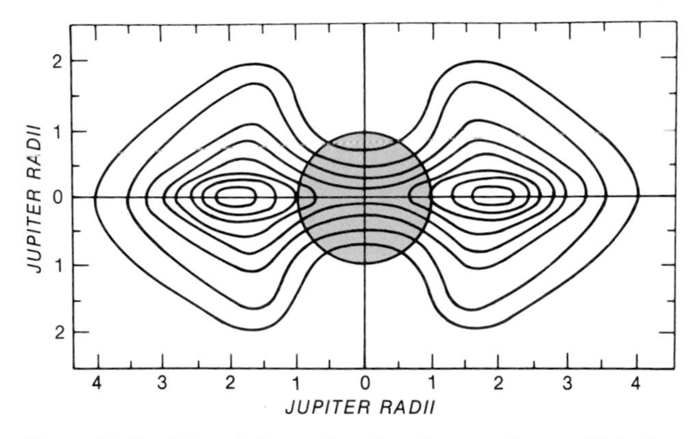

Figure 14. **Earth-based observations have been used to establish the intensity contours of Jupiter's decimetric radio emission (10.4 cm in wavelength). The colored circle represents the planet's disk; note two "hot spots" at about 1.9 Jovian radii from its center.**

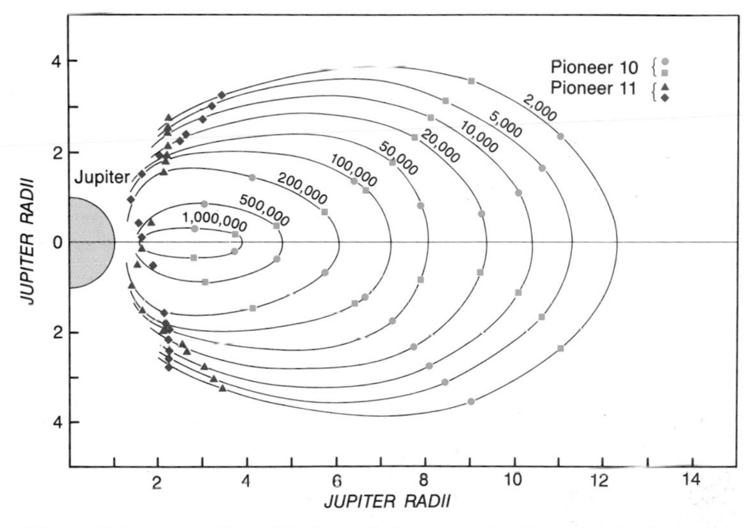

Figure 15. **A cross-section of the inner Jovian magnetosphere, showing how many electrons with energies greater than 21 million electron volts were encountered per cm² second by the author's instruments aboard Pioneers 10 and 11.**

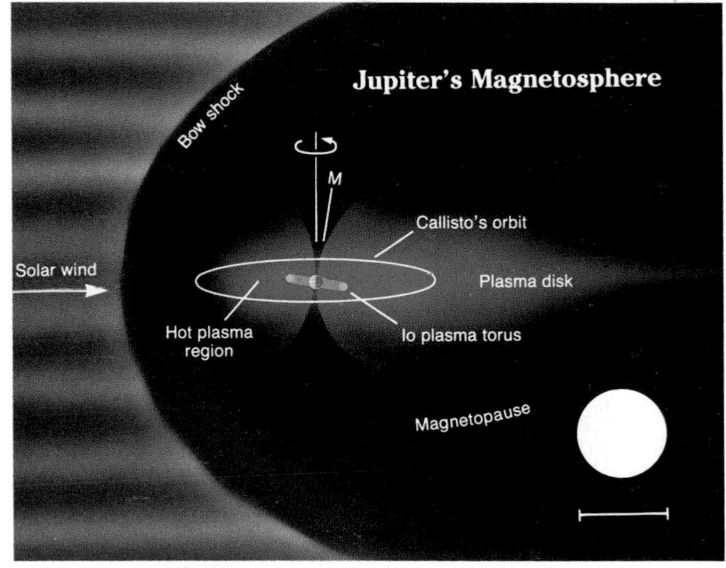

Figure 16. **Jupiter's magnetosphere is an enormous envelope much larger than the Sun (circle at lower right). *M* denotes the planet's dipole axis, which is aligned close to the rotation axis. Trapped plasma is concentrated in a disk-like region near the planet's magnetic equator. (Jupiter is shown enlarged to make it visible.)**

principally the solar wind. The Pioneer and subsequent Voyager observations have shown that Jupiter itself emits copious numbers of energetic electrons into interplanetary space. These data have gone far toward enriching our understanding of the origin of energetic charged particles in a larger astrophysical context, such as in pulsars.

SATURN

In contrast to the situation at Jupiter, prior to the arrival of spacecraft we knew nothing of the state of magnetization of Saturn or whether the planet possessed radiation belts. Our speculations favored a strongly magnetized Saturn whose ring system prevented energetic electrons from being trapped within the innermost and thus strongest part of its presumed magnetic field. (Recall that the telltale synchrotron radiation from Jupiter comes primarily from such an inner region.) Moreover, any synchrotron radiation from electrons outside of this region would be about four times harder to detect than that from Jupiter, because Saturn is nearly twice as far from Earth. On the basis of these considerations, theorists concluded that an intense radiation belt could exist outside of the outer edge of Saturn's ring A – even though it was not apparent from Earth-based radio astronomy.

The discovery of Saturn's magnetosphere came in September 1979 as instruments aboard Pioneer 11 detected the presence of a bow shock 24 Saturnian radii (1.44 million km) from the center of Saturn on its sunward side. Soon thereafter the spacecraft entered an intense, fully developed magnetosphere and later, as it passed beneath the outer edge of the planet's ring A, the instruments on board recorded a guillotine-like cutoff of the charged-particle population, as expected. All of the data returned during the encounter were new, rich in interpretable significance, and intensely interesting.

Pioneer found the magnetosphere of Saturn to be intermediate between those of the Earth and Jupiter both in extent and in the population of trapped energetic particles. The magnetic moment of the planet is 540 times greater than Earth's but 36 times less than Jupiter's. As at Jupiter, the magnetic moment vector and the angular momentum vector have the same sense. But in contrast to the situation at the Earth and Jupiter, the two Saturnian vectors are parallel to each other to within an uncertainty of about 1°. Moreover, Saturn's magnetic center is only slightly eccentric, lying north of its geometric center by only 2,400 km – just 0.04 of the planet's radius.

Beyond about 7 planetary radii, the outer magnetosphere is markedly distended on the dawn side but is much less so on the sunward side. Thus, Saturn's outer magnetosphere has some disklike properties that are again intermediate between those of Jupiter, where the magnetodisk is prominent, and those of the Earth, where it is nearly absent.

Saturn's largest satellite Titan, at an orbital radius of 1.22 million km (20.4 planetary radii), moves through the outer fringes of the magnetosphere. According to data obtained by Voyager 1 in November 1980, Titan apparently loses nitrogen gas from its upper atmosphere to the magnetosphere and produces substantial plasma and magnetic effects in its wake. The escaping gas is either ionized initially or soon becomes so, thus providing a source for the plasma trapped in Saturn's inner magnetosphere. Other apparent sources of gas are the inner satellites Enceladus, Tethys, and Dione, as well as the planet's rings and hydrogen-dominated upper atmosphere.

During their approaches to Saturn, both of the Voyager spacecraft detected sporadic emissions of radio energy with wavelengths in the kilometer range. Apparently, some electromagnetic process was generating these bursts and modulating their intensity with a period of 10 hours 39.4 minutes. The origin of the emissions has not been identified, but the periodicity is attributed to the magnetic field. Hence, astronomers have adopted this period provisionally as the internal rotation period of the planet (as distinct from that of its atmosphere).

Like their counterparts around Jupiter, the large inner satellites Rhea, Dione, Tethys, Enceladus, and Mimas are excellent absorbers of diffusing electrons and protons. Yet, remarkably, the satellites selectively permit the inward migration of electrons of specific energies. At the orbit of each satellite, those electrons drifting in longitude at the same rate as the moving satellite are able to diffuse across the orbit as though the satellite were not present, whereas electrons with other energies (and thus drift rates) will encounter and be absorbed by the satellite with varying probabilities. All of this has a selective effect on the energy spectrum of the electrons in a manner analogous to that on white light passing through a succession of overlapping colored filters. Inside the orbit of Mimas, for example, nearly all of the surviving electrons have an energy of about 1.6 million electron volts. No similar effect occurs in the magnetospheres of either the Earth or Jupiter.

But in contrast to the situation for electrons, inward-diffusing protons are strongly absorbed by Dione, Tethys, Enceladus, and perhaps the tenuous, particulate ring E. Consequently, there is a near-total absence of low- and intermediate-energy protons inward of 4 planetary radii. The protons found closer in have energies approaching or exceeding 100 million electron volts and are most common 2.7 radii from the planet's center. These potent protons could not have survived the gauntlet of inward diffusion from the

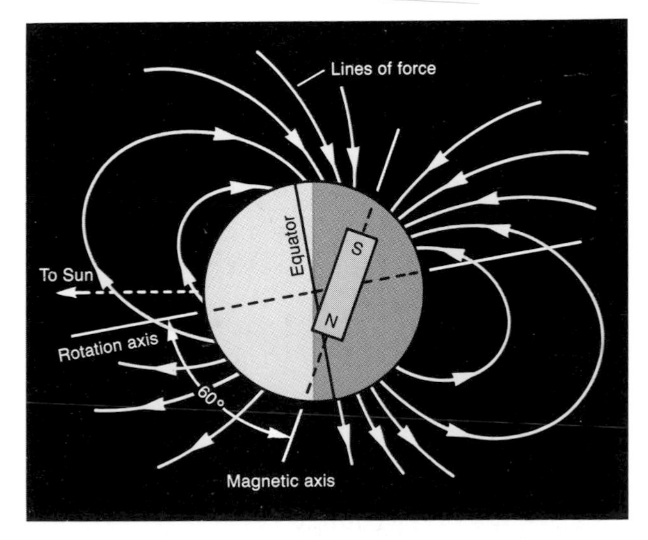

Figure 17. **A schematic diagram of the magnetic field of Uranus as observed by Voyager 2. The internal system of electrical currents that produces the magnetic field is represented by an equivalent bar magnet. Uranus' magnetic vector is tilted away from the rotation axis and offset from the planet's center more than any other world in the solar system.**

outer magnetosphere; instead, they are the decay products of neutrons released when cosmic rays pass through and interact with Saturn's upper atmosphere and ring material.

The intensity of all trapped particles drops dramatically at the outer edge of ring A, because they are absorbed by the ring material encountered there. In addition, Saturn's general magnetic field deflects away most cosmic rays before they can reach the inner magnetosphere. As a result, the region interior to the ring A cutoff is nearly free of high-energy particles – the most completely shielded region within the solar system, excluding the atmospheres and solid bodies of planets, large asteroids, and the Sun.

The imaging instrument on Pioneer 11 recorded an inner satellite first glimpsed from Earth in 1966 (since named Epimetheus) and discovered a previously unknown ring (F). It is worth noting that the spacecraft identified both of these independently by the gaps they created in the magnetosphere's electron population. Pioneer 11 picked up other distinctive absorption features, which were designated 1979 S3, S4, and S5. Fourteen months later, Voyager 1 found another new ring (G) at the location that corresponded to that of 1979 S3; the others may be the signatures of as-yet-unconfirmed rings or small satellites. The Voyagers found a number of additional small moons elsewhere in the Saturnian system.

URANUS, NEPTUNE, AND PLUTO

Voyager 2 made the first-ever encounter with Uranus in early 1986. It came closest to the planet on January 24th at a point 107,100 km (4.19 planetary radii) from Uranus' center. Prior to this encounter, the magnetic properties of Uranus were almost entirely conjectural. Pioneer 10 had already established that the solar wind extended beyond Uranus' orbit. An empirical relationship, the "Bode's law" of planetary magnetism, exists between the angular momenta and magnetic moments of Mercury, Mars, Earth, Saturn, and Jupiter; it suggested that the magnetic moment of Uranus would be about one-tenth that of Saturn. Hence, it appeared likely that the planet would be found to have a large magnetosphere. Anticipation heightened in the early 1980s, when astronomers using the International Ultraviolet Explorer satellite observed auroralike emissions from hydrogen in Uranus' upper atmosphere.

We realized ahead of time that the rotational axis of Uranus would, in early 1986, be within 8° of the planet-Sun line, whereas those of all other planets (except Pluto) are approximately perpendicular to their orbital planes. If Uranus' magnetic and rotational axes were nearly parallel, as is the case for other magnetized planets, one magnetic pole would be pointed almost directly at the Sun and a very unusual magnetospheric topology would be expected. If they were inclined markedly to one another, each magnetic pole would move in a sweeping, conical path as the planet spun – creating an even more exotic magnetosphere.

Finally, Voyager 2 ended our many years of conjecture. It found that the planet's magnetic moment was nearly the same strength as that predicted. But its orientation and eccentricity were *very* different from our expectations. Uranus' magnetic axis is tilted a huge 58°.6 from the rotational axis, and its magnetic center is offset from the core by 30 percent of the planet's radius (Figure 17). Its large magnetosphere (Figure 18) contains plasma and a large population of energetic particles. As at Saturn, the absorption of energetic particles by satellites Miranda, Ariel, and Umbriel, and by particulate matter in rings controls the development of the inner magnetosphere. But unlike the situation at Saturn, the diurnal wobble of the tilted magnetic equator creates a very complex relationship between the orbiting satellites and the magnetic field's topology. The magnetospheric rotational period was found to be 17.24 hours.

Again, a rich variety of plasma-wave phenomena and radio emissions were observed, as was an extended magnetotail. But helium and other heavier nuclei characteristic of the solar wind were conspicuously absent from the trapped-particle population. This fact favors Uranus' upper ionosphere as the primary source of such particles.

We expect the magnetic moment of Neptune to be at least the same strength as that of Uranus, based on at least two lines of evidence. First, Pioneer 10 has already established that the solar wind flows out to and beyond Neptune's orbit. Second, in 1986 radio astronomers Imke de Pater and Michael Richmond discovered the signature of what may be synchrotron radiation coming from the planet's vicinity. As at Jupiter, this energy is produced by high-energy electrons spiraling around magnetic field lines. Thus, if Voyager 2 continues to operate well, we expect to receive another rich harvest of magnetospheric phenomena. The spacecraft comes its closest to Neptune on August 25, 1989.

Thereafter, Pluto will remain the only major planet not investigated at close range, and there are no plans for a mission to Pluto within the foreseeable future. This planet and its close satellite Charon are objects of great interest in other contexts, but because of their small sizes and long rotation periods (6.39 days) it is unlikely that either is magnetic. Nonetheless, they doubtless cause numerous plasma-physical phenomena – as should the asteroids, which are also presumed to be nonmagnetic – as the solar wind flows past them.

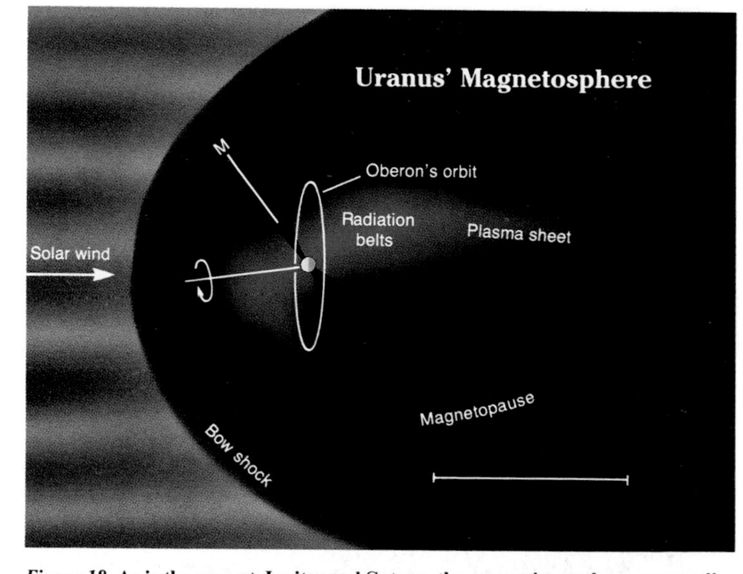

Figure 18. **As is the case at Jupiter and Saturn, the magnetic envelope surrounding Uranus is actually larger than the Sun (the scale bar corresponds the Sun's diameter). But because Uranus' magnetic axis *(M)* and rotation axis are angled far apart, its magnetosphere has an orientation unlike that of any other planet's.**

COMETS

Comet nuclei, likewise, are presumably not magnetized. But these objects create gaseous envelopes when within the inner solar system, and thus they are of interest to magnetospheric physicists. Indeed, as mentioned earlier, Ludwig Biermann's observations of the ionized-gas tails of comets provided the first evidence for the continuous nature of the solar wind.

The Sun causes both a solar-wind pressure and a radiation pressure (the latter from the "impact" of photons of light) on all objects in the solar system. On inert objects, the pressure of the impinging solar wind is about 3,000 times less than solar-radiation pressure. Both of these decrease as the inverse square of the distance from the Sun; for example, moving twice as far away decreases the pressure by four times. For a large object such as an asteroid or planet, these pressures have a negligible effect compared to gravitational forces. But for cometary dust grains, which have vastly greater area-to-mass ratios, radiation pressure is important. It is also significant even for spacecraft with large arrays of photovoltaic cells or "solar sails."

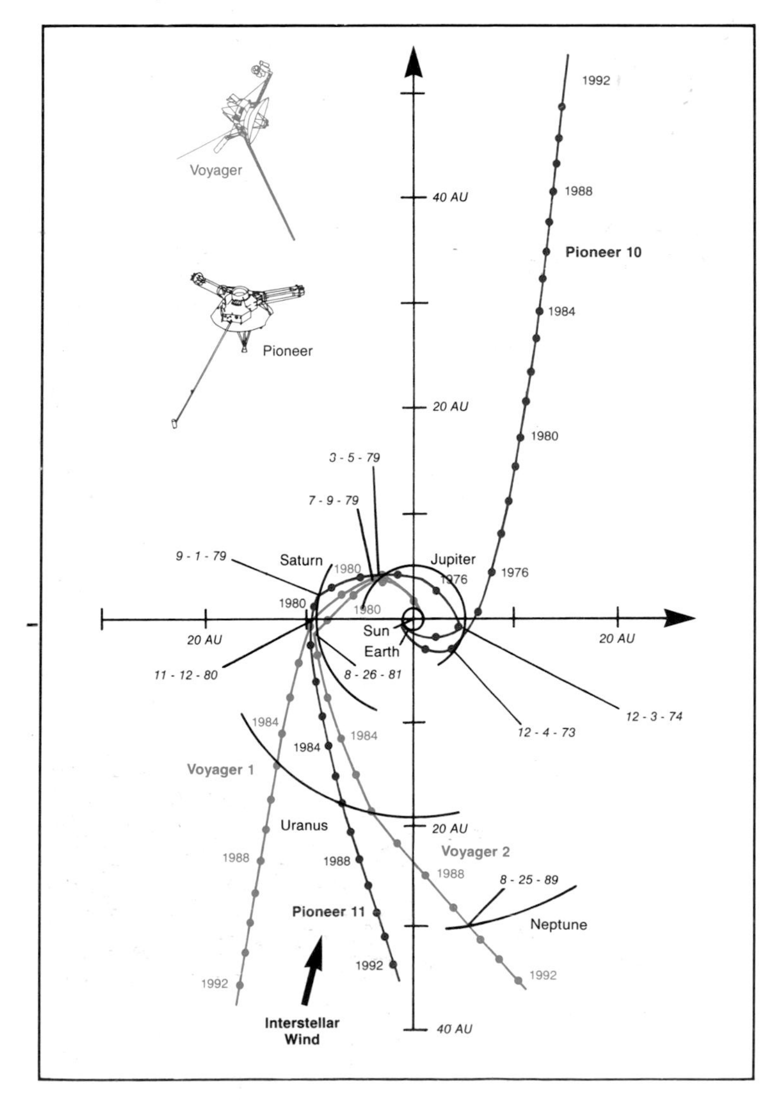

Figure 19. **Four spacecraft are now headed out of the solar system, as indicated by this plot of their trajectories (projected on the ecliptic plane). Dates of planetary encounters are indicated, and the tick marks are spaced at 1-year intervals.**

The solar wind has a much greater effect on cometary gas than does radiation pressure. Gas in a comet's coma is ionized by the Sun's ultraviolet light and by charge exchange with the interplanetary plasma moving past it. The ions are then picked up and swept along by the moving magnetic field, becoming in effect an added constituent of the outward-flowing solar wind. Many interesting plasma physical effects accompany this process. Comets' ionized comet tails are distinguished from their more familiar dust and neutral-gas tails by their form, optical spectra, and other detailed features (see Chapter 17). To test and confirm these ideas experimentally, space physicists created artificial comets in and near the Earth's magnetosphere in 1984 and 1985 as part of the AMPTE (Active Magnetospheric Particle Tracer Explorers) program. High-altitude spacecraft explosively released barium and lithium vapors, then observed (along with scientists on the ground) the consequent physical effects.

The first natural comet to be investigated by space techniques was Giacobini-Zinner. The spacecraft ISEE 3 (International Sun-Earth Explorer 3), after more than six years of operation near the L_1 Lagrangian point of the Sun-Earth system, was redirected to fly through the comet's tail in August 1985. (Months earlier, NASA renamed it the International Cometary Explorer, or ICE). In March 1986, the Soviet spacecraft Vegas 1 and 2, Europe's Giotto, and Japan's Suisei made close encounters with Comet Halley, while a number of other spacecraft observed Halley from more remote locations (see Chapter 16). The combination of data from all of these observations has added new volumes of knowledge of the plasma-physical phenomena associated with comets.

UNIVERSAL APPLICATIONS

The solar wind is an important element in the physical processes occurring in all of the planetary magnetospheres investigated so far. But it is not *essential* to the formation of radiation belts, as is amply illustrated by the Earth and Saturn. In these cases, important sources of trapped energetic particles are the neutrons produced by cosmic rays in the planets' upper atmospheres and in Saturn's rings and satellites. This process of creating charged particles by neutron decay is ubiquitous throughout our galaxy and presumably throughout the universe. Moreover, there is massive astronomical evidence that the gas pervading the universe is at least partially ionized and moving some kilometers per second with respect to any planetary object. One of our four distant robotic explorers (Figure 19) may eventually cross the heliopause to sample the interstellar medium directly and transmit its findings to Earth. Given a reasonably magnetic body lying outside the heliopause (and thus isolated from the solar wind), magnetospheric particles would not simply accumulate without limit but would strike a pressure balance with the interstellar wind.

Classical astronomy contains scarcely any mention of electrical and magnetic fields, electrical currents, or energetic charged particles. Yet comets and planetary magnetospheres provide striking examples of the importance of electromagnetic and plasma-physical phenomena. The application of knowledge from magnetospheric physics to the study of the radio, X-ray, and gamma-ray emissions from distant objects is already an active field that will likely play an increasingly important role in modern astrophysics.

4
The Moon

Paul D. Spudis

THE MOON of Earth has been a source of inspiration and curiosity throughout history. For millennia, people have gazed at its changing shape and wondered about its nature and origin. But profound changes have come within our lifetimes: we have witnessed the transformation of the Moon from a remote, passive mirror of the Sun to a planetary body with a complex history. Twelve humans have walked on the lunar surface to gather samples, take photographs, and make other scientific measurements. An even greater number of robotic explorers have scrutinized the Moon from close range.

Thanks to these and other remarkable achievements, we have begun to unravel the lunar story. Today our knowledge of the Moon is deeper and broader than for any other solar-system object save Earth. By studying the processes and evolution of this nearest planetary body, we achieve not only a deeper understanding of geologic processes in general, but a fuller appreciation of the still more complex histories of the terrestrial planets.

SURFACE FEATURES AND STRATIGRAPHY

It makes sense that we should know the Moon so well – after all, it is near enough to Earth that crude surface features can be distinguished with the naked eye. By simply looking up at the Moon, you can discern that its surface consists of two major types of terrain: relatively bright highlands (or *terrae* in Latin) and darker plains sometimes called the lunar "seas" (or *maria*).

Seen close up or through a telescope (Figure 1), the terrae resolve into an apparently endless sequence of overlapping craters, ranging in size from small craters at the limit of resolution on even the best photographs to large multi-ringed basins – some of which exceed 1,000 km in diameter. All of the basins and nearly all of the craters are the consequence of meteoritic impact (see Chapter 21). Indeed, the great number of impact scars in the lunar highlands serve to remind us that the Moon's early history was exceedingly violent. At least the top few kilometers of the crust has been repeatedly mixed by the force of these collisions.

The dark maria cover about 16 percent of the lunar surface and are concentrated on the hemisphere facing Earth. While the maria occur almost everywhere within impact basins, they are geologically distinct. Thus, it is important to distinguish between such features as the Imbrium basin (a large, ancient impact structure) and Mare Imbrium (volcanic material that later filled the basin). The maria are significantly younger than the highlands and thus have accumulated fewer craters. This difference in crater frequency is quite pronounced and easily seen through even a small telescope. Long before Apollo astronauts hopped across the lunar surface, geologists recognized that a substantial amount of time had elapsed between the heavy

Figure 1. **This photograph, obtained in 1972 by the crew of Apollo 16, is predominantly of the lunar farside – the hemisphere never seen before the space age. The large dark circle at upper left is Mare Crisium, which is on the eastern limb of the Moon as seen from Earth; below it are Mare Smythii and Mare Marginis. Innumerable craters scar the ancient, light-colored highlands, while the darker, smoother maria are younger regions flooded long ago by volcanic outpourings from the interior. These two basic terrains are distinctly visible even to the naked eye.**

System	*Age (10⁹ years)*	*Remarks*
pre-Nectarian	began: 4.6 ended: 3.92	includes crater and basin deposits and many other units formed before the Nectaris basin impact; includes formation of lunar crust and its most heavily cratered surfaces.
Nectarian	began: 3.92 ended: 3.85	defined by deposits of the Nectaris basin (a large multi-ring basin on the lunar nearside); includes almost four times as many large craters and basins as the Imbrian system; may also contain some volcanic deposits.
Imbrian	began: 3.85 ended: 3.15	defined by deposits of the Imbrium basin; includes the striking Orientale basin on the Moon's extreme western limb, most visible mare deposits, and numerous large impact craters.
Eratosthenian	began: 3.15 ended: about 1.0	includes those craters that are slightly more degraded and have lost visible rays; also includes most of the youngest mare deposits.
Copernican	began: about 1.0 ended: (to present)	youngest segment in the Moon's stratigraphic hierarchy; it encompasses the freshest lunar craters, most of which have preserved rays.

Table 1. **The basic system of lunar stratigraphy has evolved, in a relative sense, from thorough scrutiny of the Moon with telescopes and orbiting spacecraft and, in an absolute sense, from the isotopic dating of lunar rocks and soils.**

bombardment of the highlands and the final emplacement of the visible maria.

In the very best telescopic photographs, small raised lobes can be seen in some mare regions, which led to the idea that the lunar maria consist of volcanic lava flows. Photographs taken by spacecraft in lunar orbit show confirming evidence such as lava channels (sinuous rilles), domes, cones, and collapse pits. Chemical analyses made in 1967 by automated Surveyor landers – and later the study on Earth of actual lunar samples – showed that the maria are indeed volcanic outflows. They appear darker than the terrae due to their higher iron content; the lunar soil becomes momentarily molten where a meteorite hits it, and the heat produces glasses that are iron-rich and thus dark in color.

Geologists can go beyond the scrutiny of the Moon's impacts and volcanic landforms. They can assess the lunar surface in a fourth dimension, time, by determining the relative ages of geologically discrete surface "units." According to the geologic law of superposition (Figure 2), younger materials overlie, embay, or intrude older ones. This simple but powerful methodology has allowed us to make geologic maps of the entire Moon and to produce a formal stratigraphic sequence for events throughout its history (Table 1). However, stratigraphic analysis cannot by itself

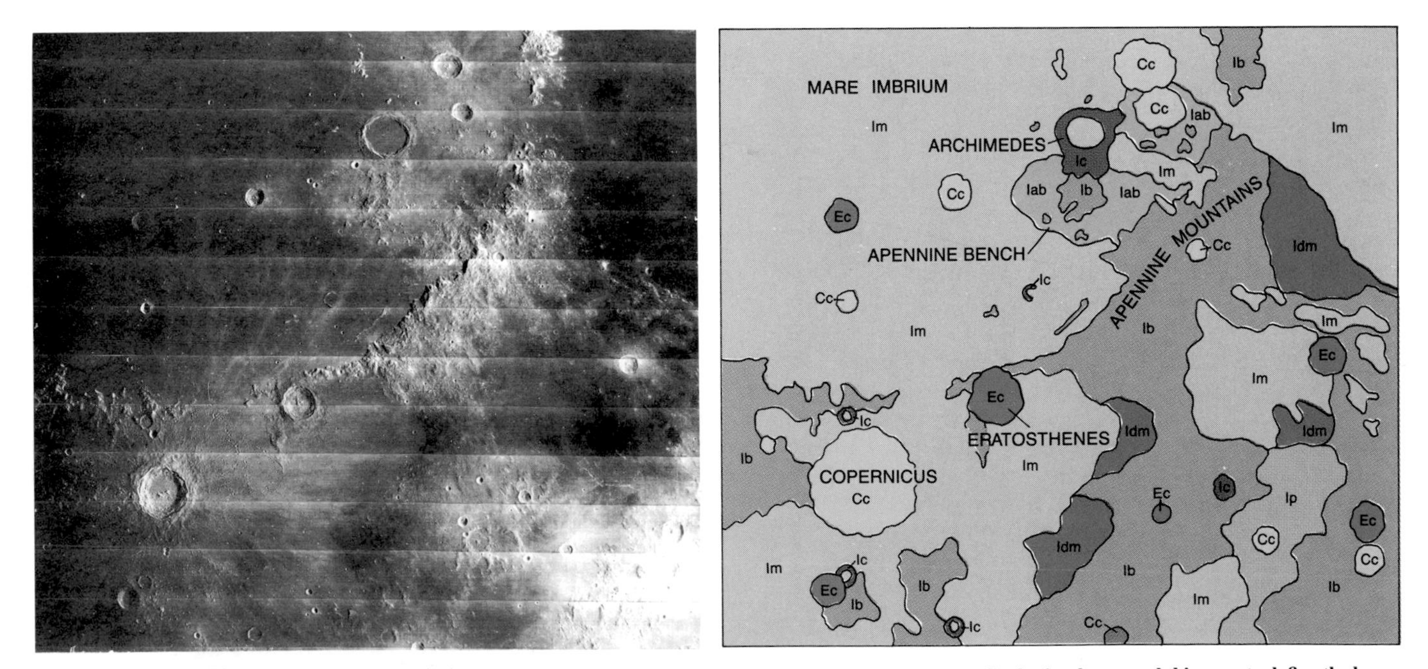

Figure 2. **At left is a photograph of the Apennine Mountain region and at right its corresponding geologic map. Geologists have used this area to define the lunar stratigraphic system (see Table 1). Debris thrown out during the formation of the crater Copernicus (yellow) lies atop all other units and is therefore the youngest. For example, faint, distant rays of Copernicus' ejecta overlie the crater Eratosthenes (green). Eratosthenes was created on top of mare basalts (pink), which in turn fill the floor of the crater Archimedes (purple). Note that Archimedes lies both on the smooth deposits of the Apennine Bench Formation (light blue) and on the Imbrium basin (dark blue); however, the Apennine Bench embays mountains rimming the Imbrium basin mountains and is thus younger. Thus, the scene's relative ages increase as follows: Copernicus, Eratosthenes, mare deposits, Archimedes, Apennine Bench Formation, Imbrium basin.**

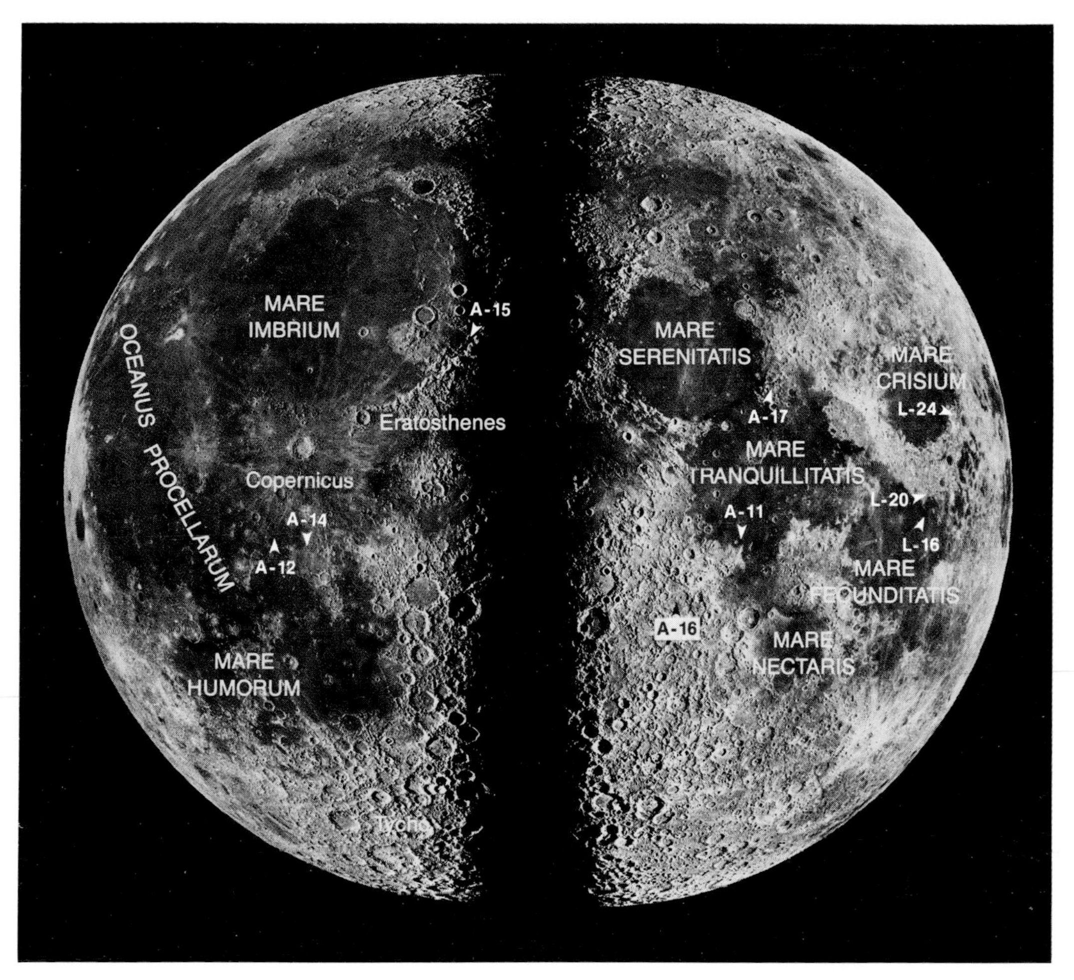

Figure 3. **A pair of Lick Observatory photographs have been labeled to show selected lunar features and the location of the nine Apollo (*A*) and Luna (*L*) sample-return sites.**

determine the *absolute* ages of surface units. Our understanding of their emplacement times, as well as their compositions and rock types, had to await the return of samples from the lunar "field trips" undertaken by the Apollo and Luna missions.

UNDERSTANDING THE LUNAR SAMPLES

From 1969 to 1972, six Apollo expeditions set down on the Moon, allowing a dozen American astronauts to explore the lunar landscape and return with pieces of its surface (Figures 3,4). The initial landing sites were chosen primarily on the basis of safety. Apollo 11 landed on the smooth plains of Mare Tranquillitatis, Apollo 12 on a mare site near the east edge of the vast Oceanus Procellarum. These first missions confirmed the volcanic nature of the maria and established their antiquity (older than 3 billion years). Later missions visited sites of increasing geologic complexity. Apollo 14 landed in highland terrain near the crater Fra Mauro, an area thought to be covered with debris thrown out by the impact that formed the Imbrium basin. Apollo 15 was the first mission to employ a roving vehicle and the first sent to a site containing both mare and highland units (the Hadley-Apennine region). Apollo 16 landed on a highland site near the rim of the Nectaris basin. The final lunar mission in the series, Apollo 17, was sent to a combination mare-highland site on the east edge of the Serenitatis basin.

Figure 4. **Apollo 17 astronaut-geologist Harrison ("Jack") Schmitt uses a special rake to collect small rock chips from the Moon's Taurus-Littrow Valley in December 1972.**

The Soviet Union has acquired a small but important set of lunar samples of its own, thanks to three automated spacecraft that landed near the eastern limb of the Moon's nearside. Luna 16 visited Mare Fecunditatis in 1970, and Luna 24 went to Mare Crisium in 1976. A third site, in the highlands surrounding the Crisium basin, was visited by Luna 20 in 1972.

Altogether, these nine missions returned 382 kg of rocks and soil (Table 2), the "ground truth" that provides most of our detailed knowledge of the Moon. While the most exhilarating discoveries came from studies completed years ago, today scientists around the world continue to examine these samples, establish their geologic contexts, and make inferences about the regional events that shaped their histories. What we've learned about the Moon's three major surface materials – maria, terrae, and the soil-like regolith that covers both – is summarized in the following paragraphs.

Regolith. Over the history of the Moon, micrometeorite bombardment has thoroughly pulverized the surface rocks into a fine-grained, chaotic mass of material called the regolith (also informally called "lunar soil," though it contains no organic matter). The regolith consists of single mineral grains, rock fragments, and combinations of these that have been cemented by impact-generated glass. Because the Moon has no atmosphere, its soil is directly exposed to the high-speed solar wind (see Chapter 3), gases flowing out from the Sun that become implanted directly onto small surface grains. The regolith's thickness depends on the age of the bedrock that underlies it and thus how long the surface has been exposed to meteoritic bombardment; regolith in the maria is 2 to 8 meters thick, whereas in highland regions its thickness may exceed 15 meters.

Not surprisingly, the composition of the regolith closely resembles that of the local underlying bedrock. Some exotic components are always present, perhaps having arrived as debris flung from a large distant impact. But this is the exception rather than the rule. The contacts between mare and highland units appear sharp from lunar orbit, which suggests that relatively little material has been transported laterally. Thus, while mare regoliths may contain numerous terrae fragments, in general these derive not from far-away highland plateaus but are instead crustal material excavated locally from beneath the mare deposits.

Impacts energetic enough to form meter-size craters in the lunar regolith sometimes compact and weld the loose soil into a type of rock called *regolith breccia*. Once fused into a coherent mass, a regolith breccia no longer undergoes the fine-scale mixing and "gardening" taking place in the unconsolidated soil around it. Thus, regolith breccias are "fossilized soils" that retain not only their ancient composition but also the chemical and isotopic properties of the solar wind from the era in which they formed.

Maria. Thanks to our lunar samples, there is no longer any doubt that the maria are volcanic in nature. The mare rocks are *basalts* (Figure 5), which have a fine-grained or even glassy crystalline structure (indicating that they cooled rapidly) and are rich in iron and magnesium. Basalts are a widespread volcanic rock on Earth, consisting mostly of the common silicates pyroxene and plagioclase, numerous accessory minerals, and sometimes olivine (an iron-magnesium silicate). But the lunar basalts display some interesting departures from this basic formulation. For example, they are completely devoid of water – or indeed any form of hydrated mineral – and contain few volatile elements in general. Basalts from Mare Tranquillitatis and Mare Serenitatis are remarkably abundant in titanium, sometimes containing roughly 10 times more than is typically found in their terrestrial counterparts.

The mare basalts originated hundreds of kilometers deep within the Moon in the total absence of water and the near-absence of free oxygen. There the heat from decaying radioactive isotopes created zones of partially molten rock, which ultimately forced its way to the surface. The occurrence of mare outpourings within impact basins is no chance coincidence, for the crust beneath these basins must have been fractured to great depth by the cataclysmic impacts that formed them. Much later, molten magmas rose to the surface through these fractures and erupted onto the basin floors.

Although they may appear otherwise, the maria average only a few hundred meters in thickness. These volcanic veneers tend to be thinner near the rims that confine them and thicker over the basins' centers (as much as 2 to 4 km in some places). What the maria may lack in thickness they make up for in sheer mass, which frequently its great enough to deform the crust underneath them (Figure 6). This has stretched the outer edges of the maria (creating fault-like depressions called grabens) and compressed their interiors (creating raised "wrinkle" ridges).

Basalts returned from the mare plains range in age from 3.8 to 3.1 billion years, a substantial interval of time. But small fragments of mare basalt found in highland breccias solidified ever earlier – as long ago as 4.3 billion years. We do not have samples of the youngest mare basalts on the Moon, but stratigraphic evidence from high-resolution photographs suggests that some mare flows actually embay (and therefore postdate) young, rayed craters and thus may be no older than 1 billion years.

A variety of volcanic glasses – distinct from the ubiquitous, impact-generated glass beads in the regolith – were found in the soils at virtually all the Apollo landing sites. They even were scattered about the terrae sites, far from the nearest mare. Some of these volcanic materials are similar in chemical composition, but not identical, to the mare basalts and were apparently formed at roughly the same time.

One such sample, tiny beads of orange glass, came from the Apollo 17 site (Figure 7). They are akin to the small airborne droplets accompanying volcanic "fire fountains" on Earth, like those in Hawaii. The force of the eruption throws bits of lava high into the air, which solidify into tiny spherules before hitting the ground. The Moon's volcanic glass beads have had a similar origin. The orange ones from the Apollo 17 site get their color from a high titanium content, more than 9 percent, and some of them are coated with amorphous mounds of volatile elements like zinc, lead, sulfur, and chlorine.

Terrae. One could easily imagine the lunar highlands to contain outcrops of the original lunar crust – much as we find in Earth's continents. But what really awaited the astronauts was a landscape so totally pulverized that no traces of the original outer crust survived intact. Instead, most of the stones collected from the terrae were breccias (Figure 8), usually containing fragments from a wide variety of rock

Mission	*Arrival date*	*Landing site*	*Latitude*	*Longitude*	*Sample returned*
Apollo 11	July 20, 1969	Mare Tranquillitatis	0°67′ N	23°49′ E	21.7 kg
Apollo 12	Nov. 19, 1969	Oceanus Procellarum	3°12′ S	23°23′ W	34.4 kg
Apollo 14	Jan. 31, 1971	Fra Mauro	3°40′ S	17°28′ E	42.9 kg
Apollo 15	July 30, 1971	Hadley-Apennine	26° 6′ N	3°39′ E	76.8 kg
Apollo 16	Apr. 21, 1972	Descartes crater	9°00′ N	15°31′ E	94.7 kg
Apollo 17	Dec. 11, 1972	Taurus-Littrow	20°10′ N	30°46′ E	110.5 kg
Luna 16	Sep. 20, 1970	Mare Fecunditatis	0°41′ S	56°18′ E	100 g
Luna 20	Feb. 21, 1972	Apollonius highlands	3°32′ N	56°33′ E	30 g
Luna 24	Aug. 18, 1976	Mare Crisium	12°45′ N	60°12′ E	170 g

***Table 2.* A total of nine "field trips" to the Moon have returned with samples of the lunar surface. In addition to these missions, the United States successfully soft-landed five automated Surveyors and the Soviet Union four other Lunas (including two Lunakhod rovers) on the Moon between 1966 and 1973.**

types that have been fused together by impact processes. Most of these consist of still-older breccia fragments, attesting to a long and protracted bombardment history.

The highland samples also include several fine-grained crystalline rocks with a wide range of compositions. They are not breccias, but they *were* created during an impact. In these cases the shock and pressure were so overwhelming that the "target" melted completely, creating in effect entirely new rocks from whatever ended up in the molten melee. Of course, the impacting "bullets" become part of this mixture, and these impact-melt rocks contain distinct elemental signatures of meteoritic material.

Based on the samples in hand, virtually all of the highlands' breccias and impact melts formed between about 4.0 and 3.8 billion years ago. The relative brevity of this interval surprised researchers – why were all the highland rocks so similar in age? Perhaps the rate of meteoritic bombardment on the Moon increased dramatically during that time. Alternatively, the narrow age range may merely mark the conclusion of an intense and continuous bombardment that began 4.6 billion years ago, the estimated time of lunar origin. To resolve the enigma, we must return to the Moon and sample its surface at carefully selected geologic sites.

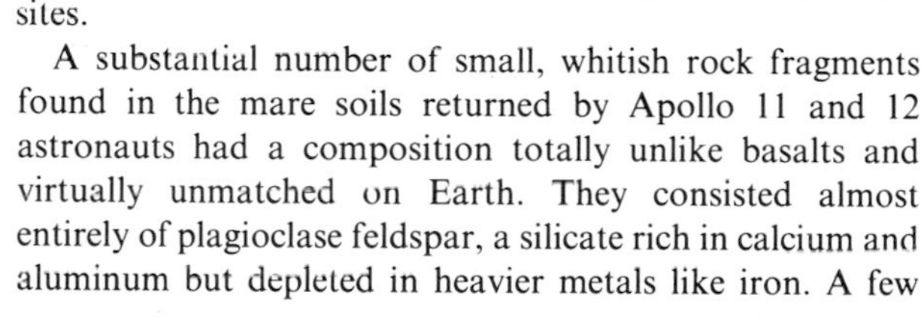

A substantial number of small, whitish rock fragments found in the mare soils returned by Apollo 11 and 12 astronauts had a composition totally unlike basalts and virtually unmatched on Earth. They consisted almost entirely of plagioclase feldspar, a silicate rich in calcium and aluminum but depleted in heavier metals like iron. A few

***Figure 5.* This mare basalt, sample 15016 from the Hadley-Apennine landing site of Apollo 15, crystallized 3.3 billion years ago. The hand specimen's numerous vesicles (bubbles) were formed by gas that had been dissolved in the basaltic magma before it erupted. By shining polarized light through paper-thin slices of a lunar rock, geologists can learn much about its crystal structure and composition. Sample 15016 exhibits the minerals plagioclase (lath-shaped black and white crystals), pyroxene (lath-shaped colored crystals), olivine (roundish, brightly colored grains), and ilmenite (opaque).**

***Figure 6.* The western edge of Mare Serenitatis, looking north, as photographed from Apollo 17. The mare's surface exhibits numerous deep rilles (bottom center) and wrinkle ridges that resulted from strain and deformation within the massive basalt sheet. Mare Imbrium is on the horizon at top left.**

prescient researchers postulated that these rocks came from the lunar highlands. The last four Apollo missions, sent to highland landing sites, confirmed that plagioclase feldspar dominates the lunar crust. The resulting implication was broad and profound: at some point in the distant past much of the Moon's exterior – and perhaps its entire globe – had been molten.

The detailed nature of this waterless "magma ocean" is only dimly perceived at present; for example, the lunar surface may not have been everywhere completely molten. But the consequences seem clear. In a deep, slowly cooling layer of lunar magma, crystals of low-density plagioclase feldspar would have risen upward after forming, while higher-density minerals would have accumulated at lower levels. This segregation process, termed *differentiation*, left the young Moon with a crust that was, in effect, a low-density rock "froth" tens of kilometers thick consisting mostly of plagioclase feldspar. At the same time, denser minerals (particularly olivine and pyroxene) became concentrated in the mantle below – the future source region of mare basalts.

It is unclear to what depth the magma ocean extended, but the volatile-element coatings discovered on some mare glasses provide an important clue. If the Moon's exterior really was once molten, the most volatile components in the melt would have vaporized and escaped into space. But the volatile-coated glasses sprayed onto the lunar surface long after the magma ocean solidified. If the glasses' compositions did not change in their upward migration from the lunar interior, they imply that volatile-rich pockets remained (and perhaps still exist) in the upper mantle. The implication, therefore, is that the magma ocean was at most only a few hundred kilometers deep.

The highland samples returned by the last four Apollo crews provided other surprises. Unlike glasses and basalts, which quench quickly after erupting onto the surface, some of the clasts in the highland breccias contained large, well-formed crystals, indicating that they had cooled and solidified slowly, deep inside the Moon. These igneous rocks sometimes occur as discrete specimens (Figure 9). At least two distinct magmas were involved in their formation. Rocks composed almost completely of plagioclase feldspar, with just a hint of iron-rich silicates, are called *ferroan anorthosites*. These appear to be widespread in the highlands. Absolute dating of the anorthosites has proved difficult, but it appears that they are extremely ancient, having crystallized very soon after the Moon formed (4.6 to 4.5 billion years ago).

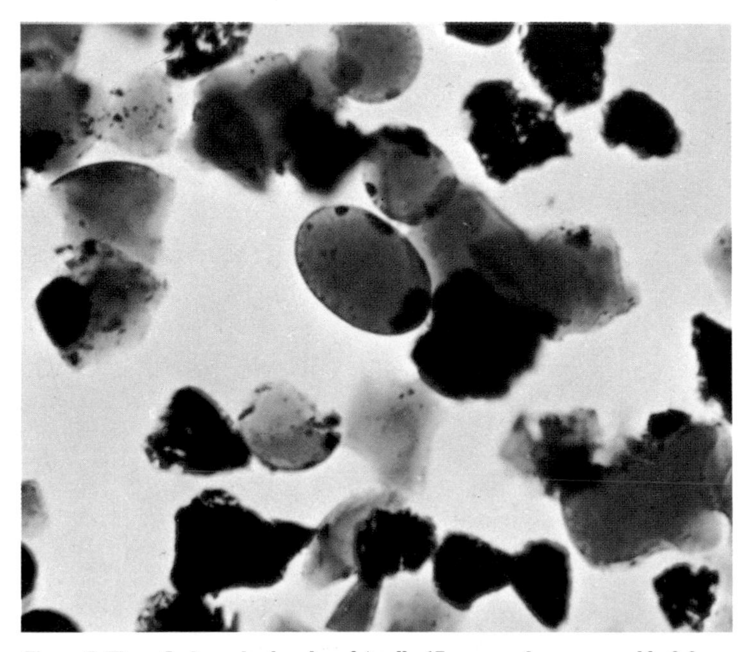

Figure 7. **These flecks and spherules of Apollo 17 orange glass are roughly 0.3 mm across. The lunar equivalent to terrestrial ash deposits, they were sprayed onto the Moon's surface about 3.7 billion years ago in erupting fountains of basaltic magmas. The black particles are pieces of orange glass that have crystallized over time.**

The highlands' other dominant rock type is also abundant in plagioclase feldspar, but it contains substantial amounts of olivine and a variety of pyroxene low in calcium. This second class of rocks is collectively termed the *Mg-suite*, so called because they contain considerable magnesium (Mg). These rocks appear to have undergone the same intense impact processing as the anorthosites, and their crystallization ages vary widely – from about 4.3 billion years to almost the age of the Moon.

The anorthosite and Mg-suite rocks could not have crystallized from the same "parent" magma, so at least two (and probably more) deep-seated sources contributed to the formation of the early lunar crust. Conceivably, both magmas might have existed simultaneously during the first 300 million years of lunar history. This would contradict our notion of the Moon as a geologically simple world and greatly complicate our picture of the formation and early evolution of its crust.

During early study of the Apollo samples, an unusual chemical component was identified that is enriched in incompatible trace elements – those that do not fit well into the atomic structures of the common lunar minerals plagioclase, pyroxene, and olivine as molten rock cools and crystallizes. This element group includes potassium (K), rare-earth elements (REE) like samarium, and phosphorus (P); geochemists refer to this element combination as *KREEP*. It is a component of many highland soils, breccias, and impact melts, yet the trace-element abundances remain remarkably constant wherever it is found. Moreover, its estimated age is consistently 4.35 billion years. These characteristics have led to the consensus that KREEP represents the final product of the crystallization of a global magma system that solidified aeons ago.

But the evidence for chemically distinct, widespread volcanic rocks in the highlands – KREEP-rich or otherwise – remains tenuous. Some highland rocks are compositionally similar to mare basalts yet exhibit KREEP's trace-element concentrations. For example, the Apollo 15 astronauts returned with true basalts that probably derive from the nearby Apennine Bench Formation, a large volcanic outflow situated along the Imbrium basin's rim. These "KREEP basalts" have a well-determined age of 3.85 billion years, so the Imbrium impact must have occurred before this date and probably just before the Apennine Bench Formation extruded onto the surface. Thus, although the extent and importance of highland volcanism remains unknown, it apparently took place early in lunar history and contributed at least some of the KREEP component observed in highland breccias and impact melts.

Apart from the Apollo and Luna samples, our knowledge of the distribution of rock types elsewhere on the Moon is limited to telescopic observations of the nearside and to a near-equatorial zone observed from lunar orbit by geochemical instruments aboard Apollos 15 and 16. These remote observations (Figure 10) provide tantalizing indications that diverse rock types beyond anorthosites exist in the highlands and that a wide variety of volcanic flows cover the maria, only about a third of which are similar to our sampled basalts. Particularly intriguing are indications that the central peaks of the crater Copernicus may contain large amounts of olivine and that pure anorthosite deposits exist in the inner rings of the Nectaris and Orientale basins.

THE LUNAR INTERIOR

What little we know about the internal structure of the Moon comes from seismic measurements made at the Apollo 12, 14, 15, and 16 landing sites. Mild "moonquakes" shake the lunar interior from time to time and were recorded by the four seismic stations. Some of these seemed to emanate from the upper mantle, while others came from the deeper within. Most important, the seismometers were able to record occasional impacts on the Moon – both natural and man-made. We derive our knowledge of the interior by measuring how the resulting shock waves of differing frequency propagate through and around the lunar globe.

On the basis of these data, geophysicists believe that, early in lunar history, an intense meteoritic bombardment shattered and brecciated the crust to a depth of a few tens of kilometers. In the aeons since, impacts have continued to pound the uppermost crust and mix it to depths of at least 2 km but perhaps down to 10 or 20 km. The seismic data also suggest that below about 25 km fractures in the crust are self-annealing, and that rocks deeper still may be largely intact (except those under large impact basins like Imbrium).

On average, the lunar crust is about 70 km thick, but it varies from a few tens of kilometers beneath the mare basins to over 100 km in some highland areas. Under some of the largest basins, the crust was weakened (and indeed partially removed) so much that the mantle has bulged upward. One manifestation of this movement is that basin floors are frequently raised and fractured. Moreover, the intrusion of dense mantle material into the crust changes the local gravity field – an orbiting spacecraft that passes over these mass concentrations, or *mascons*, experiences slight changes in velocity that can be used to map the mascons' locations.

The lunar crust varies from region to region, but does it contain stratified layers as well? The sparse seismic results do not require different rock types to exist in the lower crust. But we know that mafic (iron/magnesium-rich) rocks exist in abundance on the rims of the large impact basins Imbrium and Serenitatis – precisely where material blasted out from great depths in the crust should have come to rest. These basaltic rocks have some peculiar properties. They were formed 3.9 to 3.8 billion years ago (the age of the last basin-forming impacts) but cannot be made by melting any combination of the known highland rock types. They also contain rock and mineral clasts of relatively deep-seated origin (no soil or regolith-breccia fragments are present). If these rocks were thrown out as molten ejecta during the cataclysmic blasts that formed the basins, they provide direct evidence for a lower crust that is more mafic than the "average" upper crust.

The mantle constitutes about 90 percent of the volume of the Moon and is thought to consist of an olivine-pyroxene mixture that varies both regionally and with depth in complex ways that are not fully understood. Source regions for the mare basalts were apparently situated 200 to 400 km below the surface, so our mare samples provide tracers of the upper-mantle compositions. For example, at least some zones in the mantle must contain large concentrations of ilmenite (an iron-titanium oxide), because they spawned the titanium-rich basalts found in Mare Tranquillitatis and Mare Serenitatis.

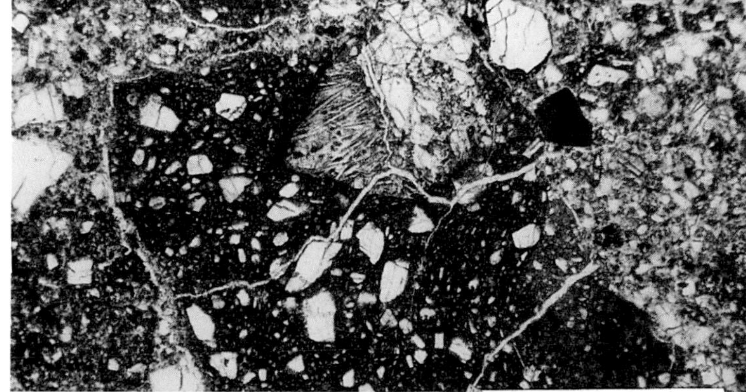

Figure 8. **A breccia from the lunar highlands, sample 67015, collected at the Apollo 16 landing site near Descartes crater. This rock is termed polymict because it consists of numerous fragments of pre-existing rocks, some of which are themselves breccias. Fused into a coherent mass about 4.0 billion years ago, this breccia demonstrates dramatically how impacts have altered rocks on the lunar surface.**

The Moon's center of mass is offset from its geometric center by about 2 km in the direction of the Earth, probably because the crust is generally thicker on the lunar farside. This may not seem like much of an offset, but it may explain why so few maria exist on the farside of the Moon. Imagine a subsurface boundary akin to a global water table, attracted toward the center of mass with equal gravitational force at every point (Figure 11). Because of the 2-km offset, this so-called equipotential surface lies farther from the top of the crust on the farside. It is possible, therefore, that basalt magmas rising from the interior reached the surface easily on the nearside, but encountered difficulty on the farside.

The Moon currently has no global magnetic field. Yet many of our lunar samples cooled in the presence of a surprisingly strong magnetic environment that was most intense 3.8 to 3.6 billion years ago (an estimate considered crude due to our sparse sampling of the crust). The "paleomagnetism" in certain lunar samples has led some researchers to postulate that the Moon once possessed a significant global magnetic field produced by dynamo motion within a metallic-iron core.

However, the size – and even the existence – of this metallic core remains unresolved. First, the low uncompressed bulk density of the Moon (3.3 g/cm^3) means that it is depleted in iron relative to other terrestrial planets and particularly the Earth (4.5 g/cm^3). Second, the best estimate of the lunar moment of inertia implies that the Moon's interior has a nearly uniform density throughout and that an iron-rich core can be no larger than about 400 km in radius. Third, the Moon's weak interaction with the Sun's magnetosphere argues that a highly conducting lunar core can be no greater than 350 to 450 km in radius. Such a core would constitute some 2 to 4 percent of the total lunar mass.

Figure 9. **Sample 76535, collected from the lunar highlands by Apollo 17 astronauts, is a coarse-grained igneous rock containing plagioclase feldspar (white) and olivine (dark). The large crystals indicate that this rock cooled slowly, well below the lunar surface. It is a single rock type, unmixed with other material by impact. Such rocks are relatively rare in the lunar sample collection.**

HYPOTHESES OF LUNAR ORIGIN

In their surveys of the solar system, astronomers have discovered dozens of satellites around other planets. Yet, of the four inner planets, only the Earth and Mars have moons (and Mars' are probably captured asteroids). Ours is remarkably large as satellites go, particularly when compared to the modest size of Earth itself. The creation of the Moon was thus an unusual event in terms of general planetary evolution, and our knowledge of the solar system – however detailed – would be profoundly incomplete without determining how our enigmatic satellite came to exist.

Traditionally, scientists have investigated three models of lunar origin. In the simplest hypothesis, termed *co-accretion*, the Earth and Moon formed together from gas and dust in the primordial solar nebula and have existed as a pair from the outset. A second concept, called the *capture* scenario, envisions the Moon as a maverick world that strayed too near the Earth and became trapped in orbit – either intact or as ripped-apart fragments – due to our planet's strong gravity. According to the third model, termed *fission*, the Earth initially had no satellite but somehow began to spin so fast that a large fraction of its mass tore away to create the Moon.

It was hoped that our astronauts would return with results that would allow us to choose decisively from among these three models. Study of the Apollo samples has indeed provided some constraints on the true lunar origin, but none of these models has proven completely satisfactory. First, the Moon's bulk composition appears to be similar, but not identical, to the composition of the Earth's upper mantle. Both are dominated by the iron- and magnesium-rich silicates pyroxene and olivine. But one important distinction is that, unlike Earth, the Moon generally lacks volatile elements. Another involves the relative dearth in lunar material of what are termed siderophile ("metal-loving") elements such as cobalt and nickel, which tend to occur in mineral assemblages containing metallic iron.

A second key constraint comes from oxygen's three natural isotopes: ^{16}O, ^{17}O, and ^{18}O. Ratios of these isotopes are identical in lunar and terrestrial materials, which suggests strongly that the Moon and Earth originated in the same part of the solar system (see Chapters 18,22). These same ratios are different in meteorites such as the eucrites (asteroidal basalts), the so-called SNC group (possible Martian igneous rocks), and various subgroups of the ordinary chondrites.

Beyond geochemical evidence, several physical properties of the Earth-Moon system provide important clues in determining lunar origin. For example, the pair possess a great deal of angular momentum. Also, the Moon's orbit

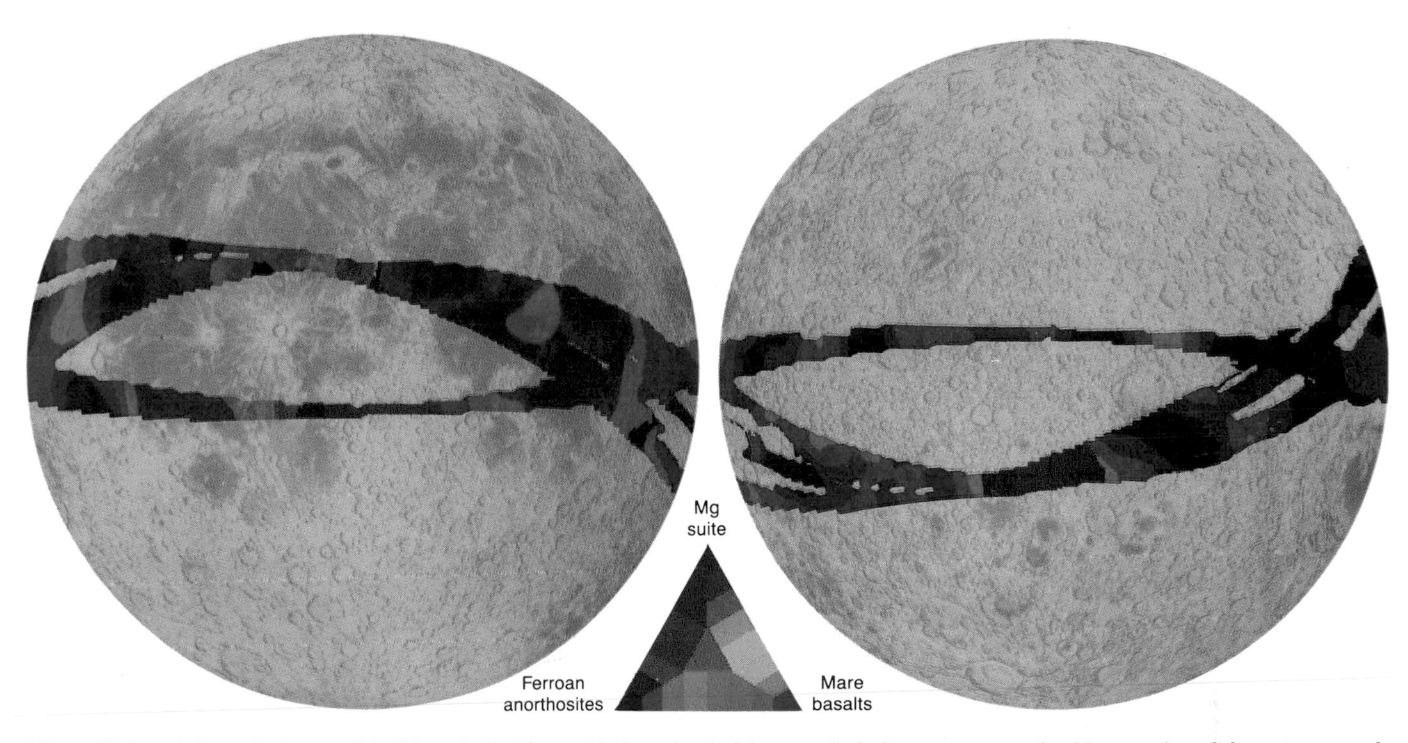

Figure 10. **A partial petrologic map of the Moon derived from orbital geochemical data, particularly gamma rays emitted by a number of elements commonly found in silicate minerals. The ratios of these emissions allows geologists to distinguish anorthosites, mare basalts, and rocks enriched in magnesium or the so-called KREEP elements on the surface below. The dominance of blue in the highlands indicates that anorthosite is the most important and widespread rock type in the lunar crust.**

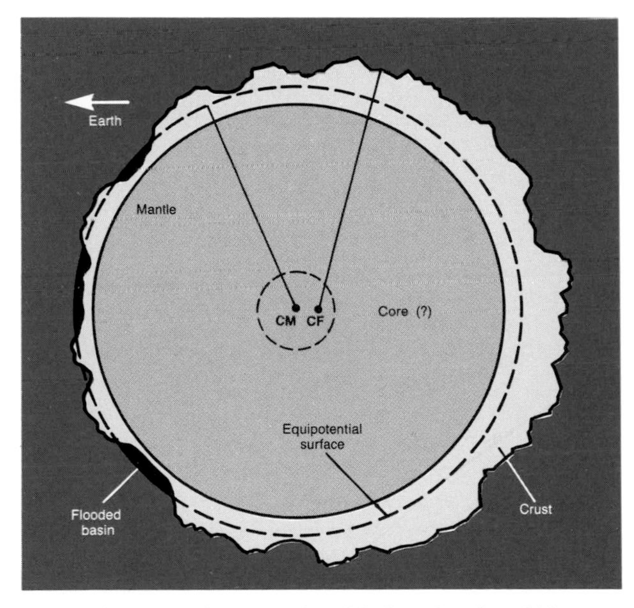

Figure 11. **A schematic cross section of the lunar interior, which may or may not include a small metallic-iron core. The Moon's center of mass (*CM*) is offset by 2 km from its center of figure (*CF*), so an *equipotential surface* (which experiences an equal gravitational force at all points) lies closer to the lunar surface on the hemisphere facing Earth. Therefore, magmas originating at equipotential depths will have greater difficulty reaching the surface on the farside.**

does not lie within in the plane either of the Earth's equator or of its orbit (the ecliptic plane). Finally, the Moon is gradually receding from the Earth at roughly 3 cm per year – a curious effect caused by the gravitational coupling of the Moon and our oceans. Tidal bulges raised in seawater do not lie directly along the Earth-Moon line but actually precede it

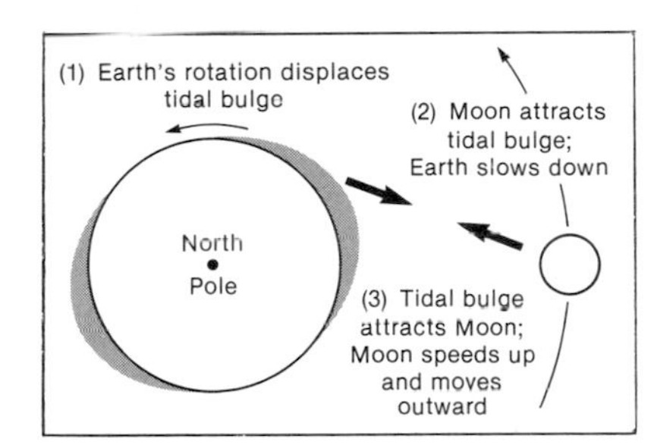

Figure 12. **Ocean tides do not actually lie along the Earth-Moon line. Over time, this misalignment slows the Earth's rotation and causes the Moon to move farther away.**

(Figure 12), because the Earth's rotation drags them along for some distance before they can adjust to the Moon's changing location in the sky. This misalignment causes the Earth's rotation to decelerate slightly; the Moon in turn is pulled forward in its orbit, speeds up, and inches farther away. Unfortunately, we cannot determine the original Earth-Moon distance because the orbital recession going on now cannot be extrapolated back to the time of lunar origin.

Recently, a new idea for the birth of the Moon has gained popularity and even something of a consensus, although all the attendant problems that it poses have yet to be resolved. This idea is that a giant object, possibly a planet-sized body as big as Mars, hit the Earth around 4.6 billion years ago (Figure 13). It may have struck off-center, thereby increasing the Earth's rotation rate. A mixture of terrestrial and impactor material would have been thrown into Earth orbit and later coalesced to form the Moon.

Because this material would have jetted into space in a predominantly vaporized state, the *giant-impact* hypothesis could explain both the Moon's dearth of volatile elements and its possible slight enrichment in refractory elements (those that remain solid at high temperature). To account as well for the Moon's depletion in metallic iron and siderophile

Factor	Co-accretion	Intact capture	Earth fission	Giant impact
Mass of the Moon	B	B	D	I
Earth-Moon angular momentum	F	C	F	B
Depletion of lunar volatile elements	C	C	B	B
Depletion of lunar iron	D	F	A	I
Oxygen-isotope match to Earth	A	B	A	B
Allows lunar magma ocean	C	D	A	A
Physical plausibility	C	D−	F	A
(Trace-element match to mantle)	(D)	(C)	(A)	(C)

***Table 3.* A "report card" for theories of lunar origin. For those unfamiliar with the U.S. educational system, *A* is the best grade, *F* is a failing grade, and *I* indicates that a grade cannot be given because the course work is incomplete. This table is by John Wood, who notes: "The line in parentheses symbolizes my doubt that similarities in trace-element abundance patterns between the mantles of Earth and Moon are a reliable constraint."**

***Figure 13.* The birth of the Moon? During the 1970's, two teams of scientists independently proposed that an object perhaps the size of Mars could have collided with Earth and thrown enough matter into orbit to create the Moon.**

elements relative to Earth, theorists must assume that the incoming object had already differentiated into a core and mantle. Their calculations show that at least half to nearly all of the lunar mass was derived from the outer layers of the colliding body (Figure 14). So to create a proper Moon, depleted in iron and siderophiles, these elements would have to be concentrated in the impactor's core, which became incorporated into the Earth shortly after the initial collision.

The giant-impact hypothesis appears to explain, or to allow for, several fundamental relations (Table 3) – not just bulk composition, but also the orientation and evolution of the lunar orbit. It also makes the uniqueness of the Earth-Moon system seem more plausible. That is, impacts of this magnitude might have occurred only rarely, rather than being a requirement for planetary formation. Part of the reason for this model's current popularity is doubtless because we know too little to rule it out: key factors such as the impactor's composition, the collision geometry, and the Moon's initial orbit are all undetermined.

Scientists realize that the advent of the giant-impact hypothesis has not "solved" the problem of lunar origin. For example, the close genetic relation of Earth and Moon (inferred from the oxygen-isotope ratios) is not an obvious consequence of a giant impact, especially if most of the lunar mass derived from the projectile. Consequently, research into the effects of such cataclysmic impacts in early planetary history continues at a brisk pace. But this model for lunar origin appears to explain the most salient features of the Moon with the minimum amount of special pleading.

A CAPSULE HISTORY

More than two decades of intensive study with spacecraft now enable us to devise an outline of the origin and geologic evolution of the Moon (Figures 15,16). However, the following scenario should be regarded only as a progress report. Many chapters in this history are still obscure, and some of the speculations here could easily be disproved by further research or, more probably, by the extensive new data that will become available when lunar exploration is resumed.

About 4.6 billion years ago, a giant object collided with the newly formed Earth, blasting a massive cloud of vaporized rock into orbit. This matter was depleted in iron, but enriched in refractory elements with respect to the terrestrial material that remained behind. Most of the entrained volatile elements were soon lost to interplanetary space. As the white-hot vapor cooled and solidified it probably formed a disk around Earth. Within some few tens of millions of years, the disk of debris coalesced to form the Moon. This assembly was quite brief by planetary standards; in fact, chunks of debris cascaded together so rapidly that the growing sphere became very hot and melted almost completely to a depth of at least a few hundred kilometers.

As the global magma ocean gradually cooled and crystallized, meteorites continued to bombard the Moon at a very high rate, fragmenting and mixing the uppermost portions of the primordial crust. On a global scale, the Moon's molten outer shell solidified by about 4.3 billion years ago, when the last residues of the original magma system crystallized as the KREEP source region. This was not the end of the Moon's magmatic life, however. Deep within the lunar mantle, radioactive heat created zones of magma that were forced upward and onto the surface as eruptions of volcanic lavas (the maria).

Meanwhile, violent collisions continued to overturn and mix the upper crustal materials thoroughly, destroying most of the original geologic formations within the primordial crust and surface outflows of volcanic rock. Some of the larger impacts created multi-ring basins that penetrated below the broken, intermixed debris layer and threw deep-seated, pristine samples of the Moon's interior onto the surface for our collection and inspection several aeons later.

The Imbrium and Orientale basins represent the last major impacts on the Moon; the Imbrium impact took place an estimated 3.85 billion years ago, and the Orientale impact probably occurred within a few tens of millions of years thereafter. At about this time the cratering rate was declining very rapidly, and more volcanic flows were being preserved from destruction. Mare volcanism may well have been more extensive before the Imbrium basin was formed, but just how much more is not known.

After about 3.0 billion years ago, the cratering rate apparently became relatively constant. The flooding of impact basins by molten basalts also began to fall off rapidly about then. Conceivably, some very small amounts of basalt surfaced onto the maria until the crater Copernicus appeared (roughly 1 billion years ago). But the dominant geologic activity on the Moon ever since has been the ongoing peppering of the surface by meteorites, punctuated by the occasional formation of a large crater. For all practical purposes, the Moon is now geologically "dead."

Despite its violent beginnings, the Moon became quiescent long ago and now affords us the opportunity to examine a "fossil" from the early solar system, a planetary body frozen in time. Its most active geologic period, from 4.6 to about 3 billions years ago, perfectly complements the observable geologic record of the Earth, for which rocks older than 3 billion years have been almost completely destroyed (see Chapter 6). Thus, the Moon holds secrets of planetary processes that we could barely imagine before the Space Age, and we see in its battered surface many of the processes observed ubiquitously throughout the solar system.

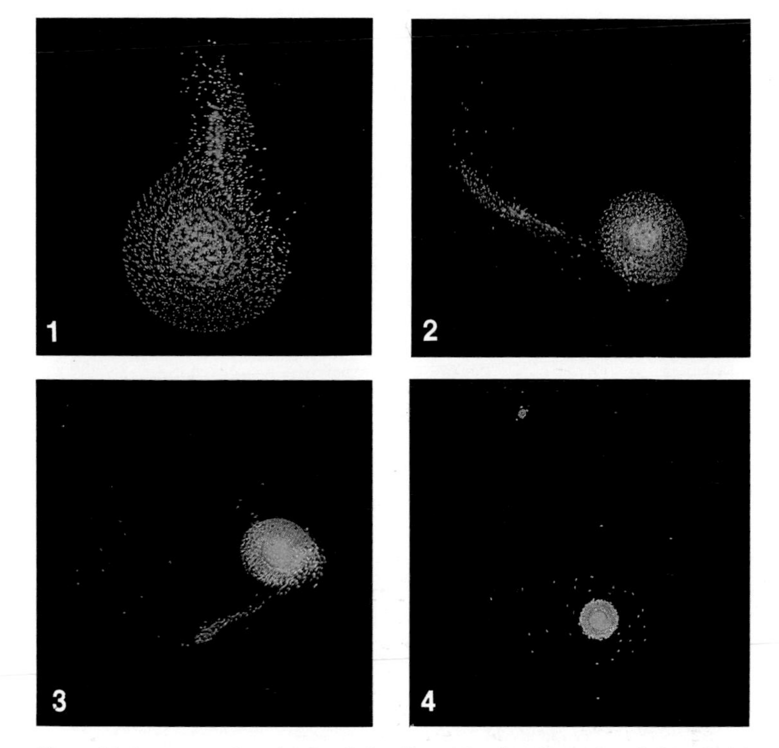

Figure 14. **A computer has simulated the effects of a giant impact on the Earth 4.6 billion years ago. The mantles of both the Earth and the impactor are vaporized; some of this material ends up in Earth orbit and forms a circumterrestrial disk from which the Moon coalesces shortly afterward.**

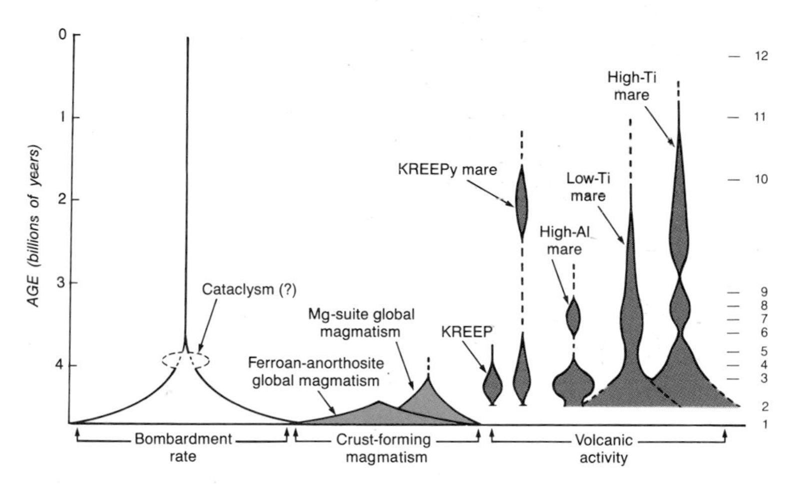

Figure 15. **The rates of impact and magmatism have varied on the Moon over time. Here the width of each envelope corresponds to the intensity of activity. Key dates, numbered at right, are: *1,* Moon forms; *2,* magma ocean solidifies; *3,* age of "average" highland surface; *4,* Orientale and Imbrium basins form; *5,* Apollo 11 and 17 maria; *6,* Luna 16 mare; *7,* Luna 24 mare; *8,* Apollo 15 mare; *9,* Apollo 12 mare; *10,* late Imbrium mare flows; *11,* Copernicus impact; *12,* Tycho impact.**

OUR FUTURE ON THE MOON

Although we have explored the Moon extensively with spacecraft, much of its surface remains *luna incognita*. Our most pressing scientific need is detailed geologic and geochemical data for the entire globe. This objective could be fulfilled by an orbiting spacecraft such as the proposed Lunar Geoscience Observer (LGO) mission. It would ideally spend at least a year in a polar orbit and systematically map the chemistry, mineralogy, topography, and morphology of the surface below. When combined with the "ground truth" of the Apollo and Luna samples, the data from such a mission should give us an unprecedented picture of the relative ages, compositions, and origins of most of the lunar crust. These new data could also be used to select a wide variety of geologic settings from which the next lunar samples are acquired, for only with actual samples can we ascertain exact chemistries and absolute ages.

Beyond getting new samples in hand, the emplacement of a global network of geophysical stations would help us learn more about the Moon's mantle and core structure, variations in its crustal thickness, and the enigmatic lunar paleomagnetism. It would also lead to a more accurate determination of the enrichment of the Moon in refractory elements through the measurement of lunar heat flow (because two of these elements, uranium and thorium, are radioactive and thus important sources of heat). A three-pronged effort – geochemical orbiter, sample returns, and geophysical network – would allow us to characterize the Moon as a global entity.

Eventually, humans will probably go the Moon to live, and the establishment of a permanent presence there opens up scientific vistas that are difficult to foresee clearly. Each Apollo mission provided some geologic surprise within its sample collection, so there is little doubt that both the variety of rock types and geologic processes that have operated on the Moon exceed by far those that we have currently deciphered. From a permanent lunar base, we could begin a detailed exploration of our complex and fascinating satellite that could last for centuries – uncovering not only its secrets, but the early history of our home planet as well.

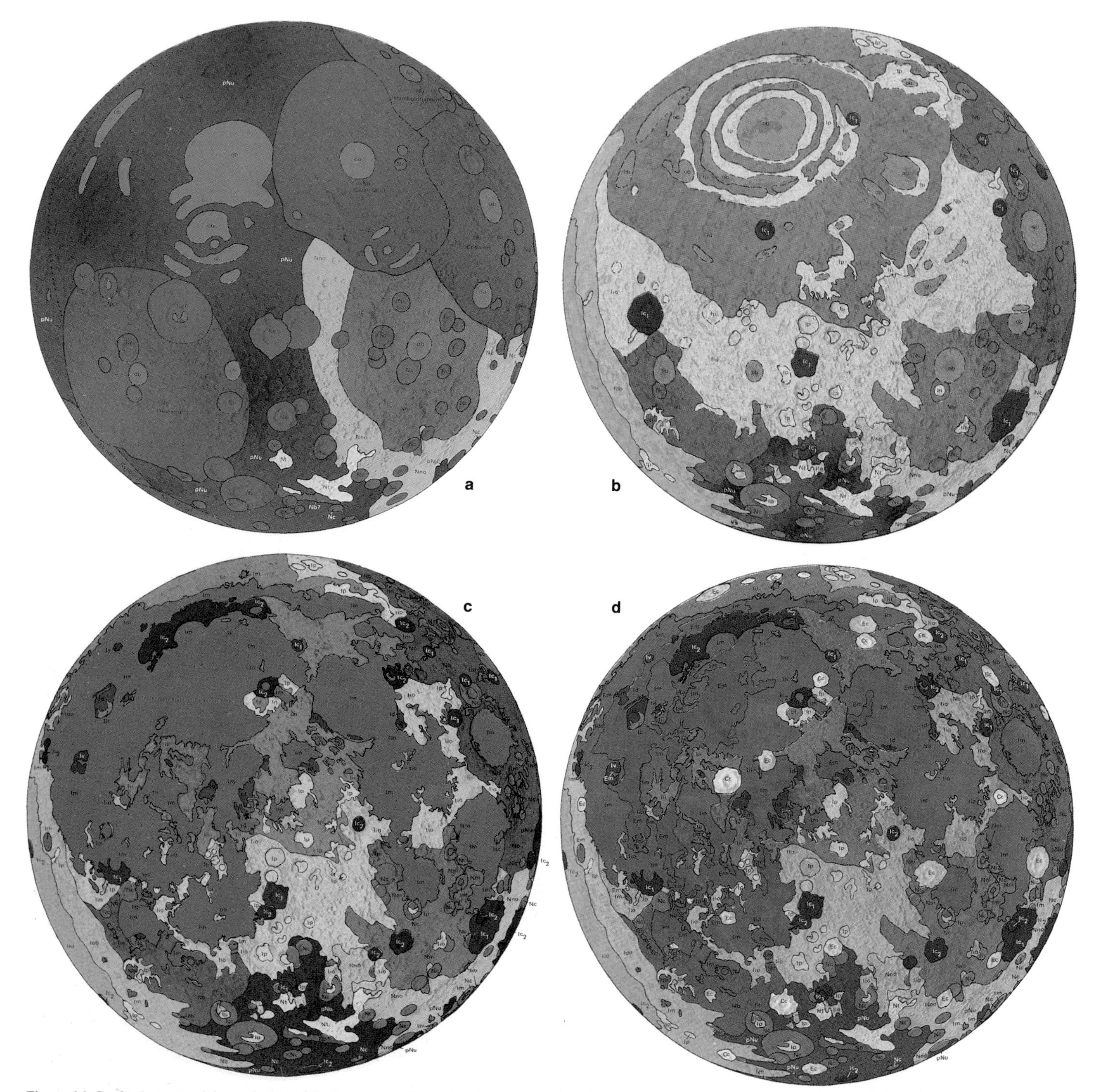

Figure 16. Geologic maps of the evolution of the lunar nearside at four key dates: (*a*) Just before the mammoth impact that formed the Imbrium basin, about 3.87 billion years ago. Brown represents pre-Nectarian and Nectarian deposits; pink is ancient mare basalts (now obliterated). (*b*) Just after the Orientale basin impact (beyond the limb at left), about 3.8 billion years ago; deposits from the Imbrium basin (blue) dominate much of the lunar nearside. Purple signifies post-Imbrium deposits. (*c*) At the end of the Imbrian Period, about 3.2 billion years ago; widespread mare basalts (red and pink) have largely covered the Imbrium basin deposits. (*d*) At present. Greens and yellows represent craters and the ejecta that surround them. The face of the Moon has remained largely unchanged for 3 billion years.

5
Mars

Michael H. Carr

WE HAVE LONG been fascinated by Mars. Its red color, periodic brightenings, and slow looping movements across the starry background make it particularly distinctive in the night sky. But in the last two centuries fascination with Mars was stimulated largely by the prospect that life may exist there and by the certainty that it will be the first planet to be visited by humans. Our perception of Mars has changed greatly over the last several decades. Early this century it was widely believed that advanced civilizations had developed there and that long linear markings, seen by many observers, were canals built to transport water from the poles to the parched equatorial deserts. As the 20th century progressed, belief in intelligent life dimmed considerably – yet even as late as the early 1960s, maps prominently portrayed canals and oases. In addition, many astronomers continued to believe that seasonal changes in the surface markings (Figure 1) could be due to changing vegetation patterns.

This perception of a planet hospitable to life changed dramatically a quarter century ago when we were able to determine Mars' surface temperature and when Mariner 4 flew by the planet and returned the first closeup pictures. These revealed an apparently lifeless, cratered landscape somewhat resembling that of the Moon. The subsequent flybys of Mariner 6 and 7 in 1969 appeared to confirm the impression of a Moonlike planet. However, our perception changed again after the Mariner 9 spacecraft arrived at Mars late in 1971. It was an orbiter, not a flyby craft like its predecessors. This proved fortunate, for when the spacecraft arrived at Mars, the entire planet was enveloped in a dust storm, and the surface could barely be seen. By early 1972, however, the dust had cleared, and Mariner 9 was then able to explore the planet systematically with its various instruments. Contrary to earlier explorations, it revealed the complex, very *un*-Moonlike planet that we know today. The Soviet Union sent four additional spacecraft to Mars in 1971, and four more in 1974, confirming the planet's geologic diversity.

The most comprehensive and detailed information we presently have about Mars came from the Viking mission, which was primarily devoted to the search for life (see Chapter 22). In 1976 NASA set two Viking spacecraft down onto the Martian surface and placed two others in orbit around the planet. All four worked flawlessly and returned data to Earth for more than four years. Although the landers' complex biology experiments did not find any life, they revealed much about the reactivity and physical properties of the soil. Other experiments on each lander observed the local scene as it changed through the seasons, recorded the local meteorology for more than two Martian years, determined the soil chemistry, and listened for marsquakes. At the same time, the two orbiters photographed the entire planet,

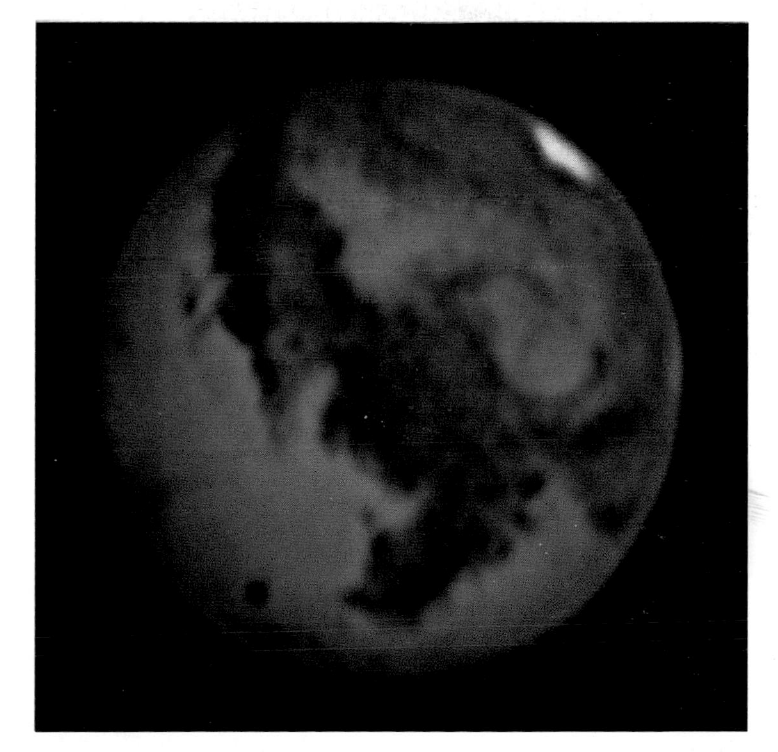

Figure 1. **During 1988, Mars was its closest to Earth since 1971 and in its best viewing location for Northern Hemisphere observers since 1875. Astronomers the world over took advantage of this fortunate geometry and made studies of the planet that were unprecedented in their detail. This spectacular image, for example, far surpasses any previous telescopic photograph of the red planet. It was obtained on September 26, 1988, when Mars was only 59.2 million km away. To make this image astronomers combined blue-, green-, and infrared-filtered images made separately with an electronic (CCD) detector and the 1.1–m reflector at Pic du Midi, an observatory high in the French Pyrenees. Mars' south polar cap is at upper right; bespeckled Mare Tyrrhenum is at center, with the prominent Syrtis Major jutting off to the north (down).**

measured the thermal properties of its surface, and monitored the water content of the atmosphere. By the mission's end, all of Mars had been photographed at a resolution of a few hundred meters, and much of it at higher resolutions that ranged down to 10 m.

We now know that Mars has had a long and varied geologic history. The planet has a global asymmetry, with most of its younger surfaces on one hemisphere and most of the older surfaces on the other. Volcanism appears to have occurred throughout the planet's history, possibly up to the present day. Deformation has taken place both on local and regional scales. The surface has also preserved a long record of impacts, despite having been extensively modified by wind, water, and ice. While many of the processes that have shaped the planet are familiar to us on Earth, the results on Mars are spectacularly different. Huge volcanoes have accumulated atop broad regional bulges. Extensive fault systems disrupt the surface. Vast canyons have formed. Several areas have been subjected to episodic floods of enormous magnitude, and closer to the poles ice has apparently caused pervasive modification of the surface.

Why do familiar geologic processes have such large-scale effects on Mars? In what ways are these processes similar to, and different from, those at work on Earth? What was been the sequence of events that led to the configuration of the Martian surface that we currently see? This chapter will try to answer these questions. The focus here will be the surface; Mars' interior is discussed in Chapter 7 and its atmosphere in Chapter 8. However, the histories of the surface, atmosphere, and interior are inextricably intertwined, and discussion of the Martian landscape leads inevitably to consideration of climate and volatile inventories. Indeed, one of the most fascinating aspects of Mars' surface is the clues it provides about former climatic conditions that were, perhaps, more hospitable to life.

Mars' geology differs from that of the Earth mainly because of its smaller size, its climate, and its lack of plate tectonism. The planet has a radius of 3,393 km, compared with 6,378 km for Earth. As a result, its interior has cooled more quickly than the Earth's, so Mars now has a lower heat flow and is less active volcanically. The Martian surface is very cold; temperatures there may reach the freezing point of water (273° K) in summer, but the average daily temperature is some 50° colder. In winter, polar temperatures fall to 150° K. The planet gets so cold mainly because the atmospheric pressure at its surface is only about 7 mb, less than 1 percent that at sea level on Earth. The atmosphere, which is almost totally carbon dioxide (CO_2), is too thin to reduce significantly heat loss from the surface. Under present conditions, the ground is permanently frozen down to depths of 1 km or more, liquid water cannot exist anywhere on the surface, and ice is stable on the surface only at the poles. Thus, liquid water, which is such an effective erosive and weathering agent here on Earth, is not available on present-day Mars. (We will see later that this was not always true.) Finally, Mars differs from Earth in that it lacks plate tectonics; that is, its crust is rigid and static. Consequently, the planet lacks analogs to many of the Earth's most prominent features that result from plate motions: mountain chains, oceanic troughs, mid-ocean ridges, and lines of volcanoes.

VIKING LANDING SITES

We have closeup views of the Martian surface only where the two Vikings touched down. The Viking 1 landing site, in Chryse Planitia at 22°.3 N, 48°.0 W, is on a plain that looks featureless from orbit except for impact craters and wrinkle ridges like those on the lunar maria. As viewed from the lander (Figure 2), the surface is gently rolling and strewn with blocks centimeters to meters across. The rocks exhibit a wide range of color, shape, and texture, probably reflecting variations in origin and length of exposure on the surface. Many are angular with coarsely pitted surfaces.

The blocks are thought to be pieces of the local volcanic rocks excavated by impact, but why there are so very many of them, as compared to lunar maria, is unclear. Perhaps the presence of an atmosphere protects the surface from microscopic erosion by small meteoritic particles. Or the Martian wind may winnow out small particles and carry them elsewhere, so that a Moonlike regolith of intermixed coarse and fine material does not develop everywhere on Mars.

Fine-grain debris is seen interspersed among the blocks and has accumulated into dunelike features that are prominent in several parts of the scene. The lander set down on an extensive patch of this dusty material – indeed, that was all the spacecraft's 3-m-long sampling arm could obtain for analysis. (The arm picked up what appeared to be rock fragments, but these all turned out to be clods of the fine-grain material.) The individual dust particles are on average extremely small, well under 10 microns across. From its chemical composition, and from simulations of the results of

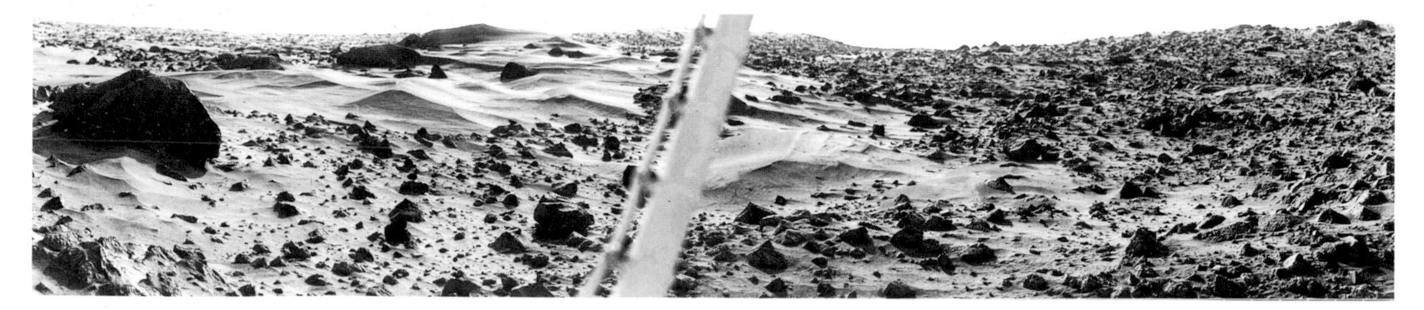

Figure 2. **Two weeks after touching down on Mars in July 1976, the Viking 1 lander returned this spectacular 100°-wide panorama from its site in western Chryse Planitia. Early morning sunlight backlights most of the scene, delineating drifts of fine-grain sediments and shadowing a pair of large rocks at left (nicknamed "Big Joe" by mission scientists) that together are about 2 m across and 8 m from the camera. Just to the right of the lander's white meteorology boom is a drift that has been scoured to the point where its internal stratification shows. The rocky area at right may contain exposed patches of bedrock.**

the biology experiments, the debris is thought to be mostly an iron-rich clay. It differs, therefore, from the lunar regolith in that it is a weathering product instead of merely ground-up rock.

Near the surface the tiny particles become cemented together to form a hardpan or duricrust, which probably results from cementation of the clays by soluble salts such as sulfates and nitrates. The soil has an additional peculiar chemical characteristic. If water is added to it, oxygen is given off. This suggests that the soil contains a small fraction of some oxidant, probably a peroxide. The soil also contains no organic compounds despite the continual rain of small amounts of organic material in meteoritic debris. The implication is that organics are destroyed by a combination of the oxidants and the ultraviolet radiation at the surface (see Chapter 22).

The Viking 2 lander set down in Utopia Planitia at 47°.6 N, 225°.7 W, part of a vast plain that extends over much of the high northern latitudes. As seen from orbit, the region is characterized by complex albedo patterns, polygonal fractures, and a variety of textures attributed to repeated deposition and removal of material by the wind. All around the lander is a level, block-strewn, but otherwise featureless plain that stretches out to the horizon (Figures 3,4). The blocks are remarkably uniform in size and appearance; most are angular and deeply pitted. No dunes can be seen, though fine-grain debris is interspersed among the blocks, commonly in their lee. We think the site got its blocky appearance when it was covered by a lobe of ejecta from Mie, a 90-km-diameter crater situated 200 km east of the landing site. The chemistry of the soil at the Viking 2 site is almost identical to that at the Viking 1 site, which suggests that the fine-grain debris is chemically homogenized over the whole planet as a result of mixing by the annual dust storms.

Figure 3. **Thousands of rocks litter the Viking 2 landing site in Mars' Utopia Planitia. Their porous texture may be the result of gas bubbles formed as the rocks cooled or of erosion by windblown dust. The largest boulder, near the center of the mosaic, measures about 0.6 by 0.3 m. A narrow trough, running from upper left to lower right, is part of a polygonal network that resembles "patterned ground" observed in Earth's polar regions. The horizon appears tilted by about 8° because the spacecraft landed with one footpad on a rock.**

Figure 4. **Patches of frost surround the Viking 2 lander in this view taken late in the spacecraft's first Martian winter. The frost probably arose from water vapor transported to the site during dust storms; dust-water aggregates precipitated when carbon dioxide froze out of the atmosphere and adhered to them. Here the frost layer is perhaps no more than a few hundredths of a millimeter thick.**

CRATERED HIGHLANDS

Mars is markedly asymmetric in the distribution of its surface features. It almost seems as if halves of two dissimilar planets have been fused together. Most of the southern hemisphere and part of the northern hemisphere (some 60° to either side of the 330° longitude meridian) are covered with heavily cratered highlands situated 1 to 4 km above the Martian equivalent of sea level. In contrast, much of the northern hemisphere is covered with sparsely cratered plains. The plains are mostly below the mean surface level, except in the volcanic regions of Tharsis and Elysium. The cause of the hemispherical asymmetry is unknown, but perhaps it is the result of a giant impact very early in the planet's history.

The abundant large craters in the highlands suggest that the terrain has survived with relatively little modification for aeons. We know that the Moon experienced a prolonged period of heavy meteoritic bombardment that ended around 3.8 billion years ago, and since that time the rate of cratering has been very low. Mars probably had a similar history, the heavily cratered highlands having formed just prior to 3.8 billion years ago and all the more sparsely cratered surfaces having formed since. Although the highlands of the Moon and Mars resemble one another, there are differences. First, the Martian craters tend to be slightly more degraded, perhaps reflecting faster erosion early in the planet's history. Second, the Martian highlands are dissected by valley networks (to be discussed more fully later). Third, the Martian highlands have extensive smooth areas between the craters. These areas probably result from relatively high rates of volcanism, and other forms of deposition, during and immediately after the heavy bombardment.

A fourth difference between the Martian and lunar highlands involves the craters themselves. Large lunar craters typically have coarse hummocky ejecta close to their rims. Farther out the ejecta has a radial "fabric," which merges outward with lines of secondary craters. In contrast, the ejecta around most Martian craters is deposited as thin sheets, each with a lobate outer margin that is clearly defined by a low ridge (Figure 5). These ejecta look like flows – indeed, where obstacles were in its path, the material clearly flowed around them. The distinctive patterns around Martian craters may result from the ejecta's interaction with the atmosphere; or perhaps the ejecta materials were water-saturated, causing them to run across the surface after being thrown out of the crater. While generally favoring the second explanation, geologists are puzzled by why such flow patterns are found high on the flanks of large volcanoes, where water-saturated ground would not be expected.

One peculiarity of the Martian highlands is of special interest because of its climatic implications. At latitudes poleward of 30° the terrain has a softened appearance, as though the surface materials had oozed and so caused a rounding of all the landforms (Figure 6). In addition, at these same latitudes, every cliff or steep slope is accompanied by flow-like features that extend up to 20 km away from its base (Figure 7). Both of these characteristics may be telling us that the cratered highlands near the poles contain abundant ice. It seems that buried ice acts as a lubricant and causes the surface materials to creep downhill. This movement is

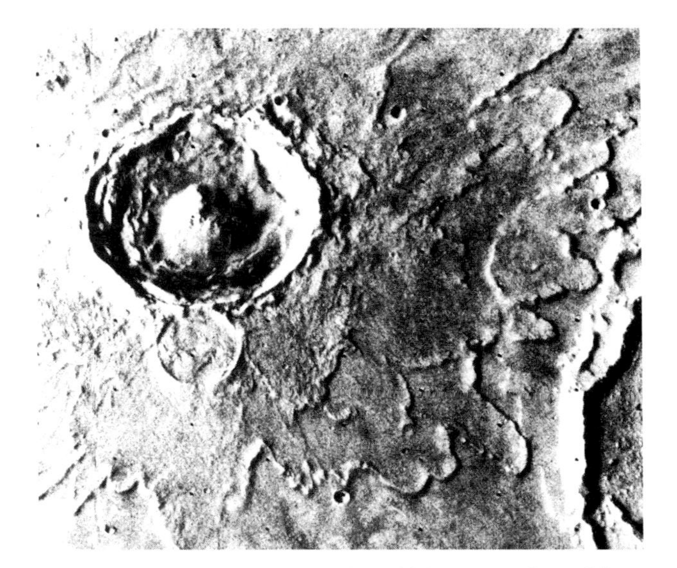

Figure 5. **Yuty, an impact crater about 20 km across, has a lobate, layered ejecta blanket. The distinctive pattern probably formed because the airborne debris was saturated with groundwater and thus tended to flow across the ground as a fluidlike mass after being ejected outward.**

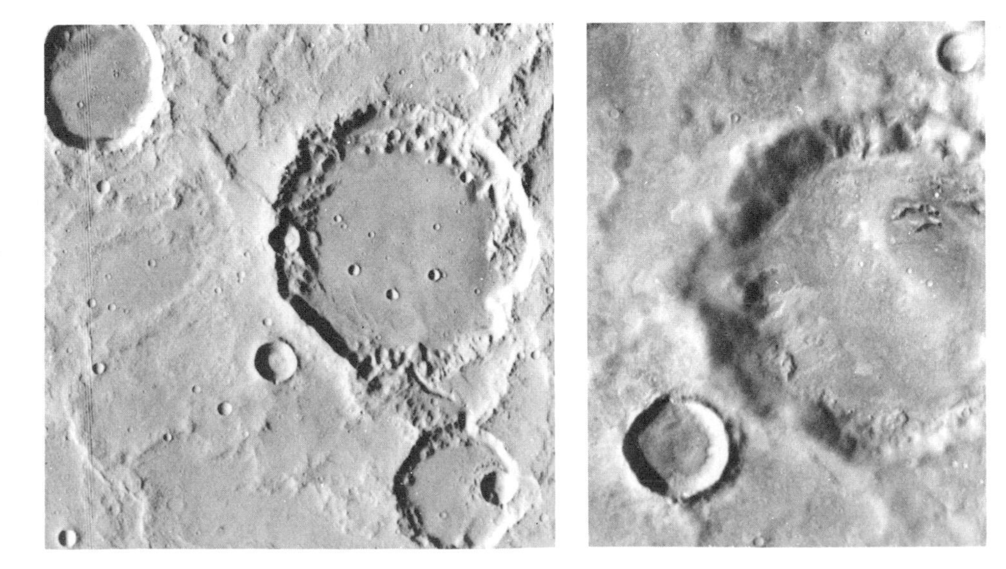

Figure 6. **Compare these Viking images of Martian craters. Those at left (the largest is 24 km across) are typical of the craters found near the equator. They have sharp rim crests, ridges and escarpments between them are well defined, and small craters are distinctly visible. Those at right, however, are typical of the heavily cratered highlands found on Mars at latitudes poleward of 35°. Crater rims are rounded, and other features are indistinct and poorly preserved. The differences seen in this comparison have been ascribed to the presence of ground ice at high latitudes, which permits the gradual creep of material near the surface.**

especially evident in the talus that accumulates on steep slopes. In contrast, relatively little ice should lie buried in the cratered highlands nearer the equator. From what we know of Martian surface conditions, at low latitudes subsurface ice is unstable and should slowly sublime into the atmosphere, whereas at high latitudes it is stable. Thus, at low latitudes, creep of surface materials cannot occur so easily; talus simply accumulates on steep slopes and protects them from further erosion.

THE PLAINS

The extensive plains that dominate Mars' northern hemisphere differ from the highlands mainly in being only sparsely cratered. They clearly postdate the period of heavy bombardment but have a wide range of ages. The most heavily cratered plains, such as those of Lunae Planum, probably formed more than 3.5 billion years ago – shortly after the end of heavy bombardment. An assortment of very sparsely cratered plains in Tharsis are probably less than 500 million years old. Some of the equatorial plains, particularly those in the volcanic regions of Tharsis and Elysium (Figure 8), clearly consist of enormous flows of lava and ash. The flows resemble those on the lunar maria; they are also very large, suggesting eruptions rates larger even than those associated with flood basalts on Earth. In many areas flows are not visible, but numerous mare-like ridges are; such plains also appear to be volcanic, though other origins are possible. On all these plains impact craters have the distinctive flow-ejecta patterns described in the previous section.

However, many plains areas, particularly those low in elevation or at high latitudes, do not fall into the two categories just described. Some have a distinctive mottled appearance because the ejecta around their impact craters is much brighter than the intervening areas. Others are crosscut by strange polygonal fractures. Still others appear to be etched or partly covered with easily eroded debris. These plains are probably of diverse origin, with volcanism, wind, and ice all participating in their formation. Since plains at the ends of large channels often bear many of these characteristics, some of them are possibly ice-rich sediments deposited by large floods.

VOLCANOES

Among Mars' most impressive features are its large volcanoes. Many occur in the Tharsis region (centered on the equator at 105° W) and Elysium (centered at 25° N, 210° W). Tharsis is situated in the middle of a huge uplift or bulge on the Martian surface that is about 4,000 km across and 10 km high at the center. On the bulge's northwest flank are three large volcanoes (Arsia Mons, Pavonis Mons, and Ascraeus Mons), and just beyond its northwest edge is Olympus Mons, the tallest volcano on the planet. All of these are enormous by terrestrial standards. The main edifice of Olympus Mons is 550 to 600 km across, rises more than 24 km above its surroundings, and is rimmed by a cliff 6 km high in some spots (Figures 9,10). To the north of Tharsis lies Alba Patera, which though only a few kilometers high is over 1,500 km

***Figure 7.* Debris flows in the Protonilus Mensae region (46° N, 311° W). The hills are isolated remnants of the cratered highlands. Material has flowed down the hills' steep slopes to form small aprons of debris, a process that was likely aided by ground ice. Such debris flows are common on Mars between 35° and 55° in latitude but rare closer to the equator.**

***Figure 8.* Overlapping lava flows cascade down the flank of Alba Patera, a huge shield volcano in Mars' northern hemisphere. The longest flows seen in these Viking orbiter images have traveled for at least 225 km, making them many times larger than eruptions on the Earth.**

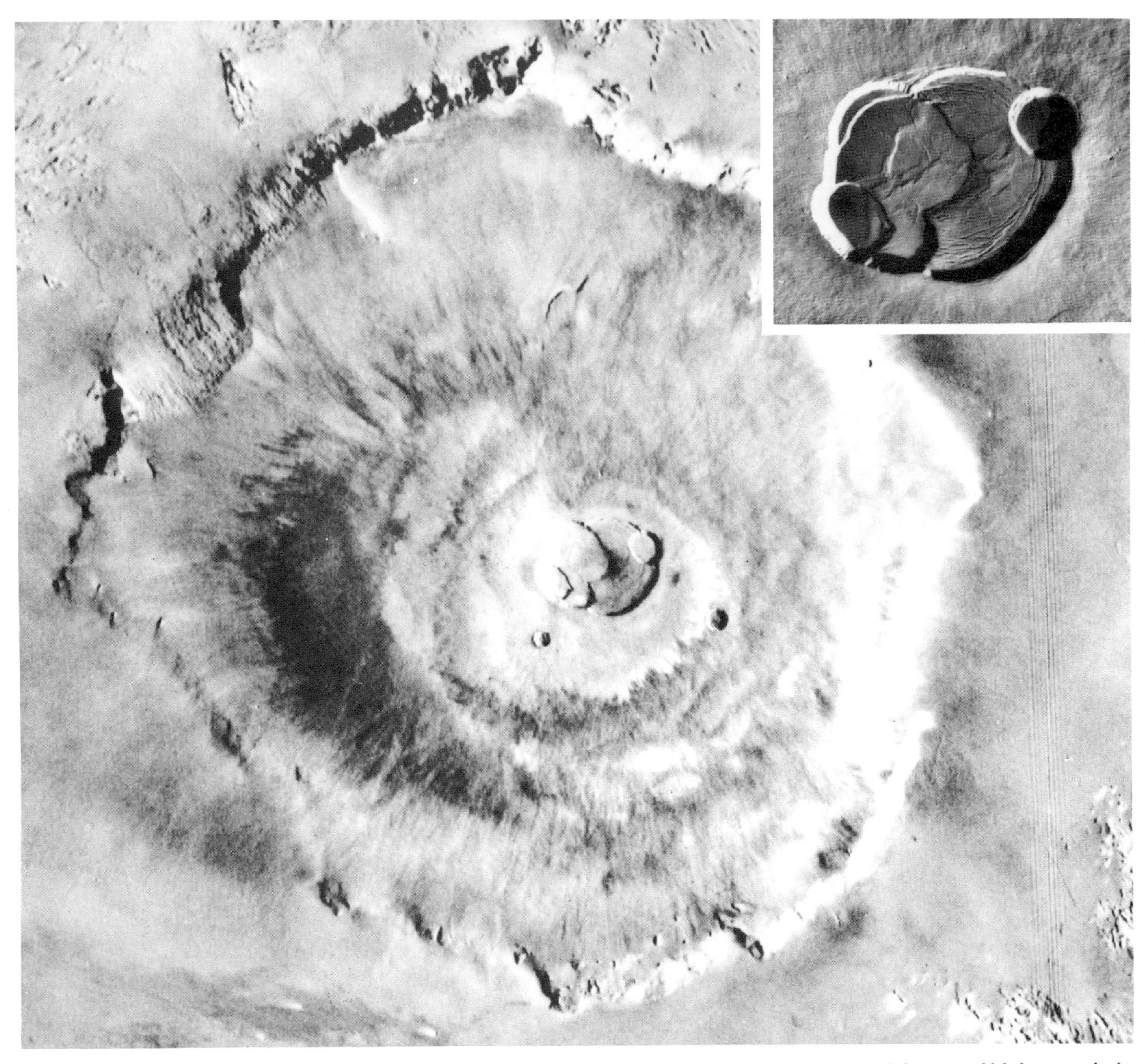

Figure 9. **The huge shield volcano Olympus Mons covers an area the size of Arizona. Its peak is roughly 24 km above the broad plateau on which the mountain sits and more than 26 km above Martian "sea level." The volcano's undulating flanks slope at an average angle of 4°. Somewhat north of its summit is a conspicuous caldera (inset), a complex nest of collapse craters 90 km across. Long before spacecraft discovered its true nature, Olympus Mons and the clouds that frequently attend it were recognized by telescopic observers as a bright spot on Mars' disk named Nix Olympica (Snows of Olympus).**

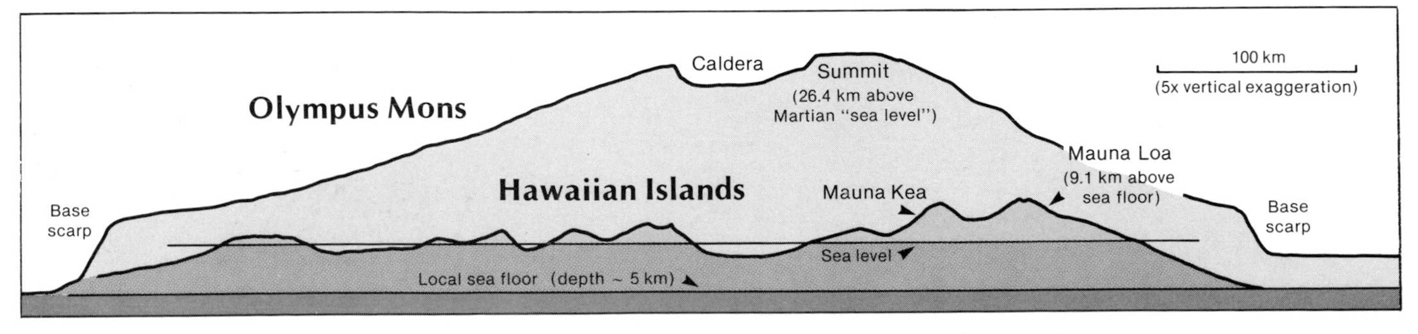

Figure 10. **Seen in cross section, Olympus Mons dwarfs the principal peaks of the Hawaiian Island chain (here depicted as if they formed in a straight line). Many parallels can nonetheless be drawn between Martian and terrestrial shield volcanoes.**

across. For comparison, the largest volcano on Earth – Hawaii's Mauna Loa – is 120 km across at its base and has a summit 9 km above the ocean floor. The Elysium province is also at the center of a bulge, but the bulge is smaller than Tharsis (2,500 km across), as are its volcanoes.

Olympus Mons and its like-size siblings resemble large terrestrial shield volcanoes such as those in Hawaii. Each has a summit pit or caldera, and numerous long flows and leveed channels cascade down its flanks. The main difference between the Martian and terrestrial volcanoes is size. Martian summit calderas are 10 to 100 times wider than terrestrial ones, probably because larger magma chambers lie beneath their summits. The flank flows and associated features are likewise 10 to 100 times longer; Mars' lower gravity may play a role here, but it appears that the rate and volume of the planet's eruptions are simply much higher than those on Earth.

The large size of the volcanoes themselves probably results from a combination of the deep source of magma that built them plus the lack of plate tectonism on Mars. Hawaii's shield volcanoes have relatively brief active lifetimes because the Pacific plate on which they stand constantly moves northwest. Once separated from its magma source, a Hawaiian volcano falls silent and a new one forms to its southeast. The result is a line of ever-older extinct volcanoes that stretches northwestward from Hawaii across the Pacific Ocean. In contrast, volcanoes on Mars remained over their magma sources and continued to grow as long as magma was available. The large height of the Martian volcanoes also implies that their magma came from very deep down and was thus under great pressure. To reach the top of Olympus Mons, magma had to squeezed up from depths of 150 to 200 km, versus about 60 km in the case of Hawaii.

Compared to elsewhere on Mars, few impact craters are superimposed on the large shield volcanoes in Tharsis and Elysium, which suggests that the mountains' current, topmost surfaces are relatively young. However, evidence from surrounding flows suggests that the volcanoes have been accumulating for much of Mars' history. Thus the planet's large shield volcanoes may actually be quite old, despite their young surfaces.

Not all Martian volcanoes have shield shapes. Some have relatively little vertical relief. Alba Patera has already been mentioned; others are found close to the large impact basin, Hellas. Tyrrhena Patera, northeast of Hellas, is very different from any of the volcanoes so far discussed. It is surrounded by deeply eroded, stratified deposits (Figure 11). Most probably, the eruptions of Tyrrhena Patera consisted primarily of ash rather than lava. In this respect the volcano more resembles Mount St. Helens than Mauna Loa. Its more explosive eruption style could arise from magma with a higher proportion of silicon than that which formed the Tharsis volcanoes, or the magma could have encountered water or ice on its way to the surface.

The Tharsis bulge has clearly played a major role in the evolution of Mars. On it are the planet's largest volcanoes and most of its youngest features. A vast system of fractures that surrounds Tharsis affects almost a third of the planet's surface. On the eastern flank is a huge equatorial canyon system, and numerous flood features scar its eastern periphery. What caused the Tharsis bulge is not known. According to one suggestion, it was formed by an upward doming of the crust, possibly in response to convection in the underlying mantle. Another theory is that the bulge is simply a thick accumulation of volcanic materials. Whatever its origin, the Tharsis bulge clearly formed very early in the planet's history, because all the visible lava flows (including very old ones) have moved downhill in directions consistent with the present topography. The surrounding fractures appear to have formed response to the huge loads on the underlying crust caused by the bulge's immense mass.

CANYONS AND CHANNELS

To the east of Tharsis, just south of the equator, is a vast system of canyons collectively known as Valles Marineris (Figure 12). The system begins in the west with Noctis Labyrinthus, at summit of the Tharsis bulge, and extends about 4,000 km eastward until the canyons merge with areas of what is termed *chaotic terrain* (to be described later). Depths range from 2 km at the east and west ends to over 7 km in the central section, where three parallel canyons merge to create a chasm over 600 km wide. The canyons appear to have formed largely by faulting. Walls typically have long, straight sections, and linear offsets are common at the base of the walls, as are triangular-faceted spurs – all common attributes of fault scarps on Earth.

Processes other than faults have also clearly played roles in forming the canyons (Figure 13). In places, particularly on the south rim of Ius Chasma, deep branching side valleys suggest fluvial (water) erosion, not by rainfall but rather by the seepage of groundwater. Elsewhere the canyons have been widened by enormous landslides. But we know that fluvial erosion did not dominate the canyons' formation, and in the west, at least, it appears to have played a minor role. There the system is poorly graded and consists largely of interconnected closed depressions. Indeed, Hebes Chasma is a large, closed canyon entirely isolated from the others. Farther east, the canyons become better graded, and at the extreme eastern end, where the canyons merge with chaotic terrain, are many water-formed features such as teardrop-shaped islands and streamlined walls. Thus the Martian canyonlands had a complex origin, with faulting and subsidence common in the west and fluvial effects common in the east.

One of the most intriguing aspects of the canyons is the presence, in places, of thick stacks of layered sediments. They

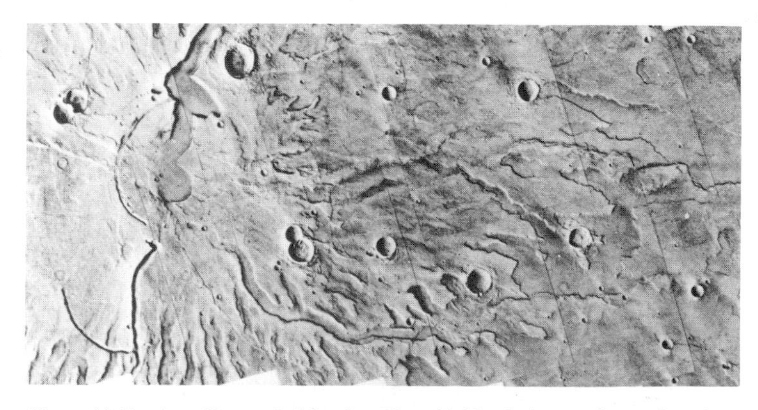

Figure 11. **Tyrrhena Patera, in Mars' southern highlands, is one of several ancient, degraded volcanic centers found on the planet. Partly eroded deposits (probably volcanic ash), with superimposed impact craters, surround a deteriorating central crater that is about 45 km across.**

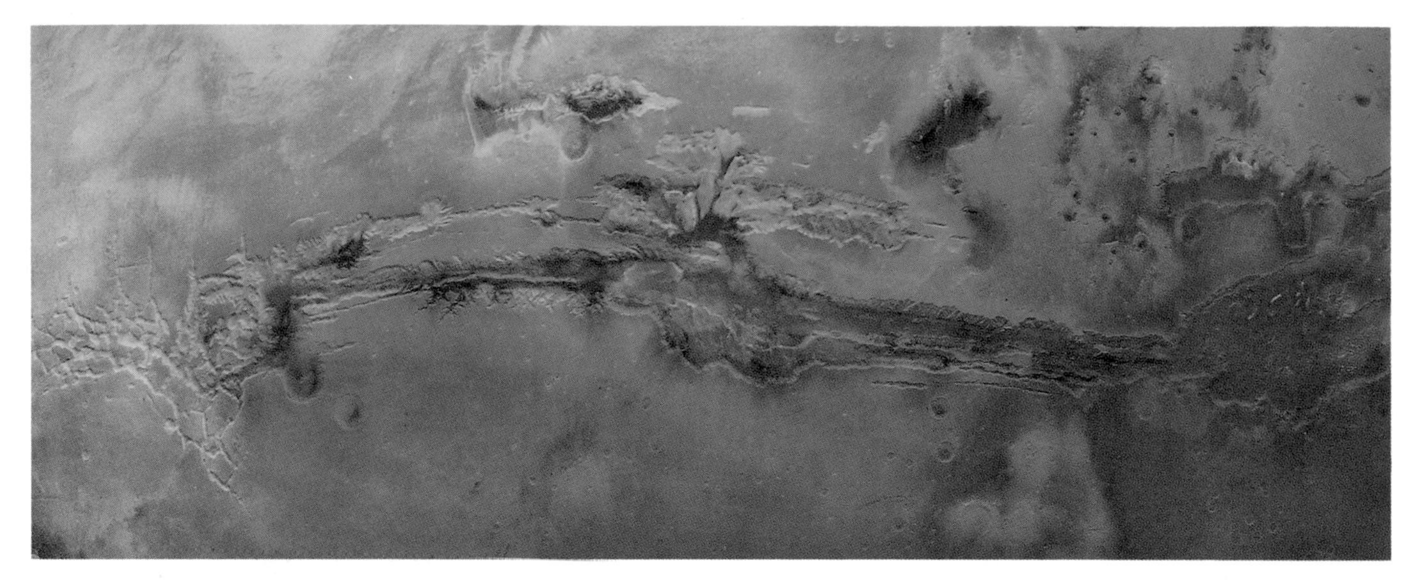

Figure 12. The vast canyon complex called Valles Marineris splits Mars equatorial region for more than 4,000 km. It begins at the crest of the Tharsis bulge with Noctis Labryinthus, the strange polygonal network at left (west), continues eastward as an interconnected series of enormous steep-walled rifts, and empties to the north at its east end in the dark-hued region known as Margaritifer Sinus. This view of Valles Marineris is part of a larger mosaic of 102 images obtained by the Viking orbiters from an altitude of 32,000 km.

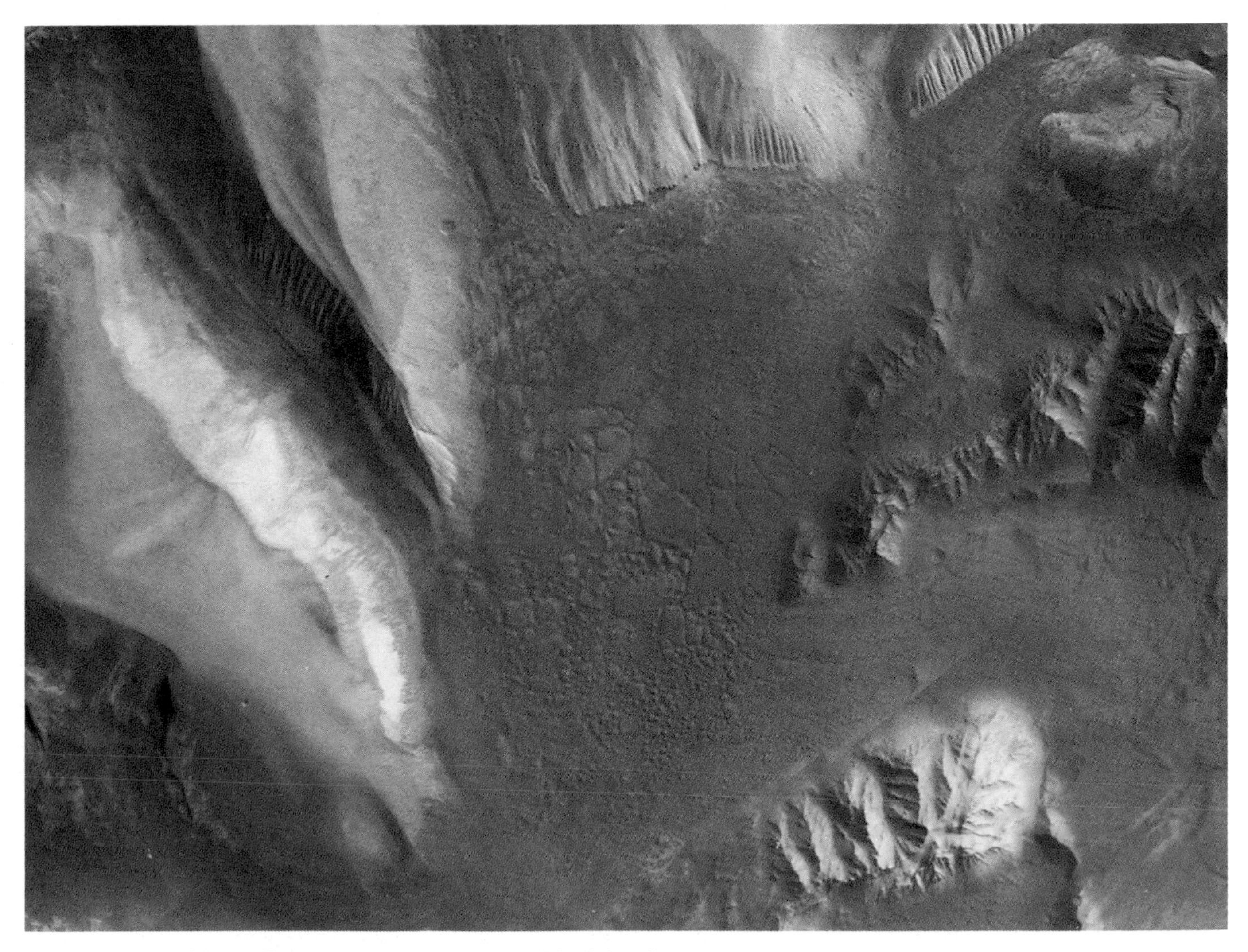

Figure 13. A portion of Candor Chasma, which lies near the middle of Mars' giant rift complex. The dark low-lying material appears to be sediment of some kind. Various geologists have speculated that it could be volcanic ash, thick layers of dust laid down early in the planet's history, or the floor of an ancient sea. Steep canyon walls exhibit deep fluting and other evidence of erosion. The area shown measures 170 by 110 km and is from a mosaic of Viking images.

are particularly common in the Hebes, Ophir, and Candor chasmata. The only satisfactory explanation is that these sediments were deposited under water. Accordingly, many geologists think standing bodies of water partly filled the canyons in times past. These lakes were probably fed by water seeping from the canyon walls. It has been suggested, further, that the lakes within Candor and Ophir chasmas were released catastrophically as the ridge dividing Candor Chasma from the main canyon to its south failed. This would explain the pattern of erosion of the canyon-floor sediments and the previously mentioned fluvial features found abundantly at the "downstream" (eastern) end.

To the east, the canyons' flat floors become rubbly, and their linear walls become inward-facing cliffs enclosing areas of chaotic terrain. Situated 1 to 2 km below the surrounding landscape, the chaotic debris consists of jostled blocks that look as though broad areas of the former surface collapsed. Large channels emerge out of these chaotic areas and extend thousands of kilometers northward across Chryse Planitia until they merge with the low-lying plains at high northern latitudes (Figure 14). Other large channels emerge from box canyons to the north of the main east-west canyon complex. All these channels emerge full-size from areas of chaos and have few, if any, tributaries. They tend to be narrow and deeply incised where they cross cratered highlands, but broad and shallow on the volcanic plains. Their paths are easily recognized by scour marks and numerous teardrop-shaped islands.

The abrupt beginnings, lack of tributaries, abundance of sculpted landforms, and strong resemblance to terrestrial flood features all suggest that these channels formed by catastrophic floods. The largest flood features known on Earth are the Channeled Scablands of eastern Washington, which appeared about 10,000 years ago. Numerous branching channels, tens of kilometers wide, are believed to have formed within a few days when a large ice dam collapsed, thereby releasing water from a large lake in western Montana. Peak discharges in the Channeled Scablands floods were an estimated 10^7 m^3 per second. The immense flows that created the Martian channels may have been 10 to 100 times larger – up to 10,000 times the average discharge rate of the Mississippi River.

The Martian floods may not all have resulted from the sudden drainage of canyon-bound lakes. Another possible cause is catastrophic eruption of groundwater under high pressure. This is implied by the sudden emergence of large channels from local areas of chaotic terrain (Figure 15). If groundwater were trapped beneath a thick permafrost zone, pressures could have become so great that the water broke through the overlying seal. Rapid escape of water from the underground aquifer would have caused the host rock to disintegrate and be carried along in the flood; the surface collapsed thereafter, creating the chaotic terrain. Relative ages determined from crater frequencies on the eroded surfaces suggest that the process was repeated episodically throughout much of Mars' history.

Branching valley networks, which are common throughout the ancient cratered highlands, probably formed in a different way. These networks differ from the flood features just described in that they are smaller, have tributaries, increase in size downstream, and lack the sculpted bedforms (Figure 16). Thus they are more akin to typical terrestrial river valleys than to flood features like the Channeled Scablands. Valley networks are found almost everywhere in the cratered highlands but only in a few places on younger terrains. It appears that whatever process created

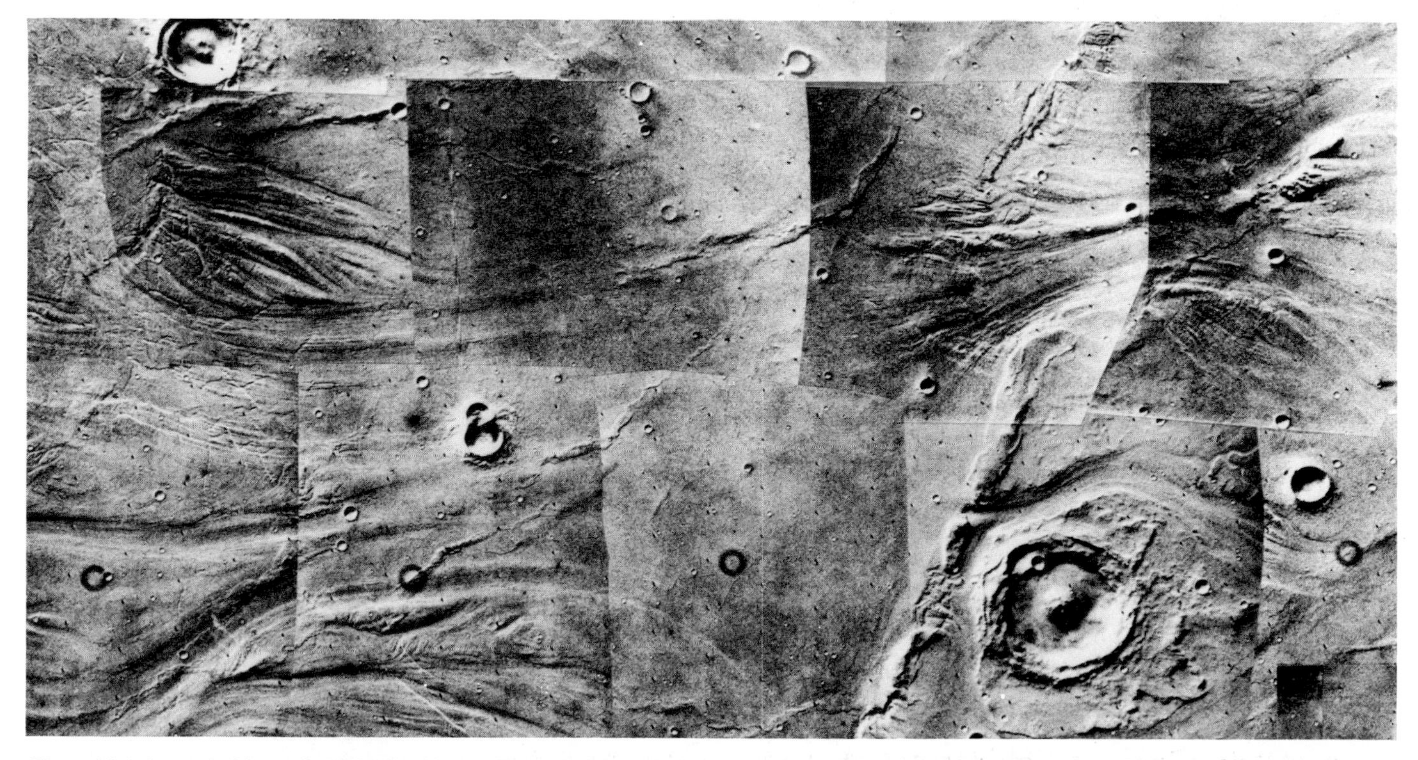

Figure 14. **As it searched for a suitable landing site, the Viking 1 orbiter looked down onto a huge flood plain in the region of Mars called Chryse Planitia (Plains of Gold). Billions of years ago, huge torrents of water deeply scoured the landscape as it raced northward (to the right). Mare-like ridges partly obstructed the flood, but the water found low points and gaps to flow through. The scene shown is 155 km across.**

Figure 15. **A large outflow channel, 20 km wide, emerges from the depression at left and continues eastward toward an even larger channel called Simud Vallis. Inside the depressed area is what geologists term chaotic terrain – an irregular jumble that may have resulted after groundwater erupted and flowed away, causing the ground above it to collapse.**

Figure 16. **A typical intermediate-size channel, located about 4,000 km northeast of the Viking 1 lander, displays characteristic sinuous form and branching tributaries with blunt ends.**

Figure 17. **Branching, dendritic channel networks, like this well-preserved example, are common in Mars' southern highlands. The largest neighboring craters are roughly 35 km in diameter.**

them was more effective very early in the planet's history. Several characteristics, such as the alcove-like terminations of many tributariues (Figure 17), suggest that the networks formed by groundwater seepage rather than runoff following precipitation. The stream beds are small, so only modest discharges were involved in their formation. Since such shallow bodies of water would freeze rapidly in Mars' current climate, it seems that warmer conditions must have existed during the era when most of the networks formed. In contrast, the floods involve such huge discharges that freezing would be insignificant even under present climatic conditions.

If warmer climatic conditions did prevail very early in the planet's history, they did not last long, for the amount of fluvial erosion is quite small and localized. The rims of craters that formed at least 3.8 billion years ago remain almost intact, and the craters themselves have not been filled with waterborne debris. Moreover, the valley patterns themselves suggest that the fluvial action was brief; long, individual trunk channels did not have time to extend themselves, capture streams from adjacent drainage basins, and so control the drainage over large areas. This territorial consolidation happens relatively quickly on the Earth, so Mars' fluvial action was either short-lived, very intermittent, or not truly analogous to what occurs here on Earth.

THE GLOBAL WATER INVENTORY

Geologists can use the channels to estimate the amount of water present near the Martian surface, but their conclusions vary considerably. The issue is of significant geologic, climatic, and biologic interest. A great deal of water could have coursed across Mars early in its history. If true, the planet probably was warm and wet then, perhaps enough so to allow life to gain a foothold there.

We have seen that floods have caused a large amount of erosion around the Chryse Basin (Figure 14). More than 4 million km^3 of material has been removed from the channels and smaller canyons in its vicinity. This would require about 70 million km^3 of water carrying its maximum sediment load (about 40 percent by volume). We have seen, further, that the floods probably formed by massive breakouts of groundwater. The artesian basin from which the water erupted can be outlined roughly from the topography. It constitutes about one-seventh of the Martian surface. But the ubiquitous presence of valley networks in the old cratered terrain suggests that groundwater was everywhere early in the planet's history. If the abundance estimate for the Chryse region applies over the whole planet, the total volume of water would cover all of Mars with a layer roughly 500 m deep. The corresponding value for the Earth is 3 km.

A global 500-m layer is considerably more water than the estimates derived from the very small amounts of inert gases now present in the Martian atmosphere. Because they are not chemically reactive, all the inert gases that outgassed from the planet's interior should still be in the atmosphere. Therefore, inert gases can be used as a gauge of how completely a planet has purged its interior of volatile compounds – including water. However, the "low-water" estimates may be in error because Mars never acquired the inventory of inert gases when it formed that Earth did, or large impacts could have driven inert gas from Mars' atmosphere early in its history. Alternatively the "high-water" values derived from erosion rates may be wrong.

One problem with these estimates is that under present climatic conditions the Martian atmosphere can contain only minute amounts of water, and ice can exist stably on the surface only at the poles (Figure 18). If Mars outgassed a 500-m layer of water from its interior, where is it all now? Much of the water that cut the large Chryse channels probably pooled and froze in lowland areas at high northern latitudes, which explains many peculiarities of these plains. Some of the water is currently trapped in the layered terrains at the poles. But most of it probably exists as buried ice and groundwater in the ancient and deeply fractured cratered terrain. This is especially evident at high latitudes, where ice has caused a general softening of the terrain and caused debris flows wherever cliffs are found.

THE POLAR REGIONS

Mars' polar regions are distinctively different from the rest of the planet. Layered deposits occur at both poles and extend outward about to the 80° latitude circle. In the south they lie on the cratered highlands; in the north they overlie plains. Cut into the deposits are valleys that curl outward from the poles, resulting in a distinctive swirl texture in spring when the CO_2 frost sublimes from the sunward-facing slopes (Figures 19,20). Layering is visible on these frost-free slopes down to the resolution of the available Viking photographs, about 20 m. In the north, the layered deposits are surrounded by a vast array of sand dunes (Figure 21), which form an almost complete dark collar around the polar regions. Dune fields of comparable scale do not occur in the south, though small ones are common within craters at high latitudes.

The polar deposits are believed to be mixtures of ice and dust. In the north, when the seasonally deposited CO_2 frost completely sublimes, a residual cap of water ice is exposed. (This has not been observed in the south because the CO_2 cap never completely dissipates there.) The relative proportions of each component and their respective accumulation rates are probably being modulated by climatic variations. For example, the 1977 dust storms observed by the Viking

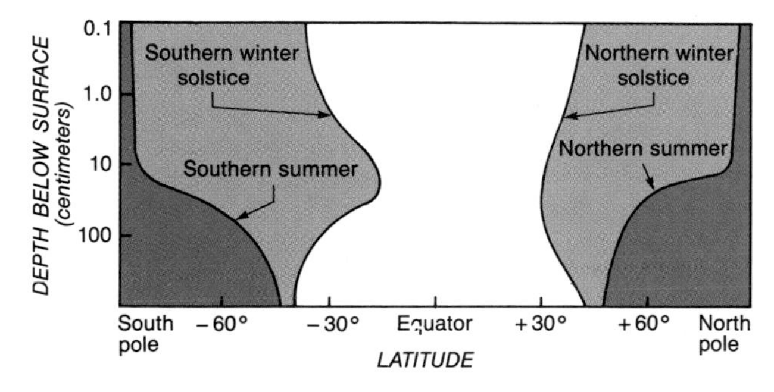

Figure 18. **Today liquid water does not exist anywhere on the Martian surface, and the stability of water ice depends strongly on its subsurface depth and latitude. For example, ice is unstable within about 30° of the equator down to at least 10 m (white); over time, the surface at these latitudes has become completely dried out to great depths. Farther to the north and south (light blue), ice could exist during local winter but would be driven into the atmosphere as vapor at other times. In the polar regions (dark blue), water ice is always stable. This calculation by Crofton B. Farmer and Peter E. Doms assumes that the Martian atmosphere is well mixed and carries only enough water vapor to create a layer on the surface 12 microns thick.**

Figure 19. **From its high-inclination orbit, the Viking 1 orbiter photographed the involved spiral of Mars' north polar cap. During local winter, atmospheric carbon dioxide freezes out onto the polar terrain and creates a thin white veneer covering a much greater area. But in summer the CO_2 returns to the atmosphere, leaving this residual cap of water ice, which is about 600 km across.**

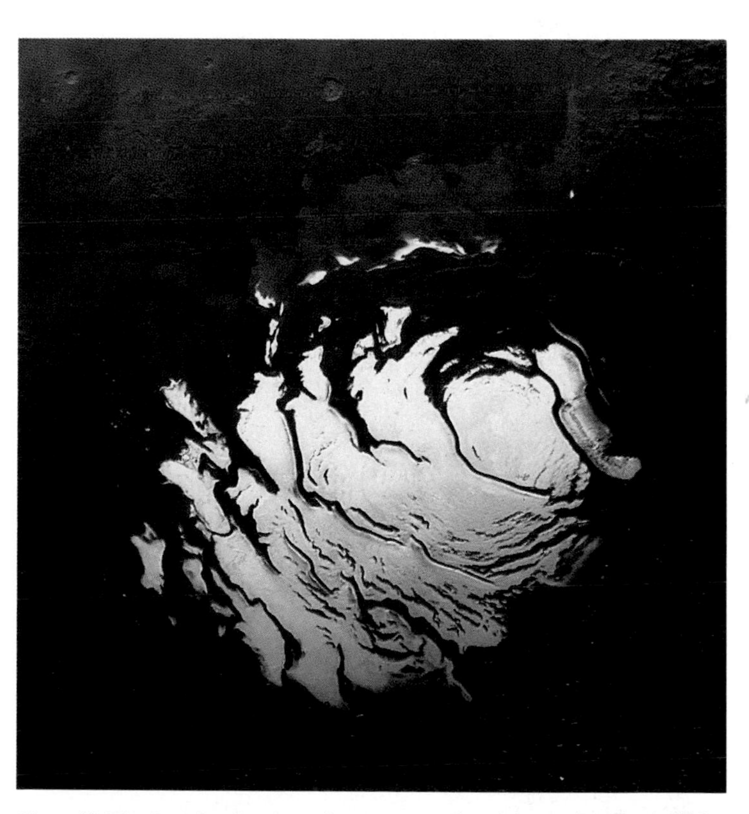

Figure 20. **Mars' south polar cap, as it appears near its minimum size (about 400 km across) during local summer. In contrast to the residual cap at the north pole, this one is believed to consist mainly of frozen CO_2. The true south pole of Mars lies just at cap's edge at lower right. The false blue shading in this U.S.G.S. mosaic arises from the dim lighting on the cap when the Viking 2 orbiter acquired these images in September 1977.**

spacecraft (Figure 22) deposited an estimated 0.4 mm of dust at the north pole. At this rate it would take 100,000 years to accumulate a 30-m layer, and 6 to 10 million years to accumulate the observed thickness of 2 to 3 km. However, precession of Mars' orbit and the planet's axis of rotation causes deposition to alternate between the poles on a 51,000-year cycle, and variations in Mars' obliquity and orbital eccentricity will alter the frequency and intensity of dust storms in an unknown way (see Chapter 8).

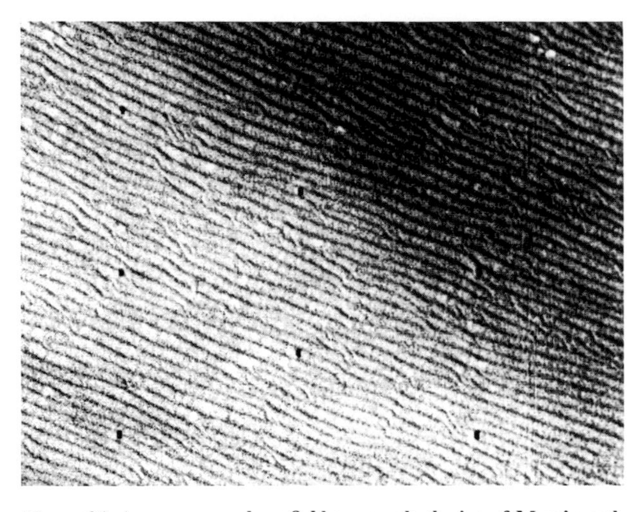

Figure 21. **An enormous dune field surrounds the ice of Mars' north polar cap. The dunes here align roughly north-south, and the vague circular forms are probably buried craters.**

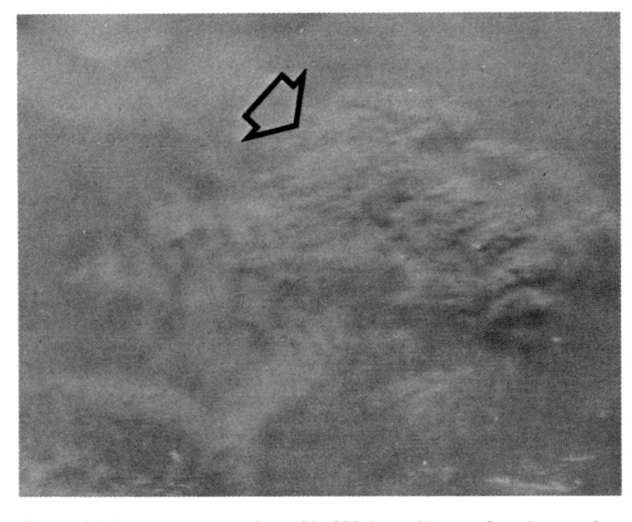

Figure 22. **Dust storms, such as this 300-km-wide one found scurrying across the floor of the Argyre basin in early 1977, carry micron-size particles high into Mars' atmosphere. Dust storms often grow so large that they cloak the entire planet – as happened twice during 1977 as the four Viking spacecraft looked on.**

The present surface of the polar deposits lacks impact craters, suggesting that they are very young. However, the materials making up the layers themselves may be rather old, because geologists suspect that the deposits are being continually reformed. Ice may sublime from the dark, sunward-facing slopes of the incised valleys and be redeposited on the intervening frost-covered flats. In this way, the valleys march across the landscape and slowly renew the surface. Alternatively, the deposits may have eroded away and reformed throughout Martian history in response to long-term climatic changes. Some investigators have noted that the uplifting of the Tharsis region may have realigned the planet's rotation axis, but for now no one knows with certainty whether polar wandering has actually occurred on Mars.

SUMMARY

We have seen that, in some ways, Mars resembles the Earth. The planet has an atmosphere (albeit a thin one) and has been both volcanically and tectonically active. Water has eroded parts of the surface and reacted with materials there to produce weathering products. Ice and wind have modified the surface as well.

Yet the two planets remain profoundly different. Because of plate tectonism, the outer layers of the Earth (down to depths of several hundred kilometers) are continually turned over, as old lithosphere returns to the upper mantle at subduction zones and new lithosphere is created along mid-ocean ridges. The Martian lithosphere, despite all its volcanism, is actually very stable. Volcanic products erupted onto the surface simply remain there, creating massive piles several kilometers high as in Tharsis and Elysium. In fact, it is the *lack* of plate tectonism that allows the Martian volcanoes to attain such enormous sizes.

An additional major cause of differences between Mars and the Earth are climatic conditions. Water plays an essential role in weathering, erosion, transportation, and deposition. On Earth these processes work to reduce surface relief, constantly competing with volcanism and deformation that are working to create relief. Throughout most of Mars' history liquid water could not exist on the surface, and while fluvial erosion has occurred on Mars, its cumulative effect as been trivial. Consequently wherever relief has been created, it largely remains. The result is a spectacular planet on which geologic features of enormous scale and a wide variety of origins and ages are preserved.

6
Planet Earth

Don L. Anderson

"It is my opinion that the Earth is very noble and admirable . . . and if it had contained an immense globe of crystal, wherein nothing had ever changed, I should have esteemed it a wretched lump and of no benefit to the Universe."

– Galileo Galilei

A PLANET'S SURFACE provides geologists with clues as to what is happening inside. But many of these clues are ambiguous because so many other processes (impacts and erosion, for example) contribute to surface characteristics. Most of the surface of the Earth is less than 100 million years old, and even its oldest rocks are less than 4 billion years old, so the record of the origin of our planet has been erased many times. Part of this erasure is due to erosion by wind and water, and part is due to the continual recycling of material back into the interior and the repaving of the ocean basins by seafloor spreading.

The solar system's other solid planets and smaller worlds have much more ancient surfaces in general, and this tells us two things that we cannot learn from the Earth itself. One is that in the early days of the solar system, violent and destructive impacts were common as larger bodies swept up and devoured smaller ones. The other is that most other worlds preserved evidence of these early happenings, while ours did not.

Geophysicists try to sidestep the mixed signals they encounter at ground level by studying the interiors of planets directly. On our home planet this effort has been rather successful: the Earth is the only body for which we have detailed information, including three-dimensional images of the internal structure, from the surface to the center. And while we realize that other planets do not all share Earth's present behavior, they may at least have been put together in similar ways. So intensive study of the Earth's interior may yield knowledge that is applicable elsewhere, just as our cognizance of the Earth will in turn be shaped by the study of other worlds.

Earlier views of Earth's origin envisioned a gentle rain of dust and small particles that slowly accumulated layer by layer. A planet growing this way would remain relatively cool, building up heat mainly by the slow decay of radioactive elements. According to this scenario, an initially cold, homogeneous Earth eventually heated up, started to melt, and formed a buoyant crust and a dense core in a way that somehow left behind (in most versions of this story) a homogeneous mantle.

However, planetary evolution has not been so simple. The energy associated with a single large impact is enough to melt, or even vaporize, much of both the impactor and the planet it strikes. If the Moon really came into being when a Mars-size object struck the Earth (see Chapter 4), the energy from that collision would have melted much of the Earth itself. Even smaller hits, of which there were many more, would have caused widespread melting where they penetrated and generated shock waves.

Did the Earth start out cold or hot? The answer depends on whether it accreted slowly (100 million to 1 billion years) or rapidly (100,000 to 10 million years). In the latter case, kinetic energy would have been delivered faster than the growing Earth's ability to conduct and radiate it away as heat, so our planet would have remained molten, at least in its outer parts, as it accreted (Figure 1). In addition, however, every giant impact is essentially an instantaneous accretion event, and planets that grew by gathering up relatively large objects experienced widespread melting over and over again.

We assume, therefore, that growing planets were molten – at least partially, and at least once. At such times, their component materials had the opportunity to separate according to melting points and densities. The "heavy" materials sank toward the interior, creating cores, and the "light" ones rose to the surface, creating crusts. This process of gravitational separation is usually called *differentiation,* and it played a key role in the early histories of the Earth and other terrestrial planets. Over time these worlds may have acquired more internal stratification than simply a light crust and a dense core; as we shall see, the mantle situated between them can itself become layered according to chemistry and density.

Seismology is the geophysicist's principal tool for probing planetary interiors. In a sense, the Earth is a huge spherical bell that is periodically "struck" by earthquakes. We learn about the interior by listening to how the Earth "rings" – that is, by noting how seismic waves move away from the source point, or focus, of an earthquake (Figure 2). Of the four types of seismic waves, two travel around the Earth's surface like the rolling swells on an ocean. A third type, called primary or

P waves, alternately compress and dilate the rock or liquid they travel through, just as sound travels. Secondary or *S* waves propagate through rock (but not liquids) by creating a momentary sideways displacement or shear, like the movement along a rope that is flicked at one end. Both P and S waves slow down when moving through hotter material, and they are refracted or reflected at the boundary between two layers with distinct physical properties.

In fact, we have relied on physical properties such as density and seismic velocity, rather than chemistry or composition, to distinguish the three principal divisions of the Earth's interior. The *crust, mantle,* and *core* account for 0.4, 67.1, and 32.5 percent of the planet's mass, respectively. Rocks' physical properties vary with depth due to increasing temperature and pressure and, in places, changes in chemistry or physical state. For example, the most common minerals in the crust and upper mantle are all unstable farther down. As pressure increases the atoms in their crystals become more tightly packed, and their density increases. These changes are gradual except at *phase transitions* (such as when carbon transforms from graphite to diamond under pressure). Phase transitions cause a rapid or abrupt change in physical properties, including those measured by seismic techniques. If the change occurs abruptly, it is called a *seismic discontinuity*.

In the early 1980s, several groups of researchers discovered that seismic waves could be used to produce three-dimensional maps of the Earth's interior, a technique known as *seismic tomography*. The word "tomography" derives from the Greek word for a cutting or section, and in effect geophysicists create a series of cross-sections of the interior at various depths (Figure 3). Seismic tomography is a very powerful technique that has revolutionized our study of the Earth's interior.

COMPOSITION OF THE EARTH

Ours is the only planet for which we can speak with some confidence about its bulk composition or chemical makeup. By combining the Earth's mass with seismic determinations of the radius and density of the core, we have deduced that the Earth is about one-third iron and that this iron is concentrated toward the center of the planet. In fact, the Earth's solid inner core, which is smaller in size than the Moon but three times denser, may be pure iron and nickel. Seismology also tells us that the outer part of the core is liquid and indicates strongly that it is molten and mostly iron. To explain its lower density and molten state, the outer core needs to incorporate a small amount of oxide, silicate, or sulfide material.

The Earth is the largest terrestrial planet and contains slightly more than 50 percent of the mass in the inner solar system, excluding the Sun. Compared to Earth, the dense planet Mercury contains proportionately more iron; Mars and the Moon contain substantially less iron, even though they may have small cores. Based on its similarity to our planet in size and density, Venus probably has an Earthlike core. But a solid inner core may be absent, because we expect Venus to have slightly lower pressures and possibly higher temperatures in its interior.

The bulk of the Earth is contained in its mantle, the region between the core and the thin crust. We can sample the top of the mantle in several ways. Fragments of it are exposed in eroded mountain belts and brought to the surface by volcanic eruptions. The major mantle minerals excavated in these ways are olivine $(Mg,Fe)_2SiO_4$ and pyroxene $(Mg,Fe)SiO_3$; thus, iron is present but only as a minor constituent.

The most abundant material we see emerging from the mantle is *basalt,* and it must exist there in vast quantities.

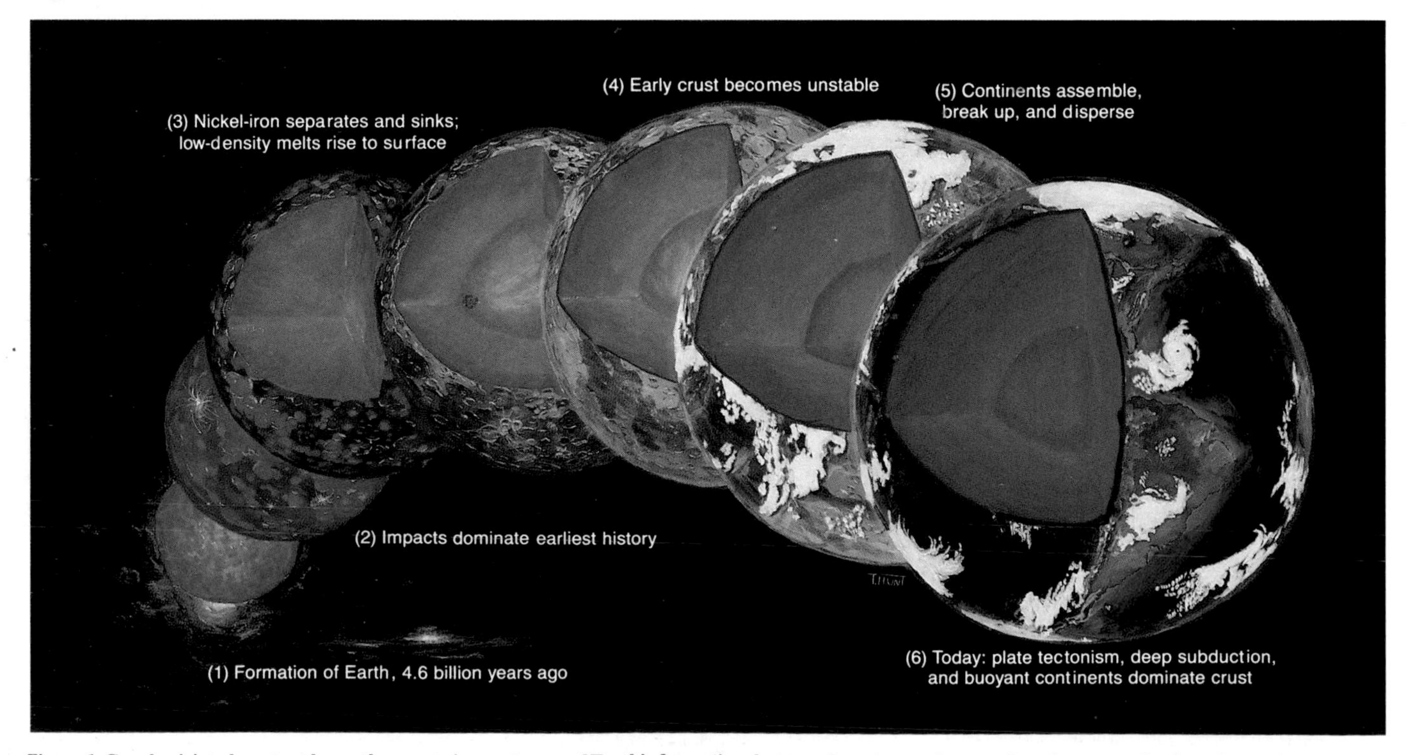

Figure 1. **Geophysicists do not yet know the exact circumstances of Earth's formation, but our planet's exterior must have been completely molten at least early in its history. Much of the energy needed to melt its outer layers came from innumerable collisions with interplanetary material left over from planetary formation.**

Basaltic magma is rich in the elements calcium and aluminum and is less dense than upper-mantle material, which allows it to erupt into or onto the crust. The ocean floor is covered with basalt. Iceland and Hawaii (Figure 4) are two examples of thick basalt piles that have accumulated on the ocean floor. Hidden from our view under seawater is a 40,000-km-long network of volcanoes – the oceanic ridge system – which generates new oceanic crust at the rate of 17 km^3 per year. In fact, the majority of the Earth's crust was made in this way.

However, at depths below 60 km in the mantle, cold (solid) basaltic material converts to a form of rock called *eclogite,* which is much denser than shallow-mantle rocks because it contains garnet, a complex, aluminum-bearing silicate mineral. Large bodies of eclogite can sink through the upper mantle, which probably explains why the crust on Earth never gets thicker than about 60 km. Inside smaller terrestrial planets, like Mars, the pressures at a given depth are lower, so their crusts can extend farther down without converting to dense eclogite. On a hotter planet, like Venus, a thick basaltic crust would melt at its base rather than convert as a solid to eclogite.

Although our direct samples of the Earth's interior are limited to the crust and shallow mantle, we know from seismic tomography that broad regions with low seismic velocities extend to a depth of at least 400 km under oceanic ridges and other volcanic terrains. Magmas and rock-magma mixtures have low densities and low seismic velocities, so it seems reasonable that the basalt source region lies below about 400 km. When a hot silicate rock or low-density magmatic mush ascends from that great depth, it eventually separates into molten liquids (which erupt at volcanoes) and crystals (which stay behind in the mantle or form new crustal material).

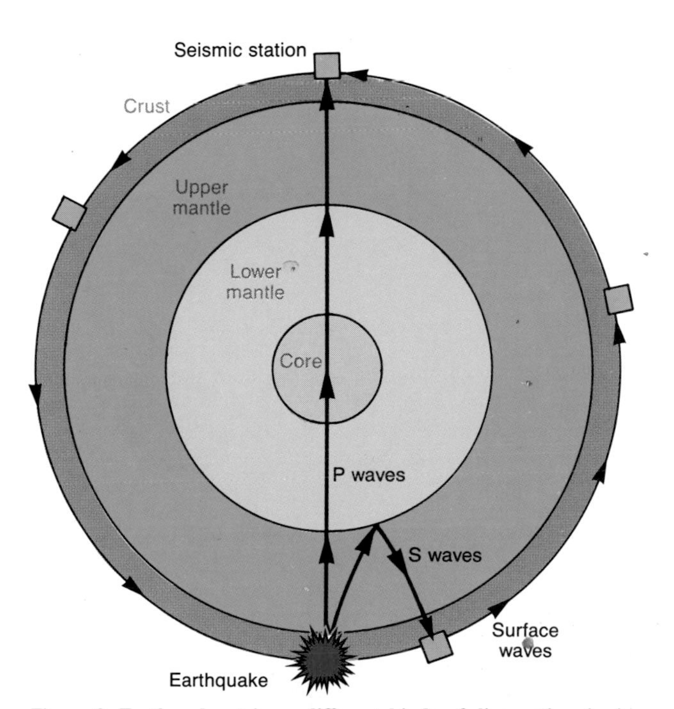

Figure 2. **Earthquakes trigger different kinds of diagnostic seismic waves that travel around the Earth and through its interior at 3 to 15 km per second. Compression (*P*) waves move almost twice as fast as shear (*S*) waves; they can also pass through the liquid outer core, which the S waves cannot.**

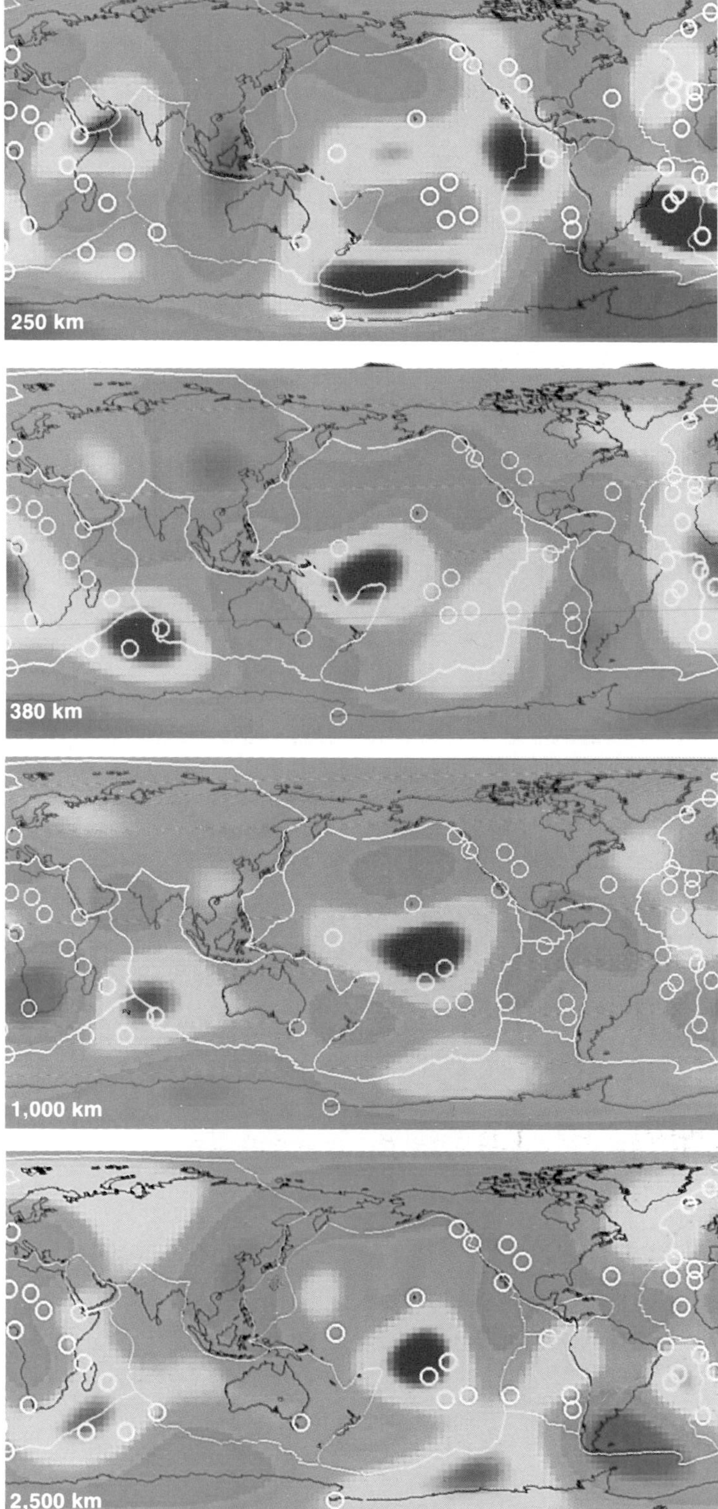

Figure 3. **This series of maps shows the state of the Earth's interior at different depths as determined by seismic tomography. In each, dark lines denote land masses, white lines show plate boundaries, and white circles mark the locations of "hot spots." The red regions have slower-than-average seismic velocities (4 percent in the upper panels, 0.5 percent in the lower ones) and are therefore hot. The blue regions have faster velocities and are therefore colder. In the 250-km map, derived from surface-wave data, notice the association of hot mantle with oceanic ridges and continental tectonic regions. The 380-km map, from shear-wave data, shows a large hot region in the central Pacific. Compression-wave velocities were used for the maps at depths of 1,000 and 2,500 km.**

Thus, we have been able to identify three outer layers in the Earth: (1) the buoyant crust, containing low-density minerals dominated by quartz (SiO_2) and metal-poor silicates called feldspars; (2) the uppermost mantle, containing minerals (primarily olivine and pyroxene) that are refractory (crystalline at high temperatures) and thus settle out of rising magma mushes; and (3) a "fertile" layer, below 400 km, that contains a large basaltic component and therefore abundant calcium and aluminum. This third layer is dense when cold, due to the garnet it contains, and buoyant when hot, because garnet and related minerals melt easily to form basalt.

Underneath all of this is the lower mantle. If the Earth has "cosmic" abundances of the elements, as deduced from their proportions in the Sun and primitive meteorites, then the lower mantle (with 70 percent of the mantle's mass) must be mainly silicon, magnesium, and oxygen. It probably also contains some iron, calcium, and aluminum. Although Ca and Al are well-represented in Earth's crust, the crust is too thin to yield Ca:Si or Al:Si ratios for the whole Earth as high as those found in the Sun, meteorites, and, by inference, the planets. Moreover, there is little calcium or aluminum in upper-mantle rocks (otherwise basalts could not rise through them en route to the surface), nor are they present in the core. So, by elimination, the bulk of Earth's calcium and aluminum must reside in the lower mantle or in the mesosphere.

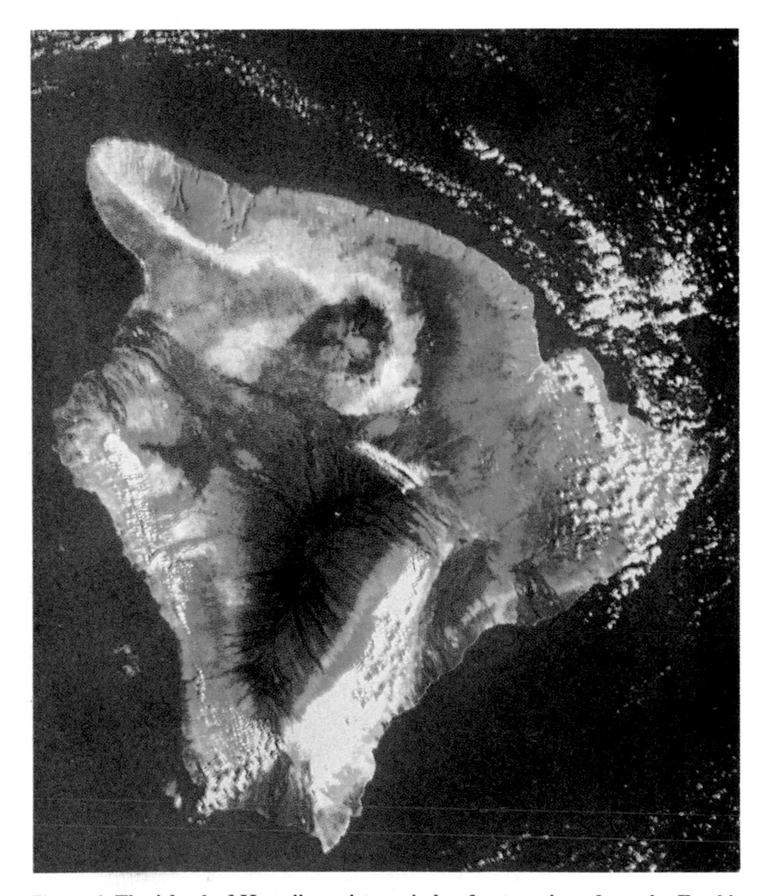

Figure 4. **The island of Hawaii consists entirely of outpourings from the Earth's mantle. The now-dormant volcano Mauna Kea, in the island's northern half, has become the site of numerous astronomical observatories. But Mauna Loa, to its south, is still quite active, especially along its southeast flank. The Hawaiian Islands are part of a long chain of peaks that formed as the Pacific lithsopheric plate slowly moved northwest over a plume of upwelling mantle material. Hawaii, at the southeast end of the island chain, currently sits almost directly over the hot spot; the exact location is marked by a small submerged peak, Loihi, to the island's south. This photograph is a composite of two Landsat images.**

The lower mantle must also be richer in silicon than the layer above it. The reasoning behind this assumption is as follows: Primitive meteorites and the Sun have about one magnesium atom for every silicon atom. In the mantle, this 1:1 ratio would favor the formation of the mineral enstatite ($MgSiO_3$, a pyroxene) over forsterite (Mg_2SiO_4, an olivine). However, we know from its surface exposures that the upper mantle is olivine-rich and has a Mg:Si ratio of about two. Farther down, at the high pressures present in the lower mantle, Mg_2SiO_4 decomposes to two new minerals. One is periclase (MgO), which has the crystal structure of ordinary table salt, NaCl. The other is an ultrahigh-pressure form of enstatite (which, incidentally, has the same crystal structure as many of the new high-temperature superconductors). This enstatite variant propagates seismic waves at much higher velocities than periclase does and matches the seismic velocities we have observed for the lower mantle. Therefore, at great depths $MgSiO_3$ would appear to be the most abundant mineral – and MgO largely absent. Some seismic evidence also indicates that the lower mantle has more iron (as FeO) than the upper mantle does; it may be similar to the mantles of the Moon and Mars, which we also suspect to be rich in FeO.

DIVIDING THE EARTH'S INTERIOR

The Earth's seismic properties have allowed geophysicists to distinguish rather distinct layering in its interior (Figure 5). In 1906, the British geologist Richard D. Oldham found that at a certain depth, compression or P waves slow sharply and S waves cannot penetrate further. It was the first evidence that the Earth has a liquid core. Only three years after Oldham's revelation, the Yugoslavian seismologist Andrija Mohorovičić discovered that the velocity of seismic waves takes a large jump about 60 km down. This Mohorovičić or "Moho" seismic discontinuity marks the crust-mantle boundary, where changes in rock chemistry and crystal structure occur. At the core-mantle boundary, averaging 2,890 km in depth, the composition of rock changes from silicate to metallic and its physical state changes from solid to liquid. This boundary is also known as the Gutenberg discontinuity, after Beno Gutenberg, who made the first accurate determination of its depth. Seismic discontinuities allow a further division of the Earth into inner core, outer core, lower mantle, transition region, upper mantle, and crust (Table 1). These regions are not necessarily all chemically distinct, nor can we assume that each of them is chemically homogeneous.

The inner core represents only 1.7 percent of the Earth's mass. It is solid, primarily the result of "pressure-freezing" (most liquids will solidify if the temperature is decreased or the pressure is increased). Probably the entire core was once molten, but over time it has lost enough heat for the inner core to solidify. It "floats" in the center of the outer core and is thus essentially decoupled from the mantle.

The outer core (30.8 percent of Earth's mass) is liquid, a result of its high temperature and the fact that iron alloys melt at lower temperatures than do common rocks. The viscosity of the outer core is very low, probably not much greater than water. We expect it to behave in general like other fluid parts of the Earth. Rapid motions of molten iron in the core are responsible for the Earth's magnetic field and for some of the subtle jerkiness in our planet's rotation. The density of the outer core is slightly less dense than pure molten iron and requires about 10 percent of some lighter elements such as sulfur or oxygen, or both. These elements are considered likely because they are cosmically abundant and would readily dissolve in the hot metallic soup.

Just above the core is a 200- or 300-km-thick layer, called D'', that may differ chemically from the rest of the lower mantle lying above it. It may represent material that was once dissolved in the core, or dense material that sank through the mantle but was unable to sink into the core. The D″ layer comprises about 3 percent the Earth's mass, or about 4 percent of the mantle.

Smaller seismic discontinuities occur at several depths in the mantle and halfway through the core. These are often attributed to phase transitions, but they may signify changes in composition. The two largest ones in the mantle, 400 and 650 km down, represent abrupt rearrangements of the atoms in the major mantle minerals. Large variations in seismic velocity have also been found from place to place. These "lateral variations" have been revealed by seismic tomography. In fact, the Earth's upper mantle exhibits as much variation horizontally as it does vertically (Figure 6).

Region	*Depth (km)*	*Percent of Earth's mass*	*Percent of mantle-crust mass*
Continental crust	0–50	0.374	0.554
Oceanic crust	0–10	0.099	0.147
Upper mantle	10–400	10.3	15.3
Transition region	400–650	7.5	11.1
Lower mantle	650–2,890	49.2	72.9
Outer core	2,890–5,150	30.8	
Inner core	5,150–6,370	1.7	

***Table 1.* A summary of Earth's internal structure, as deduced from decades of probings with seismic techniques.**

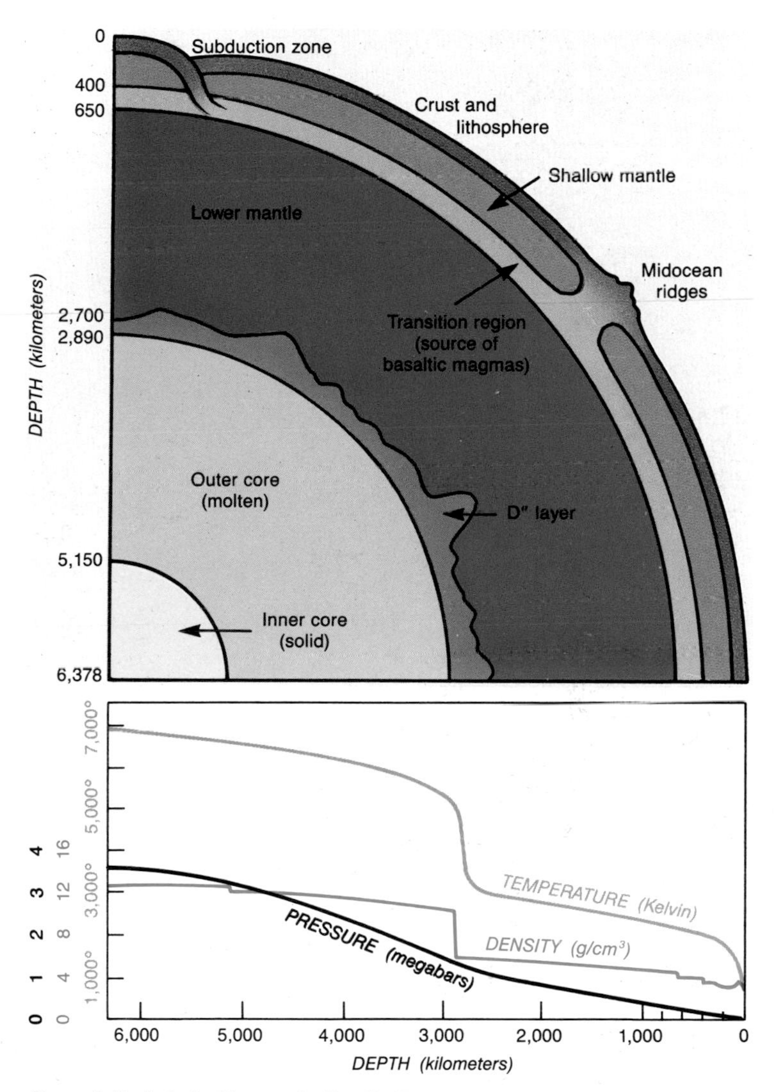

***Figure 5.* Early in its history, the Earth differentiated into a series of layers with distinct physical and perhaps compositional properties.**

WHERE ON EARTH IS THE CRUST?

Planets grow by colliding with other objects, an energetic process that results in melting or even vaporization. Most of the energy is deposited in the outer layers, except for the small number of truly giant impacts that are as likely to destroy the target object as add to its bulk; these may melt a large fraction of a planet. A global ocean of magma can segregate incoming material into solid and liquid fractions that float and refractory crystals and iron-rich melts that sink. Differentiation is akin to what takes place in a blast furnace or fat-rendering plant. By heating and boiling, the original material is reduced to frothy scum, dense dregs, and a "purified" liquid in between.

Planetary geologists have invoked such a global magma ocean to explain the Moon's anorthositic highlands (calcium- and aluminum-rich silicates that floated to the surface) and its "KREEP" basalts (which cooled from the final liquid dregs of a crystallizing magma ocean and thus became highly enriched in trace-elements). A similar process probably occurred inside the Earth, except the pressures were much higher, which caused dense garnet-bearing eclogite to form instead of buoyant anorthosite. In fact, high-grade anorthosite is fairly rare on Earth. Therefore, one key product of Earth's magma ocean did not float to the surface but sank from view. Had we not obtained actual anorthositic samples of the Moon, the magma-ocean concept might never have occurred to terrestrial geologists.

The crust of the Earth would be about 200 km thick if most of the low-density and easily melted material in the interior had separated out during Earth's formation. Yet the average terrestrial crust (20 km) is considerably thinner than the lunar crust (100 km) – even though the Moon has only 2 percent of the Earth's volume. Does this mean that the Earth did not have a magma ocean? Or has the crustal material mostly remained in or returned to the mantle?

The lunar crust is so thick and contains so much of the Moon's calcium and aluminum that it must have formed very efficiently, for example, with its light crustal minerals rising directly to the top of a deep magma ocean. However, on a larger body like Earth, the pressures far down in a magma ocean are so great that buoyant minerals never form. Instead, dense crystals such as garnet and pyroxene soak up the calcium and aluminum. These, by and large, stay in the mantle and may even sink to the base of a magma ocean, thus limiting the crust's thickness. Even so, the high concentrations of some elements in the Earth's crust (Table 2) tell us that most of the mantle *must* have differentiated

Rubidium	(Rb)	68
Cesium	(Cs)	67
Thorium	(Th)	55
Barium	(Ba)	49
Uranium	(U)	47
Lanthanum	(La)	27
Strontium	(Sr)	21
Sodium	(Na)	13
Aluminum	(Al)	2.4
Calcium	(Ca)	0.9
Silicon	(Si)	0.7
Iron	(Fe)	0.07
Magnesium	(Mg)	0.06

***Table 2.* The abundance of various elements in the Earth's crust, as a percentage of their estimated abundance in the whole Earth.**

either during accretion or shortly thereafter. These elements happen to be ones that are not easily incorporated into the high-pressure minerals that form at depth in a magma ocean. It is therefore unlikely that the Earth made its crust inefficiently but very likely that the missing crust resides somewhere in the mantle. The amount of crust now at the Earth's surface is much less than the *potential* crustal material and probably only a small fraction of the total volume of the crust that has been generated in $4\frac{1}{2}$ billion years.

There is also a good reason why the Earth cannot have a thick "secondary" crust – that is, one formed by continental collision, mountain building, or the accumulation of volcanic materials. Wherever these processes cause the Earth's crust to thicken to more than about 60 km, the low-density crustal minerals convert to denser ones, causing the bottom of the crust to "fall off" or, technically, to delaminate. But even if delamination did not occur, the great pressure present below 60 km makes the seismic velocities there so high that a seismologist would call this deep-lying material part of the mantle, not the crust. In fact, "crust" is a physical concept, and its properties and thickness are derived from seismology. However, since erosion and volcanism supply us with many samples of the lower crust and shallow mantle, we know that the crust truly is compositionally distinct. As mentioned, it is calcium-, aluminum-, and silicon-rich compared to the shallow mantle, so changes in physical properties at the crust-mantle boundary are accompanied by changes in chemistry as well.

THE LITHOSPHERE

Although most of the crust and mantle are solid, we know from seismic velocities, the abundance of volcanoes, and the rise of temperature with depth in wells and mines that much of the outer part of the Earth is near or above its melting point. In fact, the coldest part of our planet is its surface. Since cold rocks deform slowly, we refer to this rigid outer shell as the *lithosphere* (the "rocky" or "strong" layer). On the Earth the lithosphere is not a single seamless shell, but rather a patchwork of rigid, snugly fitting *plates* that ride atop the mantle (Figures 7,8). These plates – eight large ones and about two dozen smaller ones – are moving with respect to one another, and their interactions are collectively called *plate tectonics*, a subject to be discussed later.

At depths between about 50 and 100 km, lithospheric rocks become hot and weaken enough structurally to behave as fluids – at least over geologic time. This portion of the upper mantle is called the *asthenosphere* (or "weak" layer), and it may be partially molten. Seismic velocities under young seafloor and tectonic regions are so low that some partial melting is required to depths as great as 400 km. Below the asthenosphere, the temperature continues to climb, but the solidifying effects of high pressure become dominant. So at still-greater depths the Earth again becomes strong and harder to deform. The region between the 400- and 650-km seismic discontinuities is called the *transition region* or *mesosphere* (for middle mantle), and the basalts that make up midocean ridges and new oceanic crust may be derived from this region.

Within the Earth's lithosphere, rocks are so cold and their viscosity so high that they support large loads and fail by brittle fracture rather than by deforming smoothly. New lithosphere forms at midocean ridges and thickens with time as it cools and moves away from the ridge, a process termed *seafloor spreading* (Figure 9). From the way it deflects under large submarine volcanoes and enters deep-sea trenches, we

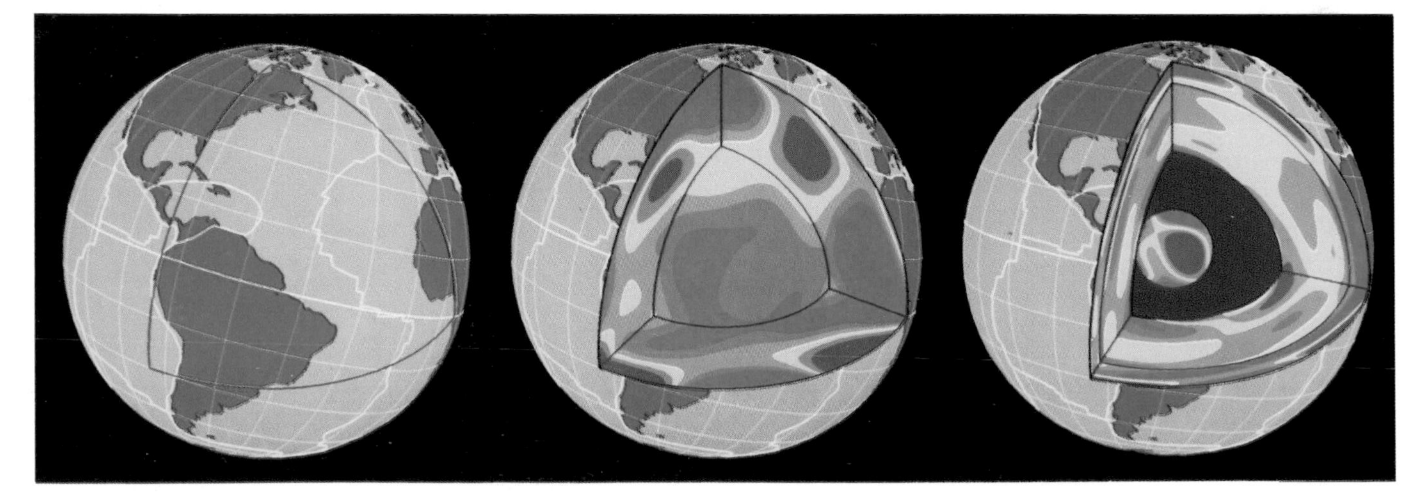

***Figure 6.* Earth's interior exhibits considerable variation with both depth and location. In these plots, red indicates the slowest seismic velocities and the hottest material, blue the fastest and coldest. The panel at left shows continents and plate boundaries (yellow lines) surrounding the Atlantic Ocean. The middle panel shows a slice into the underlying mantle to a depth of 550 km (with depth exaggerated by five times). The cutaway at right goes to a depth of 2,890 km, the mantle-core boundary; it also shows the solid inner core suspended within the outer core (red).**

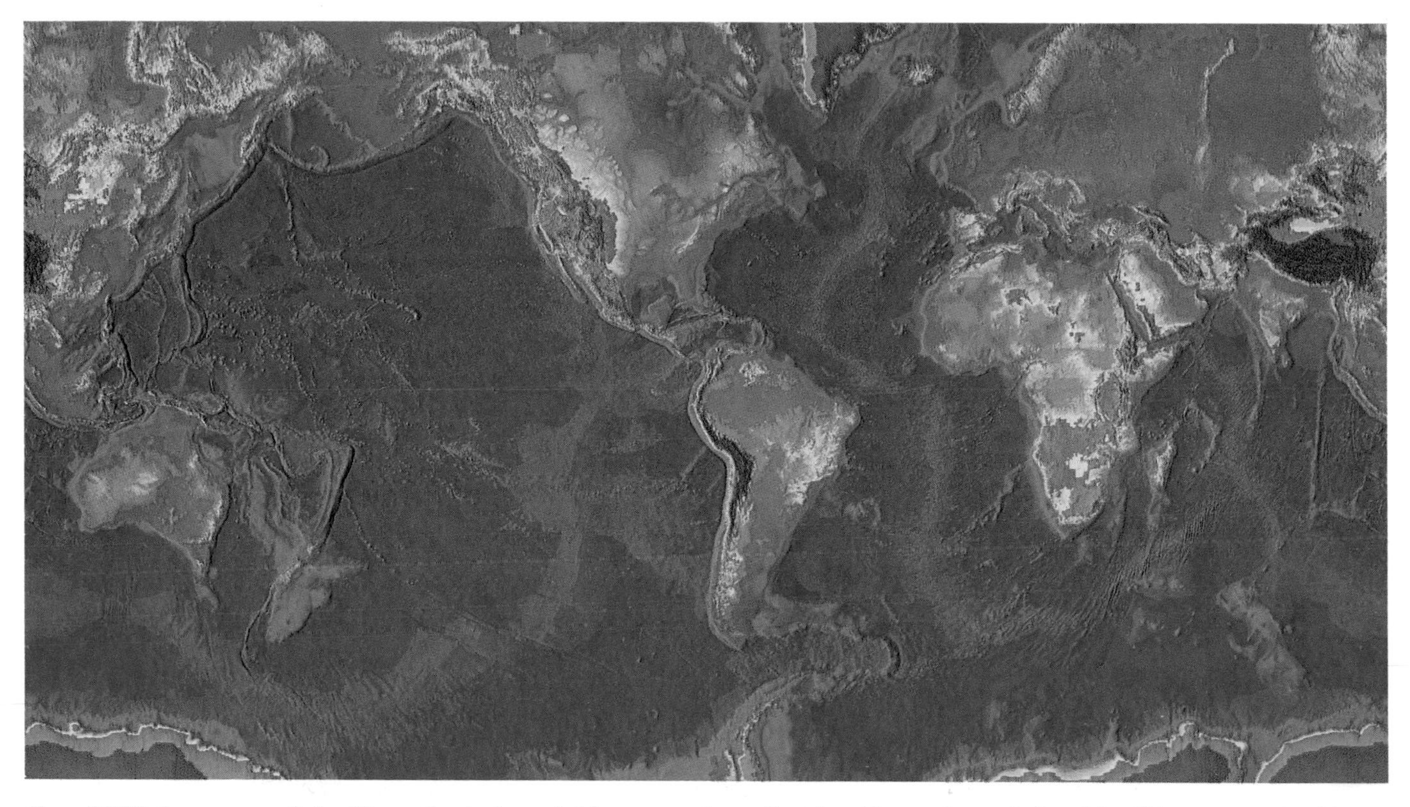

Figure 7. **With the oceans emptied and the continents obscured, this maps reveals a seafloor shaped by ceaseless geologic activity. This map uses radar altimetry from NASA's Seasat satellite of the ocean's surface, which reflects the underlying topography.**

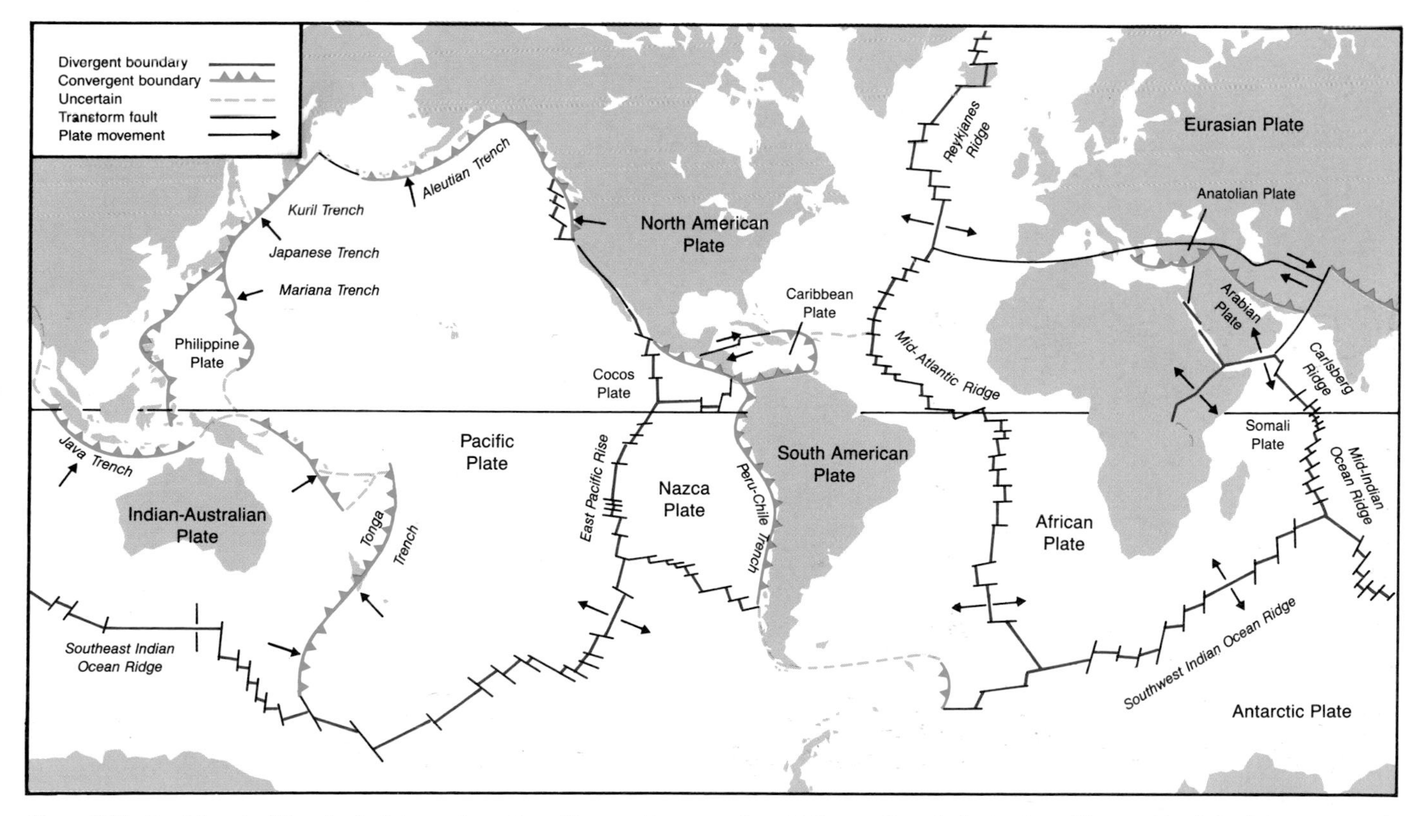

Figure 8. **The Earth's major lithospheric plates are in motion with respect to one another. At divergent boundaries (such as midocean ridges) the plates move apart, only to collide and overlap at convergent boundaries (subduction zones). Plates slide past each other along transform faults, the most famous of which is the San Andreas fault that runs the length of California.**

know that the oceanic lithosphere acts as an elastic plate whose thickness varies from zero at the midocean ridges to about 40 km under older seafloor. As it cools and moves aside, it also becomes denser and loses the high-temperature buoyancy it had initially; eventually, it tries to sink back into the interior.

Let us examine the formation and evolution of oceanic lithosphere a bit more closely. Portions of the Earth's mantle, especially the asthenosphere, behave like hot plastic and are in continuous, slow convective motion that brings heat from the interior to the surface. Upwelling mantle material partially melts in the asthenosphere, where a segregation takes place; denser refractory crystals are left behind when the lighter, easily melted material erupts upward. This buoyant, erupting melt creates the lithosphere's topmost layer, the oceanic crust, which is basaltic and averages about 6 km in thickness.

The lithospheric layer beneath this is essentially normal mantle material that has lost its basaltic component. The basalt is missing either because it rose as a melt to the crust or, at depths of roughly 60 km, it reverted to dense eclogite and sank as large blobs through the upper mantle and into the mesosphere. However, melting of eclogite-rich mesosphere material restores its buoyancy, and the resulting magma mush will rise toward the surface. If it has a clear upward path, as occurs along the midocean ridges, it will erupt as basalt onto the seafloor. The magma does not always ascend vertically, however, and may first migrate laterally through the mantle for great distances. Elsewhere its path may be blocked completely, so it pools on the underside of previously formed lithosphere. We are ignorant of the composition of the lower oceanic lithosphere, but it is probably a mixture of basalt (which ultimately converts to eclogite as the lithosphere cools and thickens) and refractory crystals. Substantial amounts of eclogite in the older lithosphere would help explain why it eventually sinks back into the interior.

In contrast, the continental lithosphere is about 150 km thick, and its crustal and upper-mantle components are both buoyant relative to the normal mantle below. Continents therefore float around as icebergs and do not directly participate in the deeper circulation currents of mantle convection. But lateral movement in the mantle can and does move these lithospheric icebergs around, and once they come to rest they can insulate the underlying mantle and cause it to warm. In the course of this *continental drift,* continents can override the thinner oceanic lithosphere along *subduction zones* – linear or arcuate features characterized by deep oceanic trenches and large volcanic cones. If the oceanic lithosphere is still young and thus hot, it tends to slide under the continent at a shallow angle; older, thicker lithosphere is denser and tends to dive steeply into the mantle.

On Earth and elsewhere, the lithosphere is an important element in planetary dynamics. If it gets too cold or too thick, it can shut off the access of hot magmas to the surface or become too hard to break and descend (subduct). If its proportion of light minerals is too great, it will stay buoyant and will not sink back into the mantle. If there are too many plates or if they are moving rapidly, they again may not become dense enough to subduct. Thus, there are a variety of ways to "choke up" the surface. In the extreme, a lithosphere may get too thick to break anywhere, creating one uninterrupted plate that can slide around as a unit on the underlying mantle. On such a "one-plate planet," a huge meteoritic impact or the mass of a large new volcano could alter the planet's moment of inertia enough to make the whole outer shell rotate with respect to the spin axis.

Several mechanisms can fragment a lithosphere. Hot mantle upwellings can both heat and deform it. Diverging mantle currents below can create extensional stresses on its base. A lithosphere moving over an ellipsoid-shaped (rotating) planet will experience large stresses due to the changing contour of the surface. Tidal despinning of a planet is a related method for generating large stresses in the surface layer and a global fracture pattern. If the lithosphere becomes too dense it may sag and break. Several subduction

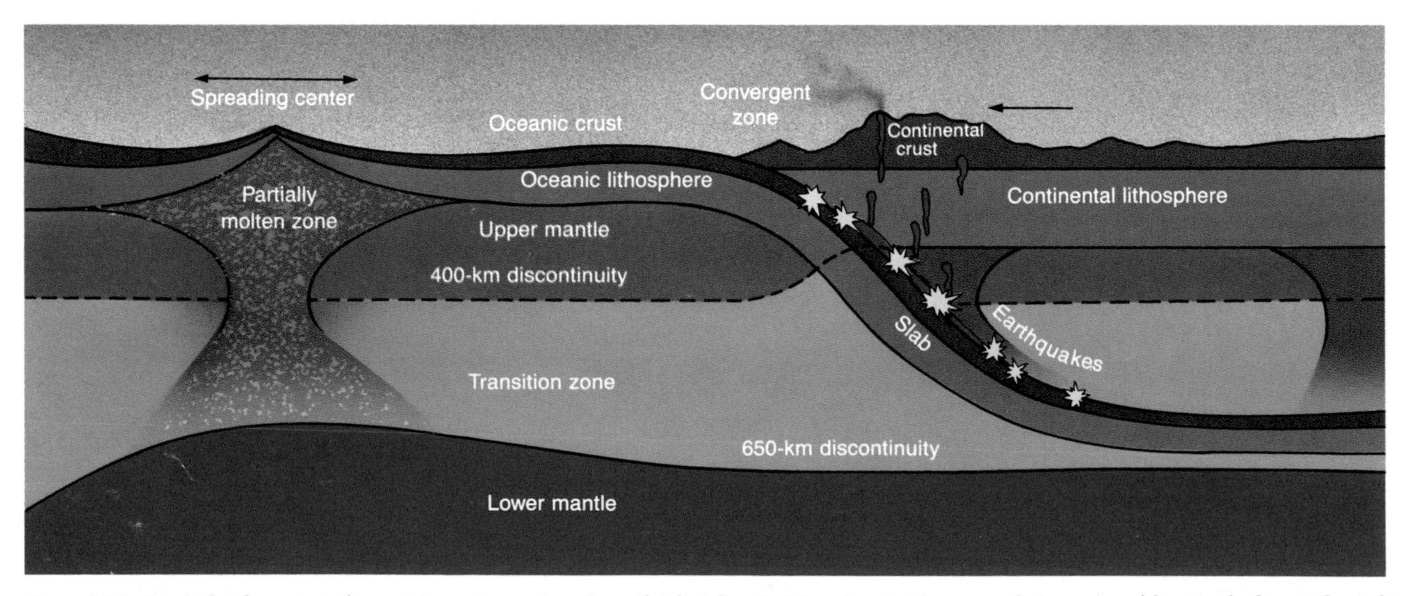

Figure 9. **The basalt that forms oceanic crust does not come from immediately below the lithosphere but from a much-deeper transition zone in the mantle. As it rises, this material decompresses and may become partially molten; finally it erupts at a spreading center. As the new oceanic crust moves away from its formation site, it cools and thickens, eventually becoming dense enough to plunge back into the mantle. This subduction occurs dramatically along zones of convergence marked by deep trenches, frequent earthquakes, and active volcanism.**

zones currently exist entirely under the Earth's oceans. But it is not clear if they are the result of an instability of the oceanic lithosphere, or if subduction started at the edge of a continent and later migrated toward the ocean. We do know that both oceanic ridges and island arcs can migrate relative to the underlying mantle and the spin axis.

The spin axis of a planet is controlled by the distribution of masses on the surface and in the interior. By analogy, the rotation of a spinning top is controlled by its shape, and its spin axis will change if bits of clay are attached to the surface. The physics of planetary reorientation is the same. If a large impact or a new volcano redistributes the mass, the planet will reorient itself relative to the spin axis so that the mass excesses lie closer to the equator. This shift is termed *true polar wander*. Both Mars and the Moon have apparently reoriented themselves to accommodate the effects of impacts or volcanoes.

On the Earth at present, true polar wander is very slight and results mainly from the rearrangement of mass due to melting glaciers. Polar motion in times past, as evidenced by magnetically aligned crystals in ancient rocks, is usually considered to be *apparent polar wander*, since we know that the continents have been drifting relative to the magnetic pole. In addition, however, Earth's rotation axis has apparently moved about 8° in the past 60 million years and 20° in the past 200 million – a period of time when the configuration of continents and subductions zones was also changing dramatically. Our extraterrestrial experiences now tell us that major shifts of the Earth's lithospheric shell relative to its spin axis might have followed convective rearrangement of mass in the interior, plate subduction, or the build-up of heat beneath large continents. These mass adjustments might be responsible for some major events in the geologic record, such as the breakup of supercontinents discussed in the next section.

Although plate tectonics, or at least "seafloor" spreading, may exist on Venus (see Chapter 7), that planet does not have the more obvious manifestations of terrestrial-style plate tectonics such as long linear ridges, subduction zones, and deep trenches. Planetary geologists have recently found evidence in Venus' equatorial highlands for some crustal extension, with short ridge segments and fracture zones, but there is as yet no hint of the subduction process. Since Venus and Earth are so similar in size, why do they differ so much in tectonic style? One reason is that Venus spins much more slowly and therefore has a much smaller tidal bulge. Its shape is thus nearly spherical, so its lithosphere experiences no large stresses while moving around. Another reason is that the surface of Venus is much hotter than the Earth's, which makes its lithosphere thinner and more buoyant (particularly if the two planets' crusts are compositionally similar). At present we do not know which of these explanations is correct, or whether other factors await discovery.

PLATE TECTONICS

Planets have various options for relieving themselves of their internal heat. The Earth chooses the plate-tectonic option, and most of our planet's interior heat is removed by this mechanism. Plate tectonics begins by the creation of new crust and upper mantle at long, globe-encircling cracks – the midocean ridges. While Arthur Holmes suggested that the oceans were a source of new crustal material as long ago as the 1920s, it was not until the early 1960s that Harry Hess (and later Robert Dietz) refined the scenario of a dynamic, self-renewing seafloor and focused attention on the midocean ridges and deep-sea troughs. The associated volcanism occurs mostly underwater, but the ridges can be traced around the world by their bathymetry and their seismic activity.

Newly formed lithosphere cools and contracts as it moves away from a ridge. Consequently, the ocean depth above it increases in a smooth and characteristic way as a function of distance from the ridge and, therefore, of age. The oceanic lithosphere also thickens with age and eventually becomes denser than the mantle material below; in response, it sinks back into the mantle at subduction zones.

Most of the ocean floor is less than 90 million years old, and nowhere is it older than 200 million years. It takes about 200 million years for the oceanic lithosphere and shallow mantle to cool to a depth of about 100 km, and when this is inserted back into the hot mantle it becomes, in effect, an ice cube in a warm drink. Subduction is the main mechanism by which mantle deeper than 100 km cools. Earthquakes have been recorded at depths as great as 670 km, and geophysicists believe the cold oceanic plates, or slabs, can sink this far into the mantle (Figure 10).

The Earth is apparently unique among the known worlds in its use of deep subduction as a cooling mechanism, and this can only occur if the lithosphere gets cold enough to cause it to become unstably dense and sink. On a planet with a thicker crust, a hotter surface, or a colder interior, the lithosphere may be permanently buoyant. In fact, on Earth the continents *are* permanently buoyant, a combination of thick low-density crust capping a buoyant upper-mantle "root" extending down to about 150 km.

Smaller planets cool more rapidly than large ones, have lower gravity, and experience less vigorous internal convection. Therefore, a lithosphere of a given thickness would be harder to break up on planet smaller than Earth. The Moon (with 1 percent of Earth's mass) and probably Mars (11 percent) are single-plate planets. Their interiors can convect heat outward, but they are never exposed to the cooling effect of subducting lithosphere because their outer layers behave as more-or-less rigid shells. Except for isolated volcanoes, they must lose their internal heat by conduction. Mantle upwellings can focus heat on one portion of the shell, weaken and thin it, and permit magmas to erupt onto the surface. This situation occurs on Earth as well; variously called midplate volcanism, hot spots, and plumes, it accounts for about 10 percent of the heat flow from the terrestrial interior (Figure 11).

The Earth actually exhibits at least three tectonic styles. The oceanic lithosphere recycles itself. The continents are buoyant; they may break up and reassemble, but they remain at the surface. A third characteristic is the way continents affect and are affected by the underlying mantle and adjacent plates. They are maintained against erosion – rejuvenated, in a sense – by compression and uplifting (mountain building) at their boundaries with other plates, by the sweeping up of island arcs at their leading edges, and by eruptions of basalt onto, into, or under the continental mass. Heat escapes from below a continent mostly by conduction, a relatively slow process. Therefore, the underlying mantle can heat up

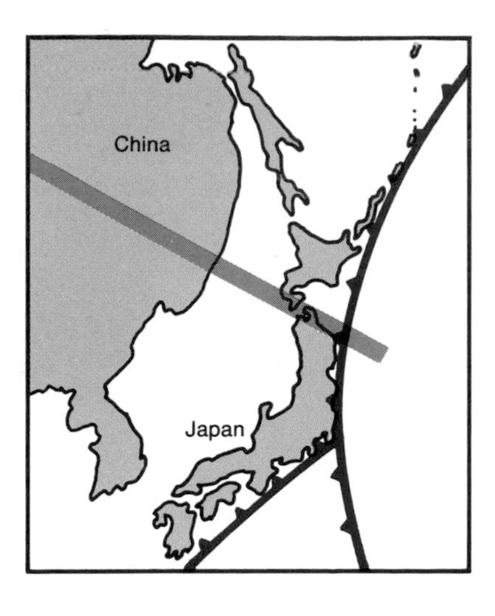

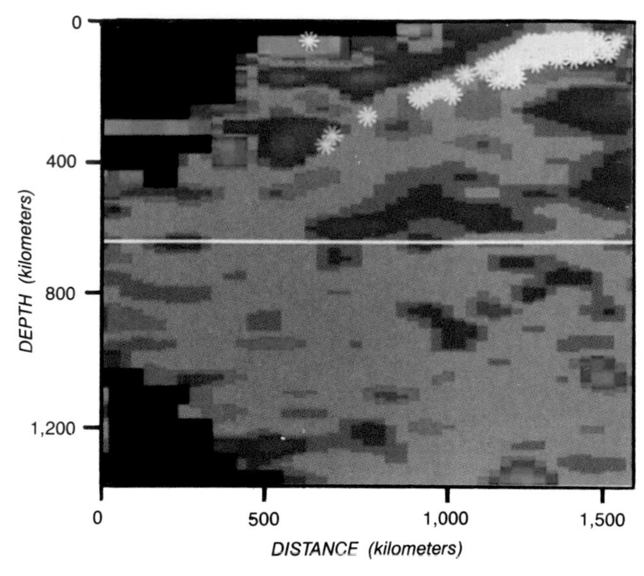

Figure 10. **Seismic tomography has been used to create a cross-section of the subduction zone of northern Japan. In the red regions, compression waves travel at relatively slow velocities, indicating hot regions. Earthquakes (yellow stars) fall near the top of a cold (blue) slab of descending lithosphere.**

enough to melt partially or, at least, to offset the cooling action of subducting oceanic lithosphere.

Since material flows from hot to cool parts of a convecting system continents will tend to drift away from hot mantle zones and come to rest over cool ones. When viewed from Africa, the continents are drifting away from each other at rates of some 5 to 10 cm per year. When this motion is traced back in time, we find that about 180 million years ago the continents were assembled into a supercontinent called Pangea (Figure 12). Moreover, for at least several hundred million years prior to that, the southern continents (Africa, South America, Australia, and Antarctica) plus India were a single assemblage, Gondwana. About 360 million years ago, Gondwana was centered on the South Pole, but it moved toward the equator just prior to its breakup. Initially, the continents' separation was rapid, but it slowed as the distances between them increased. As the continents moved apart, the Atlantic Ocean opened up and the Pacific Ocean shrank. Part of the Pacific lithospheric plate disappeared beneath the continental plates surrounding it.

Most of the continents are now sitting on or moving toward cold parts of the mantle. The exception is Africa, which was the core of Pangea. As they move around, the continents encounter oceanic lithosphere and force it to subduct into the mantle. Many active subduction zones are currently at the leading edges of continents. Perhaps all such zones formed along continental margins, after which some of them migrated to their present midocean locations.

There is another conceivable type of plate tectonics. If a large temperature difference does not exist between the surface and the interior, or if plate generation is very rapid, or if the crust-lithosphere system is completely buoyant – then deep subduction cannot happen. Consequently, the plates must remain near the surface, and their interactions will result in "pack-ice" underthrusting (much the way ice flows behave in the polar oceans). The convergence zones will be diffuse, elevated jumbles characterized by deformation, plate thickening, shallow underthrusting, and lithospheric doubling. Venus and the early Earth may have experienced this tectonic style, for we still see evidence of it in western North America, parts of western South America, and Tibet.

Figure 11. **Iceland sits in the Atlantic Ocean atop both a midocean ridge and a stationary hot spot in the upper mantle. As a result, the entire island is alive with volcanic and tectonic activity. Underground water can become superheated, reaching the surface through fractures. Shown here is the Namaskard geothermal area in northeast Iceland.**

THE EARTH'S GEOID

On an entirely fluid planet the shape of its surface – the *geoid* – is not controlled solely by rotation. Concentrations of mass in the interior (actually, pockets of anomalously high density) attract the fluid, cause it to pool above them, and make the regional surface stand high. The geoid is usually defined with respect to the perfect ellipsoid that the planet would assume if its interior were completely fluid, with density changing only with depth. The result on a real planet is a global pattern of broad undulations, with heights of some hundreds of meters and a variety of wavelengths.

On Earth, the surface of the ocean approximates the geoid, but a more accurate figure for the entire planet has been obtained by tracking the motion of low-altitude satellites (Figure 13). While these geoid, or gravity, data cannot identify subsurface structures unambiguously, they can be used to calculate the contribution from isostatically

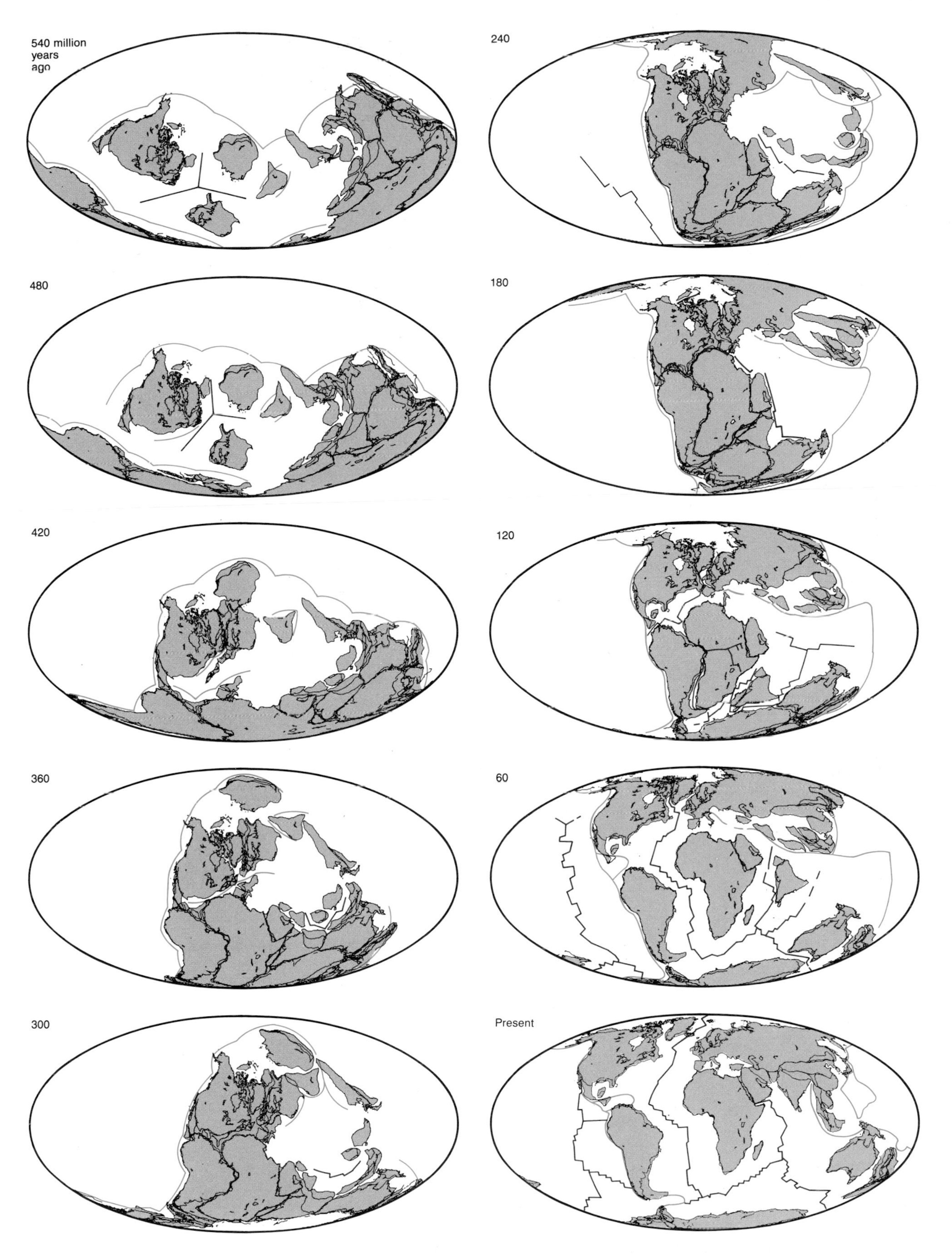

Figure 12. **The Earth's face has changed dramatically in the last half billion years, as shown here in 60-million-year intervals. Note the assembly of Gondwana at the south pole prior to its incorporation into the supercontinent Pangea. Pangea moved northward across the equator over 150 million years; its eventual breakup created the Atlantic Ocean and greatly diminished the extent of the Pacific Ocean.**

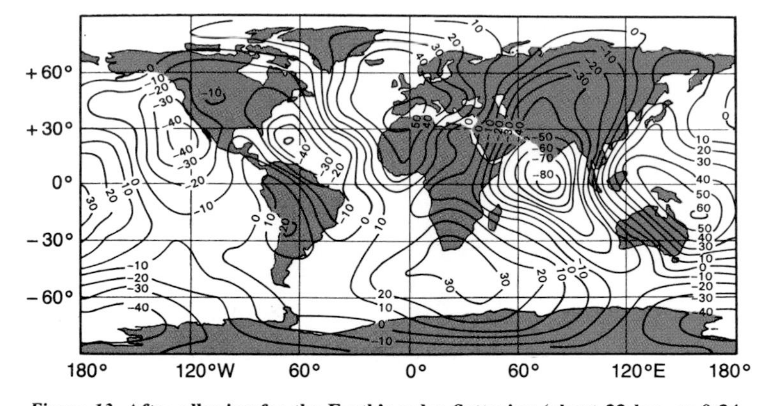

Figure 13. **After allowing for the Earth's polar flattening (about 22 km, or 0.34 percent), geophysicists have analyzed satellite-tracking data to reveal that our planet displays residual geoid highs and lows, given here in meters with respect to the ideal spheroid. The highs correspond to subsurface concentrations of dense rocks, the lows to accumulations of less-dense rocks probably within the mantle.**

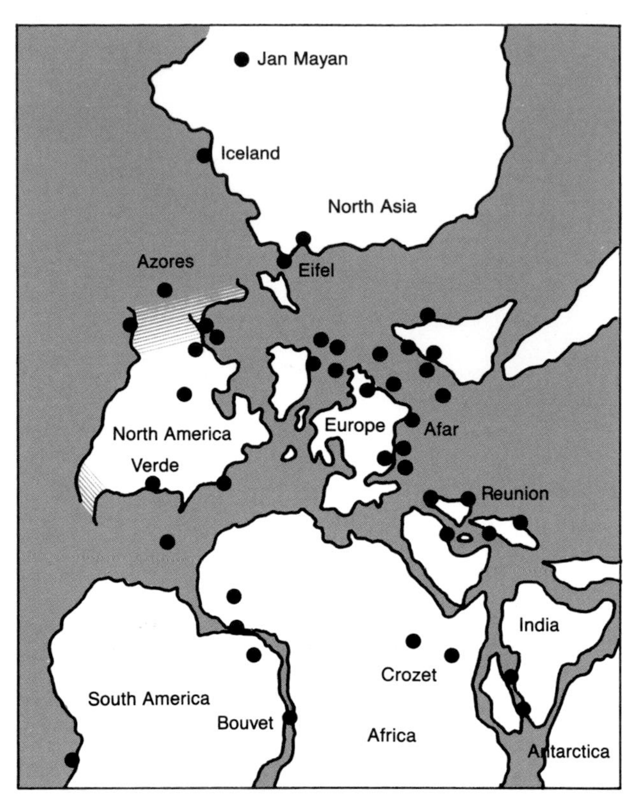

Figure 14. **The locations of hot spots about 350 million years ago. Today hot spots are primarily in the Atlantic and Indian oceans and under Africa, but most of them were originally underneath Pangea. They may have come about because the great mass of continental material kept heat from leaving the mantle, causing a buildup of thermal energy that has taken hundreds of millions of years to be released.**

compensated continents, slabs, and density variations in the lower mantle. (A continent is considered to be in a state of *isostasy* if equilibrium exists between gravity's downward pull on the mass sitting above sea level and the upward push of the mantle on the continent's low-density "root." Icebergs, in a sense, float isostatically in sea water.)

At very long wavelengths, there are equatorial geoid highs centered on the Pacific Ocean and Africa. Geoid lows occur in a polar band extending through North America, Brazil, Antarctica, Austria, and Asia. Brad Hager, Robert Clayton, and Adam Dziewonski have shown that this pattern correlates with the seismic velocity distribution in the lower mantle, as expected. The long-wavelength geoid highs arise from upwellings of hot mantle material that deform the core-mantle boundary and the Earth's surface upward. At the same time, the hot upwelling mantle is expected to be buoyant and thus relatively low in density, which seismic waves travel through more slowly than in cold material elsewhere. Except for Africa the continents are in or near geoid lows. We think they migrated into these regions as they moved away from Africa after the destruction of Pangea.

The major geoid highs of moderate wavelength are associated with subduction zones stretching from New Guinea to Tonga and along the Peru-Chile coastline. These highs, centered on the equator, undoubtedly contribute to the moment of inertia that controls the orientation of the Earth's spin axis relative to its mantle. At shorter wavelengths subduction zones show up as geoid highs, or mass excesses. This is expected as long as the descending slabs are cold, dense, and supported from below by a strong or dense lower mantle.

From Figure 13, it is apparent that Earth's present-day expressions of tectonism correlate poorly with its geoid. However, there is *good* correlation with the continental and subduction-zone configurations of the past. For example, the geoid high centered over Africa has about the shape and size of Pangea, and geoid lows correspond roughly with where regions of subduction should have existed prior to extensive opening of the Atlantic Ocean. This is an excellent demonstration of the time-scales on which planetary processes operate – the heat trapped under the super-continent of Pangea more than 100 million years ago (Figure 14) continues to escape from the mantle today. The still-hot mantle has thus elevated the continent of Africa; it represents a geoid high.

We have been studying the Earth's surface since our arrival here, but only within the past several decades have we come to appreciate the internal turmoil that continuously shapes the landscapes around us. We were learning about the roles of continental drift and plate tectonics on the Earth at the same time we realized that every other world has a unique style of operation. As far as we know the Earth is the only planet that has active plate tectonics, oceans, and life. One wonders if these facts are interrelated.

7
Surfaces of the Terrestrial Planets

James W. Head, III

"We shall not cease from exploration, and the end of all our exploring will be to arrive where we started and know the place for the first time."

– T. S. Eliot

IN THE LAST THREE DECADES, two parallel revolutions in the Earth sciences have radically altered our perceptions of planets and how they work. The first was the development of the theory of terrestrial plate tectonics. We came to realize that the Earth's geologic expressions are not an assortment of isolated puzzles to be solved individually, but rather a record of the movement and interaction of a small number of large lithospheric (crustal) plates operating in an integrated, global manner. The second revolution resulted from the unfolding view, thanks to spacecraft exploration, of planetary surface features. Almost overnight, it seems, the Moon and planets changed from astronomical objects to geological ones, and in doing so they began to provide a framework within which the Earth could be viewed not as a single data point, but as one of a family of planets.

Let us examine these parallel revolutions in a bit more detail. The decade of the 1960s was one of revelation. We began to comprehend the geology and geophysics of the Earth's ocean basins and the unorthodox processes they signified. Old and new ideas on continental drift and seafloor spreading finally melded late in the decade into a new theory of global plate tectonics. Its predictive nature gave renewed vigor to the geological sciences, which to that point had been mired by the extreme complexity of the Earth. During the 1970s our attention remained fixed on plate tectonism and how the last several hundred million years of Earth history fit into its context. By the 1980s, the emphasis had shifted somewhat, first to the study of the kinematics (motions) and dynamics (the forces producing the motions) behind the plate-tectonic theory, and then to how all these surface processes might be linked to the structure and motions of the Earth's interior (see Chapter 6). Throughout this period, terrestrial geologists wondered how the emerging picture of global tectonism during the last several hundred million years might apply to the rest of Earth history.

Meanwhile, planetary exploration in the 1960s saw Mariner spacecraft reveal the surface of Mars and Apollo astronauts begin intensive exploration of the Moon (see Chapter 1). A flood of discoveries, including rocks dating to the planets' formative years and volcanoes on Mars more immense than could have been imagined, freed us from our Earthbound view of geology and ushered in an era of comparative planetology. During the 1970s, American and Soviet spacecraft continued to course through the inner solar system, even as researchers back home struggled to understand the important factors in planetary origin and evolution. In the 1980s the Soviet Union sent a half dozen highly successful spacecraft to Venus; these dropped instrumented balloons into its atmosphere, delivered landers that revealed the nature and composition of its surface, and used radar imaging from orbit to characterize its regional terrain. Simultaneously, radar observatories on Earth were also obtaining high-resolution images that added further to our understanding of Venus.

The point is that, prior to the 1960s, geologic thought focused on specific areas of the Earth's surface and, in a rather passive sense, was analogous to the pre-Copernican, Earth-centered view of the solar system. Now we are abandoning this chauvinism, working and thinking instead in terms of a new solar system in which Earth's history is inextricably linked to those of the Moon and the other terrestrial planets. The 1990s promise to be a time of convergence of thought. Earth scientists, on one hand, will begin to apply their understanding of recent geologic history to the first 4 billion or so years of our planet's existence. Simultaneously, others will consolidate their understanding of the formative years of planetary evolution (the first $1\frac{1}{2}$ billion years) and begin applying it to the Earth. The study of Venus will provide a driving force for this convergence. It is the most Earthlike planet in terms of gross physical properties and position in the solar system, and we must learn to what extent it does – or does not – mimic the Earth in detail.

SURFACE PROCESSES

This chapter emphasizes the nature of the terrain types observed on the terrestrial planets, and what such information tells us about the geologic processes that form and modify planetary surfaces, their distribution with time, their relation to planetary interiors, and the general evolution of planets. Three fundamental processes – *impact cratering,*

Figure 1. **Standing on the outer ring of a Brazilian impact crater known as Serra da Cangalha, geologist John McHone and a guide look south toward the striking inner ring some 5 km away. Pockets of dense vegetation and a small dwelling sit in the intervening flat sediments, which are 350 m lower than the distant peaks.**

volcanism, and *tectonism* (crustal movement) – give planetary surfaces most of their characteristic features. And, as later sections will show, each terrestrial planet expresses these processes with a unique signature.

Most geologic activity occurs over long time scales, with only modest short-term changes. For example, mountains are built over millions of years and eroded down over additional millions of years. But the formation of impact craters is dramatically different. A meteoritic projectile strikes a planetary surface at velocities measured in kilometers per second. The kinetic energy concentrated at that point can easily equal the total annual heat flow of the Earth! Most of the impact energy goes into fragmenting, shock-heating, and ejecting material from the crater cavity. Less than 1 percent of it is radiated away from the target area as seismic waves, but an impact can be so energetic that it creates trains of shock waves several orders of magnitude stronger than any terrestrial earthquake on record.

Impact craters are not common on the surface of the Earth, (Figure 1), and the known ones were not commonly recognized as such until recently. Prior to planetary exploration, therefore, impact cratering was not considered a significant geologic process (see Chapter 21). Now we realize that not only are all the terrestrial planets cratered, but also that the first 600 million years of their history (for which the record is missing on Earth) was a period of incessant cratering.

Impact basins exceeding several hundred kilometers in diameter are common on Earth's neighboring worlds (Figure 2), and their characteristics reveal much about the nature of the cratering process. An initial, transient cavity is created by the fragmentation and ejection of a vast quantity of material (1 to 10 million km^3 for the Moon's Orientale basin, seen in Figure 3). The detailed geometry of the transient cavity is not known, but for a large basin it could reach depths of several tens of kilometers. Ejected debris spreads over great distances, accompanied by projectiles that can form secondary craters as large as 25 km across. When the Orientale basin was blasted out of the lunar crust, the impact energy left enough molten rock within the crater cavity to cover about 200,000 km^2 – equivalent to the great Columbia River basalt deposits in the Pacific Northwest, which took several millions of years to accumulate! With the crust below them heavily fractured, the basins then became the locus for mare lava flooding and tectonic deformation. Further study of how cratering affected the early Earth may provide clues to crustal formation, early continental growth, and the onset of plate tectonism.

Igneous and volcanic processes have also been significant in the evolution of planetary surfaces. It appears that volcanism is a consequence of radioactive decay within the interior, which heats the surrounding rocks until they melt. The amounts and rates of lava production have varied from planet to planet (and with time on each individual planet), but the influence on the surfaces of all planets has been widespread. For example, the dark lunar maria first appeared about 4 billion years ago and began to wane about $1\frac{1}{2}$ billion years later. Volcanic activity on Mars and the formation of its great shield volcanoes continued well into the most recent half of the solar-system history, and of course volcanic activity on Earth occurs frequently even now (Figure 4).

Although basaltic plains constitute the most common style

Figure 2. **The remarkable Caloris basin, situated near the terminator in this Mariner 10 mosaic, is a huge impact feature some 1,300 km in diameter. Its broad floor is laced by fractures, often in polygonal patterns, and by sinuous ridges resembling those on the lunar maria. The fractures probably arose when the basin's central area sank slightly from the weight of overlying lava.**

Figure 3. **The Moon's Mare Orientale, seen by Lunar Orbiter 4, lies within the larger, impact-created Orientale basin. The outermost ring of peaks, called the Cordillera Mountains, is some 900 km across. The impact that formed the basin about 3.8 billion years ago produced large secondary craters up to 2,000 km away.**

Figure 4. **In May 1980, Mount St. Helens in the Pacific Northwest exploded violently, devastating the surrounding region. Shortly thereafter, a NASA aircraft flew over the region and recorded this detailed radar image. About 400 m of the once-dormant volcano's beautifully symmetrical cone is now missing, leaving in its place a gaping crater that stands as a reminder that – even today – the Earth is a dynamic, evolving planet.**

of volcanism, different planets create them in different ways. The Earth's ocean basins are created by upwelling and injection of magma at divergent plate boundaries. Once cool, these outpourings are rifted apart, with new material injected in a continuous process known as seafloor spreading. On the Moon, the process was quite different. Molten rock traveled up through the thick lunar crust and emerged along extensive fractures often related to the creation of impact basins. On Mars, many of the volcanic plains came about when the great shield volcanoes erupted and sent lava flowing onto the low-lying plains surrounding them. Thus the formative circumstances of basaltic plains can provide information on the nature of major crustal provinces and environments. Discerning how much lava appeared over time helps pin down the amount and duration of thermal activity in a planet's interior.

Heavily cratered terrain generally represents the oldest surface on a planet, for it records the final stages of an object's accretion and the intense bombardment endured by its just-formed crust. Thus the degree of preservation of cratered terrain is a measure of local or global crustal stability over a planet's history. Volcanic units, on the other hand, imply melting in the interior and fracturing of the crust (to provide the magma access to the surface). Therefore, we can interpret the abundance and type of volcanic units in terms of the thermal history of the interior, the state of stress in the crust and lithosphere, and the general level of activity since the period of heavy bombardment.

Tectonic activity is also a significant indication of internal dynamics. Deformation of the outer rigid layer of a planet (its lithosphere) can occur in a variety of ways. For instance, the Moon has a thick global crust that exhibits few tectonic

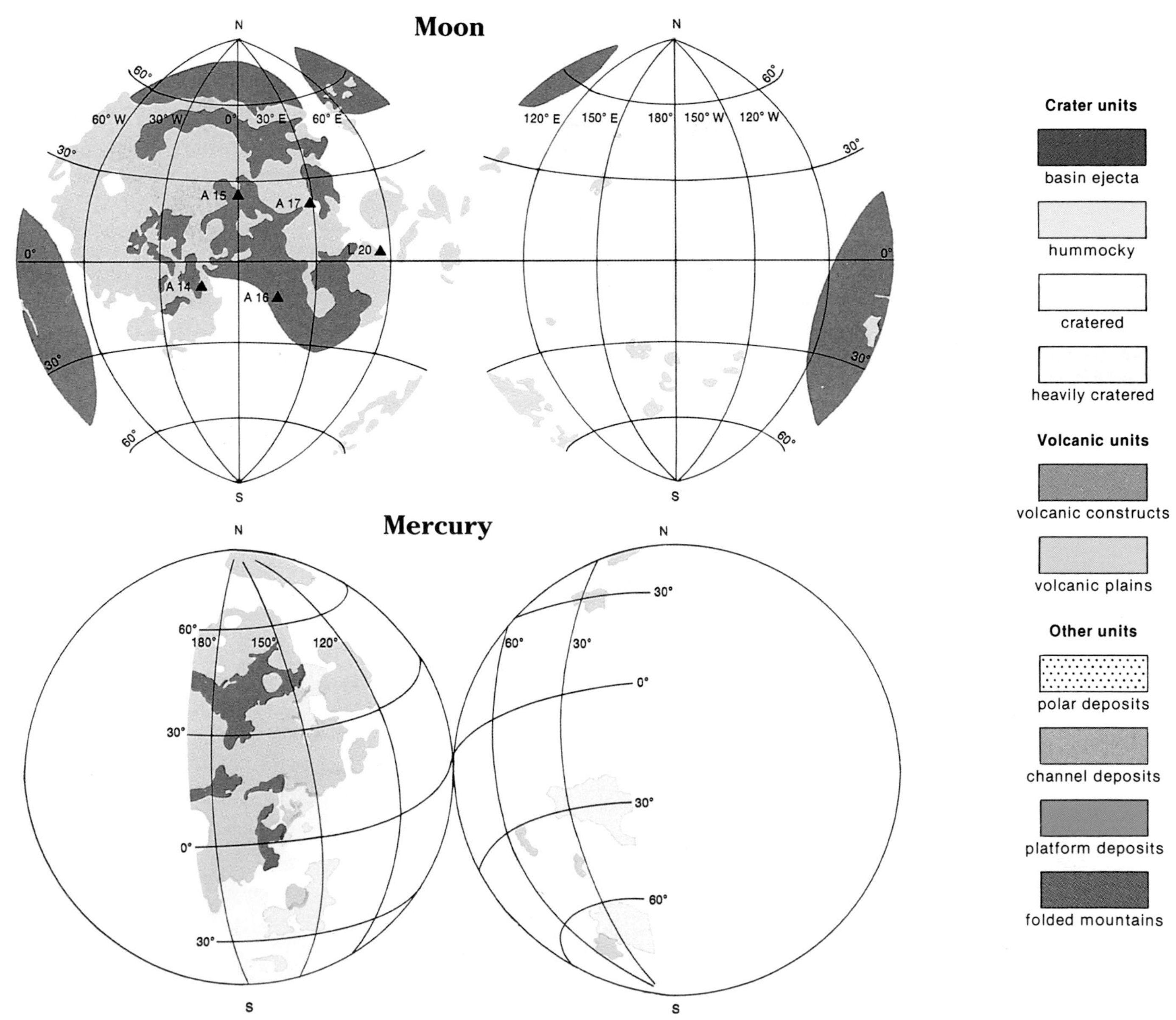

Figure 5a. **Reproduced here are maps showing the geologic terrain units on the Moon. Craters and near-side "seas" of relatively thin, solidified lava dominate the lunar landscape; absent are more complex landforms found on the Earth and Mars. Landing sites of Apollo *(A)* and unmanned Soviet Luna *(L)* spacecraft that were in or near highland areas are shown.**

Figure 5b. **Mariner 10 photographed only about 45 percent of Mercury's surface, but enough to demonstrate that the planet looks remarkably like the Moon. Cratered terrain is common, and extensive areas of smooth plains surround the Caloris basin (Figure 2), which is centered at 30° N, 190° W.**

features, and it lacks any evidence of the major lateral movement so typical of Earth's plate tectonics. Instead, lunar tectonics are characterized by downward movement, due to loading by the thousands of cubic kilometers of lava that have poured out onto the Moon's surface. The extensive lobate scarps on Mercury seem to be a manifestation of the global compression that occurred when the planet cooled and contracted. Mars has numerous tectonic features comparable to those seen on the Moon, but there is one important exception. The Tharsis region, with its concentration of tectonic and volcanic features, is a distinctive area unlike virtually any other observed on the terrestrial planets. Whatever the origin of Tharsis – be it deep-seated uplift or long-lasting volcanism – the nature of Martian tectonism is still vertical, rather than horizontal.

THE EARTH

The morphology and geology of the vast majority of the Earth's surface is concealed by liquid water and vegetation. Even though the atmosphere, hydrosphere, and biosphere are not considered terrain units, they are nonetheless significant agents in the modification of the solid crust beneath us. Earth's surface (Figure 5d) appears remarkably different from those of the other terrestrial planets. In particular, the Earth's crust is in constant motion, with the positions and relative abundances of its terrain units continually changing. Our planet possesses some units not commonly seen on neighboring worlds (like folded mountain belts and platform deposits), while certain terrain types seen

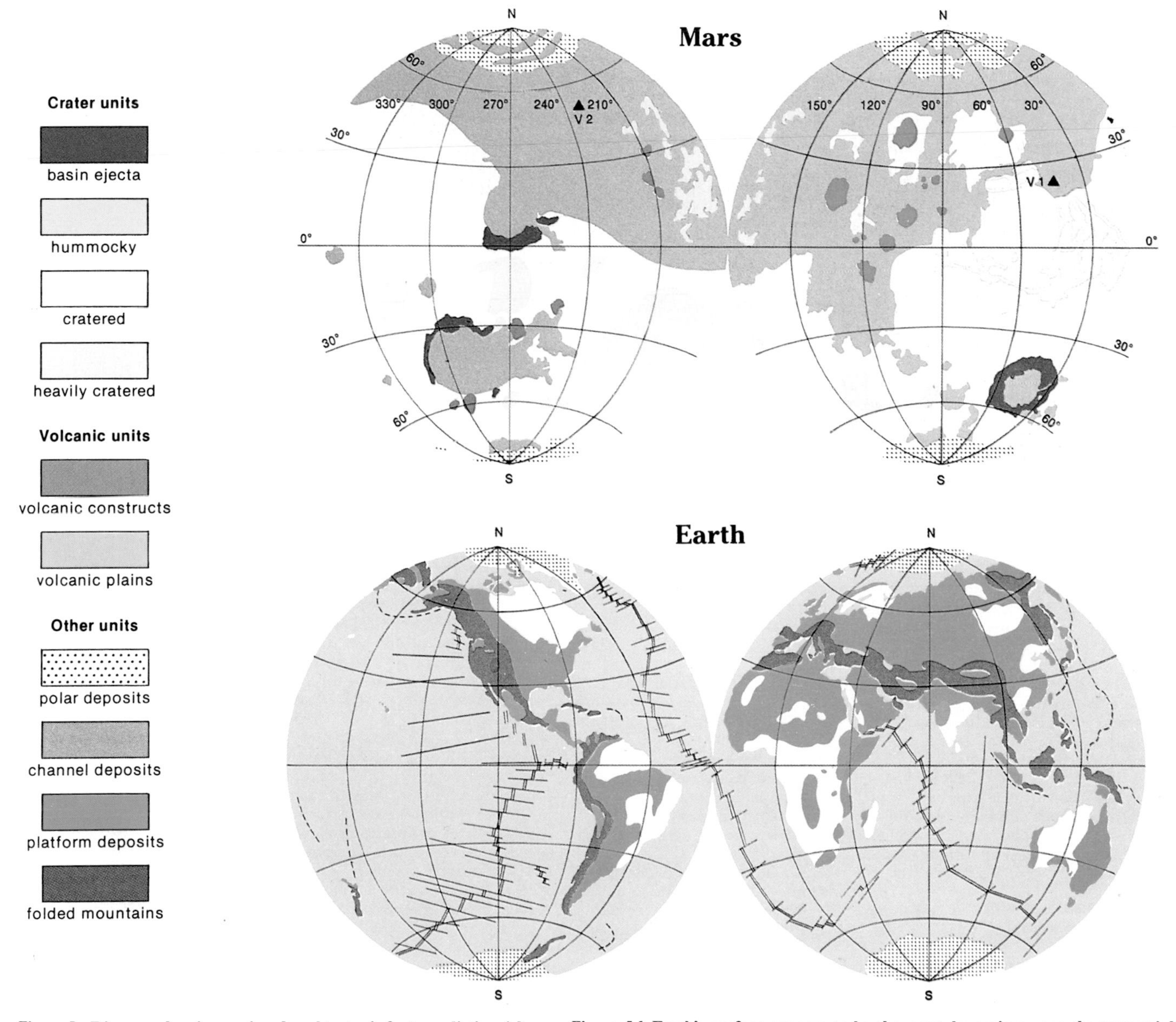

Figure 5c. **Diverse volcanic, erosional, and tectonic features distinguish Mars from the less-complex surfaces of Mercury and the Moon. The dichotomy of northern lowlands and southern cratered uplands results in a hemispherical asymmetry, which is a significant, unanswered puzzle. *V* indicates the two Viking landing sites.**

Figure 5d. **Earth's surface appears to be the most dynamic among the terrestrial planets. Only small portions of our planet are covered with aeons-old rocks; the remainder is relatively young, occurring mostly in the ocean basins. Like others in this series, the maps bring together data from numerous spacecraft investigations and a variety of other sources.**

elsewhere are either much less widespread here (cratered terrain) or much more so (volcanic plains).

Despite Earth's cover of water and vegetation, we now know that the ocean basins are geologically and morphologically dissimilar from the continents, a discovery that paved the way for the development of the theory of plate tectonics. From this theory has come the realization that the Earth is divided up into a series of rigid lithospheric plates, and that the formation, lateral movement, interaction, and destruction of these plates is responsible for most of the Earth's large-scale structural and topographic features. Plates collide to produce folded mountain belts (Figure 6) and deep trenches where plates dive down into the mantle and are destroyed. Continental rift valleys and vast plateaus of basalt accompany plate breakup. Strings of volcanoes appear where plates are consumed, and new crust forms at the gradually separating midocean ridges. Chapter 6 contains a more complete description of Earth's geology and geophysical characteristics.

Figure 6. **NASA's Seasat satellite recorded this radar image from high above Harrisburg, Pennsylvania (bright area at lower left). Conspicuous are the numerous folded hills that comprise this portion of the Appalachian Mountains. Note that the Susquehanna River, near the bottom, assumed its present course after the deformation of the crust around it. Similar techniques have been used on Venera spacecraft orbiting Venus to map the planet's surface at resolutions of 1 to 3 km (compare this image with Figure 13).**

Figure 7. **In this Mariner 10 image of Mercury's north polar region, the lower half is dominated by numerous impact craters superimposed on an intercrater plain, while the upper half shows a smooth plain with relatively few craters.**

THE MOON

As our closest neighbor in space, the Moon was first to occupy the attention of geologists studying other planetary surfaces (see Chapter 4). Eugene Shoemaker and his co-workers applied the basic principles of terrestrial stratigraphy to the surface features on the Moon. This enabled them to produce geologic maps (Figure 5b) and thus delineate the major surface processes and sequence of events in lunar history. Similar mapping techniques have been applied to each successive solid body explored over the last 25 years, and the collective maps provide the basis for understanding the history of each object and comparisons of planetary evolution.

For the Moon, the early formation of its crust was accompanied and succeeded for several hundred million years by a massive influx of projectiles impacting the newly formed surface at several kilometers a second, producing craters of many sizes. This *heavy bombardment* had three major effects: (1) the fragmentation, fracturing, and brecciation (shock-induced cohesion) of the upper few kilometers of the Moon's crust, forming a massive soil layer called the megaregolith; (2) the production of geologic units that represent the first few hundred million years of lunar history; and (3) the creation of extremely rough surface topography. The bombardment ended about 3.8 billion years ago, but not before the largest projectiles had excavated huge depressions (perhaps as large as 2,000 km in diameter) and spread ejecta over immense areas that sometimes affected an entire lunar hemisphere.

Volcanic flooding of the surface of the Moon became evident during the waning stages of the heavy bombardment. By 2.5 to 3.0 billion years ago, basaltic lavas had covered approximately 16 percent of the lunar surface, preferentially filling in the low-lying basins to form its "seas" or *maria*. Small amounts of lava may have been extruded onto the surface at even later times. Tectonic activity on the Moon stands in stark contrast to that of our own planet: instead of the Earth's multiple colliding lithospheric plates, the Moon has but a single lithospheric plate – it is a "one-plate" planet. This has significantly inhibited the formation of lunar tectonic features. The few it has occur predominantly in and near the maria, where crustal expansion has created linear rilles and grabens, and compression has uplifted sinuous ridges. The maria bear these geological undulations because outpourings of basaltic lava have loaded the lithosphere to the point of flexure. Virtually no major geologic episodes have occurred on the lunar surface in the last 2.5 billion years.

The Apollo expeditions and Luna missions returned samples from which a chronology of lunar events could be constructed (see Chapter 4). The Moon thus provides a picture of the first half of solar-system history that is characterized by impact bombardment and early volcanism, and it serves as a cornerstone for the interpretation of the records preserved on other terrestrial planets.

MERCURY

During its three flybys of Mercury in 1973–1974, the Mariner 10 spacecraft returned images of about 45 percent of the planet's surface and revealed a lunarlike terrain. Geologists found this somewhat surprising, since the planet is intermediate in size to the Moon and Mars and has a density comparable to Earth's. Detailed geologic mapping of the surface, however, has shown that Mercury differs from the Moon in several respects. Large areas of relatively ancient *intercrater plains* (Figure 7) may indicate that more extensive volcanism accompanied the heavy-bombardment period on Mercury than on the Moon. The planet's large, extensive scarps (Figure 8) attest to episodes of regional and perhaps global compression that would result from a modest decrease in Mercury's circumference (1 or 2 km) as it solidified from a molten state. Its smooth plains are nearly as reflective as its heavily cratered regions, which has led to controversy over the origin (volcanic or otherwise?) of the smooth regions; unfortunately, the modest resolution of Mariner 10's images make it difficult to resolve the uncertainty. With these exceptions, Mercury generally resembles the Moon on its surface, but its high overall density (5.4 g/cm^3) suggests that the planet's interior is more like the Earth's.

Why is Mercury's density so high? Many believe it is the result of where the planet formed in the solar nebula (see Chapter 23). Being so near the Sun, the growing planet must have experienced very high temperatures. Consequently, many elements and molecules proved too unstable to remain incorporated in Mercury's rocks, and only less-volatile elements, particularly iron, could have accumulated. Yet this explanation does not seem to account completely for the high density of Mercury. In 1987, computer simulations by Willy Benz, Wayne Slattery, and Alastair G. W. Cameron showed that a large, high-speed projectile may have struck Mercury early in its history (Figure 9), preferentially stripping off part of its silicate-rich outer layers, and thus increasing the planet's proportion of iron content and its bulk density.

MARS

Thanks to Mariner and Viking images, we now realize that Mars is more geologically diverse and complex than either the Moon or Mercury (see Chapter 5). Moreover, the red planet shows a distinct hemispheric asymmetry in the distribution of its geologic units (Figure 5c). The often densely cratered southern hemisphere stands 1 to 3 km above the topographic "sea-level," while the northern hemisphere is sparsely cratered and lies generally below that level. The boundary that separates these hemispheres is extremely complex.

Mars' southern hemisphere, in turn, has two general types of terrain: a very ancient crust that is nearly saturated with large craters, often fractured, and crossed by abundant small channels; and younger intercrater plains that also look ancient but have been modified less. The older, higher terrain of the southern hemisphere contains numerous channels, tens of kilometers wide and hundreds long, that are reminiscent of those formed on Earth by catastrophic flooding. In the northern hemisphere, volcanic flows clearly surround the largest volcanoes, but elsewhere the plains are featureless except for craters and marelike ridges. At mid-to-high northern latitudes the plains contain patterned and striped ground, scarps, and irregularly shaped mesas. Their complexity may represent the influence of volatiles and changing temperature.

A major departure from Mars' general surface geology is the Tharsis rise (Figure 10), a broad topographic bulge that exhibits ancient heavily cratered units and young shield volcanoes. The 8,000-km-wide region stands about 10 km high – and its enormous shield volcanoes rise another 15 km

Figure 8. **More than 500 km long and 3 km high in places, Discovery Rupes probably formed when Mercury's crust contracted as the planet cooled and solidified early in its history. The scarp is younger than the craters it transects.**

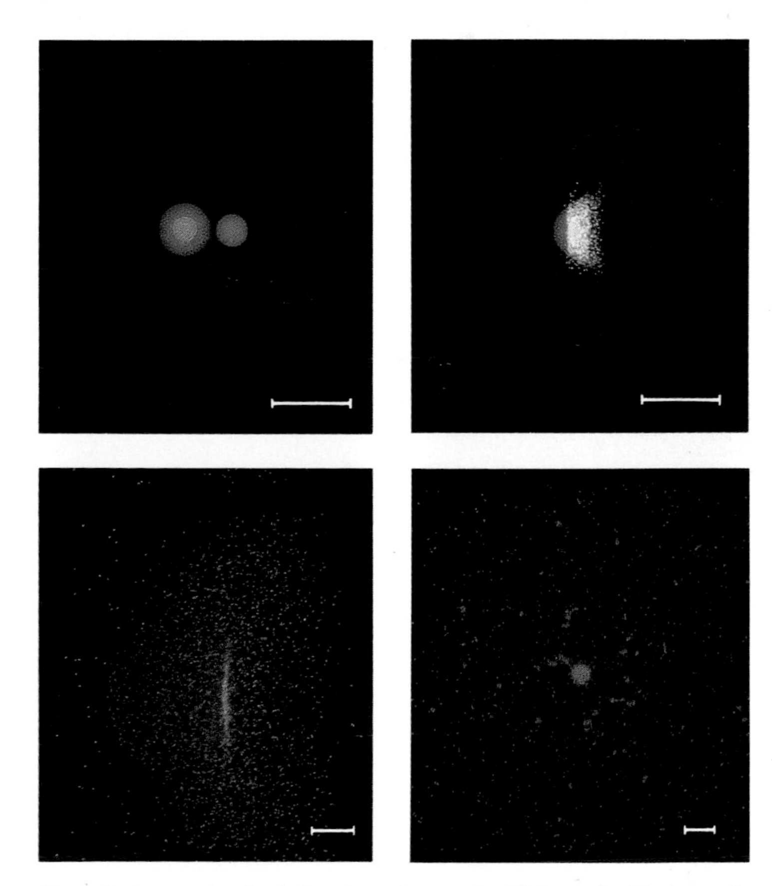

Figure 9. **A computer simulation shows how much of Mercury's silicate mantle (brown) may have been lost during a cataclysmic collision early in solar-system history. During the impact, which occurs at a speed of 20 km per second, the metallic cores (red) of both Mercury and the colliding object are also disrupted, but enough of their mass remains to recombine into a new planet. The four frames shown span 3.7 hours; scale changes are indicated by the scale bar, which corresponds to 10,000 km. While this simulation yielded a Mercury with the correct size and high density, other test runs show that the final outcome depends critically on the impactor's size, velocity, and target point.**

or more in altitude! The vast majority of Mars' linear rilles and fractures surround the Tharsis region. Valles Marineris, an enormous equatorial canyon system, extends radially away from Tharsis and is probably related to faulting that followed the bulge's formation. The origin of Tharsis is uncertain; some believe that it is predominantly a massive uplift of the crust caused by some dynamic process deep within the planet's mantle, while others propose that the lithosphere under Tharsis is thin, rendering it more susceptible than elsewhere to volcanic outpouring, topographic buildup, and related stresses.

From the interrelations of its surface units, planetary geologists believe Mars had an early history much like the Moon's and Mercury's, but volcanism (particularly in the Tharsis region) extended well into the last half of solar-system history and perhaps continues today. However, the absolute chronology is not known because of the lack of documented samples from the surface units on Mars.

One of the most exciting aspects of Mars is the abundant evidence that its climate was once dramatically different. Certain Martian craters have odd ejecta patterns that suggest a layer of ice or water existed in the upper crust where and when they were excavated. Much of Mars is covered with windblown deposits, and alternating layers of ice and dust at the poles suggest wide cyclic swings in climatic conditions (see Chapter 8). What was the early atmosphere of Mars like? Under what conditions could its channels have formed, and how did the fluids in the upper crust get there? Further exploration of Mars in the 1990s by spacecraft from the Soviet Union and United States will provide important data for understanding the nature of the atmosphere, the interaction of the atmosphere and the surface, and possible

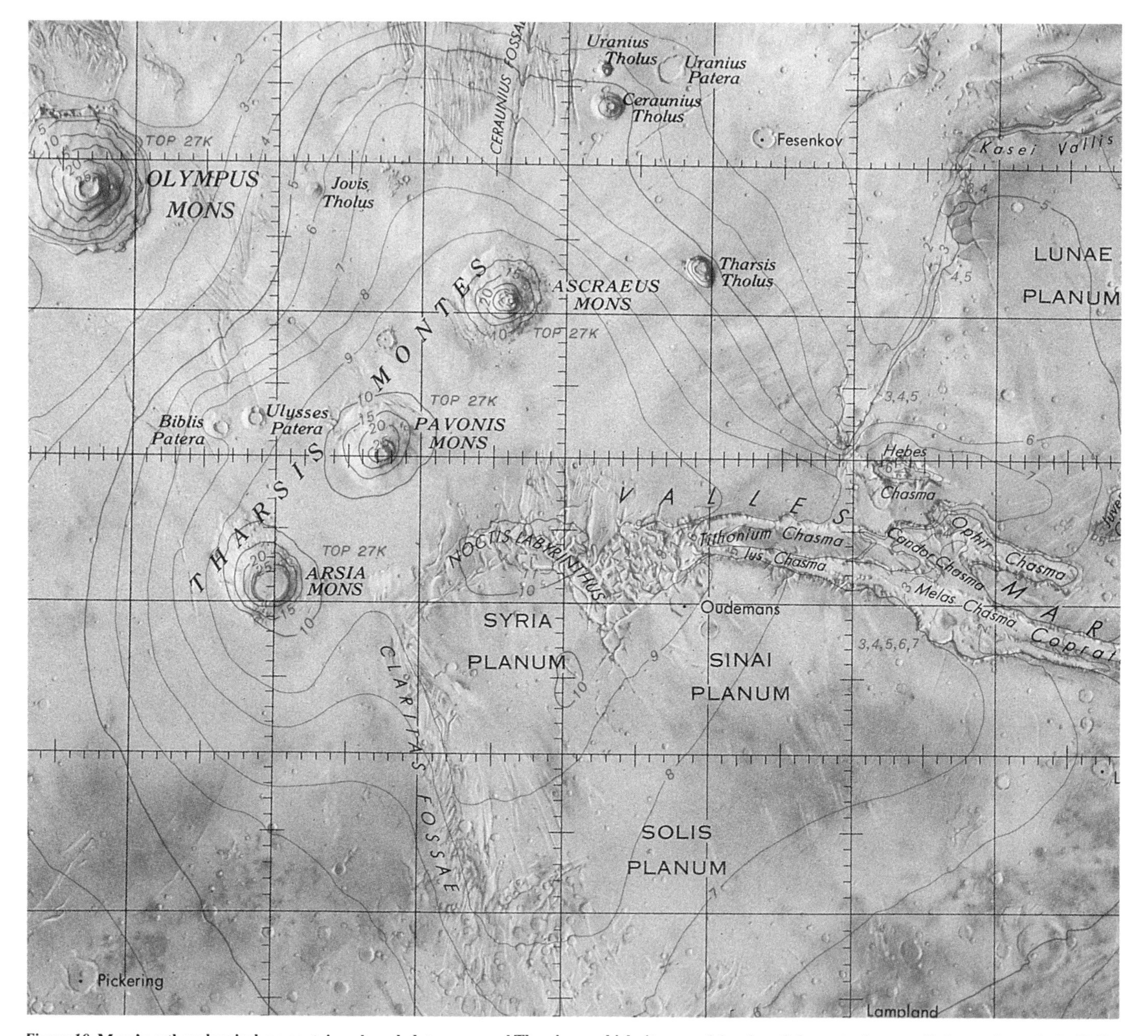

Figure 10. **Mars' northern hemisphere contains a broad plateau, named Tharsis, on which sit some of the planet's largest volcanoes. To its southeast is the Valles Marineris canyon complex, which may have a related origin. The red contour lines correspond to 1-km steps in altitude, and the map scale is 1:25,000,000.**

climatic change. A major scientific goal is to return documented samples of the planet's surface; with those in hand, we can determine not only their composition but also the history of the atmosphere and the absolute chronology for Martian geologic evolution.

Taken together, the smaller terrestrial planetary bodies (the Moon, Mercury, and Mars) are quite different than the larger Earth. Their single, globally continuous lithospheric plates stabilized early in solar-system history. Consequently, their surfaces have experienced predominantly *vertical* tectonism linked to loading and flexure, rather than the creation, *lateral* movement, and subduction of multiple plates as occurred on the Earth. Heat escaped from their interiors by conduction through a thick static outer shell, rather than by the convective overturn of the mantle and recycling of the lithosphere. Their ancient surfaces have retained evidence of the fundamental geologic processes at work early in solar-system history.

VENUS

The geology of Venus is of extreme interest because of the planet's similarity to Earth in size, density, and position in the solar system, as well as its *dis*similarity from the smaller terrestrial planets. The dense carbon dioxide atmosphere of Venus has served, however, to obscure the surface of the planet from ground-based and spacecraft cameras. Even well into the space age, our only views of Venus' topography consisted of radar images of small areas obtained from Earth. Planetary geologists longed for the answers to a host of fundamental questions, not the least of which was knowing what the surface features looked like.

The question with arguably the broadest implications is, simply, *how has Venus chosen to get rid of its internal heat?* For example, does it escape by conduction, as occurs in the smaller terrestrial planets? In this case we might expect to see an ancient lunarlike surface dominated by impact craters and volcanism. Does Venus cool itself by sending magmas directly from the interior to the surface? Then we would expect to see widespread volcanic deposits and numerous "hot spots," like those on Jupiter's satellite Io. Does Venus get rid of its heat as Earth does, predominantly through the recycling of lithospheric plates? If so, we might expect to see oceanic rises, transform faults, folded mountain belts, and long, deep trenches. Or perhaps Venus employs a combination of these mechanisms.

Beginning in 1975, the Soviet Union dropped a series of its Venera spacecraft onto the planet, and over several years their cameras returned a total of six surface panoramas (Figure 11). In the late 1970s the U.S. Pioneer Venus spacecraft returned radar soundings from hundreds of orbits around Venus. Pioneer's radar-altimetry data, covering 93 percent of the surface, allowed us to characterize Venus' global topographic provinces (Figure 12). About 60 percent of the surface lies within 500 m of the mean planetary radius (Venus' equivalent of sea level), and only 5 percent lies more than 2 km above it. This relatively even surface is in contrast to the distinctly bimodal distribution of the Earth's topography, which reflects the density differences between the lighter, high-standing continents and the heavier, low-lying ocean basins. Despite this strong clustering of altitudes on Venus, the total range of elevations – about 13 km – is comparable to that of the Earth.

The terrains that cover Venus can be subdivided into lowland plains (20 percent), rolling uplands (70 percent), and highlands (10 percent). Venus' highlands are unlike any topography seen on the smaller terrestrial planets. For example, Ishtar Terra (Figure 13) is larger than the continental United States and stands several kilometers above the average planetary radius. It is separated from the surrounding rolling uplands by relatively steep flanks, and the western portion is a vast plateau (Lakshmi Planum) some 2,000 km across. Eastern Ishtar contains Maxwell Montes, the single most dramatic topographic feature on Venus. These mountains rise up to 12 km above the average radius, a height exceeding that of Mount Everest above sea level. Ishtar Terra has an elevation comparable to that of the Tibetan Plateau but covers approximately twice the area.

The largest highland region on Venus, Aphrodite Terra, covers an area more than half the size of Africa. It extends along the equator for at least 10,000 km and appears topographically rougher and more complex than Ishtar. Several dramatic linear depressions transect east-central Aphrodite; some are up to 3 km deep, hundreds of kilometers wide, and trace across the surface for 1,000 km.

The lowland areas of Venus occupy some approximately circular areas (such as Atalanta Planitia, a large depression east of Ishtar) and broad but linear depressions (such as that

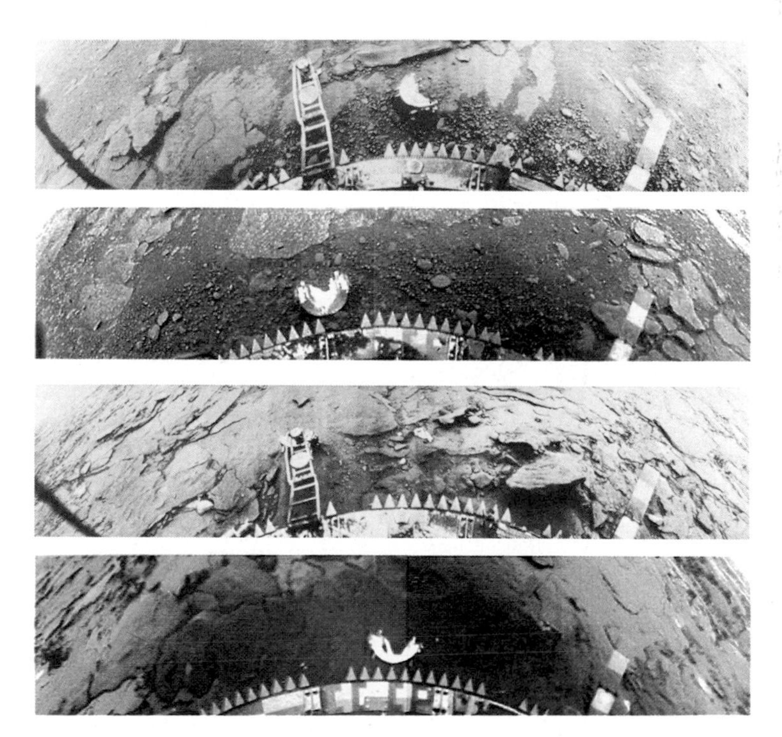

Figure 11. **In March 1982, twin Soviet landers obtained these panoramas of Venus' surface. The top one, from Venera 13's landing site at longitude 304° W and latitude 7° S, detail rocky outcrops and accumulations of small particles that extend to the horizon (upper corners). The spacecraft's shock- absorbing ring has a sawtooth lip to stabilize its descent and touchdown; also visible are a calibration standard at right and a crescent-shaped lens cover on the ground. The unprecedented color view has an orange cast because the cloudy atmosphere scatters and absorbs the blue component of sunlight before it reaches the ground. Venera 14's site, at longitude 313° W, latitude 13° S, includes flat expanses of rock that apparently lack the fine-grain material seen near the craft's twin. Samples collected by both spacecraft have compositions similar to basalts found on Earth, and the slabs seen by Venera 14 may be thin layers of lava that fractured after solidifying.**

extending along Ishtar's southern edge). Radar imagery reveals that the extensive rolling uplands contain a diversity of topographic features, including linear troughs, parallel ridge-trough systems, and shallow circular structures up to 1,700 km across that often contain central mounds.

All of these features suggest that Venus had (and perhaps has) tectonism, volcanism, impact cratering, and a complex geologic history. But despite our improving understanding of Venus' surface character, the resolution of our observations has remained insufficient to determine the exact geologic processes responsible for the diverse and tantalizing topography. Fortunately, recent high-resolution data have helped to characterize these broad features better and to begin to understand how they came to be. This, in turn, helps us deduce Venus' method(s) of heat transfer and its general tectonic evolution. The new data have come from the radar imaging and altimetry of the Soviet spacecraft Veneras 15 and 16, which began orbiting Venus in October 1983, and from radar studies conducted with radio telescopes in Arecibo, Puerto Rico, and Goldstone, California.

The Venera spacecraft mapped the high northern latitudes of Venus, about 25 percent of the surface, at a resolution of a few kilometers. Chief among their discoveries (Figure 14) were: *ovoids*, generally circular features up to several hundred kilometers across characterized by an annulus of ridges and central volcanism; *ridge belts*, complex linear deformations reminiscent of compressional mare ridges; *mountain belts*, high-altitude chains with folds and faults; *tessera*, complexly textured and tectonically dissected terrain standing above its surroundings; *volcanoes*, as shields and abundant small domes; and widespread *volcanic plains*. On the basis of these features and their distribution, Soviet investigators have tentatively concluded that the surface of Venus shares characteristics of both the Earth and the smaller terrestrial planets. The terrain imaged by the Venera orbiters has relatively few craters, indicating that the average age of the surface poleward of about 30° N is less than 1 billion years old – considerably younger than the surfaces of the Moon, Mercury, and Mars, and slightly older than the average for Earth's.

Meanwhile, the continuing analysis of Earth-based and Pioneer Venus radar data has revealed some interesting things about the equatorial highlands, south of the areas examined by Veneras 15 and 16. For example, Beta Regio exhibits a gigantic rift valley and interior faults; a major volcano, Theia Mons, sits astride Beta Regio's western boundary and has flooded the rift valley. The valley itself branches to the south and extends westward, where it connects with Venus' near-global equatorial highland system of rift valleys. The region's extensional deformation and volcanism appear to be relatively young.

Halfway around the planet, western Aphrodite Terra has numerous features that bear some similarities to Earth's system of oceanic spreading centers (Figure 15). These

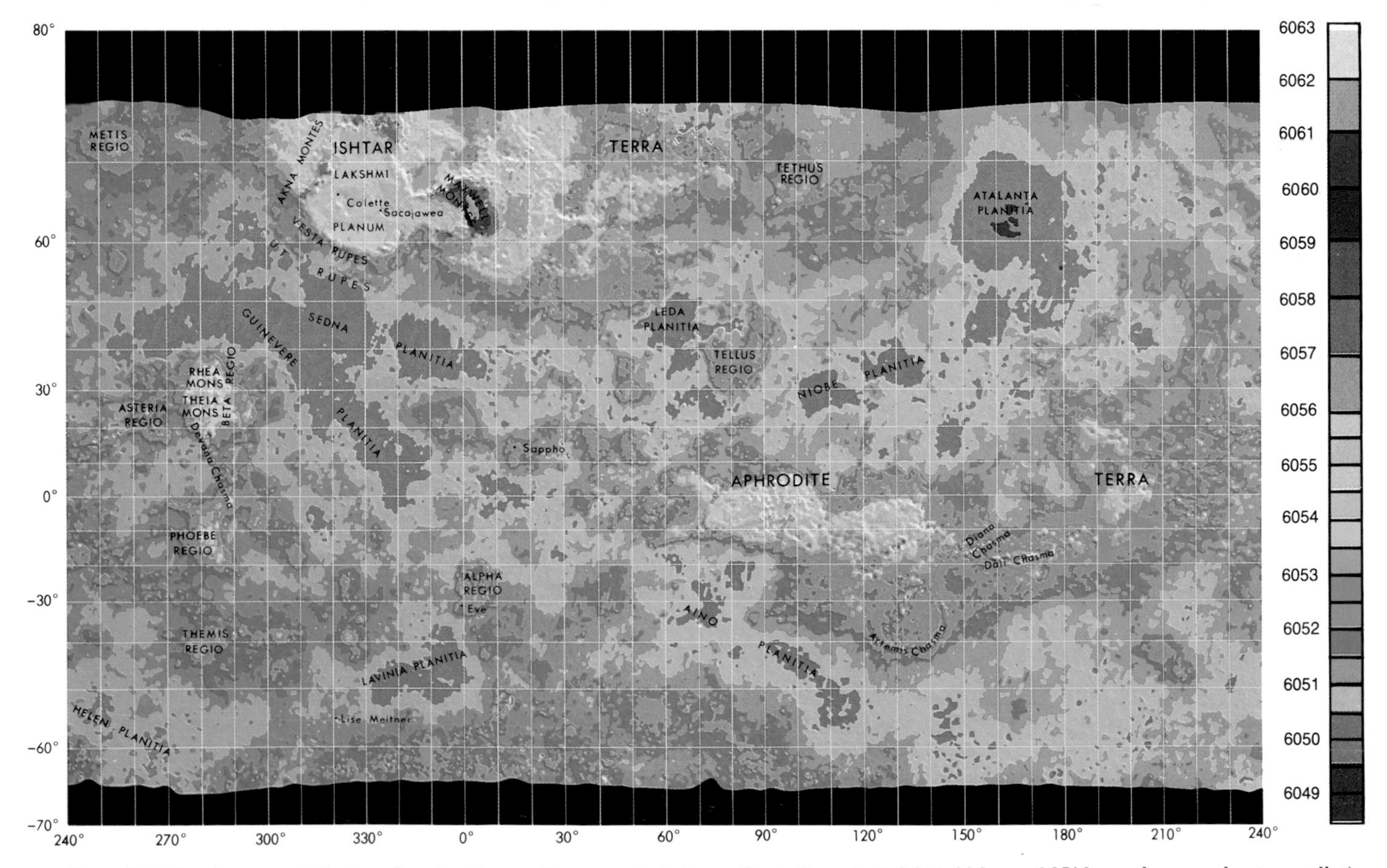

Figure 12. **Colored contours indicate surface elevations on Venus, graduated according to the scale at right (which uses 6,051 km as the mean planetary radius). Features have been named for female figures in mythology and science. Data for the map were gathered by a radar altimeter aboard the Pioneer Venus orbiter; a Mercator projection is used, making features near the poles appear larger than their true relative sizes. Lowland regions (dark blues) and highlands (reds, yellows, and light greens) together cover only about one-third of the planet's surface; the dominant remainder (light blues and dark greens) is rolling upland terrain that lies close to Venus' equivalent of "sea level."**

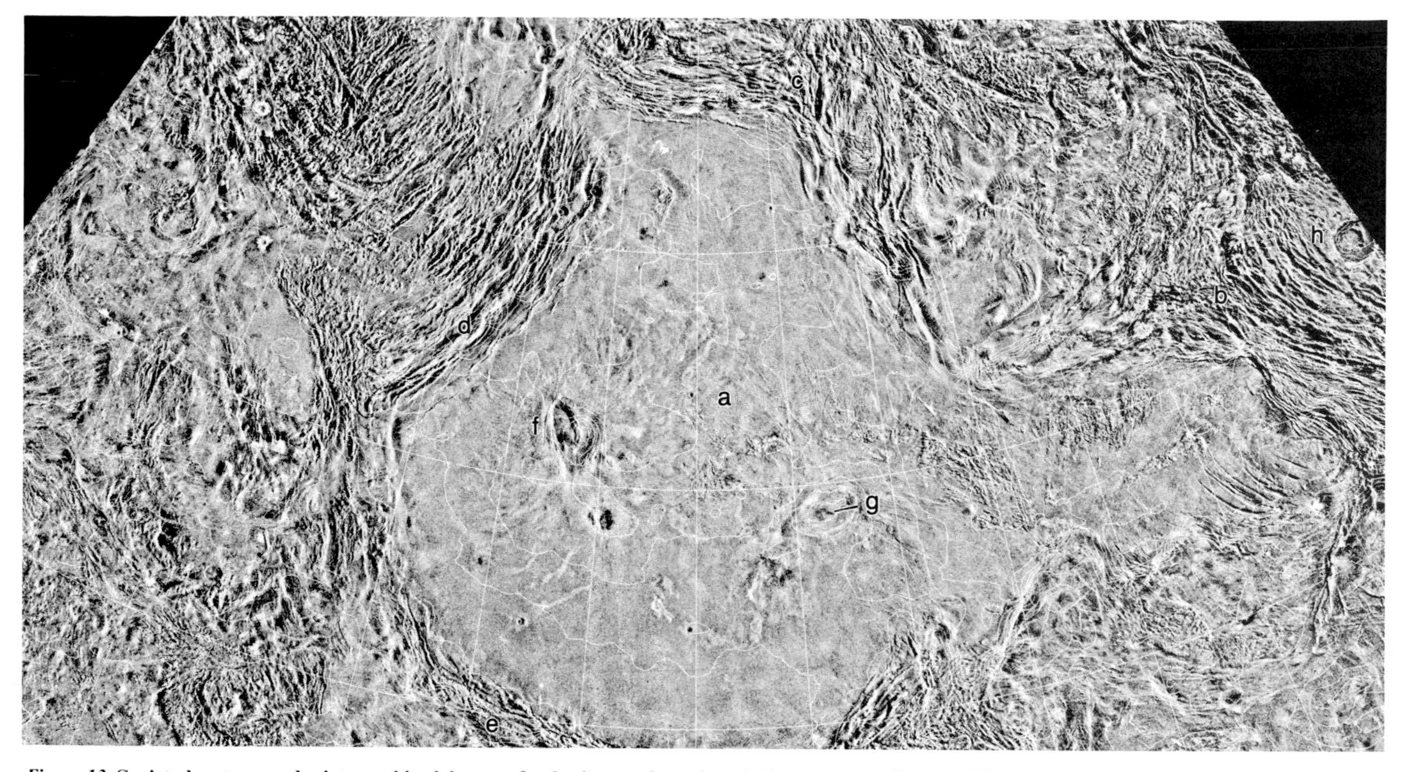

Figure 13. **Soviet planetary geologists combined dozens of radar images from the orbiting spacecraft Veneras 15 and 16 to create this mosaic of western Ishtar Terra, a "continent" the size of Australia in Venus' northern hemisphere. The vast plateau Lakshmi Planum *(a)*, about 2,000 km across, is rimmed on all sides by rugged, lineated mountain chains: the towering Maxwell Montes *(b)* to its east; Freyja Montes *(c)* to its north and northeast; Akna Montes *(d)* to the west; and Danu Montes *(e)* to the south. Two large calderas, Colette *(f)* and Sacajawea *(g)*, may be the source of much of the basalt floods that cover the plateau. Cleopatra *(h)*, the enigmatic 100-km-wide circular feature at extreme right, may be either a large impact crater or a caldera. The landscape's elevation is denoted by white contours spaced 0.5 km apart in altitude.**

include sets of ridges with troughs running between them, discontinuities similar to the fracture zones on Earth that offset linear rises in a stair-step fashion, and topographic profiles with striking bilateral symmetry. There appears to be considerable evidence that topographic features (ridges, volcanoes) have been created at rise crests, rifted and separated, and moved laterally to their present symmetrical positions. Taken together, all this argues that western Aphrodite Terra is the site of crustal spreading on Venus, with an estimated separation rate of roughly several centimeters per year. If these features are manifestations of divergent plate movement, how widespread is the process, is there any evidence for convergence and subduction elsewhere, and does it represent the Venus equivalent of terrestrial plate tectonics or some variation on that theme?

Radar studies of the third major highland region on Venus, Ishtar Terra, reveal characteristics very different from the extensional environments of Beta Regio and Aphrodite Terra. Ishtar Terra is characterized by a series of compression-generated mountain belts (Figure 13) much like the folded, uplifted rises that form along convergent plate boundaries on Earth. The parallel linear folds and faults are reminiscent of the Appalachians (Figure 6) and other mountain belts on Earth. The crust beneath Maxwell Montes and the northern edge of Freyja Montes has become thickened by flexure and underthrusting. These mountain belts and their associated topography repesent evidence for the large-scale deformation due to compression and convergence.

But do they represent true subduction zones? Does Venus

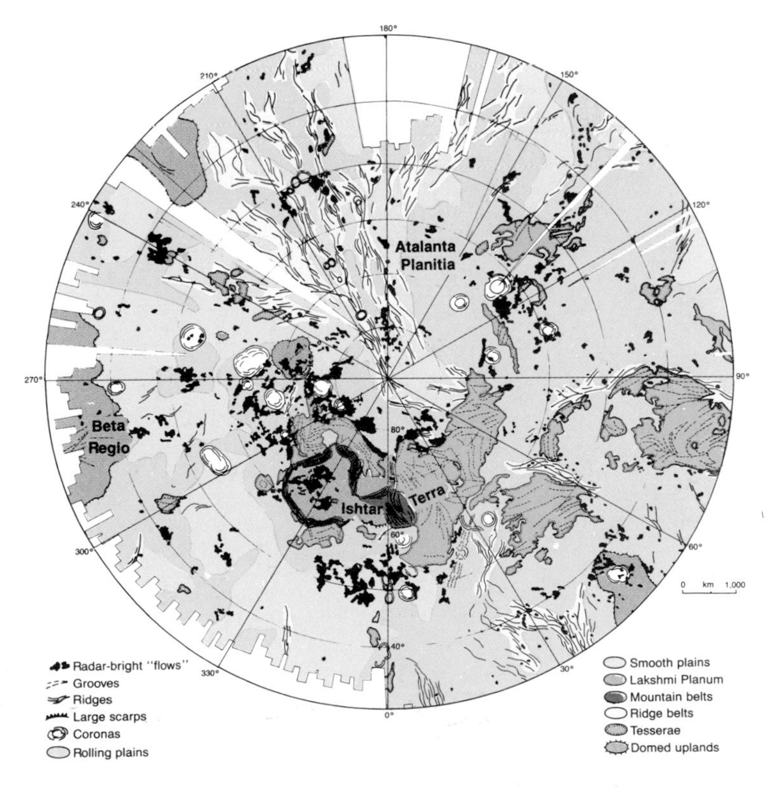

Figure 14. **Venera 15 and 16 radar data was used to prepare this geologic map of Venus' northern latitudes. It shows the distribution of various terrain units, some of which (like ovoids) are unique to Venus.**

really possess an integrated system of plate tectonism, as the Earth does? Planetary geologists suspect that the rigid outer shells of Venus and Earth are different enough, at least in detail, to cause their plate-tectonic styles to operate rather differently. Thus, although we may be seeing extension, crustal spreading, and compressional deformation in the highlands of Venus, it is too early to know if this is a global process or to what extent it mimics terrestrial plate movements. The United States' Magellan orbiter should provide radar images of nearly all of Venus at resolutions below 1 km. These data will allow us to make a more definitive assessment of the planet's dominant tectonic style and, in turn, how it transports heat out of the interior.

COMPARATIVE PLANETOLOGY

Based on the preceding brief examinations of the terrain types of the terrestrial planets, it is clear that these objects have important similarities and differences. Cratered terrain and volcanic plains are ubiquitous, but their abundances vary from world to world. On Earth, Precambrian shields cover little of the surface, while volcanic plains dominate more than two-thirds of the crustal area. More than 70 percent of the Moon's surface consists of cratered terrain and widespread volcanic plains. Cratered terrain dominates the hemisphere of Mercury seen by Mariner 10. But smooth plains cover about one-fourth of that area, and although there is disagreement as to how the Mercurian plains formed, their widespread occurrence (comparable to those on the Moon and Mars) argues for a volcanic origin.

Despite the apparent uniqueness of many aspects of Earth, several terrain units appear comparable to some seen on other planets. The Earth's most ancient rocks – the 10 percent of its surface comprising the Precambrian shields – are the nearest analog to the *cratered terrain* of other planets, even though the craters here are much more sparse and the shields' ages much younger (no more than 600 million years) than the versions of this unit found elsewhere. The basaltic plains of the ocean floor are the most pervasive terrain type on the Earth and, having formed within the last 200 million years, are among its youngest rocks. However, oceanic basalts' mode of origin is different than that of the basaltic plains on other terrestrial planets. Recurrent volcanism has built broad basaltic plateaus (such as the Ethiopian and Indian flood basalts), conical mountains (Hawaii), and dramatic craters (Mount St. Helens) over the last 65 million years. Our polar caps, like those on Mars, wax and wane seasonally.

The relative abundances of these two fundamental surface types give us a general impression of the extent of evolution of a planetary surface. For example, with more than 70 percent cratered terrain and less than 20 percent lava plains, the surface of the Moon appears extremely stable, while the Earth, with the percentages approximately reversed, has evolved significantly since its crust first formed. But to understand the full significance of these general impressions, we must both learn in detail *how* these terrain units formed and establish the absolute chronology of *when* they formed. Armed with this information, we can then turn to comparing the nature and evolution of the planets.

Thanks to decades of field and laboratory investigations, a framework of radiometrically determined true ages exists for most of the terrain units on the Earth and Moon. But since no samples have been returned from Mars, Venus, or Mercury, we can only estimate their surface ages by counting the impact craters superimposed on each terrain unit (in other words, how long have the various surfaces been exposed to incoming projectiles; see Chapter 21). Thus the ages derived for these planets depend crucially on models that predict how frequently impacts occur throughout the solar system. For Venus, the lack of high-resolution images over the whole surface complicates the problem, though age estimates of the better-imaged northern hemisphere can be derived from counts of those circular structures interpreted to be impact craters.

Figure 16 compares the estimated distribution of surface ages as a function of surface area for the Earth, Moon, Mars, and Mercury. Evolution of the surfaces of the Moon and Mercury was essentially complete by the end of the first half of solar-system history. In contrast, 98 percent of the Earth's present surface was formed since then and 90 percent within the past 600 million years. Mars' surface ages appear to be

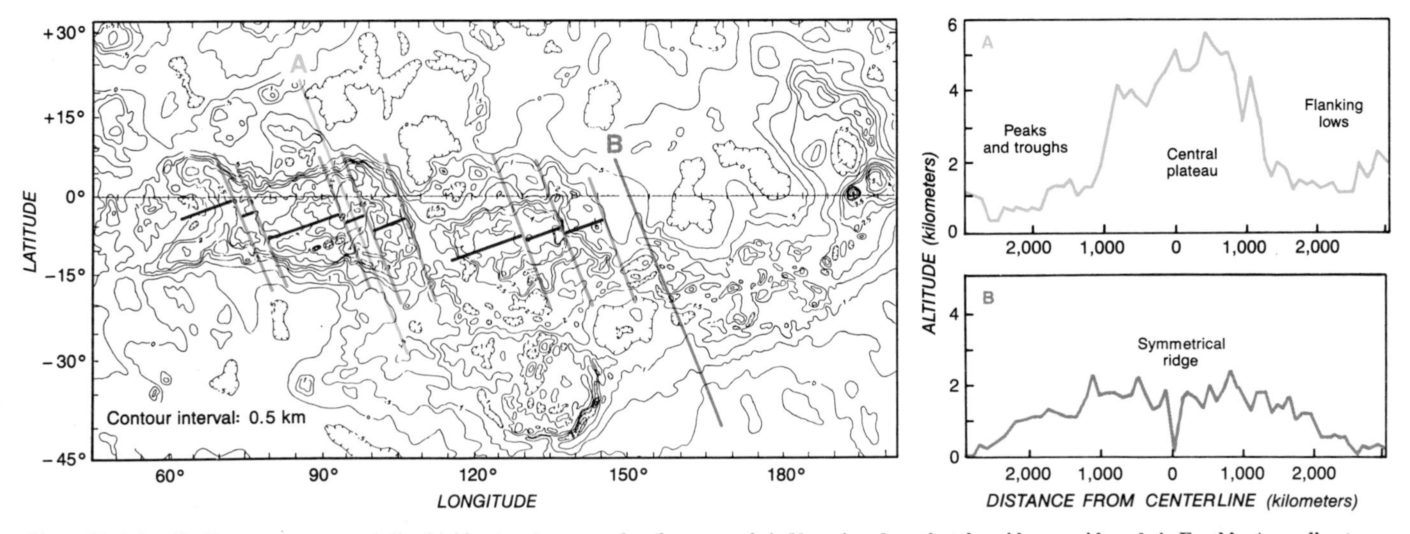

Figure 15. **Aphrodite Terra, an equator-girding highland region, may play the same role in Venus' geology that the midocean ridges do in Earth's. According to an analysis by the author and Larry S. Crumpler, Aphrodite's flanks are spreading away from its central ridge (red) and are occasionally offset by cross-strike discontinuities (blue). The cross-sections at right show the remarkable symmetry of the topography lying on either side of the central ridge – a characteristic seen as well across the Earth's midocean spreading zones.**

intermediate between those of the Moon and Earth. Preliminary evidence indicates that at least the northern high latitudes of Venus are relatively young – more like the Earth in age than the ancient heavily cratered terrain on the Moon, Mercury, and Mars. Of course, the continued exploration of

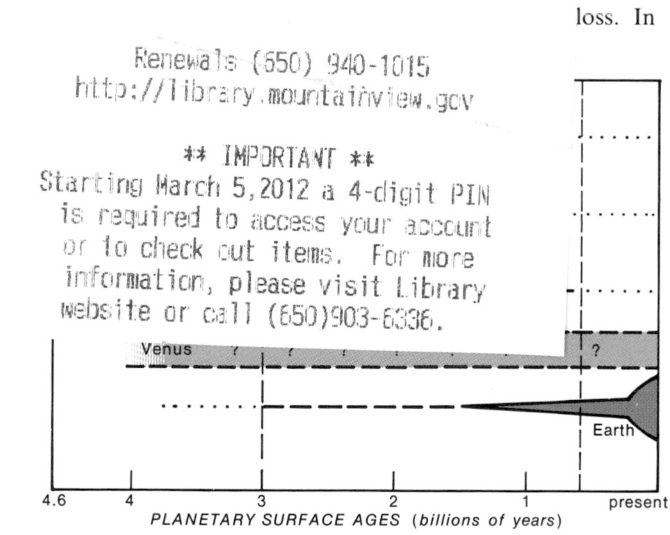

Figure 16. The ages of the terrestrial planets' surfaces, with colored shading representing the total surface area of each planet. Most regions on the Moon, Mars, and Mercury are several billions of years old, while two-thirds of the Earth's surface (its ocean basins) formed only within the last 200 million years. The age of Venus' surface is uncertain, but the paucity of craters seen in radar images from Veneras 15 and 16 indicate that the northern high latitudes are in general no more than about 1 billion years old.

particular, the advection process is often characterized by the presence of "hot spots" (such as Hawaii and Iceland on Earth). Io, the innermost of Jupiter's large satellites, is a good example of a planetary body whose heat loss is almost totally dominated by hot-spot volcanism (see Chapter 13). Although the surfaces of the smaller terrestrial planets do exhibit broad expanses of volcanic deposits, their thickness is not great, and they make up only a very small percentage of the total crust. Thus, conduction – not hot spots and advection – dominates these bodies.

The Earth rids its interior of heat yet another way. Our planet's segmented lithosphere forms along ridges under the oceans, moves laterally outward at geologically rapid rates, and is destroyed (subducted) at linear boundaries. This "conveyer belt" recycling process transfers heat out of the interior very effectively. One consequence of this relatively rapid process is that on average the Earth's surface is very young and sparsely cratered. The lithospheric recycling also creates very distinctive signatures at the surface: broad linear rises, extensive transform faults, long trenches, and linear mountain belts. Of course, we don't know for certain that the Earth has always been as it is today. For example, there is much evidence to suggest that the temperature of its interior was higher very long ago. In fact, the upper mantle of the infant Earth may have been roughly as hot as the upper mantle of Venus is today. Under such circumstances, advection and conduction may have played larger roles.

And where does Venus fit in this picture? The dearth of craters observed in radar imagery of the planet suggests that much of its surface is relatively young (less than 1 billion years old) and that heat loss involves something more than pure conduction. The abundance of plains, domes, and edifices of apparent volcanic origin means that advection and hot-spot volcanism probably play significant roles in Venus' heat transfer. The linear rises, central troughs, and transform-like features observed in Aphrodite Terra, and the

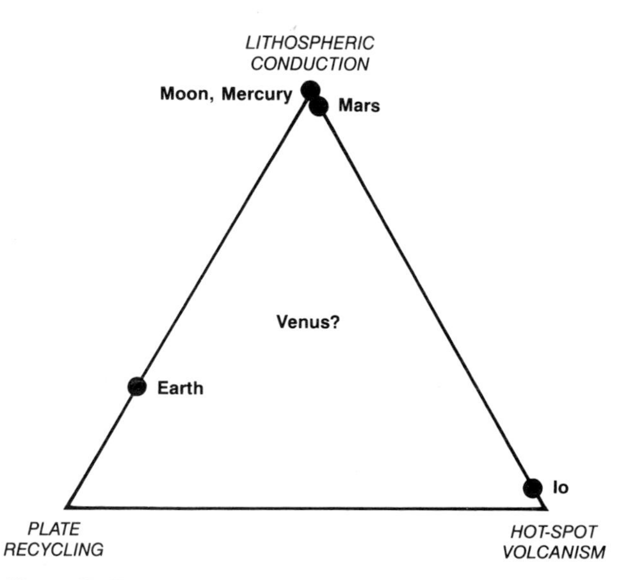

Figure 17. The geologic activity seen on planetary surfaces is a manifestation of how such bodies choose to transfer heat out of their interiors. The Moon, Mercury, and Mars simply conduct it outward through a thick solid outer shell or lithosphere. Heat generated inside the Jovian satellite Io escapes to the surface through a system of hot spots. The Earth employs a combination of passive conduction and the dynamic convection-driven recylcing of its lithospheric plates.

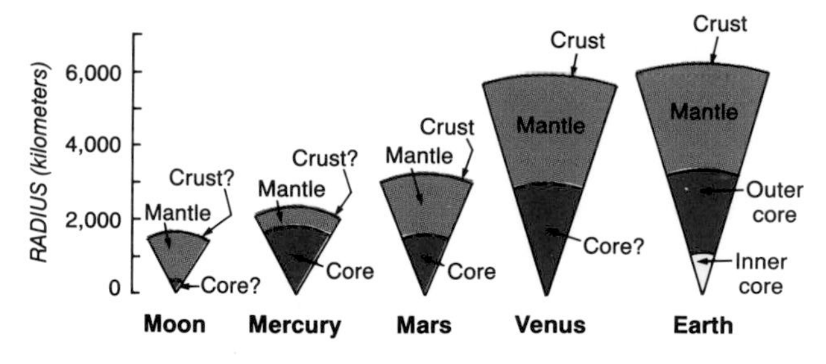

***Figure 18.* Interiors of the terrestrial planets, as deduced from a wide range of observations. The *crust, mantle,* and *core* of a planet are distinguished from one another on the basis of their geochemistry. It is not clear whether the Moon and Venus have discrete cores, nor does Mercury necessarily have a chemically distinct crust.**

long mountain belts of Ishtar Terra, suggest that both crustal spreading and convergence are occurring on Venus. But what are the relationships and relative contributions of these disparate processes? How do Venus' torrid surface temperatures (and its correspondingly higher crustal and upper-mantle temperatures) influence them? Are the zones of apparent crustal spreading and convergence arrayed in a global pattern as on Earth, or are they a regional phenomenon? Continuing radar observations from our observatories and the detailed global imagery expected from the Magellan orbiter will help to resolve these questions. We might even gain direct insight into the early history of the Earth, a period for which most of our planet's geologic record is no longer preserved.

Any comparison of the terrestrial planets must also consider their internal chemical layering (Figure 18). The internal structure of the Earth – a relatively thin oceanic and continental crust, with the rest of the interior about evenly divided between the mantle and core – is relatively well-known from seismic data (see Chapter 6). For the Moon, a crust roughly 70 km thick is indicated from various types of data, but we can deduce only an upper limit on the size of a possible lunar core (see Chapter 4). The surface geology and magnetic field of Mercury suggest, but do not demonstrate, that the planet has a core, and there is no significant information on the geometry of any crustal layer it might have. For Mars, a distinct crust and core are indicated from gravity data, but their radial extents are poorly defined. Gravity data from orbiting spacecraft and surface analyses by Soviet landers indicate that Venus has a low-density crust, but the characteristics of a possible core are speculations based on the planet's mean density and by analogy with the Earth. Obviously, we have much to learn about terrestrial-planet interiors, and future studies by spacecraft should provide significant advances in our understanding of their structure and evolution.

PLANETARY EVOLUTION

What are the fundamental processes that formed and modified the surfaces of the terrestrial planets? In terms of areal coverage, time duration, and volume, the most dominant mechanisms appear to be impact cratering and volcanism. Atmospheric processes have been influential on the surfaces of the Earth, Mars, and perhaps Venus as well. There is even evidence that polar wandering may have occurred on Mars and that relict polar deposits exist in what are now its equatorial regions. Ironically, based on Earth's surface alone, we never would have listed impact cratering as a fundamental process. Yet the records of other terrestrial planets have revealed its unquestionable significance, particularly in early planetary history. Indeed, recent theories suggest that the crater-poor Earth may have experienced the penultimate impact event very early in solar-system history: its near-destruction (and the Moon's formation) by a collision with a Mars-sized object! Similarly, Earth's ocean basins and the surface of our neighbor worlds have made us realize the significance of volcanism as a planetary process.

But do terrestrial planets share a common early history? In particular, did volatile-rich planets (like Earth) undergo evolutionary paths different from those of volatile-poor planets (like Mercury)? Analyses of the Apollo and Luna samples suggest that the Moon underwent extensive melting of its outer several hundred kilometers, with the source of heat thought to be the intense impact bombardment associated with the terminal stages of planetary accretion. The bombardment significantly affected other planetary surfaces too, but we have yet to learn whether all the terrestrial planets possessed extensive "magma oceans" early in their histories.

Given our present knowledge of surfaces and interiors, we can make three broad generalizations concerning the evolution of the terrestrial planets. The *size* of a planet is an important influence on its subsequent thermal evolution, because smaller planets may cool more efficiently due to their higher ratio of surface area to volume. Planetary *chemistry* is also important, for variations in the ratio of iron to silicate materials and in the relative abundances of volatiles and radiogenic heat sources can affect evolution significantly. The bulk chemistry of the planets may well vary as a function of distance from the Sun, a vestige of the decreasing temperature and pressure farther from the center of the collapsing solar nebula during planetary formation (see Chapter 23). A third major influence involves the *energy sources* available throughout planetary history and their relationship to one another. The possibilities include energy derived from impacts, tidal interaction between planetary bodies, differentiation of materials in the interior, and radioactivity.

No single factor stands out as a control on the evolution of the terrestrial planets. Continued study of these bodies, particularly Venus and Mars, will allow us to assess further the relative importance of each factor. From such investigations should come a more complete view of Earth's place in our solar system, and we will indeed "know the place for the first time."

8

Atmospheres of the Terrestrial Planets

James B. Pollack

THE ATMOSPHERES of the solar system's innermost planets – Mercury, Venus, Earth, and Mars – show an enormous diversity in their characteristics (Figure 1). For example, compared to sea-level pressure on Earth, the pressure at the surface of Venus is 90 times greater, whereas that on Mars is 150 times *less*. Mercury has almost no atmosphere: it is a million billion times less dense than ours. The composition of the Earth's atmosphere has been strongly influenced by the presence of life on its surface, with nitrogen and oxygen the dominant gas species. By contrast, carbon dioxide (CO_2) is the chief gas in the atmospheres of Venus and Mars. Helium and sodium are the major components of Mercury's (such as it is), and some of this is not intrinsic to the planet but rather derived from the rarefied solar wind that flows past it. Dust storms rage across Mars, creating an ever-present suspension of dust particles; a dense cloud layer of concentrated sulfuric acid shrouds the entire surface of Venus; and water clouds in all their numerous variations are present over half the area of our world at any given time. Winters in the polar regions of Mars are so cold that seasonal caps of dry ice form there; it is so hot at Venus' surface that lead, if present, would melt; temperatures on Mercury vary from very hot on its day side to very cold on its night side; and the temperatures of Earth are just right to permit the occurrence of the most precious form of matter in the universe: life.

Remarkably, there are common threads of history and processes underlying these discordant atmospheric properties. Earth's endowment of carbon dioxide, for example, is comparable to that in Venus' atmosphere, but almost all of it is now locked up in carbonate rocks, such as limestone. "Astronomical variations" in the orbital and axial characteristics of the Earth and Mars are thought to have caused quasi-periodic climatic changes on both planets, with these variations being responsible in part for the succession of ice ages that have occurred on the Earth over the last few million years.

This chapter will explore how the terrestrial planets' atmospheres are at once alike and dissimilar. It first reviews the current characteristics of the atmospheres and the factors responsible for these characteristics. Discussed next are the atmospheres' origin and long-term history and some of the major climatic changes that have occurred. The discussion concludes with a consideration of the conditions necessary for the origin and maintenance of life and the interactions between life and its environment.

COMPOSITION

Tables 1 and 2 summarize our current knowledge about gaseous and particulate components in the atmospheres of the inner planets. From Table 1, we can estimate how the absolute abundance of a gas varies among the different atmospheres using the product of the total pressure and the gas's *mixing ratio* (the fraction of the molecules it represents). Table 2 defines the major types of particles in each atmosphere and gives their vertical and horizontal locations, as well as a measure of the degree to which they interact with sunlight (termed *optical depth* or τ). When τ is much less than 1, very little sunlight is scattered or absorbed, while just the opposite is true when τ is large.

Only a little water vapor exists in the atmospheres of the Earth and Mars because water easily condenses at moderate and low temperatures. As a result, much more water resides in Earth's oceans and polar ice sheets than in the air above them. On Mars, the chief water reservoirs are ice deposits in the north polar region and, nearer the equator, water molecules adsorbed onto soil grains in the surface regolith ("soil"). When sunlight warms these reservoirs, some water vapor is driven into the atmosphere; if the air is cool enough the vapor condenses into clouds and eventually returns to the surface reservoirs. On average, the lower atmospheres of the Earth and Mars have one-third to one-half of the maximum amount of water that can be contained in an air mass at the temperature of the water reservoirs. The surface of Venus is so hot that no reservoirs of water exist there. Instead, all of the planet's water resides in its atmosphere. High above Venus' surface, where it is much cooler, the water vapor is incorporated into sulfuric acid (H_2SO_4) cloud droplets.

In all three terrestrial atmospheres, a small fraction of the water-vapor molecules is broken apart by ultraviolet sunlight into hydrogen atoms and hydroxyl (OH) radicals. Hydroxyl radicals are highly reactive and play an important role in the chemistry of these atmospheres: they help to convert sulfur gases into sulfuric acid vapor and act as a catalyst in the destruction of ozone.

Ultraviolet sunlight converts gases containing oxygen into

other oxygen-bearing species. For example, ozone (O_3) is produced in the Earth's atmosphere by a series of chemical reactions that begins when sunlight breaks molecular oxygen into two oxygen atoms. These atoms then combine with other oxygen molecules to form ozone. Ozone reaches its peak concentration in the stratosphere at an altitude of about 30 km. It helps to shield the ground from biologically dangerous ultraviolet radiation. Small amounts of other gases, such as chlorine atoms and nitric oxide (NO) act as catalysts in chemical reactions that destroy ozone. There is growing evidence that activities such as the use of fluorocarbons and fertilizers may increase the amount of these catalysts in the atmosphere, which reduce the amount of ozone and hence increase the ultraviolet sunlight reaching the surface. Considerable efforts have been made by the scientific community to predict such changes accurately and thus assess the reality of this potential problem.

Figure 1. **At left are spacecraft photographs of the four terrestrial planets: Mercury from Mariner 10, Venus from the Pioneer Venus orbiter, Earth from Apollo 17, and Mars from the Viking 1 orbiter. Venus appears in a false-color version of an ultraviolet image; its cloud details are not apparent in visible light. White clouds hide about half of Earth at any given time, while clouds of water ice, carbon dioxide ice, and dust enshroud varying portions of Mars. At right, artist Don Davis has portrayed all four terrestrial landscapes as they would appear with the Sun 20° above the horizon.**

The Earth's so-called "ozone hole" may be a harbinger of our impact on the atmosphere's fragile chemical balance. The abundance of ozone over Antarctica drops markedly during local spring, creating a vast spectral hole in the stratosphere through which ultraviolet light can stream more freely (Figure 2). In 1987 an aircraft expedition to Antarctica and laboratory simulations revealed the chemical factors responsible for the strong seasonal variation. Normally, chlorine atoms in the atmosphere reside in hydrogen chloride (HCl) and chlorine nitrate ($ClNO_3$), gases that do not catalytically destroy ozone. However, during the Antarctic winter the lower stratosphere becomes so cold that HCl condenses into tiny ice particles. Chlorine nitrate then reacts with these particles, forming condensed nitric acid (HNO_3) and chlorine gas. When spring arrives, the combination of sunlight and the chlorine destroys ozone. Thus, the drop in ozone abundance is a direct consequence of the freezing out of hydrogen chloride, a condition that occurs only over Antarctica. In this sense, the "ozone hole" has few implications for other parts of the world, except to confirm the potent catalytic powers of chlorine. What has heightened concern among scientists, however, is that the already low springtime level of ozone has declined steadily for more than a decade; when measured in October 1987 the abundance had dropped by 60 percent over that measured 11 years before.

For Venus and Mars, carbon dioxide – not molecular oxygen – is the chief oxygen-bearing gas. But small amounts of molecular oxygen and trace amounts of atomic oxygen and ozone occur in these atmospheres as a result of chemical reactions that begin when ultraviolet light dissociates carbon dioxide into carbon monoxide (CO) and atomic oxygen. In the case of Mars, ozone is most prevalent near each polar region during local winter; when the water-vapor abundance is small, few hydroxyl radicals are generated to attack the ozone.

Sulfur gases are present only in trace amounts in the Earth's atmosphere. They either dissolve readily in cloud water droplets or, in cloudless regions, react with oxygen-rich radicals, water vapor, and other gases to form particles. Both types of particles eventually fall onto the ground, taking the sulfur with them. Still, sulfuric acid and ammonium sulfate represent an important component of the non-water particles found in the lowest regions of the Earth's atmosphere, and sulfuric acid predominates higher up in the stratosphere. By injecting large amounts of sulfur-containing gases, such as sulfur dioxide (SO_2), into the stratosphere, volcanic explosions such as those of Mount St. Helens in 1980 and El Chichon in 1982 can cause a large, but temporary increase in the amount of sulfuric acid there. Human activities, like the burning of coal, have accelerated the input of sulfurous gases into the lower atmosphere, which in turn increases the abundance of sulfur-rich particles there. These particles, together with ones that result from hydrocarbon emissions,

Planet	g	P	T	Major gases	Minor gases
Mercury	395	10^{-15}	440	He (.42) Na (.42) O (.15)	H (10,000) K (2,000)
Venus	888	90	730	CO_2 (.96) N_2 (.035)	H_2O (≈100), SO_2 (150), Ar (70) CO (40), Ne (5), HCl (0.4), HF (.01)
Earth	978	1	288	N_2 (.77) O_2 (.21) H_2O (.01) Ar (.0093)	CO_2 (330), Ne (18), He (5.2), Kr (1.1) Xe (.087), CH_4 (1.5), H_2 (.5), N_2O (.3) CO (.12), NH_3 (.01), NO_2 (.001) SO_2 (.0002), H_2S (.0002), O_3 (.4)
Mars	373	.007	218	CO_2 (.95) N_2 (.027) Ar (.016)	O_2 (1,300), CO (700), H_2O (300) Ne (2.5), Kr (.3), Xe (.08), O_3 (.1)

***Table 1.* Listed above for the four terrestrial planets are: *g*, the acceleration of gravity at the surface (in cm/sec/sec); *P*, the atmospheric pressure at the surface (in bars); and *T*, the average surface temperature (in degrees Kelvin). Parenthetical values following major gases are their fractional abundances by number; values following minor gases are their fractional abundances in the lower atmosphere in parts per million. Except for the rare gases, many minor constituents vary considerably with altitude and sometimes latitude.**

are responsible for the smog that exists in many regions of the industrialized world.

The hot surface of Venus prevents sulfur from being removed from its atmosphere. As a result, SO_2 is abundant in its lower atmosphere and provides source material for the dense layer of sulfuric acid particles that exists between about 50 and 80 km above the surface. At lower altitudes, these caustic droplets evaporate into water vapor and sulfuric acid vapor; the latter ultimately decomposes into sulfur dioxide. Measurements made by the Pioneer Venus orbiter show that the amount of sulfur dioxide present near the cloud tops declined from approximately 100 parts per billion (ppb) in 1978 to about 10 ppb in 1986. There is also fragmentary evidence for similar SO_2 increases and decreases at earlier times. Such fluctuations might be due to episodic injections of SO_2 high into the atmosphere by powerful volcanic explosions. Alternatively, long-term changes in circulation may be affecting how vigorously SO_2 is brought up from below the clouds (where it is abundant) to above them (where it reacts to form sulfuric acid).

No sulfur-containing gases have been detected in the Martian atmosphere. But they may have been released from Martian volcanoes in the past and converted to sulfur-containing particles that eventually formed sulfate minerals. (Sulfates occur abundantly at the two Viking lander sites on the Martian surface.)

Planet	Particulate composition	Altitude (km)	Areal distribution	Optical depth (τ)
Mercury	None			
Venus	Concentrated sulfuric acid	50–80	Everywhere	≈25
Earth	Concentrated sulfuric acid	12–30	Everywhere	0.003–0.3*
	sulfates, silicates, sea salt, organics	0–12	Everywhere but spatially variable	0.05–3
	water	0–12	50% cloud cover	5
Mars	dust	0–50	Everywhere; large temporal changes	0.3–6**
	water ice	0–50	winter polar region; morning fog; isolated clouds behind high places	≈1
	CO_2 ice	≈60 0–25	many places winter polar region	≈0.001 ≈1

***Table 2.* The nature of particulate layers in the atmospheres of the terrestrial planets. Note that in the listings under optical depth, Earth's highest sulfuric acid values (*) occur after volcanic eruptions, and that Mars' maxima due to dust (**) occur during global dust storms.**

TEMPERATURE

What if Venus, Earth, and Mars had no atmospheres? Then their globally averaged surface temperatures would represent a balance between the amount of sunlight they absorbed and the amount of heat they radiated to space. The presence of an atmosphere affects both components of this balance (Figure 3). On one hand, highly reflective clouds allow less sunlight to be absorbed by the planet, which tends to cool the surface. But on the other hand an atmosphere absorbs some of the heat emitted by the surface; part of this is then reradiated back to the surface, warming it. This latter situation, termed the *greenhouse effect,* makes the mean surface temperature of Mars, Earth, and Venus about 5°, 35°, and 500° K warmer, respectively, than they would be if their atmospheres were completely transparent to thermal infrared radiation (heat). Earth and Mars experience only a modest warming because each has an atmosphere transparent at some infrared wavelengths, which permits much of the heat emitted by the surface to escape to space.

On Mars carbon dioxide and dust particles are the chief suppliers of atmospheric infrared opacity; on Earth that role is filled by carbon dioxide, water vapor, and water clouds. The very large rise in surface temperature for Venus stems from the complete absorption of thermal energy radiated by its surface, with carbon dioxide, water vapor, sulfur dioxide, and sulfuric acid clouds playing key roles. In part, the differences in the degree of greenhouse warming among the three planets can be attributed to differences in the masses of their atmospheres.

Particles interact much more effectively with sunlight than gases do, absorbing both sunlight and upwelling thermal radiation. Particles thus tend to warm only the atmosphere around them and to *cool* the atmosphere and surface below, by scattering and absorbing sunlight and thus reducing the

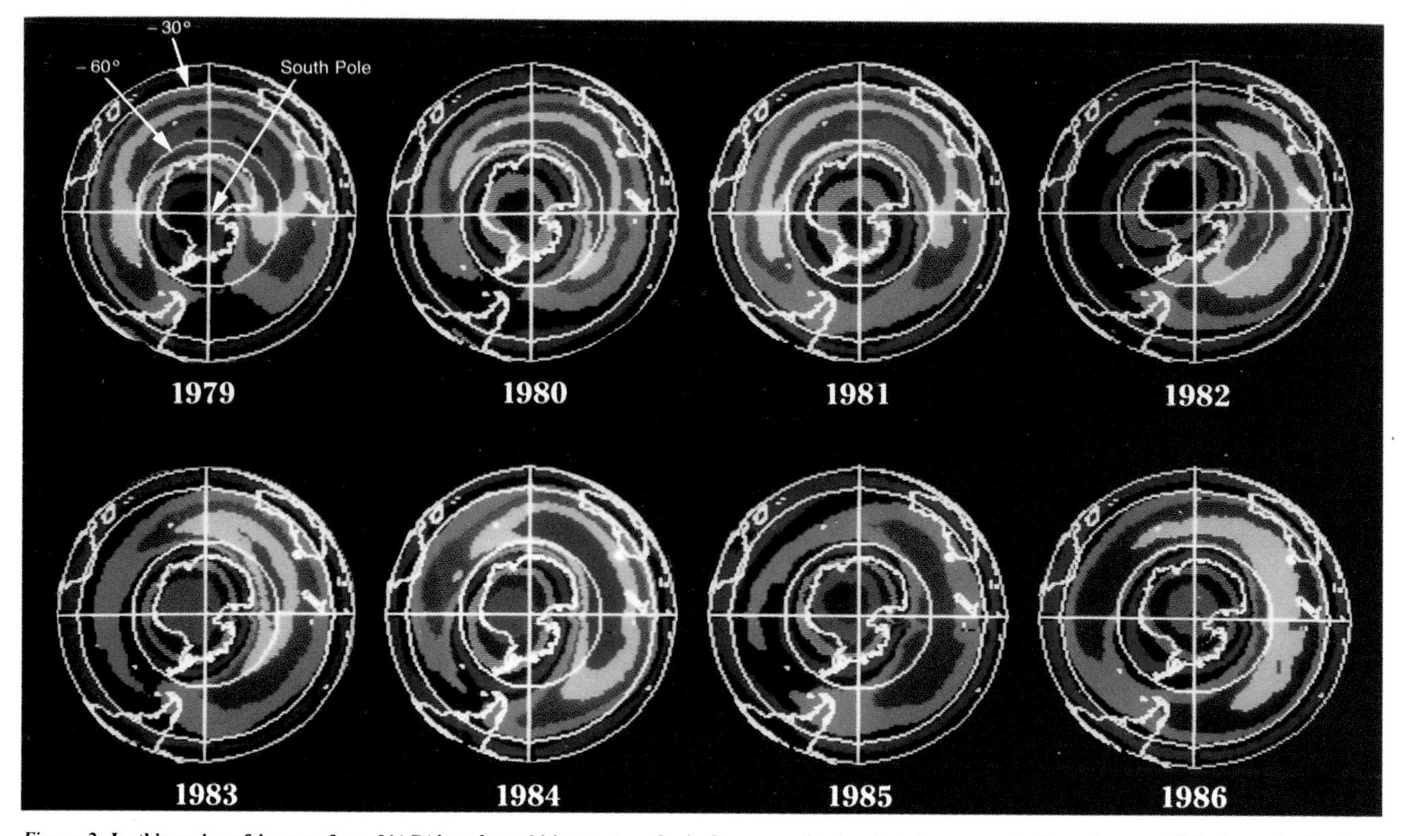

Figure 2. **In this series of images from NASA's polar-orbiting meteorological spacecraft, the changing concentration of ozone (O_3) above the Southern Hemisphere is shown for an 8-year period. Generally, ozone is most abundant at mid-latitudes (red and brown) and least abundant over Antarctica (pink and black). As spring arrives in the Southern Hemisphere, an ozone depletion or "hole" appears in the stratosphere above Antarctica.**

amount reaching deeper levels. This occurs on Earth whenever volcanic explosions drive sulfur dioxide into the stratosphere. There it reacts to form particles of sulfuric acid, which warm the stratosphere and slightly cool the lower atmosphere and surface. When global dust storms occur on Mars, the highest dust-laden levels in the atmosphere become substantially warmer. Collaborating scientists Richard Turco, Brian Toon, Thomas Ackerman, the author, and Carl Sagan suggested in 1983 that a similar but more severe temporary cooling of the Earth's continents would follow a major nuclear war. This "nuclear winter," which would be more pronounced in the Northern Hemisphere, arises chiefly from the huge quantities of smoke produced when bombs, detonated over major urban centers, create giant fires.

The density and temperature of a planetary atmosphere vary considerably with altitude. Density falls steadily with increasing height in such a fashion that, at a given altitude, a very close balance exists between gravity, which tries to pull the gas down, and gas pressure, which tries to move gas upward from places of high pressure to places of low pressure. In all three terrestrial atmospheres, the density changes by a factor of three with every 10 km of height. However, the magnitude of this *scale height* is greater (that is, the falloff of density with altitude is less rapid) in hot regions. Figure 4 illustrates how temperature varies with height for the three planets. It has been traditional to divide the Earth's atmosphere into four major altitude domains, as indicated, based on its temperature profile. For example, in the troposphere the temperature drops steadily with increasing altitude, while the opposite is true in the stratosphere.

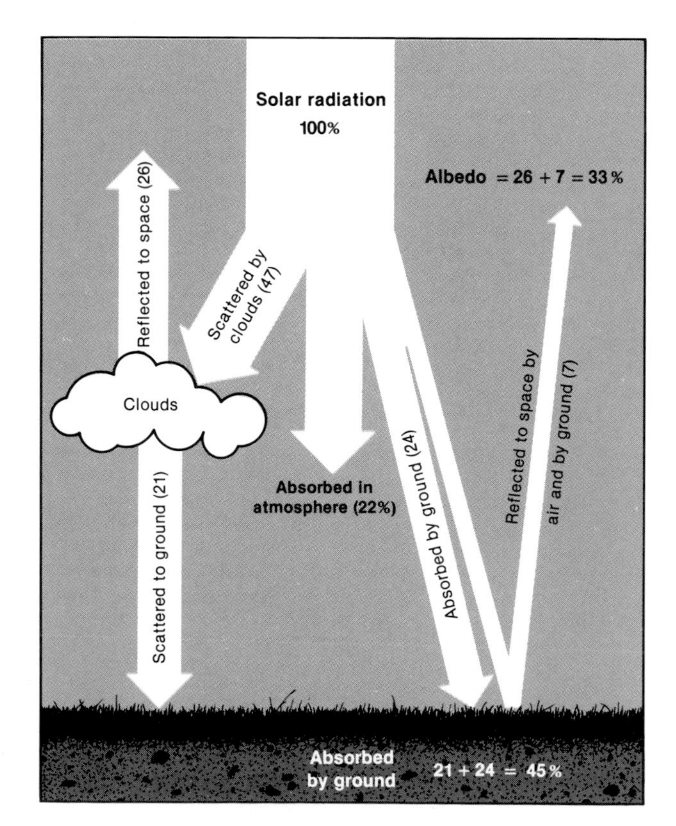

Figure 3. **The interaction of sunlight with Earth and its atmosphere. Ultimately, about one-third of the energy striking our globe is reflected away.**

Several processes can transfer heat within the atmosphere, and collectively these determine the temperature structure there. A given layer of air is warmed by absorbing sunlight and thermal radiation produced elsewhere and is cooled by emitting its own thermal radiation. Within the troposphere, heat is transferred upward from the ground by the evaporation of water vapor at the surface (which requires heat) and its condensation in the atmosphere (which releases heat). More upward transfer is accomplished by small-scale turbulence and large-scale wave motions.

Small-scale turbulence occurs whenever the temperature decreases faster with altitude than the ideal thermodynamic "norm," which is termed the *adiabatic lapse rate*. If the Earth's troposphere were completely water-free, this value would be about 10° K per km of altitude. But the observed temperature gradient in the troposphere is only 6° per km, about two-thirds of the dry adiabatic value, due to the combined effects of water-vapor condensation and heat transport by wave motion. Small-scale turbulence transports heat so efficiently that the actual temperature gradient never exceeds the adiabatic lapse rate by more than a minute amount. For Mars, the lapse rate in the troposphere is also substantially less than the dry adiabatic value, but there absorption of sunlight by ubiquitous atmospheric dust "stabilizes" the temperature profile. For Venus, the temperature variations with altitude closely match the dry adiabatic value in the atmosphere's lowest 35 km. But from 35 km up to 48 km, the location of the cloud bottoms, the atmosphere does not cool with altitude as quickly (termed a *subadiabatic lapse rate*). In this region the gas molecules are less densely packed and thus partly transparent to thermal wavelengths. This allows more thermal energy to be exchanged with adjacent layers, and so shallower temperature gradients result.

Unlike the Earth's troposphere, the stratosphere gets *warmer* with increasing altitude, a reversal caused by ozone's absorption of ultraviolet sunlight. This absorption is more than adequate to counteract cooling due to the emission of thermal radiation by carbon dioxide and ozone molecules. At higher altitudes, within the mesosphere, there is less ozone and hence the temperature gradient reverses again due to radiative cooling by carbon dioxide. Because the atmospheres of Venus and Mars have much less ozone than that of the Earth, they lack the equivalent of a stratosphere.

In the highest layer of the Earth's atmosphere, the thermosphere, temperatures again rise with increasing altitude. The numbers of atoms and molecules there are too sparse to cool the atmosphere very well by emitting thermal radiation. Atomic oxygen and nitric oxide – not carbon dioxide – produce most of the infrared radiation in the thermosphere, but they are much less efficient coolants than carbon dioxide. Consequently, the absorption of sunlight at very short wavelengths warms this region rather effectively. Much lower temperatures occur in the thermospheres of Venus and Mars because they contain much more carbon dioxide than does the Earth's and are thus better able to radiate away energy.

UPPER ATMOSPHERES

Fluid motions in the lower atmospheres of the three planets keep the inert gases well mixed. However, above about 110 km for the Earth and Mars and somewhat higher for Venus, low densities render these motions ineffective. The boundary at which fluid motion relinquishes control is termed the *homopause* (or turbopause). Above this boundary, each species behaves as if it were the sole atmospheric constituent, with the result that lighter gases thin out less rapidly with increasing altitude than do heavier ones. Below it, variations in the sunlight absorbed by the lower atmosphere over the course of a day induce vertically propagating atmospheric "tidal" motions. Air density is lower at the homopauses of Mars and Venus than at the Earth's, possibly because their tidal motions are stronger.

In the upper regions of an atmosphere, sunlight at very short wavelengths strips electrons from atoms and molecules and creates an ionized layer, the *ionosphere*. In the case of Mars and Venus, carbon dioxide is the chief victim, while for the Earth it is molecular oxygen and nitrogen. However, due to a series of chemical reactions, ionized oxygen atoms and molecules actually constitute the most abundant ion for all three. Ionized nitrogen oxide (NO^+) is another important constituent of the Earth's ionosphere.

The boundary between a planet's domain and interplanetary space is determined by how its atmosphere interacts with the solar wind – a high-velocity, fully ionized gas of very low density that emanates from the Sun. Descriptions of these interactions are presented in Chapter 3.

METEOROLOGY

Because more sunlight falls on a planet's equator than at its poles over the course of a year, the equator is the warmer place. This latitude-dependent temperature gradient sets winds into motion that transport heat from warm spots to cold ones. To some extent, dynamic atmospheric motions are self-destructive, in that they try to equalize the very temperature differences that drive them. For example, Earth's atmosphere is dense enough that its motions diminish somewhat the equator-to-pole temperature differences that would exist otherwise. The atmosphere of Venus takes this process to the extreme; it is so massive that surface temperatures vary by no more than a few degrees between its equator and poles. Conversely, the Martian atmosphere is so thin that heat transport does little to diminish latitudinal temperature gradients. As a result, Mars' polar regions

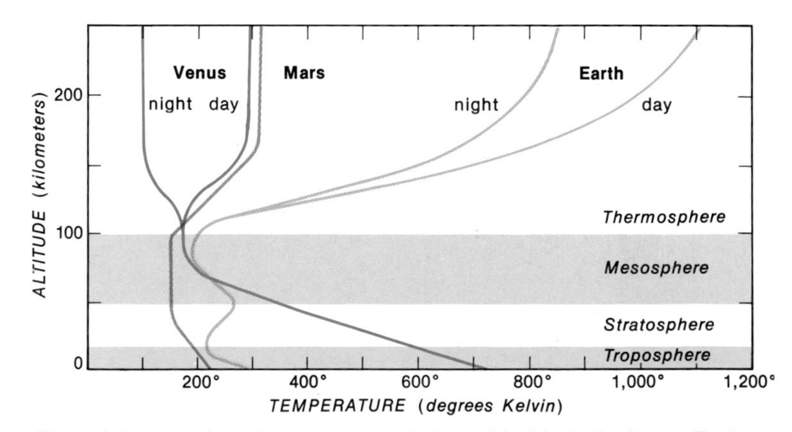

Figure 4. A comparison of temperature variations with altitude for Venus, Earth, and Mars. The day-night pairs of curves show the strong diurnal cycle in the upper atmospheres of Earth and Venus. Also indicated are the names given to regions within the Earth's atmosphere.

become so cold in winter – about 150° K – that carbon dioxide condenses on the surface as dry ice. Near the end of the winter in a given hemisphere, this polar ice deposit extends down to a latitude of about 50°. In the northern hemisphere, it totally sublimates away during the subsequent spring and summer. However, a small residual CO_2 ice cap remains throughout some years near the south pole.

The direct response of planetary atmospheres to latitudinal temperature differences is a *Hadley circulation pattern,* in which air rises over the warm regions, moves toward the cooler places at high altitude, sinks in the cooler regions, and returns back to the warmer places at low altitude (Figure 5). However, rotating planets possess what is called the *Coriolis force,* which acts to deflect air perpendicularly to its direction of motion. Consequently, this force adds an east-west component to the ideally north-south movement in a Hadley circulation pattern.

On the Earth, a single Hadley cell does not span the entire distance from the equator to the pole; it would be unstable, because even a small perturbation to the flow pattern results in a drastic change. Rather, each hemisphere has three cells, with the middle one – the Ferrel cell – circulating in a thermodynamically indirect sense; that is, air rises at the cold end and sinks at the warm end. The equatorial Hadley cell is driven primarily by heat released during the condensation of atmospheric water vapor rising from the ocean. Much of this condensation takes place in a narrow latitudinal belt called the Intertropical Convergence Zone, within clusters of cumulus clouds that draw on moisture brought into the ICZ by trade winds. (The trade winds correspond to the low-altitude, equatorward return flow of the Hadley cell there.) The Coriolis force becomes more pronounced farther from the equator, so the winds at mid- and high latitudes blow mostly in an east-west direction rather than north-south.

Other factors further subdivide this large-scale circulation into smaller-scale "eddy" motions, such as the progression of cyclones (pressure lows) and anticyclones (highs) that constitute an important component of the weather at mid-latitudes on the Earth. The perturbation responsible for these particular disturbances is called a *baroclinic instability,* which occurs when the temperature differences across adjacent latitudes become too large (Figure 5). Baroclinic waves – the eddy motions resulting from this instability – play a major role in transporting heat poleward at mid-latitudes and in transporting heat upward, thus decreasing the vertical temperature gradient or lapse rate within the troposphere. They also transport momentum toward a narrow latitudinal region in the upper troposphere and thus help to maintain the fast, eastward-moving "jet stream." Typically about six pairs of highs and lows girdle the Earth in this latitudinal belt.

Our planet's topography (large mountain ranges, for example) and the temperature differences between the oceans and neighboring continents set up another type of pattern called standing waves or *stationary eddies* (Figure 5). In contrast to baroclinic waves, these do not propagate horizontally, though they can transport heat and momentum vertically. They also tend to have longer wavelengths – typically it takes only one to three wavelengths to girdle a latitude circle. Stationary eddies in the troposphere transfer momentum into the stratosphere and so influence the circulation patterns there. In summary, then, the winds in the Earth's troposphere can be roughly divided into three major components: a longitudinally averaged or mean circulation mainly in the tropics and, at higher latitudes, baroclinic eddies and stationary eddies.

Because the length of a Martian day (24.6 hours) is very similar to ours, Mars' circulation patterns are expected to resemble crudely those of the Earth. Thus, for example, a Hadley cell is thought to be present at equatorial latitudes, and baroclinic waves have apparently been observed by the Viking lander meteorology experiment at mid-latitudes during certain seasons. Also, standing waves set up by strikingly large elevation differences (up to 25 km) are expected but have not yet been observed.

However, some fundamental differences exist between the circulation patterns in the lower atmospheres of these two planets. Because Mars has no oceans, its entire surface responds very rapidly to changes in incident sunlight. During the summer season in a given hemisphere, the hottest place on Mars is not at the equator, but displaced to the latitude where the Sun is directly overhead at noon. As a result, a separate Hadley cell no longer exists for each hemisphere, as at the Earth. Instead, warm air rises in the summer hemisphere of Mars and sinks in the winter hemisphere, spanning the equator in the process.

Another consequence of the essentially instantaneous response of the entire Martian surface to incident sunlight is that strong latitudinal differences in temperature are present only at mid-latitudes in the winter hemisphere. Thus, baroclinic waves are expected and observed only there. Baroclinic waves on Mars appear to be somewhat more regular than on the Earth in the sense of having a more well-defined period. In this regard, weather may be more predictable for Mars. As on the Earth, baroclinic waves transport momentum toward (and thus help maintain) a strong eastward-moving jet; on Mars, this is located in the winter hemisphere close to the boundary of the seasonal CO_2 ice cap.

Because CO_2 is the major constituent of the Martian atmosphere, its condensation at high latitudes in the winter hemisphere makes the pressure there somewhat lower than elsewhere. (The formation and dissipation of the seasonal CO_2 ice caps cause the atmospheric pressure to vary by about 20 percent from one season to the next.) The reduced pressure in turn sets up a strong planetwide circulation that transports heat, momentum, and mass toward the region of the forming polar cap (Figure 5). Because of this "condensation flow," the mean circulation is the dominant wind component at most latitudes. In particular, baroclinic waves on Mars play a less central role than they do on Earth in transporting heat poleward at midlatitudes.

Mars' thin atmosphere also causes temperatures just above its surface to drop significantly – by tens of degrees – after the Sun sets. These diurnal temperature variations drive "thermal tidal" winds that propagate in the direction of motion of the Sun in the Martian sky (see Figure 5). Atmospheric dust enhances these winds by strongly absorbing sunlight and acting as a second good infrared radiator. Earth's atmosphere is much more massive, so its diurnal temperature differences are much smaller and hence its tidal winds much weaker.

Other very interesting relationships exist between dust and winds on Mars. When the wind speed close to the surface exceeds a critical threshold value – about 50 to 100 m per second – sand grains about 100 microns across initiate a skipping motion called saltation. Upon striking the surface, they propel smaller dust grains (a few microns in size) upward into the atmosphere, where they remain suspended for months before settling back out. About 100 localized dust storms occur over the course of a Martian year in places where the winds are particularly strong. Particularly susceptible sites are the edges of the seasonal CO_2 caps during local spring, where large temperature differences between ice-covered and bare ground drive strong winds; mid-latitudes of the winter hemisphere, where strong winds occur during the passage of "fronts" associated with baroclinic eddies; or subtropical highlands, where standing eddies reinforce tidal winds.

During some Martian years (equal to about two of ours), a local dust storm grows to global proportions in only a few weeks, thereby enveloping much or all of the planet in a dense shroud of dust. But the storm soon begins to decay, and dust gradually settles out of the atmosphere over a period of several months. These global dust storms develop only during the spring and early summer in the southern hemisphere, when Mars is closest to the Sun and thus its winds somewhat stronger. They also appear to be self-limiting. When the atmosphere is relatively dust-free, localized but strategically placed dust storms can load the lower atmosphere with particles; these greatly strengthen tidal and Hadley winds, which in turn raise dust over a progressively larger area. However, once the dust-loading becomes too great, the temperature profile of the lower atmosphere stabilizes; strong winds aloft become poorly coupled to surface winds, dust can no longer become airborne, and the global storm begins to decay. Dust particles can be removed quickly in the winter polar regions by acting as growth sites for CO_2 snowflakes, which fall more rapidly to the ground.

As atmospheric pressure increases, the threshold wind speed needed to set particles into motion goes down. Hence, slower winds are required to raise dust into the Earth's atmosphere than into that of Mars, and even smaller speeds are needed on Venus. However, vegetation protects the soil on much of our continental terrain, so dust storms occur mostly in sparsely covered desert regions. (The "dust bowl" that occurred in the central United States during the 1930s was preceded by a loss of the vegetation cover.) Near-surface winds on Venus are quite slow – about 1 m per second – but that may be enough to set sand into motion there.

The very slow rotation rate of Venus (about 117 Earth days) would imply a nearly ideal Hadley circulation in its atmosphere, with winds moving north-south. However, space probes have shown that the winds in its lower and middle atmosphere blow primarily in an east-west direction. The Coriolis force is undoubtedly too small to cause this reorientation, but another inertial force due to the winds themselves, *centrifugal force*, may play a key role instead. The east-west winds on Venus blow in the direction of the planet's rotation at all altitudes, increasing from a modest 1 m per second near the ground to a phenomenal 100 m per second close to the cloud tops. In contrast to the Earth and Mars, where such high velocities are encountered only in jet streams, rapid winds occur over essentially the entire globe of

	Hadley Circulation	*Baroclinic Eddies*	*Stationary Eddies*	*Condensation Flow*	*Thermal Tides*
VENUS	Single cell goes from equator to pole at a given altitude; probably several cells situated at different altitudes	?	"A" eddies result from flow over large elevated regions; strong vertical motions have been found 50 km above these locations		May have affected planet's rotation rate
EARTH	Three cells per hemisphere	Occur in both summer and winter hemispheres; very irregular in duration	"A" eddies result from flow over topography; "B" eddies are generated by temperature differences between oceans and continents		Weak
MARS	Single cell from tropics of summer hemisphere to subtropics of winter hemisphere	Form only in winter hemisphere; quasi-periodic	"A" eddies result from flow over topography in winter hemisphere; "B" eddies from same in the tropics	Uniquely Martian phenomenon caused by seasonal condensation of CO_2	Very strong

Figure 5. **Major circulation patterns within the atmospheres of Venus, Earth, and Mars appear schematically above a comparison of the patterns' characteristics. Arrows on the globes indicate the direction of flow, and each type of circulation is described more fully in the text.**

Venus. This strong "superrotation" of the atmosphere – the winds move much more rapidly than the planet rotates – could be the consequence of both Hadley circulation and eddies. The Hadley circulation may transport momentum from the denser parts of the atmosphere upward to the more rarefied regions, while eddies aloft may transport momentum toward the equator, thereby spreading it over a wide range of latitudes. Key eddies may include thermal tides, barotropic eddies (produced by strong horizontal gradients in wind speed), and gravity waves (up-and-down oscillations of the atmosphere). Compared to the swift east-west air flow at the cloud tops, much more modest speeds of 5 to 10 m per second characterize the winds higher up that blow from the equator to the poles.

Important differences exist between the wind pattern near the cloud tops determined from photographs obtained by Mariner 10 and, five years later, by the Pioneer Venus orbiter (Figure 6). Certain mid-latitude regions exhibited enhanced east-west winds in 1974 but appeared much smoother in 1979. Thus, the circulation patterns on Venus may undergo natural oscillations on a time scale of years.

While thermal tidal winds are quite weak in the massive lower atmosphere of Venus, they may nevertheless have greatly influenced the current rotation rate of the planet. Not only does Venus rotate more than a hundred times more slowly than Earth does, but it also rotates in the opposite direction (sunrise there occurs in the west, not the east). Because Venus is closer to the Sun, gravitational tidal forces exerted by the Sun on the solid body of the planet may have markedly slowed the spin of Venus over its lifetime. Meanwhile, thermal tides would have acted to speed up the atmospheric rotation rate, with their strength growing steadily as Venus itself slowed down. According to Anthony Dobrovolskis and Andrew Ingersoll, the current rate of rotation may represent the balance struck between these two opposing forces, and Venus' massive atmosphere could well have prevented the planet from spinning even more slowly than it does now.

ATMOSPHERIC ORIGIN

We next consider the sources of planetary atmospheres, the factors that determine their chemical makeup, and their possible variations in composition with time. The terrestrial planets formed about 4.6 billion years ago within a giant disk of gas and dust, the solar nebula. Over time, dust grains aggregated into larger and larger solid bodies (planetesimals) that eventually became the four terrestrial planets (see Chapter 23). Magnesium, silicon, iron, and oxygen – in the form of oxides – are the most abundant elements of the terrestrial planets. Had these planets obtained elements in the same proportions present in the solar nebula and now present in the Sun, hydrogen and helium would be most abundant, followed by carbon, oxygen (mostly in the form of water), and nitrogen.

Yet only traces of H and He exist in the terrestrial planets, showing that these worlds retained at best minuscule amounts of the solar nebula's gaseous components and reinforcing the notion that they formed by the aggregation or accretion of solid matter. Also, the dearth of C, N, and water-borne O in the terrestrial planets implies that these elements were chiefly in the gas phase rather than the solid phase of the inner solar nebula. For example, for Earth as a whole, C, N, and H_2O are about 10,000 times less abundant relative to Mg, Si, and Fe than they were in the solar nebula. Nevertheless, these greatly depleted "volatiles" have played key roles by serving as the chief sources of the atmospheres of Venus, Earth, and Mars, the ice deposits on Mars, and the oceans and life-generating macromolecules on this planet.

There are two logical possibilities for the birthplace of the planetesimals that supplied volatile components to the terrestrial planets. The volatiles may have resided in the planetesimals that formed the bulk of these planets, that is, their "rocky" (Mg, Si, and Fe) component. Or they were carried in from farther out in the solar system – from either the asteroid belt or the region of the giant planets.

Consider the first possibility, a "local" source. Our knowledge of the chemical composition of meteorites suggests that the water available locally or in asteroids would have been chemically bound into the crystal structure of hydrated minerals such as serpentine. Nitrogen and oxygen would have been combined, along with carbon and hydrogen, chiefly in complex (but nonbiological) organic compounds. These may have originated in the dense molecular cloud from which the solar system originated and may have been preserved (not vaporized) as this cloud contributed material to the forming solar nebula. It is also possible that minor amounts of elemental C and N were "dissolved" in the metallic Fe phase of the planetesimals and that some H_2O was ultimately derived from the reduced H of the organic component.

Because temperatures decreased outward in the solar nebula, volatile-rich planetesimals may have been most abundant near where Mars formed and least abundant near where Venus and Mercury formed. As they began to accumulate, the terrestrial planets drew chiefly on nearby planetesimals, and so their initial content would have reflected the volatile abundance dictated by local temperatures within the nebula. However, as Earth and Venus approached their final masses, they possessed enough gravity to redirect close-approaching planetesimals onto orbits that crossed those of the other terrestrial planets. From a compositional standpoint, whatever material remained uncollected in the inner solar system would have become homogenized, and thus the terrestrial planets would have accreted comparable amounts of volatiles during their later growth phases – *if* all their volatiles originated locally.

Suppose instead that terrestrial planets obtained their volatiles chiefly from planetesimals that formed farther out in the solar system. If these came from the asteroid belt, hydrated minerals would have been the chief water-bearing phase and organic compounds would have been the major source of C and N. If the volatile-rich planetesimals came from still farther out, from near where the giant planets formed, organic compounds would have again borne the C and N, but water ice would have been the chief water-bearing phase. Gravitational effects exerted by the massive, nearly complete giant planets could then have perturbed these distant planetesimals inward so that they crossed the orbits of the terrestrial planets. (Such scattering of the planetesimals near Uranus and Neptune may have created the Oort cloud of comets inhabiting the outermost regions of the solar system; see Chapters 17,21.)

Even bodies in the asteroid belt could have been affected greatly. If the orbital periods of the infant Jupiter and an asteroid had been simple fractions of one another, the gravitational force exerted by Jupiter on the planetesimal would have become greatly amplified by keeping a common phase over many orbital revolutions. Therefore, a "distant" source of volatiles should have arrived in the inner solar system late in the growth of the terrestrial planets, or even at the very end. Comparable amounts of volatiles would have been added to Earth and Venus, but somewhat less to Mars and Mercury (because they were less capable of capturing passing objects).

All this implies that the four terrestrial planets would have ended up with similar fractional amounts of volatiles from either a local or distant source of volatiles. We can confirm this conclusion to some extent by comparing the abundances of carbon and nitrogen found on Venus and Earth. Because Venus' surface is so hot, much of its C and N resides in its atmosphere as carbon dioxide and molecular nitrogen. By contrast, almost all of the carbon once present in the Earth's atmosphere is contained in rocks – as carbonates and, to a lesser degree, buried organic compounds produced in the past by living organisms. Much of the Earth's nitrogen is present as atmospheric N_2, but some resides in buried organic material. When all these reservoirs are taken into account, we find that Earth and Venus have very comparable amounts of C and N. This similarity implies that the two planets *initially* had comparable amounts of water as well. However, today Venus' atmosphere contains 100,000 times less water than is in Earth's oceans, and possible ways that Venus may have lost an "ocean" of water will be discussed later.

The earliest planetesimals to be accreted by the terrestrial planets would have arrived at low velocities, since they came from neighboring regions of space and the growing protoplanets were not yet massive enough to accelerate the planetesimals very much as they were drawn in. Thus, the volatile content of planetesimals was not vaporized during their impacts. However, later impacts would have been much more energetic as both the planetesimals' approach velocities and their final gravitational acceleration steadily increased. Arriving volatile compounds, having been vaporized during the collisions, would have formed the planets' earliest atmospheres (Figure 7). Laboratory studies with volatile-rich meteorites imply that infalling planetesimals began to degas when the protoplanets had grown to about 0.3 percent of Earth's present mass and became totally degassed when the protoplanetary masses exceeded about 5 percent that of Earth. Thus, almost all of the volatiles in the planetesimals that formed Earth and Venus would have been released into these planets' atmospheres during accretion. But Mars-forming planetesimals would have degassed only partially and even less so for the objects that formed Mercury.

The transient heat and pressure generated during impacts drove chemical reactions between the volatile elements and the rock-forming minerals that determined the chemical composition of the gases released. For example, if metallic iron was present, as seems likely during much of the accretionary period, some of the water released during an impact would have been converted to hydrogen (losing its O to Fe). Carbon would have been released chiefly as CO and nitrogen as N_2 in this case. But if metallic iron was absent, the chief gas products would have been H_2O, CO_2, and N_2 instead.

However, the released gases did not simply remain in these primordial atmospheres (Figure 7). If the surfaces were not too hot, much of the released water vapor condensed to form oceans, deposits of ice, or both. Also, the first atmospheres were vulnerable to being lost to space – at least partially – by two mechanisms that operated very efficiently during these early times. One is the impact process itself. The arrival of planetesimals more than 10 km or so in diameter shocked and thus heated the atmosphere as they passed through, enough so to drive much of the local atmosphere into space. Thus, accreting planetesimals both contributed and eroded atmospheric gases, and in the absence of other factors each growing planet would have achieved a steady-state atmospheric mass.

The other important loss mechanism involved the Sun. Its strong ultraviolet radiation dissociates H_2O in the upper atmosphere into H, H_2, O, and OH. Furthermore, when H and H_2 absorb this energy they become heated (agitated), and these low-mass gases can then escape to space. The rate of escape is regulated both by the fractional abundance of hydrogen-bearing species present in the upper atmosphere and by the amount of ultraviolet radiation incident there. The first, "diffusion-limited" loss rate is controlled by how quickly H and H_2 can diffuse upward to the atmosphere's *exobase* (where the chance of hitting another molecule before speeding away drops to 50 percent). As the fractional abundance of hydrogen-bearing species above the tropopause approaches 1, the loss of H and H_2 depends

Figure 6. **The clouds of Venus, as photographed by the Pioneer Venus orbiter in 1979. This image was obtained in ultraviolet light to bring out details in the upper cloud layer, which appears very bland at visible wavelengths. Although Venus rotates once every 117 days (as seen from the Sun) in a retrograde direction (right to left in this view), its atmosphere near the cloud tops whirls around in only four days. In doing so, it also moves from equator to pole in a Hadley circulation pattern, giving rise to the chevron-shaped markings seen here.**

instead on the amount of solar ultraviolet radiation available to heat the gases ("energy-limited" escape). The loss may then take place so rapidly that the escaping gases form an outflowing wind that reaches supersonic speeds below the exobase. If this "hydrodynamic escape" is vigorous enough, the departing H and H_2 may drag other gas species into space as well, particularly the lighter ones. Hydrodynamic escape may have occurred and been especially strong during planetary accretion, both because water (and possibly H_2) may have been major components of atmospheres then and because the flux of ultraviolet radiation emitted by the young Sun may have been much greater than at present.

The possible importance of hydrodynamic escape for the earliest inner-planet atmospheres is indicated by the elemental and especially the isotopic contents of rare gases in their current atmospheres. These gases do not combine with other elements, so once added to an atmosphere they remain there unless they escape to space. The ratios of neon-20 to neon-22 in the atmospheres of Venus, Earth, and Mars are less than the solar ratio of 13.5 and also appear to differ from one another. The enrichment of neon-22 may have resulted from the preferential loss of lighter neon-20 through hydrodynamic escape. As Robert Pepin, Donald Hunten, and James Walker have shown, other aspects of the rare-gas patterns of the atmospheres of the terrestrial planets can likewise be attributed to elemental and isotopic fractionation during hydrodynamic escape.

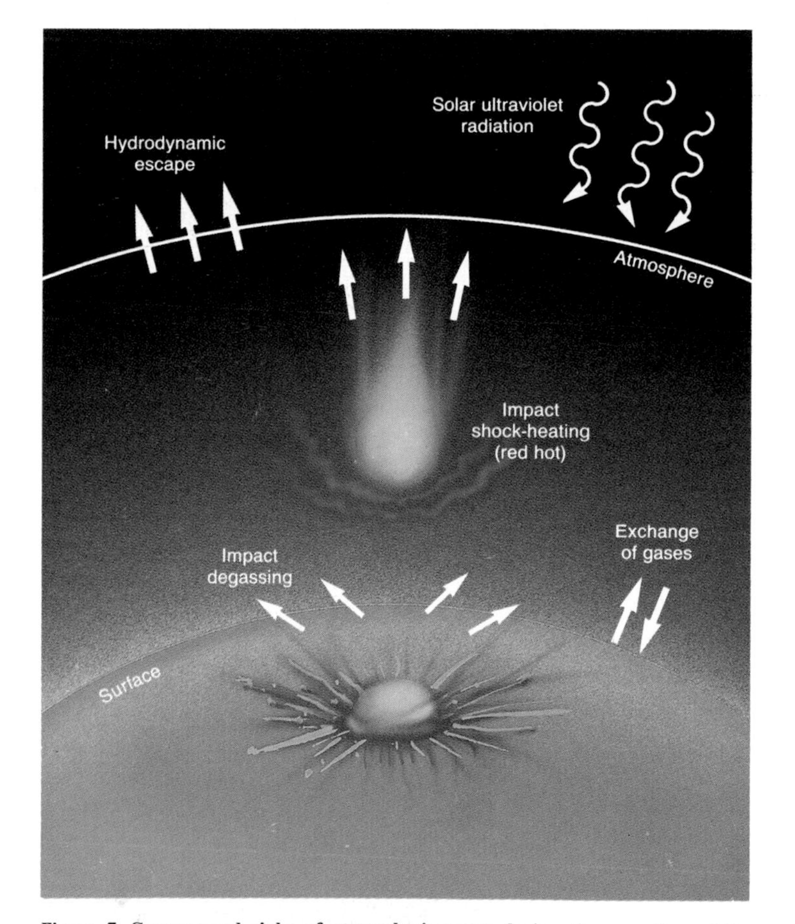

Figure 7. **Sources and sinks of atmospheric gases during the accretion of the terrestrial planets. All planetesimals partially vaporize and release their volatiles to the atmosphere if they strike the surface with enough velocity. The largest planetesimals shock heat the pre-existing atmospheric gases, causing some of them to be lost. Other gases, especially the lighter ones, are also lost to space in a hydrodynamic wind powered by solar ultraviolet light. Gases are exchanged between the atmosphere and surface through condensation, evaporation, gas-to-particle conversion in the atmosphere, dissolution of gases into surface magmas (hot molten rock), and release of gases from surface magmas.**

The partitioning of water between a planet's atmosphere and surface is very sensitive to surface temperature (Figure 8). For example, the fact that almost all of our water resides in oceans – rather than in the atmosphere or polar caps – is a consequence of Earth's distance from the Sun and the heating of its surface only by sunlight. But hypothetically, if the Earth were moved to Venus' orbit, thus receiving about two times more solar energy than at present, all the water in the oceans would evaporate and become part of the atmosphere. This climatic sensitivity to energy input is due to water vapor's strong contribution to greenhouse heating. Once there is enough water vapor in the atmosphere to plug all the infrared windows through which the surface's thermal emissions can escape to space, Earth becomes so hot that the oceans totally vaporize. Andrew Ingersoll has termed this situation "a runaway greenhouse."

Although the Sun put out less total energy while the planets were forming than it does today (see below), the heat generated when planetesimals collided with the growing planets probably more than compensated for the diminished sunlight available at that time. If the Earth and Venus accreted in 100 million years or less, as calculations indicate, and if the planetesimals contained water in any significant amount, then a runaway greenhouse may have occurred late in the accretion process. Their surface temperatures would have risen to about 1,500° K, at which point the landscape became partially molten beneath a "steam" atmosphere. However, any further heating would have been greatly inhibited, because some water vapor would have dissolved in the surface magma, thereby removing itself from the atmosphere and bounding the greenhouse warming.

As the population of available planetesimals dwindled and the growth of the terrestrial planets approached its end, accretional heating began to diminish and solar energy became the only important source of surface and atmospheric heating. Consequently, the Earth's surface temperature dropped from perhaps 1,500° to about 300° K. Almost all the water vapor present in its early atmosphere condensed into oceans. The oceans also gained mass as the partially molten surface cooled and released the water dissolved in it. Mars and perhaps Venus cooled as well as their growth neared completion.

Beyond the possible condensation of water onto planetary surfaces, gases continued to be removed from the atmosphere due to ultraviolet-driven chemistry at high altitudes and liquid-water chemistry at ground level (Figure 9). Specifically, the dissociation of water by sunlight led to the ultimate loss of H and H_2 to space and a net destruction of water; carbon dioxide, dissolved in water, reacted with surface rocks to form carbonate minerals, resulting in a net loss of CO_2 from the atmosphere. These losses were partially or totally compensated by the evaporation of water from surface reservoirs, the volcanic venting of gases from the planet's interior, and the occasional arrival of volatile-rich stray bodies like asteroids and comets. Some volcanic gases were "recycled atmosphere," released from carbonates and other surface rock that had been conveyed by tectonic motion or volcanic burial into the interior and heated to the point of chemical breakdown. Other volcanic gases were "juvenile," either trapped in the material that formed the planet or dissolved into the early magma surfaces.

Following accretion, the atmospheres of the terrestrial planets evolved, becoming more hydrogen-poor and oxygen-rich – though not necessarily abundant in free molecular oxygen. Besides the preferential loss of hydrogen to space, two other processes contributed to the transformation as well: (1) the removal of metallic iron to the planet's deep interior (the "core"); and (2) the gradual oxidation of near-surface rocks, which reacted with oxygen left over from the ultraviolet-driven destruction of water. These latter processes imply that volcanic gases contained proportionately more water and carbon dioxide, and less hydrogen and carbon monoxide, than the volatiles outgassed at earlier times.

In summary, the terrestrial planets could have obtained their water, carbon, and nitrogen from planetesimals originating in the inner solar system, asteroid belt, or the outer solar system. Regardless of the sources, these worlds obtained comparable initial endowments of volatiles. Especially in the cases of Earth and Venus, much of the volatiles were released into their atmospheres during the later stages of planetary growth. Again, especially for Earth and Venus, impact-generated heat may have engendered a runaway greenhouse during these times, making their surfaces hot enough to melt partially. The cratering process itself, along with a high ultraviolet flux from the early Sun, may have driven large quantities of atmospheric gases into space, particularly the lighter ones. At the end of accretion, the surface temperatures on Earth moderated and its oceans formed with a total mass comparable to that at present.

ATMOSPHERIC EVOLUTION

Change, not constancy, has been the rule for the atmospheres of the terrestrial planets over their 4.6-billion-year history. They have experienced not only an overall oxidation but also large swings in total mass and in the relative abundance of their gaseous constituents (Figure 10). In the process, each world's surface climate has been dramatically altered as well. We next consider changes that have occurred on time scales of tens of thousands to billions of years, and the combination of factors that promoted them.

Our Sun, a rather ordinary star, produces energy by slowly converting hydrogen to helium in its deep interior through a series of nuclear reactions. Over the age of the solar system, this conversion process has caused the Sun's total luminosity to increase by several tens of percent – a long-term trend that, by itself, would have induced dramatic changes in the climates of the terrestrial planets. Figure 8 illustrates how surface temperature varies with the amount of solar energy falling on a planet, given that it possesses a terrestrial ocean's worth of water, a fixed amount of carbon dioxide in its atmosphere, and cloudless skies. According to this figure, the Earth of today should have a mean surface temperature about 15° K above freezing, and the Venus of today should have a runaway greenhouse.

But the curve also suggests that Venus' surface temperature was just on the borderline of not being in a runaway state in the early solar system, when the Sun's total energy output was 25 to 30 percent less than it is today. The likelihood that Venus had oceans in its early history (albeit hot ones) improves further when the possible effects of water

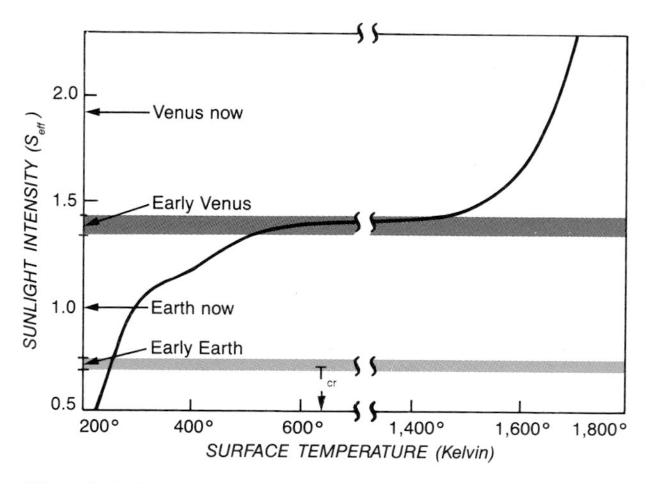

Figure 8. **Surface temperature of Earthlike planets as a function of the amount of sunlight absorbed per unit area, S_{EFF} (which has a value of 1 for the present Earth). T_{CR} is the critical temperature of water, above which it exists only as a dense gas. Also indicated are estimates of S_{EFF} for the early Earth, early Venus, and the present-day Venus endowed with Earth's complement of water and other volatiles. Early in these planets' history, the Sun put out 25 percent less energy than it does today.**

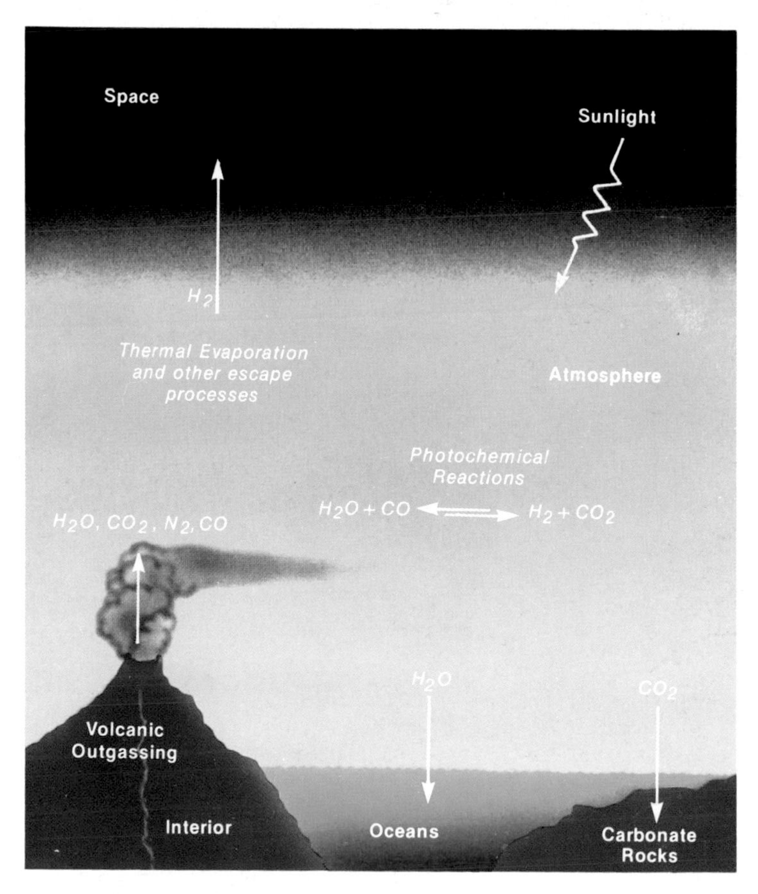

Figure 9. **Sources and sinks of atmospheric gases following the end of planetary accretion. New or "juvenile" gases are introduced by volcanoes. Nitrogen and rare gases, among others, tend to accumulate in the atmosphere over time, while certain species become depleted through various loss processes: hydrogen escapes to space, water condenses out, carbon dioxide takes part in chemical weathering, and so on. Thus the steady-state abundances of atmospheric gases may be markedly different from the mixtures initially spewed out by volcanoes.**

clouds are considered. Clouds reflect sunlight back to space better than they trap thermal radiation, so on balance they tend to cool the atmosphere and surface below them. Thus, the early Venus may have experienced a "moist" greenhouse (hot oceans) rather than a runaway greenhouse (no oceans). Ironically, this may have made it *easier* for the planet to lose virtually all of its original water over the next several billion years. Here is the reasoning behind such a counterintuitive conclusion.

As discussed earlier, the loss rate depends strongly on the amount of water vapor present high in its atmosphere and the flux of solar ultraviolet light reaching that level. When water vapor is a minor component of a planet's lower atmosphere, where the temperature falls with altitude, almost all the water condenses out there before it can reach higher altitudes. Specifically, the fractional abundance of water in the upper atmosphere is controlled by its vapor pressure at the top of the troposphere, the tropopause. For example, water vapor constitutes several percent of the gas molecules at the Earth's surface, but only a few parts per million of those in our stratosphere. However, as water vapor becomes a major constituent of the lower atmosphere, the change in temperature with height slows (because water releases heat as it condenses). Therefore, the altitude of the tropopause rises. At these higher levels the total number density of gas molecules decreases, so the fractional abundance of water vapor in the upper atmosphere dramatically increases. As shown in Figure 11, water vapor becomes a major constituent of the lower and upper atmosphere of a planet when the surface becomes about 100° K hotter than the mean temperature of present-day Earth. In this situation, the oceans would not boil since the enhanced amount of water vapor would substantially increase the atmospheric pressure. In fact, oceans could persist until their temperature reached 647° K, at what is termed the critical point of water.

If early Venus had a moist greenhouse, it would have lost water rapidly by ultraviolet dissociation and the escape of hydrogen to space. This rapid escape would have been promoted by a hot surface (which would drive a large mass of water vapor into the upper atmosphere through evaporation) and by a much higher flux of solar ultraviolet radiation in early planetary history. As long as they persisted, hot oceans would have abetted the rapid loss rate by continually replenishing whatever water vapor had been destroyed and by promoting the weathering of surface rocks by oxygen and carbon dioxide. Surface weathering was an important link in this chain of events, because it prevented O_2 and CO_2 from dominating the atmosphere and thus diluting the concentration of water vapor. When the amount of water left was but 0.1 to 1 percent of its original abundance, not enough of it remained to liquify at Venus' surface. So any carbon dioxide released from the heating of carbonates below the surface remained in the atmosphere, and the climate evolved toward what we observe today: a very hot surface and an atmosphere dominated by CO_2 (Figure 10).

The initial loss of water, which took place as long as these hypothetical oceans were present, may have spanned the first several hundred million years after Venus' formation. Thereafter, water continued to be broken down by solar ultraviolet light, but its irreversible loss hinged on the escape of hydrogen atoms into space. Venus' present water vapor abundance is so low – a mere 0.001 percent of Earth's oceans – that the rate at which it is now lost may be balanced by resupply sources (outgassing from the planet's interior and collisions with volatile-rich asteroids and comets).

Mercury should have received a significant number of volatile-bearing planetesimals scattered inward from the zones where Earth and Venus formed. If so, the early climate on Mercury would have been one of a runaway greenhouse because of its closeness to the Sun. But a combination of the runaway state, Mercury's less-than-Earthlike gravity, the early Sun's intense ultraviolet flux, and an early strong solar wind would have completely eliminated the planet's initial gas envelope (Figure 10). Today Mercury has only a very rarefied atmosphere (Table 1). Its sodium, potassium, and oxygen atoms are supplied by the vaporization of small impacting bodies and by the sputtering of atoms from surface rocks by energetic ultraviolet sunlight. All the hydrogen and some of the helium there derive from the solar wind, which occasionally penetrates the planet's magnetic field and reaches the surface. Ions in the solar wind become neutralized upon hitting the surface, and the lighter atoms

	Volatile Inventory	*The First 10^9 Years*		*The Next $3.5 x 10^9$ Years*	*Present*
MERCURY	$CO_2 \approx 5$ b $H_2O \approx 25$ b $N_2 \approx 0.15$ b	$CO_2 \approx 5$ b $H_2O \approx 25$ b $N_2 \approx 0.15$ b	*runaway greenhouse*	*all volatiles lost to space*	minute amounts of He, Na, O from solar wind, surface, and interior
VENUS	$CO_2 \approx 90$ b $H_2O \approx 450$ b $N_2 \approx 3$ b	$CO_2 \approx 1$ b $H_2O/CO_2 > 1$ $N_2 \approx 1$ b	*moist greenhouse*	*runaway greenhouse* → *almost all water lost*	$CO_2 \approx 90$ b $H_2O/CO_2 \ll 1$ $N_2 \approx 3$ b
EARTH	$CO_2 \approx 60$ b $H_2O \approx 300$ b $N_2 \approx 2$ b	$CO_2 \approx 1$ b $H_2O/CO_2 \ll 1$ $N_2 \approx 1$ b		*CO_2 level falls (in rocks) O_2 level rises*	$CO_2 \approx .0003$ b $O_2 \approx .26$ b $N_2 \approx .8$ b
MARS	$CO_2 \approx 10$ b $H_2O \approx 50$ b $N_2 \approx 0.3$ b	$CO_2 \approx 2$ b $H_2O/CO_2 \ll 1$ $N_2 \approx 0.2$ b		*loss of CO_2 and N_2 to surface and space*	$CO_2 \approx .007$ b $N_2 \approx .002$ b

Figure 10. **In the column at left are estimates of the amounts of volatile species initially present in the terrestrial planets' atmospheres. Successive columns show possible evolutionary paths, including estimated abundances for several gases at various critical times.**

soon escape into the atmosphere due to their thermal motions. Helium may also come from the interior due to radioactive decay of uranium and thorium; this gas worked its way to the surface through cracks. Mercury's atmospheric gases are lost when they are ionized by solar ultraviolet light, picked up by the magnetic field entrained in the solar wind, and convected away from the planet. This loss process keeps the atmosphere at its very low density.

Since the Sun was less luminous in the distant past, the Earth could have experienced a deep ice age during its early history and Mars an even colder climate than it has today. In fact, if no other factors intervened, the Earth's oceans should have frozen completely prior to 2 billion years ago. Yet there is sound geologic evidence for the existence of liquid oceans on Earth for the last 3.8 billion years and of life here for almost that long. Similarly, the ubiquitous networks of dry river valleys on the oldest terrain on Mars suggests that the climate long ago was more clement than at present (Figure 12). Thus there is a paradoxical relationship between our knowledge of past climates and the expected effects of decreased solar output.

Resolution of the "faint ancient Sun" paradox must lie in the recognition of other, counteracting factors. In particular, a stronger greenhouse effect could have moderated the surface temperatures on the young Earth and Mars. Since the early atmospheres of the planets may have been somewhat reducing, Sagan and George Mullen have suggested that as little as 10 parts per million of ammonia (NH_3) may have produced the desired greenhouse enhancement. However, practically no ammonia should have been outgassed by volcanoes into the early atmospheres. Moreover, NH_3 is easily broken down by ultraviolet radiation into nitrogen and hydrogen, and it is very difficult to recombine these two back into the parent molecule. Therefore, ammonia is probably not the agent involved in the early greenhouse enhancement.

A second and perhaps more plausible possibility involves elevated concentrations of carbon dioxide at that time. In order to warm Mars' surface above the freezing point of water, James Kasting and the author estimate that some 100 to 500 times more atmospheric carbon dioxide was needed. For the Earth, Robert Cess, Tobias Owen, and V. Ramanathan calculate that 300 to 3,000 times more CO_2 was necessary. Such increases are not unreasonable. First, Earth and presumably Mars were endowed with much more CO_2 than the above amounts. For example, the carbon dioxide currently locked in the carbonate rocks of Earth's crust exceeds the CO_2 now in the atmosphere by a factor of 100,000! By analogy to the Earth and Venus, Mars possesses perhaps 2,000 times the CO_2 in its current atmosphere. Second, factors such as enhanced volcanism and tectonism during Earth's and Mars' early histories probably kept a significantly larger fraction of their CO_2 inventories in their atmospheres than is the case today, as discussed further below.

Over periods of hundreds of millions of years, carbon dioxide levels in the Earth's atmosphere may have been determined by a balance between its enrichment by volcanic outgassing and its removal by the formation of carbonate rocks (Figure 13). The continual shifting of large blocks of the Earth's surface eventually transports these carbonates into our planet's hot interior, where they decompose and release carbon dioxide back into the atmosphere. In other words, CO_2 is not permanently lost by weathering. Thus, an enhanced abundance of carbon dioxide in the Earth's early atmosphere could have resulted from increased volcanic activity and a smaller continental landmass (thus exposing less rock to weathering) at the time. Also involved, or probably so, was the inverse relationship that exists between surface temperature and the concentration of atmospheric CO_2. As pointed out by Walker and Kasting, the weathering rate of rocks, and hence the rate at which carbon dioxide is removed from the atmosphere, may decrease as the planet gets cooler.

On Mars, the weathering of surface rocks could have converted massive amounts of carbon dioxide to carbonates in some millions to tens of millions of years, a time much shorter than the billion or so years over which valley networks formed and a strong greenhouse is thought to have

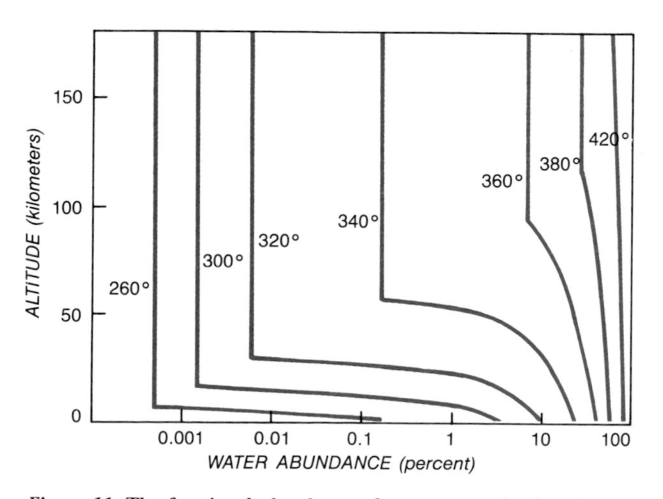

Figure 11. **The fractional abundance of water vapor is plotted as a function of altitude for various surface temperatures, T_S. As T_S increases, water vapor becomes a more dominant component of the gases residing above the tropopause.**

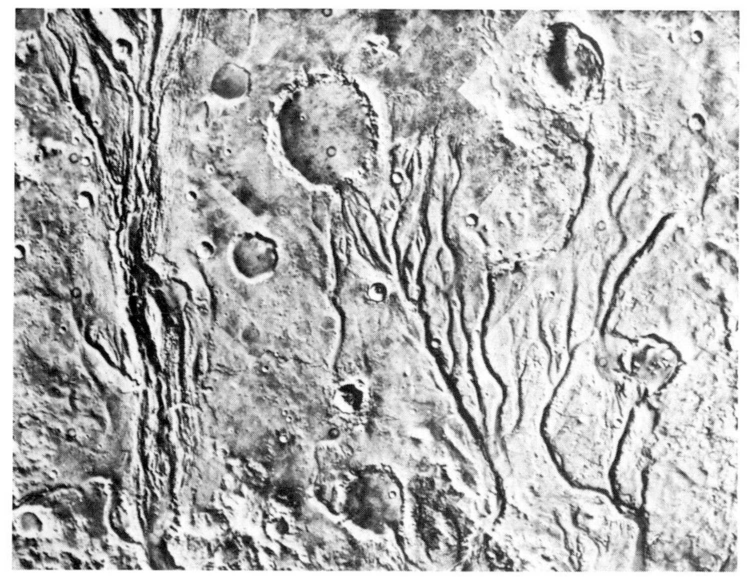

Figure 12. **A Viking photograph of channels on Mars. Bearing strong resemblance to Earth's river beds, these features may have been carved by running water – even though no liquid water now exists anywhere on the Martian surface.**

existed. The role of liquid water is key to this loss of atmospheric carbon dioxide. In this sense, a massive CO_2 atmosphere on Mars contains the seeds of its own destruction. However, intense global volcanism early on may have greatly delayed the demise of Mars' putative massive CO_2 atmosphere. According to calculations made by Kasting, Steven Richardson, and the author, lava could have buried carbonate rocks deep enough to trigger their decomposition to CO_2, with the gas subsequently working its way back into the atmosphere. Michael Carr has suggested that the debris thrown out during frequent cratering events could also have contributed to the burial and decomposition of carbonate rocks on the young planet. This recycling process may have continued throughout much of the period of water-aided weathering.

However, after the first billion or so years of Martian history, the rate of volcanism declined and became confined to localized areas. (By contrast, the more massive Earth cooled more slowly and is still capable of recycling carbonate rocks.) Also, by then meteoritic impacts were occurring much less often. Once most of its original carbon dioxide was locked permanently in carbonates and liquid water no longer flowed across the Martian landscape, much of the remaining atmospheric carbon dioxide may have been removed by adsorption on the surfaces of tiny grains in sediment layers as deep as several hundred meters. Fraser Fanale estimates that the CO_2 adsorbed by this regolith amounts to anywhere from 10 to several dozen times that now in the Martian atmosphere.

Due to the gravitational pull of other large solar-system bodies, the *orbital eccentricity* of Mars and Earth (a measure of how much their orbits deviate from perfect circles) and the tilt or *obliquity* of their rotation axes (with respect to a line perpendicular to their orbital planes) undergo quasi-periodic variations. These variations occur on time scales of 10,000 to 100,000 years for the Earth and 100,000 to 1,000,000 years for Mars, with the departures from the mean eccentricity and obliquity being much larger for Mars. In addition, both planets' rotation axes precess continuously – much like a spinning top (Figure 14).

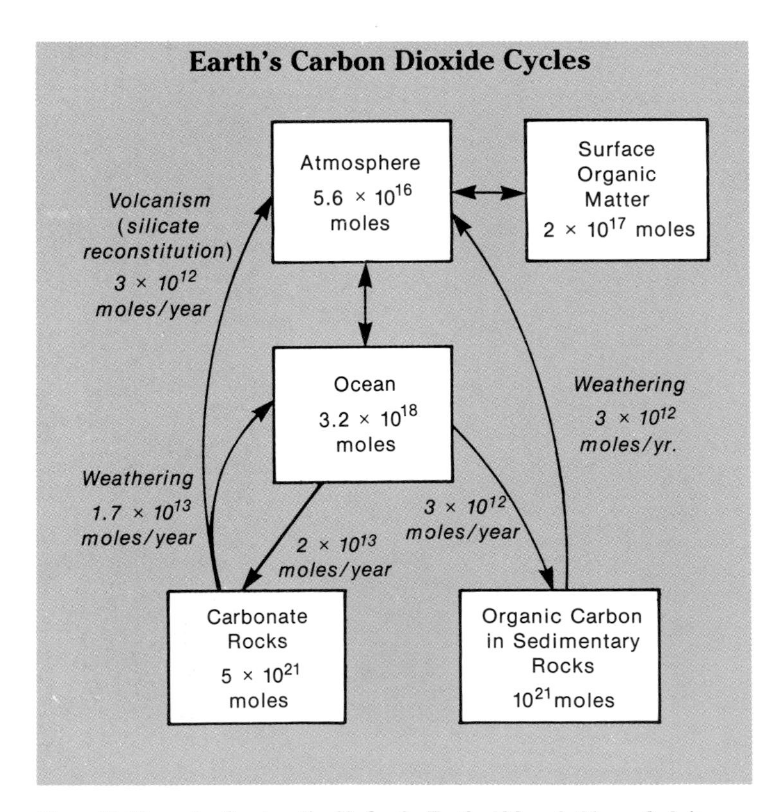

***Figure 13.* The cycle of carbon dioxide for the Earth. Although this gas finds its way into the atmosphere, oceans, biosphere, and crustal rocks over hundreds of millions of years, at any given time almost all of it is confined to the last three of these reservoirs.**

These changes in orbital and axial characteristics cause the seasonal and latitudinal distribution of sunlight to vary, which can lead to climatic changes. For example, the polar regions receive more sunlight at large obliquities than at low ones. Also, large eccentricities increase the sunlight falling on one hemisphere during summer. Studies of sedimentary deposits in the Earth's oceans show that the succession of ice ages and ice-free epochs over the last million years are not random. Rather, these climatic cycles closely match the periodicities expected from orbital and axial variations. When the obliquity is small, high latitudes receive less sunlight over a year. Moreover, if summer in the Northern Hemisphere occurs when the Earth is farthest from the Sun, less sunlight is available during the northern summer to dissipate snow that fell during the previous winter. These two conditions, occurring simultaneously or even separately, tend to initiate ice ages. When these happen, giant ice sheets form at near-polar latitudes on continents of the Northern Hemisphere and expand to mid-latitudes (Figure 15). Because only Antarctica is situated at high southern latitudes and has always been covered with snow over the last tens of millions of years, there has been little change in land area covered by ice in the Southern Hemisphere.

Studies of gas bubbles trapped in ice cores indicate that at the height of the last ice age our atmosphere contained several tens of percent less carbon dioxide than it does now. Such swings in atmospheric CO_2 abundance could have played an important role in inducing or reinforcing the ice-age cycle through the greenhouse effect. Most likely, this change reflects a biological component of the carbon dioxide cycle, which could have varied significantly over the comparatively short time scale of the ice ages.

Prior to several million years ago, no major ice advances occurred in the Northern Hemisphere for almost 300 million years, even though the astronomical cycles continued over that entire period. Thus, they are not the only stimuli to the icing-over of our planet. Another factor might be continental drift, the slow shifting of Earth's continents relative to the poles and to each other in response to internal forces. When a land mass ventures near the poles, as Antarctica and Greenland have, massive sheets of ice cover them continuously. The positioning of continents near the poles may be necessary for the onset of ice ages, for once they become capped with bright ice, the Earth as a whole will absorb less sunlight and soon become cooler.

The layered or "laminated" deposits discovered in the polar regions of Mars (Figure 16) imply that periodic climatic changes have also occurred there and are almost certainly the result of Mars' pronounced astronomical variations. This layering may reflect quasi-periodic cycles in the deposition of dust, water ice, and carbon dioxide ice, with the first two materials constituting the permanent deposits we observe today. Consider, for example, the effect of changing obliquity on dust deposition. During times of high obliquity, latitudes near the poles experience higher temperatures. As a result, less carbon dioxide is adsorbed onto the surfaces of particles in the subpolar regolith. The atmospheric pressure increases, meaning slower winds will set particles into motion, and therefore more dust is driven into the atmosphere. So at times of high obliquity, dust may be deposited in the polar regions at a faster rate. Conversely, at times of low obliquity the dust deposition rate may slow and in fact even stop, since supersonic winds (a very unlikely possibility) may then be required to set sand into motion.

The layered deposits occur just in the polar regions of Mars because water ice, a key ingredient, is stable only at high latitudes. It helps to cement the dust particles together and makes them more resistant to erosion. Astronomical variations also modulate the amount of water ice deposited in the polar regions. The layered appearance of the laminated terrain may reflect different proportions of dust and ice (or their deposition rates) over the period of the astronomical changes.

Because its axis of rotation is almost perpendicular to its orbital plane, and because its atmosphere has a large heat capacity, Venus does not experience any appreciable "seasons." Hence, orbital variations have little effect on the planet's climate.

ZONES OF HABITABILITY

As best we know, Earth is the only planet in the solar system on which life exists and has persisted over almost all of its 4.6-billion-year age. Why is the Earth unique in this regard? The two key factors seem to be our planet's distance from the Sun and its size or mass. Venus is in some sense a twin of the Earth, in that it was endowed with a similar mass, bulk composition, and perhaps initial volatile inventory. However, if the early Venus had hot oceans, temperatures may have been too high to permit life to evolve there. Perhaps more important, such hot oceans would have lasted for at most a few hundred million years. Had the Earth formed 5 percent or more closer to the Sun, it too would have lost all its water by now, and life here would no longer exist. Our good fortune is not permanent, however, for in another billion years the Earth will reach this regressive stage due to the steady brightening of the Sun.

Carbon dioxide and water vapor play key roles in maintaining life-sustaining temperatures on our planet. Fortunately, the Earth has a large inventory of carbon dioxide, and enough of it is accessible for the atmosphere to moderate the surface temperature over a wide range of incident solar fluxes. Mars was not so lucky. The recycling of carbonate rocks back into atmospheric CO_2 may have been possible only during Mars' first billion or so years of existence, because the planet formed with only one-tenth the mass of the Earth. This means that Mars has developed a very thick, rigid outer crust, and its volcanism during the last several billion years has been much less vigorous than at the outset and confined to localized areas. Life could have arisen on *early* Mars (indeed, this is a key question for future explorations there!) but could not have sustained itself.

Maintaining life on Earth over so much of its history has not been easy. Numerous extinction events – some sudden, some gradual – punctuate the geologic record. For example, the Cretaceous geologic period, during which dinosaurs roamed the Earth, came to an abrupt end 66 million years ago; about 75 percent of all the species alive during the last part of the Cretaceous did not survive into the next (Tertiary) geologic period, the age of mammals. Many, though perhaps not all, of these extinctions occurred very suddenly.

Luis and Walter Alvarez and their colleagues have maintained that the Cretaceous-Tertiary extinctions occurred precisely when the Earth was struck by a large, stray asteroid or comet (see Chapters 18,21). Their evidence comes from chemical analyses of sedimentary layers formed around the time of this extinction event. They studied the abundance of the metal iridium, which is about 10,000 times less abundant in the Earth's crust than in most meteorites. In sediments from diverse locations around the world, all laid down exactly at the time of the extinction event, the Alvarez's research team found a very sizable increase in the iridium concentration. They concluded, therefore, that the enrichment was due to the impact of a large extraterrestrial

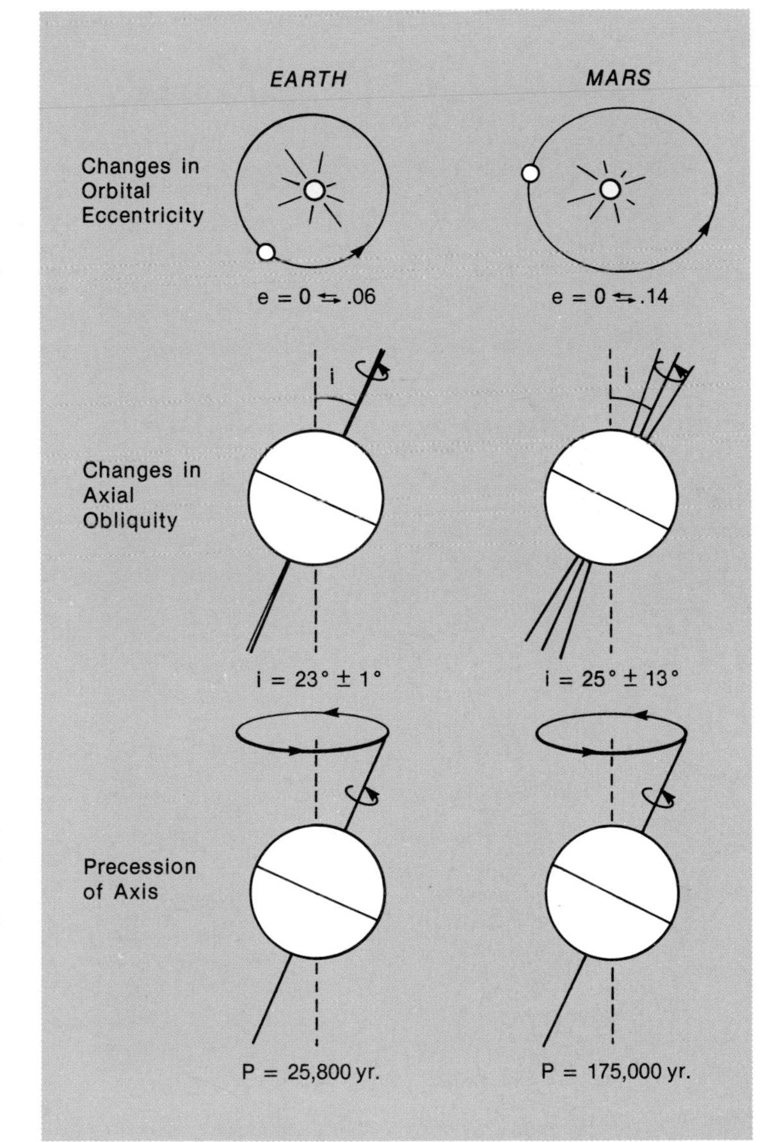

Figure 14. **This diagram illustrates quasi-periodic changes in the orbital and axial characteristics of Mars and Earth. Note how much greater the variations are (in most cases) for the latter. Such fluctuations are far smaller and less important for Venus.**

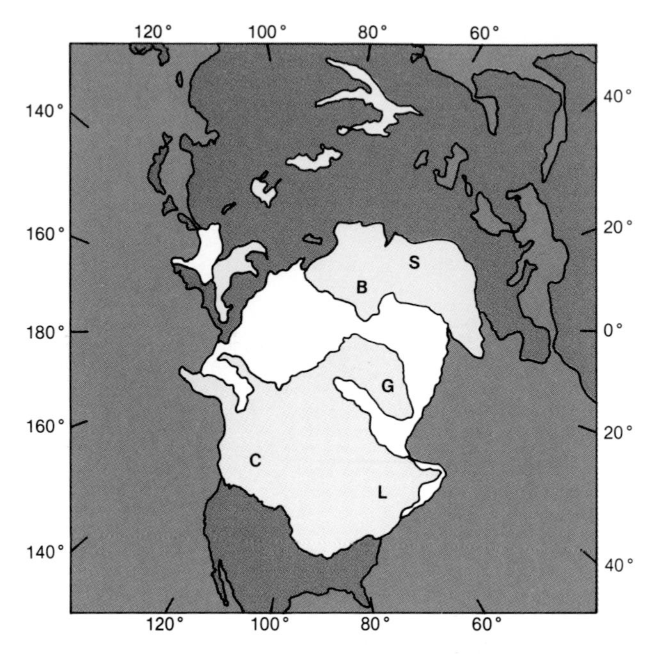

Figure 15. **The maximum coverage of the Earth's Northern Hemisphere by ice during the last million years. Continental ice sheets, indicated by colored areas, are: *B*, Barents Sea; *S*, Scandanavian; *G*, Greenland; *L*, Laurentide; and *C*, Cordilleran. Sea ice is indicated by white. It is apparent that all of Canada and much of the northern United States were covered at some time. The last glacial maximum, which occurred only about 18,000 years ago, covered about 90 percent of the area mantled by ice here.**

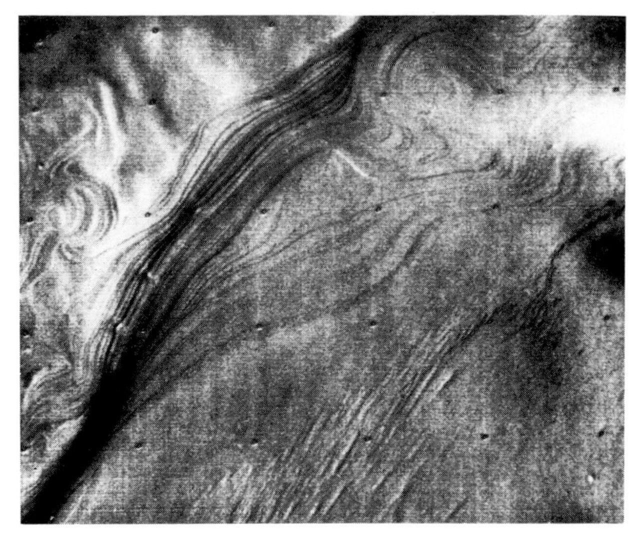

Figure 16. **A detailed photograph of layered or "laminated" terrain near Mars' south pole, taken by the Mariner 9 spacecraft. This landform lies exposed on steep slopes, appearing as a series of stripes (due to illumination) that suggests terraced topography.**

object. If so, they reasoned, the widespread extinction of life could have been caused by the large amount of very fine debris thrown into the stratosphere by the impact. For a period of months, much less sunlight would have reached the surface of the Earth, conceivably causing photosynthesis to cease suddenly over the entire world. Plants would have died in vast quantities, and with their food supply interrupted, many marine species (for which the carbon cycle is relatively brief) would have perished along with the plants. The suspended dust may also have blocked enough sunlight to lower temperatures sharply on the continents and for a time to prevent animals from seeing. These two factors may have contributed to the extinctions of land animals, including the dinosaurs. Finally, shock heating of the atmosphere as the incoming object passed through it could have produced copious quantities of nitrogen oxides (NO and NO_2) and ultimately nitric acid (HNO_3). Once in the ocean, the HNO_3 may have caused a lethal change in the ocean's acidity.

The chemical origin of life on Earth during its earliest history may have been delayed by the much higher rate of giant impact events than at present. Studies of the Moon's craters and rocks indicate that impacts occurred a thousand times more often 4.2 billion years ago than they do now. By extrapolation, the impact rate may have been even higher during the first several hundred million years of the Earth's history. If frequent enough, large impact events might have extinguished the earliest life forms by creating continuous clouds of opaque dust, polluting the oceans with nitric acid, and perhaps even heating the oceans to lethal temperatures.

Despite these early hazards, as well as later ones, life arose during the early history of the Earth and has persisted until today. Gradual and sudden extinction events have not eliminated life but rather led to its evolution – simple organisms were replaced by more complicated ones. The earliest life forms, which were probably single celled, perhaps obtained their energy and food from chemicals rather than sunlight. Environmental change and evolution led to the development of multi-celled organisms, photosynthetic species, and ultimately intelligent creatures. It is perhaps no coincidence that the rise of humans coincides with a period of ice ages.

Life has been affected by – but in turn also has affected – the environment. Perhaps the most obvious manifestation of the latter is the presence of large quantities of oxygen in the atmosphere, a waste product of plants' conversion of water and carbon dioxide into food. Oxygen is intrinsically a "poison" to organic matter, so life needed to evolve ways to isolate and, later, to utilize O_2 as the gas became more abundant. Yet oxygen is crucial to the way humans and many other organisms derive energy from organic food. More generally, biology participates in and moderates many of the key geochemical cycles on Earth that move chemical species from the atmosphere to the oceans to the rocks and back to the atmosphere. However, we have reached the point where the activities of the most advanced life form on our planet – human beings – represent potential hazards to Earth's climate. It is perhaps fortunate that this coincides with our first overtures to explore the solar system. The histories of these other bodies offer forceful reminders of how drastically planetary atmospheres and climate can change. Those atmospheres also constitute huge natural laboratories that can be used to test our understanding of the basic processes at work here and our ability to predict the changes that could affect this oasis of the solar system.

9

The Voyager Encounters

Bradford A. Smith

"Our sense of novelty could not have been greater had we explored a different solar system."

– Anonymous

THIS REFLECTION on Voyager 1's reconnaissance of the Jupiter system was expressed by a member of the imaging-science team shortly after our historic encounter with the giant planet on March 5, 1979. It conveyed for all of us our astonishment with the thousands of scenes we had witnessed of this assemblage of bizarre, yet beautiful worlds – worlds that had previously belonged only to the astronomer or to the occasional watcher of the night skies. Changed forever was our distant view of Jupiter and its Galilean satellites; the pale yellow disk surrounded by its starlike points of light had suddenly become real, and even if we failed to understand all that we saw, we had nonetheless become curiously familiar with a distant planetary system. Over the next seven years this intimacy would extend first to Saturn and then Uranus.

JUPITER: THE PRE-VOYAGER VIEW

Jupiter is the largest of the planets, possessing more than the combined mass of everything else in the solar system except the Sun. Yet, at a mean distance of nearly 800 million km from Earth, it is a difficult object for telescopic observation. The disk of Jupiter displays a series of alternating light and dark bands, parallel to the equator and shaded in subtle tones of blue, brown, and orange (Figure 1). In addition to these features, there are numerous bright and dark spots, the largest and best known being the Great Red Spot. By carefully tracking the more conspicuous of such localized features, astronomers – both professional and amateur – were able to map the global circulation patterns of Jupiter's tropospheric winds. The meridional profile showed a pattern of alternating easterly and westerly (zonal) winds that seemed to be related to the banded cloud structure. At the equator was a 30,000-km-wide jet traveling eastward at up to 400 km per hour. The only winds *not* seen strictly confined to latitude zones were in the anticyclonic flow around the Great Red Spot, first noted by Elmer J. Reese and the author in 1966. Centuries of visual and, more recently, photographic records show that the contrast and color of the Jovian cloud system is in a continuous state of change. Dramatic increases or decreases in the reflectivity of a single band have been observed over only a few months, and the entire planet's appearance can change in just a few years.

Among Jupiter's satellites (13 were known prior to Voyager), the four Galilean satellites stand in a class by themselves; Io, Europa, Ganymede, and Callisto are each planet-size bodies in their own right and can actually be seen as tiny disks in a large telescope. Vague, dusky markings have been reported by visual observers, and when combined with records of periodic brightness variations during the course of an orbit, these observations show that each Galilean satellite always keeps the same face toward Jupiter.

Although other types of telescopic instrumentation have told us much about Jupiter and its satellites, information based on direct observation has left much to be desired. Observers attempting to make systematic observations of Jupiter's clouds have been hampered by unpredictable

Figure 1. **One of the most detailed photographs of Jupiter ever taken from Earth, this image acquired by Stephen Larson in December 1975 with the University of Arizona's 1.5-m telescope hints at the dynamic turbulence present in the planets's upper atmosphere.**

weather and other difficulties; variable observing conditions give rise to sporadic data of uneven quality. When the objectives involve time-variable phenomena, such as the morphology and motions of atmospheric features, the inability to obtain uniform and consistent records is a severe impediment to an observer.

The space age came to the outer solar system with the arrival of Pioneer 10 at Jupiter on December 3, 1973. This extraordinary accomplishment was repeated one year later when its twin, Pioneer 11, flew by the giant planet on December 2, 1974. The colorful images transmitted back by these spacecraft (Figure 2) gave us our first appreciation of the complex nature of the Jovian cloud system. Much of this intricate structure was located at the boundaries between light and dark bands, where telescopic observers suspected large wind shears to exist. At high latitudes the morphology of individual features, some as small as 500 km across, suggested convective cloud systems. Structure seen within the Great Red Spot and elsewhere implied complex atmospheric motion; unfortunately, because of the brief duration of the Pioneer encounters, repetitive coverage was seldom obtained and few cloud motions could be measured. Thus, although Pioneer gave us a new and captivating insight into the detailed structure of the global cloud system on Jupiter, it provided little in the way of dynamical information. Pioneer also recorded several images of the Galilean satellites, and while the resolution was superior to that obtained from telescopic observations, it was insufficient to give even a hint of any geologic sculpturing of their surfaces.

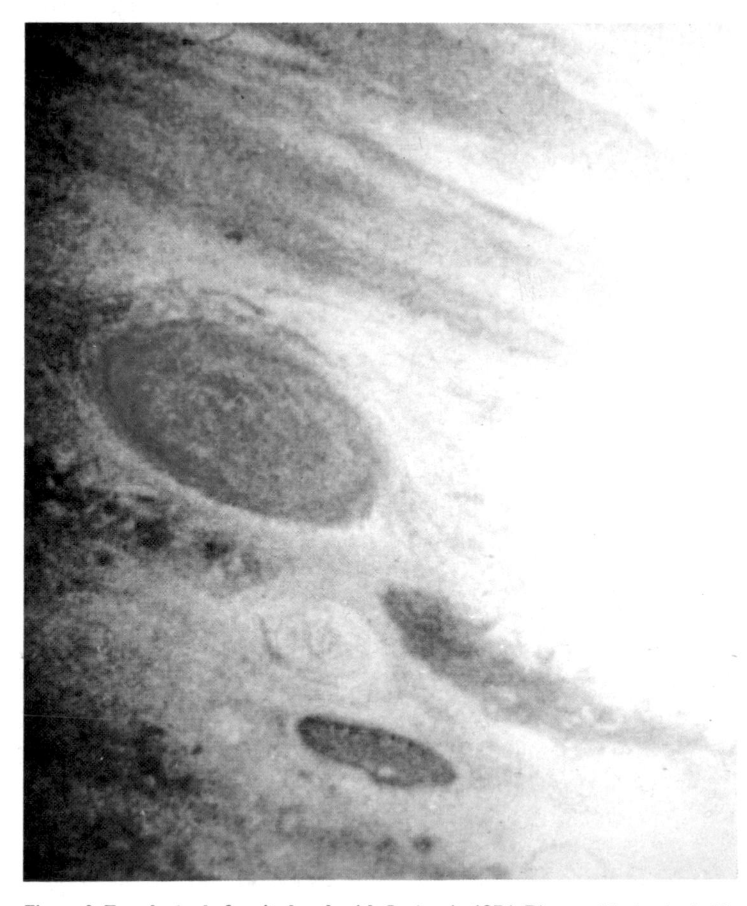

Figure 2. **Four hours before its brush with Jupiter in 1974, Pioneer 11 obtained this close-up of the Great Red Spot and neighboring features in the Jovian southern hemisphere. At that point, the spacecraft was as close to Jupiter as the Moon is to the Earth.**

VOYAGER: THE APPROACH

The Voyager 1 approach phase at Jupiter began on January 4, 1979, just 60 days before encounter. At this time its narrow-angle (telephoto) camera began systematic imaging of Jupiter, taking a multicolor sequence every two hours. More than a month earlier Voyager test images had already exceeded the resolution of the best telescopic pictures taken from Earth; now we were beginning a systematic series of observations designed to collect data on the structure and dynamics of the Jovian atmosphere. Even at that great distance of 60 million km, the pictures were spectacular and getting better every day (Figure 3). From the very earliest images, it became clear that the structure of the Jovian clouds was even more complicated than we had thought. As the data continued to accumulate and cloud motions became apparent, we could see that the detailed dynamical picture was going to be equally complex. The model that we had developed from telescopic observations was wrong – rarely did we see simple zonal motion, but instead there was a planetwide pattern of small-scale vortices. The interaction of zonal flow with the anticyclonic vortex of the Great Red Spot (Figure 4) produced a complicated, time-variable maelstrom in which clouds were torn apart and often swept into the swirling anticyclone. Elsewhere, interacting currents caused clouds to consume their neighbors, only to disgorge them a few tens of hours later. Others engaged briefly, circled each other in a *pas de deux*, then separated and continued on their way. Later we would learn that the vorticity, drawing its energy from the interior of Jupiter, was the source of momentum for the well-known zonal flow.

During the approach phase, we watched certain cloud systems with a special interest. These were the features selected from a ground-based photographic patrol, operated by team member Reta Beebe, as potential targets for close scrutiny at encounter. The continued monitoring of these atmospheric systems by Voyager was essential if we were to predict their precise location and extent at the time of encounter. Final coordinates had to be provided one month before the high-resolution pictures were taken, thereby placing a requirement on the imaging team to make a 30-day forecast of Jovian weather. That we were successful with nearly every target is testimony to the fact that not all Jovian phenomena are unpredictable.

Some 30 days before encounter Voyager's images exceeded the highest spatial resolution achieved by Pioneer. We had entered *terra incognita*. Jupiter loomed before us in a spectacle of bizarre cloud formations and dazzling variegated color that went beyond the sum of our collective imaginations. The immensity of what we were seeing was staggering: the vortex of the Great Red Spot alone could swallow several Earths. As the encounter approached, Jupiter grew larger than the camera's field of view (Figure 5). Features barely visible from Earth filled an entire Voyager picture. Later would come the analysis and the scientific significance (see Chapter 11), but at the moment we were caught up in the sheer wonder of it all.

Throughout the approach the images of the Galilean satellites had grown in size as they passed back and forth across our television monitors. Among all of the planetary bodies visited thus far by exploratory spacecraft, none elicited less distinct images in our minds than these four large moons of Jupiter. Now they were becoming individually recognizable bodies. Within a few days we would see them as clearly as astronomical telescopes view our own Moon – and we were about to make one of the most important discoveries of the space program.

THE ENCOUNTER

While Jupiter was the star of the show during the approach, we were actually *too* close to the huge, cloud-covered planet at encounter. The images were low in contrast and fuzzy. As we passed behind on the night side, however, our cameras recorded a number of interesting phenomena such as lightning, a meteor, and auroras.

Just as we passed through Jupiter's equatorial plane, about 17 hours before closest approach, our cameras turned to the side for a single, 11-minute-long exposure. In that picture was a faint ring. The discovery was not accidental. Everything we then understood about the formation and stability of planetary ring systems had told us that Jupiter could not have a ring. Nevertheless, one does not travel nearly 1 billion km

Figure 3. **Voyager 1 recorded this view of Jupiter from a range of nearly 33 million km. From that distance, cloud features as small as 600 km across can be resolved. The Great Red Spot and whorls of turbulence to one side dominate the planet's southern hemisphere. Farther to the south are several white ovals that first appeared about 50 years ago. The white puff in the equatorial zone north of the red spot marks one area where bright clouds originate, then stream westward. Voyager color photographs are actually composites of several black-and-white frames taken through colored filters.**

Figure 4. **By the time Voyager 2 reached Jupiter in mid-1979, the turbulence west of the Great Red Spot viewed by Voyager 1 had broken into smaller cloud parcels and vortices. A region of white clouds cap the GRS's northern boundary, preventing smaller puffs from circling the immense cyclonic feature as they had done four months before.**

into space without at least a quick look. Even though the image had been a part of the encounter sequence (primarily at the insistence of team member Tobias Owen and experiment representative Candy Hansen), most of us felt the chance of actually finding anything was nil and had actually forgotten about it. The ring, it turns out, was seen much better by Voyager 2 in back-illumination (Figure 6), but it was a major discovery of the first encounter. More is said about Jupiter's ring (and those of other planets) in Chapter 12.

Even with our camera's 1,500-mm focal length (granted us by NASA to our everlasting gratitude), only the moon Amalthea and the four Galilean satellites would show significant disks. Amalthea, so small that it is seldom seen even with large telescopes, was certain to present a serious targeting problem; we did not know exactly where it would be. Fortunately, it was identified by Voyager navigation specialist Stephen Synnott in the approach images. This made it possible to aim our cameras accurately and during encounter several high-resolution images were obtained. As team member Joseph Veverka pointed out, Amalthea turned out to look a little like a dark red potato, complete with eyes. Synnott would later discover two very small satellites of Jupiter in the Voyager images. A third would be found by David Jewitt, then a graduate student working for team member Edward Danielson. Those satellites and Amalthea are described further in Chapter 20.

Our preconceptions of what the surfaces of the Galilean satellites would be like varied from one team member to the next. Many of us thought that Ganymede and Callisto, with their thick ice mantles, would be smooth; ice would flow as a plastic, we reasoned, reducing all impact or tectonic structures to something approaching an equipotential surface. Io, being without ice, would look like a reddish twin of Earth's Moon, covered with sulfur-coated impact craters. Io, in fact, would be the Rosetta stone for meteorite fluxes in the outer solar system, allowing us to translate the impact records, and therefore the ages, of all other satellite surfaces. Europa probably had a thin coating of ice and, therefore, a hybrid surface – but then, we considered Europa to be the least interesting of the four, anyway.

I don't think we could have been more wrong in predicting what we would see on the Galilean satellites. What we failed to consider is that ice becomes as rigid as steel in frigid space 5 AU from the Sun; also, we could not have known that processes were occurring within Io and Europa that had not yet been encountered in planetary exploration.

The first of the ice-mantled satellites to be seen at close range was Ganymede. We had survived the radiation

***Figure 5.* This is one of the most striking images from Voyager 1, which captures ruddy Io and pearl-like Europa against the vast backdrop of Jupiter, here about 20 million km from the spacecraft. Circulation patterns are evident in the planet's atmosphere, especially the intriguing eddies located within and between cloud bands. Even from this distance, Io shows color variations on its surface.**

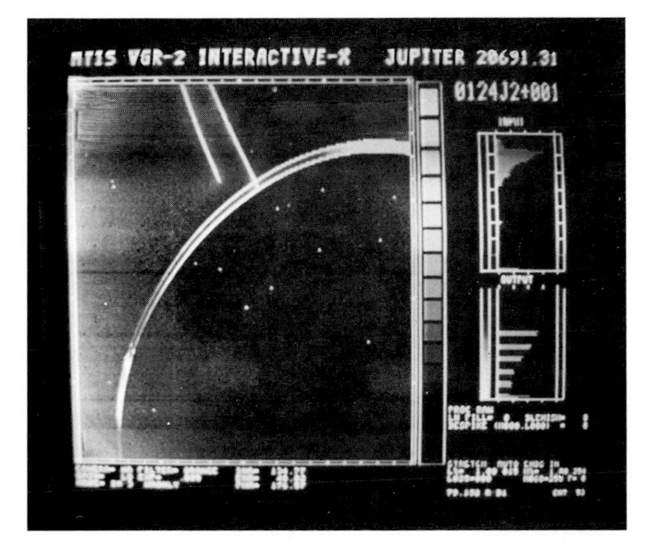

***Figure 6.* Soon after passing Jupiter, Voyager 2 turned its television eyes back on the planet, now backlit by the Sun, from within its shadow. Monitors at the Jet Propulsion Laboratory in California flashed with images of the tenuous Jovian ring, which gleamed with unexpected brightness when viewed from this perspective.**

environment at our close pass by Jupiter and the whole team was now standing around, drinking champagne and watching Ganymede on the television monitors. Would there be any topography at all on these icy bodies? The geologists were nervous. Then, deputy team leader Larry Soderblom saw a well-defined impact crater, jumped up, and pointed at the screen; the team cheered, but the expression on Soderblom's face conveyed more relief than jubilation. Several hours later, we were looking at Callisto's battered surface, and craters of all sizes were everywhere.

On both Ganymede and Callisto there was ghostly evidence of impact features that had lost all of their relief, presumably a result of plastic flow in the warm ice of an earlier time. But these were relatively rare; the surfaces of both satellites were sculpted by topography that had held its form over aeons in the 120° K ice. The geologists were ecstatic – an evolutionary record had been preserved. We soon realized that the surface of Callisto is very old, perhaps dating back to the late torrential bombardment period more than 4 billion years ago. It is a world long dead and its aspect vaguely familiar to geologists experienced in studying the silicate bodies of the inner solar system. Some regions on Ganymede also date back to that ancient epoch, but others are younger and show a strangely grooved landscape that is suggestive of internal activity. This was *not* familiar terrain, and discussion of the responsible mechanisms would go on for months.

Europa was not seen well by Voyager 1, though what we did see was fascinating. No craters were evident, but the satellite's surface was interlaced with a remarkable pattern of linear features, looking curiously like a Lowellian drawing of Mars. We suspected that global tectonic activity had come into play, and this greatly enhanced the anticipation of our more favorable Voyager 2 encounter with Europa – this "least interesting" Galilean moon.

At a distance of 8 million km, one week before encounter, Io was beginning to show circular features that we assumed were craters. Each passing day, however, brought increasing doubts about our Rosetta stone; the "craters" really didn't look like craters. As we approached, the surface of Io began

***Figure 7*. This full-disk image of Io from Voyager 1 reveals a variety of features that appear to be linked with the satellite's intense volcanic activity. For example, the circular "doughnut" in the center has been matched with an erupting plume seen in other images. Despite the wildly variegated colors on its surface, Io seems to be covered entirely by only sulfur and sulfur dioxide. No impact craters have been found at all, leading scientists to conclude that the eruptive vents constantly deposit new material on the surface, thus burying the cratering evidence.**

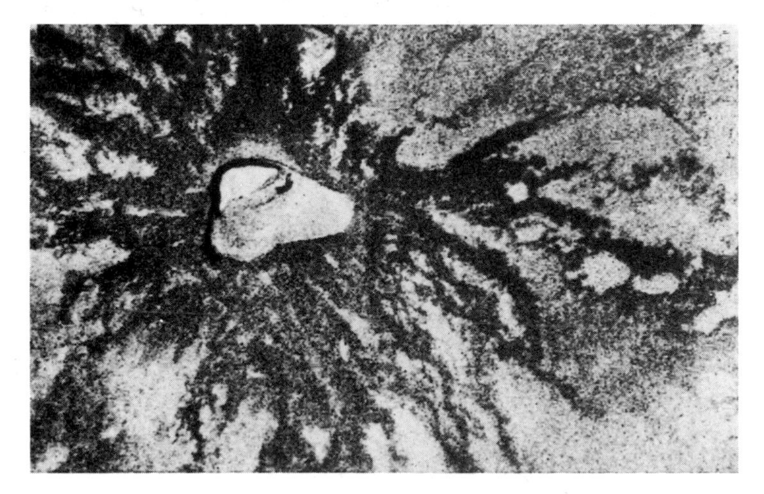

***Figure 8*. Voyager 1 passed Io at a distance of only about 30,000 km and recorded intricate details like the volcanic feature seen here. Dark flows of basaltic or perhaps sulfurous lava radiate from a complex caldera that measures about 50 km in diameter.**

to take on the most bizarre appearance of any object yet seen in the solar system. Great splotchy regions of yellow, orange, white, and black gave its surface a grotesque "diseased" appearance (Figure 7). The "craters" were actually albedo features; no true impact craters were to be found. Their absence had profound implications: since Io could hardly have escaped meteoritic bombardment, some very active process had to be destroying or burying the craters. The surface of Io was, therefore, relatively young. Our first estimates of 100 million years seemed to be incredible, but as resolution improved and impact features continued to be absent, our estimates of the surface age became even younger. At 10 million years it was obvious that volcanism must be responsible; in fact, features that looked like volcanic landforms were already becoming recognizable. At 1 million years we realized that volcanism must be as extensive as on Earth and that it was only bad luck that had prevented us from seeing an actual volcanic eruption taking place as Voyager flew by.

The nature of the volcanism was debated within the team. Red, orange, and yellow sulfurous material seemed to be associated with the volcanic features, but there were closely related black deposits as well. Silicate volcanism, indicated by the black deposits, was proposed by Harold Masursky and Michael Carr. Eugene Shoemaker and I suggested sulfur-driven volcanism. Carl Sagan then pointed out that allotropes of sulfur can take on many colors, including white and black. Long sinuous flows could be seen emanating from volcanic calderas in images that showed features less than 1 km across (Figure 8). Still, no impact craters could seen. Voyager had made an important and surprising finding, but the biggest surprise was three days away.

On March 8th, navigation team member Linda Morabito made her historic discovery. On an image of Io taken to evaluate the spacecraft's trajectory was a bright volcanic plume rising hundreds of kilometers above the limb. Encounter-weary imaging scientists, home for a weekend's rest, quickly returned to JPL. Within hours several more volcanic plumes were found in the encounter images. The active volcanoes had been there all along, but special contrast enhancement of the pictures was required to make the relatively faint plumes stand out against the black sky. A final count showed that eight volcanic eruptions were occurring during the flyby of Voyager 1. Our guess that Io might be as volcanically active as the Earth was far too conservative. Team member Torrence Johnson calculated that the volcanic resurfacing rate on Io cold be as much as 10 mm per year. Io's surface was as young as yesterday.

Voyager 2 arrived at Jupiter on July 9, 1979, four months after Voyager 1, giving us a different view of the planet's global wind patterns and a new look at phenomena found by the first spacecraft: the ring system, Europa's smooth surface, and Io's volcanoes. As team members Garry Hunt and Jim Mitchell had predicted, Jupiter's weather had undergone a number of changes. Europa was found to be the smoothest world in the solar system; no surface relief greater than a few hundred meters was evident. The enigmatic linear features continued to puzzle us, though they obviously resulted from some active process in Europa's thin ice crust. Team member David Morrison remarked that the pattern looked like a cracked eggshell. Shoemaker thought it reminded him of packed sea ice. Crater experts Robert Strom and Joseph

Boyce remained silent; not a single well-defined impact crater was seen, but a dozen or so circular features may actually turn out to be badly eroded craters. The surface of Europa was relatively young, though not nearly as young as Io's.

One volcano on Io had died in the four months between the encounters, one had turned out of our view, and two new eruptive vents had formed. From these crude statistics it appeared that eruptions may last for several years. The driving volatile for the plumes seemed to be sulfur dioxide (SO_2) or sulfur, but were we witnessing sulfur-enriched silicate volcanism or a new type based entirely on sulfur and sulfur compounds? The issue remains unresolved to this day.

Even as our cameras looked back at the receding crescent of Jupiter (Figure 9), we had all become aware that this giant planet would never again be the mystical, wandering star of the ancients – or even the changeable banded disk of the modern astronomer. The secrets of Jupiter and its satellites had been revealed; these astronomical bodies belonged now to the atmospheric physicist, the geophysicist, and the geologist. Meanwhile, ahead of us lay Saturn, and it too would soon become known.

SATURN: THE PRE-VOYAGER VIEW

If the surprises found at Jupiter were an inevitable consequence of our prior ignorance of that planetary system, Voyager was certain to treat us to a host of unexpected and exciting revelations as we approached Saturn. Of those planets known to the ancients, Saturn is the most remote and the dimmest. Moving slowly in the cold, faintly lit fringe of our solar system, it is more than 1.4 billion km from the Sun – nearly twice as far away as Jupiter. At such a great distance, Saturn receives scarcely more than 1 percent of the solar energy that falls on Earth.

The planet's great distance and dimness had severely limited what we knew of the planet, its satellites, and its rings prior to the arrival of spacecraft – this despite a century of photographic records and more than 350 years of visual observations. Saturn's disk is marked by a series of light and dusky bands, similar to those on Jupiter, but with far less contrast in both brightness and color (Figure 10). The planet is unblemished by the light and dark spots that give Jupiter its characteristic appearance, yet it was known that this generally featureless appearance could not be due entirely to Saturn's greater distance. In all, fewer than a dozen well-documented spots had ever been observed in Saturn's atmosphere, but those few that could be followed gave us an approximate value for the planet's rotation period and showed that the zonal wind jet at the equator was considerably stronger than the one on Jupiter. Little more could be said about the dynamics of Saturn's atmosphere.

Among Saturn's satellites, only Titan can be seen as a disk, even through large telescopes. Several visual observers had reported dusky shadings that seemed to vary with time, but such observations of the tiny disk (1.0 arc-second across) were suspect. Saturn's other satellites are seen only as points, though they have been noted to display variations in brightness with their orbital longitude; the most striking example is Iapetus, whose brightness during one orbit varies by a factor of six. In 1966 Audouin Dollfus discovered a small satellite orbiting just outside the bright ring system; later, after reexamining their 1966 observations, John Fountain and Stephen Larson reported still another satellite in the vicinity of Dollfus' object. Both satellites were confirmed in 1980 and found to be co-orbital. In March 1980, Jean Lecacheux reported the discovery of yet another faint

Figure 9. **Before moving on to Saturn, Voyager 2 captured this dramatic view of Jupiter's receding crescent. The Great Red Spot appears near the limb.**

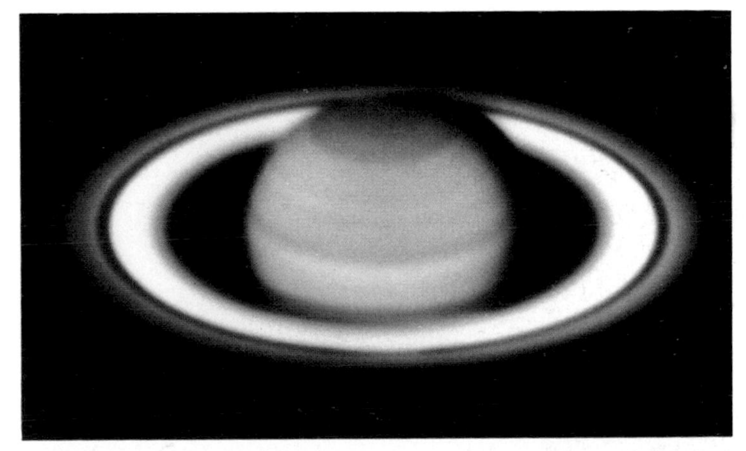

Figure 10. **Before Voyager, this was the best existing color photograph of ringed Saturn, taken by astronomer Stephen Larson in March 1974. At that time the planet's axis was tipped 27° toward Earth, allowing terrestrial observers to view the rings fully open. Despite this exquisite detail (the result of combining 16 separate transparencies), Larson's photograph gives no hint of the hundreds of fine ring divisions revealed by Voyager imagery.**

satellite near the Lagrangian point 60° ahead of Dione in its orbit. This brought to 12 the total number of known Saturnian satellites prior to the arrival of Voyager 1 in November 1980.

Saturn is observed, telescopically, to have three bright rings. The more conspicuous A and B rings are separated by the easily seen Cassini division; the "Crepe" or C ring is too faint to be recorded by the limited dynamic range of most photographic emulsions, but is not difficult for a visual observer. Several brightness minima in the A and B rings, most notably the Encke division, were reported by various visual observers, but these features had never been obvious photographically. The existence of the Encke division, however, was confirmed in 1977 during an eclipse of the satellite Iapetus by the Saturnian rings. In addition to the three bright rings, two fainter ones have been reported by ground-based observers: a faint band (subsequently designated D) interior to ring C was reported in 1969 by Pierre Guerin, and an even fainter ring extending outward from the bright rings to beyond the orbit of Dione was photographed in 1966 by Walter Feibelman. During three passages of the Earth through the Saturn ring plane in 1979 and 1980, several observing groups confirmed that this faint outer band (designated ring E) was real and that it extended nearly to the orbit of Rhea.

The Pioneer 11 encounter with Saturn on September 1, 1979, contributed greatly to our knowledge of the magnetic and charged-particle environment of Saturn and of the thermal, photometric, and polarimetric properties of the atmospheres of both Saturn and Titan. In most respects the imaging results, however, were disappointing. Pioneer's views of the cloud tops of Saturn showed the same banded pattern seen from Earth, but little else. No clearly identifiable discrete clouds were seen, dashing our hopes that a Jupiter-like spottiness might lie just below the resolution of the better telescopic photographs. Titan, the only Saturnian satellite recorded by Pioneer as anything more than a point of light, was frustratingly featureless. However, several new and intriguing features were evident within the ring system (Figure 11): we saw the Encke division clearly for the first time and discovered structure within the Cassini division. The spacecraft discovered a narrow ringlet, designated the F ring, approximately 4,000 km outside the outer edge of A. Although the actual width of the new ring appeared to be less than the images' resolution limit of 500 km, a hint of clumpiness was suspected. Not all of Pioneer's contributions to rings and satellites involved direct imaging, however; the instruments that measured charged-particle fluxes noted a number of minima that implied the existence of several as-yet-undetected satellites or rings located between the new F ring and Mimas, the innermost of Saturn's bright satellites.

Figure 11. **Pioneer 11 saw Saturn's rings from their unlit side, which created a kind of negative effect because sunlight filtered through each ring differently, according to the density of particles within it. Seen from left to right are the C ring (very bright), the nearly opaque B ring (almost black), Cassini's division (thin and bright), the A ring (edged in brown), and a faint, detached arc of the previously unknown F ring. At far right is a small satellite.**

THE APPROACH TO SATURN

With this and little more to prepare us, we awaited the Voyager 1 encounter with Saturn during the late summer and autumn of 1980. Systematic imaging was to begin on August 25th, 80 days before the November 12th encounter, but test frames taken in early summer had already exceeded the best images obtained with telescopes. Those of us interested in atmospheric dynamics were greatly concerned over the lack of discrete cloud features in the telescopic, Pioneer, and early Voyager images; without such cloud systems to track, it would not be possible to map the global circulation of Saturn's atmosphere. Fortunately, our fears were put to rest at the start of the approach phase; contrast-enhanced images revealed a subtle bright spot in Saturn's North Tropical Zone. Within a few weeks several more discrete clouds came to our attention, including a relatively large oval feature in the southern hemisphere that was first seen by experiment representative Anne Bunker. "Anne's Spot" was observed throughout the approach and bore certain similarities to the Great Red Spot on Jupiter. As resolution improved, more atmospheric features – bright and dark – became visible, and their morphology was encouragingly familiar. We had seen such clouds dozens of times only 16 months earlier at Jupiter, though these were extremely low in contrast. Our atmospheric enthusiasts relaxed, as computers on the ground safely recorded the data needed to define Saturn's global circulation.

Preliminary measurements, made within a few weeks of encounter, confirmed the extraordinary equatorial wind current that had been inferred from ground-based observations; its maximum speed was 1,800 km per hour, four times greater than Jupiter's equatorial jet. But this was only the first hint that Saturn's zonal wind system is strangely dissimilar to that of Jupiter. Saturn's equatorial jet (if "jet" is still the proper term) is very wide, more than 80,000 km, extending to 40° latitude. Poleward of 40° there appear to be several alternating easterly and westerly zonal currents, similar to those on Jupiter, but the relationship of maximum velocities to the locations of bright and dusky zones is decidedly different. Why should the zonal wind system at the cloud tops on Saturn be so different from that on Jupiter? To a first approximation both planets have the same dimensions, the same bulk composition, and similar cloud morphology; both have internal heat sources. The answer may be found in the physical structure of Saturn's interior, as was suggested by team member Andrew Ingersoll, or in the strong seasonal effects that are geometrically amplified by the presence of the ring system.

There are, of course, other atmospheric questions to be answered. Why are the clouds of Saturn so low in contrast and unvariegated in color? Saturn is, after all, a colorful planet (Figure 12); it's just that it is nearly all the *same* color – the *chromophores*, or coloring pigments, are well mixed. Is the clouds' low contrast an intrinsic characteristic or due to an overlying haze? The evidence seems to point toward

Figure 12. **Three weeks away from its encounter with Saturn, Voyager 2 had this view of its majestic target, 21 million km distant. The planet's surface is crossed by light and dark atmospheric bands. Within the rings are the now-familiar "spokes" (whose multiple images in this color composite show their motion between exposures). The icy moon Tethys is just below the planet; Dione is at bottom. Tiny Mimas appears to the upper left of Tethys, inside Saturn's disk.**

intrinsically low contrast, consistent with the idea that the chemistry within most of the convective cells is similar.

During the Jupiter approach phase, we were content to do little more than wait patiently for the accumulation of atmospheric dynamical data. With two Jupiter encounters behind us, the imaging team was even more relaxed as the spectacle of the ringed planet grew even larger on our television monitors (Figure 13). It was team member Richard Terrile who broke the quiet routine on October 6th with the discovery of dark, spokelike features extending radially outward across ring B and revolving around Saturn with the ring particles. A hastily prepared "movie" from early October's time-lapse sequence of Saturn's rotation gave a dramatic visual rendition of Terrile's findings. Working quickly with Voyager project personnel, we were able to modify a similar, higher-resolution sequence scheduled for October 25th, pointing the camera off Saturn and toward one ansa (tip) of ring B. The sequence was carried out flawlessly and the results spectacular. Although the data were everything that we could have hoped for, the interpretation of the spokes was to become a headache that would continue to persist even during the Voyager 2 encounter. Spokes were readily apparent in two specially designed time-lapse sequences made during the approach. As one frustrated team member commented, "I wish they'd just go away."

This was not the only surprise to come out of the approach phase; team member Andy Collins and Terrile, while tracking the motion of the co-orbital satellites, found two new satellites just outside and inside the F ring. Although very small (some 100 km across), it was immediately evident that these "shepherding" satellites play a crucial role in the Saturn ring system, stabilizing ring F against disruptive nongravitational forces. Just before encounter, Terrile found still another small satellite (20 by 40 km) orbiting only 800 km beyond the outer edge of ring A. This tiny object might be considered insignificant anywhere else, but it appears that in its unique location the satellite provides stability against outward diffusion for the entire bright ring system of Saturn.

Figure 13. **Press conferences during the encounters were often spur-of-the moment events prompted by the appearance of knowledgeable project personnel. Here chief scientist Edward Stone offers fresh insights to eager reporters in the press room.**

As the final days of the approach phase were upon us, anticipation grew. The satellites were finally being seen as individual disks and there was something very peculiar about the bright rings – they were daily showing more and more structure, far more than could be accounted for by simple satellite resonance theory. We were prepared for an exciting encounter, but even bigger surprises were to come.

THE ENCOUNTER

The Saturn encounter would be more intense than either of those at Jupiter; the compact system of satellites and rings and the rapidly changing viewing geometry meant that the most interesting data would all be transmitted within just a few hours of Voyager's closest approach to Saturn. Eighteen hours before passing Saturn, Voyager 1 would encounter Titan and cross over to the unlit side of the rings. Five hours after closest approach, the spacecraft would cross back to the illuminated side of the rings and look sunward on a back-illuminated planet with a phase angle of 150°. During this brief interval would occur the closest approaches to Titan, Dione, Mimas, the co-orbital satellites, and Rhea. The best views of Tethys, Enceladus, and the outer satellites would have to await the arrival of Voyager 2.

Our first brush with a Saturnian satellite was with Titan, considered by many to be one of the most important targets of the entire Voyager program. But after so many years of expectation, the imaging results were disappointing to even the most loyal Titan enthusiasts. On the evening before encounter, team members watched silently as dozens of images of the cloud-enshrouded satellite paraded across JPL's television monitors with monotonous, featureless repetition. Invoking various enhancement tricks via our interactive computer terminals didn't help much; the only details that emerged were a slight hemispheric difference in cloud brightness, displaying a curiously sharp demarcation at Titan's equator, and a global haze layer that thickened near the north pole into a polar hood; 10 months later Voyager 2 would find that the hood had evolved to a pole-girding collar. With none of the hoped-for holes in the clouds evident and no surface features to be seen, James Pollack wore the only smile – he was interested in Titan's atmosphere. (Later enhancement of the images revealed a series of low-contrast bands parallel to Titan's equator, giving the satellite an appearance similar to that of Jupiter and Saturn). Although the imaging results were a disappointment, other Voyager instruments had been reassuringly successful (see Chapter 14), and Titan still retains its high position on any list of interesting bodies in the solar system.

From the very beginning we knew that we might not see the surface of Titan, but now we were approaching the bright, icy satellites on whose surfaces we would surely see features, in some cases as minuscule as 1 km across. This group of bodies represents a class of objects never before encountered. Intermediate in size between the Galilean satellites and Amalthea, they are composed largely (in some cases, perhaps entirely) of water ice. After our experience at Jupiter, we were prepared for almost anything. What we saw turned out, at first glance, to be rather commonplace: in detail the icy surfaces of Dione and Rhea looked very much like the Moon or Mercury, though the morphology of the individual craters seemed slightly different. There were puzzling features, however. The poorly resolved hemispheres of both satellites showed bright, wispy streaks that might have been caused by ejecta from impacts, but some of the team members believed they could have been created by the condensation of volatiles leaking from the interior. Were these small bodies exhibiting internal activity? Furthermore, both Dione and Rhea had regions in which craters seemed to be too few in number, as though many had been obliterated or covered up. Was this still more evidence for internal activity at some time in the distant past? Mimas seems to be more normal – except for a giant impact crater (later named Herschel) that is more than a third the diameter of Mimas itself. As the image of Mimas and this absurd crater first appeared on our television monitors, there was a sense of *déjà vu*. Of course! It was George Lucas' "Death Star!" The resemblance (Figure 14) is uncanny. Aside from its strange appearance, Herschel's very size seemed improbable. It is difficult to understand how Mimas could have received such an impact and survived.

Iapetus, Hyperion, and Tethys, seen poorly by Voyager 1, were recorded at much higher resolution by the second spacecraft. Icy Tethys displayed an enormous impact feature, proportional in size to the large crater on Mimas, and a battered, irregular Hyperion was revealed to have a spin orientation and rate that was and remains completely unpredictable – a characteristic dynamicists term "chaotic rotation." Iapetus was seen well enough to show us one bright, heavily cratered hemisphere and another very dark, apparently featureless hemisphere. Unfortunately, the images did not show enough detail for us to rule out any of several competing hypotheses offered to explain the satellite's dichotomous character. The external mechanisms proposed by astronomers Sagan and Morrison, internal ones championed by geologists Masursky and Shoemaker, and hybrid models suggested by Soderblom all remain viable (see Chapter 15).

Enceladus also was poorly seen by Voyager 1, but not so poorly that its highly anomalous appearance escaped notice. At the limit of resolution (10 km) its surface appeared to lack topography of any kind! Since there is no way that Enceladus could have escaped the meteoritic or cometary bombardment suffered by its neighboring satellites, some mechanism must be destroying its craters at an astonishing rate. Enceladus is in an eccentric orbit forced by Dione, suggesting to team member Allan Cook that tidally dissipated energy, similar to that experienced by Io and Europa, is also responsible for the rapid resurfacing of this Saturnian satellite. Voyager 2, however, revealed a heterogeneous surface, cratered in some places but devoid of impact features elsewhere. The very existence of topography was a great relief to the team geodesist, Merton Davies, and the complex nature of the surface has delighted (if not confounded) the geologists.

Several observations were planned to observe the two co-orbital satellites, the Lagrangian companion to Dione, and two tiny new moons (recognized from the ground after the Voyager 1 flyby) that occupy the Lagrangian points preceding and following Tethys. The irregular shapes of the co-orbital satellites and the proximity of their orbits implies that they are now the remaining pieces of a single satellite that received an impact having rather more serious consequences than the one that formed the big crater on Mimas.

Meanwhile the rings continued to draw our attention. Even when Voyager was still several weeks away from Saturn, the three bright rings were starting to show far more structure than could be accounted for by a simple application of resonance theory (described in Chapter 12). Now, as resolution improved, the trio seemed to be breaking down into hundreds of individual ringlets. Some indeed were located where major resonances with satellites should occur, but they were everywhere else as well. In some regions, the structure appears to have organization, with a spacing pattern that follows some sort of arithmetic or geometric progression. Elsewhere the ringlets' spacing and widths are apparently random. In all, there are more than 1,000 ringlets at the resolution of the Voyager cameras and probably still more that we are unable to see. The Cassini division alone contains at least 100 individual ringlets. Some of these, like others found in ring C, are eccentric; even the outer edge of ring B is out of round. Eccentric ringlets within the bright ring system were not high on our list of expected phenomena; in fact, such a suggestion prior to their discovery would probably have been met with verbal abuse. Although the dynamic mechanisms responsible for some of the ring structure have so far been elusive, there now exists an enormous data base in which we hope to find the answers. It was team member Jeffrey Cuzzi who summed it up when he said that we now have far more information about the Cassini division alone than previously for the entire ring system.

In the final hours before encounter with Saturn our attention was fixed on the F ring. Resolution was improving rapidly, and the apparent decreasing width of the ring kept pace. Already we knew that it could not be more than 100 km wide and, furthermore, it was exhibiting a curiously nonuniform brightness that we thought might be due either to variations in width or particle accumulation. The most detailed images of the F ring to be obtained by Voyager 1 were received just 10 hours before closest approach to Saturn; those of us watching the monitors were stunned. If, by our third planetary encounter, we had become somewhat jaded to the unpredictability of the outer solar system, our sense of astonishment was brought back in an instant. In some tabulation of ring phenomena that we least expected to see, the observed structure of the F ring would have been somewhere off the top. Staring back at us from the television monitors were three individual strands, each approximately 20 km wide and separated by a few tens of km; they appeared to be knotted, kinked, and braided (Figure 15). To me, it was the most improbable picture yet sent back by either Voyager spacecraft.

The F ring's (or rings') apparent deviations from simple Keplerian motion must be due to complex gravitational interactions with the two shepherding satellites or to interactions with nongravitational forces such as those generated by charged particles moving in Saturn's magnetic field. The latter become important when particles are small, and the enhanced brightness of the F-ring when viewed under forward-scattering conditions suggests that a large fraction of its particles have diameters of only a few times the wavelength of light. Additional Voyager images have shown that the F ring's structure varies along the circumference and probably changes with time. At the moment, however, all of our explanations are likely to be little more than an exercise in vigorous arm-waving.

Without doubt the biggest surprises of the Voyager encounters with Saturn have come from the rings. Ironically, prior to the Voyager 1 encounter, it was with the rings alone that we really felt comfortable; the astonishing discoveries, we thought, were certain to be found elsewhere. As somewhat of an anticlimax we found the D ring, but it was extremely faint, too faint. This ring could not possibly have been seen from the Earth. Various observers (including me) had reported it, and Voyager had photographed it. But it was all just a coincidence – another of Voyager's ironies.

As the Voyager spacecraft left the Saturnian system, their cameras gave us memorable views (Figure 16). On December 19, 1980, the two cameras on Voyager 1 were turned off, probably forever. But the mission of Voyager 2 was not yet complete. Still ahead lay its lengthy journey to Uranus, an odyssey that took longer than the entire flight to Saturn.

URANUS: THE PRE-VOYAGER VIEW

Barely visible to the naked eye, Uranus was the first planet to be discovered within modern times. In 1781, Sir William Herschel, observing from his home in Bath, England, spotted the planet's tiny, fuzzy disk, thinking at first that he had found a comet or "nebulous star." However, the bluish-green object was quickly recognized as a new, trans-Saturnian planet. Averaging nearly 2.9 billion km from the Sun, Uranus receives only 0.27 percent of the sunlight that falls on our planet, and despite a diameter nearly four times that of Earth, its telescopic image is so small it could fit within that of Jupiter's Great Red Spot. Moreover, Uranus' disk appears virtually featureless in visible light.

It is, therefore, not surprising that we have learned precious little about this remote body since its discovery

Figure 14. **An enormous impact crater on Mimas (right), one of Saturn's inner satellites, gives the moon a surprising resemblance to the Death Star spaceship fron *Star Wars* (left).**

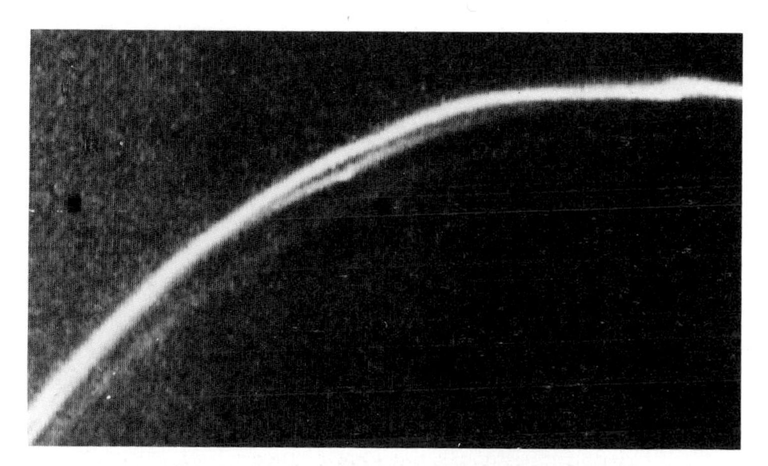

Figure 15. **The enigmatic F ring of Saturn, as photographed from its unilluminated side, shows two knotted and "braided" components, plus a more diffuse third strand. The narrow rings are about 10 km across and have occasional bright knots. These knots retained their integrity for 15 orbits and perhaps mark the location of the ring's very largest members.**

more than two centuries ago, and what we knew of it prior to Voyager's arrival can be summarized rather briefly. Spectroscopists had found a surprisingly large amount of methane in Uranus' atmosphere. Methane gas in such abundance selectively absorbs the longer (redder) wavelengths of sunlight, thus giving Uranus the dim, ghostly bluish-green appearance first noted by Herschel. Five satellites had been discovered, orbiting in planes tipped nearly at right angles to the ecliptic – a consequence of the large tilt of the planet's rotation axis to its orbital plane. Finally, Uranus curiously seemed to lack the large infrared thermal excess found in Jupiter, Saturn, and Neptune.

In the spring of 1976, I, along with James Janesik and Larry Hoveland, placed a new type of detector called a charge-coupled device (CCD) on the University of Arizona's 1.54-m telescope. CCDs have excellent near-infrared sensitivity, so we were able to examine Uranus through an optical filter transparent only to wavelengths near 8900 angstroms, where a deep absorption band of methane lies. But when we examined those first CCD images, we thought the telescope was out of focus – Uranus had the appearance of a doughnut with a dark center (Figure 17). Our astonishment ended when we realized that we were seeing thin clouds of ice or photochemical particles so high in the planet's atmosphere that they could reflect near-infrared sunlight back to space before it could be absorbed by the methane deeper down. Geometric projection made this particulate haze appear thicker toward the edge of the disk, a phenomenon known as limb brightening. In the regions not covered by haze, methane absorption caused the disk of Uranus to be blacker than coal.

In the decade thereafter, other observers and I struggled without success to achieve the combination of cloud asymmetry and observing continuity necessary to derive a rotation period for the planet. Furthermore, the clouds seemed to be getting thinner with time, and by the mid-1980s, as we looked almost straight down on the planet's south pole, they had nearly vanished entirely.

Meanwhile, in 1977 a team led by James Elliot had unexpectedly discovered a suite of narrow rings around Uranus while watching the planet occult a star (see Chapter 12). Five rings were apparent at first, but other astronomers observing later occultations brought the total to nine. The widest ring, called Epsilon, was found to be both eccentric and variable in width. Had it been made up of icy particles, as are the rings of Saturn, it would have been visible in our early CCD images. It was, but only marginally (Figure 18). I concluded, therefore, that Epsilon's ring particles had to be

***Figure 16*. Four days after its dazzling and flawless encounter with Saturn, Voyager 1 looked back on the planet from a distance of more than 5 million km. This view of a crescent Saturn – unobtainable from Earth – is but one of 20,000 pictures obtained by the spacecraft.**

very dark indeed, with a reflectivity of only a few percent. What could these rings be made of to appear so very dark? There was so much about Uranus that we did not know.

THE APPROACH TO URANUS

And so, during the final months of 1985, Voyager 2 was racing toward a planet whose rotation period was unknown, whose thinnest of clouds had nearly disappeared, and whose exceedingly narrow rings just happened to be jet black. The imaging team watched and waited with a collective emotion that could best be described as "anticipation well-mixed with anxiety."

Voyager 2 began to collect data systematically on November 4th, producing seemingly endless views of Uranus surrounded by the little white dots (its brighter satellites) but otherwise unblemished by any features of potential scientific interest. Only the long exposures taken for navigation purposes broke the monotony; Uranus itself was burned out in these intentional "overexposures," but the surrounding rings could be seen with ever-growing clarity. Cuzzi had calculated beforehand that long exposures would successfully record the rings, but it was nevertheless comforting to see them there in the images – even if only slightly brighter than the dark sky beyond.

There was certainly no comfort yet for Ingersoll, Beebe, and their atmospheric colleagues; the disk of Uranus continued to be as featureless as an airbrushed Ping-Pong ball. Boredom was setting in. Some early-arriving news reporters and a few of the imaging scientists argued the pointless question of whether the pictures proved that Uranus was bluish-green or greenish-blue. Color prints, of course, could provide only an approximate representation of the true color of the planet's atmosphere, but it was at least something to talk about.

For more than six weeks the greenish (bluish?), featureless eye of Uranus stared back at us. Then, in late December, the planet began to reveal its face. Computer-enhanced composite images started to show a banded pattern of clouds (Figure 19) not unlike a very-low-contrast version of the Saturn atmosphere. The south polar region of Uranus was dark, but it was surrounded by a broad cloud band with a distinct reddish-brown hue. From our unfamiliar perspective, viewing Uranus from above its south pole, the banded disk gave the appearance of a bloodshot bull's-eye.

We were at last seeing structure in the atmosphere, but to measure atmospheric wind speeds, one needed to discern individual clouds or groups of clouds. As early January slipped past, the frustration grew as the hoped-for features continued to elude us. Then, just 10 days before encounter, several discrete clouds popped out of the murky atmosphere. Too little time remained before the encounter to perform the kinds of comprehensive studies we would have preferred, but in the days that followed Uranus' winds were indeed measured and we learned how the planet's global circulation compares with those of Jupiter and Saturn.

But the euphoria of finding discrete clouds on Uranus soon gave way to perplexity: the winds appeared to be blowing the wrong way! Because of the orientation of Uranus' spin axis, the polar regions receive more sunlight on average than does the equator and should in theory be warmer (which proved to be true, according to Voyager's infrared sensors). Therefore, according to the "thermal wind equation" – the Golden Rule of atmospheric dynamicists – atmospheric winds on Uranus should slow down with increasing altitude. But instead we found that the clouds' tops were streaking out ahead of their bases, indicating that wind speed was *increasing* with altitude. The Uranian winds continued to confound the imaging team for several months before a satisfactory model emerged (see Chapter 11).

Since Voyager had added significantly to the satellite inventories of both Jupiter and Saturn, we fully expected that it would do no less for Uranus. However, we had not anticipated that the number of known Uranian moons would triple. The first was spotted on December 30th, and team member Robert H. Brown promptly dubbed it "Puck." (In keeping with established guidelines, Uranian satellites are named after characters appearing in the writings of Shakespeare and Pope.) Shortly thereafter, new moons began popping out quickly. One pair was found to be

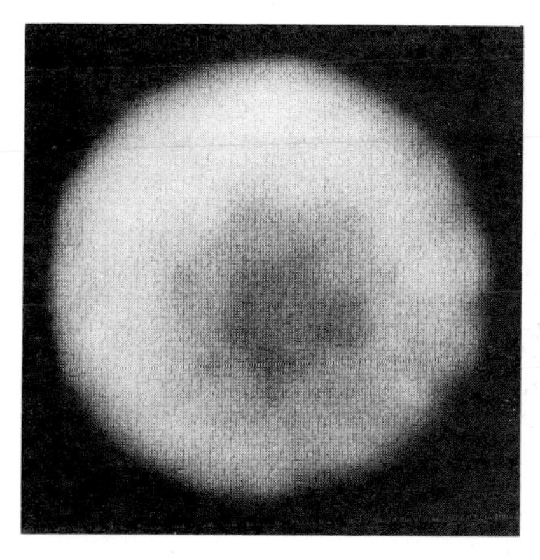

Figure 17. **Uranus, as seen from Earth at the infrared wavelength of 8900 angstroms. This image, taken in 1976, was the first known use of a charge-coupled device (CCD) on an astronomical telescope.**

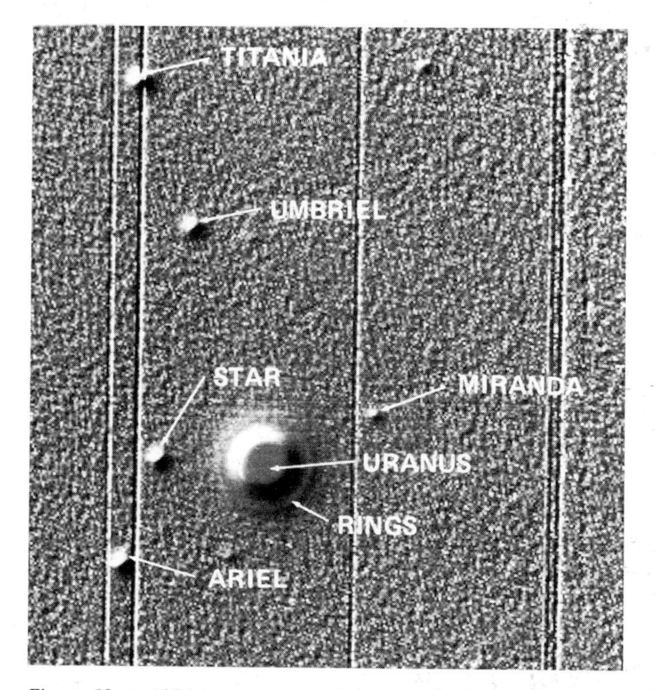

Figure 18. **A 1984 image of the Uranian system shows the planet surrounded by a faint band – its ring system. The individual rings were not resolved, but all five of the satellites know at that time were resolved. (Oberon lies beyond this reproduction's left edge). Vertical lines are missing data, and computer processing has added the three-dimensional "shadow" effect.**

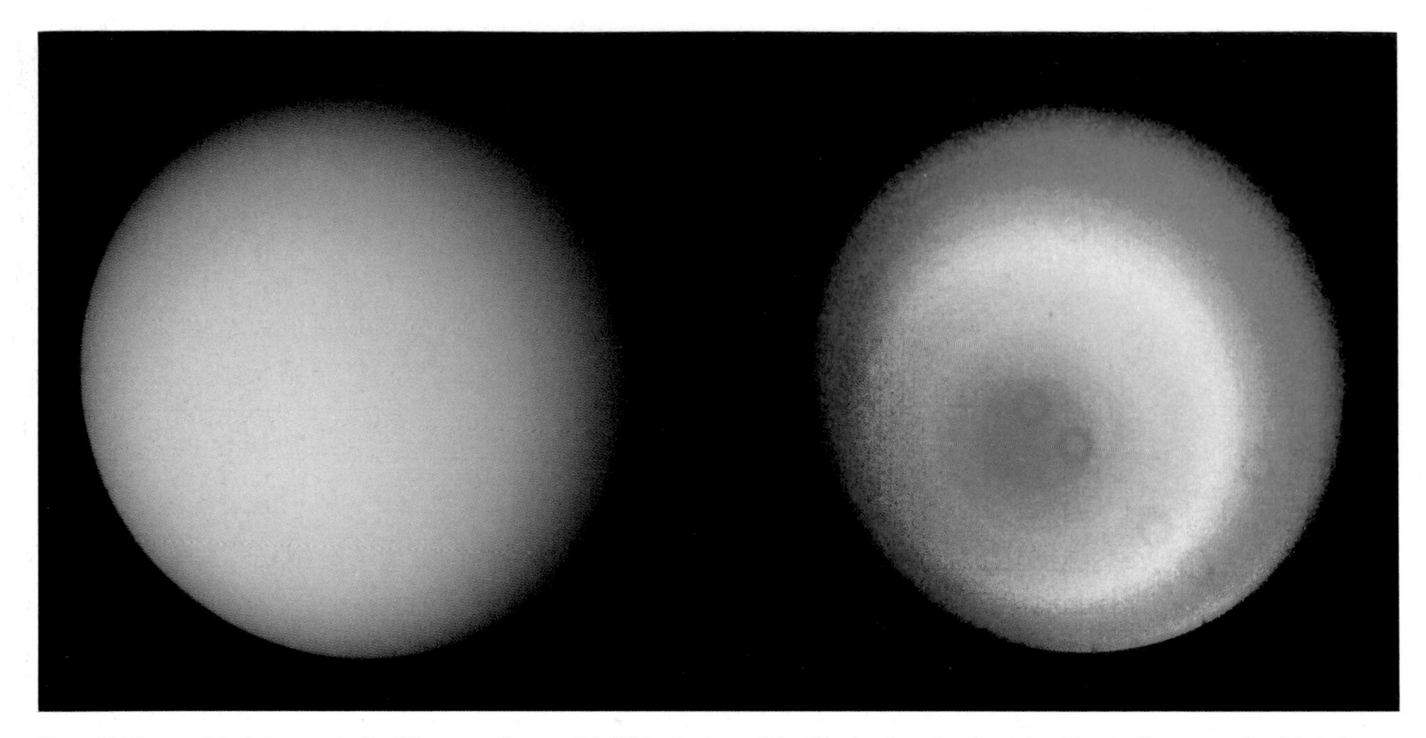

Figure 19. **Voyager 2 took these portraits of Uranus on January 17, 1986, when it was 9.1 million km from the planet. Each is actually a composite of three frames made through color filters. The planet appears at left as it might to the human eye; at right the color and contrast have been greatly exaggerated to bring out subtle details. The orange-hued polar region is covered by a thin haze of what may be hydrocarbon molecules like acetylene and ethane.**

Figure 20. **Craters dot the surface of Puck, one of several small moons discovered around Uranus by Voyager 2. This image, taken from a distance of 500,000 km, is the only close-up acquired of the 170-km-diameter satellite.**

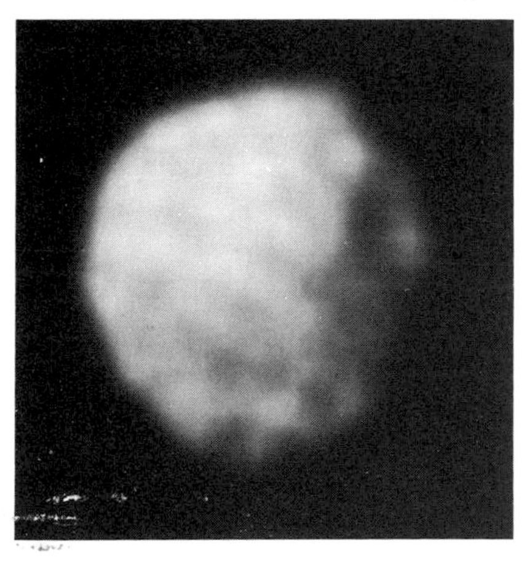

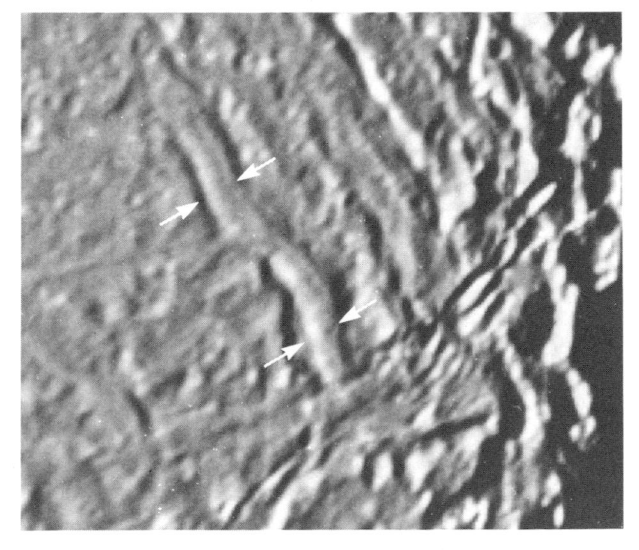

Figure 21. **According to researchers David Jankowski and Steven Squyres, this 20-km-wide canyon on Ariel (arrowed) is filled with solid ice that oozed up from cracks in the surface. This Voyager 2 image provides the first clear evidence for ice volcanism in the solar system.**

shepherding the Epsilon ring, thus creating a jurisdictional crisis between ring and satellite enthusiasts. Anxious to please all, Voyager 2 proceeded to find a tenth ring of Uranus.

Fortunately, Puck was discovered early enough to be targeted for closeup observations at the time of encounter (Figure 20). Like Saturn's Phoebe, it turned out to be very dark. With black rings and black moons, Uranus was generating its own individual mystique. But more was to come. Even from afar, the appearance of Uranus' five large satellites was quite different from our early views of the Galilean satellites and the icy moons of Saturn. Furthermore, each had characteristics distinct from all others. Umbriel, in the midst of the five, was the darkest, reflecting only a little more sunlight than Puck. Miranda, closest to Uranus, seemed to look stranger with each passing day (by the time of encounter, its peculiar appearance would truly defy description).

It should be noted that as Voyager had moved ever-farther from the Sun, the illumination on its target objects became correspondingly weaker. This demanded longer exposure times from our cameras, which, in turn, increased the smear caused by unavoidable spacecraft motion. Before Voyager 2 reached Saturn, JPL's engineers had taken a big step in solving this problem by commanding the whole spacecraft to rotate at just the rate needed to compensate for the apparent motion of the target whizzing by. This technique permitted us

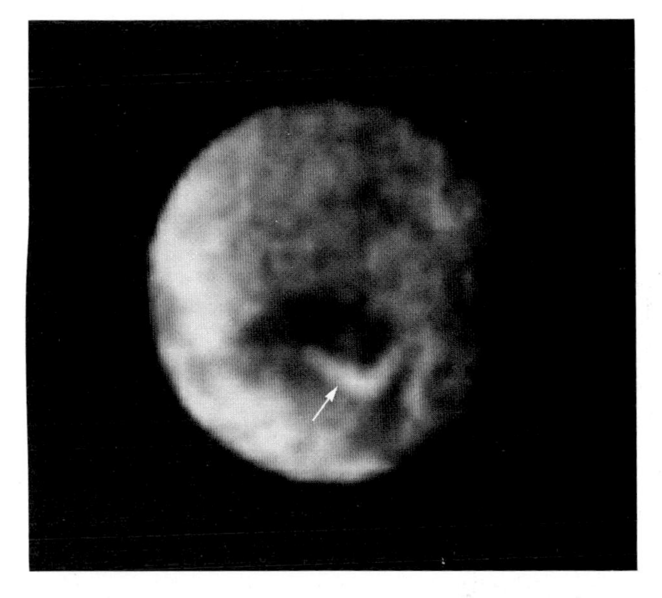

Figure 22. **From 1.4 million km away, Miranda showed Voyager scientists some strangely shaped features that were seen with much higher detail in later images.**

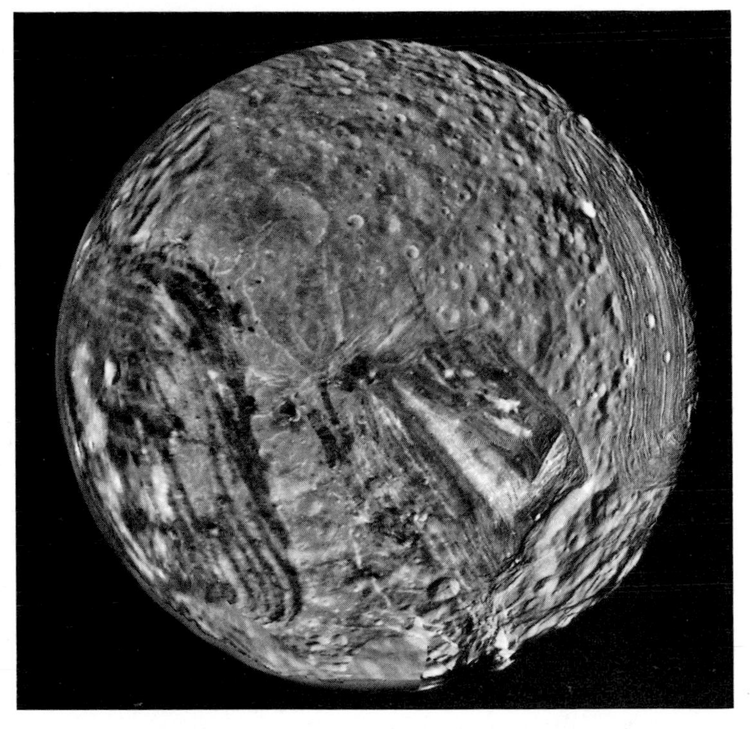

Figure 23. **Only 485 km across, Miranda exhibits a bizarre collection of terrains. The pattern to the lower-right of center has been nicknamed "the chevron," which lies at the center of a larger trapezoid-shape structure. Along the limb at left and right are "ovoids," concentric sets of ridges and grooves. These three major features perhaps represent different evolutionary stages of the same process. Miranda's tortured surface may also reflect the final settling out of material that reassembled after the satellite was shattered by a collision with another object long ago.**

to take a half dozen pictures of Dione with greatly reduced smear. However, the flyby of dimly lit Uranus was another matter entirely. To remain on course for Neptune as it passed Uranus, Voyager 2 would have to skirt very close to Miranda – so close that each image of the satellite would require its own compensation rate, an extraordinarily difficult task to perform with a spacecraft nearly 3 billion km away. But the JPL engineers, headed by William McLaughlin, set out to work miracles, and well before reaching Uranus the spacecraft was prepared to obey their complex commands. As a direct consequence of this effort, the resolution of our Miranda images would be the highest of any object since launch.

We continued to watch the dark rings from the sunward side of Uranus, but learned little in doing so. One early computer-enhanced color image of the rings produced some striking results and a brief flurry of speculation within the team. Carolyn Porco, however, warned us that the processing had been too hurried and that the results were almost certainly spurious. Later images proved that she had indeed been right; Uranus' rings showed no measurable color whatsoever. Several hypotheses and a few egos vanished when the color did.

Voyager 2's cameras did reveal some structure in the Epsilon ring, and other instruments were able to discern much finer detail by noting how the brightness of stars changed as they passed behind the rings. But our time would come. We knew that as the spacecraft had crossed over to the dark sides of both Jupiter and Saturn, forward scattering of sunlight suddenly brought into view the planets' very smallest ring particles, in much the same way that a photographer highlights a model's hair by placing a bright light behind her head. So if the Uranian rings had a sizable component of very small particles, the dark rings would light up like a halo when we looked back toward the Sun from the dark side of the planet. They did. Fine ring material seemed to fill the space between the rings and even extend inward from the innermost ring.

Team member André Brahic pointed out that such material would have a very short orbital lifetime, but that collisions between the larger ring particles might be a likely source of replenishment. At the time, none of us realized just how short those lifetimes would have to be. Observations by Voyager's ultraviolet spectrometer soon suggested that Uranus has a greatly extended hydrogen atmosphere and that, though very thin, it continues outward far enough to envelop the rings themselves. The resulting drag on the tiny particles made their lifetimes very short indeed – describable not in eons nor millennia nor even years, but just a few tens of days! So in order to maintain the rings at all must involve a lot more grinding among the larger particles than Brahic had at first thought. The ring group had developed its own dynamical headache.

Meanwhile, our geologists had no reason to feel complacent. The five largest satellites were too small, we thought, to have experienced any significant surface modification due to internal processes. But as Voyager closed in and image resolution continued to improve, it began to look as though the satellites had not enjoyed a very quiescent history. Oberon was showing evidence that something very black had oozed out onto the floors of some of its impact craters, and Titania appeared to have suffered even more from internal activity. Nearer the planet, Ariel also showed clear signs of some kind of liquid having flowed on its surface (Figure 21). Old theories were rapidly being replaced by newer ideas, which themselves did not last but a few hours.

Then the long-awaited, highest-resolution images of Miranda came in, and everything began to unravel.

Miranda was a geologic *bête noire*. With each new picture, we saw new and often unfamiliar types of terrain. Our team geologists, temporarily at a loss for appropriate terminology, were resorting to such highly technical terms as "chevron," "race track," even "layer cake" to describe the various geological features exhibited by the satellite (Figures 22, 23). As the pictures were hastily assembled into a global mosaic, we became aware that Miranda had various regions resembling just about every solid body we'd seen in the solar system, along with a few twists none of us had *ever* seen. And we all knew that, in just a few hours, Soderblom would have to get up before several hundred members of the press to explain the whole thing with authority and conviction. Only later – much later – would come erudite explanations of a body frozen in a partial, but incomplete, state of differentiation. But, for those who were there, the memory of those outlandish images as they first appeared on our television monitors will remain one of the outstanding highlights of the American space program.

VOYAGER AT NEPTUNE, 1989

On August 24–25, 1989, Voyager 2 swept through the Neptunian system for its fourth and final planetary encounter. Thanks to Earth-based observations made before the distant rendezvous, scientists had expected the planet to show more atmospheric activity than neighboring Uranus. But as the spacecraft closed in, Neptune's atmosphere proved surprisingly active. It displayed zonal bands and giant storms reminiscent of Jupiter. This dynamic climatic activity is probably powered in part by Neptune's internal heat, which provides over twice as much energy as the feeble sunlight that the planet receives.

Equally surprising was the planet's electromagnetic environment. Although relatively weak, Neptune's magnetic field is inclined to the axis of rotation by 47°, making the surrounding magnetosphere remarkably similar to that of Uranus. Voyager 2 also recorded periodic pulses of radio energy, which imply that Neptune's deep interior is turning once every 16 hours, 7 minutes.

Voyager confirmed that Neptune has a ring system – not the partial "ring arcs" that Earth-based occultation data had suggested, but a set of complete (if tenuous) rings relatively near the planet. Voyager also spotted six small moons that range in diameter from 50 to 400 km. Not seen, however, were any of the numerous "shepherding" moons believed to be orbiting near or within the rings. Distant little Nereid was not seen at close range but appeared unremarkable when the spacecraft looked its way.

Voyager's encounter climaxed with a close (40,000-km) approach to Neptune's largest moon Triton. The surface of this icy world proved dazzlingly complex. Its crust appears to have been shaped by widespread eruptions of fluid that probably began in the moon's infancy. Because of Triton's strange orbit – retrograde and well inclined to Neptune's equator – many scientists believe that this satellite was originally a huge planetesimal that strayed too close to the planet and was captured by its gravity. Then, as Triton's orbit evolved from elliptical to circular, the resulting tidal stresses could have completely melted the moon's icy interior and triggered voluminous outpourings of water at the surface. There are even indications that nitrogen-driven volcanism continues there today.

The Neptune flyby brought a fitting climax to Voyager 2's historic 12-year trek through the outer solar system, and the following pages display some photographic highlights from that remarkable scientific harvest.

Voyager 1 and Voyager 2 are now both sailing outward to interstellar space. Their cameras will never again capture the magnificence of another Jupiter or Saturn or Neptune, but they will continue to send us information about the interplanetary medium. Voyager 2, which culminated in the spectacular encounter with Neptune in 1989 is widely regarded as the most successful planetary mission ever undertaken.

***Figure 24.* Speeding away from Neptune following its historic encounter in August 1989, Voyager 2 recorded this view of the planet's southern hemisphere as a slender crescent.**

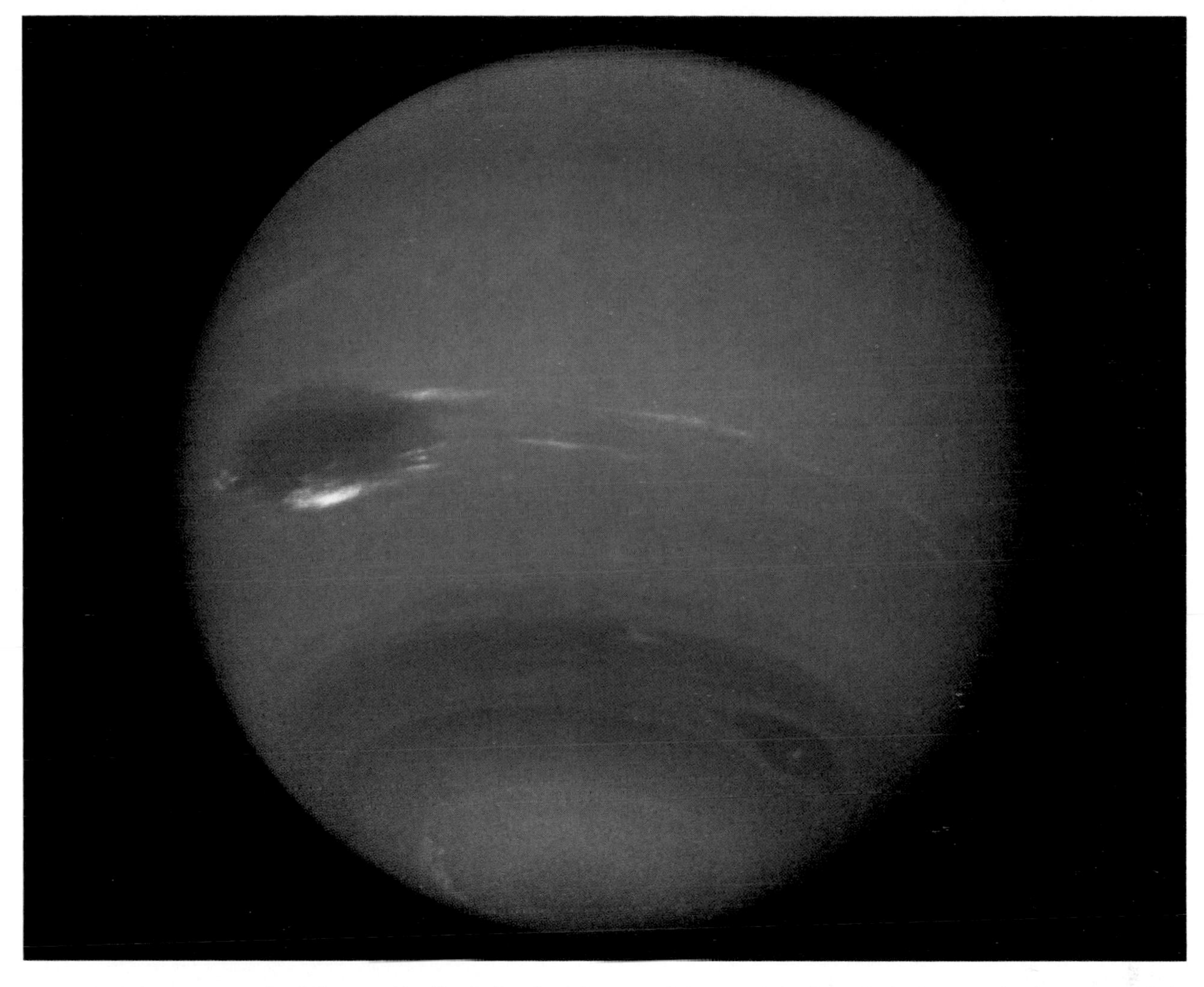

Figure 25. A Voyager portrait of Neptune. The Earth-size cloud feature at left, named the Great Dark Spot by scientists, is overlain by bright clouds of methane ice crystals. A second dark oval appears at lower right within a dark band of clouds that rings the planet's south pole. Neptune's blue color results from absorption of red light by the small amount of methane in its atmosphere, which is dominated by hydrogen and helium.

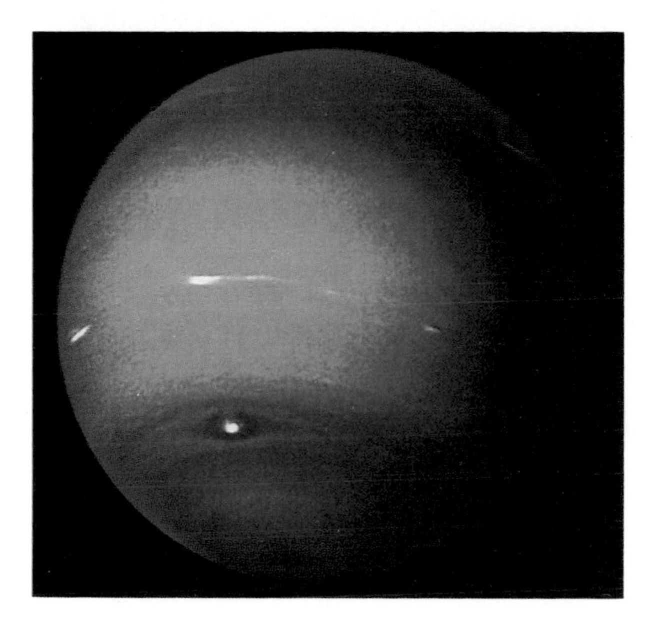

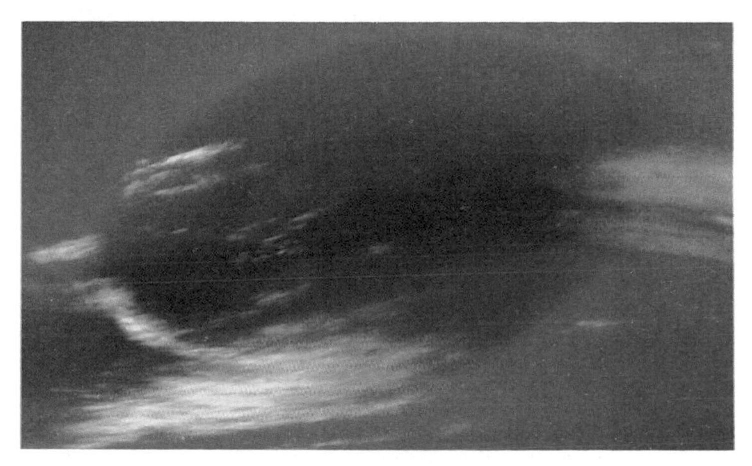

Figure 26. Above: The Great Dark Spot's pinwheel edge and the distribution of overlying clouds suggest counterclockwise rotation. *Left:* A false-color view of Neptune, made by combining images taken through blue, green, and methane-band filters, shows high-altitude clouds (white) and a planet-wide translucent haze (red) that lie above most of the methane.

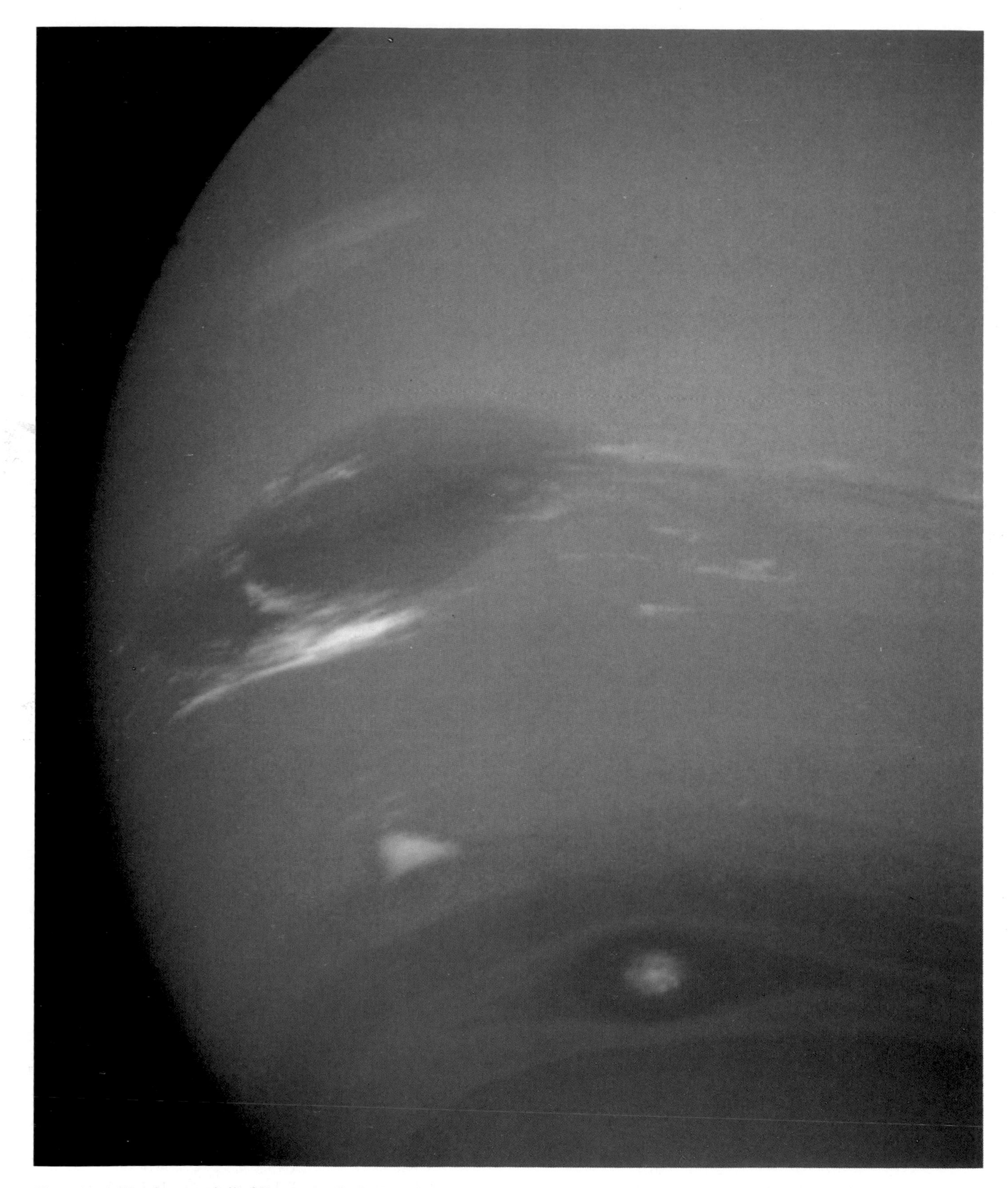

Figure 27. A "family portrait" of Neptune's principal cloud features. These spots and streamers move around the planet at different rates because wind velocity varies greatly with latitude. The Great Dark Spot, left of center, moves around the planet every 18.3 hours at a latitude of 22° south. The winds around it blow westward at an amazing 2,200 km per hour with respect to the planet below them. At bottom center, near 55° south, is the eyelike Small Dark Spot, or "D2." It takes only 16.0 hours to circle Neptune, which virtually matches the interior's rotation rate. In between, at 42° south, is an isolated cirrus-type cloud patch dubbed "the Scooter," because its 16.8-hour trips around Neptune are faster than those made by other bright cloud features. As Voyager 2 watched, the Scooter's shape changed from round to square to triangular.

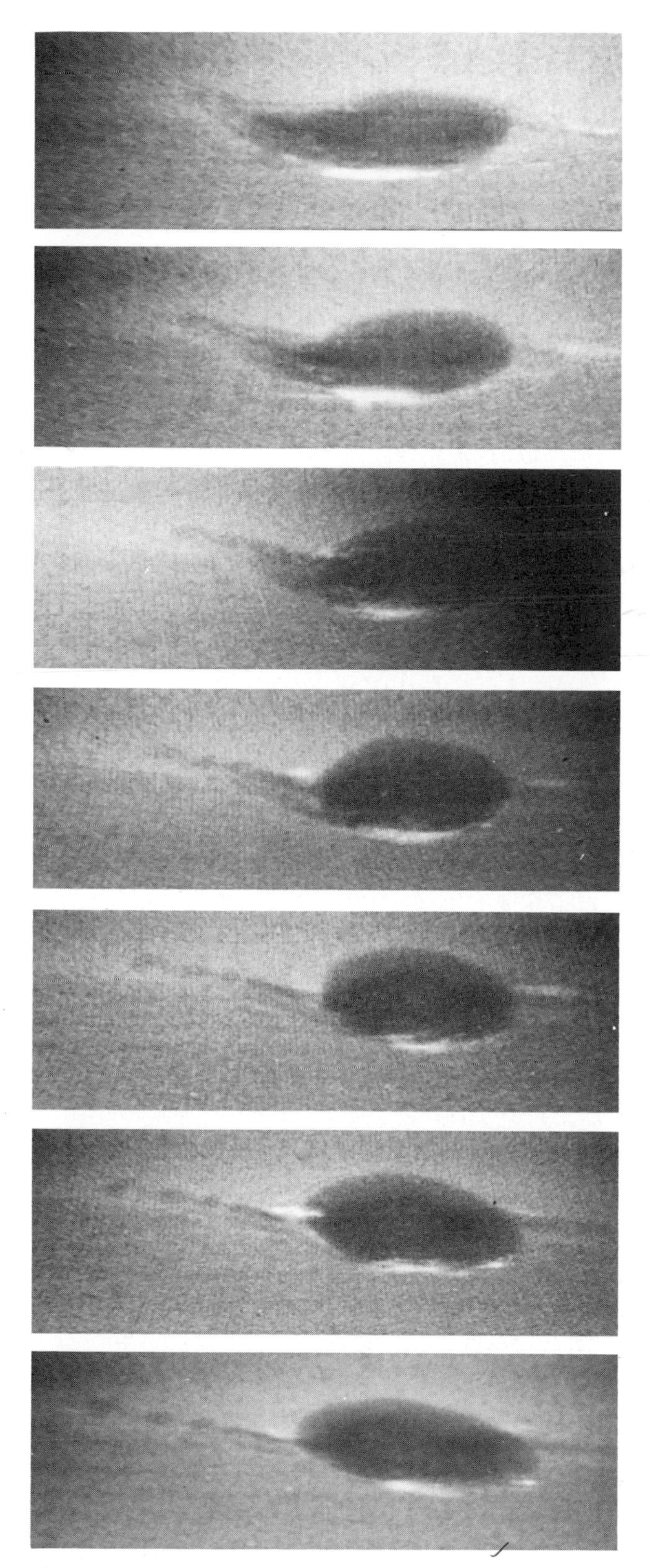

Figure 28. This sequence shows changes in Neptune's Great Dark Spot that occurred over a 4½-day period. A dark extension on the spot's western (left) edge breaks up into a string of smaller features. Despite the spot's rotation, bright clouds above its southern edge do not change position over time.

Figure 29. A close-up of Neptune's Small Dark Spot shows discrete features down to 20 km across. Bright clouds have formed atop an upwelling in the spot's center. Because this feature appears to rotate clockwise, in contrast to the Great Dark Spot, material within the dark oval is inferred to be descending.

Figure 30. High-altitude cirrus clouds at Neptune's terminator cast shadows on the vast blue cloud deck some 50 km below. The clouds are between 50 and 200 km wide. Voyager 2 recorded this scene just 2 hours before passing a mere 4,905 km from the planet's cloud tops. Because sunlight at Neptune is about 900 times dimmer than at Earth, Voyager's cameras utilized exposure times that ranged from a few seconds to 15 minutes. For close-ups like this one, the spacecraft or its cameras had to pan precisely to prevent the images from being hopelessly smeared.

Figure 31. A composite portrait of Neptune's ring system. Two main rings orbit 38,000 and 27,500 km from the planet's cloud tops. A diffuse sheet of material at least 4,000 km wide extends outward from the inner of these two rings. There is also a very tenuous third ring 17,000 km from Neptune. The outer two rings were most clearly visible in backlit views such as this, indicating that they are dominated by very small (micron-size) particles. The rings' narrowness probably means that unseen ''shepherding'' satellites are orbiting among them and controlling the rings' extent with their weak gravity.

Figure 32. Above: Voyager 2 photographed the curving shadow that Neptune cast on the rings, which does not quite extend to the outermost ring. In this view the diffuse sheet located between the two main rings displays a distinct brightening along its outer edge. *Left:* Before Voyager 2's encounter, astronomers had detected what they thought might be partial rings around Neptune, or ''ring arcs.'' This image shows that the suspected features are actually clumps of material within complete rings. Each of the three clumps within the outer ring is 6° to 8° in extent. Neptune's highly overexposed crescent is at lower right.

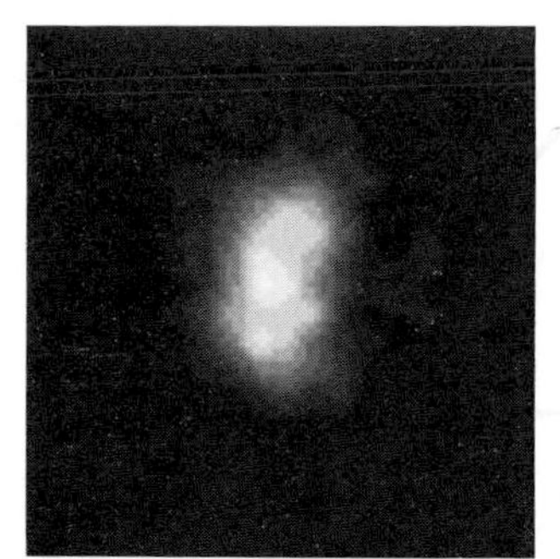

Figure 33. **Voyager 2 came no closer than 4.7 million km to Nereid, the smaller of Neptune's two previously known satellites. This image reveals that Nereid is about 340 km across and relatively dark. The Sun is to the left, so the moon's featureless disk appears only half illuminated.**

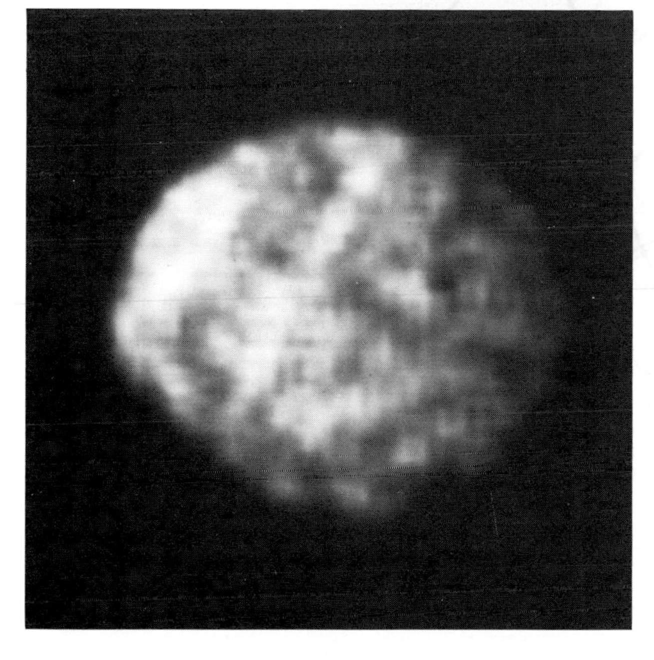

Figure 34. **Voyager found six previously unknown moons circling Neptune. The largest, designated 1989N1 *(above)*, is 400 km across; 1989N2 *(left)* is roughly half that size and was probably spotted from Earth during an occultation in 1981. Both moons are very dark and heavily cratered.**

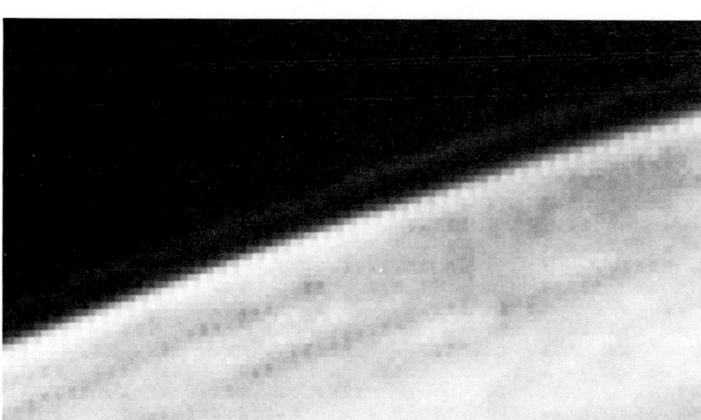

Figure 35. **Triton proved to be somewhat smaller than scientists expected, 2,705 km across. Voyager detected an exceedingly rarefied atmosphere consisting of nitrogen and a trace of methane, and having a surface pressure 70,000 times lower than Earth's. At roughly 38° K, Triton's surface is the coldest known place in the solar system. Triton's complex seasons are described in Chapter 15. *Above:* Voyager photographed a layer of haze over Triton that may consist of ice particles or photochemical smog. *Right:* Much of Triton's southern hemisphere, where it is now summer, is coated with a polar cap of very bright frost. The cap's pinkish hue may result from alteration of methane ice by cosmic rays into complex hydrocarbons. Near Triton's equator, which runs roughly along the terminator, is a darker region that probably is an exposure of the water-ice crust. The bluish cast (exaggerated here) may be due to the preferential scattering of blue light by microscopic particles of frost.**

Figure 36. A mosaic of 14 images of Triton. At left is the moon's bright south polar cap, which is crisscrossed by somewhat darker streaks. To the right, beyond the cap's margin, is Triton's equatorial region; the crust displays a stunning variety of terrains, indicating a complex history of volcanic activity. Specific features are shown in greater detail on the opposite page.

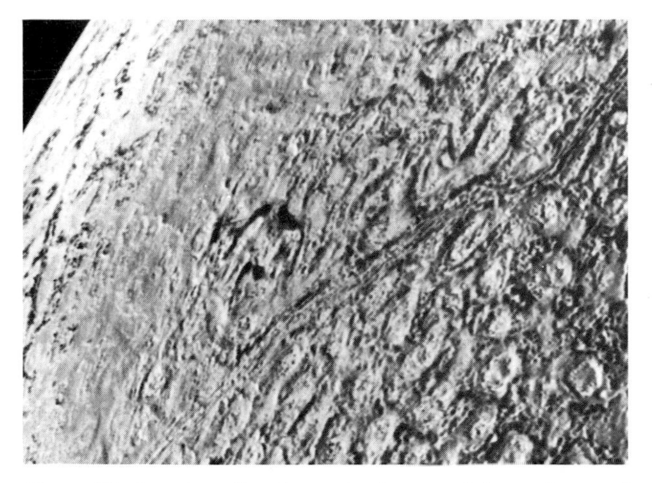

Figure 37. A region of curious, closely spaced depressions and ridges, nicknamed "cantaloupe terrain," may have formed by repeated episodes of localized melting and collapse of Triton's icy crust. Impact craters are scarce, indicating a surface younger than a few billion years old.

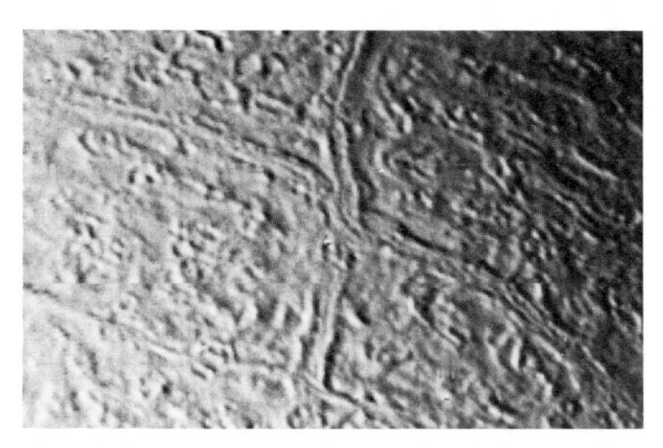

Figure 38. Vast cracks in Triton's crust, some 35 km across, are filled with relatively fresh ice that welled up from the interior.

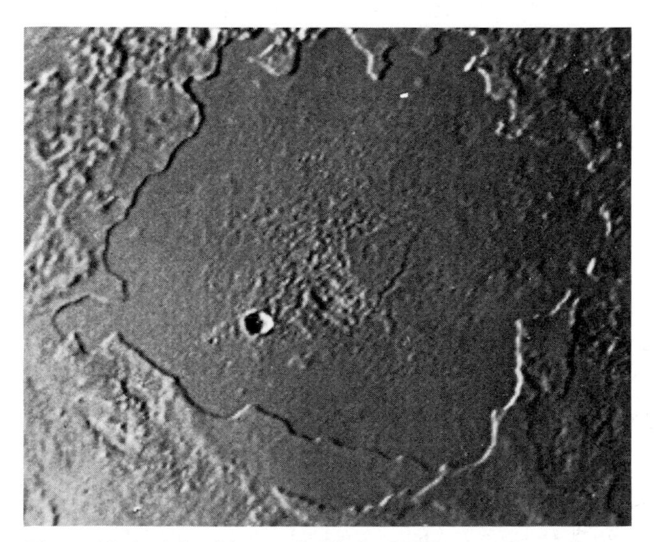

Figure 39. A plain of ice, probably formed by eruptions of water or a water-ammonia slurry, fills what may be the remains of an ancient impact basin. Ledges at the basin's margin may be remnants from earlier volcanic outpourings, and the outlying terrain has also apparently been modified by volcanism.

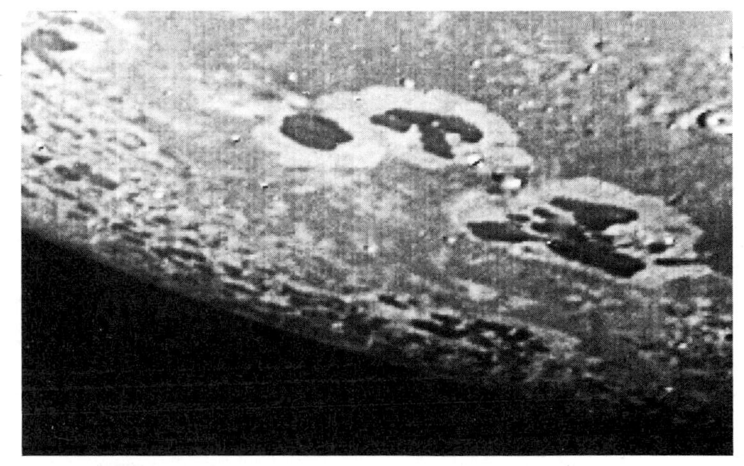

Figure 40. Strange features with dark centers and bright collars lie in one of the most heavily cratered areas of Triton. The largest clump of features, at right, is about 320 km across.

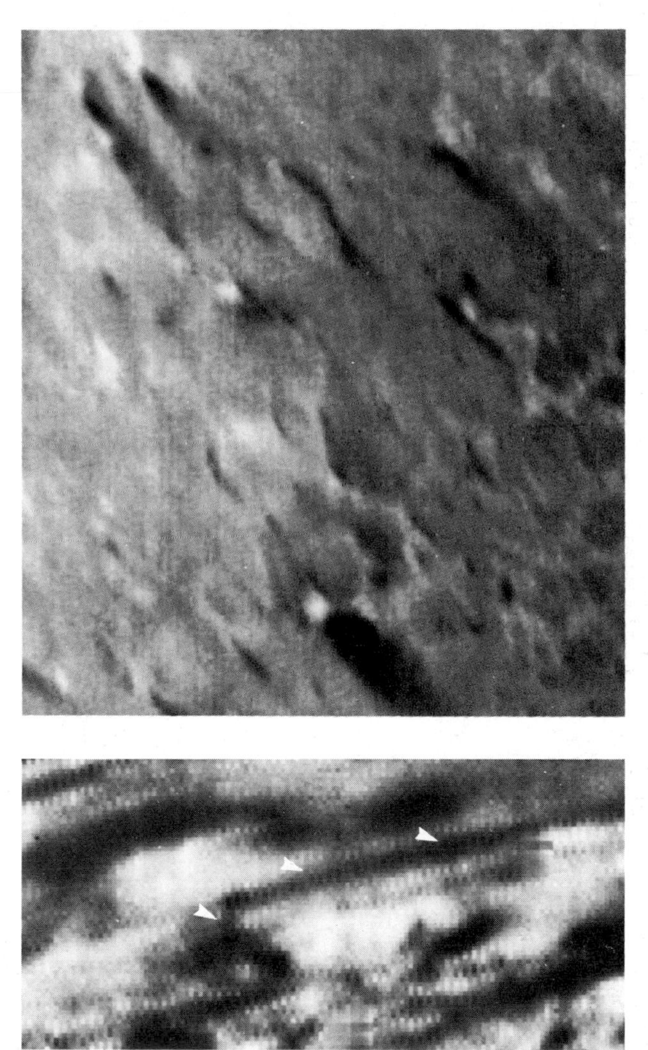

Figure 41. Top: Dark streaks superimposed on the south polar cap emanate from what may be small volcanoes. The streaks, up to 75 km long, may consist of nitrogen frost mixed with hydrocarbon-rich material that has been ejected by explosive eruptions. *Bottom:* Triton, like Jupiter's moon Io, is still volcanically active. A dark, geyser-like plume (arrows), seen in profile, rises vertically 8 km and forms a wind-blown trail 150 km long.

Figure 42. **A little more than three days after its encounter with the Neptune system, Voyager 2 looked back at Triton's slender crescent rising beyond the planet's limb.**

10

Interiors of the Giant Planets

William B. Hubbard

ALL TOGETHER, the five worlds closest to the Sun, the terrestrial planets and the Moon, amount to only two Earth masses of material. In contrast, the four giant planets – Jupiter, Saturn, Uranus, and Neptune, along with a small contribution from their numerous satellites – represent 445 times the Earth's mass! Jupiter alone amounts to 318 Earth masses, and Saturn another 95. Uranus (14.5 Earth masses) and Neptune (17.2) complete the known inventory of major bodies. Pluto's mass is not great, but undetected cometary material orbiting far beyond it could possibly represent a great deal of matter as well. The four giant planets likewise greatly surpass the Earth in size (Figure 1). They can be logically divided into two subclasses, with Jupiter and Saturn forming one like-sized pair, and Uranus and Neptune a second, smaller pair. One key objective in studying these planets' interiors is to understand how sharp chemical divisions arose between the two types of giant planets, and how these are expressed through the great differences in mass and size.

As we shall see, the giant planets differ further from the Earth in their chemical make-up. To begin, consider the average composition of the Sun not as it is today, but as it was when the solar system was just forming, before nuclear reactions had begun in the Sun's core. Further, let this protosolar matter be at 100° to 200° K, like the cold temperatures that prevailed in the outer solar system where the giant planets formed.

Under these circumstances, pairs of hydrogen atoms combine to form molecules of gas, whereas hydrogen exists as fully ionized protons and electrons throughout most of the Sun's interior, which is at temperatures of many millions of degrees. After hydrogen, the second most abundant constituent is helium, present in atomic form as a noble gas. (So are other noble gases such as neon and krypton, but in such low abundances as to contribute insignificantly to the total mass.) By number, the cool nebular gas contains about six hydrogen molecules for every helium atom; by mass, these two elements contribute 74 and 24 percent, respectively. Thus, hydrogen and helium together account for a little more than 98 percent of the mass of this primordial gas (Figure 2).

Other elements are present, such as carbon, nitrogen, and oxygen. Because of the low temperatures, these three atoms bond with the readily available hydrogen to form methane (CH_4), ammonia (NH_3), and water (H_2O), respectively. In the language of the builders of model giant planets, these are termed "ices," because within a cold nebula they condense to their solid phases. As Figure 2 shows, taken together they dominate the 2 percent of the condensable matter from which giant planets could initially form. It is possible that some of these molecules exist as solids and others as gases in the cold giant-planet zone. Initially, some carbon might even be found in carbon monoxide (CO) and carbon dioxide (CO_2) rather than methane. Most of the remainder of the planet-building matter, the metal-and-silicate "rock" out of which the terrestrial planets formed, represents only about a quarter of the condensable nebular material.

***Figure 1.* A size comparison of Jupiter, Saturn, Uranus, Neptune, and Earth. The rotational flattening of these planets is also shown to scale. Note how nearly identical Uranus and Neptune are in size.**

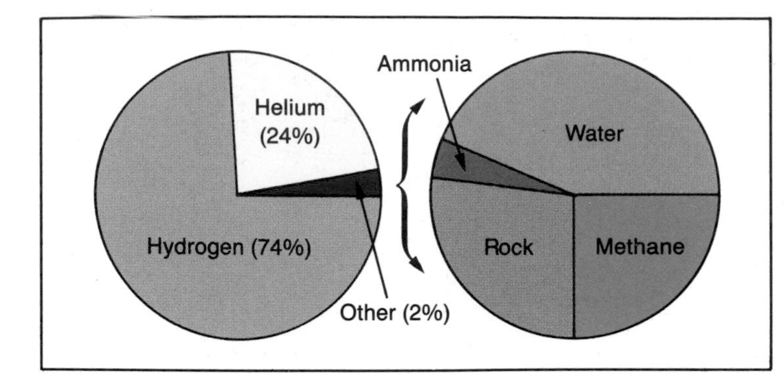

Figure 2. **The distribution of molecules, by mass, in the cool gas of the primordial solar nebula. Hydrogen and helium, the most dominant and volatile components, account for 98 percent of the nebula's mass (upper circle). Yet these gases were never in a condensed state in the primordial solar system. Condensable solid materials (right circle) account for only 1.5 to 2 percent of the solar system's mass.**

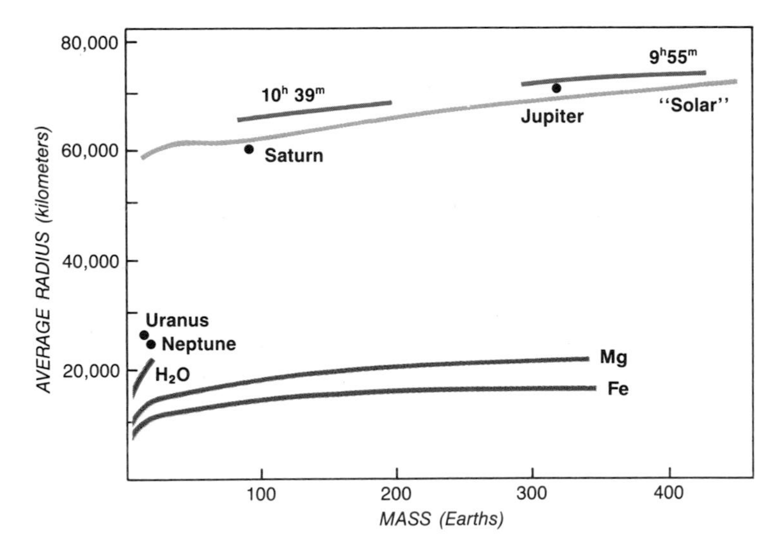

Figure 3. **Theorists can calculate just how big Jovian planets would be assuming various masses and possible compositions. The longest curve is the radius-mass relation predicted for a liquid body of solar composition with its interior in equilibrium. The short arcs, labeled with the rotation periods of Jupiter and Saturn, show how much spinning planets having this composition would be flattened at their equators by centrifugal force. At the bottom are curves for pure magnesium and iron ("terrestrial") planets, as well as a curve for liquid water based on its experimentally determined characteristics under pressure. Dots show the actual values for the four giant planets.**

Moving outward from the Sun, the giant planets prove to be progressively less-complete samples of this primordial nebular matter. Jupiter resembles a large aggregation of material close to solar composition, highly compressed but at much lower temperatures than in the Sun. Saturn is similar, except that more of the primordial hydrogen and helium is missing. Uranus and Neptune appear to be primitive samples of the nebula's ice and rock, but they possess only a small fraction of the primordial allotment of hydrogen and helium. In contrast, the planets of the inner solar system retain only the rocky component of the initial solar composition; the gas and ice are almost entirely missing.

These assertions about bulk composition can be verified most directly by comparing the giant planets' radii with their masses. Figure 3's lower two curves, for pure magnesium and iron, are similar to the curves that would apply to planets with Earthlike composition but masses comparable to the Jovian planets. Such mythical giant terrestrial planets would be much smaller than the actual Jovian planets. The figure's upper half shows how the solar-composition curve shifts when allowance is made for the expansion of the equatorial radius due to the planet's rotation (Jupiter spins once every 9^h 55^m and Saturn once every 10^h 39^m).

	Equatorial radius (km)	*Polar radius (km)*	*Oblateness* $(R_e - R_p)/R_e$
Jupiter	71,492	66,854	0.0649
Saturn	60,268	54,364	0.0980
Uranus	25,559	24,973	0.0229
Neptune	24,800	24,300	0.020

Table 1. **Radii and oblateness of giant planets. Since these objects have no solid surfaces, their radii are given here at the 1–bar pressure level.**

Intuition suggests that as planets grow more massive, they should also get substantially larger. But, as Figure 3 shows, the increase in size with mass is actually quite slow for huge objects like the giant planets. Specifically, although Jupiter is more than three times more massive than Saturn, it is only slightly bigger. This behavior is due to the very great compression at work deep in these planets' interiors. Pressures at their cores, in excess of 10 million bars, cause the molecular and atomic structure of matter to break down, and the compression curve "softens." Thus, as more matter is added to a large planet, its interior begins to collapse somewhat at the atomic level, and its radius increases more and more slowly. As we shall see, eventually the radius actually *decreases* as the mass increases, such that there exists a maximum radius for a cold body of a given composition. For an object with solar composition, this maximum is about 80,000 km, and it is reached at a mass about three times that of Jupiter. The Sun exceeds this maximum radius by nine times and contains 1,000 times the mass of Jupiter. But the radius-limiting relationship does not apply to the Sun, which is a hot, gaseous body inflated primarily by its very hot interior.

The wide separation in Figure 3 between the curves for solar composition and those for other abundant materials shows, as Wendell DeMarcus concluded in 1958, that Jupiter and Saturn are indeed composed mostly of hydrogen and helium. Conversely, Uranus and Neptune can contain little hydrogen or helium, for their radii are much too small to be consistent with solar composition. But since Uranus and Neptune lie slightly above the curve for water, it is not unreasonable to suppose that water contributes substantially to their content.

OBLATENESS AND THE INTERIOR

Radius-mass relationships may give important clues to the bulk compositions of the giant planets, but they do not tell us whether dense cores are present, nor do they reveal interior structures in more detail. Such information comes from the way these worlds accelerate natural and artificial satellites in their vicinity. From a gravitational standpoint, a perfectly spherical planet would act as if all its mass were concentrated in a single, central point. In such cases the motions of a

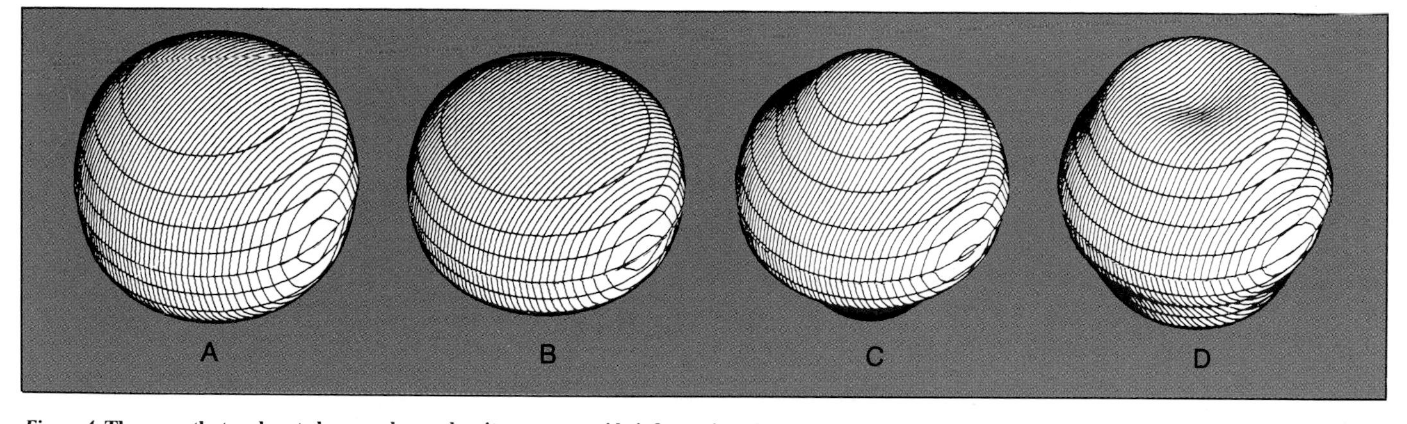

Figure 4. **The ways that a planet changes shape when it rotates provide information about its interior structure. These responses can be quantified by numerical values called harmonic coefficients. When not rotating, *(a)*, a planet will be perfectly spherical. Saturn spins very rapidly, and its distortion due to the harmonic coefficient J_2 is shown approximately to scale in *(b)*; Saturn's J_4 distortion is exaggerated by about 10 times in *(c)*, and its J_6 distortion is exaggerated by about 100 times in *(d)*.**

nearby satellite or spacecraft would obey the classical laws of Kepler.

But to be perfectly spherical (Figure 4a), a planet must not rotate and its interior must be in a state of *hydrostatic equilibrium*, in which all forces and pressures are in balance. Once the planet begins to rotate, its shape and the shape of its external gravitational potential, or field, deform and become oblate (Figure 4b). Astronomers express the extent of the oblateness as a value, J_2, called a harmonic coefficient.

Since all of the giant planets rotate very fast, more rapidly than the Earth does, they are decidedly *not* spherical. Their liquid interiors (discussed more fully later) respond readily to the strong centrifugal forces induced by rotation, producing substantial and rather obvious equatorial bulges (Table 1). The size of the bulge, when compared to the magnitude of the rotational perturbation that causes it, gives us insights into the distribution of density in the planet's interior.

In addition to the basic response J_2, rotation induces higher-order perturbations on the planet's figure and gravitational potential, which are called J_4, J_6, and so on. These progressively more-delicate distortions reflect the distribution of mass and density in the planet's outer layers, but they can only be detected and measured by "gravity probes" very close to the planet's surface. The close flybys made by the Pioneer and Voyager spacecraft have provided such data, as have the elegant studies of the motions of the rings of Saturn and Uranus carried out by Peter Goldreich, James Elliot, and their students and associates. To date the values of J_2, J_4, and J_6 have been determined for Saturn, and the planet's true figure is derived from the sum of these distortions.

Analyzing the harmonic components of the giant planets' gravity fields helps us to understand the relationships between density and pressure in their interiors. But to interpret these relationships properly, we must also understand the behavior of one of the main constituents, hydrogen, at various combinations of temperature and pressure (Figure 5).

In the observable regions of giant-planet atmospheres, at a pressure of about 1 bar, hydrogen forms a molecular gas at temperatures ranging from 165° K (Jupiter) to 135° (Saturn) to 76° (Uranus). Theory and observations show that deeper

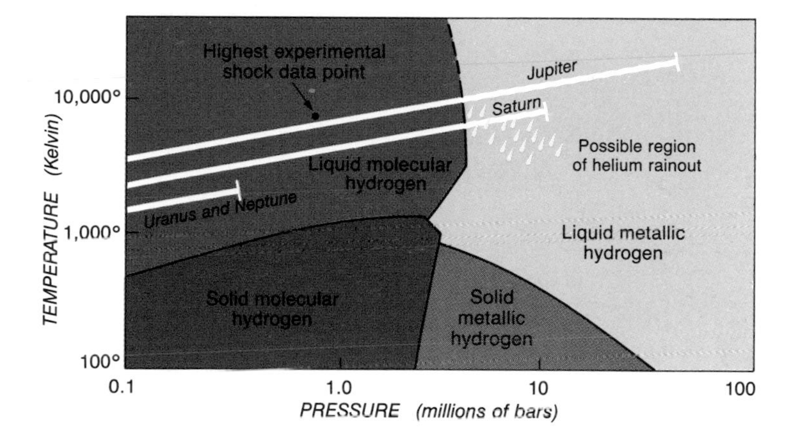

Figure 5. **A phase diagram for hydrogen, showing the domains of liquid metallic and molecular hydrogen (the main components in Jupiter and Saturn), as well as approximate interior temperature profiles for the Jovian planets. Color shading indicates where helium "rain-out" may occur, a zone where a solar-composition mixture of metallic hydrogen and helium cannot exist in equilibrium.**

down, where pressures are higher, the gas increases in temperature and density according to the thermodynamic law for *adiabatic compression* (compression of a gas without the loss or gain of heat). Still deeper, at pressures exceeding 100,000 bars, the gas gradually begins to resemble a hot liquid. The transition is gradual because temperatures are everywhere well above hydrogen's critical point at 13 bars and 33° K. If the atmospheres of the giant planets were cooler than 33°, the transition would be distinct, with layers above the 13-bar pressure level gaseous and those below it liquid.

About 20,000 km below Jupiter's cloud tops, liquid hydrogen reaches a pressure in excess of 4 million bars (4 Mbar) and a temperature of about 10,000° K. At that point, according to theoretical predictions, hydrogen's molecular and atomic bonds abruptly disappear, yielding an entirely new phase: liquid metallic hydrogen. The metallic hydrogen consists of ionized protons and electrons, as in the Sun's interior, but the temperature of matter with a given density inside Jupiter is about 1,000 times lower than it would be inside the Sun. Thus, the hydrogen acts not at all like a gas, but more like molten metal. This exotic form of hydrogen, like other metals, is both an electrical conductor and opaque to visible radiation. Most of the interior of Jupiter is in this state, which so far lies tantalizingly just beyond the limits of modern high-pressure laboratory experimentation.

Saturn is less massive than Jupiter, so its internal pressures are not as great. Even so, Saturn should also possess liquid metallic hydrogen in its deep interior. Uranus and Neptune, on the other hand, can only contain a small fraction of hydrogen, roughly 15 percent by mass (see Figure 3), and apparently the hydrogen is mostly limited to their outer layers. Thus, hot liquid hydrogen probably exists in these planets, but it is confined to the molecular phase (with the possible exception of small pockets of liquid metallic hydrogen buried deep within the "ice" layers).

Helium is an important tracer of internal processes in giant planets, because it is unlikely to have become separated from the hydrogen as these planets were forming. However, the abundances of helium in the giant planets' atmospheres are not all the same, and not all equal to the solar value. The Voyager spacecraft determined that Saturn's atmospheric helium is about four times less abundant than would be expected, while Jupiter and Uranus' atmospheric helium fractions are about equal to the solar value. Prior to these flybys, theorist David Stevenson predicted that at sufficiently low temperatures helium atoms become almost insoluble in metallic hydrogen. We think that the critical temperature for the onset of this hydrogen-helium separation may be close to those in the interior of Saturn. But accurate calculations are difficult, because while hydrogen is ionized in the deep interiors of Jupiter and Saturn, helium is not. High-pressure interactions between bare (fully ionized) hydrogen and helium nuclei are relatively simple to describe theoretically, and theorists agree that these lead to phase separation. But there is still uncertainty as to whether this relatively straightforward model can be applied to the real interactions between ionized hydrogen and neutral helium in Saturn's interior. If separation *has* occurred inside Saturn, the helium may have migrated into the metallic-hydrogen core, where it "rained out" and formed a helium-rich inner region.

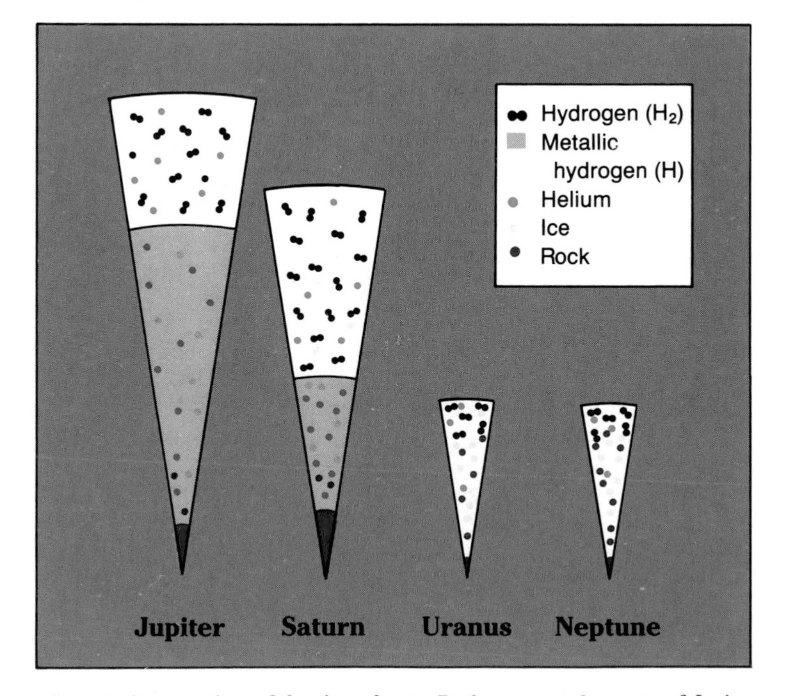

Figure 6. **Cross-sections of the giant planets. Rocky cores at the center of Jupiter and Saturn may be as massive as Uranus and Neptune.**

We can create cross-sections of the four giant planets (Figure 6) from our knowledge of theoretical compression curves (Figure 3), the gravitational harmonics of these planets (Figure 4), and the phase relationships of hydrogen (Figure 5). The hydrogen-rich envelopes of Jupiter and Saturn appear to contain substantial amounts of other elements in addition to helium, rather more than would be the case for strictly solar compositions. Both planets have dense cores of about 10 to 15 Earth masses that probably consist of rock or rock-ice mixtures. The outer "surface" of Jupiter's core is about 20,000° K, and Saturn's is about 12,000° K. The corresponding pressures there are 42 and 12 Mbar, respectively, yielding hydrogen-helium mass densities of 4.3 and 2.6 g/cm^3.

Interestingly, the core masses deduced for Jupiter and Saturn are very similar to the *total* masses of Uranus or Neptune. The latter two planets do appear to be mainly "cores," composed of mixtures of rock and ice in roughly solar proportions, with rather indistinct layering. Both are topped by hydrogen-rich atmospheres accounting for about 1 to 2 Earth masses. Although we use the term "ice" to denote a chemical mixture of water, methane, and ammonia under the high temperatures and pressures deep within a giant planet, this mixture actually will be a hot, liquid soup of various chemical species derived from these molecules.

HEAT FROM WITHIN

All planetary bodies generate *some* heat (thermal-infrared energy) deep in their interiors. For the small terrestrial planets, this energy comes mostly from the slow decay of radioactive isotopes that were incorporated into their rocks when the planets formed. Such a modest upwelling of heat would be difficult to detect from space, because it is overwhelmed by sunlight's warming effect on inner-planet surfaces.

But in the outer solar system, the situation is different. There solar heating contributes much less to a planet's atmospheric energy budget (Figure 7). Consequently, heat driven to the surface from the deep interior should be relatively easy to detect. And it was. Frank Low originally discovered Jupiter's intrinsic infrared glow in 1966, and subsequent observations from aircraft, ground-based telescopes, and the Pioneer and Voyager spacecraft have yielded progressively better heat-flow data for all four giant planets. After taking careful account of the fraction of solar energy absorbed in the planet's atmosphere and then re-emitted into space, observers concluded that Jupiter, Saturn, and Neptune have detectable internal sources of heat. This has proved to be one of the most interesting revelations of modern planetary science.

Some clues to the origin of the giant-planet heat flows can be obtained by computing the specific luminosity of each planet, that is, the average power released per unit of planetary mass (Figure 8). The luminosity of the Sun, powered by the thermonuclear fusion of hydrogen, is enormous compared with any other solar-system object. At the other extreme are primitive rocky bodies, such as the carbonaceous chondrite meteorites, whose internal heat

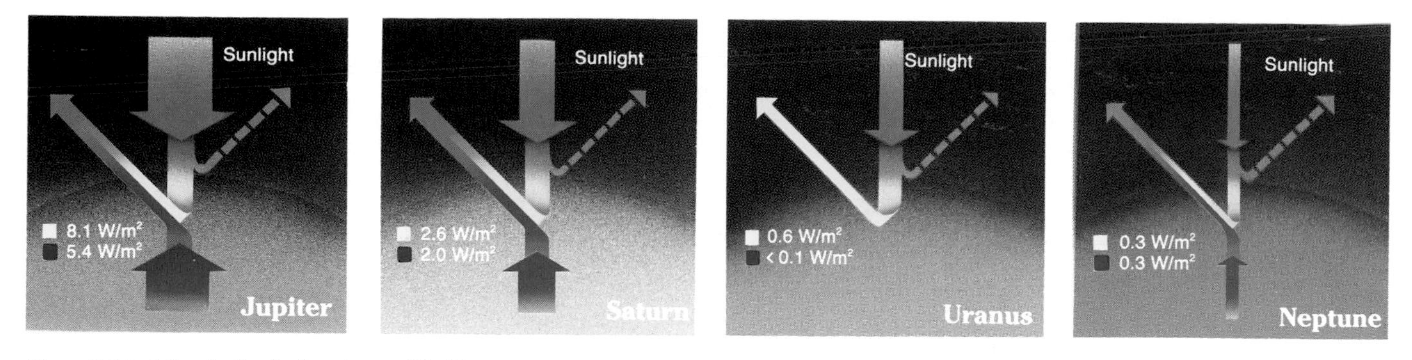

Figure 7. **Heat flow in the Jovian planets. The blue rays denote incident solar energy, some of which (dashed) is scattered back into space by the planet's atmosphere. The remaining fraction is absorbed and converted into infrared energy (yellow) deeper in the atmosphere; this, together with the infrared energy flowing out from the planet's deep interior (red), is the heat observed escaping at the surface of each planet. No interior heat has been detected coming from Uranus, but it must surely be present. With more sensitive instruments, astronomers would expect to detect the escape of heat generated by radioactive decay in the planet's possibly chondritic (rocky) core.**

comes only from the radioactive decay of what little uranium, potassium, and thorium they contain. The Earth's specific luminosity lies very close to the carbonaceous-chondrite value, suggesting that much of its internal heat derives from radioactive decay.

Among the giant planets, only the anomalous Uranus could lie close to the very low power-to-mass ratio of carbonaceous chondrites. Jupiter, Saturn, and Neptune release far more energy per kilogram of mass, so they must derive their luminosities from something other than radioactive decay. Thermonuclear fusion of hydrogen or deuterium, as in the Sun, is in principle possible, but calculations of the central temperatures of the giant planets show that they are nowhere near the required 10,000,000° K for this to occur.

Therefore, by elimination, only one process could be responsible for the luminosities of Jupiter, Saturn, and Neptune. Energy is liberated when mass in a gravitationally bound object sinks closer to the center of attraction, that is, when the object becomes more centrally condensed. In effect, potential energy becomes kinetic energy. This process causes the temperature within a gaseous body to rise. Gaseous stars like the Sun become so hot as they contract that hydrogen fusion begins, at which point they cannot radiate the accumulating heat into space fast enough and the contraction stops. But in a liquid body, like a giant planet, contraction is instead accompanied by cooling, because internal pressures depend only slightly on temperature. The giant planets probably are contracting, but only as fast as they can radiate the heat thus generated to space. This slow, self-regulated coupling of contraction and cooling, called the *Kelvin-Helmholtz mechanism*, probably represents the last phase of the giant planets' violent, high-temperature origin. By itself, the Kelvin-Helmholtz mechanism may be insufficient to explain Saturn's luminosity. The extra energy may be coming from the sinking of dense components (such as helium, or perhaps ice) toward the planet's center.

Although the escape of primordial heat from Jupiter, Saturn, and Neptune is very slow, it appears to be enough to destabilize the hydrogen layers through much of these planets' interiors, causing the hot, liquid hydrogen to convect at the rate of a few centimeters per second. Thus, the convective overturn seen at cloud tops of Jupiter and Saturn probably has deeper "roots" and may extend virtually to these planets' centers!

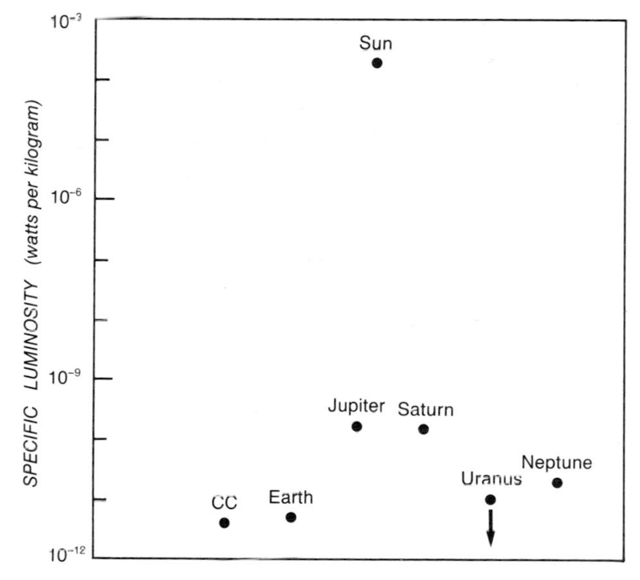

Figure 8. **Specific luminosities (energy emission per unit of mass) for the Jovian planets, compared with other solar-system bodies. The value labeled *CC* (carbonaceous chondrite) is that expected for a body of approximately chondritic (rocky) composition.**

Deep inside Jupiter and Saturn, convection creates and sustains powerful dynamos in the metallic-hydrogen zones, which in turn generate immense magnetic fields (see Chapter 3). Uranus appears to lack the heat flow required to induce such deep-seated convection, so perhaps it is no coincidence that the planet's enigmatic magnetic field differs markedly in its geometry from those of Jupiter and Saturn. William Nellis and his associates have attempted to simulate the conditions inside Uranus and Neptune through shock-compression experiments on water and other materials considered likely to exist in the planets' interiors. Their simulations show that high pressure and temperature cause a modest increase in the materials' electrical conductivity. Therefore, even though Uranus has no metallic-hydrogen layer, slow convection within its interior fluids may be reason enough for the global magnetic field observed by Voyager 2. Moreover, it seems virtually inevitable that Neptune should also have a substantial global magnetic field.

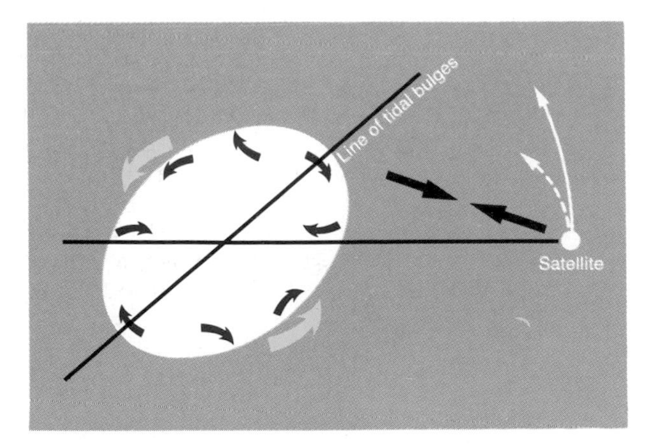

Figure 9. **The gravitational attraction of a nearby satellite raises symmetric tidal bulges on a giant planet. Because the planet is rotating with respect to the satellite, fluid currents (red arrows) occur in its interior. This motion results in a net loss of energy, as dissipated heat, that very gradually slows the planet's rotation. A fraction of this energy also serves to move the satellite's orbit outward. (The bulges' size and their angular displacement, or phase lag, from the planet-satellite line are shown much exaggerated.)**

One clue to dynamic processes at work inside a giant planet comes, oddly enough, from noting how its satellites' orbits evolve over time. The connection is as follows: A moon's gravitational field creates small, periodic tidal distortions in the shape of its parent body (as the Moon does on Earth). The planet's tidal bulges are not precisely underneath the satellite, because all the giant planets rotate at different rates than their moons orbit them (Figure 9). As the bulges slide backward in longitude to try to remain with the satellite, they encounter resistance and generate "friction" in the form of fluid currents in the planet's interior. Consequently, a displacement or phase lag develops between where the bulges should be and where they are, and its magnitude is proportional to the amount of energy being dissipated in the internal currents.

Meanwhile, the planet's altered shape (and thus its gravity field) can, over a long time, alter the motions of nearby satellites. If Jupiter were an ideal fluid body, it would have no internal dissipation of energy, the tidal bulges induced by Io would have no phase lag, Jupiter's corresponding effect on the orbit of Io would be zero, and we would be unlikely to see volcanism on that satellite today (see Chapter 13). However, based on the tidally generated heat now seen coming out of Io, we can calculate the rate of tidal evolution of its orbit and, working backward, set some interesting limits on Jupiter's effective internal viscosity. It turns out that the fraction of tidal energy converted into heat in Jupiter's fluid interior is several hundred times smaller than the corresponding fraction of the tidal energy generated in the Earth's interior by the Moon. The situation is much the same for the other Jovian planets.

The interesting question is not why the Jovian planets dissipate tidal energy at such a low rate (after all, they have liquid interiors), but rather why the dissipation is so *large*. Our estimates of ordinary fluid viscosities within the giant planets turn out to be similar to that of water at room temperature. Such values are many orders of magnitude too small to account for the tidal effects we observe among giant planets and their satellites. Apparently, some additional source of dissipation needs to be acting. Stevenson has suggested that sluggish phase transitions in the planet's deep interior may be responsible for the required tidal phase shift, but it is difficult to identify a specific region inside Jupiter which might be responsible for the evolution of Io's orbit.

MAKING GIANT PLANETS

Many aspects of the process of forming giant planets remain unclear, but a coherent picture has started to emerge. Because Jupiter and Saturn contain so much hydrogen and helium, they must have formed when the entire solar system was still enveloped in these gases. Astronomers who observe very young solar-type stars estimate that their gaseous envelopes start to dissipate after about 10 million years. Figure 10 depicts the early solar system at this point. Because the solar system contracted out of a rotating cloud, angular momentum spread the shrinking system into a disk. In the cooler parts of the disk, away from the primordial Sun, solid particles aggregated, settled into the plane of the disk, and

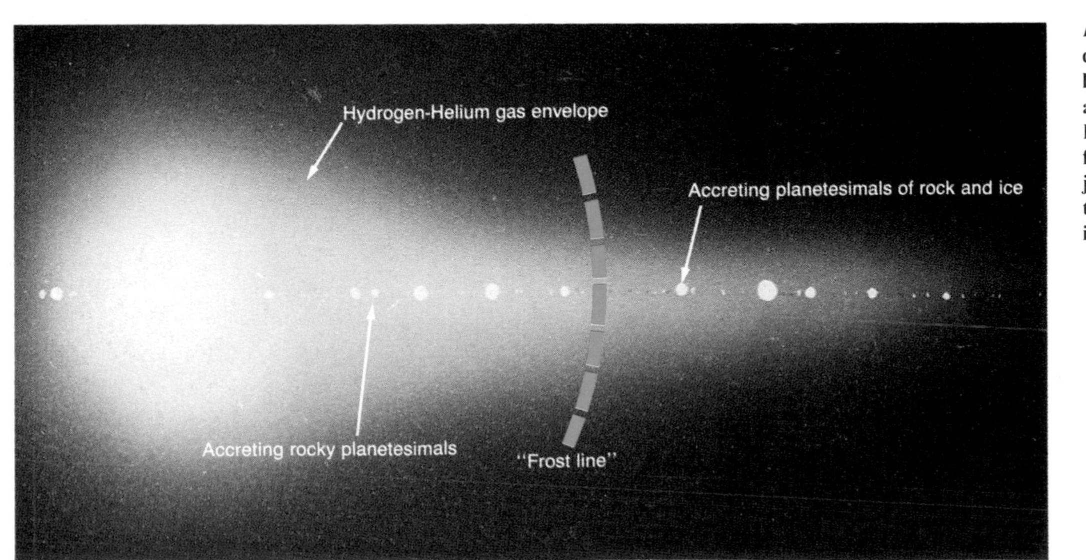

Figure 10. **This is how the primordial solar system might have looked (in cross-section) at an age of about 10 million years. Rock-ice planetesimals were forming in the outer solar system just before the hydrogen envelope that surrounded the protosun and its nascent family dissipated.**

formed asteroid-sized *planetesimals*. Outside a "frost line" at a temperature of about 170° K, not very far from the present orbit of Jupiter, the solid particles could become much more massive because they were cold enough to incorporate the abundant water ice in their vicinity (see Figure 2). Farther out, at still lower temperatures, additional ice components may have been incorporated in the solids.

According to Hiroshi Mizuno's model for formation of the Jovian planets, the icy planetesimals accreted into "trigger" nuclei for eventual Jovian planets. Once a nucleus had grown to 10 or 20 times the mass of the Earth, the gas in its vicinity could no longer remain dispersed in the primordial nebular disk. Instead, the gas succumbed to the gravitational pull of the massive nucleus and was "vacuumed up" into the growing protoplanet (Figure 11). Such captured gaseous envelopes would be hot by virtue of their sudden cascade into the primordial planets, thus providing a source for the Kelvin-Helmholtz cooling of Jupiter and Saturn that is feebly observable today. The rocky remnants of the trigger nuclei may form the dense cores at these planets' centers.

Somewhat farther from the infant Sun, at the orbits of Uranus and Neptune, the situation became a race between the accretion of the trigger nuclei and the dissipation of the gaseous nebula. Solid particles were more dispersed out there and orbital periods much longer, so it took longer to aggregate a sizable nucleus. Apparently, the nuclei lost the race, and little gas was left to capture by the time Uranus and Neptune formed.

As appealing as this picture may seem, there are many details still to be accounted for. The formation of a giant planet was probably not such a well-defined two-stage process. Large, icy bodies undoubtedly continued to splash into a protoplanet's gaseous envelope after it drew together, and a certain amount of nebular gas was probably entrained initially during formation of the trigger nucleus.

Also understood poorly are chemical reactions occurring deep in the interior. These reactions, acting over the age of the solar system, may have altered the atmospheric composition. For example, an excess of methane observed in all giant-planet atmospheres has yet to be accounted for satisfactorily. Molecules such as carbon monoxide and carbon dioxide have been observed in the upper atmosphere of Jupiter yet are highly unstable there. According to models developed by Ron Prinn and his associates, these molecules apparently were created at deeper levels under the influence of higher temperature and pressure, then were wafted upward by convection into the observable layers. Francis Ree and Marvin Ross have speculated that methane gas, abundant in the atmospheres of all four giant planets, may decompose at the high temperatures and pressures found deeper down – yielding diamond grains! But if these grains sink to the center of the planet, some mechanism (perhaps cometary infall) must replenish the methane over the planet's lifetime.

Some of the intricacies of the relations between the chemistry of molecules observed in the giant planets' atmospheres and the composition of their interiors are depicted in Figure 12, which represents a "typical" cross-section of such worlds. We now believe that the composition of a giant planet is unlikely to be uniform throughout its interior; its atmosphere, envelope, and core are roughly analogous to the crust, mantle, and core of a terrestrial planet, and they may have been similarly modified by chemical processes over geologic time.

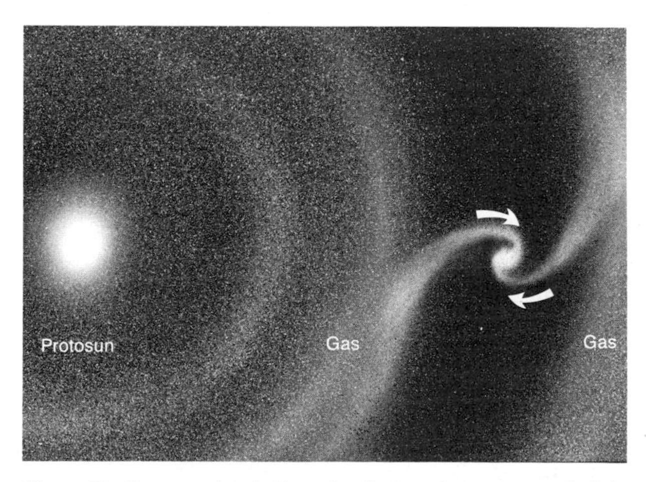

Figure 11. **Cosmogonists believe the Jovian planets assumed their immense size when nebular gas collapsed onto rock-ice cores of perhaps 10 or 20 Earth masses. The growth of the cores, and the subsequent sweeping up of gas in their vicinity, must have occurred before the primordial nebula dissipated – an interval of only some 10 million years.**

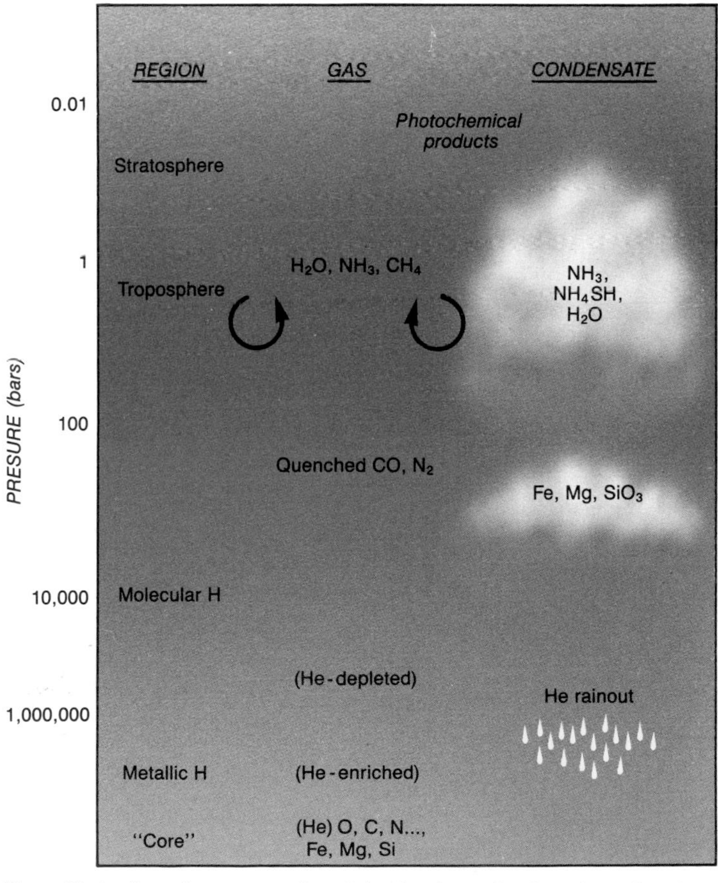

Figure 12. **A schematic representation of the chemistry of a giant planet, based on calculations by Jonathan Lunine. The dominant species is hydrogen, which exists in molecular form from a pressure level of about 1 million bars (log $P = 6$) up into the stratosphere. Regions of precipitation of various species are shown in the third column. At temperatures colder than about 1,000° K, the time for chemical reactions to reach equilibrium can become longer than the local convective mixing time. Thus, "quenched" species that are only stable at high temperatures and pressures (such as CO and N_2) are nevertheless observable higher in the atmosphere.**

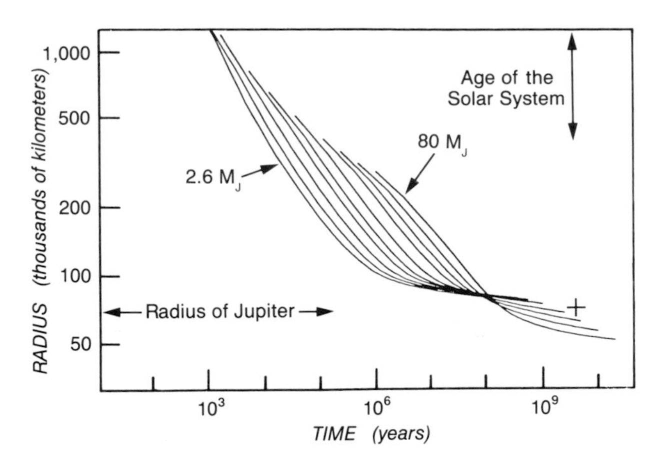

Figure 13. **The evolution of super-Jovian planets, substellar objects, or brown dwarfs ranging from 2.6 Jupiter masses up to 80 Jupiter masses. A cross denotes the radius of Jupiter and the age of the solar system (presumably the age of the giant planets as well).**

Much of our present uncertainty should be eliminated in the next decade and next century. Spacecraft being built or under serious consideration will dispatch probes directly into outer-planet atmospheres, providing compositional details and physical characteristics that we can only guess at now. This information will be key to deciphering the evolution of these planets' interiors. Meanwhile, high-pressure experiments conducted here on Earth may ultimately succeed in creating metallic hydrogen and provide further data on other light elements. All this will help us to determine how these huge planets came to be and how they have changed in the billions of years since their formation.

Finally, far beyond the solar system, there may exist a family of substellar objects or "superplanets," sometimes termed *brown dwarfs*, that have evolutionary histories similar to those of Jupiter and Saturn. As explained earlier, once a collapsing cloud of gaseous solar-composition matter becomes dense enough to act as a liquid instead of a gas, the temperature at the center stops rising and instead drops as contraction continues. Thus there is a maximum central temperature for collapsing clouds that contract with no energy sources other than gravity. Clouds of less than about 80 Jupiter masses never get hot enough in their center to start thermonuclear hydrogen fusion; thus they never become stars.

There also exists a maximum *radius* for cold hydrogen-helium objects, a consequence of the softening of the hydrogen-helium compression curve (Figure 3). Figure 13 shows the calculated contraction histories of a sequence of ever-larger brown dwarfs. The lowest-mass objects initially shrink rapidly, then very slowly cool and contract over billions of years, and end up with radii very close to that of Jupiter. Massive brown dwarfs, those close to the hydrogen-fusion limit, shrink much more slowly, but ultimately end up with *smaller* radii than Jupiter!

No brown dwarfs exist in the solar system – though Jupiter and Saturn might be considered smallish members of the class. So astronomers are looking intensively at the region of our galaxy within a few light-years of the Sun, seeking to detect these distant, elusive relatives of the solar system's giant planets either by the faint infrared glows from heat released by their slow contraction, or by their gravitational influence on companion stars. The results of these searches will give important clues to the physical conditions needed to form the largest planets.

11

Atmospheres of the Giant Planets

Andrew P. Ingersoll

THE GIANT PLANETS – Jupiter, Saturn, Uranus, and Neptune – are fluid objects. They have no solid surfaces because the light elements of which they are made do not condense at solar-system temperatures. Instead, their deep atmospheres grade downward into hot interiors where the distinction between gas and liquid becomes meaningless (see Chapter 10).

The chemical soup in these planets' cool outer layers precipitates clouds in a variety of colors. The cloud patterns are organized by winds, which are powered by heat derived from sunlight (as on Earth) and by internal heat left over from planetary formation. Thus the atmospheres of the Jovian planets are distinctly different both compositionally and dynamically from those of the terrestrial planets, making them fascinating objects for study.

Naturally, atmospheric scientists are interested to see how well the principles of their field apply beyond the Earth. For example, the Jovian planets are ringed by multiple cloud bands that move rapidly, yet somehow remain fixed in latitude. Such structures are found not only on Jupiter and Saturn (Figures 1,2), which are heated by the Sun most strongly around their equators, but also on Uranus (Figure 3). On Uranus the heating patterns are quite different because the planet spins on its side. During the Uranian year the Sun shines down on one pole then moves over the other, depositing on average more energy in the polar regions than at the equator. Although the cloud bands are hard to see, they follow latitude circles quite accurately, making Uranus look like a tipped-over version of Jupiter and Saturn. On Earth the bands are disrupted by large cyclonic storms at temperate latitudes and by long pressure waves that are anchored to the continents. Earth-based views of Neptune (Figure 4) reveal high-altitude clouds but no banding. However, even from a great distance, Voyager 2's cameras recorded bands in Neptune's deeper-lying clouds, in addition to a high altitude spot first seen from Earth (Figure 5).

Figure 1. **Jupiter as viewed by the Voyager 1 spacecraft from a distance of 54 million km. The overall color is reasonably accurate, though color contrast has been enhanced.**

Figure 2. **Voyager 1 recorded Saturn and its rings from a distance of 18 million km. The color enhancement is similar to that in Figure 1. Note that cloud features on Saturn are fewer and lower in contrast than on Jupiter.**

Chemistry provides another set of challenges for atmospheric scientists. On Earth, the composition of our air is determined largely by reactions with the solid crust, the oceans, and the biosphere. Even before the advent of life, photodissociation (the splitting of molecules by the ultraviolet component of sunlight) and the escape of hydrogen into space were causing our atmosphere to evolve. The Jovian planets have neither crust nor oceans nor life, and their gravitational fields are so strong that they retain all elements including hydrogen. Yet unstable compounds are formed by a variety of processes, including rapid ascent from the hot interior, condensation (cloud formation), photodissociation, charged-particle bombardment (auroras), and electrical discharge (lightning; see Figure 6). As is evident from their mutlicolored clouds and abundant molecular species, the atmospheric chemistry of the Jovian planets is just as rich as on Earth.

With their internal heat and fluid interiors, the giant planets can be compared with the Sun and stars. On the other hand, we can compare their atmospheric phenomena with those on Earth. Our understanding of terrestrial atmospheric phenomena was inadequate to prepare us for the revelations made in recent years about the giant planets, but by testing our theories against these observations, we can understand the Earth in a broader context. The preceding chapter delved into the interiors of the Jovian planets. This one focuses on their atmospheres, especially the observable layers from the base of the clouds to the altitude of the rings and satellites.

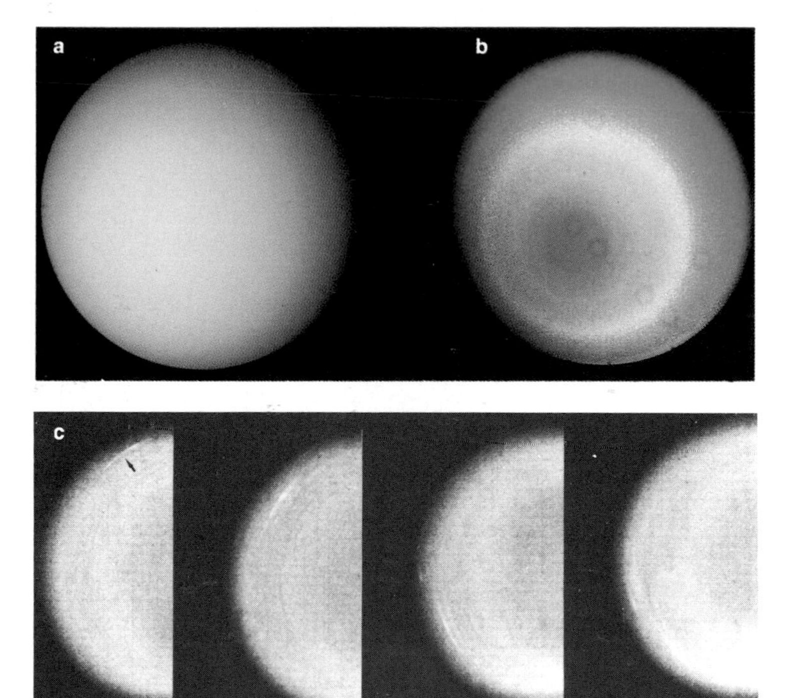

Figure 3. **Voyager 2 views of Uranus. The color in *a* has been adjusted to simulate the view the eye would normally see. The phase angle (Sun-Uranus-camera) is 14°, and features as small as 160 km would be visible – if Uranus had them. In *b*, computer processing has eliminated the globe's tendency to appear darker away from the subsolar point. Both the color and contrast have been greatly exaggerated. The small doughnut-shape features are artifacts caused by dust on the camera optics. In *c*, a convective plume at latitude 35° S is shown at several positions on the disk. It is visible only because contrast has been greatly exaggerated.**

OBSERVATIONS

Our knowledge of the Jovian planets is derived from Earth-based telescopic observations begun 300 years ago, modern observations from high-altitude aircraft and orbiting satellites, and the wealth of data returned by interplanetary spacecraft. Between 1973 and 1981, four unmanned probes flew by Jupiter (Pioneers 10 and 11, Voyagers 1 and 2) and three of them flew by Saturn. Voyager 2 continued past Uranus in January 1986 and Neptune in August 1989.

Basic characteristics like mass, radius, density, and rotational flattening were determined during the first era of telescopic observation. Galileo's early views revealed the four large Jovian satellites that now bear his name. Newton estimated the mass and density of Jupiter from observations of those satellites' orbits. Others, using ever-improving optics, began to perceive atmospheric features on the planet. The most prominent of these, the Great Red Spot, can be traced back 300 years with near certainty, and it may be older still.

Beginning in the late 19th century, astronomers made systematic measurements of Jovian and Saturnian winds by tracking features visually with small telescopes and, later, with photographs made through larger ones. Most recently, tens of thousands of features on Jupiter and Saturn have been tracked with great precision using the Voyager imaging system (Figure 7). The constancy of these currents over the centuries spanned by classical and modern observations is one of the truly remarkable aspects of Jovian and Saturnian meteorology. Uranus, on the other hand, is nearly featureless. Although the planet was discovered in 1781, the first definitive evidence for markings came from Voyager 2's

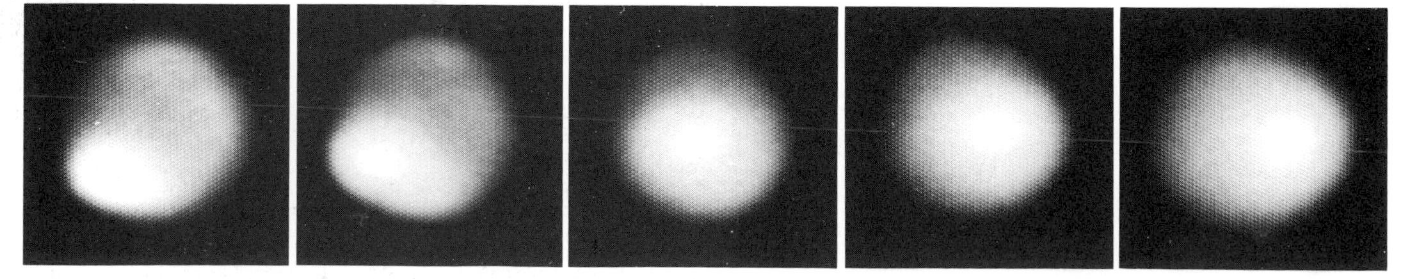

Figure 4. **These images of Neptune were taken through a filter that transmits only light near 8900 angstroms (in the near infrared), where methane gas has a strong absorption band. Therefore, the bright spots are clouds of ice crystals high above deeper layers of methane (which appear dark). Heidi Hammel obtained these images over a 5.4-hour period in July 1988 with a 2.2-m telescope on Mauna Kea, Hawaii. From them she determined that the planet's upper atmosphere rotates with a period of 17.7 hours.**

images. Neptune is slightly more photogenic than Uranus and exhibits faint markings that are visible from Earth with modern infrared detectors.

Astronomers began to decipher these planets' atmospheric compositions in the 1930s with the identification of absorptions by methane (CH_4) and ammonia (NH_3) in the spectra of sunlight reflected from their banded clouds. The detection of molecular hydrogen (H_2) followed around 1960. Because hydrogen is a simple, symmetrical molecule, its vibrational and rotational absorptions are weak. Fortunately, hydrogen occurs above the clouds in such abundance that its absorption lines are nevertheless detectable in the spectra of all four planets. Observers quickly verified that the proportions of hydrogen, carbon, and nitrogen in Jupiter's atmosphere were roughly consistent with a mixture of solar composition. Similar inferences have been made for Saturn, although actual observations of these compounds (especially ammonia) are extremely difficult. Recent observations suggest a number of intriguing departures from solar composition that are not fully understood.

During the 1970s, as infrared detectors improved, observers recorded absorptions by other gases present in extremely minute amounts (one part per million or even per billion by volume). Ethane (C_2H_6), acetylene (C_2H_2), water, phosphine (PH_3), hydrogen cyanide (HCN), carbon monoxide (CO), germane (GeH_4), and compounds with deuterium (^2H) and isotopic carbon (^{13}C) all were detected in Jupiter's atmosphere in this way (see Table 1). The list for Saturn is shorter than for Jupiter, partly because Saturn is a fainter object and partly because it is colder; many of these compounds freeze at the level of its cloud tops and become harder to detect.

Because Uranus and Neptune are so cold, only extremely volatile hydrocarbon compounds are detectable in their atmospheric spectra. Methane is particularly abundant, and its tendency to absorb red light gives the planets their blue-green color. Methane also condenses into clouds in the Uranian atmosphere, and Voyager scientists were able to determine the ratio of methane to hydrogen there by observing how these clouds affected Voyager 2's radio signal as the spacecraft passed behind the planet. It turns out that Uranus' C:H ratio is some 30 to 40 times the value for a solar-composition atmosphere.

Helium, the second most abundant element in the Sun, is presumably an important constituent of the Jovian planets as well. Unfortunately, it has no detectable spectral signature to make its presence known. However, collisions between helium and hydrogen molecules alter the latter's ability to absorb infrared light, an effect that the Pioneer and Voyager instruments could detect and thus give us a kind of "back-door" identification of helium. And we can also combine infrared measurements with the radio-occultation data obtained as the spacecraft passed behind the planets to determine the molecular weight of the mixtures in their atmospheres. When paired together, these two methods yield helium abundances with an uncertainty of only a few percent. The value thus determined for Uranus, 15 percent by number of molecules, is consistent with a gas mixture of solar composition. However, the values derived for Saturn (3 percent) and Jupiter (10 percent) suggest that helium has been depleted from Saturn's upper atmosphere and probably Jupiter's as well, a result that provides intriguing clues to these planets' internal structure and evolution.

Even though the temperatures and energy budgets of the giant planets have been studied from Earth for decades, the best determinations have once again come from spacecraft. Their infrared and radio-occultation experiments probed from the stratosphere (where temperature increases with height) down into the troposphere (where the clouds are). Ultraviolet instruments covered the much greater altitudes of

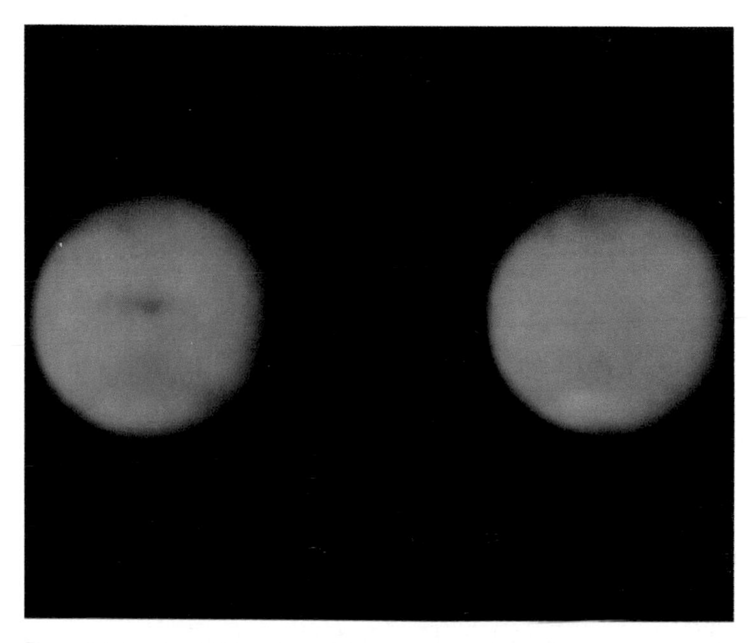

Figure 5. **In late April 1989, while still four months and 176 million km away from its encounter with Neptune, Voyager 2 recorded these views of the planet. The color and contrast have been enhanced to reveal such details such as a pattern of faint banding and the dark spot at left. (Curiously, the bright spot seen in Figure 4 was absent.) The image at right was acquired 5 hours later, during which time Neptune rotated 100°. Note the bright spot near the south pole.**

Figure 6. **The night hemisphere of Jupiter displayed surprising evidence of lightning (bright patches below center) and auroras (curved arcs at top) in this Voyager 1 image. The planet's north pole lies roughly midway along the auroral arc.**

the thermosphere and ionosphere. We find that pressures near the cloud tops range from hundreds of millibars (mb) to 1 bar, as on Earth. But the temperatures at these altitudes, where infrared radiation escapes to space, are much colder – about 124° K for Jupiter and 95° K for Saturn (Figure 8).

From these observations, we infer that Jupiter radiates about 1.7 times the amount of heat it absorbs from the Sun, and that Saturn radiates about 1.8 times its absorbed heat. Uranus and Neptune both emit at temperatures of about 56° K, even though Neptune is 1.6 times farther from the Sun. This implies that Uranus has at most a small internal heat source (its estimated ratio of emitted to absorbed heat is below 1.1), but that Neptune has a relatively large one (its ratio is around 1.8). Such infrared excesses, or lack thereof, provide us with additional insight into internal structures and histories (see Chapter 10).

The ways that temperature and infrared emissions vary with latitude on the giant planets (Figure 9) indicate how effective the winds are in redistributing heat away from the subsolar zone. Recall that on Jupiter, Saturn, and Neptune the Sun heats the equator more than the poles, and on Uranus the Sun heats the poles more than the equator. But in all four cases the temperature differences with latitude are small, less than what they would be without winds. This does not automatically mean that winds *within* the cloud layers are carrying the heat from warm spots to cold ones – an interesting question (dealt with more fully later) is whether the redistribution of heat takes place instead in the planets' fluid interiors.

The Voyager scientific team explored the uppermost atmospheres of these worlds using three main techniques. First, when the spacecraft passed behind the planet, we used its radio signals, which are affected by electrons in the planets' ionospheres; the decrease of electron density with altitude is a measure of temperature. Second, the ultraviolet spectrum of sunlight and starlight that has passed through the upper atmospheres contains information about temperature and composition. And third, by observing how and where the upper atmospheres give off light, we learn about the energy sources for chemical reactions. For instance, electrically charged particles striking the atmosphere from above cause auroras near the magnetic poles, and energetic solar photons cause airglow on the daytime hemisphere. These excitation processes can alter the chemical composition of deeper layers by triggering the formation of stable compounds that ultimately are convected downward.

ATMOSPHERIC STRUCTURE AND COMPOSITION

Temperatures, pressures, gas abundances, cloud compositions – all as functions of altitude and horizontal position – are the principal components of atmospheric structure. Atmospheric dynamics, discussed in the next section, concerns the wind and its causes and effects. However, the distinction between structure and dynamics is not always clear. Atmospheric circulation alters the structure by carrying heat, mass, and chemical species from place to place. The structure, in turn, controls the absorption of sunlight, its re-emission as infrared radiation, and release of latent heat during condensation – all of which cause the gas to heat up, expand, and set winds in motion.

Atmospheres are self-supporting, in that the increase of pressure with depth provides the upward force needed to balance the downward pull of gravity. This balanced state is known as *hydrostatic equilibrium*. The degree of compression of the gas is proportional to the gravitational acceleration. The Jovian atmosphere, therefore, is the most compressed and also the thinnest – that is, the altitude range over which pressure doubles, for example, is less on Jupiter than on the other gas giants.

The variation of temperature with depth (and thus with pressure) is more subtle, and it is controlled by different processes at different altitudes. At the deepest levels measured, pressures of about 1 or 2 bars, rapid convective

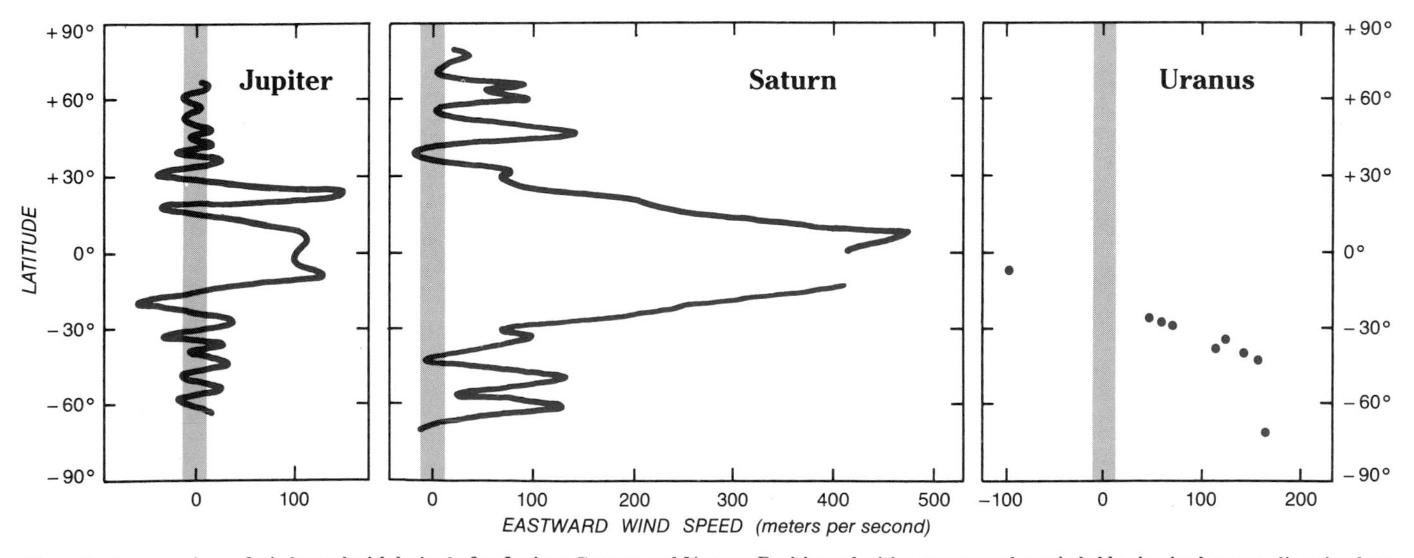

Figure 7. **A comparison of wind speed with latitude for Jupiter, Saturn, and Uranus. Positive velocities correspond to winds blowing in the same direction but faster than the planets' internal rotation periods (9 hours 55.5 minutes for Jupiter, 10 hours 39.4 minutes for Saturn, and 17 hours 14.4 minutes for Uranus), which are based on observations of periodic radio emissions. Negative velocities are, therefore, winds moving more slowly than these reference frames. Saturn's winds are strongest and blow mostly to the east. Uranus has the only slower-moving equatorial jet, whose existence was revealed by Voyager radio-occultation data; the other data points for Uranus correspond to cloud features tracked in Voyager images.**

motions result in an *adiabatic gradient* of temperature. There, temperature decreases with altitude at the same rate as for a rapidly rising parcel of air. The parcel expands and does work at the expense of its own internal energy (temperature). Mixing creates a state where all air parcels resemble one another and all follow the same adiabatic gradient.

Convection is driven by heat from below, and an adiabatic gradient is either a sign of an internal heat source or deep penetration of sunlight. As we have seen, all the giant planets but Uranus have detectable internal heat sources. At the deepest observable levels (pressures about 1 bar), Jupiter and Saturn have adiabatic gradients. Neptune is probably similar, and for all three planets we can assume that the trend continues downward indefinitely. Jupiter, Saturn, and Neptune, therefore, are inferred to have hot interiors. Uranus is a less-certain case, since the adiabatic gradient could give way to a constant-temperature interior immediately below the lowest level to which sunlight penetrates, at pressures of 10 or 20 bars.

Heat is carried upward by convection until it reaches a level where the overlying atmosphere is no longer opaque to infrared radiation. For the giant planets, this occurs at pressures of 100 to 300 mb, at which point the atmosphere can radiate its heat directly to space. Not surprisingly, infrared cooling drives the temperatures down to their lowest values in this altitude range (Figure 8).

Above the 100–mb level, the temperature increases with height because the atmosphere there is absorbing sunlight. Atmospheric gases alone probably cannot absorb energy efficiently enough to account for the rise in temperature. But

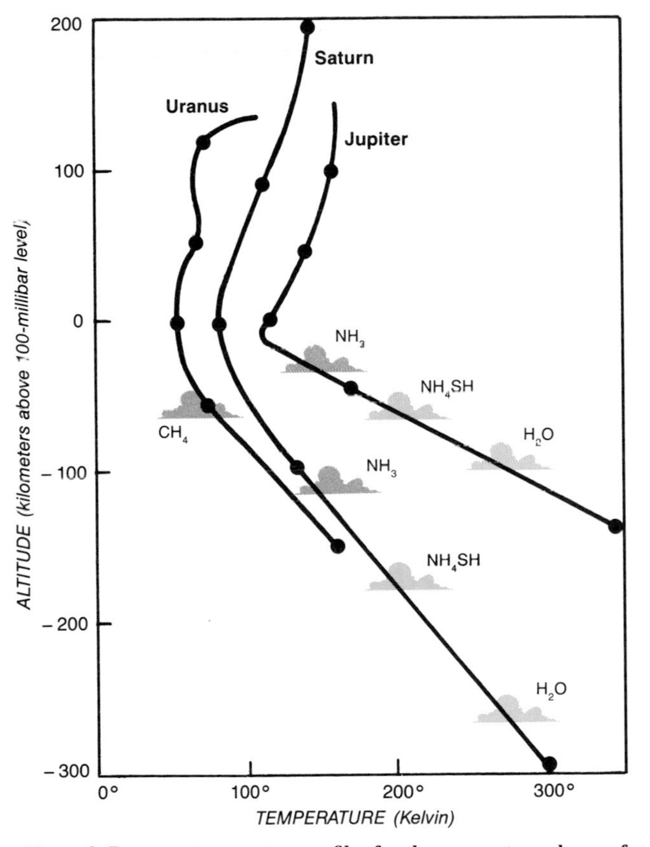

Figure 8. **Pressure-temperature profiles for the upper atmospheres of three giant planets, as determined by Voyager measurements at radio and infrared wavelengths. Each dot marks the point where atmospheric pressure is 10 times greater than the next dot above it; the total range indicated runs from 1 millibar at the top to 10 bars near the bottom. Colored bands show the altitudes at which various clouds should form. For Jupiter and Saturn the cloud altitudes are based on a gaseous mixture of solar composition. For Uranus, solar composition grossly underestimates the amounts of condensable gases, and only the methane cloud detected by Voyager 2 is shown. Temperatures are generally lower on planets farther from the Sun, except for Neptune (not shown), whose internal heat source makes it as warm as Uranus. The range of cloud altitudes is narrower on Jupiter because the stronger gravity compresses the atmosphere more than on the other planets.**

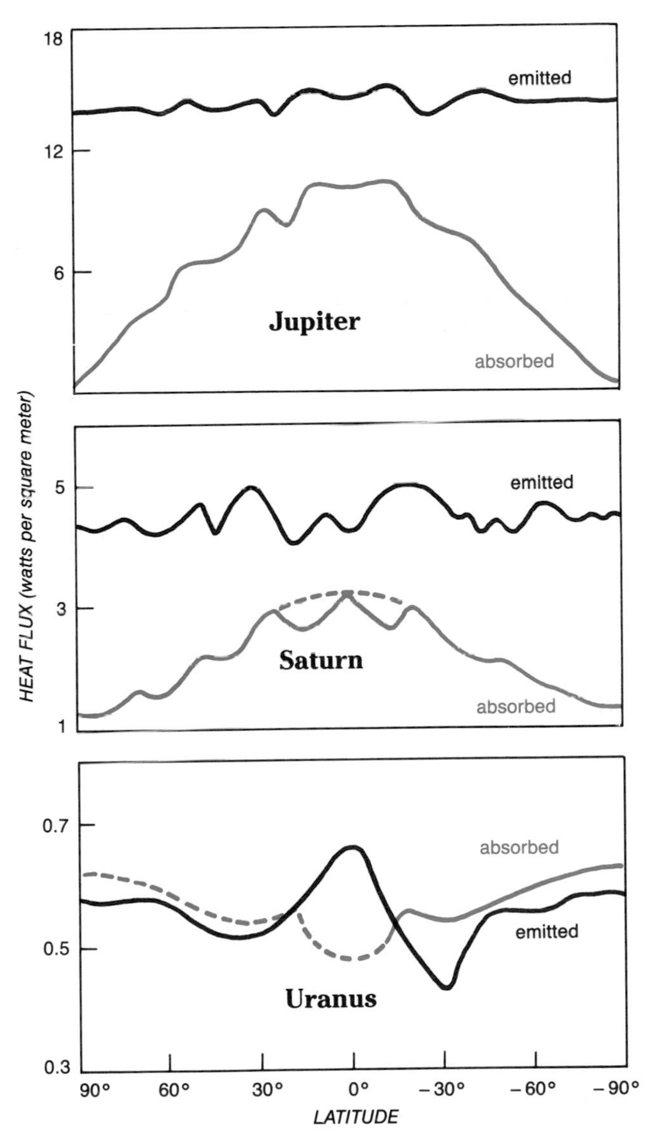

Figure 9. **A comparison of absorbed solar energy and emitted infrared radiation for Jupiter, Saturn, and Uranus, averaged with respect to longitude, season, and time of day. The dashed curve for Saturn corresponds to the planet's heat absorption calculated without the shadowing effect caused by its ring system. The dashed curve for Uranus is a northward extrapolation of Voyager measurements, which assumes that the planet's absorption of heat is symmetric with respect to latitude over the course of a Uranian year. Small bumps are due to temperature and brightness differences between adjacent latitude bands. The infrared radiation emitted by Jupiter and Saturn exceeds the amount of sunlight they absorb, implying an internal heat source. This radiation is also distributed more uniformly than the absorbed sunlight, which suggests heat transport across latitude circles at some depth within the planet.**

a haze layer, produced photochemically in the stratosphere, could absorb the extra sunlight required to heat its surroundings. Higher up, at pressures less than 0.001 mb (the thermosphere), even small energy inputs can change the temperature significantly. For example, the auroral emissions that Voyager discovered on Jupiter were especially intense at latitudes where magnetic field lines from the orbit of the satellite Io intersect the atmosphere. Spread over the planet, the energy of the incoming charged particles causing the auroras is equivalent to only 0.001 percent of the total incident sunlight, but that is a large input at these altitudes. Solar photons alone would produce temperatures only near 200° K. But, according to Voyager data, the cascade of charged particles from Io seems capable of heating the Jovian thermosphere to 1,000° K. The dissipation of electrical currents and upward-propagating waves might also be contributing to the high temperatures, but there are large uncertainties associated with all of these energy sources. The thermosphere sheds its heat principally by conducting it downward to the cooler layers below.

The thermospheric temperatures of Jupiter (1,000° K), Saturn (420° K), and Uranus (800° K) do not suggest an obvious pattern, and the exact balance of energy sources and sinks is unknown. An interesting consequence of the relatively high temperatures and low gravity of Uranus is that the atmosphere actually extends outward into the rings. Thus the ring particles experience a small amount of atmospheric drag, which shortens the lifetimes by causing them to spiral into the planet. Current estimates are that some of the Uranian rings are much younger than the age of the solar system – that is, they must have been re-supplied with material many times since the planet formed (see Chapter 12).

Once the pressure-temperature curve for an atmosphere has been determined, its vertical cloud structure can be inferred. One first computes the altitudes at which the different atmospheric constituents can condense by assuming that the gas mixture has a uniform, known composition (Table 1). A particular gas will condense when its partial pressure (its absolute abundance in the atmosphere) exceeds the saturation vapor pressure (the abundance when the gas is at its dew point). Since the latter falls rapidly with temperature, clouds will form at the coldest layers in the atmosphere. Convective currents may carry the condensates to higher altitudes, unless they rain out completely near the cloud base.

When these calculations are worked for Jupiter and Saturn, we find three distinct cloud layers (Figure 8). The lowest is composed of water ice or possibly water droplets. Next are crystals of ammonium hydrosulfide (NH_4SH), which is basically a combination of ammonia (NH_3) and hydrogen sulfide (H_2S). At the top we expect an ammonia ice cloud. At Uranus and Neptune, these three cloud layers should lie deeper – at higher pressures – than at Jupiter and Saturn. This is because the low temperatures necessary for cloud formation extend to greater depths when the atmosphere is colder. Even methane condenses if the temperature gets low enough, especially if the gas is abundant. Apparently, such conditions are met in the atmospheres of Uranus and Neptune, where a layer of condensed methane overlies the other clouds. Voyager detected this cloud on Uranus from its effect on the

Molecule	*Sun*	*Jupiter*	*Saturn*	*Uranus*
H_2	85	90	97	83
He	15	10	3	15
H_2O	0.11	0.0001	—	—
CH_4	0.06	0.2	0.2	2
NH_3	0.016	0.03	0.03	—
H_2S	0.003	—	—	—

Table 1. **Molecular abundances (as a percentage of all molecules) for a solar-composition atmosphere and for the giant planets' atmospheres near the cloud tops. Uranus' methane abundance refers to a depth below the methane cloud. The list contains seven of the 10 most abundant elements in the Sun (and the universe). The other three – silicon (Si), magnesium (Mg), and iron (Fe) – are believed to reside in the cores of the giant planets. Dashes indicate unobserved compounds, and all values are uncertain in the least significant figure.**

spacecraft's radio signal. The signal probed down to the 2-bar level, which is below the methane cloud at 1.3 bars but not yet into the ammonia cloud.

The layered-cloud model has not been fully tested and may be too simple. For one thing, vertical mixing could carry particles from the lower clouds upward, thereby changing the composition of the upper clouds. Moreover, hydrogen sulfide has not been detected on any of the giant planets, possibly because it precipitates out below the cloud tops, and possibly because it changes to sulfur when exposed to sunlight. The existence of water clouds is also uncertain. Water vapor's abundance on Jupiter has been measured as only one part per million, even though a match to solar composition would imply an abundance 1,000 times greater. The paradox is that we can see through holes in the clouds (Figure 10) to warm levels with temperatures of at least 275° K where both H_2O and H_2S should be abundant. Perhaps these "hot spots" are the deserts of Jupiter, where dry air stripped of its condensable gases is descending into the well-mixed interior. The observations nevertheless cast some doubt on the solar-composition model of Jupiter's atmosphere. The opposite situation occurs on Uranus and Neptune, where the heavy elements appear to be enriched relative to hydrogen in a solar-composition atmosphere. This is implied by the greater densities of these planets, and it is confirmed by the abundant methane measured by the Voyagers (Table 1). Jupiter and Saturn are also somewhat enriched in carbon, making oxygen even more of a mystery.

If the atmospheres were in chemical equilibrium, with hydrogen the dominant species, virtually all the carbon would be tied up in methane, all the nitrogen in ammonia, all the oxygen in water, and so on. But in Jupiter's case, we have identified gases such as C_2H_6, C_2H_2, and CO, which imply disequilibrium. Both C_2H_6 and C_2H_2 are relatively easy to account for, since they are formed in the upper atmosphere as by-products of methane photodissociation. In these reactions, solar ultraviolet photons at wavelengths shorter than 1600 angstroms knock hydrogen atoms off the methane (CH_4) molecule, leaving free radicals that can react with each other and with CH_4 molecules to form C_2H_6 and C_2H_2. These gases are seen on Saturn and Uranus as well.

The existence of CO in the upper atmosphere of Jupiter is more interesting, since oxygen is not readily available (water, the principal oxygen-bearing molecule, tends to condense in the lower clouds). Several explanations have been proposed.

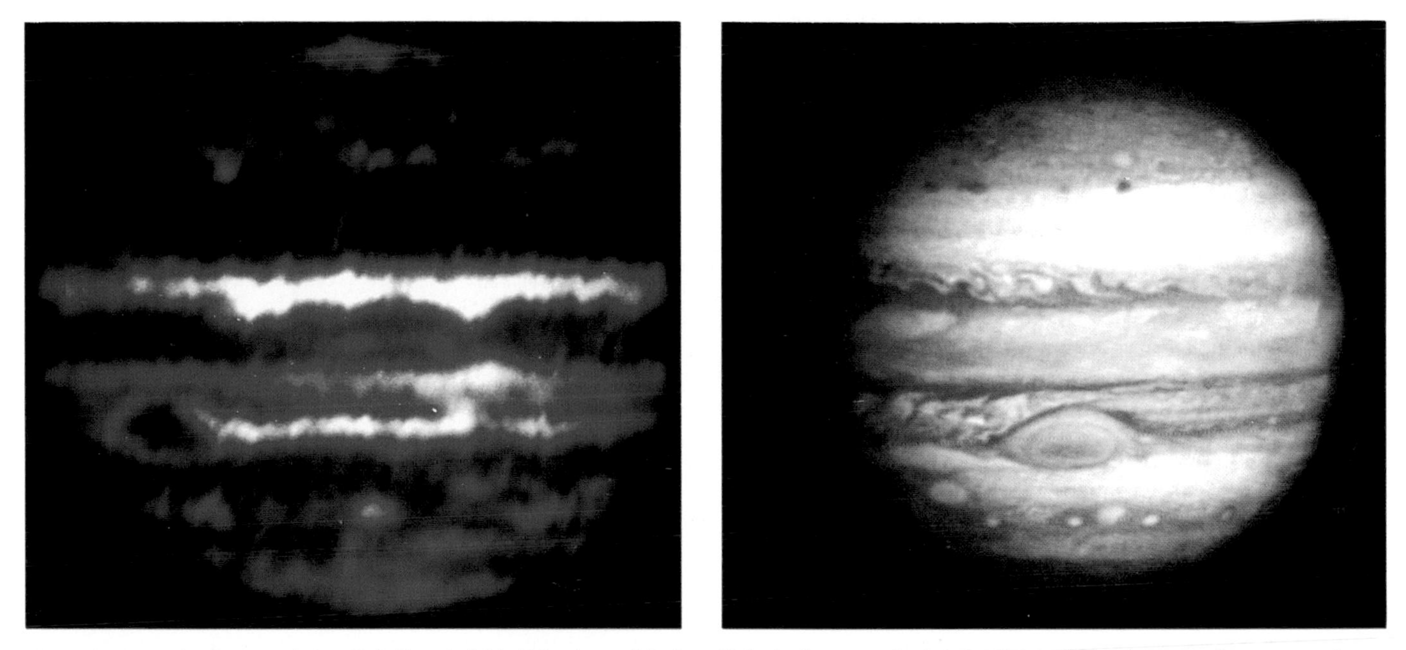

Figure 10. **Near-simultaneous infrared (left) and visible-light views of Jupiter. Holes in the upper cloud deck allow radiation to escape from warmer layers underneath; these spots appear bright in the infrared. Richard Terrile and others obtained the 4.8–micron infrared image in March 1979 with the 5–m Hale reflector at Palomar Observatory; the other photograph is from Voyager 1.**

Figure 11. **Io transits the southern hemisphere of Jupiter in this Voyager 2 image, which has a resolution of about 200 km. Note the similarity in color between the planet's clouds and Io's sulfur-covered surface.**

One is that oxygen ions enter the upper atmosphere from above after being injected into the magnetosphere (as SO_2 molecules by Io's volcanoes. Another explanation is that CO, which forms and remains stable deep in Jupiter's atmosphere, gets transported upward before it can react with hydrogen and thus be destroyed. If this latter scenario is correct, phosphorus should likewise be transported upward in an oxidized state. But instead we observe only phosphine (PH_3), which is something of a mystery. A critical unknown in these theories is how fast the high-temperature species are convected upward before they are destroyed by chemical reactions.

One insight into vertical mixing comes from the fact that molecular hydrogen (H_2) exists as two species, one in which the spins of the two nuclei are parallel and one in which they are opposed. The two types react and tend to reach an equilibrium value that is a function of temperature. This is the distribution seen when vertical motion is weak. However, when it is strong, the ratio of the two species is "frozen" at the high-temperature extreme, reflecting conditions at or above 300° K. By observing how the proportions of the two types vary from place to place, atmospheric scientists have found evidence of vertical motion in the equatorial zones of the giant planets. Not surprisingly, the rate of mixing is greatest for Jupiter and least for Uranus.

The spectacular colors of Jupiter and more muted colors of Saturn provide further evidence of active chemistry in the atmospheres. These colors correlate with the cloud's altitude (Figure 10). Blue regions have the highest apparent temperatures, so they must lie at the deepest levels and are only visible through holes in the upper clouds. Browns are next highest, followed by white clouds, and finally red clouds, such as those in the Great Red Spot, which is a very cold feature judging from its infrared brightness.

The trouble is that all cloud species predicted for equilibrium conditions are white. Therefore, color must arise when chemical equilibrium is disturbed, either by charged particles, energetic photons, lightning, or rapid vertical motion through layers of varying temperature. One possible coloring agent is elemental sulfur, which forms a variety of colors depending on its molecular structure. Sulfur is definitely present on Io, which exhibits many of the same colors as Jupiter (Figure 11). Some scientists believe phosphorus gives the Great Red Spot its color, while others have proposed organic (carbon-bearing) compounds to explain almost all of the hues seen in Jupiter's clouds.

Figure 12. **These views of Jupiter and Saturn were constructed by piecing together segments of Voyager images, then projecting them as if the spacecraft were looking down onto the planets. In the Jupiter composite, the Great Red Spot is at upper left and the corners are at a latitude of roughly 10° N. When David A. Godfrey prepared this polar view of Saturn, he discovered an unusual hexagon centered on the north pole. (The dark "starfish" is a processing artifact.) The hexagon is actually a six-lobed wave in a zonal jet at 76° N. latitude. It remained visible throughout Voyager 2's August 1981 flyby, and its rotation period matched that of Saturn's interior (it did not drift eastward or westward in longitude), but its origin remains a mystery. The dark circular band is at latitude 68° N.**

Both the blue-green color and the lack of contrasting features on Uranus may have a simple explanation. Cold causes its sulfur and ammonia clouds to lie much deeper than those on Jupiter and Saturn, so light reflected from the clouds has to pass through more atmospheric gas on its way to our eyes. The gas acts as a translucent filter, scattering the blue wavelengths and absorbing the red ones – the latter because of methane. Any features in these clouds appear washed out, and more abundant photochemical haze (due to the lower Uranian temperatures) could also obscure cloud features lying deeper down. It is less clear why the methane cloud, which is not very deep, should display so few visible features.

Thus coloration is a subtle process, involving disequilibrium conditions and trace constituents. The correlation with altitude presumably reflects the pressure-temperature conditions needed to drive specific chemical reactions. For example, higher altitudes receive more sunlight and charged particles. Some regions may provide locales for intense lightning and others for intense vertical motion. A different question is why these processes should be organized into large-scale patterns that last for years and sometimes for centuries. This question involves the dynamics of the atmospheres, to which we now turn.

ATMOSPHERIC DYNAMICS

The dominant dynamic features seen in the giant planets' atmospheres are counterflowing eastward and westward winds – the *zonal jets* (Figure 7). In some respects, Uranus resembles the Earth, which has one westward air current at low latitudes (the trade winds) and one meandering eastward current at mid-latitudes (the jet stream). Jupiter has five or six of both kinds in each hemisphere. Saturn seems to have fewer sets of such currents than Jupiter, but they move faster. In fact, the eastward wind speed at Saturn's equator is about 500 m per second – about two-thirds the speed of sound there. On all the giant planets the jets seem to be straighter and steadier than their counterparts on Earth. However, a curious feature in Saturn's atmosphere is the north-polar "hexagon" formed by a meandering jet near latitude 76° (Figure 12). In all cases, these winds are measured with respect to the planets rotating beneath them. For fluid planets, which have no solid surfaces, the rotation of the interior is deduced from that of the magnetic field, which is generated in the metallic core. Even supersonic wind speeds would be small compared to the equatorial velocity imparted by the planet's rotation.

Without continents and oceans, the planet's rotation tends to produce a pattern of zonal (east-west) currents and banded clouds. Rotation has two effects. On Jupiter, Saturn, and Neptune it smears the solar heat into a band around the equator. And on all planets, not just the Jovian ones, it creates the *Coriolis force*, which acts at right angles to the wind.

Uranus is perhaps the best example of the importance of rotation. During the Voyager encounter in 1986, the Sun was almost directly overhead at Uranus' south pole. The equator was in constant twilight, and the north pole had been in darkness for 20 Earth years (one-fourth of the Uranian year). Without the Coriolis force, circulation would have been dictated solely by the need to redistribute heat: winds in the upper atmosphere blowing away from the sunlit south pole and converging on the night side at the north pole, with a return flow underneath. But the Coriolis force apparently is at work, redirecting the flow at right angles to this hypothetical, thermally driven circulation. Even so, the problem remains that over a Uranian year the planet's poles receive more energy from sunlight than they can emit as infrared radiation. Therefore, to maintain thermal equilibrium the atmosphere *must* transport heat equatorward, via some unseen part of the circulation that cuts across latitude circles. Small-scale eddies, below the resolution of the Voyager images, are one possibility; transport below the clouds or within the clear atmosphere above them are others.

The infrared instrument on Voyager measured the temperature of the gas above Uranus' clouds (Figure 9). But instead of a smooth gradient from equator to pole, the curves are bumpy. On Uranus, the equator is the hottest place. Direct solar heating would not produce such a distribution, nor would circulation that simply brings heat from warm areas to cold ones. Rather, Voyager investigators argue that heat is being pumped from cold places to hot ones, as in a refrigerator. The power source for this "global refrigerator," they claim, is the kinetic energy of the zonal jets. The steepest gradients of temperature should then be located at the latitudes of the zonal jets (Figure 7), and so they are. However, the energy source needed to maintain the jets themselves remains unspecified.

The giant planets' zonal winds are remarkably constant in time. Ninety years of modern telescopic observations reveal no changes in the east-west jets of Jupiter and Saturn. During the four months between the two Voyager encounters of Jupiter, the zonal velocities changed by less than the measurement error (about 1.5 percent). This is remarkable for several reasons. First, although the zonal jets on Jupiter usually correlate with the latitudes of the colored bands, the bands sometimes change their appearance dramatically in a few years, while the jets do not. Second, as revealed in the superbly detailed Voyager photographs, an enormous amount of eddy activity accompanies the zonal jets. Small eddies appear suddenly along the boundary between eastward- and westward-moving streams. They last only about 1 to 2 days, which is the time required for such structures to be sheared apart by the counterflowing winds. Larger eddies, including the long-lived white ovals and the Great Red Spot, manage to survive by rolling with the currents. Voyager repeatedly observed smaller spots encounter the Red Spot from the east, circulate around it in about 7 days, then partially merge with it (Figure 13). How the zonal jets and the large eddies can exist amid such activity is something of a mystery.

We learned an important fact about the eddy motions from a statistical analysis of Jupiter's winds. During one 30-hour "cloud watch" by Voyager 1 and another 30-hour watch by Voyager 2, the motions of tens of thousands of 100-km features were tracked, usually at 10-hour intervals. On average the flow is zonal, with alternating bands of eastward and westward velocity (the jets) at different latitudes. Localized departures from the zonal flow (eddies) move at about one-fourth the mean speed, but these motions are not entirely random. Fluid parcels moving *into* a jet from another latitude are typically headed in the same direction (eastward or westward) as the jet itself. Parcels moving *out of* a jet are typically headed the other way and thus carry the opposite

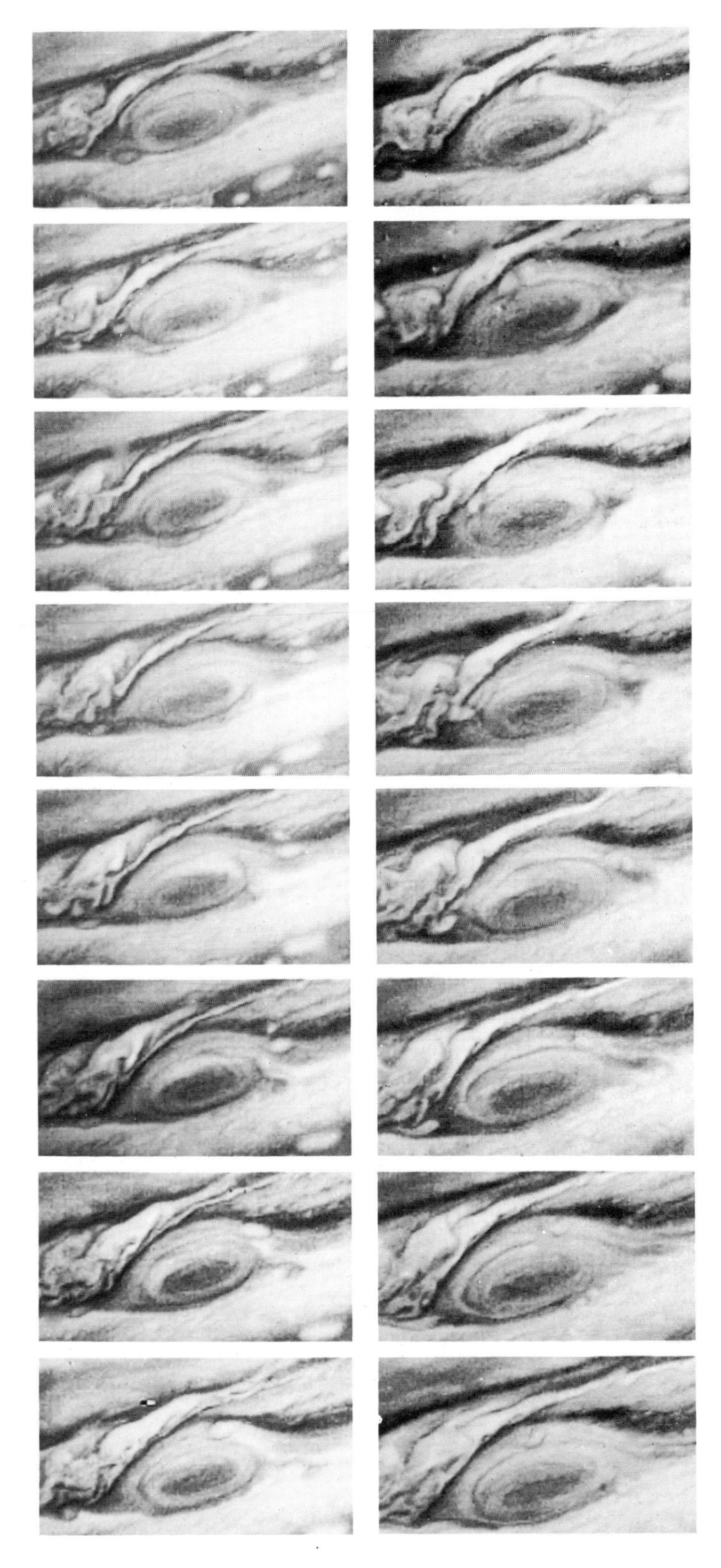

Figure 13. **A blue-light image of the Great Red Spot was taken every other rotation of Jupiter over a period of about two weeks in this Voyager sequence, which begins at upper left, continues down each column, and ends at lower right. Note the small bright clouds that encounter the Red Spot from the east, circle counterclockwise around it, and partially merge with it along the southeast boundary.**

sign of momentum. The net result is that eddies we observe are helping to maintain the jets, not vice versa. Yet the jet speeds do not change, so there must be other unseen eddies – small-scale turbulence, perhaps – that draw their energy from the jets and thus keep the system in balance.

If one regards the jets as an ordered flow and the eddies as chaotic, observations suggest that (in this case) order arises from chaos. Similar interactions among eddies and mean flows occur in the Earth's atmosphere and oceans. In all cases, the eddies get their energy from buoyancy (hot fluids rising and cold ones sinking), which releases gravitational potential energy. Most of this is dissipated as heat, but apparently some manages to power the jets. The fraction of buoyant energy converted to eddy motion is less than 1 percent on Earth, but it exceeds 10 percent on Jupiter. What causes this fundamental difference is unknown.

The constancy of Jupiter's zonal jets is remarkable in light of all this eddy activity. If the eddies and zonal jets occupy the same thickness of atmosphere, the eddies could double the kinetic energy of the jets in about 75 days at the observed rate of energy transfer. If the zonal jets extend much deeper than the eddies do, the doubling time will be much longer. Perhaps the zonal jets have remained stable for nearly 90 years because the mass involved in the jet motions is many times greater than that of the eddies. The jets might even extend through the planet and out the other side! Such behavior is not as unlikely as it first seems. In a rapidly rotating sphere with an adiabatic fluid interior, there exist two types of small-amplitude motions (relative to the basic rotation). The first are rapidly varying waves and eddies. The second are steady zonal motions on coaxial cylinders (Figure 14), with each cylinder moving about the planet's rotation axis at a unique

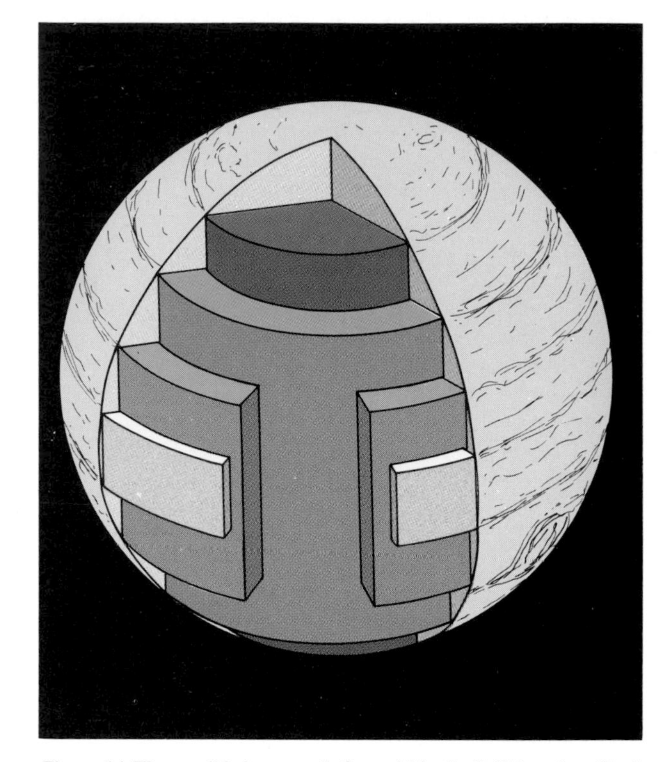

***Figure 14.* The possible large-scale flow within the fluid interiors. Each cylinder has a unique rotation rate, and zonal winds may be the surface manifestation of these rotations. The tendency of fluids in a rotating body to align with the rotational axis was observed by Geoffrey Taylor during laboratory experiments in the 1920s and was applied to Jupiter and Saturn by Friedrich H. Busse in the 1970s. Such behavior seems reasonable for Jupiter and Saturn if their interiors follow an adiabatic temperature gradient.**

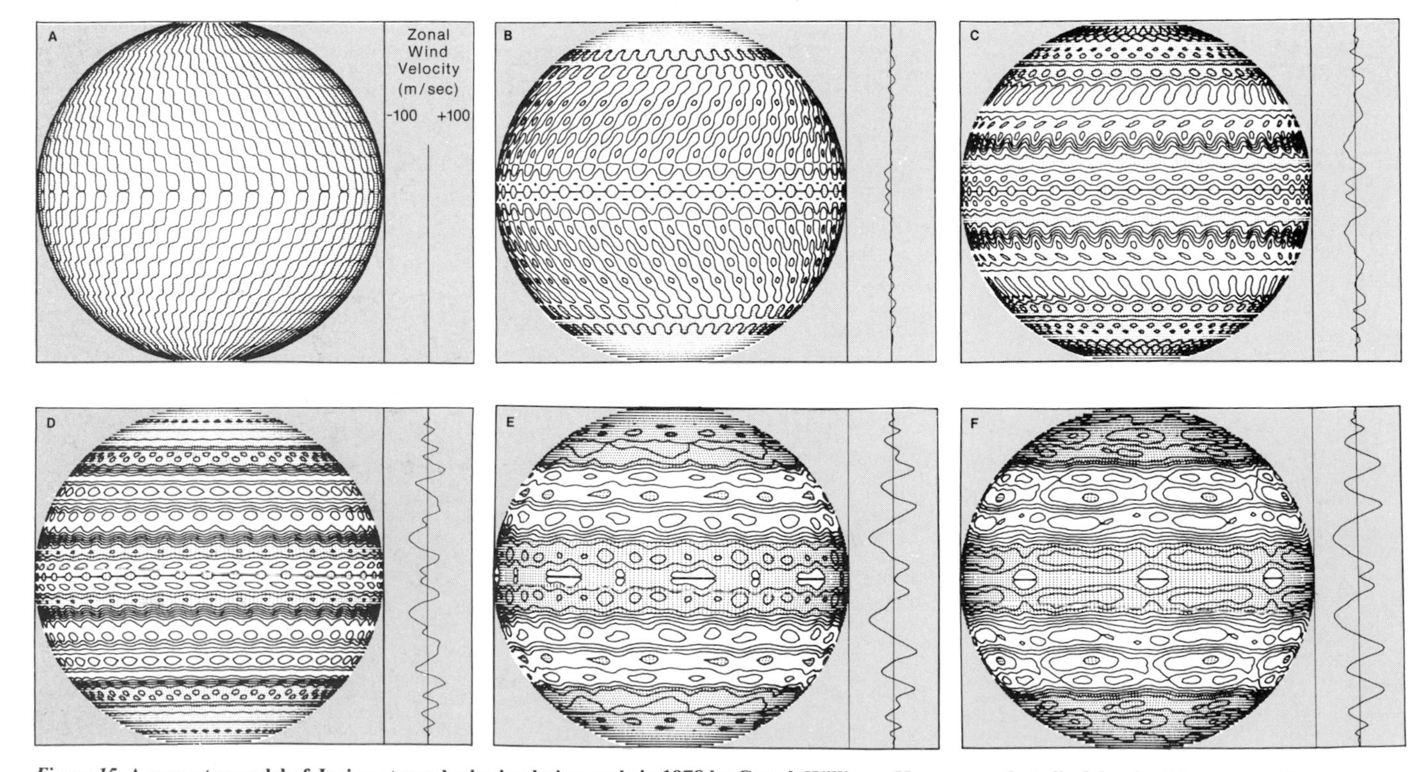

***Figure 15.* A computer model of Jovian atmospheric circulation made in 1978 by Gareth Williams. He assumes that all of the clouds' energy exchanges occur within a narrow layer of the atmosphere (as is the case on Earth); the eddies get their energy from sunlight, not from the interior. Williams' model first produces small-scale eddies *(a)*, which become unstable *(b)* and give up their energy to zonal jets *(c,d)* that eventually dominate the flow *(e,f)*. Eddies driven by internal heat demonstrate the same behavior, so remaining questions center on the depths of both the eddies and the zonal flow.**

rate. The observed zonal jets, as far as we know, could be the surface manifestation of these cylindrical patterns.

The problem with such speculation is that we lack information about winds below the visible cloud tops. A totally different approach is to treat Jupiter simply as a larger version of the Earth. In fact, several years before the Voyagers arrived at Jupiter, computer models designed for the Earth's atmosphere had also yielded realistic zonal wind patterns when applied to the Jovian situation (Figure 15). This is somewhat surprising, since the models assume an atmosphere less than 100 km thick, a flat lower boundary (the Earth's surface), and no outward heat flow from the interior. On the other hand, the process of eddies driving zonal flows is a very general one and seems to occur in a wide range of situations on rotating planets.

Clearly, the lack of data on vertical structure has allowed a wide range of models to develop explaining the zonal jets of both Jupiter and Saturn. Ultimately, we may learn which is correct by comparing the secondary effects predicted by each model with actual observations.

THE GREAT RED SPOT

There is a similarly wide range of theories concerning long-lived oval structures, which are well known on Jupiter and were discovered on Saturn by Voyager (Figure 16). As already mentioned, Jupiter's Great Red Spot (GRS) is probably more than 300 years old. It covers 10° of latitude (Figure 17), which makes it about as wide as the Earth. The three white ovals slightly south of the GRS (named BC, FA, and DE) first appeared in 1938; other spots have come and gone, lasting for months or years. These long-lived ovals, which drift in longitude but remain fixed in latitude, tend to roll in the boundary region between opposing zonal jets. The circulation around their edges is almost always counterclockwise in the southern hemisphere and clockwise in the northern hemisphere, indicating that they are high-pressure centers. They often have cusped tips at their east and west ends. Finally, during the Voyager encounters both the GRS and the white ovals had intensely turbulent regions extending off to the west.

Any theory of the GRS and white ovals must explain their longevity and isolated nature. Longevity involves two problems. The first concerns the hydrodynamic stability of rotating oval flows. If these spots were unstable they would last only a few days, which is roughly the circulation time around their edges (or the lifetime of smaller eddies that are pulled apart by the zonal jets). The second problem concerns energy sources. A stable eddy without an energy source will eventually run down, although on Jupiter this dissipative time scale could be several years.

How do theorists account for the long lifetimes of Jupiter's ovals? The "hurricane" model postulates that these structures are giant convective cells extracting energy from below (the latent heat released when gases condense). The "shear instability" model holds that they draw energy from the zonal currents in which they sit. Still another model argues that they gain energy by absorbing smaller, buoyancy-driven eddies, much as the zonal jets gain energy. In principle, these hypotheses can be tested by running simulations on high-speed computers. But the surprising result is that all three seem to work. One can input a variety of energy sources, density distributions, initial conditions, and lower

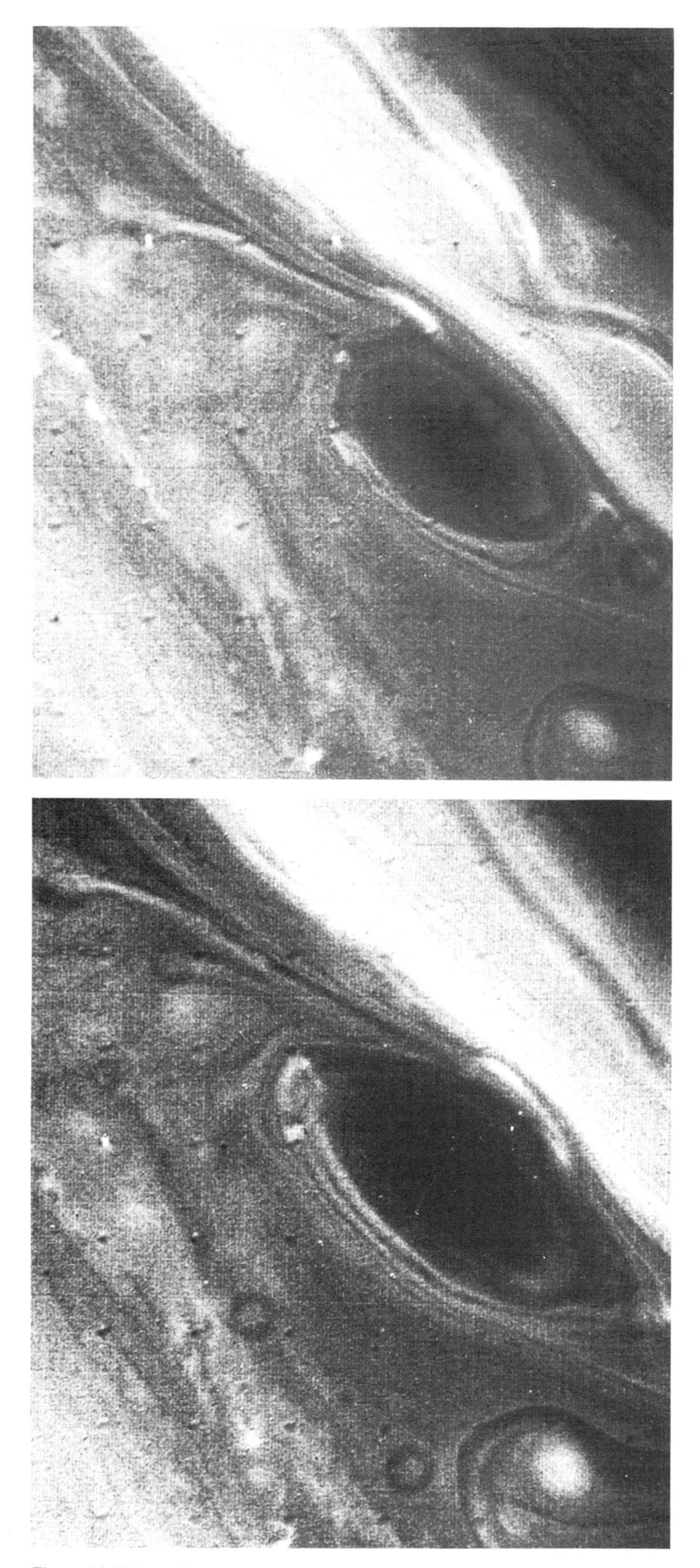

Figure 16. **This oval spot, photographed on two successive Saturn rotations, shows anticyclonic rotation (clockwise in the northern hemisphere) around its periphery. North is to the upper right. The spot's latitude is 42°.5 N, and it drifts to the east at 5 m per second. The spot is 5,000 km long across its major axis.**

boundary. But all these models have their problems, and having wind data at only one level is a serious drawback. The models themselves allow for only one or two vertical levels. This saves computer time, but it makes the dynamics less realistic. Also, Jupiter's zonal velocities are two times faster than those considered stable by the theoretical models. When a computer simulation is run with the observed zonal velocities imposed as initial conditions, little waves and eddies appear (Figure 18). Larger vortices form, merge, and take up some of the excess kinetic energy that was in the original flow. The flow settles down finally when there is only one large vortex and a weakened zonal flow that no longer fully resembles the one on Jupiter. To make the model match the observations, one has to add artificial forces to maintain the zonal flow. Thus in principle the observations already in hand will help us determine what "works" by eliminating those models that do not. But no model developed to date really fits all the data.

MOTIONS IN THE INTERIOR

In closing this discussion of the giant planets, let us speculate on possible roles that the fluid interior may play in atmospheric dynamics. Consider, for example, whether the interior could help maintain the poles and equator at the same surface temperature, even though the input of sunlight with latitude is uneven. Vertical convection is normally very effective at producing an adiabatic temperature distribution within an atmosphere, especially when the heat sources are located below the heat sinks. Once an adiabatic state is reached, it is maintained by the buffering effect of the internal circulation. This is the arrangement for Jupiter and Saturn, whose interiors generate a great deal of heat and whose atmospheres lose it (via infrared radiation) to space. An adiabatic atmosphere has constant temperature at each altitude level, with no variations in latitude.

Actually, we need to know these atmospheres' *net* heat loss: the infrared energy they emit to space minus the sunlight they absorb. Because more sunlight falls on the equators of Jupiter and Saturn, the net heat loss is greatest at the poles and least at the equator (but positive at all latitudes; see Figure 9). To compensate, heat must somehow be transported poleward. But this transport does not appear to be taking place in the cloud layers – that would disrupt the well-ordered banding. So we expect, but cannot yet know for sure, that small departures from a strictly adiabatic state have arisen in the interior to drive the internal heat flow poleward. These departures are too small to manifest themselves, but the result is that the atmosphere is effectively short-circuited by the interior.

The situation is very different on Earth. Here substantial poleward transport of heat *must* take place in the atmosphere. Absorption of sunlight by the oceans and the subsequent release of latent heat from the condensation of water in the atmosphere above them leads to a net heat gain in the tropics. Radiation to space leads to a net heat loss in the polar atmosphere. This combination sets up appreciable temperature gradients with latitude, and obvious mixing takes place across latitude circles. Consequently, Earth's atmosphere is decidedly *not* adiabatic.

Another key speculation is how the Jovian planets' zonal velocities change with depth. We have no direct measurements of this profile, but there are theoretical limits. According to the thermal wind equation of meteorology, the change of velocity with depth is proportional to the change of temperature with latitude. Since no significant temperature gradients were found on either Jupiter or Saturn by the Pioneer and Voyager instruments at the deepest measurable levels, the velocity change with depth must therefore be extremely small, and so the winds measured at the cloud tops must persist below. A rough estimate is that the eastward wind speeds (relative to the internal rotation rate) persist well below the region that is affected by sunlight, which is mostly absorbed in the clouds. This leaves internal heat as the obvious source for deep motions.

A complete theory of these deep motions is still a long way off. The problem is similar to modeling the motions in the Sun's interior, but the computational challenges are greater. The giant planets are cold objects that radiate their energy slowly, so their interiors take a long time to reach thermal equilibrium. On the other hand, certain atmospheric phenomena (turbulent waves and eddies, for example) change very rapidly. Small eddies can be quite energetic as well. Even the largest computers cannot simulate both the slow, large-scale processes and the fast, small-scale ones together. To cope at all, they must use approximations for the fast processes that convey their net effect on the slow ones. Developing such models even for the Earth is something of a black art. Terrestrial meteorologists can at least study rapid, localized phenomena at close range – a luxury those of us who model the giant planets' interiors do not have.

Here are but a few of the other questions still awaiting answers: Does the outward convection of internal heat trigger only small-scale, disorganized motion, or does it ultimately drive a large-scale pattern? How are the systems of zonal winds observed on Jupiter and Saturn linked to the interior, if at all? At what depth does the electrical conductivity of the fluid become important, and what are the effects of the magnetic field below that depth? What accounts for the spacing of the zonal jets, and why is that spacing larger for Saturn than for Jupiter? Why is Saturn's equatorial jet so much faster than Jupiter's, and why are there so few westward-moving features in Saturn's atmosphere? What novel effects occur on Uranus and Neptune, whose interiors are mostly methane, ammonia, and water (condensable gases) rather than hydrogen and helium?

Again and again, atmospheric scientists find themselves faced with pronounced differences between the gas-giant planets and the Earth. And usually we speculate that the cause of these differences may lie in the fluid interiors beyond our reach. Of course, some of our speculations may turn out to be wrong – the entire circulation may be taking place in the giant planets' outer atmospheres after all. Further analysis of Voyager data and theoretical work should resolve many of our questions. But ultimately, the best answers will come from spacecraft that actually probe these atmospheres directly.

12
Planetary Rings

Joseph A. Burns

"I DO NOT KNOW what to say in a case so surprising, so unlooked for, and so novel," confessed Galileo Galilei when Saturn's rings apparently vanished in 1612. At the time, only two years after he had discovered them, the rings were actually just presenting a slim edge-on view (as they do every 15 years or so when Earth passes through the ring plane). But since their nature as a flat disk encircling Saturn had not yet been realized, Galileo was confounded. The great Italian scientist was actually never sure of their precise form – indeed at first he believed the rings to be separate bodies, only to think six years later that they were two great arms or handles stretched toward Saturn. Many such interpretations were put forth by skywatchers in the early 17th century (see Figure 1) as the poorly viewed Saturnian system varied its configur-

Figure 1. **Our view of Saturn's rings has improved remarkably during the past $3\frac{1}{2}$ centuries: *a*, a drawing by Galileo (1610); *b*, a sketch by Gassendi (1634); *c*, Fontana's view (1646); *d*, a drawing by Riccioli (1648); *e*, Huygens' sketch (1655); *f*, Cassini's drawing (1676), showing his division; *g*, first photograph of the rings by Common (1883); *h*, Lyot's diagram (1943); *i*, S. Larson's photograph (1974); *j*, Pioneer 11's (1979) view of scattered light through the rings; *k*, Voyager 1's (1980) contrast-enhanced image of back-scattered light.**

ation due to changing Sun-planet-Earth orientations. Not until 1659 did Christiaan Huygens correctly infer the disklike nature of Saturn's "appendages."

Saturn stood as the only planet with a known ring for more than $3\frac{1}{2}$ centuries. But in 1977, nine narrow rings were detected about Uranus when the planet occulted a star. Shortly thereafter, in 1979, Voyager 1 discovered a faint band circling Jupiter, and in the early 1980s stellar occultations revealed a set of seemingly incomplete ring arcs around Neptune. Thus, in but a few years, astronomers had replaced the question of why does Saturn alone have rings with another query: why are planetary rings so commonplace, yet so different, in the outer solar system?

Thanks to this rapid growth in our knowledge of planetary ring systems, we have begun to address which properties are fundamental to all planetary rings and which are specific to particular systems. This chapter will first outline some dynamical processes that govern the overall structure of planetary rings, then describe the ring systems encircling Jupiter, Uranus, and Neptune before turning to Saturn's. Along the way I will speculate on the possible origin of planetary rings and their long-term stability, raising the issue of which other planets have had – or will have – rings. Throughout we will find that these planetary ornaments continue to be just as surprising and novel as they were to Galileo.

DYNAMICAL CONCEPTS

Through a modest telescope Saturn's rings appear to be a continuous sheet of matter, broken by at most a single division. Their smooth, nearly opaque appearance motivated Pierre Laplace and Jean Cassini, as well as other early astronomers and mathematicians, to wonder whether the rings were solid, liquid, or particulate. The answer came when James Clerk Maxwell demonstrated theoretically that the rings had to be "comprised of an indefinite number of unconnected particles," a revelation that won him the Adams Prize in 1857. Four decades later James Keeler confirmed Maxwell's idea by noting that sunlight reflected off the rings was Doppler-shifted such that the ring particles must occupy individual orbits about Saturn, with the innermost ones moving faster than their outer counterparts by some tens of percent.

How might such a system have evolved dynamically? A cloud of particles moving about a central mass will develop in three stages. First, they will swiftly flatten to a thin disk in a specific plane (Figure 2). To understand why this happens, consider the fate of two nearby particles orbiting an isolated, spherical planet on separate inclined and elliptical paths. Slight additional forces, caused for example by the planet's oblateness or by a perturbing body like a satellite, will force the orbital planes of the two particles to drift (or precess) gradually relative to a mean plane called the *Laplacian plane*. When only oblateness is a factor, the Laplacian plane is an extension of the planet's equator; it lies in the satellite's orbital plane. As their orbits gradually drift the two particles can occasionally collide, which reduces the relative velocity between them. In particular, the bodies' motions out of the Laplacian plane are lessened until both objects lie essentially in that plane.

For a system of many particles, out-of-plane orbital motion is damped rapidly and continually, since any particle not moving "in step" with the others will pass through their mean orbital plane twice every orbit and likely strike another member of the planetary ring. The frequency of these collisions depends on what fraction of the disk's area is filled with particles; for a nearly opaque system like Saturn's main rings, a renegade particle will collide with another during virtually every passage through the plane, or on the average once every few hours. This process is apparently very effective, for the vertical thickness of Saturn's ring system – just tens of meters – is but a tiny fraction of its vast breadth (more than 200,000 km); Saturn's rings are as thin as a sheet of tissue paper spread across a football field.

The mean plane mentioned above depends on the combined influence of planetary oblateness, nearby satellites, and the Sun. Since the strength of each gravitational perturbation depends on the distance to the perturber, the mean plane is thus actually a slightly warped surface, looking like a snapped-down hat brim extending out from the

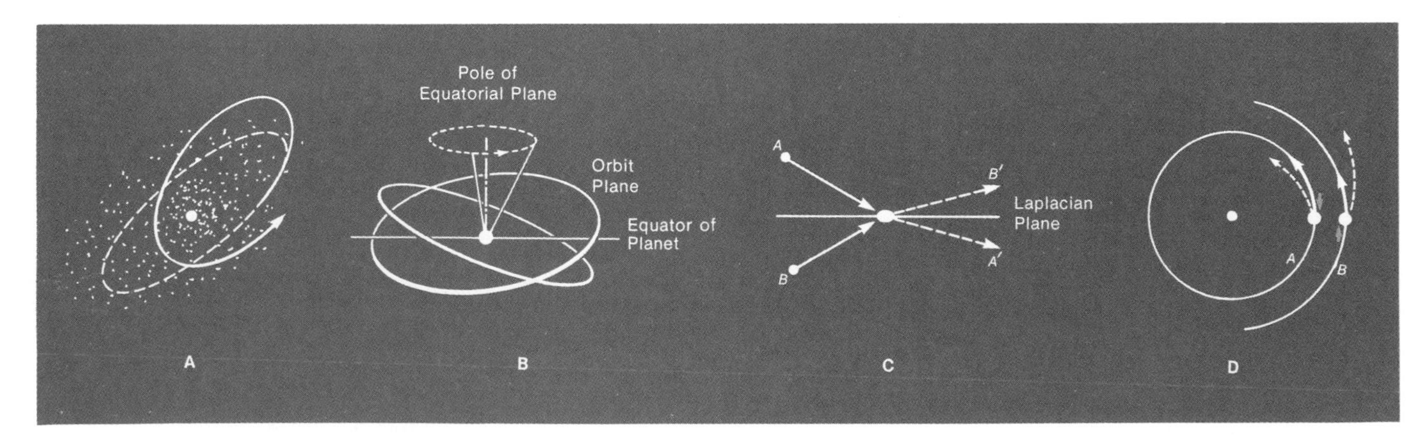

Figure 2. **The gross evolution of a circumplanetary cloud. *a*, In a swarm of particles shown surrounding a central planet, the dotted line represents the inclined, elliptical orbit of an undisturbed particle. The solid line depicts the same particle as its nearly planar path is gradually modified by perturbations. *b*, As seen in a side view, the pole of an inclined orbit precesses about the pole of an oblate planet's equator; the inclination remains constant, but the orbit's orientation continually rotates. *c*, Because of relative drifts, the orbits of two nearby particles *A* and *B* can intersect to allow collisions, which preferentially occur near the Laplacian plane. These collisions reduce out-of-plane velocities so that the cloud slowly flattens (dashed lines). *d*, As seen in this plan view, particle *A*, nearer the planet, moves faster than *B* and therefore passes it. If they collide, equal and opposite forces (colored arrows) push on the particles; *A* becomes slowed, causing it to drop along the (dashed) path closer to the planet, while *B* moves farther away.**

planet's equator. The warp amounts to less than 1 km for known planetary rings, since the disturbing satellites lie near their planets' equatorial planes. However, if Neptune proves to have an extended faint ring, it could be pulled dramatically off the equator by massive Triton, which occupies a highly inclined orbit.

Even after the first stage of dynamical evolution ends and the cloud of particles has collapsed to a flattened sheet, individual ring members will continue to interact with one another in a systematic manner (Figure 2d). To counteract the increased tug of gravity, particles close to a planet always move faster than those farther away. Thus, inner particles gradually overtake and sporadically rub against their outer, slower-moving neighbors, producing a gradient in velocity termed *differential shear* or *Keplerian shear*. The jostling that results tends to retard the inner particles, causing them to drop nearer the planet, while the outer ones are driven away from it. So, after flattening to a thin disk, planetary rings slowly spread. Unless their boundaries are restrained in some way, this radial diffusion should lead to a smooth disk of material. The third and final stage in a ring's internal evolution, which is never fully achieved, would occur when the ring particles are so widely separated that they no longer collide; thereafter, the ring does not change dynamically unless external forces act upon it.

Actually, the initial out-of-plane damping never produces a perfectly thin layer, because collisions cause particles to move out of the plane. In particular, the differential shear motion brings particles of radius r together with a typical velocity of ωr, where ω is the orbital angular velocity about the planet. Any collision reorients the relative velocities of the particles involved as well as damping them slightly. Therefore, the vertical component of ωr causes a ring at least several particles thick. The scattering of particles by one another's gravitational fields can also inflate the ring if the objects are large enough, at least 10 m. Obviously, for a ring containing particles of various sizes, collisions of large and small objects may deflect the latter vertically by many times their radii, while the former may be confined to a relatively thin layer near the central plane.

Low-velocity collisions also cause an unbounded ring to diffuse radially inward and outward, as mentioned earlier, at a rate proportional to the square of the particles' radii. Ring particles also drift gradually under the influence of both *Poynting-Robertson drag* (a drift toward the planet due to impacts with photons of sunlight) and *plasma drag* (a drift away from synchronous orbit due to collisions with magnetospheric plasma). But the rates of orbital evolution caused by these two processes are *inversely* proportional to particle size; that is, bigger objects are affected less. Consequently, if the rings we observe today formed aeons ago, the particles comprising them can be neither too small nor too large.

The model of a flattened, nearly opaque disk of particles orbiting a central object pertains to more than just planetary rings. Probably it also is a good representation for one phase in the evolution of the protoplanetary nebula from which the planets supposedly developed (see Chapter 23). But unlike the planets' accumulation, large objects cannot grow within planetary rings because the latter almost always lie close to planets, within what is known as the *Roche limit*. The Roche limit is 2.456 $R(p'/p)^{\frac{1}{3}}$, where p' is the planet's density, p that of the orbiting object, and R the planet's radius. Anywhere inside this distance a *fluid* satellite can no longer remain intact but instead gets torn apart by planetary tides. Solid satellites, because of their material strength, can exist inside the Roche limit; but above a certain size, they too will be broken apart by tidal forces. If a large satellite were to rupture, the resulting fragments would occupy similar orbits and continually chip away at one another until the largest members were no larger than some tens of kilometers.

THE GENERAL CHARACTER OF KNOWN RINGS

The realization that virtually all planetary rings lie within the Roche limits of their planets has naturally prompted the long-held popular speculation that ring systems arise when comets or satellites stray too close to a planet and become tidally disrupted. However, these same powerful tides might also have prevented material initially within the Roche limit from accreting in the first place. In other words, if a disk of debris extended to great distances from a just-formed planet, material far away could have accumulated into satellites, whereas that within the Roche limit would not have grown so easily. In addition, whatever existed very close to the planet might even have been lost to its surface by Poynting-Robertson and atmospheric drag. Thus a primordial circumplanetary disk should ultimately develop into a system with exterior satellites, rings, and then a gap just above the planet's surface – much like what we see today around Jupiter and Saturn.

As just stated, when far enough from the central planet, a disk of individually orbiting objects will gradually aggregate into a few larger bodies. However, Stuart J. Weidenschilling and Donald R. Davis have proposed that, near the Roche limit, the large aggregations neither fully coalesce nor entirely disrupt. Rather, groups of particles cluster together temporarily; but once large enough or spinning fast enough, they are torn asunder by tides. In this view, Saturnian ring members are "dynamical ephemeral bodies," changing their individual characters every few weeks.

Another key orbital distance is the *synchronous radius*, R_S, at which a satellite orbits in the same time as its planet rotates. It can affect the evolutionary paths of ring particles in two ways. First, planetary tides push satellites outward when they orbit beyond R_S but inward when they are closer; thus the orbits of large nearby objects can gradually collapse toward planets. Second, *electromagnetic forces*, which may overpower gravity for small particles (say, those much less than 1 micron across), reverse direction in a relative sense at R_S and, as such, may cause material either to accumulate there or move away from it.

Satellites undoubtedly play a pair of fundamental roles in sculpting the planetary rings we see today. First, they gravitationally perturb the orbits of ring particles, especially at resonances (to be discussed next). In this way satellites account for much of the understood gross form of Saturn's rings, confine the Uranian and Neptunian rings, and perhaps influence the structure of the Jovian ring. Second, the largest objects embedded within a ring (which could themselves be called satellites, though we will coin the name *mooms*) can serve as either sources or sinks for ring particles. Moreover, collisions with high-speed interplanetary projectiles may drive debris off these mooms, a process that over time creates many new small particles at the expense of the large ones. In combination these effects influence the particles' total surface area and may account for some of the observed brightness structure of planetary rings.

Resonant orbits are conspicuous features of the solar system's makeup today, in that matter is often either absent from, or preferentially present in, such locations. Resonances between two objects occur at those positions where the orbital period of one of them is an integer fraction of, or "commensurate with," the other's period (Figure 3).

Numerous examples of resonances exist. For instance, the *Kirkwood gaps* in the asteroid belt contain substantially fewer minor planets than do adjacent locations; these gaps occur where the orbital period of Jupiter (the primary perturber of the asteroid belt) and local orbital period form a simple ratio (like 2:1, 3:1, 5:2, and so on). Numerical studies by Jack Wisdom have shown that orbits at the 3:1 Jovian resonance become chaotic and develop large eccentricities; over time some asteroids traveling along such elongated ellipses can collide with a planet like Mars or the Earth. Resonant positions are not always vacant, however, and indeed sometimes they may be unusually populated: a family of asteroids called Hildas have orbital periods two-thirds that of Jupiter, while literally thousands of Trojan asteroids match Jupiter's period and share its orbit.

The Cassini division, which separates the two major classical rings of Saturn, is like a Kirkwood gap, since a particle at its inner edge (that is, the B ring's outer boundary) would have a period one-half that of the Saturnian moon Mimas. Many low-density regions in Saturn's A ring are located at positions resonant with the rather small satellites Janus, Epimetheus, Pandora, and Prometheus, all of which lie just beyond the ring system.

Resonance patterns are also seen in satellite systems. For example, the orbital periods of Jupiter's satellites Io, Europa, and Ganymede have a ratio of 1:2:4. In any resonance, given configurations are repeated periodically; for the just-mentioned Laplace resonance of the Galilean satellites, these alignments continually reinforce (through gravity) the eccentricity of Io's orbit and ultimately play a fundamental role in creating its vigorous volcanism (see Chapter 13). Resonances and near-resonances are likewise common among the Saturnian and Uranian moons, where they are suspected to have played a role in heating the interiors of enigmatic Enceladus and Miranda.

Despite these many examples, however, Voyager pictures of Saturn's rings show that the situation is more complicated: there is *not* a simple one-to-one correspondence of resonant locations with ring structure; even Cassini's division, while generally containing less material than its surroundings, is filled with many ringlets.

In the past the dearth of particles in resonant ring gaps was thought to be caused by a buildup of perturbations at such positions, as the involved particles and satellites periodically repeated their relative configurations. In a sense the mechanism is much like rhythmically pushing a swing until its occupant arcs high above the ground, whereas randomly timed pushes would accomplish little. Since orbits are most disturbed near resonances (so the argument goes), particles there run into one another more frequently and more violently than average, which inhibits the accumulation of material. But this is probably only a small part of the story. Except in a very narrow band to either side of the exact resonance, adjacent particles are almost equally perturbed and march side-by-side along similarly affected orbits.

A more likely cause was put forth in the late 1970s by Peter Goldreich and Scott Tremaine, who predicted that *spiral density waves,* like those thought to operate in galaxies, would originate at resonant locations in circumplanetary disks. Their mechanism begins much like that just described, but there is an additional twist. As before, the gravitational attraction of a satellite in the ring plane alters the orbits of ring particles, making them elliptical. In a populous disk this perturbation causes particles to bunch up, especially near resonances (Figures 4,5). However, say Goldreich and Tremaine, the combined gravity of these clumped particles pulls on the rest of the disk, producing condensations and rarefactions elsewhere and ultimately giving rise to a spiral-shaped density wave that propagates radially. If the perturbing satellite lies outside the ring, this wave moves outward. An entrained ring particle, if allowed to proceed

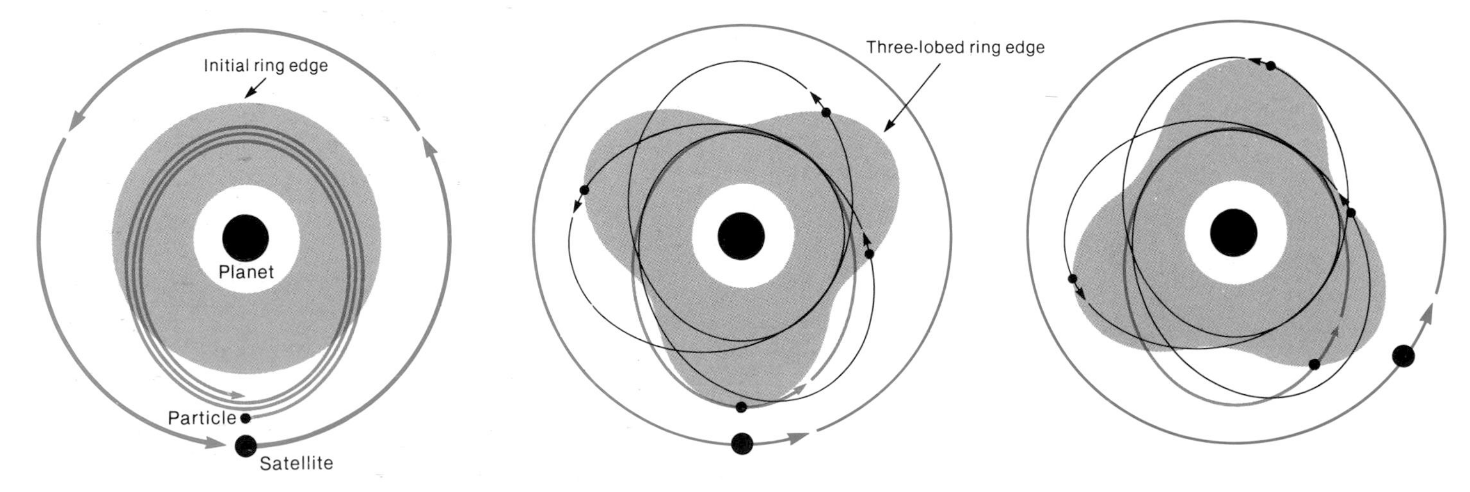

Figure 3. **These diagrams show the effect of a hypothetical 3:1 resonance between a satellite and a ring's outer edge. A particle at the ring edge (left) completes exactly three orbits in the time the satellite takes to make one. The satellite's repetitive gravitational influence forces the particle into an elliptical orbit. All particles at the resonance are also forced into elliptical orbits, which become most distorted as they pass beneath the satellite. Because of the threefold faster motion of the ring particle, three peaks will occur around a circumference (that is, the edge assumes a three-lobed shape, as at center). A short time later (right), the particles continue to move in elliptical tracks under the satellite's influence. But the net effect of their orchestrated motion is that the three-lobed edge appears to rotate with the satellite.**

without interference, would eventually return to its original position. But the rings are dense enough to make collisions probable, which robs the particles of energy and angular momentum, causing them to "fall" inward toward the planet. Over time, this process can clear a wide gap just outside the resonance position.

The physics deduced by Goldreich and Tremaine also cause a satellite and a nearby planetary ring to repel each other. In fact, their "shepherding model" was originally proposed to account for the narrow rings surrounding Uranus, because small satellites on either side of a ring could constrain its edges from spreading (Figure 6). The shepherding mechanism received substantial vindication when Voyager 1 discovered two satellites herding the narrow outer F ring of Saturn, and then again when Voyager 2 pinpointed yet another pair astride Uranus's ε ring. The spacecraft also discovered some four dozen spiral density wave trains, primarily in Saturn's outer A ring. Through very

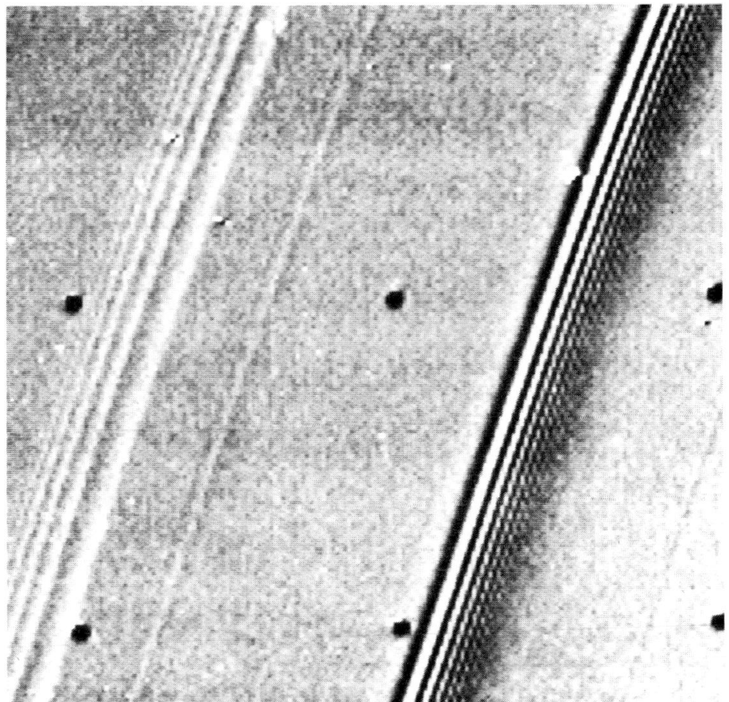

Figure 5. **Recorded by Voyager 2 in Saturn's A ring, these two wave features are very tightly wound and resemble watch springs. Both are caused by the satellite Mimas. The outer (left) one, a spiral density wave, is a series of particle-density fluctuations propagating outward (toward left). The inner one (right), a spiral bending wave, propagates inward as a train of vertical corrugations. This image is roughly 1,000 km across.**

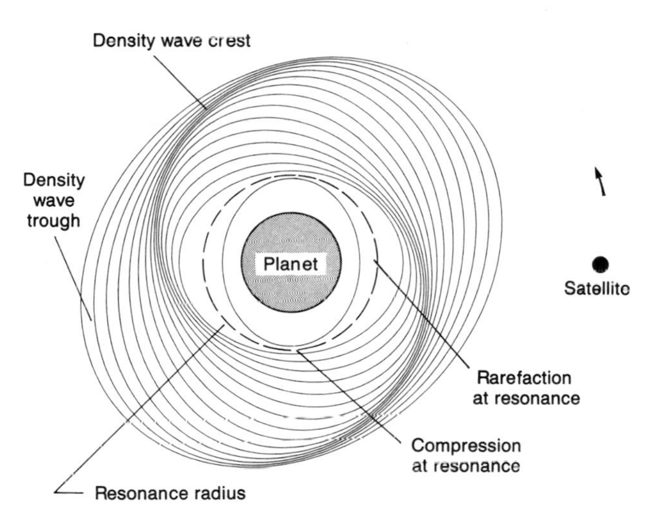

Figure 4a. **This diagram shows the orbital perturbations that induce a spiral density wave. A two-armed spiral wave is excited at the 2:1 resonance (shown dashed) with the exterior moon. The ovals represent particle streamlines as seen in the frame rotating with the orbiting satellite. Their long axes become less and less well aligned with the satellite's direction at greater distances from the resonances. The clustering that occurs in the orbital paths induces coherent oscillations in neighboring particles as they drift past by Keplerian shear. Actual spiral waves are much more tightly wrapped than shown here.**

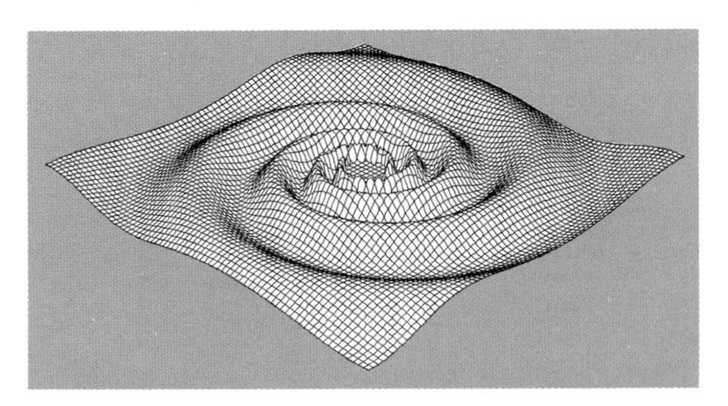

Figure 4b. **If resonant perturbations are induced in rings by a satellite on an inclined orbit, the out-of-plane tugs induce a vertical bending wave, as shown in this computer schematic of a two-armed spiral. The vertical displacement of these waves (about 1 km) has been greatly exaggerated but is 10 to 100 times greater than the rings' physical thickness.**

Figure 6. **The Goldreich-Tremaine model for constraining narrow rings employs "shepherd satellites" that force a group of particles to travel along narrow paths. Moving slower than the ring inside it, the outer satellite attracts particles going by. This force is slightly greater after particles have passed the satellite, because in slipping past it their paths have been pulled somewhat closer to the shepherding object (not shown). This extra force causes them to lose energy and "fall" closer to the planet, as shown by the red dashed lines. Conversely, the faster-moving inner satellite adds energy to nearby particles and kicks them onto higher orbits. Together, these forces herd the particles into a narrow ring. The same process can explain how an embedded satellite could cause a ring system to spread apart.**

similar processes, inclined satellites can pull ring material up out of the ring plane, producing a local rippling of the surface that propagates away from the resonance location. However, only a few of these *bending waves* have been found in Saturn's system (Figure 5).

We can gain considerable insight into the nature of the planets' ring systems by matching actual ring characteristics against the dynamical concepts described above, and also by comparing the known systems with each other (Figure 7, Table 1). For example, most rings are very thin relative to their extent. Essentially all of them lie within their planet's re spective Roche limits and, for the most part, the synchronous-orbit radius. Rings can be almost opaque in places, though Jupiter's (among others) is diaphanous throughout. Almost certainly, a range of particle sizes exists in each ring. The material in Saturn's rings is highly reflective, as are most of Saturn's icy moons. Yet particles in the other ring systems are dark, which suggests that silicate- or carbon-rich material – not ice – dominates their surfaces. Perturbing satellites border (and sometimes mingle with) the rings of Jupiter and Saturn, and a few small moonlets are adjacent to Uranus' rings.

As described in the following sections, all of these characteristics challenge our understanding of the rings' origin, evolution, and current state. Their study has been helped tremendously by the Pioneer and Voyager spacecraft, whose flybys in the 1970s and 1980s provided detailed observations unobtainable from Earth.

THE RING OF JUPITER: FAINT YET DISTINCT

As Pioneer 11 traversed Jupiter's magnetosphere in 1974, its counts of high-energy particles decreased when the

	Jupiter	*Saturn*	*Uranus*	*Neptune*
Width of main and narrowest structure (km)	7,000	20,000	100	≈15
	<100	<0.01	<0.01	?
Thickness (km)	<30 (halo 10^4)	0.01 – 0.1	0.01 – 0.1	?
Optical depth	$1-6\times10^{-6}$	0.1 – 2	0.1 – 2	0.1 – 0.4
Albedo	≈0.05	0.2 – 0.6	0.03	low (?)
Particle sizes	10^{-3} mm	cm – 5 m	10 cm – 10 m	cm – m (?)
Surface mass density (g/cm²)	$10^{-10}-10^{-3}$	10 – 100	10 – 100	?
Total mass (g)	$10^{11}-10^{16}$	$10^{20}-10^{21}$	$10^{18}-10^{19}$	?

Table 1. **Each outer planet has a ring system whose characteristics are distinctly different from the others'. *Optical depth* refers to the fraction of light able to penetrate an obstructing layer; for small values, it is essentially the fraction of photons that cannot pass through. The typical *particle sizes* listed are based on numerous assumptions and should not be taken too literally. *Surface mass density* is the optical depth multiplied by the average particle's size and density, and *total mass* is the ring's surface mass density times its area. The extensive ring system around Saturn, if compressed to a single object, would form a small satellite no more than 100 km across.**

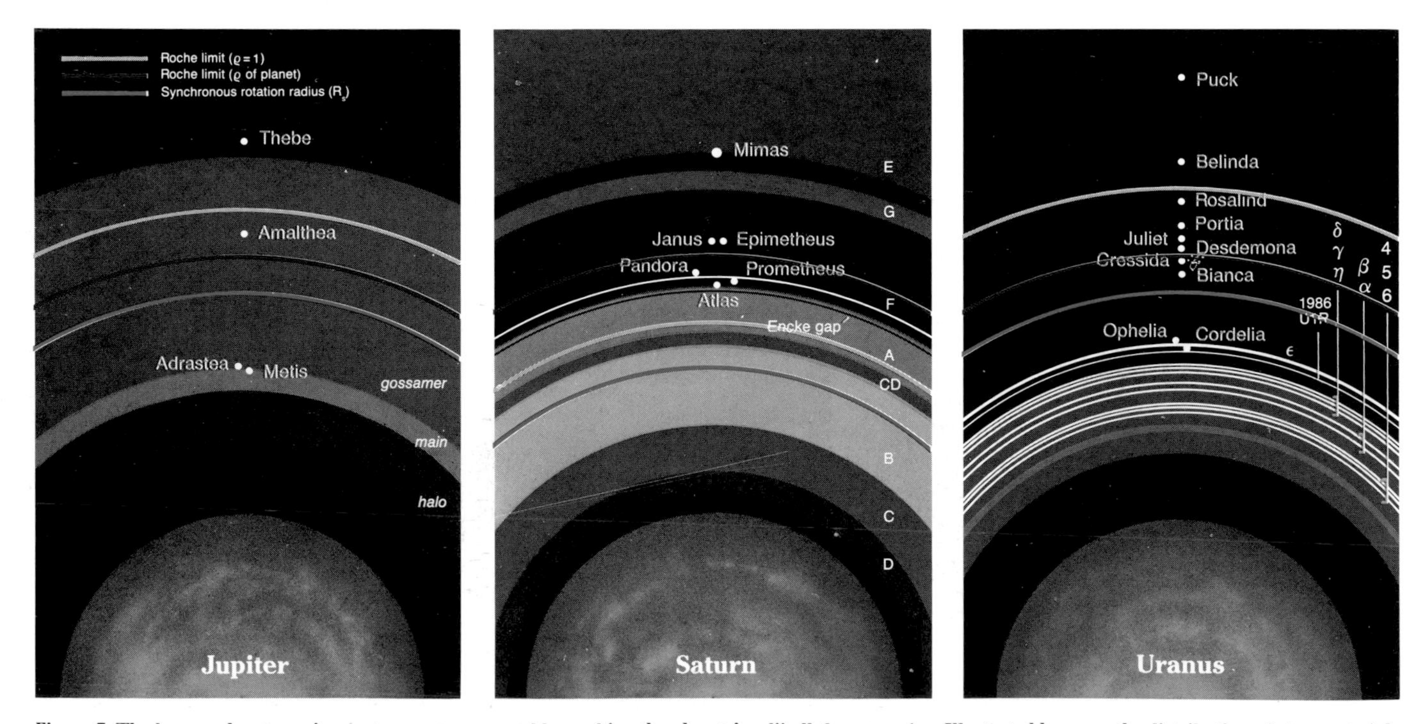

Figure 7. **The known planetary ring systems are compared by making the planets' radii all the same size. Illustrated here are the distribution of ring material, nearby satellite locations, the synchronous-orbit radius R_s, and the Roche distances for satellites having either the density of 1 g/cm³ (that of water) or the density of the planet.**

spacecraft came its closest to the planet's center, a distance of 1.6 Jovian radii (R_J) or about 114,000 km. This unexpected trend prompted investigators Mario Acuña and Norman Ness to hypothesize that a hitherto undetected satellite or a ring might have been in Pioneer 11's vicinity, which would absorb charged particles that it encountered. But most scientists dismissed this possibility, since ground-based observations had not revealed any such object, and other explanations of the reduced counts were available. Five years later Voyager 1 was programmed to look along Jupiter's equatorial plane while traversing it. In doing so, the spacecraft detected a faint disk of material encircling the planet.

In analyzing observations taken by Voyager 2 four months later (Figures 8,9), Mark Showalter and his co-workers distinguished three components for the Jovian ring. The main band starts abruptly at 1.81 R_J and ends more gradually at 1.72 R_J, 7,000 km closer to the planet. Its thickness is no more than 30 km. Even on the brightest part of the ring, matter covers a very small fraction of space (its optical depth is a few times 10^{-6}). At least partly because of the image smearing, the Jovian main ring appears unusually smooth and featureless compared to other planetary rings; only three bright lanes are visible in the main band, and two of them seem to be associated with the known satellites Adrastea and Metis. A toroidal halo (with a comparable optical depth) arises at the main ring's inner edge and rapidly expands inward to a full thickness of roughly 20,000 km; it is faintest where closest to the planet and farthest from the equatorial plane. Beyond the main band is the so-called gossamer ring, which is even more tenuous. Its brightness decreases linearly outward from the main ring until it fades into the background at about 3 R_J. The only feature in this ring is a slight enhancement at, or near, the synchronous-orbit radius.

The Voyagers found that the Jovian ring is more than 20 times brighter in forward-scattered light (which is redirected mostly by diffraction away from the Sun) than in reflected (back-scattered) light. This implies that a significant number of the ring particles are only 1 or 2 microns across. Such small objects have correspondingly short lifetimes because (1) Poynting-Robertson drag, plasma drag, and perhaps other electromagnetic effects cause them to leave the system within roughly 1,000 years; and (2) in a comparable time they are destroyed by micrometeoroid impacts or eroded more gradually by the energetic particles that are abundant in Jupiter's magnetosphere.

Given its brief existence, the dust around Jupiter must be regenerated continually if the ring is a permanent feature. Most likely, fast-moving projectiles from outside the Jovian system are bombarding boulder-size objects within the ring, chipping off flecks of dust in the process. The impacting projectiles are probably interplanetary micrometeoroids but could even be volcanic dust drawn out of Io's rarified atmosphere and transported inward.

The magnetospheric plasma that surrounds Jupiter continually sweeps past the ring particles and creates a negative electric potential on their surfaces. Because of this charging, Jovian ring particles of micron size feel a force due to the planet's magnetic field that is about 1,000 times weaker than the planet's gravity. At most places in this ring, such small perturbations will induce comparably small oscillations about a circular path. However, plasma drag causes particles from the main ring to evolve inward, where the Jovian magnetic field and its consequent effects on charged particles intensify. As Les Schaffer and I have found, when grains reach the inner regions of the main ring they experience electromagnetic forces that vary in time with the same periods as their orbits. At such *Lorentz resonances*, an initially circular orbit becomes substantially eccentric and inclined. Grains that drift farther inward encounter another Lorentz resonance near the halo's interior boundary, where their inclinations and eccentricities are increased yet more.

Throughout this evolution the grains are continually being ground down and thus subject to larger and larger electromagnetic perturbations. Eventually they become minuscule

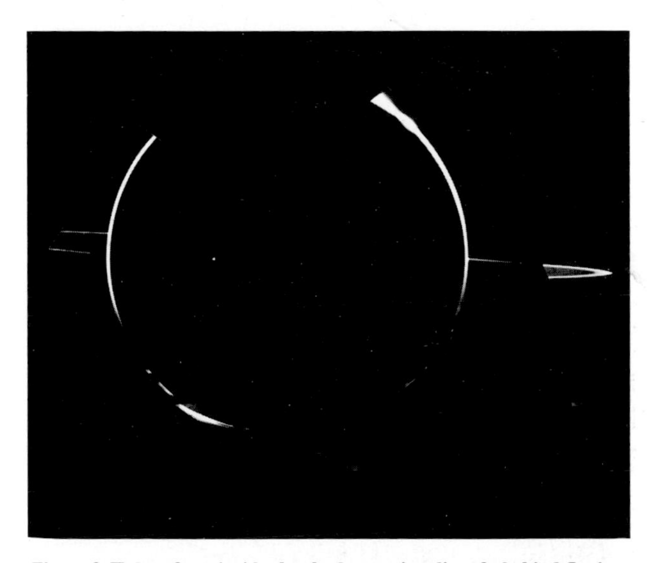

Figure 8. **Taken from inside the shadow region directly behind Jupiter, this mosaic of Voyager 2 frames in forward-scattered light shows a faint ring to either side of the planet. Haze in Jupiter's upper atmosphere accounts for the brightness of its limb.**

Figure 9. **One arm of Jupiter's ring system, as seen by Voyager 2 when in the planet's shadow and color-coded according to brightness. The main ring (the nearly uniform white strip) is the brightest; the halo (outlined in red-yellow) begins at the main ring's inner edge and extends vertically above and below the ring. The gossamer ring (blue-green band) extends outward from the main ring. The mottled background is image noise.**

enough (about 0.03 micron across) that electromagnetic forces acting on them overpower the effects of gravity. Once this point is reached, the particles are swiftly yanked totally out of the ring plane and travel along field lines into the planet's atmosphere. Because the Jovian ring is relatively uncrowded, this microscopic dust gets pumped into the halo before interparticle collisions can dampen its motion. Thus, the Jovian halo does not flatten as other rings do.

Jupiter's main ring contains a variety of particle sizes. Very tiny grains are by far the most abundant, and these are almost surely impact debris. Their unseen parent bodies are responsible for the absorption of energetic particles observed by passing spacecraft; dark and red (much like the nearby small satellites), these larger objects are probably contaminated by the material from Io that pervades Jupiter's inner magnetosphere. The largest fragments may be the outcome of the catastrophic breakup of a small satellite, or may just represent uncompleted accretion of a satellite located within the Roche limit. Whatever their origin, they have gradually spread apart from differential shear and drag, filling the main region populated by dust. The satellites Adrastea (an object 25 km across that skirts the ring's outer boundary) and Metis (somewhat larger in size but located a little further into the ring) may be merely the largest of these mooms.

The Galileo mission, which is to survey the Jovian system in the mid-1990s, will be able to address many of the questions that we now have as to the character of the Jovian ring. Despite its insubstantial nature, the tenuous band around Jupiter has already illuminated our understanding of planetary ring systems because its low optical depth and the smallness of its constituent particles accentuate certain phenomena. In particular, it is apparent that relatively large mooms must surely be embedded in Jupiter's ring. Accordingly, we wonder whether the other ring systems are also fed from as-yet unseen reservoirs of material.

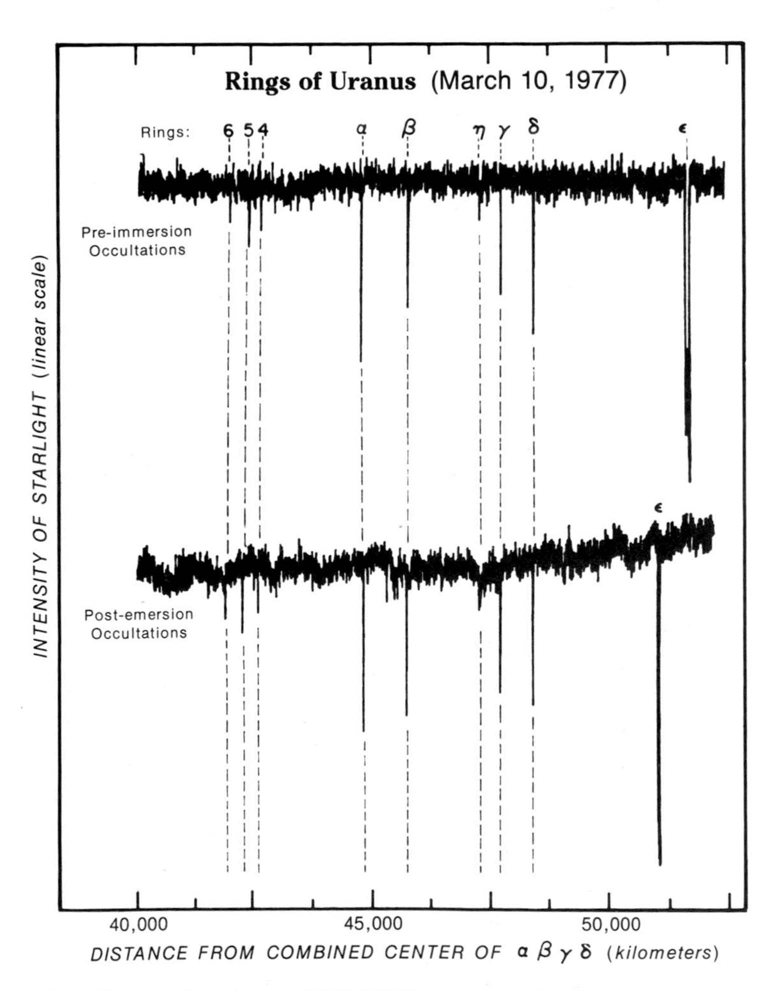

Figure 10. **Light from the star SAO 158687 was recorded before (top) and after (bottom) its occultation by Uranus' disk in 1977. Dips in the tracing are due to each of the nine Uranian rings (labeled individually); apparent fluctuations in the star's brightness are caused by system "noise." Note that the starlight is diminished at nearly the same distances on either side of the planet, implying the rings are nearly circular. Most rings obscure the star very well and rather abruptly, so they must be quite opaque and have well-defined edges. But to have these properties, small satellites must exist nearby to prevent the rings from spreading. Ring ε is the most eccentric and shows different widths on the two sides of the planet.**

URANUS' RINGS: NARROW AND NEAT

The way in which a star's light is extinguished as a planet passes in front of it tells much about the planet's atmospheric properties and also permits measurement of a chord across the planet's disk. With this in mind, several teams of astronomers trekked to observatories surrounding the Indian Ocean to witness the occultation of a star by Uranus on March 10, 1977. They received an unexpected bonus when a set of slender rings revealed its presence about the planet by interrupting the starlight. The best observations (Figure 10) came from James Elliot's group, who watched the occultation high above the Indian Ocean through the Kuiper Airborne Observatory's 91-cm telescope.

During the 1977 occultation the rings attenuated starlight substantially at approximately symmetric points about the planet; thus observers concluded that the Uranian rings are densely packed and are almost circular. The observations revealed a total of nine rings, which were named in order of increasing distance from Uranus: 6, 5, 4, α, β, η, γ, δ, and ε. Over the years numerous stellar occultations have been observed from the ground and from Voyager 2 during its flyby of Uranus in January 1986, and we now know the shapes of these rings – which are 3 billion km from Earth – to a precision of few hundred meters! Compared to their circumferences (some 250,000 km), the rings are remarkably narrow; most do not exceed 10 km in width, and only the outermost one, ε, spans as much as 100 km.

At least six Uranian rings are inclined with respect to the planet's equatorial plane, but typically by no more than a few hundredths of a degree; they are also slightly out-of-round, with eccentricities of 0.001 to 0.01, and have variable widths. The shapes of several rings seem to pulsate slightly; one "breathes" in and out, while another changes its eccentricity somewhat. The ε ring, one of the few not inclined, is by far the most massive and eccentric one in the system. Its distance from Uranus varies by about 800 km and its width changes accordingly: from 20 km (where closest to Uranus) to 100 km (where farthest from it).

A planet's oblateness causes elliptical and inclined rings to precess slowly about the planet. In the Uranian case, the rings precess as rigid structures, which is surprising since oblateness should drive a ring's inner edge more rapidly than its outer boundary, turning each ring into a circular band in no more than a few hundred years. Apparently, therefore, the rings' eccentricities must be continually induced, either by satellites or by the ring material itself. However, while models that employ this concept seem generally correct, they still fall short of matching our observations completely.

Uranian ring boundaries, like many features in the Saturn system, are remarkably crisp, suggesting the ineffectiveness of the differential shear that should force particles to collide and thus diffuse smoothly. Sharply defined rings can only be explained if satellites bound the edges in some way, either by shepherding or by trapping ring particles along so-called horseshoe orbits that almost precisely duplicate the satellite's path. If nearby, such moons need only be a few km across and could thus have escaped detection by Voyager 2's instruments and ground-based telescopes. However, the spacecraft did confirm, in remarkable detail (Figures 11,12), our impressions of the rings gained from ground-based observations. In addition, Voyager 2 discovered two faint rings of its own: 1986U1R, a narrow strand between the ϵ and δ rings, seems to contain predominantly small particles and may be longitudinally variable; 1986U2R begins about 1,500 km interior to ring 6 and has a radial extent of about 3,000 km. The spacecraft also found a pair of small satellites that

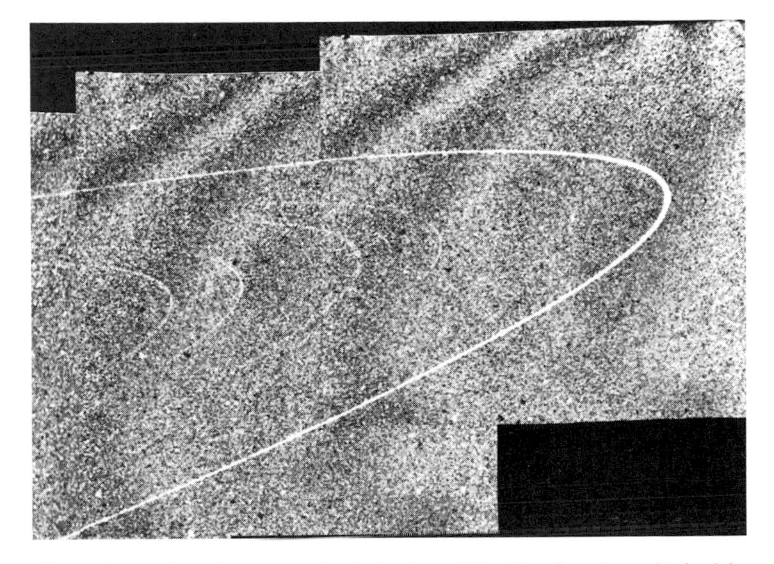

Figure 11. **A three-frame mosaic of the thread-like Uranian rings obtained by Voyager 2 as it neared the planet's equatorial plane. A short arc of the new ring feature designated 1986U1R can be seen between the bright ϵ ring (outermost) and the δ ring (the next one in).**

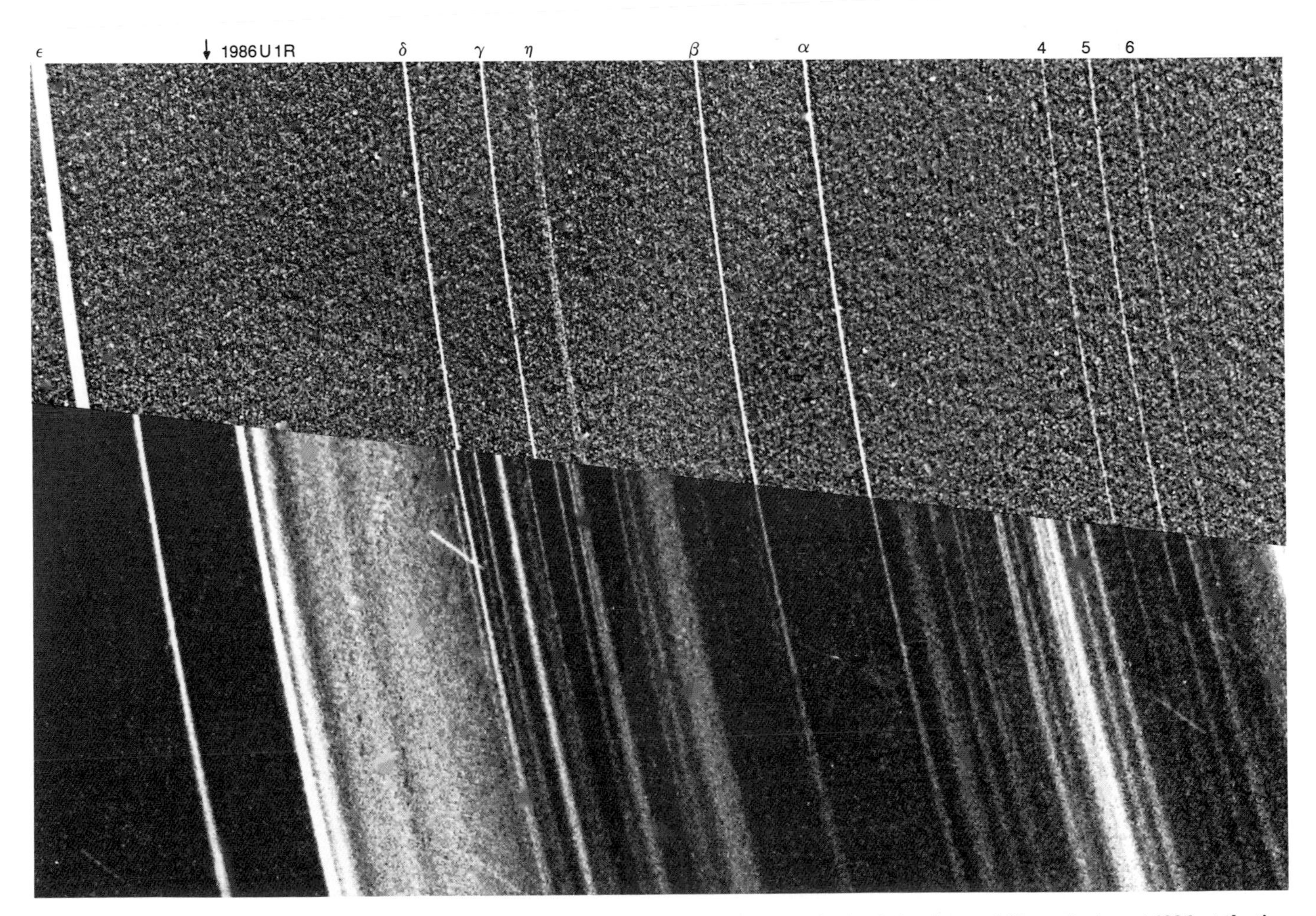

Figure 12. **Compare these images of the Uranian ring system. Voyager 2 acquired the upper one the day before it passed Uranus in August 1986; at the time sunlight striking the ring particles was reflected back toward the camera. Computer enhancement makes the rings appear much brighter than they really are. (To see the faint ring 1986U1R, put your eye close to the page and look along the arrow.) In the lower image, taken somewhat later, Voyager 2 was looking almost directly in the Sun's direction (the phase angle is 172°.5). This backlighting of the rings dramatically enhanced the visibility of the micron-size dust particles they contain. Voyager 2's motion during the exposure caused some smearing of the detail, especially near the bottom edge. The nine rings discovered from Earth can be discerned with relative ease (note the poor segment match for the markedly eccentric ϵ ring). But many other ringlets visible here do not correspond to known features in the system.**

shepherd the ϵ ring (Figure 13) and likely also constrain edges of rings δ, γ, and perhaps 1986U1R.

According to visual and radio observations, the nine "classical" rings are composed mainly of meter-size boulders and contain scarcely any dust at all. These strands are among the darkest objects in the solar system, reflecting just a few percent of the feeble sunlight that strikes them, and they have neutral colors. Such dark color, which the rings share with the numerous small moons that Voyager found nearby, may result from the magnetospheric bombardment of surfaces that contain some organic molecules.

While in the shadow of Uranus, Voyager 2 briefly peeked sunward toward the Uranian rings, which presented quite a startling image in forward-scattered light (Figure 12). The entire region inward of 1986U1R was filled with faint dust (with optical depths of 10^{-5} to 10^{-4}) organized into many slender ringlets, among which the nine rings are unexceptional members. Most of the ringlets are remarkably narrow, which is surprising since Poynting-Robertson and plasma drag should be driving small grains into Uranus relatively fast (in no more than a million years for a micron-sized grain). But the actual lifetimes are probably much shorter, because the outermost vestiges of Uranus' atmosphere should extend into the ring region and thus hasten the orbital collapse of particles near the planet. Mechanisms like those at work in the Jovian ring system probably account for the continued presence of Uranian dust. A major puzzle is what causes the ringlet structure, with theorists now concentrating on the roles that embedded moonlets play as both sources and sinks for dust particles.

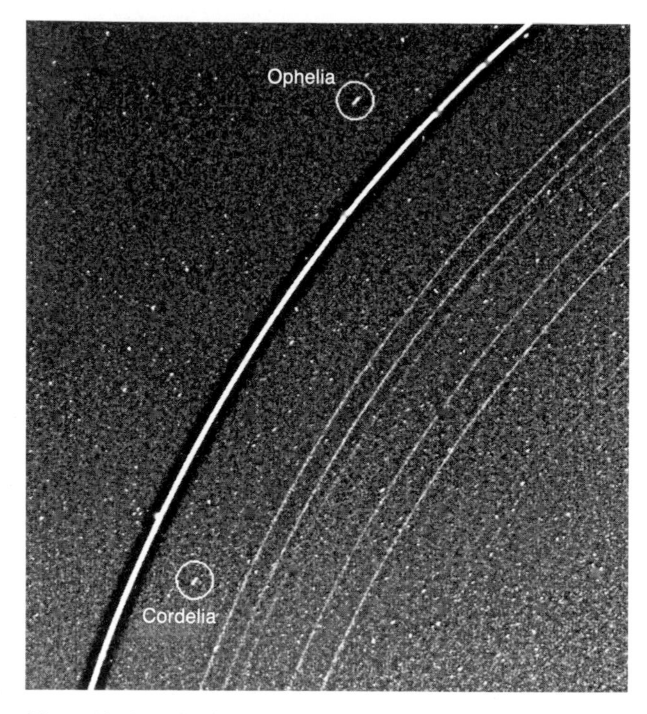

Figure 13. **Cordelia and Ophelia, a pair of shepherding satellites on each side of Uranus' ϵ ring, keep the ring particles in place through resonant gravitational forces. No other shepherds for the other rings were spied.**

NEPTUNE'S ENIGMATIC RING ARCS

Following the discovery of rings around Jupiter and Uranus during the late 1970s, planetary astronomers naturally expected to find material about Neptune, the most distant giant planet. Rather than awaiting Voyager 2's encounter with Neptune in August 1989, campaigns were mounted in the early 1980s to detect rings whenever the planet occulted a star (Figure 14). After the first several attempts proved unsuccessful, the scientists generally believed that no material existed about Neptune. However, during an occultation on July 22, 1984, teams of observers led by André Brahic and William Hubbard found good evidence for a strip of obscuring material about 50,000 km from the cloud tops. In many regards, the characteristics implied by this event and a few subsequent successful occultations bear

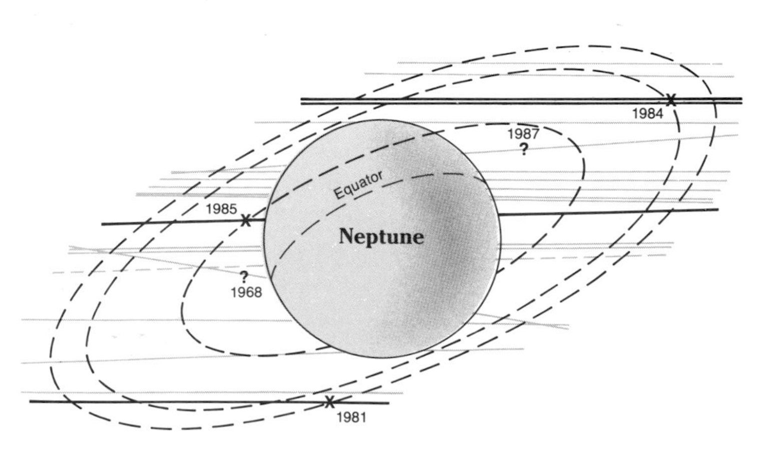

Figure 14. **Here, from Earth's perspective, are the paths of many stars that have been occulted by Neptune over the past decade. Few of these events yielded detections of ring arcs (crosses); question marks denote reported, but doubtful, events. Some tracks (dashed) have a north-south ambiguity because only one telescope observed the occultation. Dashed ellipses represent circular "ring arc" orbits in the plane of Neptune's equator with radii of 56,500, 67,000, and 70,000 km.**

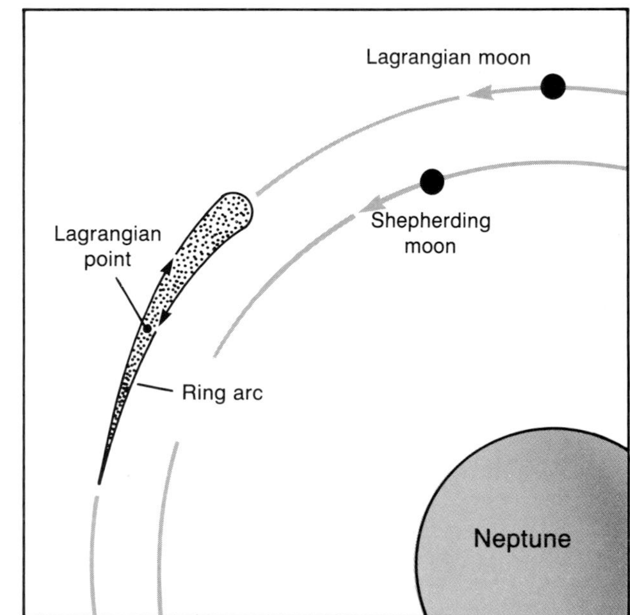

Figure 15. **Soon after it became obvious that something unusual was orbiting Neptune, theorist Jack Lissauer proposed a scheme in which two satellites could create ringlike arcs of particles at stable Lagrangian resonances. (The arc's width is shown much exaggerated.) According to Lissauer, the satellites need be no larger than a few hundred kilometers.**

similarities to the Uranian rings: moderate optical depths, abrupt edges, and widths on the order of 10 km.

But they differ dramatically in one very important aspect. Neptune's rings apparently do not encircle the planet – from the hit-and-miss nature of occultation data, they instead only fill about 10 percent of the orbital circumference! Moreover, the precise radial locations of these so-called "ring arcs" vary, implying the existence of several separate arcs each many hundred kilometers in length (their absolute radial positions are undeterminable because of the large uncertainties in Triton's mass and the orientation of Neptune's rotational pole). Finally, based upon an observation in which only one member of a binary star was occulted, the arcs terminate abruptly at their ends.

These observations pose huge challenges for theorists. One of their models holds that ring material is trapped around a series of discrete resonances corotating with an inclined satellite that, even though about 200 km or more across, cannot be seen from the ground. In another, material is confined at one moon's triangular Lagrangian point by a smaller shepherding object (Figure 15). In each case, interactions with satellites prevent diffusional spreading. Voyager 2 should be able to spy these odd slivers of material – and if it succeeds it should expand on the little we currently know about their bizarre nature.

SATURN'S SYSTEM: RINGS GALORE

Ground-based observations at infrared, visible, and radio wavelengths have provided considerable information on the properties and overall form of Saturn's celebrated rings. But these results have been all but eclipsed by the findings of the Voyager spacecraft, which resolved previously unimaginable architecture and taught us much about the size, shape, and distribution of particles in the system.

Saturn's rings are much more elaborate and complex than any of the other systems (Figures 16-19). The bright,

Figure 16. **As seen by Voyager 1 from 1.57 million km above and behind Saturn, the planet's bright, overexposed limb is visible through the rings. The C ring is relatively faint as it crosses the planet's limb. The material in parts of the B ring is so closely packed that it entirely screens the planet in places; radial spokes near the top center of the B ring are bright in forward-scattered light. The thin F ring displays its characteristic brightness variations.**

Figure 17. **The rings of Saturn, in a computer-processed image from Voyager 1. The threadlike F ring is accompanied by one of the two shepherding satellites that skirt its edges. Interior to this lies the bright A ring (the darkest gap within it is called the Encke division). As seen here, the Cassini division consists of four ringlets, each about 500 km across. The broad B ring does not exhibit as much structure because it is more uniformly opaque – though at higher resolution (as in Figure 22) it is very complex. The inner C ring shows many narrow ringlets up to the point where it ends abruptly.**

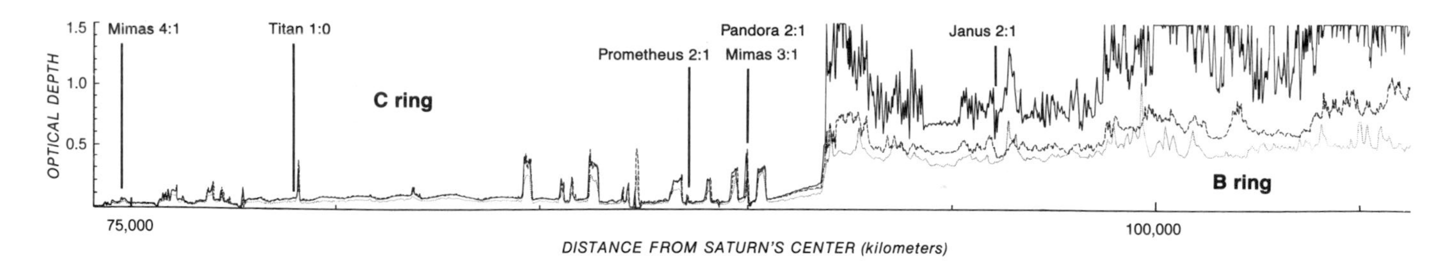

Figure 18. **The optical depth of Saturn's rings, as determined by occultations at three wavelengths. Various named features and resonance locations are labeled. In visible light (solid line), which is affected by particles larger than 1 micron, the rings' opacity can vary greatly even over short intervals. The dashed and dotted curves were obtained when the rings interrupted the Voyager spacecraft's radio signals at two wavelengths sensitive to particles larger than about 1 and 4 cm,**

Figure 19. **The narrow and enigmatic F ring of Saturn can appear either as a smooth and featureless ribbon or, as here, a set of knotted and braided strands. This clumpy appearance is thought to be caused by the shepherding satellites Pandora and Prometheus (inset) discovered by Voyager.**

classically recognized components A and B are separated by Cassini's division, which was first noticed in 1675. Optical depths average 0.5 to 0.7 for the A ring but rise to between 1 and 2 for the B ring. Interior to these is the crepe or C ring, recognized in 1850, with an average optical depth of roughly 0.1. Voyager 1 detected some material, including a few narrow, widely spaced bands, inside the C ring that extends at least halfway to the planet; however, this D ring is essentially undetectable from Earth.

Outside the traditional system lies the E ring, which is so tenuous (optical depth: about 10^{-6}) that it is scarcely more than a slight concentration of debris in the satellite orbital plane. It becomes visible only when the ring system is viewed approximately edge-on, as in 1979–80, when astronomers found that material extends to at least Rhea's orbit and thickens to perhaps 30,000 km at its outer edge. Its particles must be predominantly micron-size, like those in Jupiter's ring, so they have correspondingly short lifetimes and thus were only recently injected into the system. Since the E ring's density peaks near Enceladus' orbit, this uniquely smooth and bright satellite is the suspected source of the ring's material.

In 1979, Pioneer 11's crude imaging system located the slender F ring just 4,000 km beyond the A ring. But the spacecraft also found that charged particles were intermittently absent from this region (see Chapter 3), which hinted that undiscovered moonlets or rings might also lurk in the vicinity as well. When the Voyagers arrived in 1980 and 1981, they indeed found several new satellites skirting the edge of the rings and confirmed the existence of two more that had just been properly identified by terrestrial astronomers. The Voyagers also glimpsed a ring of their own, named G, which is located about 2.8 Saturnian radii out, surprisingly narrow (8,000 km wide), and even more tenuous than the E ring. And the F ring (Figure 19) turned out to be a contorted tangle of narrow strands.

When viewed edge-on, Saturn's main rings remain faintly visible and appear to be about 1 km thick. This is not their true thickness, which several lines of evidence have shown to be some tens of meters, but instead it may correspond to the height of localized vertical ripples (bending waves). Other possible contributors include an overall warping of the ring's Laplacian plane due to Titan and the Sun, an "atmosphere" enveloping the rings, and the F ring having some slight inclination.

The rings also exhibit what is termed the *opposition effect* – they surge suddenly in brightness whenever the phase (illumination) angle approaches 0°. That is, whenever we observe the system to one side of the direction of incoming sunlight, we must be seeing some of the ring material in shadow. The particles themselves may have rough surfaces, but more likely they occasionally shadow one another. Calculations show that in the latter case the particles must be separated by 5 to 10 times their size and be at least several layers thick. However, it is possible that the largest objects (which represent only a small fraction of the total surface area) do in fact reside in a more confined, central layer.

As described earlier, collisions or gravitational scattering by such large ring members can thicken a ring. Scattering might also be responsible for an interesting brightness variation that has been noticed primarily in the A ring. When the rings are fully open (that is, when one of Saturn's poles is tipped 26° toward us), the A ring appears about 10 percent brighter in the two quadrants approaching the Earth-Saturn line. This brightening reaches 20 percent when the rings have

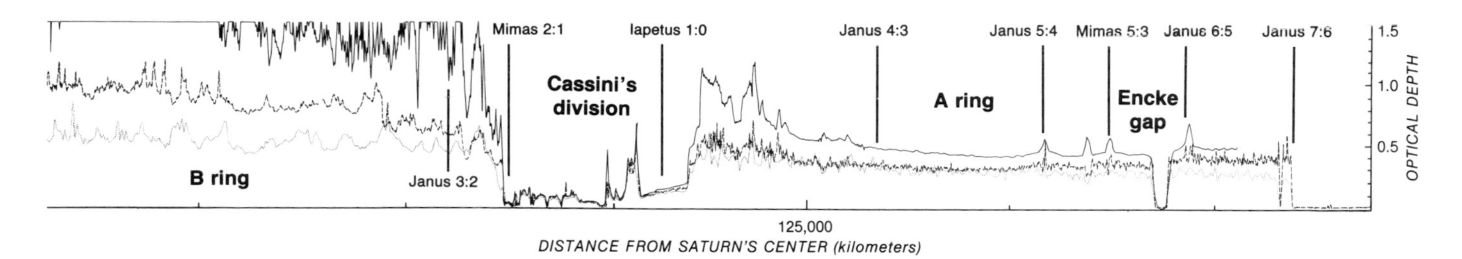

respectively. Separations in the three signals' strengths indicate the area covered by particles in the intervening size range. Differences in the structural character of the A, B, and C rings are even more apparent in these occultation profiles than in the images.

intermediate tilts (about 10°) but then decreases as the rings approach their edge-on configuration. Perhaps the particles are not spherical and have become rotationally locked by tides, such that their long axes all point toward Saturn.

But the particles are almost certainly too small for tides to exert such control, and in any event interparticle jostling would disrupt any well-ordered orientation. Instead, the largest ring members probably act to cluster particle orbits. When viewed from above, the pattern would contain a series of short arcs like the blades of a turbine. These structures appear "broadside" to our view in the morning and evening quadrants, and thus reflect more light. This clustering mechanism would be most effective when the rings are only moderately filled, as the A ring is.

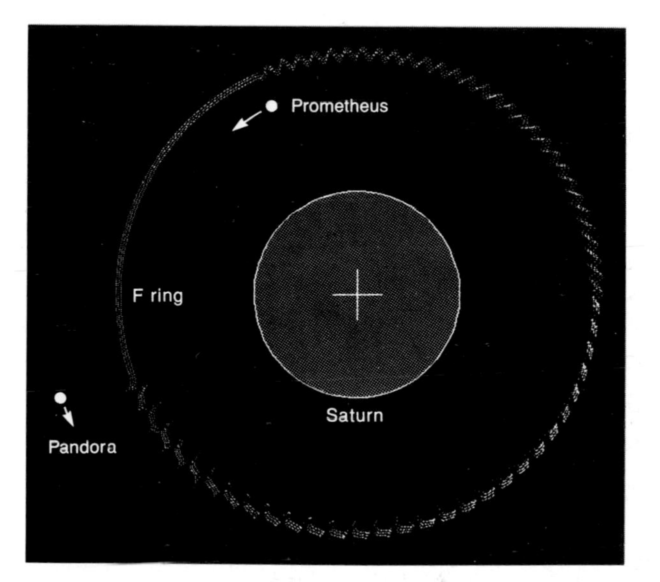

Figure 20. **A computer simulation by Mark Showalter and the author of the ring perturbations produced by shepherd satellites. The faster moving inner satellite has a circular orbit and creates a trail of smooth, sinusoidal waves. The sluggish outer moon, in an eccentric orbit, creates waves that contain discrete clumps and travel ahead of it. The spacings of some structures in Saturn's F ring compare favorably with those that should develop due to the shepherd satellites assumed in this model.**

THE COMPOSITION OF SATURN'S RINGS

During the 1970s, Gerard Kuiper, Carl Pilcher, and their colleagues used infrared spectroscopy to determine that water ice dominates the ring particles' outer layers. But a slight reddening of the rings in visible light indicates some surface contamination, perhaps from trace impurities, micrometeoroid debris, or radiation-induced modification of the ice's crystalline lattice. Clathrate compounds (in which methane or ammonia molecules are packed into ice's crystal lattice) may be present as well, because these are presently indistinguishable from water ice in near-infrared and visible spectra. Calculations that assume the optical properties of ice also match our observations of how the rings vary in brightness with changing wavelength and phase angle.

Based on infrared and microwave measurements, the ring particles' interiors (as well as their surfaces) could be water ice. Oddly, ice-coated *metallic* particles would also satisfy these measurements. But, cosmochemically speaking, metal just doesn't make sense (see Chapter 23); not only should water ice be the dominant solid condensate in the outer solar system, but also the densities of nearby Saturnian satellites and of the particles themselves are not far from 1 g/cm^3. Therefore, Saturn's ring particles are thought to be dirty snowballs throughout. Curiously, somehow the icy exteriors of Saturn's rings and satellites remain relatively pristine, especially compared to the darkened rings of other planets and the black cometary nuclei common in the outer solar system.

Particles in the C ring and within Cassini's division have noticeably different properties than do those elsewhere in the system (Figure 21). They are generally somewhat bluer, larger, darker, and do not scatter light forward as well as the average material in the A and B rings. The eccentric C ringlets have distinct optical properties as well. Such segregation of particle compositions – and its maintenance over geologically long times – is a further indication that diffusive processes are less effective than once believed. Abrupt changes in properties across ring boundaries suggest that differences among the system's components may have survived from primordial time.

We can estimate particle sizes using several observational techniques. The cooling rate of Saturn's ring particles as they enter the planet's shadow implies that, if solid, they are bigger than about 1 or 2 cm; in fact, they more likely even larger than this. Moreover, much of the ring population must exceed a few centimeters in size because the system as a whole is remarkably reflective to ground-based radar transmitting at the two frequencies most sensitive to particles larger than about 1 and 4 cm, respectively. This observation, combined with the rings' darkness at radio wavelengths, led James Pollack and Jeffrey Cuzzi to conclude that a many-particle-thick layer with a "power law" size distribution (containing many centimeter-size particles for every meter-size one) would satisfy the observations. According to G. Leonard Tyler and Essam Marouf, who analyzed radio signals transmitted through the rings by the Voyager spacecraft, particles in the A and C rings fit a power-law size distribution with a lower limit of about 1 cm and an upper limit of about 5 to 10 m. These cutoffs seem to vary somewhat from region to region (Figure 18).

However, this model requires several caveats. First, while little dust is evident, undoubtedly some particles in the A and B rings are quite small, particularly those continually being generated by micrometeoroid impacts. Second, to account for all the intricate ring structure, a few massive objects hundreds of meters or even 1 km across are likely present. Furthermore, because the unilluminated sides of the rings are extremely cold (about 55° K), some scientists argue that a single layer of material is called for because it would shadow portions of particles more effectively. Finally, however unlikely, the existence of metallic objects larger than a few centimeters cannot be ruled out.

SATURN RING DYNAMICS

The Voyagers' high-resolution images disclosed Saturn's rings to be much more finely divided than anticipated. Rings lay within rings: upwards of 1,000 have been counted in images (Figure 22), but the Voyager occultation results revealed that many more "ringlets" are present at scales less than 10 m. Even the so-called gaps are crammed with material – Cassini's division, for example, contains perhaps 100 ringlets. Therefore, such classically identified divisions merely denote *relative* absences of matter, in comparison to adjacent portions of the A and B rings.

Brightness variations in the A and B rings are due principally to the spacing of ringlets; Saturn's disk can be seen through most of the "opaque" B ring (Figures 23,24), and only a few places are truly impenetrable to light. Surprisingly, the C ring exhibits a more organized structure than either A or B. This arises in part from its lower optical depth, so there is little multiple scattering to mask details from our view. But this region is also structurally different, as can be seen in Figure 18.

As mentioned earlier, some prominent features of the rings appear to be associated with satellite resonances – especially the strongest ones, which lie mainly in the outer A ring. Nevertheless, much of the radial structure found in rings B, C, and elsewhere does not seem to be correlated with the simple perturbations of known satellites. Some strongly perturbed locations contain material, while others do not;

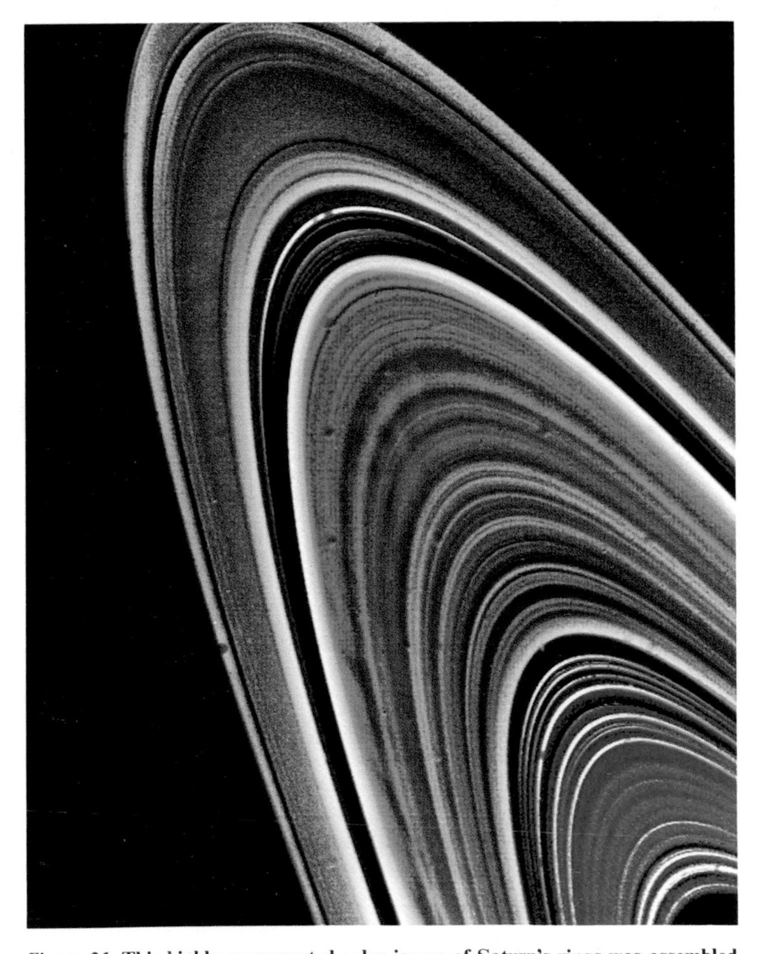

Figure 21. **This highly exaggerated color image of Saturn's rings was assembled from clear-, orange-, and ultraviolet-filtered frames taken by Voyager 2 from 8.9 million km away. The "blue" of the C ring and the Cassini division are similar, but color mismatches exist between the inner and outer B ring, and between these and the A ring. These color differences are partly explained by multiple-scattering effects and partly by intrinsic color differences.**

Figure 22. **A 6,000-km-wide section of the outer B ring. Most of the structure visible in this region consists of features several hundred km wide. But finer structures down to 15 km in width can be seen throughout the image, and these vary with time and longitude. The relatively vacant Cassini division is at upper left.**

some gaps occur at resonant positions, but most do not. Perhaps the intricate structure of the rings is due to large objects (mooms) within the disk, which serve as sources and sinks of ring matter and perturb nearby material. It has also been suggested that the multitude of ringlets in the B ring is caused by a viscous instability of the disk material. The highly organized structure seems to conflict with the notion of diffusional spreading, which should produce a smoother overall distribution. Eccentric rings may also be telling us that large objects reside nearby in the disk. Such ringlets occur in the C ring and Cassini division, and the F ring itself is out of round.

Saturn's ring system is a dynamic place, and our understanding of it has been helped by theories and observations of galactic interactions. Spiral density waves have been identified across the outer A ring, and a few are present in the outer Cassini division and in parts of the B ring. These in-plane clumpings are due primarily to the co-orbital moon Janus and the shepherd satellites. A few vertical bending waves, produced by the out-of-plane forcing of Mimas, are seen in the A ring with amplitudes of a few hundred meters; another due to Titan is in the C ring. Ironically, these elaborate wave features are among the few things in ring morphology that scientists understand at all.

The boundary between the B ring and Cassini's division is located near the 2:1 resonance of Mimas, the largest inner satellite. Carolyn Porco and her co-workers have shown that the B ring's perimeter forms an ellipse centered on Saturn, just as it should if forced by Mimas' action, with a variation in radial position of about 140 km. Numerous transient, noncircular features, roughly 20 km wide, occur in the same region and are presumably caused by waves and instabilities. The outer edge of the A ring seems to have a six-lobed petal shape that, along with its location, shows it to result from a 7:6 resonance with Janus. It is not surprising that the major outer boundaries of rings A and B coincide with the locations of the two strongest satellite resonances in the system. However, no explanation can yet account for the inner edges of the A, B, and C rings.

Satellites Pandora and Prometheus herd the particles in the narrow F ring (Figure 19) and may cause its clumpy appearance. Because their orbits are slightly eccentric, these two satellites periodically perturb ring particles as they mutually drift past one another (Figure 20). But whether perturbations from the same satellites account for the "braiding" of the F ring is still an open question. (The strands making up the F ring may appear interwoven, but the paths of individual particles in the separate braids do *not* actually pass through one another; rather the entire structure orbits as a whole and only gradually distorts by Keplerian shear motion.) The F ring forward-scatters visible light very efficiently, just as Jovian ring particles do, suggesting that it too consists principally of micron-size grains. These are perhaps generated in the collisional turmoil produced within the strands by the shepherds, as well as by unseen (but suspected) embedded moonlets.

While moonlets in the F ring and Cassini division are only suspected, there is clear evidence that one exists in the Encke gap, because the gap's inner and outer edges are scalloped. The physics of this scalloping is much like that illustrated in Figure 20. These edges, plus density undulations called gravitational wakes in the nearby ring, implicate an embedded satellite several kilometers across. If it is large enough, the resident moonlet should be able to pry open a disk of material by the Goldreich-Tremaine mechanism. Furthermore, in the middle of the Encke gap is a kinky, perhaps discontinuous, ringlet quite reminiscent of the knotted F ring.

Narrow rings and "empty" gaps are not the only places to find unexpected behavior. Curious, sporadic features called "spokes" occur near the densest part of the B ring, about 104,000 km from Saturn's center, and extend outward almost to the edge of the Cassini division (Figure 25). Each spoke appears as a pair of opposing triangles, like an hourglass, with the narrowest part located near the synchronous-orbit radius. Spokes form swiftly as radial strips. As they develop, one boundary always remains roughly radial, while the other assumes the differentially sheared profile of the initial radial strip. From their reflective properties, we know that these enigmatic features consist of microscopic grains. This fact and the long-lasting radial edges (which rotate at the same rate as the planet's magnetic field does) suggest that electromagnetic effects are at work. For example, small particles may become charged and thus levitate off larger ring bodies. This notion is reinforced by the spokes' location near the synchronous-orbit radius and by their approximate symmetry about that position.

It is fitting that Saturn's rings, as the first discovered set, remain the most ornate and interesting system. Fortunately they are also the best observed and, as the Voyager findings become more fully assimilated, are sure to provide many fundamental insights into the nature of all planetary rings.

Figure 23. **Eight hours after its closest approach to Saturn, Voyager 1 looked back upon the ring system and recorded this mosaic of images. Clearly visible is the F ring, seen as a thin, detached band along the outer edge; its brightness is probably due to minute particles that preferentially scatter light in the forward direction. The broad, shimmering B ring contains bright spokes (which also forward-scatter strongly) from the center to the lower right.**

Figure 24. **Voyager 2 was able to examine Saturn's rings from their unilluminated side. This eerie false-color view is a composite of images taken through green, clear, and violet filters. In most places some sunlight peeks through the ring system, but the densest part of the B ring is visible primarily because of Saturn-shine.**

The Cassini orbiter, part of a mission that is to arrive at Saturn early next century, will scrutinize the intriguing and elegant structures of this complex system over a four-year period and should teach us much about the important processes going on there.

THE ORIGIN OF PLANETARY RINGS

Two possible modes for the origin of planetary rings – the tidal breakup of a satellite and incomplete accretion – were mentioned briefly at the outset of this chapter. The first of these was espoused by Edouard Roche in the mid-19th century. If applied to a satellite drawn in by planetary tides, Roche's mechanism could explain only ring systems located within the synchronous-orbit radius but would not apply to the outer reaches of Saturn's rings. However, the rings could also be the disrupted remains of a comet or other interplanetary body that strayed inside the Roche limit. Either way, some fragments would still be tens of kilometers across – even after aeons of collisional grinding. Such large remnants could be creating the elaborate structure in Saturn's rings or fashioning the narrow Uranian ribbons.

New planetary ring systems may soon develop elsewhere in the solar system. The Martian moon Phobos, residing well inside the planet's synchronous-orbit radius, is being drawn inexorably inward due to tides. If Phobos remains intact, it should strike Mars in about 50 million years. More likely, it will fracture or, at the very least, be denuded of its loose surface covering and thereby produce another faint ring system. An even more rarified dust belt probably circles Mars today, though it has not been detected: this would be debris blasted off Phobos and Deimos that remains in orbit about Mars. The orbit of Neptune's giant satellite Triton is also collapsing, though dynamicists do not expect Triton to come dangerously near Neptune for billions of years. Given enough time, therefore, planetary rings will be commonplace and not at all the rarity they were once thought to be. Indeed, since the discovery of rings about Uranus, Jupiter, and Neptune, some researchers have maintained that rings may have – or may still – encircle Venus, Earth, and even the Sun itself!

A close-in satellite may be shattered not by tides but instead by the impact of a large projectile. Since a planet's gravity acts to draw in interplanetary material in the vicinity, the frequency of collisions will be higher near the planet. By extrapolating how often craters form on distant Saturnian and Uranian satellites into the ring locale, Eugene Shoemaker has argued that any Mimas-sized satellites within 2 planetary radii were pummeled into pieces many times early in the solar system's history. At Mimas' distance, the debris ring created by these cataclysmic collisions swiftly re-accumulated back into a single object, but the debris from a shattered satellite closer in (within the Roche limit) cannot reaccumulate and would form a permanent ring.

The second general hypothesis for the origin of planetary rings was first proposed by Laplace and the metaphysicist Immanuel Kant at the end of the 18th century. In their view, Saturn's rings formed from the same circumplanetary nebula as the planet's satellites – much as the planets themselves accumulated in the primordial solar nebula. In fact, the essential properties of their nebular model of *planetary* origin were derived by observing Saturn's resplendent entourage. In the modern version of nebular theory (see Chapter 23), giant planets form in the outer parts of the circumsolar cloud wherever local instabilities become sufficiently large and dense. The end result is a large gaseous protoplanet at the center of its own flattened disk of gas and dust. It is interesting to note in this regard that only the gaseous outer planets – but not the rocky terrestrial ones – have ring systems.

In regions beyond the Roche limit, the composition of accumulating objects may have depended upon their distance from the protoplanet; its collapse generated considerable heat, which governed where specific volatile compounds could have condensed out of the nebula. Pollack and his colleagues have shown that high temperatures in the neighborhood of today's planetary rings would have prevented the condensation of water for several million years after the systems first formed. Apparently, they conclude, Jupiter and Uranus lost their gaseous disks before cooling completely, leaving only less-volatile materials like silicates from which to assemble their rings. In contrast, Saturn cooled earlier, which allowed water vapor to condense into the magnificent rings we see today.

Material that accretes inside the Roche limit cannot grow without bound. Particle sizes will represent a balance between disruptive tidal stresses and the attraction of gravity and interparticle "stickiness." In this regard, the apparent absence of many large particles in Saturn's rings could be meaningful. Indeed, the identification of many objects larger than 1 km or so in Saturn's rings would be very unsettling to the nebula model.

Finally, the very notion that planetary ring systems formed billions of years ago must be viewed with some skepticism, because several lines of evidence argue that the present-day rings – or at least parts of them – are quite young. (Clearly, the *dust* in all these systems is youthful.) To make this case for

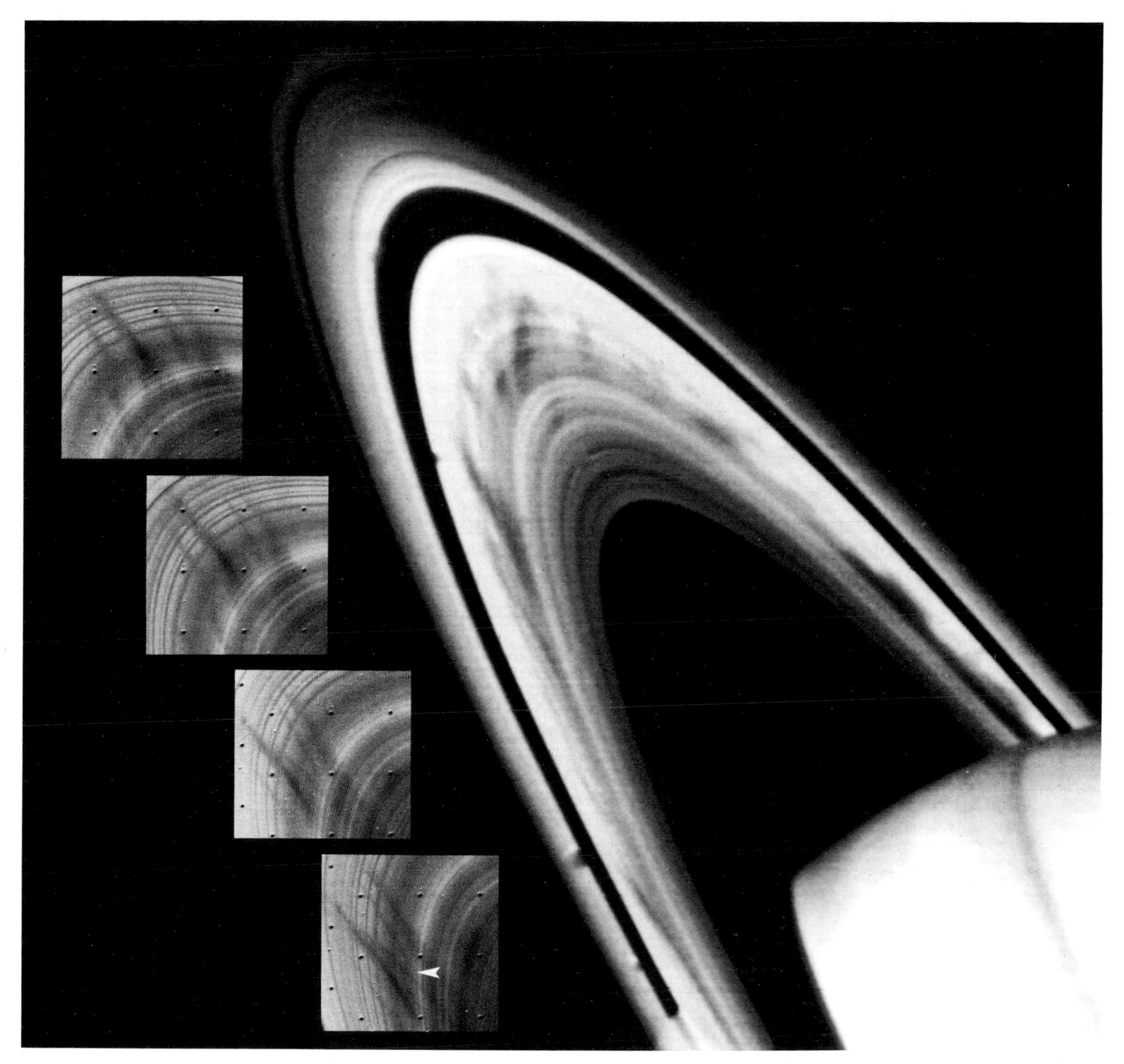

Figure 25. **The dark, shadowy fingers known as "spokes" in Saturn's B ring occur sporadically in time, with a preference toward the morning ansa. Depending on illumination and viewing geometry, their brightness changes relative to their surroundings (for example, seen here in back-scattered light they appear dark); this indicates that spoke regions are areas of enhanced dustiness. The insets show the swift formation of a new, radially aligned spoke (arrow in last panel) among a number of already-existing ones in the center of the B ring over a span of 35 minutes. (The regularly spaced black dots are reference marks used for geometric calibration of the imaging system.)**

Saturn's rings, we can use the angular momentum transferred in spiral density waves to compute the times needed for nearby satellites to evolve away from the ring edge and for the A ring to be dragged down into the B ring. In each case ages are much less than that of the solar system. The lifetimes of ring particles undergoing erosion are invariably brief (although quite model dependent), and calculations demonstrate that the Uranian ϵ ring cannot be restrained by shepherds for the age of the solar system. If all this is accepted at face value, then planetary rings must indeed be young and are perhaps the consequence of recent catastrophic events. However, the number of interplanetary intruders is continually decreasing, as more and more of them either strike planets or are completely ejected from the solar system. If catastrophic events have become less likely over time, how can the rings have had a recent origin? Perhaps our models are wrong or incomplete – explanations equally unpalatable to those of us who contrive them!

A RING RECAP

The austerely beautiful rings encircling Jupiter, Saturn, Uranus, and Neptune remain the "most extraordinary marvel" that Galileo called Saturn's in announcing their discovery. As our knowledge of them has improved, the systems have been seen as perhaps more individualistic than the planets they surround. Uranus' narrow bands, made of dark boulders, reside within an extensive dusty disk and display an intriguing dynamical structure. The icy snowballs of Saturn's rings are baroque in their organization and variety. Jupiter's ring is a mere wisp that must be continually generated from unseen parent bodies. And scientists now have Neptune's rings to study, thanks to the images from Voyager 2.

The dynamical processes now occurring in planetary rings may provide an appropriate analog for events in the early solar system or in distant spiral galaxies. The rings' detailed structure could be a fossil record of an intermediate stage in the accretion of orbiting bodies. Thus, planetary rings are more than just striking, exquisite phenomena – in a very fundamental way, they may represent the solar system's ancient beginnings.

13
The Galilean Satellites

Torrence V. Johnson

FOR MORE THAN three and a half centuries following their discovery, the four largest satellites of Jupiter remained tantalizing points of light in astronomers' telescopes, tiny disks barely discernible even under the best atmospheric seeing conditions. The discovery of Io, Europa, Ganymede, and Callisto, announced in 1610 by Galileo, provided strong support for the Copernican solar system. Observations of their relatively rapid motions around Jupiter have fascinated the amateur and intrigued the professional astronomer ever since. Ole Roemer used their eclipses to determine the speed of light, Albert Michelson measured their diameters with his stellar interferometer, and mathematical analyses of the satellites' motions in the earlier parts of this century emphasized the importance of resonant phenomena in celestial mechanics.

During the 1970s, new observations and techniques resulted in a renewed awareness of the importance of the Galilean satellites to solar-system studies. Telescopic observations of the satellites' spectra led to knowledge of their surface compositions; more accurate measurements of their diameters were made; clouds of neutral sodium and ionized sulfur related to Io were discovered; the Pioneer 10 and 11 flybys provided the first on-the-spot measurements of the satellites' magnetospheric environment. And finally, in 1979, the Voyager spacecraft transformed our view of these objects from dots of light into places – new worlds seen clearly for the first time.

GENERAL CHARACTERISTICS

The Galilean satellites, together with tiny Amalthea, a trio of moons discovered by Voyager (Metis, Adrastea, and Thebe), and Jupiter's tenuous ring, form one of the three known regular satellite systems, the others being those of Saturn and Uranus. These systems are characterized by satellites and rings with circular, coplanar orbits in the planet's equatorial plane. All of these systems thus resemble the Sun's planetary system and are frequently referred to as "mini solar systems" (Figure 1).

The Galilean satellites lie deep within the Jovian magnetosphere. Instead of being continually exposed to the solar wind, as our Moon is, they are immersed in the Jovian plasma environment and are bombarded by high-energy charged particles trapped in Jupiter's strong magnetic field. The satellites, particularly Io, appear to interact strongly with this environment in several ways. Io, at least, supplies significant material – in the form of atoms and ions – to the magnetosphere, and the continuous "rain" of charged particles onto the satellites' surfaces has been suggested as a cause of coloration and even significant erosion over geologic time.

The satellites are all relatively massive, roughly comparable to the Moon, and they thus perturb each others' orbits significantly. This situation, together with a resonant relationship in the orbital motions of the inner three

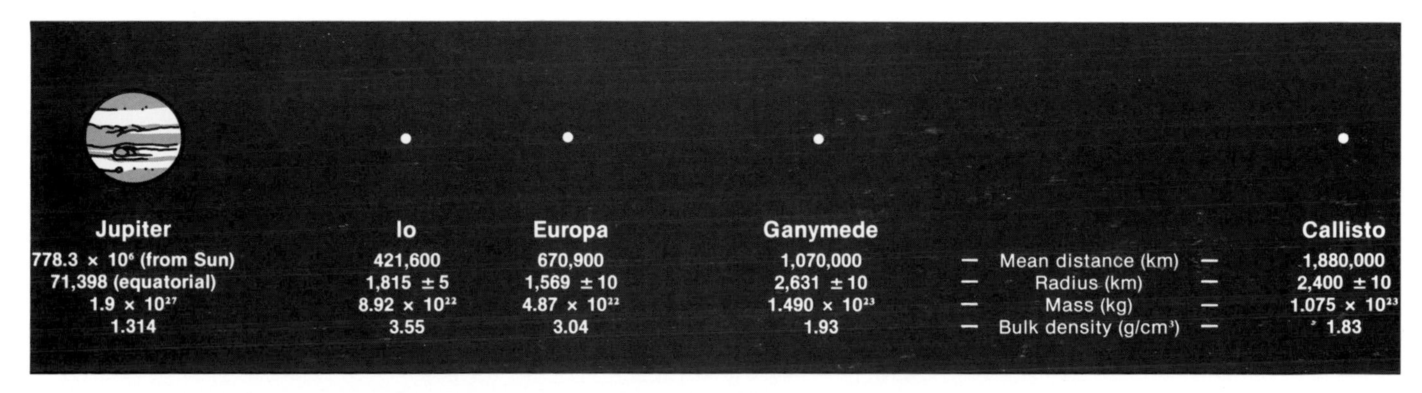

Figure 1. **Jupiter and the distances (but not sizes) of its Galilean satellites are shown to scale. The Galilean satellites are planet-size worlds. Two are roughly the size of Earth's Moon, while Ganymede is bigger than Mercury and the largest satellite in the solar system.**

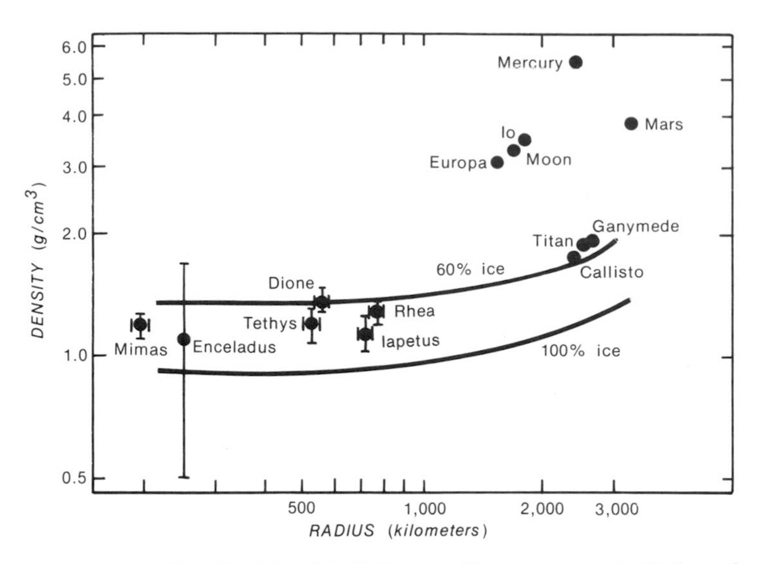

Figure 2. **The radii and densities of the Galilean satellites are compared with those of other bodies; solid lines denote models calculated by Mark Lupo and John Lewis.**

satellites, permitted DeSitter and Sampson to make reasonably good determinations of the satellites' masses in the 1920s. Much more accurate values are now available from the tracking of spacecraft as they pass nearby, but the old values were in general accurate to 20 percent or better. The resonant relation in the orbital motions of Io, Europa, and Ganymede (studied by Laplace) also appears to play a major role in powering the volcanoes on Io, which will be discussed later.

The diameters of the satellites were very uncertain prior to 1970 due to their small angular size as seen from the Earth, but stellar occultations and spacecraft images have improved this situation considerably. The satellites' masses and diameters yield their densities (Figure 2), which provides the most direct evidence of basic differences in their bulk compositions. The lunarlike sizes and densities of Io and Europa mark them as essentially rocky, silicate-rich bodies, Io being slightly more and Europa slightly less dense than the Moon. The two may be similar to the Moon in bulk properties, but they differ from it greatly in surface

Figure 3. **What vistas await would-be travelers to the Galilean satellites? Artist Don Davis has rendered these landscapes based on Voyager results. From left to**

appearance and evolution (Figure 3). Meanwhile, the low densities of Ganymede and Callisto (less than 2 g/cm^3) strongly suggest much of their mass is something other than rock, most likely water in some form. Ganymede, Callisto, and Titan form a new class of large (Mercury-size) objects with much lower density than the more familiar terrestrial planets.

John Lewis, in pioneering work done in the early 1970s, pointed out that densities like those of Ganymede and Callisto are precisely what one would expect from condensation of solar-composition gas at temperatures where water ice is stable. Such a body would have approximately equal proportions of silicates and water (see Chapter 23). He also argued that such a composition would almost inevitably lead to early melting and differentiation as a result of accretional, radiogenic, and possibly tidal heating, combined with the relatively low melting temperature of water. Other scientists then considered the possible interior structures of these differentiated, water-rock planets. Their models studied how warm, slowly deforming ice might transport heat outward from the satellites' interiors; these suggest either that complete melting of the water may never have occurred or, if it did, that convection in an ice crust could have frozen the remaining water in a short time. Figure 4 illustrates current interior models for the satellites. Note that Ganymede and Callisto both have similar structures: ice crust, convecting water-ice mantle, and silicate-rich core.

Why are there two classes of Galilean satellites? The fact that the higher density objects Io and Europa are closer to the planet than their low-density siblings leads naturally to a comparison with the solar system itself. In the early 1950s, Gerard P. Kuiper suggested that Jupiter was very hot after it formed, perhaps enough so to prevent lighter elements from condensing or cause them to "boil off" the inner satellites. Recent investigations suggest that, indeed, Jupiter's starlike infancy could have produced enough heat to make the environment of the inner satellites significantly warmer than that of the outer satellites. James Pollack, Fraser Fanale, and their co-workers have modeled early Jovian history using the planet's present-day output of energy (more than twice what

right are the surfaces of Callisto, Ganymede, Europa, and Io. Each scene shows Jupiter at its correct relative size.

it receives from the Sun) as a boundary condition. For the most plausible cases, they identified a period of some 100 million years when conditions would have permitted the formation and retention of water ice where Ganymede and Callisto are now, while allowing only higher-density (though probably water-enriched) silicates to exist at Io and Europa. As with more general models of solar-system formation, the dynamics of accretion may have influenced the satellites' composition and energy balance by mixing materials within the region of satellite formation and by bringing in material from outside the system. Theorists do not yet agree on all aspects of satellite formation, but the striking variation in the amount of condensed volatile material (mostly water ice) with increasing distance from Jupiter is strong circumstantial evidence that the early Jupiter, with its immense gravitational field and central heat source, played a dominant role in determining the nature of its satellite bodies – just as the Sun must have on a vaster scale.

Voyager data provide us with a wealth of new information concerning the satellites' surface processes and evolution. However, most of our knowledge of surface composition comes from analysis of the remarkably diagnostic reflection spectra acquired with ground-based and airborne telescopes. The first hints of infrared structure in these satellites' spectra came from observations by Kuiper and Soviet planetologist Vassily I. Moroz during the early 1960s. A decade later, Carl Pilcher and his co-workers, and independently Uwe Fink and Harold Larson, discovered strong absorptions in the infrared spectra of Europa and Ganymede from 1.0 to 3.0 microns; later work showed weak absorptions in Callisto's spectrum as well, but they are absent in Io's.

Geometric albedo is an object's brightness compared with a perfectly diffusing disk under normal illumination; Figure 5 shows the satellites' variations of albedo with wavelength from 0.3 to 5.0 microns – effectively the entire range dominated by reflected solar radiation. The deep absorptions near 1.4 and 1.8 microns in the light reflected from Europa and Ganymede are due to water ice; in fact, the strength of these absorptions and the high albedos suggest large amounts of relatively "clean" ice. Callisto's weaker absorptions and low albedo, on the other hand, indicate a "dirty" surface, though debate continues on the degree to which frozen hydrated soil can account for the spectra compared to patches of uncontaminated frost seen against a dark background.

Io's high albedo, yellowish color, and lack of diagnostic

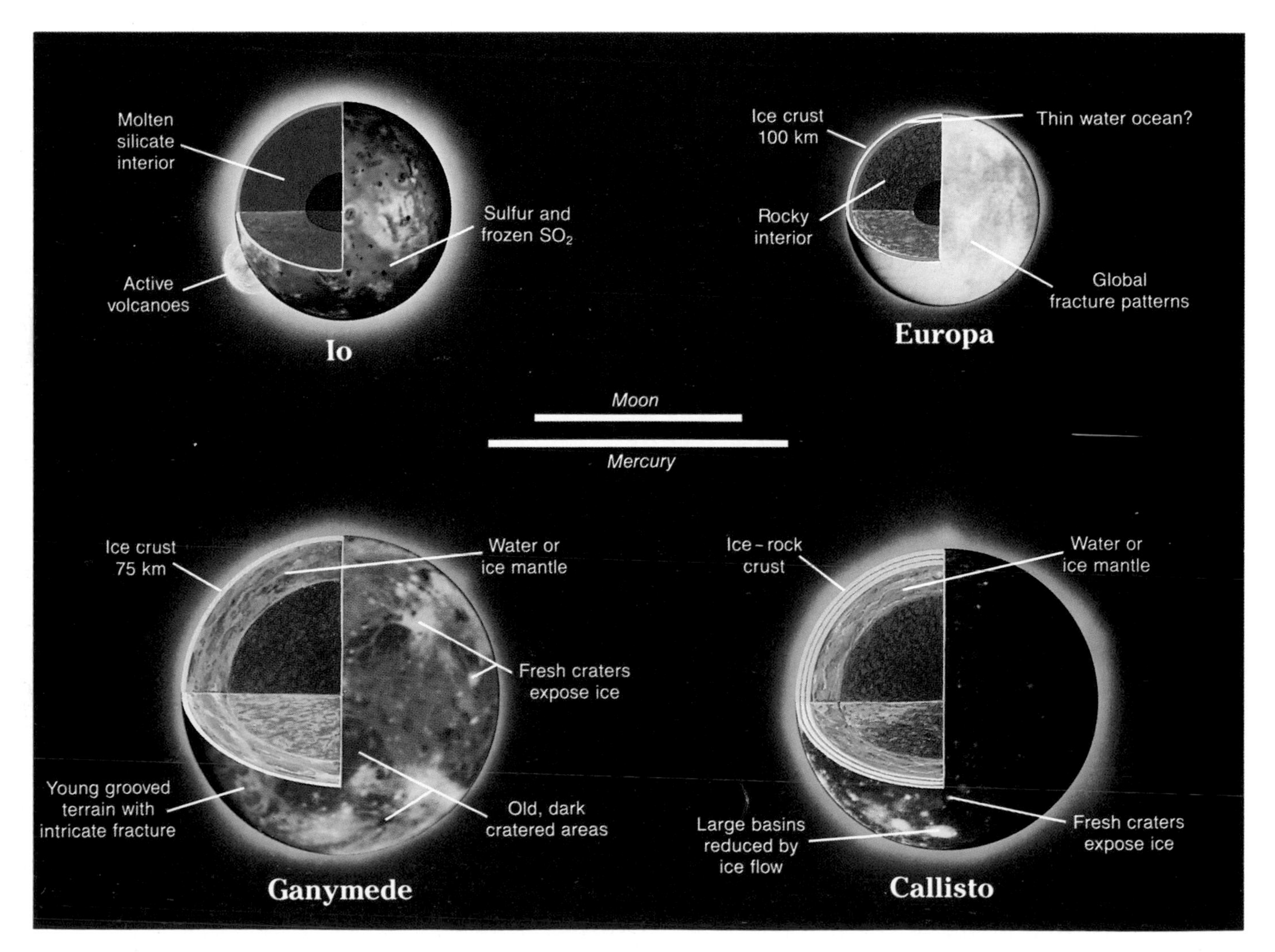

Figure 4. **These schematic illustrations portray each satellite's interior as presently understood. Earlier models, based primarily on telescopic observation, were generally confirmed in light of the Voyager findings, though it had been thought that Io's interior was solid and that Callisto's density was somewhat lower. Horizontal scale bars indicate diameters of the Moon and Mercury.**

absorptions in the near infrared posed a difficult problem to planetary scientists. Many suspected that the sharp drop in Io's blue and ultraviolet reflectance was in some way connected with sulfur or sulfur compounds (a feeling reinforced by the discovery in 1976 of ionized sulfur concentrated around Io's orbit), but the source of the sulfur and the nature of the non-sulfur component, if any, remained subjects of speculation.

In 1978, observations made from the high-altitude observatory at Mauna Kea and from NASA's Kuiper Airborne Observatory showed that Io's spectrum had a deep infrared absorption near 4.1 microns. This feature was not identified immediately, although a number of candidate materials (including most silicates) were ruled out. Spurred on by the prospect of the approaching Voyager encounters, a number of laboratories worked diligently to identify the substance causing the feature and find a consistent explanation for Io's optical properties. As often happens in science, a number of clues fell into place nearly simultaneously. In the weeks following the Voyager 1 Jupiter encounter in March of 1979, the discovery of volcanism on Io, the observation of abundant sulfur and oxygen ions in the Jovian magnetosphere, and the discovery of sulfur dioxide (SO_2) gas in the vicinity of one of Io's volcanoes (Loki) coincided with laboratory data from two independent groups identifying frozen SO_2 as the source of the 4.1-micron absorption feature.

As the preceding example demonstrates, the combined analysis of the ground-based spectra and a variety of Voyager observations has given us a much better picture of Io's surface composition than either data set alone provided. Although many problems remain in understanding the moon's chemistry, it now appears that the materials dominating the upper surface layers are forms of sulfur and sulfur compounds, including at least SO_2, brought to the surface continually by volcanic activity. What other materials may be there, either mixed with the sulfur-bearing components or buried beneath them, remain to be discovered. This includes the sodium- and potassium-bearing phases inferred from the presence of neutral "clouds" of these elements escaping from Io and of sodium ions in the magnetosphere.

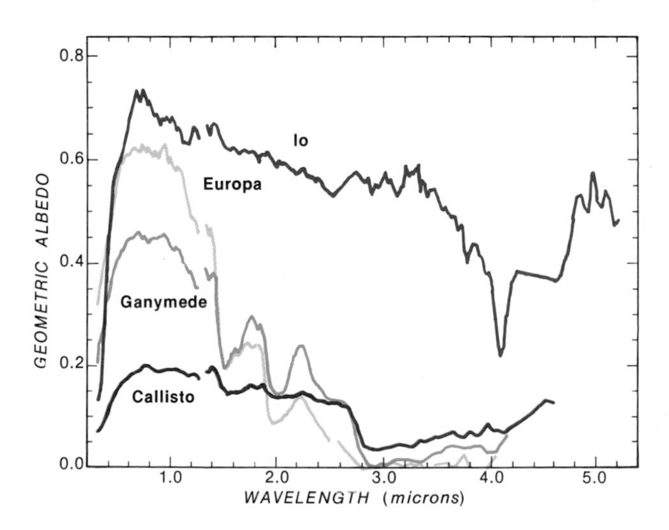

Figure 5. **Studies of Jupiter's largest satellites from Earth show that they have distinctive spectral signatures. Absorptions in the near-infrared reveal the presence of water ice on the surfaces of Europa, Ganymede, and Callisto. The reflectivities change quickly through the visible spectrum.**

CALLISTO

Impact scars dominate the surface of the outermost Galilean satellite. In fact, Voyager images show that large craters, standing nearly shoulder-to-shoulder, cover virtually its entire surface. Callisto is unique among the solid-surface worlds investigated to date in having no "plains" units where craters have been obliterated by more recent processes. The Moon, Mercury, and Mars, for instance, all have significant portions of their surfaces covered by material of volcanic origin. The only places on Callisto where the crater density appears to be significantly lower than average are near the centers of the several large ring structures believed to be the scars of even larger impacts (seen dramatically in Figure 6).

Figure 7 illustrates the degree to which impacts dominate the surfaces of the Galilean satellites and, for comparison, selected lunar and Martian regions. The curve for Callisto's average surface shows a crater population nearly as dense as that of the lunar highlands. In order to use such data to estimate the age of a satellite's surface, a number of factors must be taken into account. They include differences in the

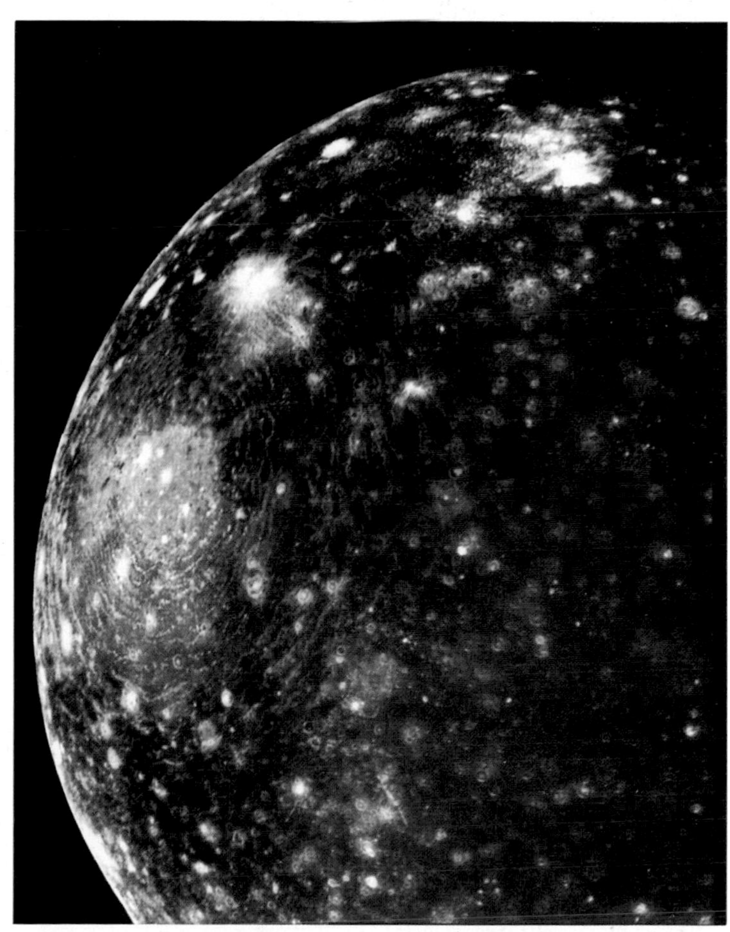

Figure 6. **This mosaic of Voyager images shows the heavily cratered surface of Callisto. Most prominent is an extensive ring structure known as Valhalla, which is similar in many respects to large, circular impact basins that dominate the surfaces of the Moon and Mercury. Valhalla's bright central area is about 300 km across; sets of discontinuous concentric ridges extend out to some 1,500 km from the center.**

flux of crater-producing objects at different places in the solar system, possible differences in the impacting bodies themselves (for instance, were they asteroids or comets?), the effects of Jupiter's gravity, and the response of ice-rich target surfaces during impacts – all of which await detailed evaluation for the Galilean satellites. But everything we know now suggests that the very heavily cratered lunar highlands and portions of Mercury, Mars, and Callisto cannot have been produced recently in solar-system history. Impact specialists generally agree that the cratering flux (at least in the inner solar system) declined very rapidly in the first half billion years following planet formation about 4.6 aeons ago (see Chapter 21). Despite the uncertainties, it seems very likely that Callisto's surface dates back to this early period and has been little modified since (except by an ongoing but lower intensity rain of comets and asteroids).

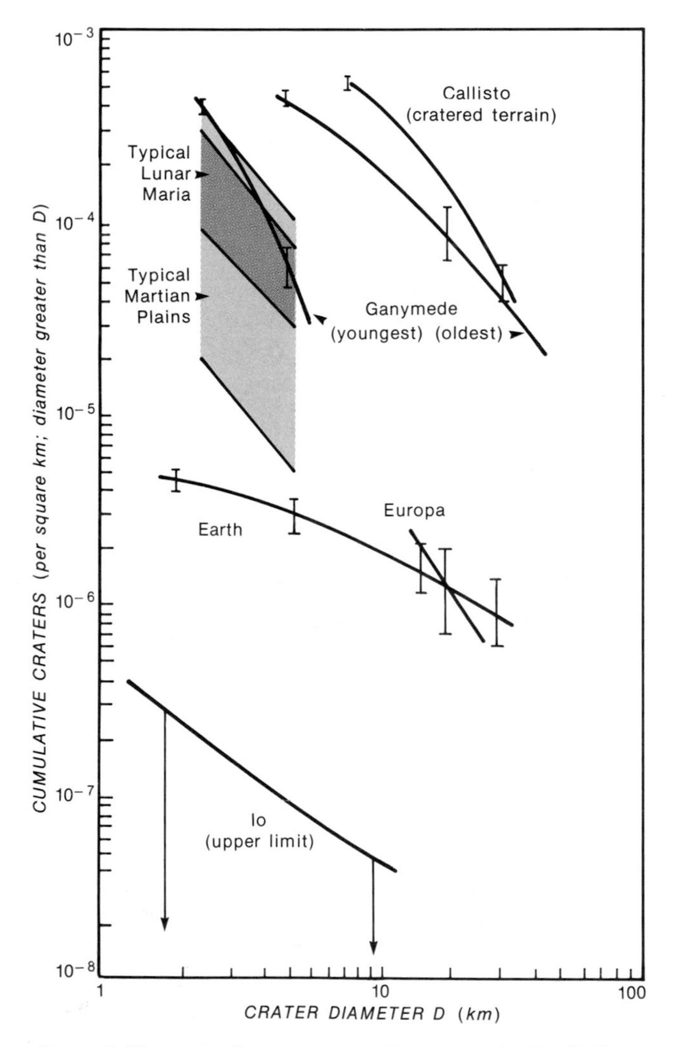

Figure 7. **Size-versus-frequency curves for craters on the Galilean satellites are compared with those for the Earth, Moon, and Mars to show these objects' relative ages. The oldest (most heavily cratered) terrains on Ganymede and Callisto are comparable in age to the oldest surfaces on the Moon, Mars, and Mercury. The youngest terrain on Ganymede plots within the range covered by the lunar maria and Martian plains. The curve for Europa is very uncertain because few craters can be identified on it in Voyager images. Io's surface is so fresh that no craters were identified down to the 1-km resolution of the Voyager images; a theoretical upper limit for their distribution has been given.**

At first glance, the surface of Callisto shows no obvious evidence of its icy nature. Most of the craters look superficially like those on the rocky terrestrial planets. But examining the details of crater shapes and of the largest impact features, we find a number of striking differences from similar features on rocky objects. Craters of all sizes on Callisto are much flatter than their counterparts on terrestrial planets. In addition, the largest impact basins lack the central depressions and surrounding ring mountains common to large basins on the terrestrial planets (like the Imbrium and Orientale basins on the Moon and the Caloris basin on Mercury). Instead, Callisto exhibits a number of large, bright, circular features with or without concentric rings and radial structure (Figure 6 shows the largest of these). Such features are similar in scale to traditional basin structures but show little or no topographic relief.

Surface temperatures on Callisto range from equatorial noontime highs of perhaps 150° K to subsurface averages of 100° K or less. Ice this cold acts more as a rock than the relatively volatile substance we are familiar with. However, even at Callisto's temperatures, ice should "flow" – albeit at a glacierlike crawl – when subjected to stress over geologic time. Deeper in the crust, tens or hundreds of kilometers below the surface, temperatures will be even higher due to an outward conduction of heat from radioactive decay deep in the interior. Thus, large impact structures, which create stresses at these warmer depths, may deform even faster. Furthermore, the material properties of ice probably play an important role in the initial impact event and subsequent deformation of the newly formed crater.

This combination of initial deformation followed by viscous flow is believed to be responsible for the flat crater shapes and the large ring structures on Callisto. Giant impact events, which would have produced classic impact basins on the terrestrial planets, probably never formed a real topographic depression for any length of time on any of the ice satellites, particularly if the cold, brittle crust was thinner and weaker early in the satellite's history. Smaller impact events, involving only the upper few kilometers or so of the crust, formed distinct craters, but these probably looked different from ones in rock even from the beginning. Subsequent glacial flow further reduced their topographic profile, again more effectively if the early crust was thinner and "softer" than at present.

GANYMEDE

The ice giant of the Jovian system, Ganymede is also the largest of all the solar system's satellites. It displays many of the characteristics of Callisto's surface but also has a baffling array of tectonic features unlike anything previously seen on the terrestrial planets. Seen at low resolution, Ganymede looks deceptively like the Moon, with irregular dark regions on a brighter background (Figure 8). The dark regions are probably the basis for Ganymede's mottled appearance in maps drawn from telescopic observations and in low-resolution Pioneer 10 imagery obtained in 1974. These were occasionally referred to as "mare," even though Ganymede's overall reflectance is four to five times higher than the Moon's. High-resolution Voyager images show how fundamentally different Ganymede and the Moon are. The dark regions, far from being younger flows of lava-like material, are instead the oldest parts of Ganymede's surface, very heavily cratered, and resemble Callisto's surface in many respects. While not cratered quite as heavily as Callisto (see Figure 7), dark areas on Ganymede are probably also quite ancient. Two of the largest ones now bear the names Galileo and Simon Marius (who independently discovered Jupiter's four giant satellites).

Conversely, the light regions are clearly younger than the darker areas and exhibit a variety of features suggestive of tectonic activity. They are dominated by parallel sets of ridges and troughs, kilometers to tens of kilometers wide and with perhaps a few hundred meters of vertical relief. These sets or "bands" of grooves wander for thousands of kilometers across Ganymede's surface and form intricate patterns (Figure 9), particularly when different sets intersect or interact with one another. They border the dark terrain everywhere and in some places seem to extend into it. The grooves have relatively fewer craters, so apparently they developed later than dark regions and probably grew at their expense, destroying previously existing dark crust. However, craters seen on the grooved areas are still plentiful in absolute terms, occurring in greater numbers than on the lunar maria, for instance.

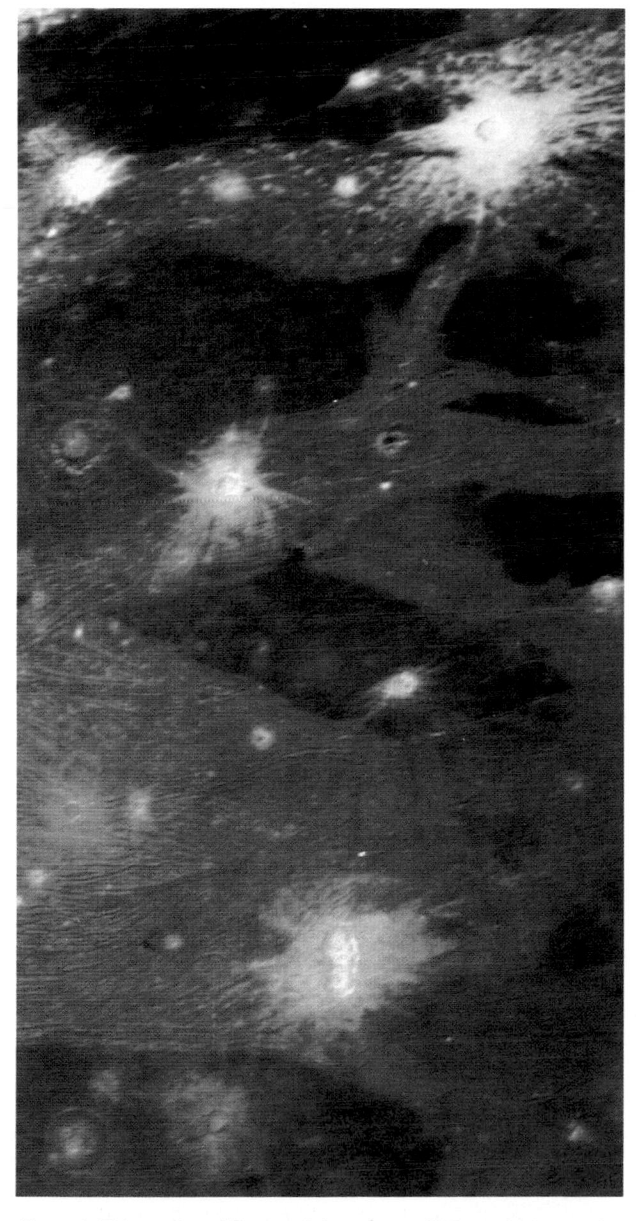

Figure 8. **This portion of Ganymede's surface exhibits typical examples of both dark cratered and light "grooved" terrains. The dark regions, frequently angular or polygonal in shape, are somewhat more reflective than Callisto's surface but considerably lower than Ganymede's global average; their color also matches Callisto's closely.**

Although the history of crater-producing events at the Galilean satellites is not yet firmly established, Eugene Shoemaker believes that to bear so many craters the surface must be quite ancient. He suggests that, as with the Moon, the two different terrains are both at least 3 to $3\frac{1}{2}$ billion years old. We know that the production of grooves was still proceeding when many large craters were being formed, because large craters both overlie and are crosscut by sets of groove systems (Figure 10).

The exact nature of the immense groove systems crisscrossing Ganymede's surface is not yet known. They appear to result from tensional rather than compressional mechanisms, analogous to the formation of graben valleys on Earth. Moreover, their high albedo (about 40 percent) indicates that these areas are "less dirty" in some ways than the darker areas. Whether this results from the upwelling of new material from below during their formation or from reworking of the existing crust is not clear. In addition, there

Figure 9. **A detailed Voyager image along Ganymede's terminator reveals complex ridge and groove systems that crisscross the entire scene. Such terrain presumably results from deformations in the moon's thick, icy crust. The smallest features seen are about 3 km across.**

are indications that the grooves' rates of formation may have varied and that many parts of the crust were mobile to a degree. Features similar to Ganymede's grooves have been observed on other icy satellites, such as Saturn's moon Enceladus and Uranus' Miranda and Ariel. Grooved terrain may be a relatively common expression of tectonism on icy bodies with active geologic histories.

On the terrestrial planets, particularly the Moon and Mercury, fresh craters are brighter than their surroundings and exhibit bright, radiating, ray patterns (the Moon's Tycho is a classic example). Bright ray craters result from the exposure of freshly pulverized crustal material that has not had time to darken from "gardening," or reworking, of the upper soil layers by impact ejecta (see Chapter 4). But *dark* ray craters, which are relatively common on Ganymede, demonstrate that other processes may be important in ice-dominated soils. It is possible, for instance, that the material thrown out during a crater's formation becomes contaminated by dark, subsurface units or even by material from the projectile itself, and this mixing may be more important on Ganymede than on the Moon or Mercury.

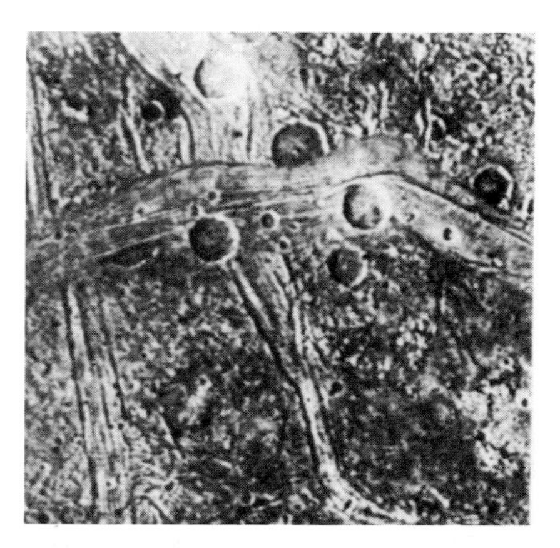

Figure 10. **This section of Ganymede, measuring about 300 km on a side, shows a group of craters that were formed before, during, and after the grooved terrain around them developed.**

Figure 11. **Nearly an entire hemisphere of Ganymede is seen in this Voyager 2 image. The prominent dark area, called Galileo Regio, is about 3,200 km in diameter and contains light-colored, closely spaced bands that resemble those seen on Callisto. Bright spots are relatively recent impact craters, while light-brown circular areas are probably the remains of older impacts. Part of Galileo Regio may be covered with a bright frost.**

Craters on Ganymede are similar in form to those on Callisto. The large, old craters (particularly on the ancient, dark terrain) are very flat for their size, showing the effects of the icy crust and viscous flow. These ancient scars have been dubbed *palimpsests,* which in archaeology refers to parchment that has been scraped clean and written over again. Numerous palimpsests occur on both Ganymede and Callisto with diameters of several hundred kilometers, usually recognized by a light, circular patch (Figures 11,12). This central patch may represent terrain covered by a continuous sheet of debris thrown from the original crater. Or it may be a more complicated phenomenon marking "ground zero," where cleaner ice was either generated during the impact itself or, sometime later, transported there as water from a less-contaminated subsurface layer.

The central part of the Valhalla ring structure on Callisto may be the largest member of the palimpsest family. No comparable feature is present on Ganymede; indeed, there is no piece of undisturbed dark crust large enough to contain such a feature. However, a series of parallel furrows run across Galileo Regio (also discernible on some nearby dark crust) that strongly resemble the ring structures on Callisto. If this is the remnant of a similar structure, the original ring system would have to have been even larger than Valhalla, affecting nearly an entire hemisphere on Ganymede. Unfortunately, the center of this hypothetical structure is no longer apparent; the area where it should lie has been extensively modified by grooved terrain. If the furrows on Galileo Regio and the surrounding dark terrain are part of one large ring structure, some movement and rotation of the dark blocks of cratered terrain seem to be necessary to account for its current configuration.

Ganymede's groove systems strongly suggest that the entire satellite experienced a dynamic early history. Tectonic forces broke up dark crustal blocks, created offsets of groove systems along faults, and perhaps rotated whole sections of dark areas containing the furrows. This activity may bear some resemblance to the global-scale plate tectonics at work on Earth, though there are obvious differences such as the composition of crust and mantle and the lack of any identifiable subduction zones (regions of crustal sinking) on Ganymede. The satellite's crustal grinding may not have lasted long, for craters on the younger, grooved terrain do not seem to be degraded as much as those on the dark regions; perhaps the crust beneath the grooves became stiffer after their formation. Voyager 2's images of the southern hemisphere show an impact basin (Figure 13) more similar to terrestrial-planet basins than to Callisto's ring structures. This basin has many large craters superimposed on it and so must itself be reasonably old. Thus, the combined evidence suggests that Ganymede's crust had stiffened to a point where it retained even large impacts at an early stage in its history, perhaps $3\frac{1}{2}$ to 4 billion years ago.

This general picture of the early evolution of Ganymede and Callisto fits the available data. Both satellites, in the later stages of accretion, probably possessed icy, heavily cratered crusts darkened by silicates and carbon-rich material present in incoming debris. Their interiors were quite likely molten water (or at least warm, convecting ice), and the combination of internal activity and occasional large impacts kept disrupting the dark crusts and bringing fresher, brighter material to the surface. As the satellites cooled, the crusts became thicker and more rigid, less susceptible to disruption. Callisto's crust may have stiffened somewhat earlier than Ganymede's and thus continued to collect the final dark debris of accretion. Ganymede, by contrast, probably remained active longer; by the time its crust stiffened, less material was available to rain down upon it. We do not yet know whether this difference in history resulted from chance (did several large bodies strike Ganymede but not Callisto late in the accretion stage?), or from small differences among their early sources of heat (radioactive decay, rapid-fire impacts during accretion, and tidal effects).

There could easily be variations to this general picture. Perhaps Callisto's crust, for some reason, was never thin or soft enough to be disrupted or that subsequent darkening by the infall of meteoritic material has complicated the situation. More detailed studies of the two satellites will be needed to understand more fully the differences in evolution between them.

EUROPA

For many of us who watched early images of Europa arriving from Voyager 1 (Figure 14), the aspect of this distant satellite was eerily reminiscent of Percival Lowell's perception of canals crossing ruddy Mars. The higher-resolution views from Voyager 2 (Figures 15,16) dispelled any hint of a relationship to the Mars of Lowell's imagination, but they revealed instead a surface unlike that of any other planet (though large fields of sea ice in the Earth's polar regions come close). A decade after the Voyager encounters, Europa remains perhaps the most enigmatic of the Galilean satellites. Our best views resolve features no smaller than about 4 km across, and these cover but a small fraction of the globe. Such relatively poor resolution, when combined with Europa's billiard-ball-smooth surface, makes normal geologic interpretation very difficult.

Among other points, we'd like to know the age and thickness of Europa's unique crust. A primary observation from Voyager data is the absence of craters larger than 50 to 100 km across. Both Ganymede and Callisto exhibit many impacts in this size range, so Europa's surface must be younger than, say, the grooved terrain on Ganymede. A key issue is *how* much younger, but our data have supplied only partial answers. Several small craters about 20 km across can be spotted near Europa's terminator. While quite sparse, this crater density could nonetheless indicate a surface as ancient as the lunar maria (3 to $3\frac{1}{2}$ billion years old) – if objects have struck the satellite only one-tenth as often as they have the Moon. However, recent estimates suggest much younger ages, perhaps as low as 100 million years. If these younger ages are correct, then significant resurfacing must be going on even at present on Europa.

The thickness of the ice covering the satellite is linked to the formation of the surface stripes and ridges. Europa's density, 3.0 g/cm^3, places some loose limits on the ice thickness. If the rocky portion of the satellite has a density between the Moon's (3.3) and Io's (3.5), and if all Europa's water has been driven from the interior and now lies frozen on the surface, then an ice crust 75 to 100 km thick would be needed to reduce the mean density to 3.0. A thicker layer would be possible if the rest of Europa has an unusually high-density composition – for instance, one more iron-rich than either the Moon or Io. A lower limit is more difficult to estimate, since even a light frosting of high-albedo ice would be enough to obscure underlying dark material. However, if Europa's thin ridges and other surface features result from processes operating in ice, then the crustal thickness must be at least comparable to the height and width of these features (at least a few hundred meters).

If Europa's icy crust is relatively thick (50 km or more) then early heating could have melted all or part of it, in a manner similar to that suggested for early models of Ganymede and Callisto. Even today radiogenic heat is probably adequate to keep temperatures at the base of a thick ice layer above the melting point of ice (if conduction through the outer layers were the only heat-loss mechanism). As with the large icy satellites nearby, the slow, glacierlike convection of Europa's ice crust causes heat from the interior to reach the surface and radiate away to space, so any subterranean "ocean" should have cooled and frozen long ago. However, Europa may have been able to retain a liquid layer in spite of convection. Io and Europa are locked in an orbital resonance that produces strong tides and, in turn, considerable heat inside Io (discussed later). Europa is farther from Jupiter and thus much less affected by tides than Io is. However, it still experiences some tidal heating. Steven Squyres and his colleagues have calculated that under some conditions this

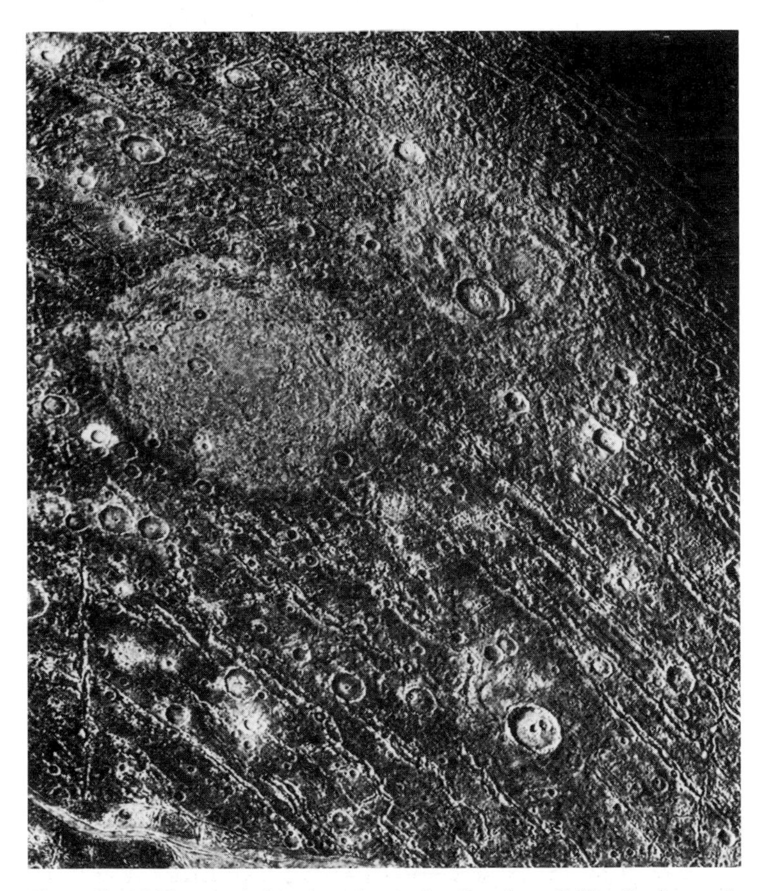

Figure 12. **A higher-resolution view of central and southern Galileo Regio reveals impact craters in various stages of degradation. Nearly all craters appear quite flat, and two prominent light-colored patches show only traces of crater rims, now almost completely erased by slow plastic flow in the icy crust. These features have been dubbed *palimpsests*.**

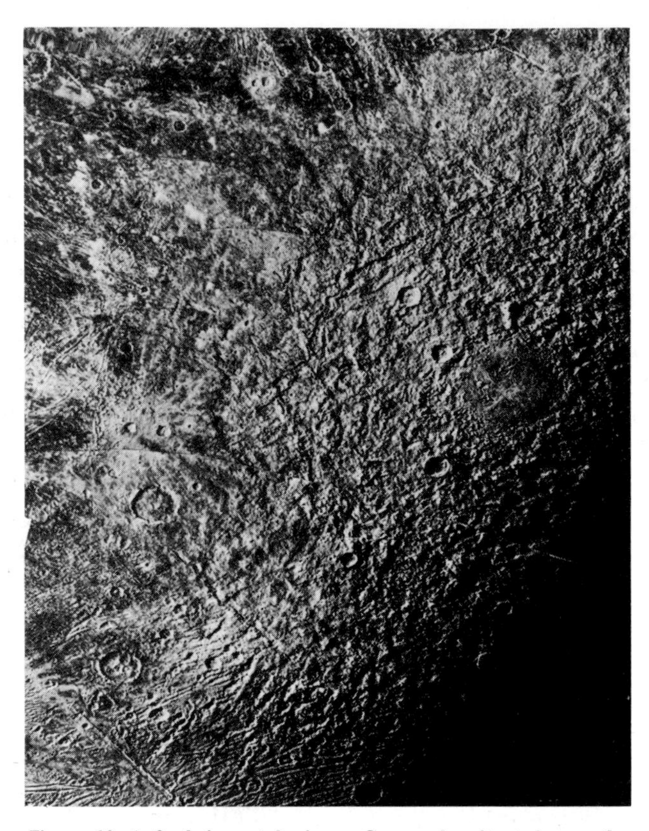

Figure 13. **A fresh impact basin on Ganymede, situated near the terminator when photographed, is surrounded by ejecta. Some 175 km across, the basin shows considerable relief and differs markedly from older craters. Evidently it formed when Ganymede's crust became rigid enough to sustain major topographic features.**

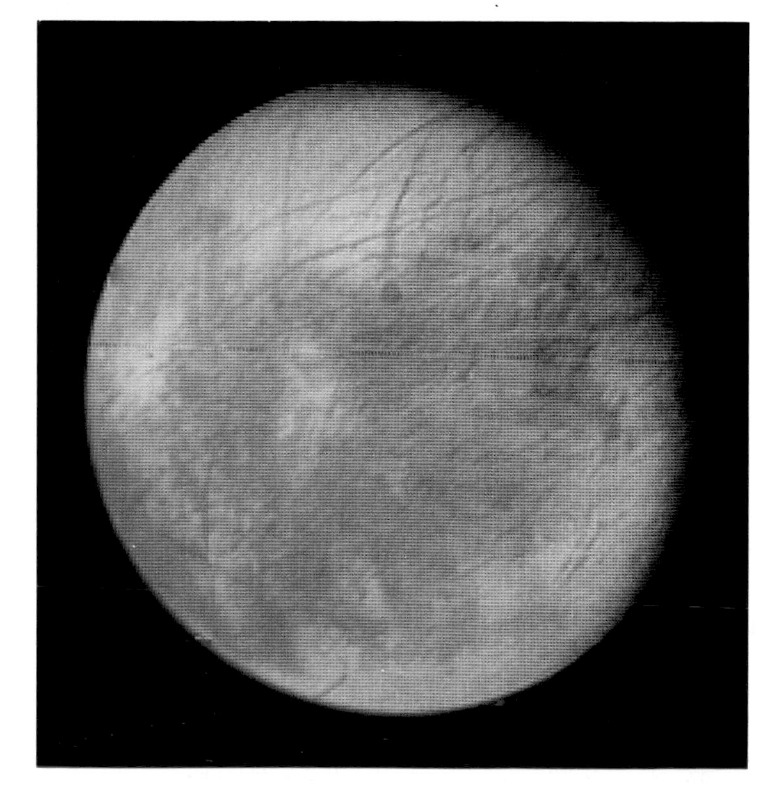

Figure 14. **The best Voyager 1 image of Europa shows a bright, low-contrast surface with darker mottlings and a tantalizing network of lines that crisscrosses much of its globe.**

tidal energy might be sufficient to keep Europa's crust from freezing completely, thus maintaining a liquid layer under a thin ice crust. The explanation for the strange cracked appearance of this satellite's surface may well lie in the history of its frozen ocean, when it formed, how deep it was, how long it remained liquid, and what forces acted upon it.

The stripes themselves suggest tensional forces and, taken together (in the absence of any major compressional zones), they indicate an increase in global surface area of perhaps 5 percent. Several models of Europa's crust associate this expansion in some way with the "frozen ocean." In the simplest scenario, virtually all of Europa's water outgassed immediately following accretion, creating a liquid layer topped by an ice crust. If radioactive decay and tidally generated heat were not sufficient to maintain the liquid layer, the freezing of this water would significantly expand its volume and increase its surface area – but not enough to explain all the "extra" area now occupied by the stripes. This model also suggests that we are looking at a very old surface, contrary to some of the estimates based on cratering. A similar model holds that water may be continually coming out of the interior and resurfacing the satellite.

Another major concern is the stripes' global distribution. Even though the Voyagers recorded only a fraction of the surface at high resolution, the major linear patterns can be traced around the globe in the lower resolution pictures. Has Europa undergone fracturing on a global scale? Did tidal forces or a wrenching rotational slowdown cause these patterns? Further analysis of Voyager data will certainly help constrain this problem better. For instance, we can now discern at least two stratigraphically distinct fracture systems and therefore can infer that they did not occur all at once.

Also awaiting explanation are numerous dark, mottled regions that may have been "dirtied" by the upwelling or mixing of silicate-laden water and ice. Although some pictures suggest that these areas may be somewhat rougher than average, they do not appear to be blocks of older crust, as on Ganymede. The Voyagers' color data also indicates that there may be at least two distinct types of this material, with different ultraviolet reflectances. This has been attributed to the implantation of sulfur ions from the Jovian magnetosphere or to other external mechanisms that have affected these color properties as well as the chemistry of the original material – particularly since the region that is darkest in the ultraviolet lies on the side of the satellite continually being overtaken and bombarded by plasma in Jupiter's rapidly rotating magnetic field. We will probably have to await further spacecraft observations to find out how young and active Europa's surface really is.

IO

Even before Voyager 1 reached Jupiter, Io was already known to be one of the strangest bodies in the solar system. It was, therefore, doubly surprising that Io surpasses even the wildest expectations of planetary scientists. The combination of intense volcanic activity, exotic chemistry, and complex interactions with the Jovian magnetosphere makes Io one of the most active and interesting planetary bodies yet explored. Most obvious to the eye are the satellite's unusual yellowish color (due primarily to sulfur, as discussed earlier) and the absence of "classical" impact features like ray craters, impact basins, or indeed craters of any size. Instead, the landscape is dominated by volcanic landforms.

The highest-resolution images of Io were of the Jupiter-facing hemisphere (much of the left half of Figure 17). These exquisite pictures show a wealth of surface detail and volcanic landforms – some only 500 m across. Sulfur, or at least sulfur-bearing compounds, are apparently responsible for Io's coloration, and some planetary geologists speculate that some of the flows themselves may have been molten sulfur rather than silicate lavas. Sulfur flows are rare but not unknown even on the Earth. The varying shades of color in and around Io's outpourings may correspond to different forms (or allotropes) of sulfur that occur at different temperatures. The stability of these allotropes and their colors under actual Io conditions have been challenged, however. An opposing view is that the flows are more like "ordinary" lavas, basaltic or otherwise, that have been colored by a high sulfur content and superficial deposits. Even in this case, however, deep-seated deposits of sulfur may still exist. Rendered molten by heat from silicate magmas, they may be the source of some of the riverlike features that snake across Io's surface.

Why should sulfur, a relatively rare element on Earth, be so prominent on Io? Actually, sulfur *is* fairly abundant here, cosmically speaking. It's just that much of our planet's allotment is tied up in iron sulfide (FeS) and lies hidden deep in the Earth's core. Studies of primitive meteorites show that sulfur is probably a common constituent of smaller solar-system bodies.

Volcanic calderas, with and without accompanying flow structures, abound on Io. Perhaps 200 of them with diameters greater than 20 km pock the globe (Figures 18,19); for comparison, the Earth, with $3^1/_2$ times more surface area, has only 15 or so in this size range. Unlike many terrestrial and Martian volcanoes, however, the eruptive centers on Io apparently do not build up large constructs of lava (such as the Hawaiian Islands, for example). Moreover, the flows from some of Io's calderas are very long, stretching for hundreds of kilometers. These observations suggest that the volcanic fluids that erupt across Io (whether molten rock or sulfur) have very low viscosities.

Io may not have craters or huge shield volcanoes, but regions of considerable topographic relief do occur there. Some calderas are several kilometers deep and there are numerous mountains up to 10 km high (Figure 20), in contrast to the flat, icy surfaces of the other satellites. The origin of these mountains is still a puzzle. They do not appear to be typical volcanic constructs; for instance, no calderas dot any of their summits. Nor are they aligned in chains or ridges in ways that suggest terrestrial mountain-building or plate tectonics. Perhaps such evidence lies buried under thick blankets of young volcanic material, making the mountains we see just the tops of an even more variegated topography. In any case, the very existence of deep calderas and mountains means that some parts of Io's upper crust are strong enough to support and maintain these features. Since much of the upper crust must be quite warm, some geochemists have argued that the required strength can only be supplied by a silicate-dominated surface over much of the satellite, with sulfur relegated to minor, shallow deposits.

The widespread volcanic landforms and lack of impact craters alone would probably have qualified Io for the title

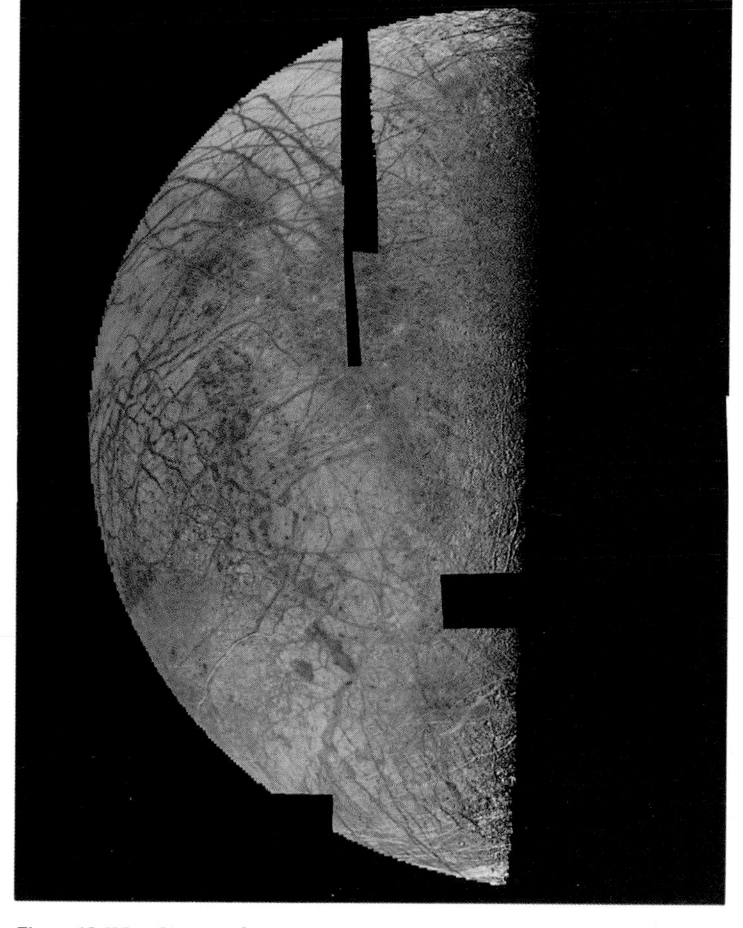

Figure 15. **When Voyager 2 observed Europa from 241,000 km away, the satellite's Lowellian maze of streaks resolved into a vast tangle of light and dark markings, tens to hundreds of kilometers across, that suggest filled-in cracks. Despite their superficial resemblance to fractures, the stripes have very little vertical relief, either positive or negative. They are primarily albedo markings, perhaps controlled by underlying fracture patterns in the crust. (Black indentations are gaps in Voyager 2's coverage.)**

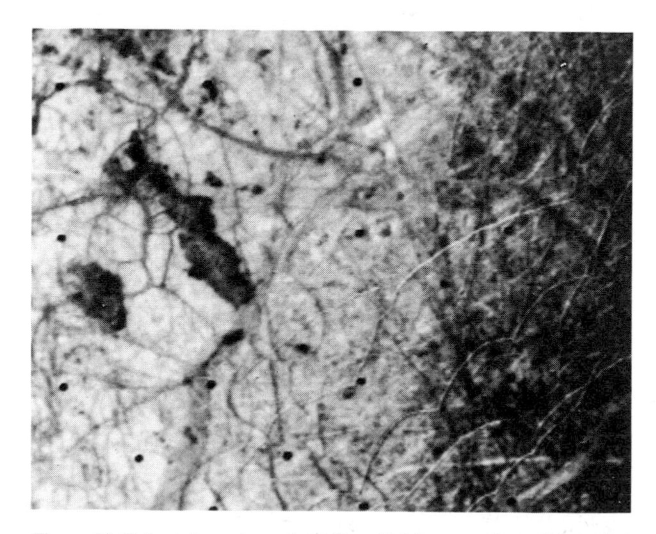

Figure 16. **Only under extremely oblique lighting was it possible to find any topographic relief on Europa. This near-terminator view is an enlarged section of a Voyager 2 image. Many narrow ridges, perhaps a few hundred meters high, loop across the surface in bright, arcuate patterns. Other ridges occasionally run down the center of the dark, broader stripes.**

"Most Volcanic Planet," but the discovery of eruptions in progress during the Voyager flybys (Figures 21-23) confirmed the title unequivocally. In all, nine eruptions spewed material more or less continuously as the spacecraft passed by. These fountainlike plumes are among the most impressive and beautiful sights in the solar system. The satellite's low gravity (about one-sixth that of Earth) and lack of appreciable atmosphere let volcanic gas and dust rise unimpeded to great heights, then fall slowly back to the surface, frequently creating symmetrical mushroom-shaped plumes.

In most cases, the plumes discovered by Voyager 1 were observed several times over a number of days; these showed little or no change during this period. Voyager 2, arriving four months later, was reprogrammed to make a study of the volcanic activity found by the first spacecraft. It found no new plumes, and most of the eruptions had remained remarkably constant. Of the eight volcanic plumes seen

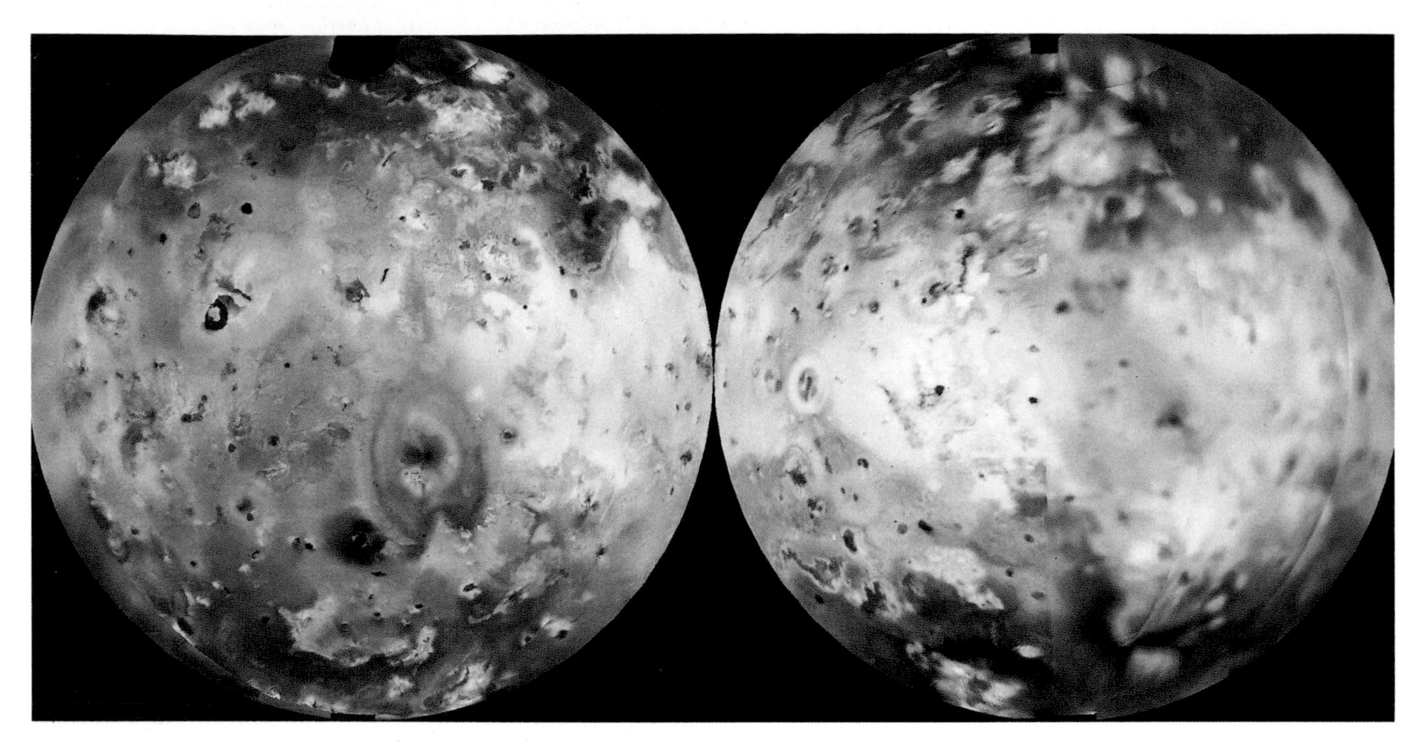

Figure 17. **Many volcanic features are evident in these mosaics of Io, though they were not initially recognized as such. For example, the large, heart-shaped marking just below center in the left half is a ring of volcanic ejecta being thrown out by the eruptive plume Pele inside it. Numerous other dark spots mark volcanic calderas, and flow features appear all over the surface.**

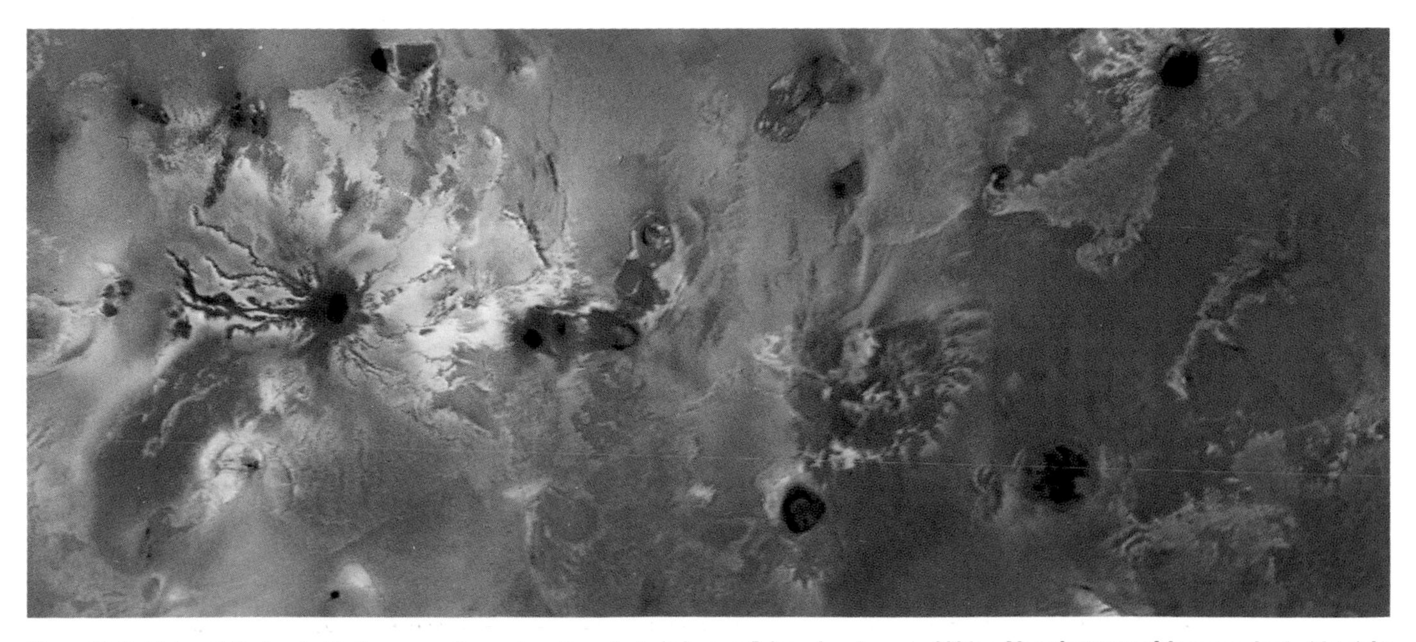

Figure 18. **Ra Patera (left of center) is the source of many long flows that wind across Io's surface for up to 200 km. Note that some of them are edged with bright halos, perhaps a frost of sulfur dioxide driven from the ground as the hot flows came by. An ongoing debate concerns whether the range of colors exhibited by Io (computer-enhanced here) matches those assumed by various phases of elemental sulfur.**

during the first flyby, Voyager 2 reobserved seven of them and found six still erupting (Pele being the exception). But several marked changes had occurred on Io itself during the interim (Figures 24,25). Thus, Io's volcanic activity may vary with time scales of months.

Most plume characteristics can be matched by ballistic ejection of material at vent velocities of 0.5 to 1.0 km per second (Figure 26). Such high speeds imply that we are not dealing with terrestrial-style explosive eruptions, for on Earth vent velocities rarely exceed 0.1 km per second. Also, the relatively constant, "sprinkler-head" appearance of Io's plumes calls to mind something more akin to a geyser than an explosive volcano. (Geologist Susan Kieffer notes that Yellowstone Park's "Old Faithful," if erupting under Io's low gravity into a vacuum, would send a plume of water and ice to an altitude of more than 35 km.)

Rapid phase changes in water (from liquid to steam, for instance) drive terrestrial geysers, but Io lacks water both on its surface and in its tenuous atmosphere. A different working fluid is required. The apparently ubiquitous nature of sulfur compounds on Io's surface and the discovery of sulfur dioxide gas near Loki imply strongly that such compounds are pivotal to the eruptions on Io. If estimates of the internal energy necessary for all this activity are correct, both sulfur and SO_2 will be molten at depths of no more than a few kilometers. One model developed by Bradford Smith, Shoemaker, Kieffer, and others uses sulfur dioxide as the principal propulsive ingredient. Models of this type can explain the major characteristics of the eruptive plumes, consistently high exit velocities, fine particle "halos," and nearly continuous eruption.

In a major study of the thermodynamics of geyser eruptions, Kieffer finds that many combinations of reservoir temperature, composition, and vent configuration can match the various aspects of Io's plumes. Most of the smaller plumes can be explained using variations of SO_2 geyser models. Some eruptions like Pele require even higher velocities, which can be achieved if the SO_2 has a higher initial temperature or if sulfur is the driving gas. Based on Kieffer's work and a study of the Voyager images, Laurence

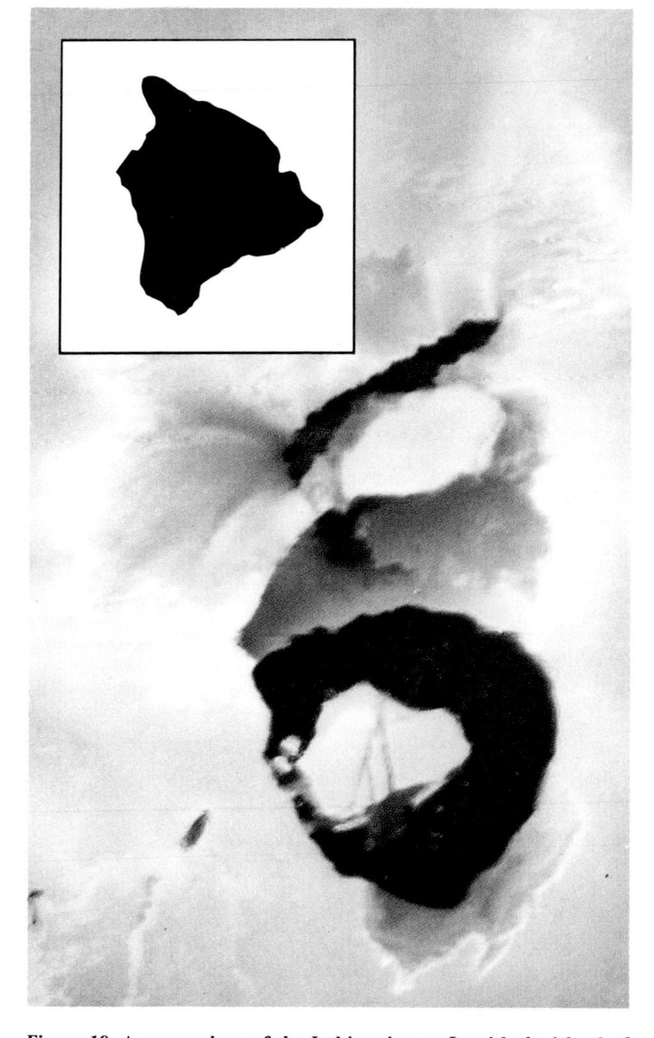

Figure 19. **A comparison of the Loki region on Io with the island of Hawaii (inset). Both are the result of volcanic activity, but Loki is considerably larger. The large black area may be a lake of liquid sulfur 250 km across. A fractured "raft" (solid sulfur?) appears to be floating inside it, surrounded by small, bright spots ("icebergs"?). Northeast of the lake is Loki's large, elongated fissure, spouting a gray plume from its left end. Computer enhancement makes the fissure and lake look black, even though they are actually more reflective than the Moon.**

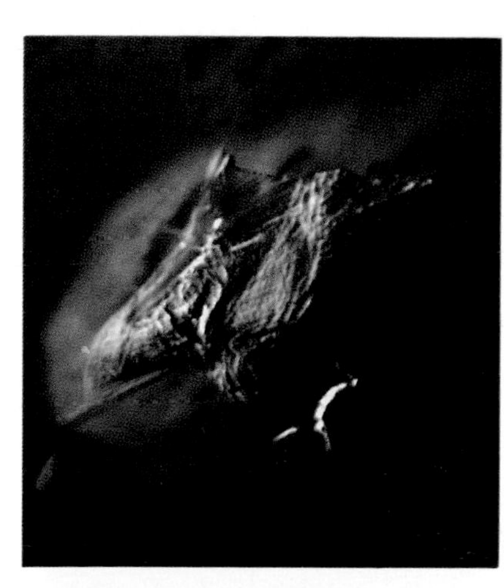

Figure 20. **Haemus Mons, near Io's south pole, has a base measuring about 200 by 100 km. It is an example of the topography that can be glimpsed on Io under favorable lighting conditions (near the terminator or against the limb). Such peaks can be up to 10 km tall.**

Figure 21. **The dome-shaped eruptive plume of Pele rises more than 300 km above the eastern limb of Io in the "discovery picture" that first alerted navigation engineer Linda Morabito to the satellite's dynamic nature. Many other plumes were quickly identified on additional Voyager images. A second plume, overlooked at first, is the bright spot at center.**

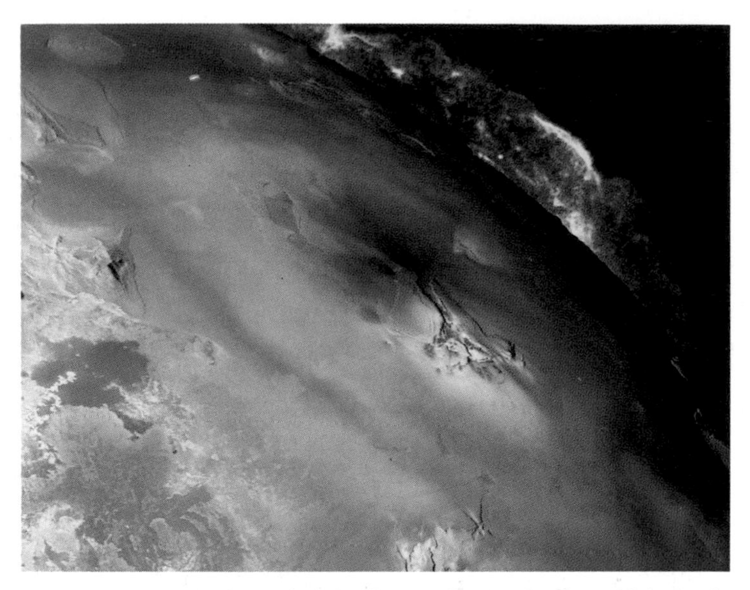

Figure 22. **Pele was the first active volcanic event discovered on Io, and it is also the largest yet found. The immense but tenuous plume of dust rises some 300 km above the satellite (the plume's contrast has been enhanced above the satellite's limb). It created a set of concentric yellow and brown rings on the surface roughly the size of Alaska (up to 1,400 km across). The source of Pele's outpourings is the hill-and-valley complex at its center. Ironically, although the volcano was erupting vigorously when Voyager 1 acquired the images used in this mosaic, it had become inactive four months later when Voyager 2 passed by.**

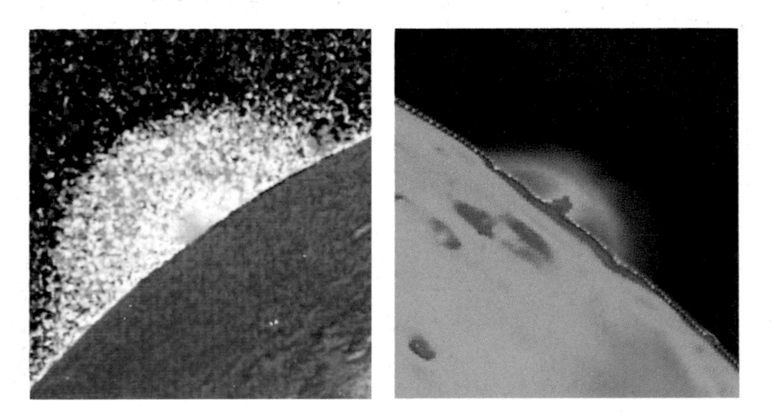

Figure 23. **All of Io's volcanoes are not alike. Compare the enormous plume of Pele (Figure 22) with those of Prometheus (left) and Loki (right), shown here in false color at the same scale. These are only 300 and 400 km across, respectively, and represent a smaller, cooler, and longer-lived class of eruptions.**

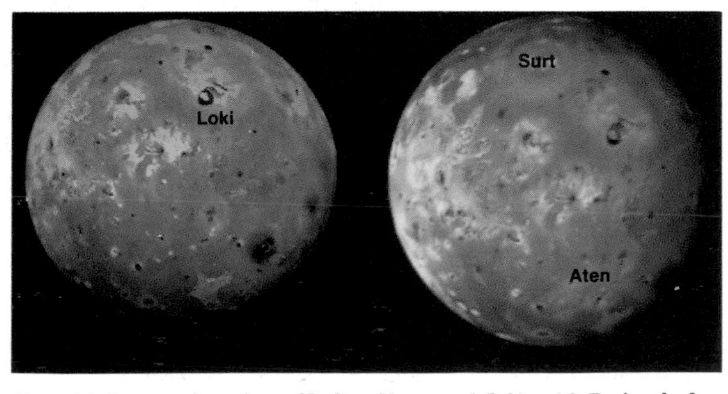

Figure 24. **Compare these views of Io from Voyagers 1 (left) and 2. During the four months separating the two visits, the deposits around Loki changed, and two new, enormous volcanic features appeared around the calderas Surt and Aten.**

Soderblom and Alfred McEwen conclude that Io's plumes fall into two categories: (1) relatively low-velocity, long-lasting, Prometheus-type eruptions that are powered by SO_2; and (2) higher-velocity, shorter-lived Pele-type events powered by sulfur.

More subtle geyserlike activity may explain another feature on Io's surface. Figure 27 shows a series of what looks like erosional scarps. Yet there is no evidence for any wind or fluid erosion on Io, so these features have proved puzzling. Perhaps fluid SO_2 has sapped (leaked) from their bases. Since this substance should be in liquid form at very shallow depths, scarps no higher than 100 m or so might suffice as sources. The result would be a small geyser of liquid SO_2 vaporizing in the near-vacuum at Io's surface.

Based on the discussions above, we can form a general idea of the likely structure of Io's upper crust (Figure 28). The interior of the satellites is probably partially molten, kept hot by tidal forces (discussed later). The rigid upper crust is primarily composed of silicate minerals. Silicate lavas probably make their way to the surface in many places, producing eruptions, lava flows, and calderas. As already mentioned, the mountains almost certainly consist of rocky material, but their origin is unclear; some may be heaps of low-density rock "floating" atop a denser silicate mantle (as Earth's continents do). The surface must also contain reasonably large amounts of sulfur and SO_2. How far down these deposits extend is one of the major uncertainties in our understanding of Io. If the sulfur layers are locally relatively deep – say, 10 to 100 m – then heat from deeper-lying silicate magmas should create molten sulfur "lakes" and flows. Should the sulfur and SO_2 liquid actually come into contact with hotter silicate magma bodies, geyser eruptions would be the likely result.

Given all these various forms of volcanic activity, it is obvious that new surfaces are being created on Io far more rapidly than are impact craters. Eruptive plumes alone are estimated to add a new layer at least 10 microns thick every year. If cratering rates at Io are similar to the Moon's, then an annual deposition of 1 mm or more may be required to erase the impact craters fast enough. This level of activity implies that Io's crust and upper mantle have been recycled by volcanic activity many times over during geologic history – a process that undoubtedly has affected Io's chemistry profoundly. The absence of water, carbon dioxide, and other volatiles on or around Io probably means that these compounds were driven from the interior and escaped to space long ago. Hydrogen, created during the dissociation of water, would have been particularly vulnerable. Only the heavier volatile species, such as sulfur and SO_2, remain today, apparently concentrated in the crust and tenuous atmosphere.

The level of volcanic activity that Io exhibits is extraordinary by any standard, but it is particularly surprising for such a small planetary body. Io has essentially the same size and density as the Moon. Models of the thermal history for small planets and satellites show that the early heat from accretion and perhaps short-lived radionuclides such as aluminum-26 can melt all or part of the body (see Chapters 22,23). But this heat is lost relatively quickly. Other sources of internal heat (longer-lived radionuclides like uranium, for instance) diminish gradually but still lose much of their effectiveness over 3 to 4 billion years.

Fortunately, just prior to Voyager 1's encounter, a new energy source was identified for Io that revised our perceptions of its thermal history. Stanton Peale and his colleagues proposed that tidal heating of Io is supplying as much or more energy than radioactive decay. This potentially large source of energy had never been seriously considered before, because Io's average orbit is almost exactly circular; there was no reason to suspect that tides (raised by Jupiter) in Io's crust could pump energy into the satellite's interior. However, Peale and his co-workers noted that orbital resonances among the inner Galilean satellites force Io into an orbit that is always somewhat eccentric – though over time this eccentricity averages to zero. This means that Io changes its distance from Jupiter as it orbits, forcing the tides to vary in height. In addition, Io's orbital velocity is not constant. But Io's spin *is* constant (one rotation per revolution). So the spot on the surface directly facing Jupiter oscillates in location, and Io is forced to flex methodically as the tidal bulges raised by Jupiter move around (Figure 29). The flexure induced by both of these effects creates friction, and the interior heats up. In fact, Peale and his colleagues calculated that the tidal-heating process could melt most of Io, and they went so far as to suggest that Voyager 1 might see evidence of volcanic processes. Rarely has such a prediction been so swiftly and overwhelmingly confirmed!

Another potential source of energy for Io may be its unique location in Jupiter's magnetosphere. Since Jupiter's rotation period is only about 10 hours compared to Io's orbital period of 1.77 days, the magnetic field of Jupiter is continually sweeping past the moon at about 57 km per second. If Io or its sparse atmosphere has a reasonable conductivity, the moving magnetic field generates an electric potential of roughly 600,000 volts across Io's diameter. According to several magnetospheric theories, currents driven by this potential difference should flow along the magnetic field lines that connect Io to Jupiter, and such currents could carry close to 1 million amps. Evidence for currents of this magnitude was provided by the magnetic-field measurements made by Voyager 1 when it flew directly under Io's south pole. These currents may control or modulate Jupiter's radio emissions. They could conceivably heat the satellite if all the current's power (about 1 trillion watts) were dissipated in the interior. However, even this seemingly huge source of energy is insufficient to explain the amount of power emanating from Io.

When Voyager 1 passed Io, it detected a number of anomalously warm regions associated with volcanic areas. For example, the infrared energy emanating from a dark region of Loki was far in excess of our expectations; there the spacecraft recorded a temperature of about 300° K, compared with typical noontime "highs" along Io's equator of perhaps 120° K. Could this feature, some 250 km across, actually be a lava lake – possibly liquid sulfur with a somewhat cooler skin or crust? Perhaps. Although not nearly as large as this one, lava lakes are a common feature of basaltic eruption on the Earth, such as those in the Kilauea caldera on the island of Hawaii.

Voyager's infrared experiment also detected numerous smaller hot spots, including one reading of 500° K over the very center of Pele. If such hot spots cover a significant fraction of the satellite's surface, their combined effect should be evident even in whole-globe infrared measurements made from Earth. And, indeed, Io is anomalously bright at some infrared wavelengths, and puzzling cooling trends occur when the moon is eclipsed by Jupiter. These could be explained by emissions from hot spots covering about 1 percent of Io's surface; a similar result has been derived from Voyager data. Another Earth-based observation in early 1979 recorded outbursts of 5-micron energy far above expectations that lasted for only hours or days. The only reasonable source appears to be small regions of hot (500-600° K) surface material.

All of these infrared measurements indicate that Io emits

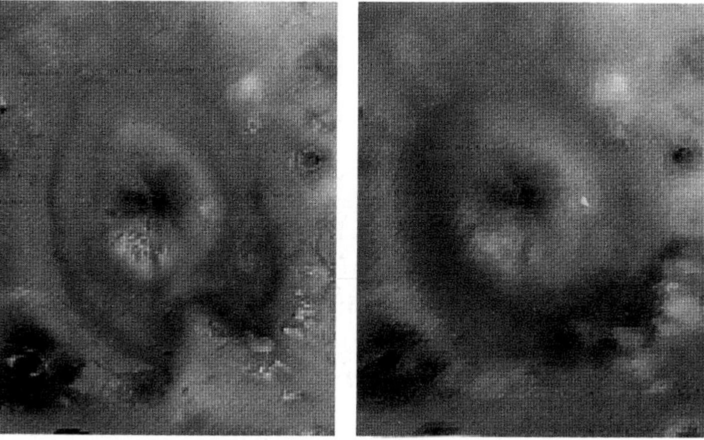

Figure 25. **Another comparison shows how the shape of Pele's "hoofprint" deposit changed between the flybys. The indentation seen by Voyager 1 (left) was probably caused by a partial obstruction of the central vent.**

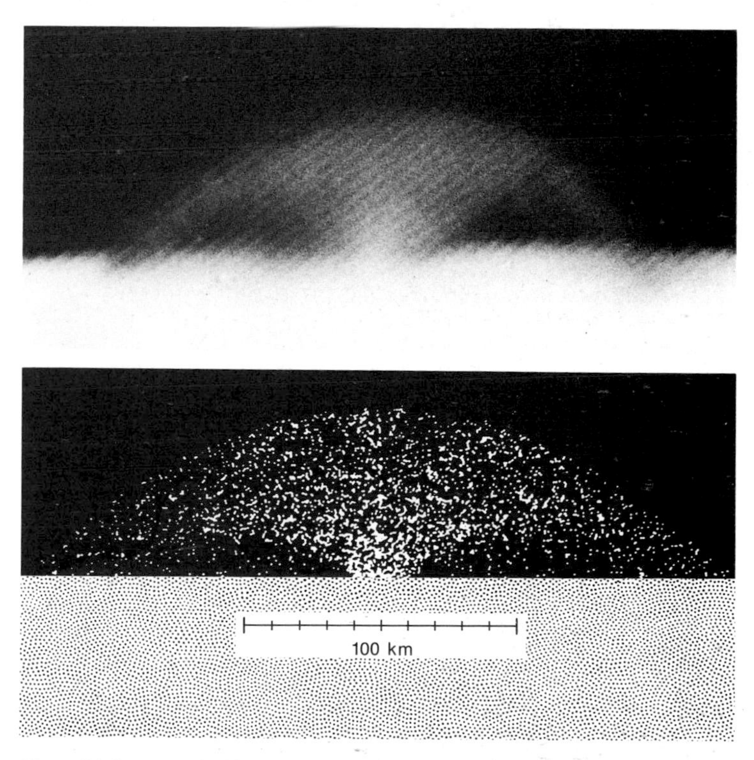

Figure 26. **Compare the Voyager 1 image of Prometheus' plume along the limb of Io with the computer simulation below it. The latter assumes that material is ejected along ballistic trajectories at a speed of 0.5 km per second and at angles of at least 55° from horizontal. The plume's actual base in both cases is 7 km below the "surface" because the source vent is turned 5° toward us from the limb.**

an impressive 100 *trillion* watts of energy in this wavelength region – an average of about 2 watts per square meter over its entire surface. Table 1 relates this heat flow to those of the Earth, the Moon, and an active terrestrial geothermal area. Note that Io's average emission is comparable to the most active areas on the Earth, making it truly the most volcanic planetary object (including the Earth) yet studied.

Since the Voyager flybys, Io has been the subject of intensive observation from ground-based telescopes. Although individual features on its surface cannot be resolved even by the largest instruments from Earth, quite a lot has been learned by using the fact that the small volcanic hot spots on Io radiate far more infrared energy at some wavelengths than does the rest of the surface added together. This gives astronomers, armed with the knowledge from Voyager, a way to distinguish the emissions of the volcanic centers from the background created by the rest of the satellite (Figure 30). Based on the results from a variety of techniques, the region around Loki appears to be the most active volcanic area on Io and accounts for much of the moon's total radiated energy. Loki also appears to be long-lived: today it radiates essentially the same amount of infrared radiation as when the Voyagers passed by in 1979. Finally, the ground-based data provide some insight into the frequency of the briefer, high-temperature eruptions like Pele's. One event, observed in 1986, seems to require a temperature of about 900° K – some 200° higher than the hottest spots observed by the Voyagers. This is significant because it is well above the temperatures thought necessary for sulfur-driven volcanism and thus probably represents our first direct evidence that silicate magmas are erupting on the satellite's surface.

Io is radiating too much energy for magnetospheric currents to be a significant heating mechanism. In fact, even potent tidal energy must be converted very efficiently within the interior to match the observed thermal output. The tidal mechanism would be most effective if the interior is partially molten. However, the rate at which the energy can be dissipated is linked to properties of the Jovian interior and to the rate at which Io's orbit changes with time in response to tidal forces (just as the orbit of Earth's Moon is changing). According to an analysis by Jay Lieske, the energy now being emitted by Io corresponds to a more rapidly evolving orbit than has actually been observed over the last 300 years. So several theorists have proposed that the amount of tidal energy dissipated inside Io (and thus its rate of volcanism) may vary with time, and that what Voyager observed is not typical of the last $4\frac{1}{2}$ billion years. As should be clear by now, Io's volcanoes continue to puzzle planetary scientists.

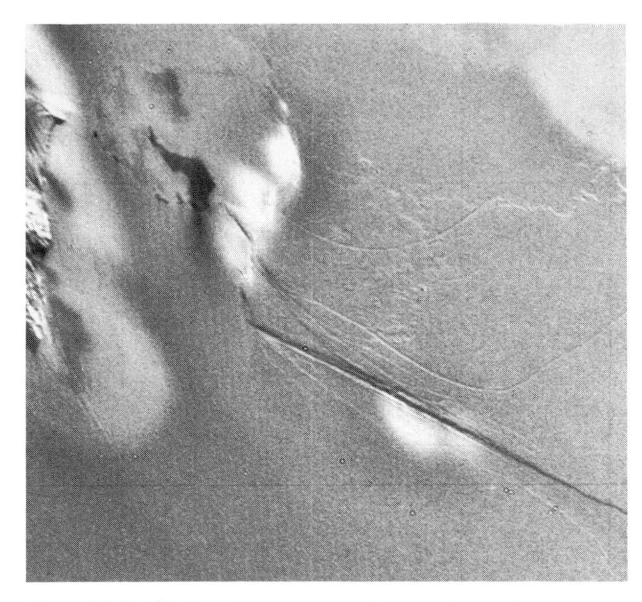

Figure 27. **Bright patches accompany what may be erosion scarps on the surface of Io. In locations such as these, sulfur dioxide may be escaping to the surface from below as blocks of crust slump or break away from the scarp's bases.**

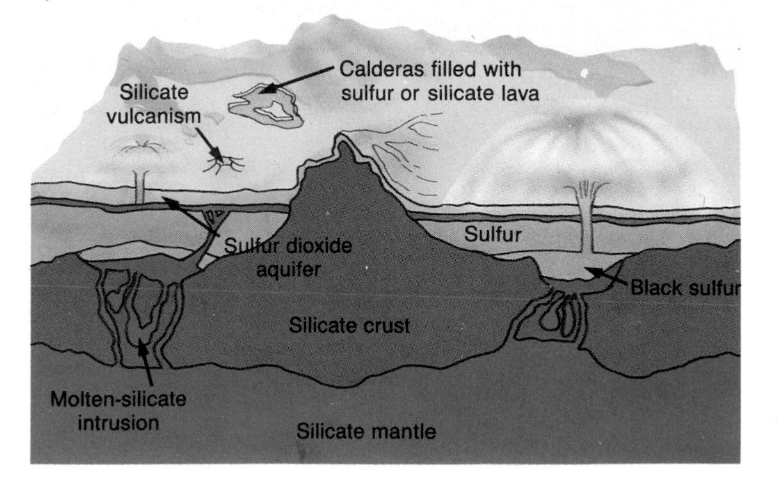

Figure 28. **A schematic depiction (not to scale) of most of the major phenomena found on Io. As described in the text, at least three distinct types of active volcanism appear to be reworking the satellite's surface and outer layers of crust.**

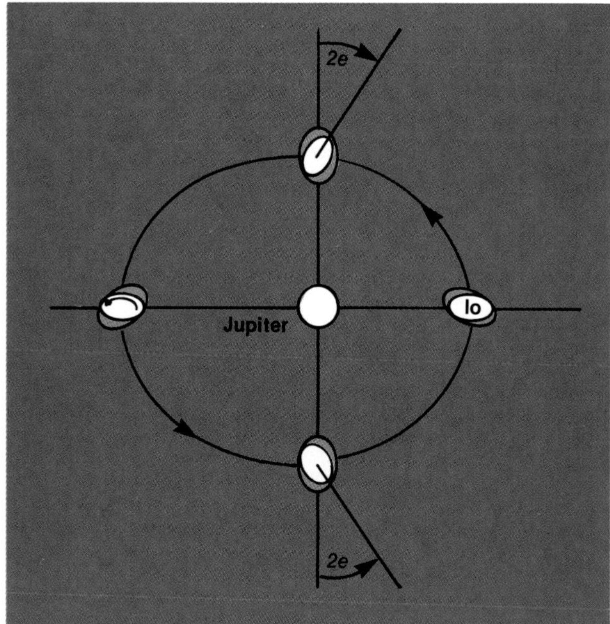

Figure 29. **Io's dynamic activity stems from an orbital resonance with nearby Europa that forces Io into a slightly eccentric orbit. Ordinarily, Jupiter's strong gravity would keep one hemisphere of the satellite facing the planet at all times. But the forced eccentricity makes Io move at different velocities along its orbit, and the side facing Jupiter nods back and forth slightly as seen from the planet. Tidal forces develop inside the satellite that generate heat through friction, and much of the interior remains partially molten as a result.**

AN ATMOSPHERE, TOO

Our current knowledge of Io's atmosphere suggests that it is "lumpy," with more gas over warm areas and active vents. Sulfur dioxide is the major known constituent, but other as-yet unidentified gases (such as hydrogen sulfide) could also be present. The first firm indication of an atmosphere surrounding Io came from Pioneer 10's discovery that the satellite possesses a significant ionosphere, with electron densities of 10,000 to 100,000 per cm^3. At least one important component, sulfur dioxide, was later identified on and above the surface, but spectroscopic studies have ruled out a number of other potential gases like water and carbon dioxide. The virtual absence of heavy ions other than sulfur and oxygen in the surrounding magnetosphere also limits the contributions from nitrogen and inert gases like argon.

Sulfur dioxide on Io has a number of interesting properties. First, new gas can be abundantly supplied by volcanic vents. Second, SO_2 can condense anywhere on Io's surface; virtually all of it will be frozen out in the polar regions and at night. This suggests that, to some degree, cold regions regulate the supply of gas, similar to the situation with condensed CO_2 in the polar caps of Mars. Under these conditions, surface temperature controls the amount of SO_2 that can sublimate from solid to gas and thus the local atmospheric pressure. In fact, the atmospheric pressure detected by Voyager near Loki is close to the value expected above the cold (roughly 120° K) regions surrounding the caldera. This simple equilibrium model is probably not correct in detail, however, because a number of mechanisms can add or subtract gas on a local or regional level. Major atmospheric variations may also be driven by changes in volcanic activity, the amount or composition of released gases, or the rates at which atmospheric species escape into space.

MAGNETOSPHERIC INTERACTIONS

Io, the Galilean satellite closest to Jupiter, lies deepest in the planet's magnetosphere and plays host to a complex array of electromagnetic interactions. One of the earliest "connections" found between Io and the magnetosphere concerned radio bursts from Jupiter with decameter wavelengths (tens of meters long). The bursts occur far more frequently when Io, Jupiter, and Earth align in a certain way. Still not completely understood, this phenomenon is related to electrical currents generated by Io's motion through the magnetosphere (discussed earlier in connection with the satellite's heating).

Another striking indication of Io's magnetospheric involvement came in 1974 when Robert A. Brown discovered the distinctive emission of sodium emanating from the vicinity of Io. This emission was found to be coming from a "cloud" of neutral sodium surrounding Io and extending for tens of thousands of kilometers along its orbit (Figure 31).

Why is sodium, a relatively minor element, so obvious in Io's spectrum? The answer is that sodium is relatively easy to observe under these conditions. Its electronic structure vibrates in a strong resonance with light at the D-line wavelengths of 5890 and 5896 angstroms, and thus scatters sunlight at these wavelengths. Other atoms have similar but generally weaker emissions; collisions with electrons in a plasma can generate still others. Shortly after sodium's discovery, spectroscopists detected potassium and then singly ionized sulfur (Figure 32). With the Voyager 1 encounter, the full extent of Io-related atomic and ionic species in Jupiter's magnetosphere became evident. Measurements by the various plasma, charged-particle, and ultraviolet detectors showed that heavy ions, particularly those of sulfur and oxygen, were important throughout the magnetosphere from the low-energy plasma to high-energy cosmic rays. All of these magnetospheric species (Table 2) appear to originate in the dense plasma torus at Io's orbit, and ultimately from Io itself.

A major question lingering since the discovery of the sodium cloud concerns the process or processes that eject material from Io into the magnetosphere. Io has about the

	Watts/m²	*Remarks*
Io	2 ± 1	Global average
Earth		
average	0.06	Global heat flow through crust
geothermal area	1.7	Wairakei, New Zealand
Moon	0.02	Average of Apollo 15 and 17 landing sites

Table 1. **Tides raised within Io by Jupiter generate (through friction) prodigious amounts of heat – enough to power its vigorous volcanoes.**

Species	*Remarks*
Na	Exists in banana-shaped cloud leading Io along its orbit; first observed from Earth in 1972
Na^+	Detected in small amounts by Voyager instruments
K	Discovered soon after neutral sodium; probably associated with it in cloud
K^+	Probably detected by Voyager's charged-particle sensors
S^+, S^{2+}, S^{3+}	Sulfur and oxygen ions exist throughout the magnetosphere, with low-energy ones concentrated in a torus centered roughly on Io's orbit; S^+ and O^+ were discovered from Earth.
SO_2^+, SO^+, and related ions	Probably discovered by Voyager's plasma experiments

Table 2. **Before the 1970s, all four Galilean satellites were thought to be inert, airless bodies. But, as the Voyagers' instruments revealed, Io bristles with activity and has a tenuous, transient atmosphere with an unusual composition.**

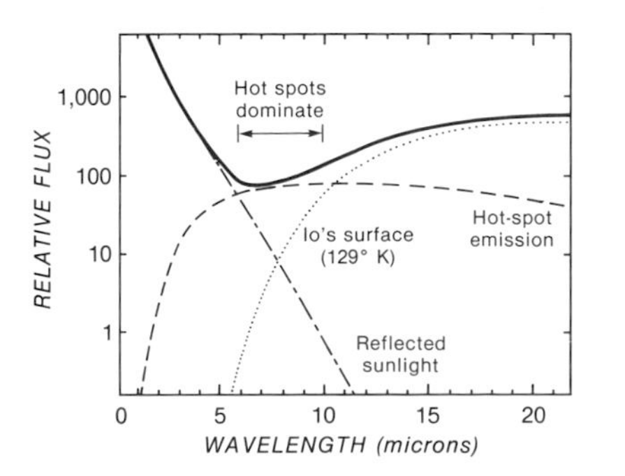

Figure 30. **Io's spectral signature involves contributions from three major components, but hot-spot emissions dominate from 6 to 10 microns. Observations in this range can be used to monitor the hot spots from Earth as they rotate into and out of view.**

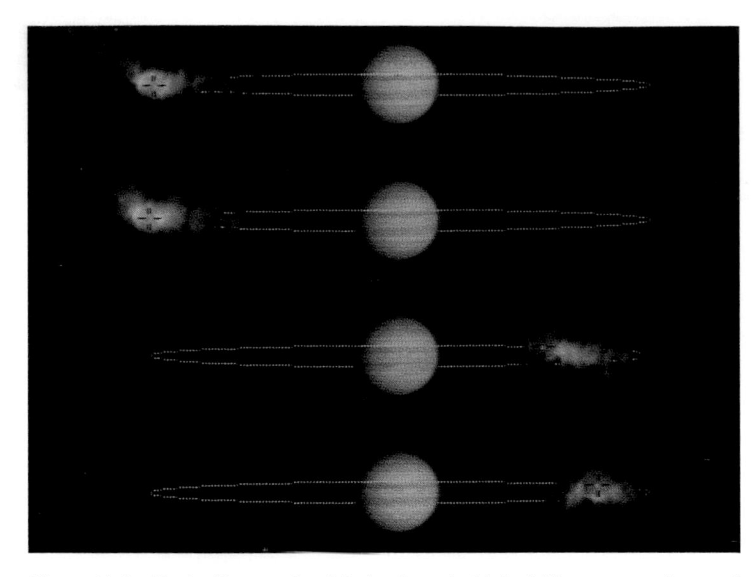

Figure 31. **Io (dot inside cross bars) is deeply embedded within a swarm of neutral sodium atoms sputtered from its surface by energetic charged particles. Most of the sodium is confined to a banana-shaped cloud that precedes the satellite for more than 100,000 km along its orbit (yellow ellipse). In these 1981 images, changes in the cloud's apparent size and shape are largely due to viewing perspective. North is up, and a telescopic view of Jupiter has been added to help visualize the system's geometry.**

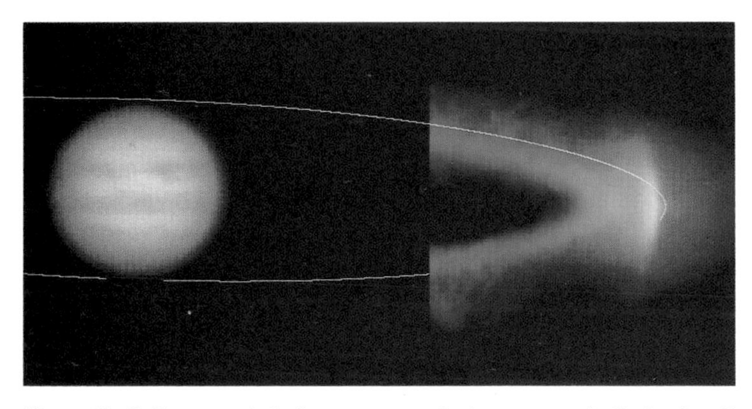

Figure 32. **Sulfur around Jupiter was recognized spectroscopically by Israeli scientists in 1976, and this false-color image from May 1983 clearly shows the three-part sulfur "nebula" that we now know to be associated with Io. Purple denotes a hot outer torus dominated by doubly ionized sulfur ions emitting in the near-infrared at 9531 angstroms. Green denotes a cooler, inner torus of singly ionized sulfur glowing at 6731 angstroms (and, not shown, at 6716 as well). Emission from both species peaks in the white vertical band, termed the hot inner torus. The yellow line shows where the ions are most concentrated around Jupiter; this trace is inclined about 7° to Io's orbit. North is up, east to the left, and an image of Jupiter has been added for clarity.**

same surface gravity as the Moon, and velocities of at least $2\frac{1}{2}$ km per second are needed to escape from Io's influence and go into independent orbit about Jupiter. Some sodium atoms have been observed leaving Io at well over 10 km per second. Various other processes for escape of material have been suggested. Pictures of the large volcanic plumes may suggest that material is "blowing off" directly to space, but it would be difficult or impossible for the volcanic gases to achieve the required escape velocity. Similarly, the escape of SO_2 or its components solely because it is hot is possible, but such thermal escape may not be efficient enough to supply the amount of material populating the magnetosphere. Ionization and sweeping by the magnetosphere is yet another possibility, though the existence of neutral oxygen in the magnetosphere indicates that this species as well as sodium and potassium may escape as neutral atoms.

Dennis Matson and his colleagues suggested in 1974 that sputtering (ejection by the impact of charged particles on a surface) might account for the escape of sodium, a theory reinforced by subsequent studies. The sputtering of material from Io's tenuous atmosphere is another important mechanism. The abundance of heavy S and O ions in the magnetospheric plasma provides a ready source of sputtering ions, and this idea remains one of the best candidates for removing material from Io. In fact, recent laboratory studies indicate that sputtering may also erode the icy surfaces of the other Galilean satellites. We now believe sputtering is a multistage process in which cascading charged particles first drive surface material into Io's atmosphere at low velocities and subsequently eject it into the surrounding Jovian magnetosphere.

All of these ejection theories have problems, so it seems likely that more than one of them is involved. Clearly, however, Io is the ultimate source of these materials, and volcanic activity must play a major role in supplying fresh gas and volatile material to the surface and atmosphere. The total injection of ions into the magnetosphere seems to lie between 10^{27} and 10^{29} ions per second. This rate of loss, maintained over geologic time, would correspond to the erosion of only a few hundred meters to perhaps 1 km of Io's surface – far less than the volcanic eruption rates seen today (millimeters of new material per year) can supply. Io seems to be in no danger of eroding away before our eyes.

FOUR NEW WORLDS

Despite more than three centuries of telescopic scrutiny, the new data about the Galilean satellites acquired by Voyager spacecraft represent an advance in knowledge almost as profound as the satellites' original discovery. During those hours of close observations in March and July of 1979, these four worlds, with a total land area equivalent to the Earth's, went from dots of light on our mental horizons to geologically and geographically known places – *terra cognita*. Totally different from previously studied planets, they present us with a stunning diversity of surface features, colors, compositions, and geophysical and geochemical processes. From frozen Callisto to bubbling Io, the satellites are a veritable laboratory for the study of planetary evolution. And, as always, we are left with a wide range of new theories, questions, problems, and controversies to fuel further exploration, such as that planned by Project Galileo when it reaches Jupiter in 1995.

14
Titan

Tobias Owen

FIRST GLIMPSED as a tiny "star" accompanying Saturn (Figure 1), Titan was discovered by the Dutch astronomer Christiaan Huygens in March of 1655. Huygens was working at the newly founded Paris Observatory, using an extremely long-focus telescope of his own design that represented the cutting edge of the day's optical technology. However, once the satellite's orbit was worked out, its size determined, and its brightness measured, little more of substance was learned about Titan for nearly 300 years. In the 1940s, Gerard P. Kuiper (another Dutch astronomer!) began a systematic spectral survey of the planets and their satellites with the 2.7-m reflector at the McDonald Observatory in Texas. He discovered that Titan's spectrum exhibits the same absorption bands of methane that were by then well-known in spectra of Jupiter and Saturn. Evidently, Kuiper concluded, this satellite has an atmosphere (Figure 2).

The possibility that the largest satellites in the outer solar system might have atmospheres had been explicitly suggested some 20 years earlier by Sir James Jeans, who had studied how thermally energized gases can escape from the planets. Jeans realized that to maintain an envelope of gas over the lifetime of the solar system, an object simply needed sufficient condensed mass to provide a strong gravitational field and a temperature low enough to keep the gas molecules from reaching escape velocity. Thermal velocities are higher for gases of low molecular weight like hydrogen and helium, so these are lost to space with greater ease. However, methane (CH_4) was quite a reasonable candidate for the atmosphere of distant Titan. This gas consists of cosmically abundant carbon and hydrogen, has a molecular weight of 16, and condenses only at extremely low temperatures.

But how much methane did Titan have, what other gases were present, and how did they get there? Unlike Jupiter and Saturn, Titan has a relatively weak gravitational field and was certainly not expected to retain large amounts of hydrogen and helium. Were there clouds on Titan like those on the giant planets? And what was the surface like? Earth-based observers struggled with these and other questions, finding various, often contradictory answers. Titan is more than 1 billion km away and appears as a barely resolvable disk even in our best telescopes – a serious handicap in trying to unravel its mysteries. By 1980, two quite different models for Titan's atmosphere were in vogue, suggesting surface pressures of 20 millibars and 20 bars!

Fortunately, help was on the way. The Voyager 1 spacecraft flew past Titan in November 1980. Its trajectory had been optimized for a close encounter, since it was already clear that we could learn much of lasting value if we could understand Titan better. The spacecraft sailed past the satellite at a distance of only 4,000 km, about one-tenth the altitude of a geosynchronous satellite above the Earth's surface. The full array of instruments on board was brought into play, and we learned more about Titan in those few days than in all the preceding 325 years. Voyager's flood of information is still being analyzed, and new details continue to emerge. This chapter, therefore, summarizes what we know as of early 1989 and includes recent results from observers who, like Huygens and Kuiper, still have their feet firmly planted on our own planet.

ORIGIN AND INTERIOR

Titan must be recognized as but one of several large, icy objects in the outer solar system. Other members of this class include Ganymede and Callisto at Jupiter, Triton at Neptune, and the Pluto-Charon system (see Chapters 13,15). Their similarities and differences (and their relationship to the smaller icy satellites, the "asteroid" Chiron, and comet nuclei) are far from clear at present. What we do know is that

Figure 1. **British amateur astronomer Paul Doherty made this drawing of Saturn and Titan on August 26, 1966, when the planet's rings were nearly edge-on as seen from Earth.**

all of these larger objects are made of a mixture of water ice and rock, with the ice apparently containing ever-greater amounts of frozen, trapped, and adsorbed gases as the distance from the Sun increases.

But it is not just heliocentric distance that determines the final composition of one of these objects. Where the object formed with respect to its parent planet is also extremely important. Each giant planet had a major influence on the gas and dust in the solar nebula surrounding it. We can thus speak of planetary sub-nebulae, in which conditions varied as a function of distance from the primary planet in a manner analogous to how the solar nebula varied at distances from the embryonic Sun. In both cases, temperature was a key variable. The formation of the giant planets produced large amounts of heat; gravitational potential energy was turned into kinetic energy when the dispersed matter in the solar nebula "condensed" into the much smaller volume of a planet. Calculations show that, early in its history, Jupiter radiated so much thermal energy that it must have glowed. Some of this planet's primordial heat is still escaping today (see Chapter 10).

We see the consequences of this early, hot phase of Jupiter in the variation of density now found among its Galilean satellites: Io and Europa, situated closest to the planet, are much denser than the more distant Ganymede and Callisto, which have bulk compositions much like that of Titan. Thus, the Jovian system provides an analogy to the inner and outer planets. The bodies richest in highly volatile compounds (hydrogen and helium for the planets, water ice for the satellites) formed farther from their primary and thus in cooler regions. In the case of Jupiter, this trend produced dramatic results. For the Saturnian system the effect is more subtle, since Saturn has less than one-third the mass of Jupiter and therefore the heat generated during its formation was correspondingly less. Instead of a distinction between rocky inner satellites and icy outer ones, we find icy bodies throughout the Saturn system, from the ring particles to Phoebe. But other volatiles, borne by the ices in these bodies, should reflect the fact that the planet at the system's center was hot when it formed. Thus, we expect substances such as methane, ammonia, nitrogen, carbon monoxide, and argon to have been more prevalent in the materials that formed the more distant satellites.

Similarly, the generally lower temperatures in the Saturnian system compared with Jupiter explain why Titan has an atmosphere, while Ganymede and Callisto do not – despite the fact that the three satellites' similar densities suggest similar compositions. Laboratory studies have shown that crystallizing water ice rapidly loses its ability to trap and retain gases as temperature increases. In particular, above 135° K, gases that have either been trapped in the crystal structure of ice or merely adsorbed on icy surfaces leave the ice with ease. In other words, if the ice that now composes Ganymede and Callisto solidified at very low temperatures but was later heated above this critical value, its gas content would have been markedly depleted. This is apparently what happened to the materials that served as the building blocks for these satellites; prior to accumulating they were, in effect, "baked out." But evidently the environment at Titan's distance from Saturn remained well below 135° K, and the gases were not released until the satellite formed. Heat generated during the accretion process would have been more than sufficient to liberate gases from the infalling ices. It could also have driven reactions among them, forming new species. These gases then became the satellite's earliest atmosphere.

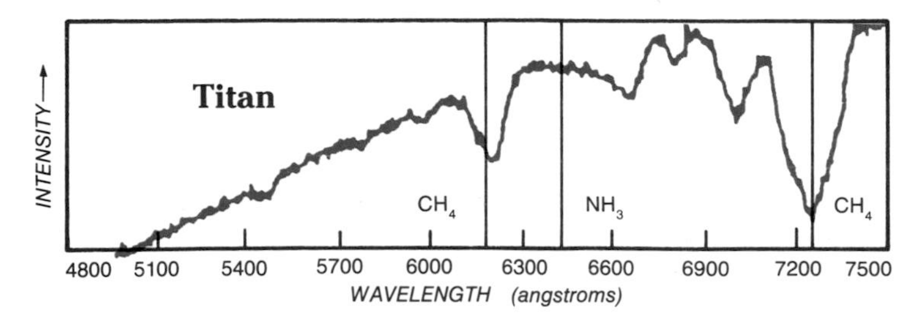

Figure 2. **The light reflected from Saturn's moon Titan bears the characteristic spectral signature of methane (deep absorption bands near the wavelengths of 6200 and 7300 angstroms). An absorption at 6400 angstroms due to ammonia (NH_3) is weakly evident in the planet's atmosphere but absent from Titan's. This spectrum was obtained in 1978, using the 2.7-m telescope at McDonald Observatory in Texas.**

Additional heat was generated later from the decay of short-lived radioactive elements in the rocky component of Titan. Just as Earth and the other terrestrial planets melted and differentiated, so too this distant satellite must have formed a core of dense, rocky material that became surrounded by a mantle of ice. Thus, Titan obtained a secondary, outgassed atmosphere from the same general mechanism that produced the early atmospheres on Mars, Venus, and Earth (see Chapter 8).

Since ice assumes different crystalline structures corresponding to the ambient temperature and pressure, we expect Titan's icy mantle to consist of layers of these different forms of ice. Depending on the amount of heat generated in the satellite's interior and the rate of its release, a layer of liquid water might exist even now beneath the icy crust. A cross-section of one possible model for Titan's interior that incorporates many of these features is shown in Figure 3.

In discussing Titan's origin, we have not considered where the material that formed the satellite came from or how it got organized in such a way that one large satellite arose instead of several smaller ones. These are problems whose solutions are still unknown. As stated above, the current consensus is that the giant planets and their satellites formed from "sub-nebulae" much as the Sun and planets formed from the primordial solar nebula. But there is at least one significant difference. It is commonly assumed that no material came from outside the solar nebula to affect the forming Sun and planets. In contrast, material from outside the planetary sub-nebulae must certainly have been incorporated into individual planets and satellites. Hence, the composition of these sub-nebulae may have been closer to that of the main nebula than one might otherwise expect. This relationship is one of the many unresolved problems that require a new mission to the Saturn system for their resolution.

THE ATMOSPHERE: COMPOSITION AND EVOLUTION

Before Voyager, ground-based astronomers had amassed considerable evidence that Titan's atmosphere contained particles in addition to gas. Polarization measurements, Titan's brightness as a function of wavelength, and large variations in the strength of its methane absorption bands all indicated that the atmosphere was not clear. But what were

Major constituent		*Percent*
Nitrogen	(N_2)	82–99
Methane	(CH_4)	1–6
(Argon)	(Ar)	0–12
Minor constituent		*Parts per million*
Hydrogen	(H_2)	2,000
Hydrocarbons		
Ethane	(C_2H_6)	20
Propane	(C_3H_8)	20
Ethylene	(C_2H_4)	0.4
Diacetylene	(C_4H_2)	≤0.1
Methylacetylene	(C_3H_4)	0.03
Nitrogen compounds		
Hydrogen cyanide	(HCN)	0.2
Cyanogen	(C_2N_2)	≤0.1
Cyanoacetylene	(HC_3N)	≤0.1
Oxygen compounds		
Carbon monoxide	(CO)	50–150
Carbon dioxide	(CO_2)	0.0015

Table 1. **In addition to nitrogen, Titan's atmosphere contains an appreciable amount of methane that varies with altitude and is still poorly determined. The presence of argon has been deduced only indirectly – there may be none at all, in which case the nitrogen fraction would increase.**

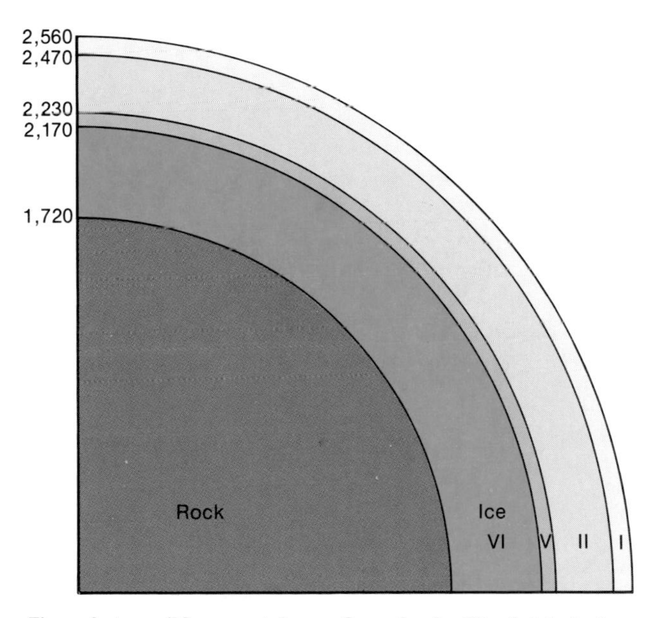

Figure 3. **A possible present-day configuration for Titan's interior has been proposed by Claudia Alexander and Ray Reynolds. Outside the rocky core, variations in pressure and temperature have produced ice layers with differing crystal structures. The rocky, ice VI, V, II, and I sections have densities of 3.52, 1.43, 1.28, 1.18, and 0.94 g/cm³, respectively.**

these particles and how were they distributed? As Voyager 1 approached in the fall of 1980, it quickly became apparent that layers of opaque material in Titan's atmosphere extend completely around the satellite (Figure 4). No hint of the surface of Titan could be seen, even in the many high-resolution pictures obtained during the spacecraft's closest approach. Yet the reflectivity of Titan showed a sharp demarcation, with the northern hemisphere noticeably darker than the southern. Just how those airborne particles know where the satellite's equator is remains an unsolved problem of Titanian meteorology. The aerosols appeared layered along the limb, and there was a dark northern cap.

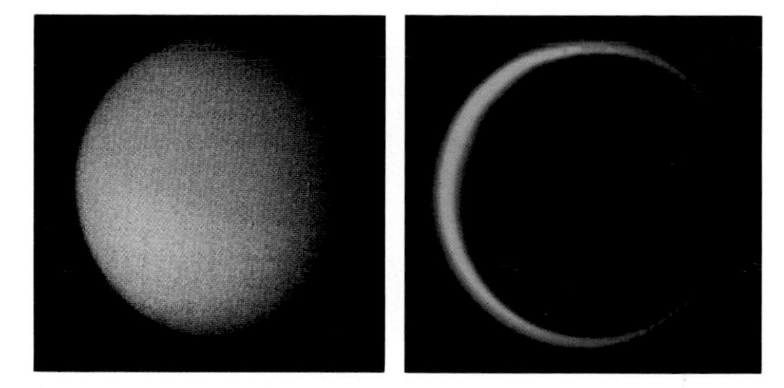

Figure 4. **Opaque layers of particles in Titan's atmosphere (left) prevented Voyager 1 from seeing the satellite's surface during its 1980 flyby. Note the lighter color of clouds over the southern hemisphere and the dark hood at the north pole. Voyager 2 looked back at a crescent Titan the following year (right); the extension of blue light around the moon's night side is due to scattering by smog particles in the sunlit portion.**

More surprises came from the experiments designed to study atmospheric structure and composition. The dominant gas surrounding Titan is molecular nitrogen, just as on the Earth. In fact, the satellite is enveloped by about 10 times more nitrogen than we are, yielding a surface pressure 1.5 times greater than at sea level on Earth. Thus, neither of the two extreme pre-Voyager models for the atmosphere proved correct. Yet Donald Hunten's high-pressure, nitrogen-rich model fits the run of pressure and temperature in Titan's atmosphere remarkably well. It was simply necessary to set the satellite's surface at the 1.5-bar level rather than at 20 bars.

Methane, the one gas identified with certainty before Voyager arrived, turned out to be a minor constituent with an abundance of a few percent. As much as 10 to 15 percent of the total atmosphere may be argon, an inert gas that cannot be detected spectroscopically. But argon may also be present only in tiny quantities; our determination of the mean molecular weight of Titan's atmosphere is uncertain enough to allow both extremes. The Voyagers' infrared spectrometers revealed a rich variety of other compounds, principally hydrocarbons like ethane (C_2H_6) and acetylene (C_2H_2), and nitriles like hydrogen cyanide (HCN). A list of all the compounds detected so far is given in Table 1.

You may be surprised to find molecular hydrogen listed there since, as mentioned, this gas can escape from the satellite easily. But actually the presence of a little H_2 is expected, given the other listed species. Some nitrogen and methane molecules high above Titan are being broken apart by the Sun's energetic ultraviolet photons and by the bombardment of electrons from Saturn's magnetosphere. The fragments then recombine to make minor constituents. So even though hydrogen escapes easily from Titan, it is being continuously generated as well, and the relative abundance of H_2 found in Table 1 represents the balance achieved between its production and escape.

Usually Titan resides within Saturn's magnetosphere, but it sometimes lies in the boundary (magnetosheath) region or even "outside" in the undisturbed solar wind. The satellite has no appreciable magnetic field, so ions and magnetic fields can interact directly with its upper atmosphere. Neutral gases are particularly susceptible, as they become ionized through impacts with high-velocity ions and electrons. Once that happens, the magnetic field picks up the newly ionized gases and sweeps them away. Thus, Titan may also be losing gases other than hydrogen through such interactions.

At Titan we have found an atmosphere that is still evolving from a primitive hydrogen-rich state. This is not too surprising, given the extremely low temperature at the satellite's surface. Voyager measured a value of 94° K, only a few degrees warmer than that expected for a body that far from the Sun with Titan's reflectivity and no atmosphere at all. At this low temperature, the vapor pressure of water is vanishingly small (which explains why the icy rings and satellites of Saturn have survived for billions of years without completely subliming away). What this means for Titan is that water ice on the surface cannot participate in the atmospheric photochemistry, providing oxygen atoms (through its dissociation by sunlight) that would interact with the methane, hydrocarbons, and nitriles. If we could magically move Titan closer to the Sun, say, to the orbit of Mars, the character of its atmosphere would immediately change. Warmer surface temperatures would drive plenty of water vapor into the atmosphere, and the resulting supply of oxygen would rapidly convert methane and its by-products to carbon dioxide, exactly the dominant carbon-carrier now found on Mars.

Meanwhile, back on the real present-day Titan, some carbon dioxide is in fact present, as is carbon monoxide (Table 1). How is this possible? There are at least two solutions to this apparent paradox. First, the carbon monoxide may be primordial – it may have been trapped in the ices that formed the satellite, then released into the atmosphere as Titan formed. Electrons from Saturn's magnetosphere, which continually bombard Titan's atmosphere, can break carbon monoxide apart, leaving the oxygen in an excited state. If one of these oxygen atoms encounters a methane molecule, it reacts to yield a hydroxyl (OH) radical; this can in turn combine with another CO molecule to make CO_2, while the hydrogen escapes. In other words, if CO was present on Titan from the beginning, the formation of a small amount of CO_2 is predicted even in the absence of water vapor.

Alternatively, water vapor could be supplied to this low-temperature atmosphere from the outside, as showers of icy debris from neighboring satellites' collisions with meteorites and comets or, more directly, from ice-rich material gravitationally captured by Saturn and swept up by Titan as it circles the planet. Then the small amounts of both CO and CO_2 observed today could be produced from photochemical reactions between water vapor and methane.

Actually, both of these alternatives may be correct. We cannot be certain, but recent evidence suggests that at least some primordial carbon monoxide may be present. Microwave (radio) observations of CO in Titan's stratosphere indicate this gas is not nearly as abundant there as it is lower down in the troposphere, where it was detected at infrared wavelengths. In other words, on Titan this gas is apparently most abundant near the surface, which is consistent with its formation there by outgassing and its destruction at high altitudes by photochemical reactions. Since both sets of CO observations were made from Earth and not by Voyager, we hope to improve the quality of these data during the next few years and find an answer to this enigmatic but basic puzzle.

Meanwhile, one more piece of evidence can be brought to bear concerning the origin of Titan's atmosphere: the relative abundances of the principal isotopes of hydrogen. The most common hydrogen atom consists of a single proton and a single electron in orbit about it. The next heavier isotope, called deuterium, has a neutron in addition to the proton, but again just a single electron. Chemically speaking, the two isotopes are nearly identical, since reactions usually involve only the electron. But the physical behavior of hydrogen and deuterium can be very different, since the mass of deuterium is twice as great. Therefore, a study of the relative abundances of these two isotopes can help us to unravel various physical and chemical processes that have taken place in a planetary atmosphere.

On Titan, most of the hydrogen is in the methane, and it has proved possible to detect the deuterated form of this molecule (CH_3D) in addition to CH_4. It turns out that deuterium occurs in Titan's methane molecules some eight times more frequently that it does in the methane in the atmospheres of Jupiter and Saturn. Only about one-fourth of this enrichment can be attributed to processes acting on the satellite's atmosphere since it formed. For example, the escape of hydrogen from Titan's upper atmosphere tends to concentrate the more-massive deuterium.

But most of the observed enrichment must be caused by something else. One interesting possibility is that Titan has been this way all along and thus that an interstellar ratio of deuterium to hydrogen has been preserved in its ices. Even higher D:H ratios have been observed in hydrogen-containing molecules in dark interstellar clouds. They are produced by ion-molecule reactions in cold, low-density environments. In other words, the ices that formed Titan and the gases these ices contained never became hot enough to reach equilibrium with either the gas in the proto-Saturnian nebula or in the solar nebula itself. Since carbon monoxide is at least 100 times more abundant than methane in the interstellar medium, the apparent preservation of deuterium-enriched interstellar methane in Titan's atmosphere lends support to the idea that at least some fraction of the CO found there today is likewise a primordial remnant.

THE ATMOSPHERE: VERTICAL STRUCTURE, HAZES, AND CLOUDS

If hydrogen is continually escaping from the top of Titan's atmosphere, what happens to the material that is left behind? Why is there any methane at all if this gas is constantly being broken apart? To answer these questions, we must consider the atmosphere from a different point of view, focusing on the composition of its aerosols. To begin this process, we need to know how the temperature in the atmosphere varies with height – a trend determined by both the radio occultation experiment and by the infrared spectrometer on the Voyager spacecraft. The combined results (Figure 5) show that temperature decreases with altitude from the surface until reaching the tropopause, about 60 km up. At this point, a reversal occurs, and temperature begins to increase with height, eventually peaking at 175° K, some 80° *higher* than at the surface. This temperature turnaround results primarily from the ultraviolet sunlight entering Titan's upper atmosphere. Its absorption not only contributes to the photochemical processes described in the previous section, but also excites the resulting molecules and the dark material making up the aerosol.

On Earth, urban smogs usually form within 1 km of its surface. But the particulate haze around Titan extends to an altitude of roughly 200 km, with a detached and much more tenuous second layer about 100 km higher. Over the dark north-polar cap the main cloud is more extended, and in Voyager images the two layers appear to blend together. Actually, Titan's atmosphere contains relatively few particles per unit volume of gas, but the haze is so deep that it completely hides the surface from our view. We now realize that some of this aerosol material is simply condensed forms of the gases shown in Table 1. With the exception of hydrogen and CO, all of the minor gases will condense at the tropopause temperature. Indeed, all of these species (except CO) were discovered in Titan's stratosphere – not in the lower atmosphere, where they would not exist as gases.

But the aerosol layers seen in Figure 4 are more than just simple condensation products, which would be white or gray. The dirty-orange color suggests that some additional chemistry is occurring, transforming simple molecules into more complex substances. Some of these end-products are probably polymers, structures in which the same molecular configuration repeats again and again. Both hydrogen cyanide and acetylene form dark polymers that could certainly contribute to the observed effects. Laboratory experiments that combine the principal ingredients of Titan's atmosphere with a variety of energy sources to drive the reactions have little difficulty in producing a long list of suitably dark organic compounds. But it has not yet proved possible to achieve a *close* match between the laboratory simulations and what is found in the satellite's atmosphere.

A SATELLITE WITH OCEANS?

Thus, what we see around Titan is a thick photochemical smog whose exact composition remains unknown, though it is likely a rich mixture of many compounds. One thing is clear, however: this aerosol cannot stay suspended indefinitely. Based on Voyager observations, the airborne particles appear to average about 0.2 to 1.0 micron in diameter. As the particles grow larger, they precipitate out of the atmosphere and end up on Titan's surface. What happens there depends on the local temperature and topography. It could be quite remarkable! For example, methane could in principle condense at the low temperature on Titan's surface. But a careful analysis of how temperature changes with altitude just above the surface indicates that this is unlikely. However, thin clouds of methane crystals may form in the lower atmosphere (Figure 5).

The story for ethane is rather different, however. One can deduce how much ethane has been produced over Titan's entire history (it is the most abundant by-product in the photochemical destruction of methane). As condensed ethane falls through Titan's lower atmosphere it may form clouds or hazes, but at the surface it will form a liquid. The result is that Titan may be covered by a global ocean of ethane with an average depth of 1 km! This ocean would contain nitrogen and methane dissolved within it, along with all of the other products shown in Table 1. The bottom could be lined with a mixture of insoluble aerosols (only the very "fluffy" ones will float) and CO_2 ice. Any surface topography protruding above the ethane sea would also be coated by this mixture (Figure 6).

In the absence of oceans, Titan's underlying landscape should be similar to those found on Ganymede and Callisto: some mixture of impact craters and their debris, plus the effects of a primitive kind of plate tectonics (present on Ganymede but not on Callisto). On the other hand, Enceladus and Miranda have taught us that it is very risky to extrapolate the geologic history of one body to another. There may have been internal sources of energy on Titan, such as resonance-induced tides from earlier configurations of the Saturnian satellites. Giant impacts may have left huge basins. But whatever the landscape, it should now be coated with precipitated smog particles – if not by the global ocean.

One other factor, climatic "seasons," probably do not affect what goes on at the surface. A year on Titan (the time it takes Saturn to circle the Sun) is 30 of ours. Since we suspect that the satellite's axis of rotation is aligned with Saturn's, its equatorial plane will be tilted some 27° with respect to the ecliptic plane. At the time of Voyager 1's encounter, the northern hemisphere had just ended a $7\frac{1}{2}$-year-long winter. With less sunlight falling on the winter hemisphere than the summer one, the photolysis rate of methane and thus characteristics of the aerosols such as their size may also change with the season. This may cause the hemispherical brightness variation Voyager observed in the ubiquitous smog layer. In fact, an apparently cyclic variation in the total brightness of Titan (observed over the past 15 years) may well be caused by seasonal variations in this hemispheric asymmetry. But an atmosphere this dense can transport heat very efficiently, so at ground level we expect the temperature to vary by no more than a few degrees throughout the year.

What would it be like to sit in a boat, rocking in the hydrocarbon seas of Titan? Dark and cold, for sure. The atmosphere is so dense that the surface and lower portions of the atmosphere experience little change in temperature with varying latitude, time of day, or even at night. But the horizontal visibility might be very good if you weren't trapped in an ethane fog bank. Models calculated by Brian Toon and his colleagues suggest that the lower atmosphere will be quite clear; the aerosol layers are concentrated where they are produced, at high altitudes (Figure 5). Titan absorbs

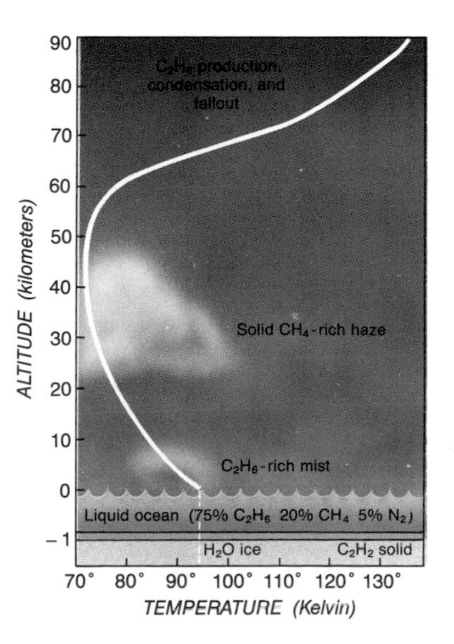

Figure 5. **The temperature and physical character of Titan's atmosphere vary markedly with altitude. Note in particular that its temperature reverses near the tropopause, about 45 km from the surface. The lower atmosphere is just barely too warm to allow the condensation of liquid nitrogen, but clouds of methane ice and a drizzle of liquid ethane are distinctly possible.**

Figure 6. **Although scientists did not glimpse the surface of Titan during Voyager 1's brief flyby, they learned enough to speculate on its appearance. Somewhere in the frozen desolation near Titan's north pole, the Sun's dull glow along the horizon marks the arrival of spring. Opaque clouds hang ceilinglike 40 to 50 km above the ground, trailing a constant drizzle of methane ice laced with brownish organic compounds. As they fall, the methane crystals evaporate in the warm lower atmosphere, coagulating the organics into drifts of cindery debris that rain onto the surface. Once ethane nears the surface, however, it changes to droplets. Here and there puddles of liquid ethane make a gradual, steamy return to the atmosphere, though a global ocean is also very likely. Muted crater forms bear witness to occasional meteoric collisions long mantled by the continual deposition from the clouds. One large pit, excavated recently, reveals fresh exposures of water ice.**

about 80 percent of the sunlight incident upon it, but the smog particles and methane gas absorb nearly all of this – only 5 to 10 percent ever reaches the ground. At noon the distant Sun's light, filtered through the smog, casts an orange glow over the local scene similar to what one experiences on a moonlit night on Earth.

Will there be waves? We have no information yet about winds on Titan's surface, so the question remains unanswered. Indeed, we are not completely certain of the existence of the global ocean itself. Ethane may undergo further reactions, producing more complex substances so efficiently that there is not enough of it left to cover Titan. Yet ponds or seas do seem likely, and cosmic chemists are especially attracted to the idea that a large ocean can resupply the atmosphere with methane as this gas is broken down and lost. What we lack is a definitive observation that tells us how much liquid is present, what its composition is, and how it is distributed over Titan's surface.

THE NEXT STEP

This account of Titan and its atmosphere should indicate just how much this satellite can tell us, if only we can learn more about it. Titan is a member of a class of volatile-rich icy planetesimals represented by everything from comet nuclei to the cores of the giant planets. In particular, comets are capable of delivering these volatiles to the atmospheres of the inner planets, so the relevance of studying Titan may be closer to home than one might think. In Titan, we can examine a secondary atmosphere produced by the degassing of the volatiles trapped in such a planetesimal. The similarities and differences between these gases and those found in the class's other members – especially comets – will tell us something about the conditions and processes involved in their formation. As we have seen, there appear to be opportunities to reach back in time to study the chemistry in the natal cloud of gas and dust in the interstellar medium.

Finally, this time travel also allows us to investigate an evolving, oxygen-poor atmosphere in which complex chemical reactions are occurring today that may resemble some of the first steps along the path from chemistry to biology on the early Earth. (It is perhaps worth stressing, however, that there is a long and as-yet uncharted path from chemical evolution to biology. *Life* on Titan seems ruled out by the exceedingly low surface temperature, which must slow chemical reactions to unproductive rates.)

This is a rich harvest indeed, and to reap it we must return to Titan. To this end, the European Space Agency (ESA) and NASA are now pursuing a joint mission called Cassini. Its orbiter, NASA's contribution, has been designed to carry a radar experiment for mapping the surface of Titan right through the smog layers, much as American and Soviet radar systems have mapped the surface of Venus. The orbiter would also carry cameras and infrared and microwave spectrometers much more capable than the aging equipment aboard Voyager. But the big news would surely come from tne ESA's probe. Descending slowly through the atmosphere the probe will determine directly the composition and isotopic ratios of the gases and aerosols around it. A sensitive camera would record the scene below. The descent is sufficiently slow (nearly 3 hours) that the probe might even survive its landing – be it a splash or a thud – and make additional measurements at the surface.

If everything goes according to plan, all of this information plus volumes more about the rest of the Saturnian system will be sent back to Earth sometime early in the next century. At that point, we will know much better just what it is that this mysterious moon is trying to tell us about our early history.

15
Icy Bodies of the Outer Solar System

Dale P. Cruikshank and David Morrison

THE OUTER SOLAR SYSTEM is the home of a remarkable variety of exotic worlds. Here are found two giant planets (Jupiter and Saturn) composed primarily of hydrogen, and two lesser giants (Uranus and Neptune) with compositions dominated by oxygen, carbon, nitrogen, and silicon as well as the ubiquitous hydrogen. There are in addition more than 50 satellites, four ring systems, and perhaps trillions of comets. Many of these objects are discussed in other chapters, including the five largest satellites of the outer solar system (Ganymede, Titan, Callisto, Io, and Europa). But that still leaves numerous slightly smaller objects, all composed at least in part of water ice, which will be described here.

Icy objects beyond the orbit of Jupiter are relatively primitive in their composition, preserving at least some of the volatile materials that condensed in the original solar nebula. Since they have experienced less modification than their cousins from the inner solar system (including even the satellites of Jupiter), these objects have the potential to reveal information about the formation and early history of the solar system that has been erased on better-known worlds situated closer to the Sun. In addition, they turn out to be fascinating objects in their own right, expanding our appreciation of the variety to be found within the Sun's family.

The subjects of this chapter include the planet Pluto and its moon Charon, the regular satellite systems of Saturn and Uranus (excluding Titan, which is discussed in Chapter 14), the irregular satellite system of Neptune, and the few asteroids that are known beyond Jupiter (Figure 1). Of these, the satellites of Saturn and Uranus are best understood, thanks largely to flybys of the Voyager spacecraft. For all of

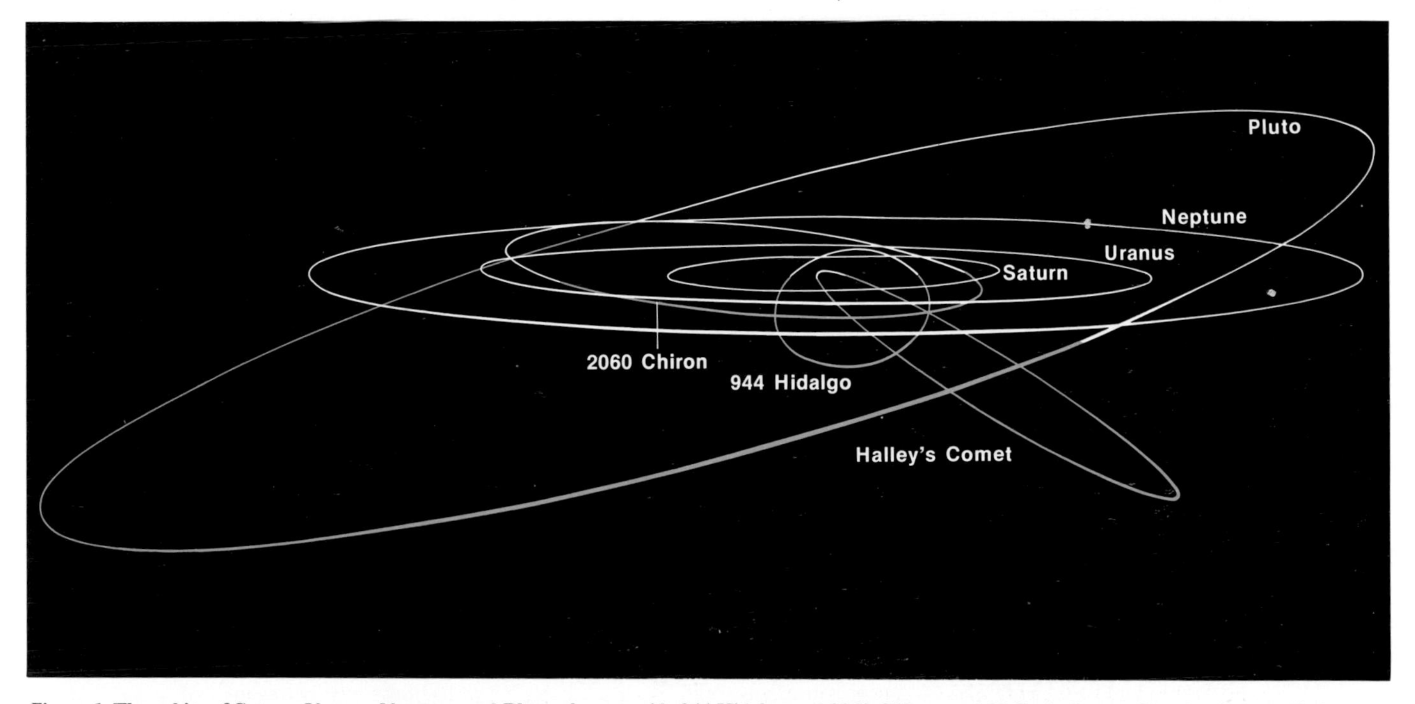

Figure 1. **The orbits of Saturn, Uranus, Neptune, and Pluto; the asteroids 944 Hidalgo and 2060 Chiron; and Halley's Comet. The view is from 5° above the ecliptic and 6.7 billion km from the Sun. Saturn, Uranus, and Neptune travel nearly within the plane of the ecliptic, but Pluto, Hidalgo, and Chiron have orbits that are both noticeably inclined and eccentric. The extremely elongated path of Halley's Comet brings it above the ecliptic only near perihelion.**

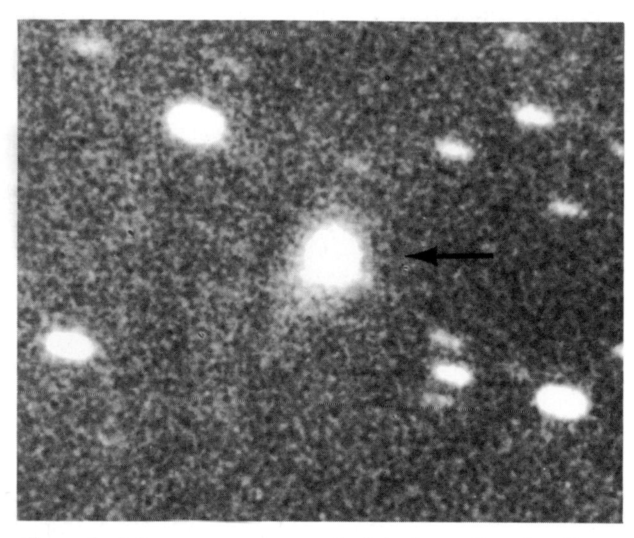

Figure 2. **Astronomers were surprised to learn that the distant "asteroid" 2060 Chiron (arrowed) can display a cometary coma, which extends faintly toward lower left in this image taken April 9, 1989. Chiron's estimated diameter is about 180 km.**

the rest, we are dependent on the feeble light collected with large telescopes and laboriously interpreted by a small cadre of planetary astronomers.

THE MOST DISTANT ASTEROIDS

In Chapter 18, the asteroids are described as rocky objects, found primarily between the orbits of Jupiter and Mars. But the term has been extended to any small object orbiting the Sun that is not a comet, and there is no reason to expect that such debris will be absent from the outer solar system. Nevertheless, only two asteroidal objects are known that travel outward as far as the orbit of Saturn. The first, 944 Hidalgo, was discovered by Walter Baade in 1920. Hidalgo's orbit carries it from a perihelion of 2.0 AU (just beyond Mars) to an aphelion of 9.7 AU (near Saturn), similar to that of a number of short-period comets. We have only a rough estimate of its diameter – about 40 km. Its spectral reflectance is unusual; perhaps Hidalgo is an extinct comet, though one of very large size.

The "asteroid" 2060 Chiron, discovered by Charles Kowal in 1977, is more unusual. A true inhabitant of the outer solar system, Chiron has a perihelion just inside Saturn's orbit at 8.5 AU and an aphelion near the orbit of Uranus at 18.9 AU. Dynamicists consider this to be an unstable or "chaotic" orbit; that is, Chiron's ultimate fate will be either collision with a planet or ejection from the solar system. There is little certain information about the physical properties of this object. But its brightness, combined a measurement of the heat it emits, suggests that its surface albedo (reflectivity) is about 10 percent and its diameter about 180 km.

In 1988 astronomers were surprised to discover that Chiron's brightness had more than doubled. They suspected that the object was exhibiting some low level of cometary behavior, perhaps similar to that of the peculiar Comet Schwassmann-Wachmann 1 (in a circular orbit just beyond Jupiter). Then, in April 1989, Karen J. Meech and Michael J. S. Belton used the Kitt Peak 4-m telescope to record a faint coma around Chiron (Figure 2). This cometary activity – occurring 11.2 AU from the Sun – implies that the object must contain some very volatile material. Perhaps most notably, Chiron appears by far to be the largest comet currently residing among the plantets. What a spectacular show it could produce if it ventured near the Sun! Chiron might also be an escaped satellite of Uranus or Saturn, but for now no one can say with certainty.

SATELLITES OF SATURN: A FIRST CLOSE LOOK AT ICY WORLDS

Seventeen regular satellites orbit the planet Saturn, and others are suspected to lie embedded in its rings (see Chapter 12). Seven of these are greater than 400 km in diameter and thus large enough to be studied in some detail with ground-based telescopes. During the decade preceding the Voyager encounters, astronomers applied a variety of photometric and spectroscopic techniques to measure their sizes and determine the composition of their surface materials. An airless object reveals its surface composition by creating absorption bands in the visible and infrared parts of the sunlight reflected from its surface. Short of directly landing and sampling the material, such remote spectroscopic methods provide our most powerful tools for compositional analysis.

Early in the 1970s astronomers at the Universities of Hawaii and Arizona used new NASA-financed telescopes to probe the chemistry of the Saturnian satellites. All of the larger ones (except Titan) exhibit the spectral signature of water ice. These measurements also revealed that the satellites have bright surfaces, a result likewise consistent with the presence of water rather than silicate rock on their surfaces. (Water is expected to be the most common volatile in regions cold enough for it to be stable.) The only exception was the outermost large satellite, Iapetus, about which more will be said later. By the eve of the first Voyager encounter, we had a fair idea of the sizes, albedos, and densities of the larger satellites and had firmly established water ice as their dominant surface material. But we knew nothing at all about their appearance or geology. They remained faint, unresolved points of light in the telescope, difficult to imagine as individual worlds with unique histories.

The cloak of mystery concerning the icy bodies of the outer solar system began to part during the second week in November 1980, when the Voyager 1 spacecraft sailed through the Saturnian system and passed relatively close to its satellites Mimas, Tethys, Dione, and Rhea. Nine months later Voyager 2 traced a similar path, concentrating on Enceladus, Iapetus, and Hyperion. Striking images returned by the twin spacecraft confirmed that these objects have diameters ranging from 400 to 1,600 km, similar to the largest asteroids. But perhaps the most important revelation is that these previously mysterious worlds really are icy, inside and out. The densities of Mimas, Tethys, Dione, Rhea, and Iapetus, as computed from the masses and diameters measured by Voyager, fall between 1.1 and 1.4 g/cm^3 – values suggesting that 50 percent or more of their interiors must be ice.

Voyager views of the Saturnian satellites also show their surfaces to be heavily cratered. At the temperatures below 100° K that prevail so far from the Sun, ice is nearly as strong as rock and behaves similarly when struck by a meteorite or comet. Thus, the high-resolution pictures of Mimas, Tethys, and Rhea (Figures 3–5) look remarkably similar to spacecraft images of Mercury or the lunar highlands, though the surfaces are brilliant white ice rather than dark brownish rock. The story is more complicated than one of simple impact cratering, however. Dione (Figure 6) and Tethys both have long, branching valleys tens of kilometers wide and hundreds of kilometers long, as well as regions of reduced crater density that indicate resurfacing by internal processes early in their history. These have not always been dead worlds; at some period they experienced geologic activity and evolution on a global scale.

Enceladus is even more remarkable, displaying a surface dominated by what appears to be water volcanism. As seen by Voyager 2 (Figure 7), parts of the satellite show impact craters no larger than 35 km in diameter. Even its most cratered regions have been bombarded less and are thus younger than the surfaces of Saturn's other satellites. In fact, broad swaths of Enceladus have no visible craters at all, indicating major resurfacing events in the geologically recent past (since the age of the dinosaurs on Earth, for example). In addition, the satellite's fresh, uncontaminated ice surface is more reflective than that of any other known planetary body. From these unexpected revelations, Voyager scientists have concluded that the interior of Enceladus continues to churn and probably remains liquid even today – even though this little world should have frozen solid aeons ago. Therefore, some presently unknown process must be supplying heat to Enceladus and maintaining its high level of geologic activity.

Iapetus, which orbits Saturn far beyond Titan and the icy inner satellites, is an equally bizarre and unique body that has been an enigma since its discovery in 1671. Spacecraft pictures only accentuate the strangeness of this 1,440-km-diameter world (Figure 8). Its trailing hemisphere (that which always faces where the satellite has just been in its orbit) has a surface not so different from Saturn's icy inner satellites; in fact, this hemisphere and the polar regions resemble the cratered surface of Rhea. But Iapetus' leading face is dramatically different. With an albedo of only about 4 percent, it is 10 times darker than the bright, trailing hemisphere. We now know that Iapetus has a low density, indicating a bulk composition dominated by ice; by implication, then, the leading hemisphere is not rocky but instead coated with a thin veneer of dark material, perhaps originating from elsewhere. Furthermore, the dark region is oriented almost exactly toward the apex of motion – a fact that strongly implies an external origin for the veneer. However, other details in the Voyager images, such as blackened crater floors within the bright hemisphere, appear to indicate that dark material may have come from inside Iapetus. Moreover, the dark hemisphere is almost

Figure 4. **Two Voyager 1 images of crater-pocked Mimas reveal different hemispheres of this inner, 400-km-diameter satellite of Saturn. The high density of craters indicates that the surface we see is an old one. One mammoth pit, roughly 135 km across, is the remaining scar from a cataclysmic impact that probably came close to shattering the entire satellite.**

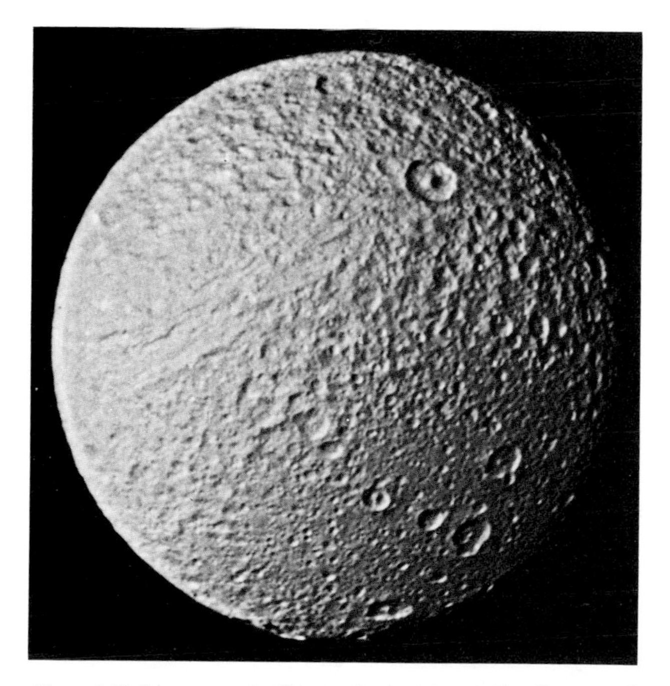

Figure 3. **Tethys, as seen by Voyager 2, appears more heavily cratered at upper left than at lower right, an indication that part of its surface has been modified by internal activity. A large, relatively fresh crater lies near part of an immense trench that here stretches from above center to extreme left.**

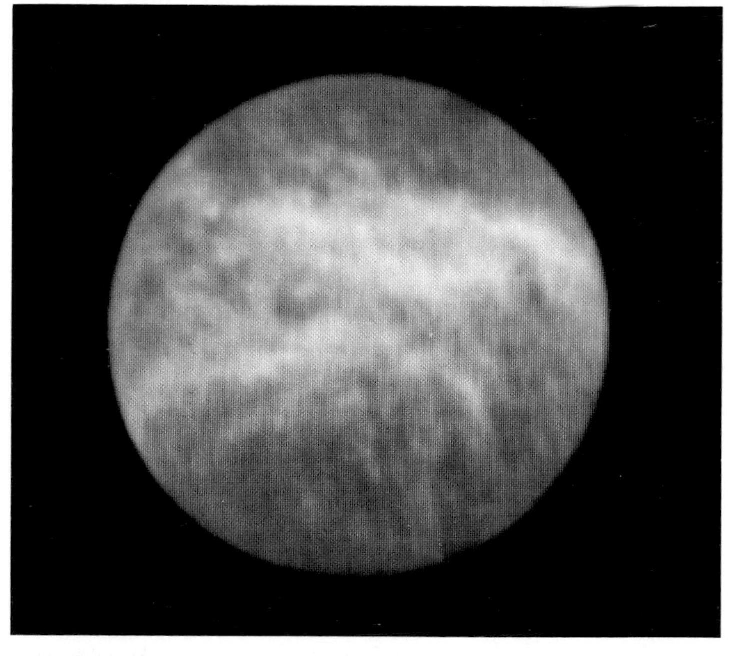

Figure 5. **Wispy white streaks, perhaps deposits of snow exuded from fractures in its crust, crisscross the surface of Rhea in the pattern seen here. They are probably not crater rays, which would appear in radial patterns. Like Mimas, Rhea has a heavily cratered surface (though it is not evident in this view). The satellite's diameter is about 1,530 km.**

Figure 6. **Thousands of craters pepper the surface of Dione, about 1,120 km in diameter. Bright streaks similar to Rhea's appear on the other hemisphere, and a fracture can be seen here near the terminator. This evidence suggests that the satellite has had an active evolution since its formation.**

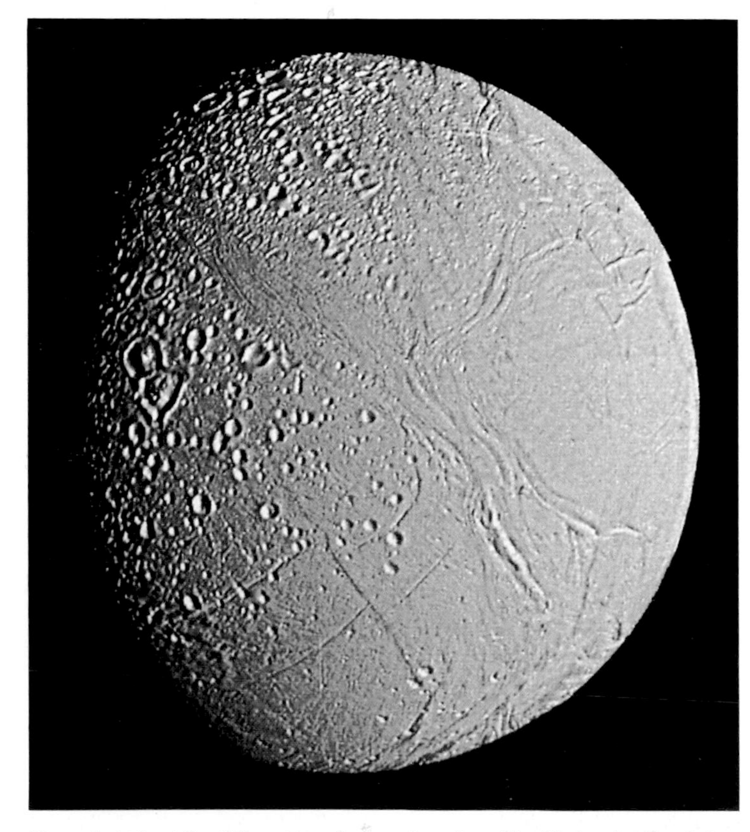

Figure 7. **At least five different terrain types have been identified on icy Enceladus, based on this false-color mosaic of Voyager 2 images. Features as small as 2 km across are visible. Crater counts give widely varying ages for different surface regions, so geologists believe Enceladus has undergone several episodes of resurfacing due to internal activity.**

featureless, with no evidence of the bright crater ejecta that would be expected if the veneer were thin (unless it forms so quickly that even the youngest craters are quickly darkened into obscurity).

We think of the dark material as black, but measurements show that it is actually reddish. The low albedo and red color suggest that this material might be a thin layer of organic material – macromolecular carbon-bearing compounds – perhaps similar to the complex substances found in the most primitive meteorites. On icy Iapetus such material may have derived from methane trapped in the water when the satellite formed, which became darkened from reactions with cosmic rays and the Sun's ultraviolet light. More likely, it was deposited on Iapetus' leading hemisphere from another source. Comets, for example, appear to have a large quantity of dark red organic matter in their nuclei. Until future observations (from Earth or spacecraft) reveal the nature and origin of the dark material on Iapetus, this body will remain one of the more baffling places in the outer solar system.

One interesting Voyager result is that the differences among Saturn's satellites show no obvious pattern. Thus, Rhea and Iapetus have almost identical sizes and masses but present entirely different appearances, perhaps related to their very different orbits. Another closely sized pair, Mimas and Enceladus, have little in common in terms of internal heat sources and resulting extent of geologic activity. Relationships involving mass, composition, orbital position, and evolution may exist in the Saturnian system, but at present they remain obscure.

ON TO URANUS

Before the Voyager flyby, astronomers had found five satellites encircling Uranus (Figure 9). The smallest of these, Miranda, was discovered in 1948 by Gerard P. Kuiper. The five satellites orbit Uranus in its equatorial plane and thus share the planet's peculiar tilt (its rotational pole lies nearly in the plane of the solar system rather than at a right angle to it).

Figure 8. **Voyager 2 viewed enigmatic Iapetus at 20-km resolution. This moon has a bright, heavily cratered hemisphere facing away from its direction of travel and a much darker, apparently smoother opposite hemisphere. (The large crater on the terminator is near the north pole.) Dark material is also visible within some craters on the bright side.**

Voyager 2 discovered 10 more Uranian satellites, all with dark surfaces and all much smaller, orbiting between their five large siblings and the planet's rings. Thus, the Uranian system consists of 15 satellites, all with regular orbits. This chapter, however, concerns just the five largest ones, which have the same general size range (400 to 1,600 km) as the satellites of Saturn covered in the previous section; Chapter 20 addresses, in part, the smaller bodies not included here.

From Earth the satellites of Uranus appear 3 to 4 magnitudes fainter than their counterparts in Saturn's system, due to their greater distance and darker, less reflective surfaces. It was not until the early 1980s that advances in astronomical detectors and the availability of larger telescopes brought these objects within reach. Robert H. Brown carried out an extensive pre-Voyager study of these bodies. Working jointly with us, he measured for the first time the diameters of four of them, and he followed up on the discovery by one of us (Cruikshank) that these objects, like the satellites of Saturn, consisted in part of water ice. However, Brown found that the ice spectral bands from Uranus' satellites were subdued, indicating a larger admixture of "dirt" and helping to explain the generally lower albedos in the Uranian system.

When Voyager 2 reached Uranus in January 1986, the planet, its rings, and its satellites formed a kind of bull's-eye pattern in the sky, so the spacecraft could make a close approach to only one satellite on the way through (in contrast to the multi-object trajectories achieved at Jupiter and Saturn). Given a choice, the Voyager scientists would have opted to pass near Titania or Oberon, the planet's two largest satellites. But there *was* no choice – to get the gravitational boost it needed to reach Neptune in 1989, Voyager 2 was forced to fly so close to the planet that only little Miranda could be seen at close range. For the four larger satellites, the project's scientists would have to settle for images that resolved details no smaller than 2 to 3 km across. Miranda, in contrast, would be photographed at a resolution of a few hundred meters, comparable to the best images of Io at Jupiter and Rhea at Saturn.

During a hectic 48-hour period, Voyager 2's cameras and other instruments scrutinized all five of the previously known moons, as well as the rings and the small inner satellites it discovered (see Chapter 9). From these observations, the project team determined the larger moons' sizes and masses, and in turn their densities. To almost everyone's surprise, these densities fell between 1.5 and 1.7 g/cm^3, significantly higher than the values found for the satellites of Saturn. Rather than being more volatile-rich than the Saturnian satellites, as had been expected, these objects possess *less* water ice and very little if any of the still lighter ices of methane and ammonia. Torrence V. Johnson, Brown, and other Voyager scientists calculate that water ice constitutes 40 to 50 percent of the Uranian moons' mass, as compared with 60 to 65 percent for those of Saturn. As we will see later, the H_2O fraction for Pluto is even lower. Thus, the expectation that water ice becomes compositionally more dominant at greater distances from the Sun is a trend that reverses beyond Saturn. Perhaps this unexpected reversal results from the increasing availability out there of solid forms of carbon, a material that joins water and silicates as an important component of solid bodies in the outer solar system.

In appearance, the four largest satellites of Uranus (Figures 10,11) generally mimic Rhea and the other heavily cratered objects in the Saturnian system. The surfaces of Umbriel and Oberon are so heavily pocked that they must date to early in the history of the solar system, when cratering rates were much higher. Ariel and Titania have fewer craters and appear to have been resurfaced by internal activity, though their present surfaces are still old (much older than the relatively smooth plains of Enceladus, for example). Ariel and Titania also possess extensive systems of broad interconnected valleys, indicating that long ago the moons expanded and split their crusts. Presumably this global tension resulted when their liquid interiors froze, but the details of such processes are obscure. In any case, all four satellites are now probably frozen throughout and geologically dead. Their surface temperatures never venture above about 80° K, and their interiors may be equally frigid.

Geologically, the real surprises in the Uranus system were provided by Miranda, an object with a diameter of only 470 km, smaller even than Enceladus in the Saturnian system. By good fortune, however, the object most closely visited by Voyager 2 was also the most remarkable, with a surface unlike anything seen before in the solar system. Miranda exhibits a bizarre variety of features jumbled together in a seemingly haphazard fashion (Figures 12,13). We may well be seeing the exposed fragments of the interior of a satellite that was broken apart after it formed and cooled. Impact specialist Eugene Shoemaker has calculated that early in its history Miranda may have been shattered by impacts as many as five times, after each of which the moon reassembled from the remains of its former self. The disrupted fragments fell back together in a jumbled configuration, with portions of the core exposed at the surface and some of the original surfaces now buried deep in the interior. Such an explanation is somewhat ad hoc, in that it explains everything and thus perhaps nothing, but we have no better theory for the peculiar surface forms revealed on Miranda by the Voyager cameras.

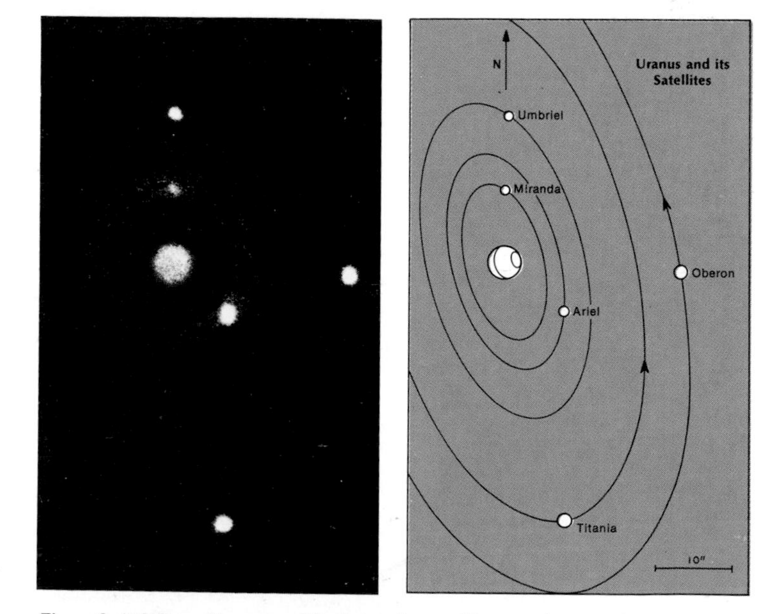

***Figure 9.* At left are Uranus and its five major satellites, as photographed at infrared wavelengths with a telescope atop Mauna Kea in Hawaii. At right, the satellites' orbits are shown at the same scale.**

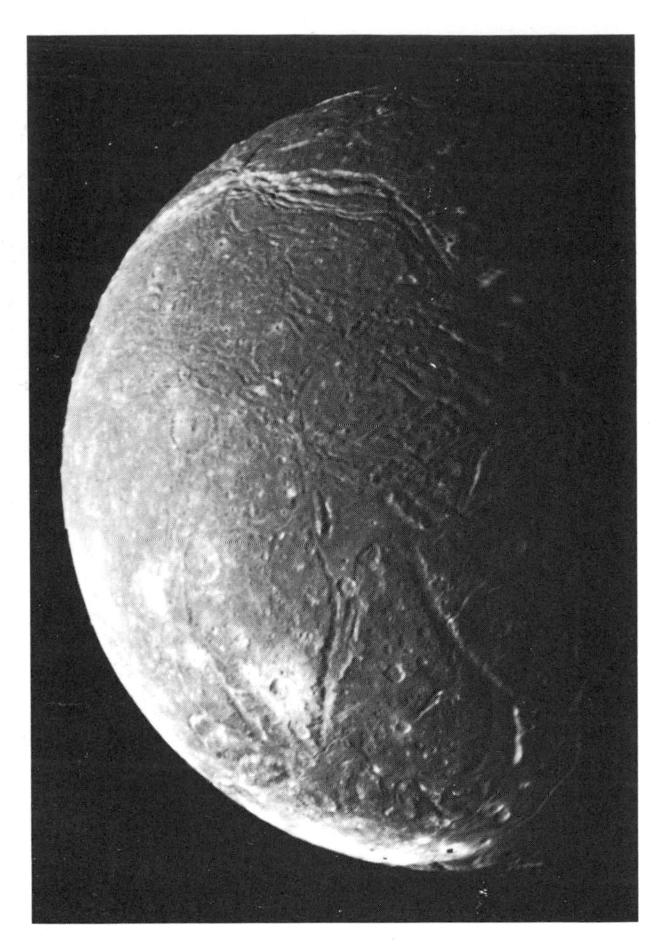

Figure 10. **This mosaic of four images shows Ariel at the highest resolution (about 2 km) obtained by Voyager 2. Networks of deep grooves crisscross some parts of the moon's icy surface, while smooth areas elsewhere appear to have been resurfaced. Ariel's exterior is younger and geologically more complex than those of Uranus' other satellites.**

SATELLITES OF NEPTUNE

The planet Uranus was discovered in 1781 by the British musician and amateur astronomer William Herschel, observing with a homemade 16-cm telescope in his backyard. Within a few years, however, positional astronomers began to have difficulty reconciling the planet's calculated and observed location in the sky, and they soon began to speculate on the existence of an unknown planet massive enough to perturb Uranus' motion. A visual search was undertaken based on positions calculated by John Couch Adams in England and Urbain Leverrier in France, and in 1846 the experienced German observer Johann Gottfried Galle spotted Neptune in its predicted location.

Within a few weeks of Neptune's discovery, William Lassell (another British amateur astronomer) found a 14th magnitude satellite about 16 arc-seconds from the new planet. Named Triton, the new satellite was quickly seen to be an unusual object because it revolves around Neptune in a retrograde orbit (Figure 14). We now know that it is locked in synchronous rotation such that its period of rotation is 5.88 days, the same as its orbital period. Furthermore, Triton is probably comparable in size to the Galilean satellites, roughly 3,000 km across. It is the largest satellite of the most distant large planet. Dynamical studies show that Triton and Neptune likely did not form together. Instead, the satellite was probably captured into an elliptical orbit that was quickly made circular by tidal interaction with Neptune. The tidal process must have produced an enormous amount of heat, which melted the satellite's interior and no doubt affected its properties as we now see them.

But why is Triton in a *retrograde* orbit? During the 1930s, astronomers in Japan and England suggested that Pluto once orbited Neptune yet somehow managed to escape the planet's gravitational grip, reversing Triton's motion in the process. However, careful calculations have since shown that this scenario simply could not have occurred without ejecting

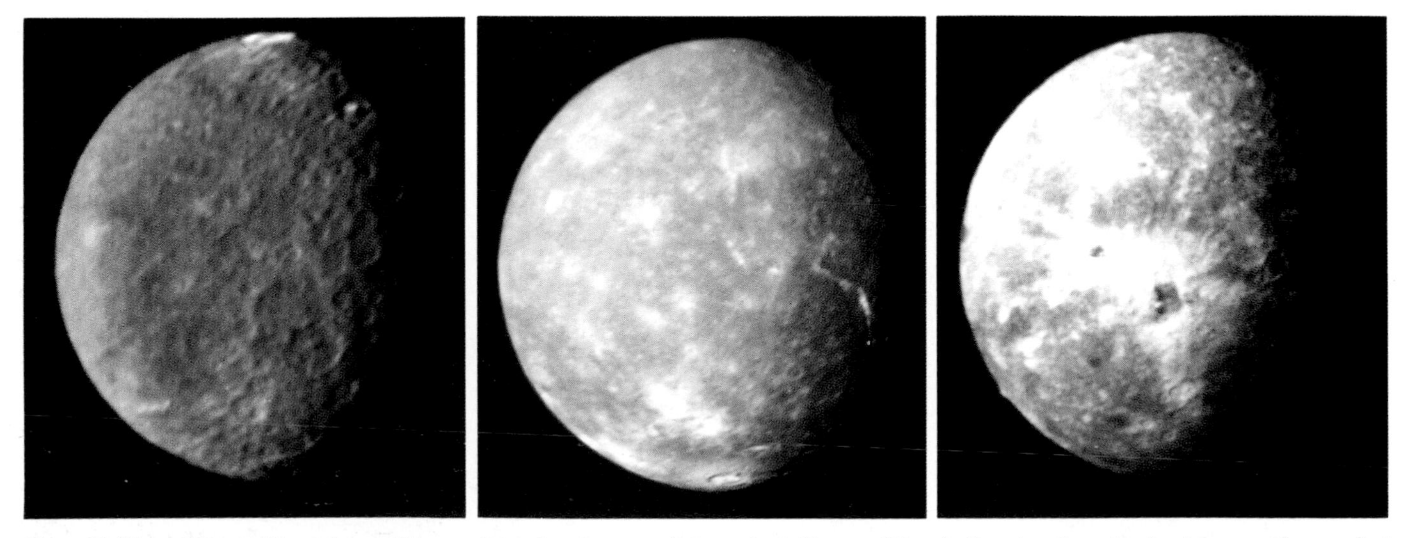

Figure 11. **Three of Uranus' five major satellites are pictured at the same relative scale. As Voyager 2 found, all are heavily cratered and thus must have ancient surfaces. Umbriel (left) appears rather uniformly dark, except for the single, bright-rimmed crater at top (near the equator) that is 110 km in diameter. (At present the south poles of Uranus and all its satellites point toward the Sun, so the equator of each body lies near its day-night terminator.) Titania (center), with a diameter of 1,600 km, is the planet's largest moon. The prominent fault valleys to the right of center extend for up to 1,500 km, and the large crater along the terminator at upper right is roughly 300 km across. Voyager observations in polarized light suggest that Titania has a porous surface. Uranus' outermost moon, Oberon (right), displays numerous impact craters surrounded by bright rays of ejected material. Dark patches within the craters probably formed when dirt-laden water erupted from within the satellite's interior.**

Pluto from the solar system entirely. We have no good explanation for Triton's retrograde orbit, but we can calculate its future consequences. The satellite is massive enough to raise tides in Neptune's "fluid" sphere, just as the Moon does in Earth's oceans. In our case the tidal bulge precedes the Moon, pulling it forward in its orbit and causing it to move gradually farther away. But because Triton's motion is retrograde, the bulge raised in Neptune lags behind the satellite and is gradually robbing it of orbital energy. At one point dynamicists believed Triton would spiral into Neptune in only 10 to 100 million years. However, more recent work suggests that the orbit is more stable and should not decay for billions of years – hardly an imminent event.

Beyond its orbital properties, however, faint and distant Triton has resisted attempts to establish even the simplest facts about it. Based on telescopic studies, Triton may be no bigger than Earth's Moon, if its albedo is about 40 percent. Triton's mass is completely unknown, and in the absence of this basic information we know nothing of its interior composition and structure because the bulk density cannot be calculated. Yet *some* important information has been garnered from the feeble light of this cold, remote world. Spectroscopic studies in the infrared, begun in the late 1970s, showed for the first time evidence of methane (CH_4) around Triton. Although it was long realized that the object is large enough to possess an atmosphere of heavy gases, photographic observations had revealed nothing. Only in the infrared, where the absorption bands of methane are very strong, is the evidence clear.

In 1978, when one of us (Cruikshank) and Peter Silvaggio first recorded a methane band in Triton's spectrum, a tenuous CH_4 atmosphere seemed the most likely cause. But

Figure 12. **When Voyager 2 swooped past Uranus in January 1986, scientists were amazed by the jumbled, chaotic surface displayed by Miranda. The smallest and innermost of the planet's five major moons, Miranda is clearly the most interesting. Some dynamicists believe the satellite has been shattered and reassembled several times since the Uranian system formed. Here several Voyager images have been combined into a seamless mosaic by a computer, which also added a coordinate grid with 10° increments.**

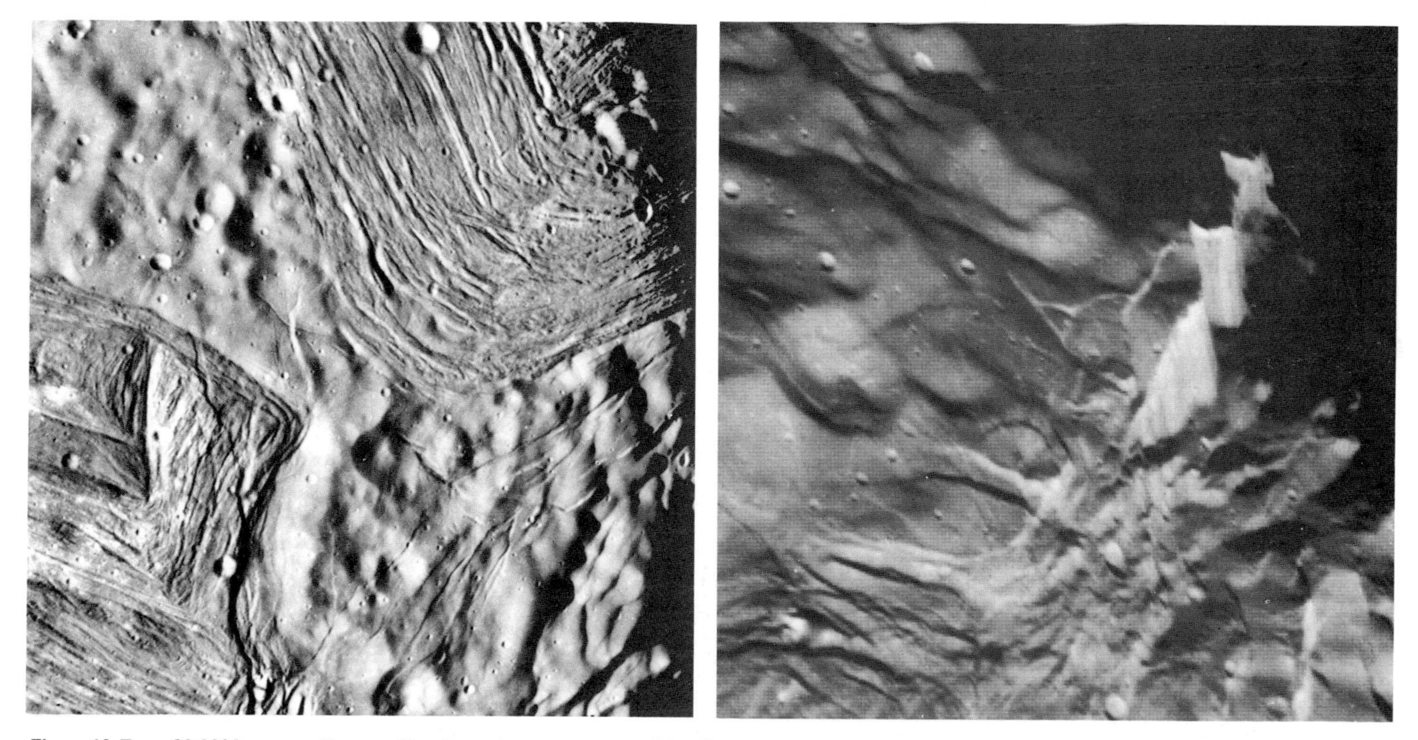

Figure 13. **From 30,000 km away, Voyager 2's telescopic camera captured details on Miranda as small as 600 m across – the best resolution for any object studied by the spacecraft. At left is a close-up of the "chevron" that shows well the light and dark grooves within its sharp boundaries; to its upper right is a region of uniformly dark grooved terrain. The area shown is about 150 km on a side. At right is the stark image of a 15-km-high cliff seen near Miranda's terminator during the Voyager flyby.**

subsequent work has shown that methane is probably only a minor constituent of an atmosphere that may be quite substantial. One reason is that another absorption band has been tentatively identified as that of nitrogen. Normally, molecular nitrogen has no infrared spectral signature, but certain bands can appear if enough of it is present. At Triton's expected temperature (at the subsolar point) of about 64° K, most of the nitrogen should be condensed; it could be in liquid form or frozen solid. In either case, a substantial atmosphere of nitrogen gas, with about 0.1 percent methane mixed in, probably surrounds Triton. It is even likely that this atmosphere is hazy, though discrete clouds are not expected to form. The state of the nitrogen on Triton depends critically upon the temperature, which we know only to within about 5°.

Figure 14. **Neptune and its two known satellites, as photographed by Gerard P. Kuiper on May 29, 1949. In the long exposure to record Nereid (arrowed), Neptune appears greatly overexposed; the spikes result from the diffraction of light within the telescope. Triton is visible at Neptune's top edge, between the upper pair of diffraction spikes.**

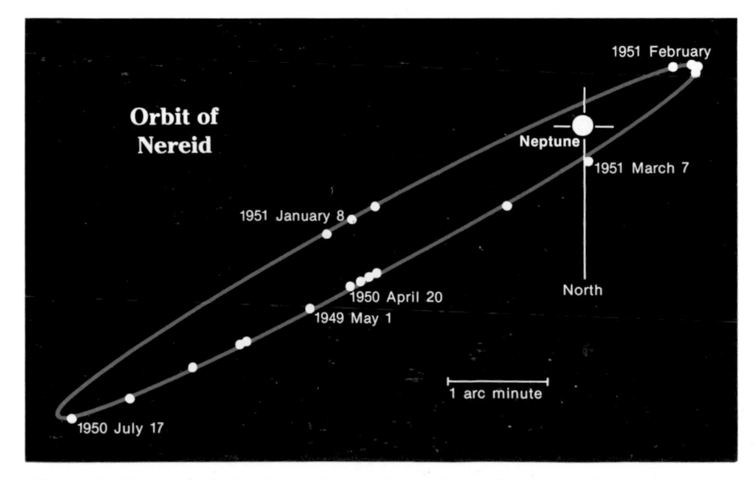

Figure 15. **The orbit of Nereid, as derived from telescopic observations made earlier this century. The satellite moves in an inclined, highly eccentric orbit with a period of about 360 days. At its closest Nereid is 1,345,500 km from Neptune, and 9,688,500 km at its farthest.**

Because of the orbital geometry of both Triton and Neptune, we know that Triton experiences extremes in a seasonal cycle some 680 years long. The corresponding range in temperature must have a profound effect on its highly volatile atmosphere and surface. Nitrogen and methane must evaporate from the sunny latitudes, migrate to the colder regions, and precipitate as fresh frost or ice – only to be mobilized again as the season changes. The wholesale migration of surface materials to a depth of several centimeters, or even meters, virtually ensures a dynamic "icescape," perhaps with very interesting structural features akin to those in the polar regions of Earth and Mars.

On August 25, 1989, the Voyager 2 spacecraft will fly past Neptune and Triton in the last planetary encounter of its long and wonderfully revealing 12-year trek through the outer solar system. From a minimum distance of 40,000 km, the cameras of Voyager should record Triton with resolution better than 1 km, and changes to the spacecraft's motion under Triton's gravitational influence will allow a good determination of the moon's mass. Will there be a glint of sunlight reflecting from a frozen expanse of arctic-like ice? Or will the surface be invisible below a hazy atmosphere, like that of Titan? In any event, data returned by the extensive complement of instruments aboard Voyager should transform Triton from a pinpoint of light in the dim, distant reaches of the planetary system to yet another world revealed by space exploration.

In his photographic search for new satellites during the late 1940s, Kuiper discovered not only Miranda, as previously mentioned, but also Nereid, the second satellite of Neptune. Nereid is a world about which nearly nothing is known. It orbits Neptune along a highly inclined and elliptical path that takes a year to complete (Figure 15). At magnitude 19, Nereid is very difficult to study, and we can only guess as to its characteristics. If, for example, the surface albedo is 40 percent (typical of slightly dirty snow or ice), the diameter is then about 330 km.

PLUTO AND ITS MOON

Emboldened by the successful prediction and discovery of Neptune in 1846, astronomers began to sift through the orbital irregularities of Uranus and Neptune for evidence of a still more remote planet. Several suggestions of new planets were made in the late 19th century, but it was primarily the efforts of two proper Bostonians – William Pickering and Percival Lowell – that gave respectability to this effort. Using predicted positions by these astronomers, unsuccessful searches began in 1905 at Lowell and Mount Wilson observatories. Before his death in 1916, Lowell urged that a special wide-field camera be constructed to carry out a systematic search for Planet X, but not until 1929 was the survey with that camera begun. A Midwest amateur astronomer named Clyde W. Tombaugh had joined the Lowell Observatory staff expressly to carry out this project. On February 18, 1930, after scrutinizing some 90 million star images on photographic plates, Tombaugh spotted a tiny, 15th-magnitude dot that had shifted with respect to the background stars. The image, though indeed that of a distant planet, showed no measurable disk and was unexpectedly faint for something supposedly massive enough to perturb Neptune and Uranus.

The new planet, named Pluto after some debate, was too

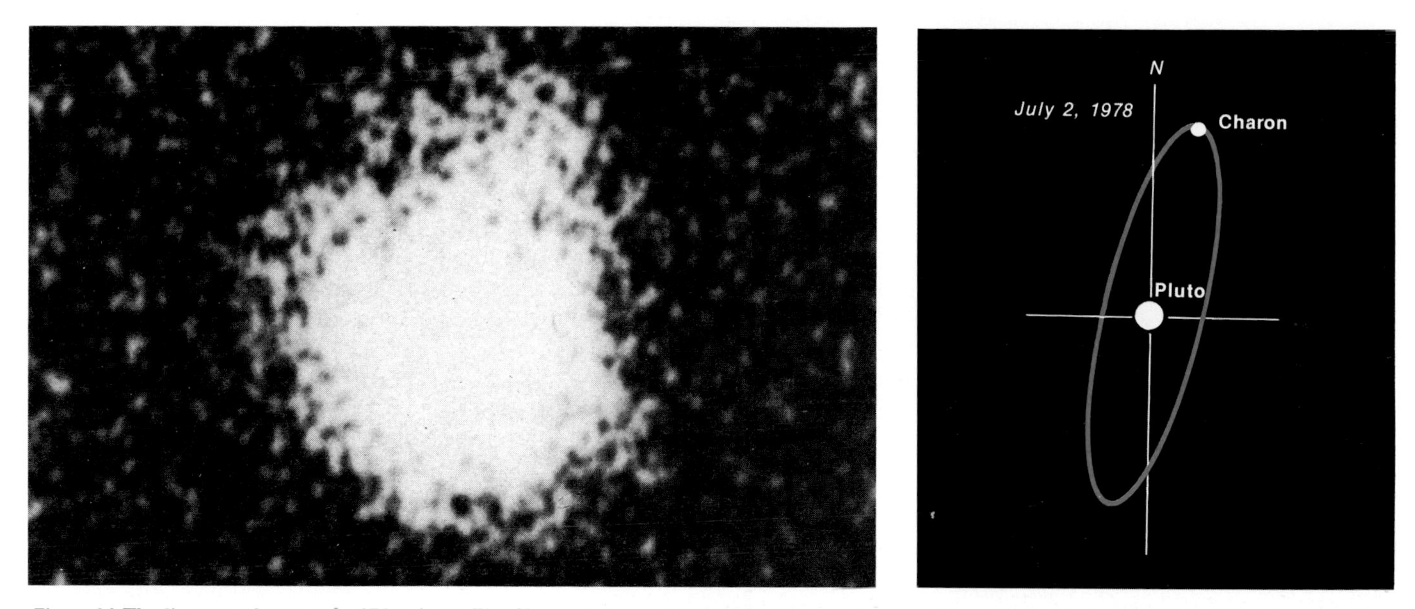

Figure 16. **The discovery photograph of Pluto's satellite Charon, taken on July 2, 1978. Charon appears as a bump on the upper-right edge of Pluto's smeared-out image. The accompanying diagram shows Charon's position in its orbit at the time, as well as the true relative sizes of the two objects.**

dim to permit the kind of observations during the 1930s and 1940s that might yield information on its physical nature. This little planet is an excellent example of those astronomical bodies that wait patiently for Earth-bound astronomers to invent and perfect devices, techniques, and strategies to study them. We scurry between laboratory and telescope in the pursuit of developing ever-better detectors, spectrographs, photometers, and other instruments. After years of iterative effort, we eventually emerge with a new device that can turned on a favorite planet, satellite, star or galaxy to wrest from it yet another secret. Take Pluto's orbit, for instance, which is more highly inclined (17°.2) and eccentric (0.25) than that of any other planet. The orbits of Pluto and Neptune actually cross one another, and every 248 years, as Pluto approaches perihelion, it comes closer to the Sun than Neptune does. (The two planets themselves can never collide because they are locked in an orbital resonance relationship that keeps them far apart.) Recently, theorists Gerald Sussman and Jack Wisdom discovered that Pluto's motion is "chaotic" – its orbit's inclination, eccentricity, and orientation are changing constantly. They conclude that the ninth planet may well have formed in a more nearly circular orbit that over time has become "pumped up" to its current elliptical shape.

Regarding composition, Pluto held nearly all its mysteries intact until March 1976, when we and colleague Carl B. Pilcher used the Kitt Peak 4-meter telescope to make the first near-infrared photometric measurements of it. Our plan was to search not only for frozen water, but also to distinguish water from the other cosmochemically plausible frozen ices of methane and ammonia. The signature of methane was readily identified from our measurements and later confirmed by other astronomers. Highly reflective methane frost on Pluto's surface meant that the planet had to be relatively small to appear so faint from Earth. In fact, we reasoned that it is smaller than the Moon, and because its bulk composition was likely to consist largely of frozen ices, its mass was surely too small to perturb Uranus and Neptune. This conclusion was soon confirmed by an unexpected discovery made just two years later.

As he sat at his measuring machine at the U.S. Naval Observatory on June 22, 1978, methodically determining the positions of Pluto on astrometric photographs, James Christy noticed that several images of the planet were all uniformly distorted in one direction, as though there was a very close star interfering with Pluto's image (Figure 16). After inspecting other astrometric plates, Christy quickly concluded that Pluto must have a satellite. A colleague called in from another room confirmed the startling discovery. The newly discovered satellite was named Charon, after the boatman in Greek mythology who operated the ferry across the River Styx to Pluto's realm in the underworld. By happy coincidence, the nickname of Christy's wife Charlene is Char.

In the discovery plates the planet and satellite were not clearly resolved from one another, but even so astronomers could estimate crudely the radius and period of the satellite's orbit. Armed with this information, Robert Harrington made the first direct calculation of Pluto's mass, which was indeed much less than that of our Moon.

Two other curious points also became obvious soon after the satellite's discovery. First, Charon's orbital period is 6.4 days, matching the rotation of Pluto (determined decades earlier from its light curve; see Figure 17). Second, the satellite moves north-south in the sky, indicating that the pole of its orbit lies essentially in the ecliptic plane (Figure 18). Dynamicists believe Pluto's rotation axis must be parallel to this pole and thus also in the ecliptic. They reason that the synchronism between Pluto's rotational period and the satellite's orbital period means that the system has become dynamically locked. As a consequence, Charon can be seen from only one hemisphere of Pluto, and vice versa. Furthermore, for such locking to occur, the satellite's mass must be quite large relative to Pluto's. Previously, the Earth and its Moon have always been considered the planet-

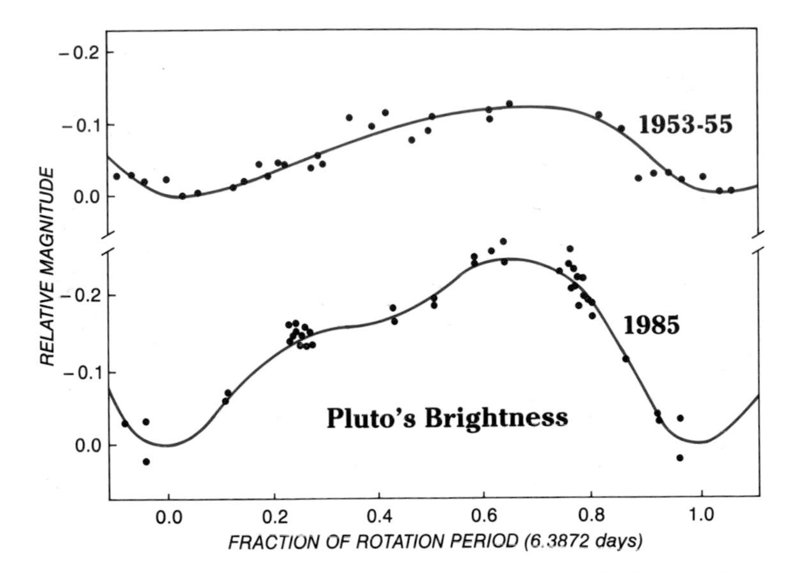

Figure 17. **Pluto's rotational light curve has changed dramatically over three decades. While light reflected from Charon adds a small contribution, virtually all of the differences between these two curves is due to bright and dark markings on the surface of Pluto itself. Also, since its discovery in 1930, the planet has moved in its orbit such that its (frost-covered?) pole no longer points towards us; instead, we are now looking down on Pluto's equator, so bright and dark surface markings in the planet's midsection affect its overall reflectivity much more at present as they rotate into and out of view.**

satellite pairing most like a "double planet," that is, with the most comparable masses. But now that distinction appears to have passed to Pluto and Charon.

Even with the pair's images smeared together in photographs, careful measurements showed that sometime in the mid-1980s Charon's orbital plane would be seen edge-on from Earth. That is, the satellite would appear pass in front of Pluto (a transit) and, half a revolution later, pass behind it (an occultation). Planetary astronomers realized at once their good fortune, for not only does this alignment occurs only twice during Pluto's long trek around the Sun, but we had chanced upon the satellite just before these "mutual events" began. (Had Charon been discovered in, say, 1993, we would have been forced to wait until the 22nd century to witness the next series of overlappings.) Thus, with hope and anticipation, a small cadre of photometrists and spectroscopists began to monitor the feeble light from Pluto and Charon with unprecedented precision and sensitivity.

A transit or occultation in progress would be evident as a drop in the pair's combined light, as either part of Pluto's surface or Charon's became obscured (Figures 19,20). But the exact onset of the event "season" depended critically upon the dimensions of the two bodies, which had been only crudely estimated, and on the uncertain orbital geometry of Charon. Richard Binzel made the first certain observation of a grazing occultation on February 17, 1985; the combined light from Pluto and Charon diminished only 3 to 4 percent as the edge of Charon slowly passed in front of Pluto. Within a few days two other teams of astronomers confirmed that the eagerly awaited mutual events had begun.

Throughout 1985 and 1986 the occultations and transits were only partial, but in 1987 the first total events occurred as Charon passed fully behind and in front of Pluto. Observations of complete occultations and transits, which contain the most telling information about the distant pair, continued during 1988. From a growing body of precise photometric observations, combined with refined estimates of Charon's orbital radius, David Tholen deduced that Pluto's diameter is 2,300 km, with an uncertainty of less than 40 km, making the planet just two-thirds the size of Earth's Moon. To the same precision, Charon is 1,190 km in diameter, or about 20 percent bigger than the largest asteroid, Ceres. With the dimensions and period of Charon's orbit now well established, the total mass of the objects can be derived and, in turn, their bulk density – a quantity vital to the question of internal composition. The combined density of the planet and satellite is 2.0 g/cm^3, indicating that, unlike the satellites of Saturn and Uranus, their interiors must contain a substantial amount of rocky material. Water ice should be abundant as well, but the methane identified spectroscopically on Pluto is surely only a minor constituent.

Deducing the state and make-up of Pluto's surface would be easier if we knew the exact surface temperature, but the planet is so cold – the best guesses range from 53° to 61° K – that Earth-based telescopes have been unable to record the feeble heat radiating from it. Fortunately, the Earth-orbiting Infrared Astronomical Satellite (IRAS) has succeeded where ground-based telescopes have failed. When IRAS scanned Pluto and Charon in 1983, it found that the level of thermal emission was less than expected (Figure 21). Perhaps the planet's thin atmosphere (discussed later) is conducting heat to its night side, or perhaps we do not yet fully understand

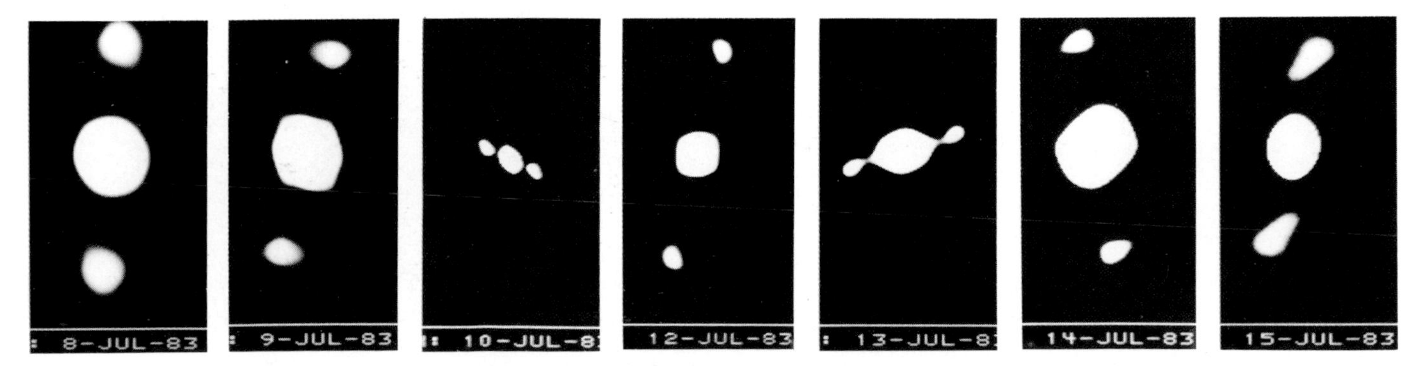

Figure 18. **On seven nights in 1983, German astronomers G. Baier and Gerd Weigelt used speckle interferometry to record the Pluto-Charon system. This technique involves obtaining a great number of extremely short photographs, which "freeze" turbulence in Earth's atmosphere, that are later combined by computer processing. The resulting *autocorrelations* are not actual photographs, and they each yield two "Charons." But they do show the objects' correct angular separation and how the planet-satellite line is oriented. Charon orbits Pluto every 6.3872 days.**

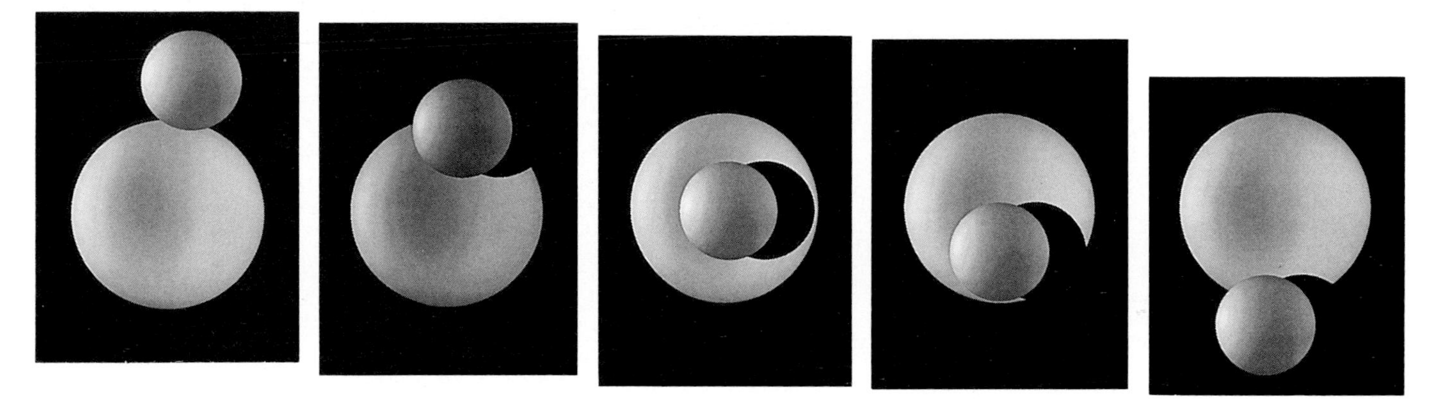

Figure 20. **Using a supercomputer to assimilate several years of carefully timed "mutual events" between Pluto and Charon, astronomers have created crude maps of these two far-flung worlds. This series represents how the pair appeared from Earth during a 4-hour transit of Charon across the planet's disk on March 19, 1987. Celestial north is up, but Pluto's north pole is toward the right. Note the suggestion of bright (frost-covered?) polar regions on both bodies, and how the shadow of Charon covered a considerable fraction of Pluto's disk during the event.**

how such remote, icy objects absorb and reradiate sunlight.

And what of Charon's surface composition? Intuition suggests that it is probably the same as that of Pluto, but the extraordinary mutual events have made it possible to test that intuitive guess. When Charon is completely hidden by Pluto's disk (or in its shadow), the light reaching our telescopes is that of the planet alone. By observing Pluto and Charon together before an occultation and then only Pluto when Charon was entirely hidden, observers could subtract one spectrum from the other and thus derive the spectrum of Charon alone. When they performed this test in 1987, independent observing teams led by Robert Marcialis and Marc Buie were astonished to see a clear indication of the spectrum of water frost – and no suggestion of methane – on Charon. It is possible that, over time, the satellite's allotment of methane has escaped into space, only to be captured in part by the stronger gravity of Pluto and dragged onto the planet's surface.

On average Pluto receives sunlight 1,600 times weaker than that falling on Earth, but the great eccentricity of its orbit carries Pluto over a wide range of distances from the Sun during each 248-year revolution. These swings result in a changing surface temperature, which in turn affects the ices of methane and whatever other volatiles may occur there. Depending critically upon the exact temperature, a certain amount of methane should form a tenuous gas envelope around Pluto, and as the temperature changes, the pressure exerted by this atmosphere on the surface could vary by up to 500 times as the methane alternately evaporates from and redeposits onto the surface. Currently the Sun is directly over Pluto's equator, so methane is presumably being driven from the equatorial regions and accumulating at the poles. But due to Pluto's polar inclination of nearly 90°, during one orbit the Sun shines down on both of the planet's poles as well. Over billions of years, the gas should have been slowly lost to interplanetary space, and some have questioned why any methane remains on the planet now at all.

Theoretical arguments and circumstantial observational evidence for the existence of an atmosphere around Pluto were borne out in mid-1988. On June 9th of that year, teams of astronomers scattered across the South Pacific watched Pluto briefly hide an obscure 12th-magnitude star from view (Figure 22). During this occultation the star seemed to

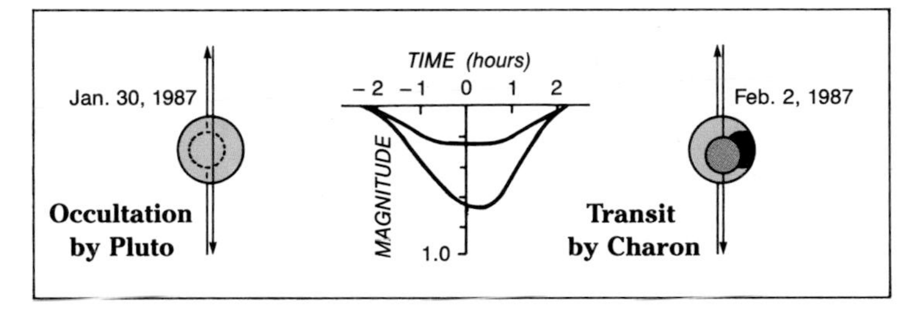

Figure 19. **When Charon transits Pluto (right), the drop in total light is greater than when it slips behind the planet (left). This difference can exceed 0.4 magnitude (as plotted at center) and is due to the combined effects of Charon's shadow and darker surface. Although the two objects were not seen separately during such events, astronomers could monitor subtle changes in their combined brightness to derive the dimensions, separations, and albedos of this remarkable planet-satellite pair. In particular, both objects have bright, ice-covered surfaces, but Charon is significantly darker than Pluto.**

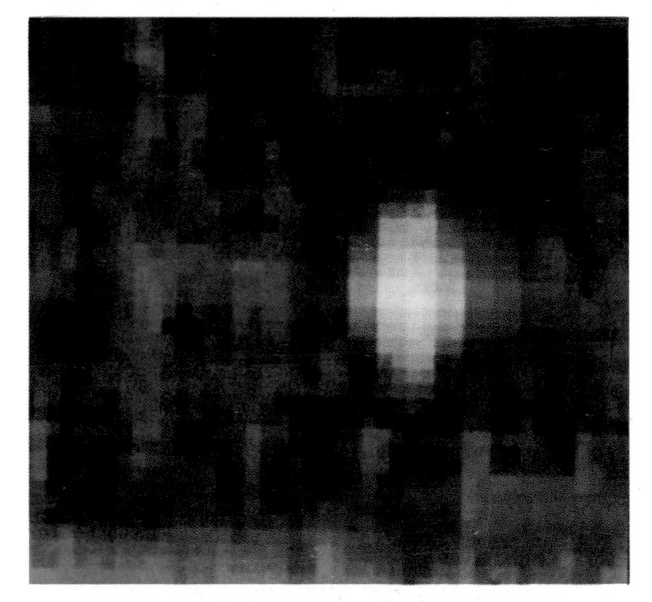

Figure 21. **Heat from Pluto and Charon. On August 16, 1983, the Infrared Astronomical Satellite (IRAS) was directed to examine the planet and its satellite, and the result is this image. Charon does not show here, and Pluto's apparently resolved disk is a processing artifact. Faint mottling is radiation from distant background sources. The IRAS data have helped determine the pair's temperature and size, and whether Pluto has an atmosphere.**

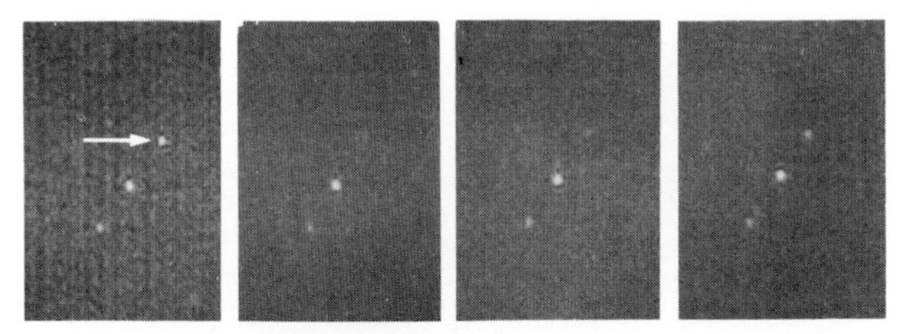

Figure 22. **In 1988, Pluto occulted a faint star in the constellation Virgo. Astronomers saw the star and planet as a single merged point of light (arrow), which dimmed as the star was covered, then brightened as it reemerged from behind the planet. These images were acquired a research team aboard NASA's Kuiper Airborne Observatory, a telescope-equipped jet transport that viewed the 2-minute-long event from a remote spot over the South Pacific Ocean.**

disappear behind the planet and reappear gradually, not abruptly, signaling to the astronomers that Pluto is surrounded by some kind of atmosphere. According to James Elliot and Edward Dunham, Jr., whose research team observed the occultation, this atmosphere might be an extremely tenuous envelope of pure methane with a surface pressure some 100,000 times less than at sea level on Earth. Their observations also suggest that a haze of Titan-like photochemical smog is suspended far above the planet's surface. An alternate interpretation, by William Hubbard and others, is that Pluto's tenuous atmosphere may contain instead as-yet-unidentified gases like carbon monoxide or nitrogen. These would give the lower atmosphere characteristics to match the occultation observations without invoking the existence of any haze.

Pluto is a cold and miserable place, yet in a sense it is dynamic. With a perihelion date of September 5, 1989, the slow-moving planet has actually been closer than Neptune to the Sun since early 1979 and will remain so until the spring of 1999. At perihelion Pluto comes within 30 AU of the Sun, but at aphelion (occurring next in 2113) it is nearly 50 AU away. Thus, we live in an opportune time to study the surface and atmosphere of Pluto – not only is it closest to the Earth, but now the sunlight there is strongest, driving activity on the planet's surface and in its atmosphere that will not be repeated until the 23rd century.

16

The Halley Encounters

Rüdeger Reinhard

DESPITE THE FACT that astronomers have observed and studied comets for thousands of years, our real knowledge of these interplanetary wanderers remains woefully limited. The most widely held theory of their nature, set forth by astronomer Fred L. Whipple in 1950, envisions comets as balls of ice and fine dust particles, which he aptly described as "dirty snowballs" (see Chapter 17). Far from the Sun, this solid nucleus remains inert. But as it ventures into the inner solar system, solar heating causes the ices and snows to evaporate. This expanding envelope, the coma, interacts with sunlight to form distinct tails of plasma (ionized gas) and dust.

But what we see as "a comet" is in fact sunlight scattered by the dust particles in the coma and the glow from fluorescing ions. The nucleus itself cannot be resolved from Earth, even in the largest telescopes. Astronomers, therefore, had no way of knowing whether Whipple's theory was correct. But cometary science made a gigantic leap forward in 1986, when five spacecraft encountered the most famous dirty snowball of them all – Halley's Comet.

Space scientists had envisioned many kinds of spacecraft missions to comets. In order of increasing complexity, these include flybys, either fast or slow; a lengthy rendezvous at close range; a landing on the nucleus; and missions to bring samples of the coma or nucleus back to Earth. Even the simplest flyby could answer many nagging questions about comets' origin, evolution, composition, and whether in fact they have central nuclei. The challenge for scientists in the early 1980s was to select a comet suitable for such a mission. With 750 known comets to choose from, and five to 10 more discovered each year, how could they single one out for study?

"New" comets, passing through the inner solar system for the first time, are the brightest and most active of their species. Visiting one of these would be very desirable, but unfortunately they are unsuitable for a spacecraft mission. Scientists could chart a rendezvous only with a comet whose orbit is well known. This rules out all but about 85 periodic comets that have been observed during at least two apparitions. Almost all of these have periods of less than 20 years. Unfortunately, comets with such brief orbital periods are not very bright; on average, they produce 100 times less gas and dust than their long-period siblings, presumably because many passes near the Sun have created thick dust layers on their surfaces.

There is only one exception: Halley's Comet. Named for the English astronomer Edmond Halley, who first predicted its return, it is by far the most famous comet. Skywatchers have noted every one of its apparitions since 240 BC – an impressive total of 30 – and therefore its orbit is very well known. It is also quite active, producing gas and dust at rates comparable to those of new comets. In addition, it is among the easiest comets for a spacecraft to reach. And finally, Halley would be seen from Earth when the spacecraft arrived there, thus allowing scientists to compare the spacecraft data with ground-based observations.

In short, Halley was an outstanding target for a first cometary mission, one that would be literally a "once-in-a-lifetime" opportunity. This comet visits the inner solar system only every 76 years and was last here in 1910, when it approached to within 0.15 AU of Earth. That spring its magnificent tails stretched across one-third of the night sky. In fact, Earth even passed through the comet's plasma tail (with no ill effects). As Halley disappeared into the outer reaches of the solar system, scientists calculated that its next perihelion would occur on February 9, 1986, on which date it would be well inside Earth's orbit at 0.59 AU from the Sun.

MISSIONS TO HALLEY'S COMET: AN INTERNATIONAL ADVENTURE

The exploration of Halley's Comet was very much an international affair. To optimize their efforts, four of the world's spacefaring powers – the United States (represented by NASA), the Soviet Union (Intercosmos), Japan (Institute of Space and Astronautical Science), and Europe (the European Space Agency) – agreed in 1981 to form the Inter-Agency Consultative Group for Space Science (IACG). Its task was to coordinate informally all matters related to the Halley missions and any other observations of the comet from space. IACG members exchanged information on spacecraft and experiment design, mission planning, and models for the anticipated behavior of dust and gas near the nucleus. The participants also agreed to exchange the results of their missions once the encounters were over.

The International Halley Watch (IHW), which coordinated all ground-based observations of the comet,

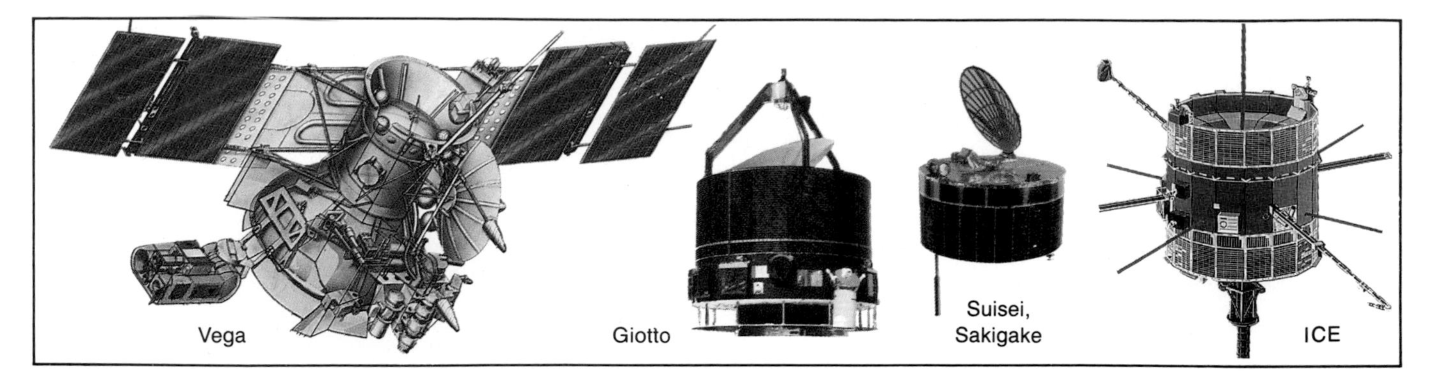

Figure 1. **The "Halley armada" of interplanetary spacecraft (from left): Vegas 1 and 2 (flown by the U.S.S.R. and Intercosmos); Giotto (European Space Agency); Suisei and Sakigake (Japan). At far right is NASA's International Cometary Explorer, a preexisting spacecraft that was redirected to Halley's vicinity.**

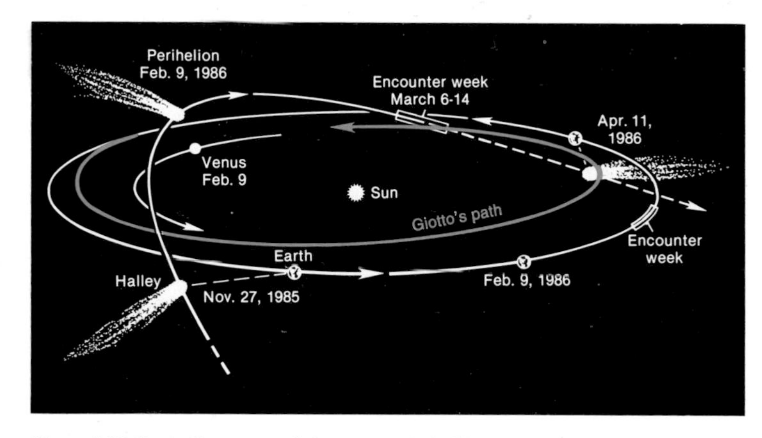

Figure 2. **Halley's Comet travels in retrograde fashion along a highly elliptical orbit inclined 18° to the plane of the solar system (the ecliptic). Consequently, it was easiest to reach by spacecraft near the times it crossed the ecliptic: on November 27, 1985, when it was 1.85 AU from the Sun; and again on March 10, 1986, when 0.89 AU away. Since the comet would be much more active during the second crossing, mission planners in the Soviet Union, Europe, and Japan all chose to have their space probes meet the comet within a few days of that date. Giotto's orbit, confined to the ecliptic, is representative of the paths of the various flyby spacecraft. At perihelion, Halley was too near the Sun to be observed from Earth. But it could be seen much better from the vicinity of Venus, from which it was observed by NASA's Pioneer Venus orbiter.**

participated in all IACG meetings, and in turn the space agencies were represented at all general meetings of the IHW The communication and coordination achieved through this "cross-representation" would prove highly efficient.

As 1986 opened, five spacecraft were headed for encounters with Halley (Figure 1). The Soviet-led Intercosmos group had dispatched Vegas 1 and 2; Japan, through ISAS, had launched Suisei and Sakigake; and the European Space Agency (ESA) had sent Giotto. Notably absent was an American spacecraft; NASA had studied several options but decided in the end not to approve a Halley mission.

However, a few of NASA's existing deep-space satellites were called into service. For example, one of the International Sun-Earth Explorers (ISEE 3), which had resided at a near-Earth libration point since 1978, was redirected by ground controllers to fly through the short-period Comet Giacobini-Zinner in September 1985, and then to monitor the solar wind "upstream" from Halley six months later. (In recognition of its new mission, ISEE 3 was renamed the International Cometary Explorer, or ICE.) The Pioneer Venus orbiter mapped ultraviolet emissions in the comet's hydrogen coma from its vantage point around Venus. And by luck Pioneer 7, equipped with plasma sensors and launched into solar orbit *20 years* before, happened to be near Halley's Comet in March 1986 and was re-activated to add its observations as well.

Meanwhile, from its orbit above the Earth, the International Ultraviolet Explorer satellite periodically recorded Halley's spectrum throughout much of 1985 and 1986. The Solar Maximum Mission satellite also provided imagery for about six weeks near perihelion, a period when the comet appeared too near the Sun to be photographed from Earth. The Dynamics Explorer 1 satellite was enlisted as well to make ultraviolet images and brightness measurements during the period after perihelion.

Because Halley's orbit is inclined some 18° to the ecliptic, we had to time the spacecraft encounters carefully. It would require the least launch energy to reach the comet at the two points where it crossed the plane of Earth's orbit (the ecliptic plane). The first instance occurred prior to perihelion on November 9, 1985, but on that date Halley's Comet was still 1.8 AU from the Sun. During the second crossing, on March 10, 1986, Halley would be closer to the Sun and much more active. Mission planners ultimately chose to have the flybys occur during the latter opportunity (Figure 2). The various spacecraft would encounter Halley over a 9-day period, providing the chance to view more than three rotations of the nucleus and the corresponding variations in its activity. The spacecraft would also fly past or through the comet at complementary distances that ranged from 600 km to 28 million km from the nucleus. All would view the comet's sunward side (Figure 3, Table 1).

To accomplish their objectives, the spacecraft would face considerable risks. The sole major disadvantage of choosing Comet Halley for a spacecraft mission is that its orbit is retrograde. That is, it moves around the Sun in the direction opposite that of the Earth – or of any spacecraft launched from the Earth. Thus, Halley and the spacecraft would approach each other head on at roughly 70 km per second, and the comet's dust would strike the spacecraft at that velocity. The spacecraft would traverse the entire coma, roughly the distance from the Earth to the Moon, in just $1\frac{1}{2}$ hours. The most interesting region, within 10,000 km of the nucleus, would be crossed in only a few minutes.

Especially for the designers of the two Vegas and Giotto,

Spacecraft	*Agency*	*Launch date (UT)*	*Flyby date and time (UT)*	*Velocity (km/s)*	*Closest point to comet (km)*	*Distance to Sun (AU)*
Vega 1	Intercosmos (U.S.S.R.)	15 Dec 1984	6 Mar 1986 7:20	79.2	8,890	0.79
Suisei	ISAS (Japan)	19 Aug 1985	8 Mar 1986 13:06	73.0	151,000	0.82
Vega 2	Intercosmos (U.S.S.R.)	21 Dec 1984	9 Mar 1986 7:20	76.8	8,030	0.83
Sakigake	ISAS (Japan)	8 Jan 1985	11 Mar 1986 4:18	75.3	6,990,000	0.86
Giotto	ESA (Europe)	2 July 1985	14 Mar 1986 0:03	68.4	596	0.89
ICE	NASA (U.S.A.)	12 Aug 1978	25 Mar 1986 10:44	64.9	28,100,000	1.07

Table 1. **Key data for missions to Halley's Comet, in order of the flyby dates. To date only two comets have been observed at close range by spacecraft. The International Cometary Explorer (ICE) had earlier encountered Comet Giacobini-Zinner; on September 11, 1985, ICE passed through the tail at a point 7,800 km from the nucleus.**

which would penetrate deeply into the coma, these fast-flyby conditions presented great challenges. The most critical objective was an essential one: survival of the spacecraft. To keep from being destroyed by hypervelocity collisions with cometary dust particles, the Vegas and Giotto carried special dust-protection shields. Another concern was getting crisp imagery during the flybys. Like a photographer trying to take pictures from a fast-moving car, the spacecraft would require sophisticated pointing mechanisms to compensate for their own motions, and they would need to acquire and transmit their images and other data very rapidly.

THE VEGA ENCOUNTERS: NUCLEUS IN A DUST COCOON

The Soviet Union seized a unique chance to combine a mission to Comet Halley with one to Venus, which for a decade had been the focus of its planetary-exploration program. This double-purpose mission was called Vega, a contraction of the Russian words *Venera* (Venus) and *Gallei* (Halley). It was conducted by the U.S.S.R. in cooperation with a number of other countries.

The Soviets chose to use the same basic design as that of their many Venus landers and orbiters. The Halley-bound portion of each spacecraft carried 14 experiments, among them a CCD imaging system; gas, ion, and dust mass spectrometers; two infrared spectrometers; dust impact detectors; and several plasma instruments. The Vegas employed three-axis stabilization and were powered by solar-cell panels with a wingspan of about of 10 m.

In order to photograph Halley's nucleus, the spacecraft required a steerable platform (Figure 4) that could track the nucleus with an accuracy of roughly 2½ arc minutes. The most difficult problem was locating the nucleus and orienting the camera toward it automatically (since the spacecraft would be 170 million km from Earth, too far away for fast direct control). The only way was to program the camera to search for the brightest spot in its field; in theory, this would be the cometary nucleus or, more precisely, a region near its surface thick with highly reflective dust. This would allow the Vegas to produce images of the innermost parts of the coma and, scientists hoped, the nucleus itself.

Two spacecraft, Vega 1 and Vega 2, were launched by Proton rockets from the Baikonur Cosmodrome on December 15 and 21, 1984, respectively. Each 4.5-ton craft carried a spherical shell, 2.5 meters in diameter, that contained the Venus-bound payload. On June 11 and 15, 1985, the large spheres plunged into the planet's atmosphere

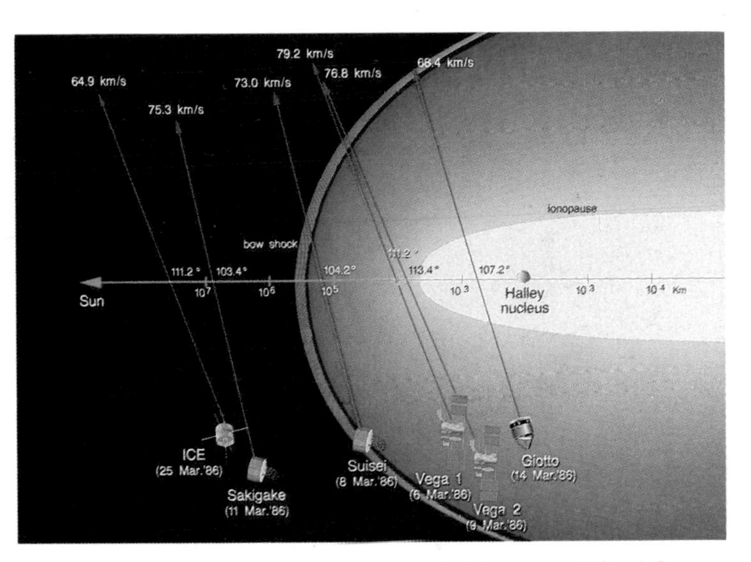

Figure 3. **The trajectories of the six flyby spacecraft with respect to Halley's nucleus, with distance represented logarithmically. Because of its retrograde orbit, all the spacecraft encountered the comet at very high velocities (see also Table 1).**

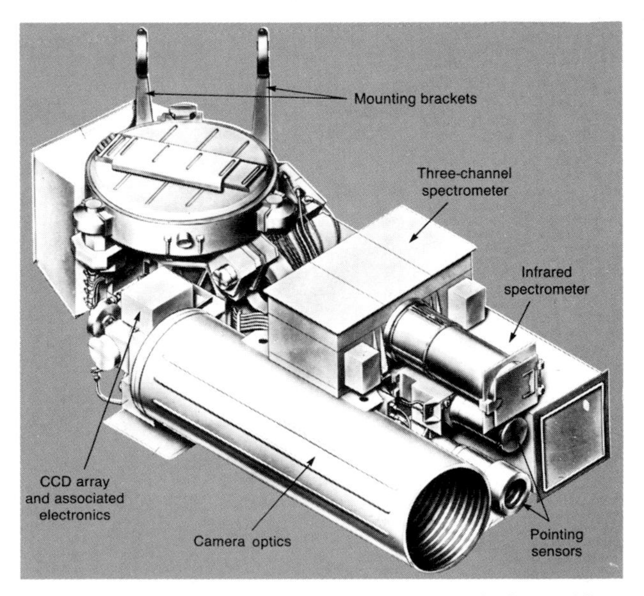

Figure 4. **The pointing platform on the Vega spacecraft, shown without its protective thermal blanket, compensated for the nucleus' rapid relative motion as the Vegas sped past it. During the flyby the camera exposure times ranged from 0.08 to 1.2 seconds. Three instruments were mounted on the platform: a two-camera television system (one for imaging and the other for pointing the platform), an infrared spectrometer, and a three-channel spectrometer.**

and, during their descent, released helium-filled balloons that carried a quartet of instruments and a tiny transmitter. During their two-day lifetimes, the novel, balloon-borne payloads relayed data on the planet's dense atmosphere and thick clouds directly to waiting scientists on Earth. The descent capsules also delivered instrumented landers to Venus' surface. Meanwhile, high overhead, the planet's gravity swung the main Vega craft onto a new trajectory toward their next target, Comet Halley.

On the morning of March 6, 1986, scientists gathered for Vega 1's flyby at the Space Research Institute (known by its Russian initials, IKI). Here the data radioed by the spacecraft would arrive $9\frac{1}{2}$ minutes later by way of receiving antennas at Evpatoria in the Crimea and Medvezy Ozera near Moscow. The Vega experimenters were an international group, including American participant John Simpson, who had been invited to provide a dust impact detector for each spacecraft. Not far from where the experimenters would receive and analyze their data were about 100 invited guests: politicians, well-known scientists (with Whipple among them), representatives from the other Halley missions, and a few Western science journalists.

All experiments aboard Vega 1 had been switched on two days before closest approach, and each day they acquired measurements and transmitted them to Earth. In between these intensive 2-hour sessions, a limited amount of data were also acquired and periodically played back slowly from an on-board tape recorder. Vega 1's path would take it past the nucleus 8,890 km from the sunward side, with closest approach predicted to occur at 10:20 a.m. Moscow time (7:20 Universal time).

At 6:20 a.m. Vega 1 crossed the comet's bow shock. Inside this turbulent boundary the solar-wind plasma is decelerated and deflected around the comet's ionosphere, much like the bow of a ship moving through the water. The bow shock has the shape of a hyperboloid and was closest to the nucleus (400,000 km away) in the direction of the Sun. Vega 1 crossed the shock to one side, where the distance to the nucleus was 1.1 million km. In their models of the bow shock, plasma physicists had correctly predicted this standoff distance by assuming that the comet would produce 10^{30} gas molecules per second.

High-speed telemetry began arriving 3 hours before the time of closest approach. In the conference room at IKI, two large television screens displayed the incoming data and images. Vega scientist Alec Galeev stood before them, offering the audience spontaneous interpretations of the incoming results. At the other end of the room, in a small enclosure converted into a television studio, well-known astronomer Carl Sagan prepared to broadcast his own commentary to the U.S. (where it was late evening) for the "Nightline" television show.

About an hour before the flyby, somewhat earlier than predicted, Vega 1 recorded the first impacts by dust particles. But these were very small, less than 10^{-13} g in mass. Closer to the nucleus, larger grains of up to 10^{-6} g also registered. As Vega 1 continued its approach, the concentration of cometary ions (mostly derived from water), neutral gas molecules, and dust particles continued to rise ever faster – as did the audience's suspense. Our main question was, of course, what would we see? Would the images show a single nucleus in the center of the coma, as Whipple predicted, or the "gravel bank" envisioned several decades ago by Raymond Lyttleton? Would the nucleus have a bright, icy surface or a dark coating of dust? Would it be spherical or irregular in shape? Another possibility, suspected by many, was that we would not see the nucleus at all; it might be hidden by a dense swarm of outward-streaming dust particles.

Meanwhile, the pictures on the big screens at IKI steadily improved. The processing computer had displayed them so that regions of equal brightness had the same color. The result was like a bull's-eye (Figure 5), with the bright center of the comet seen as a red shape surrounded by multicolored rings. A debate arose about what we were actually seeing. At 9:50, when the image resolution had shrunk to 2.4 km, some pointed out a small dot, presumably the nucleus. Then came a comment that the image was too overexposed to show the

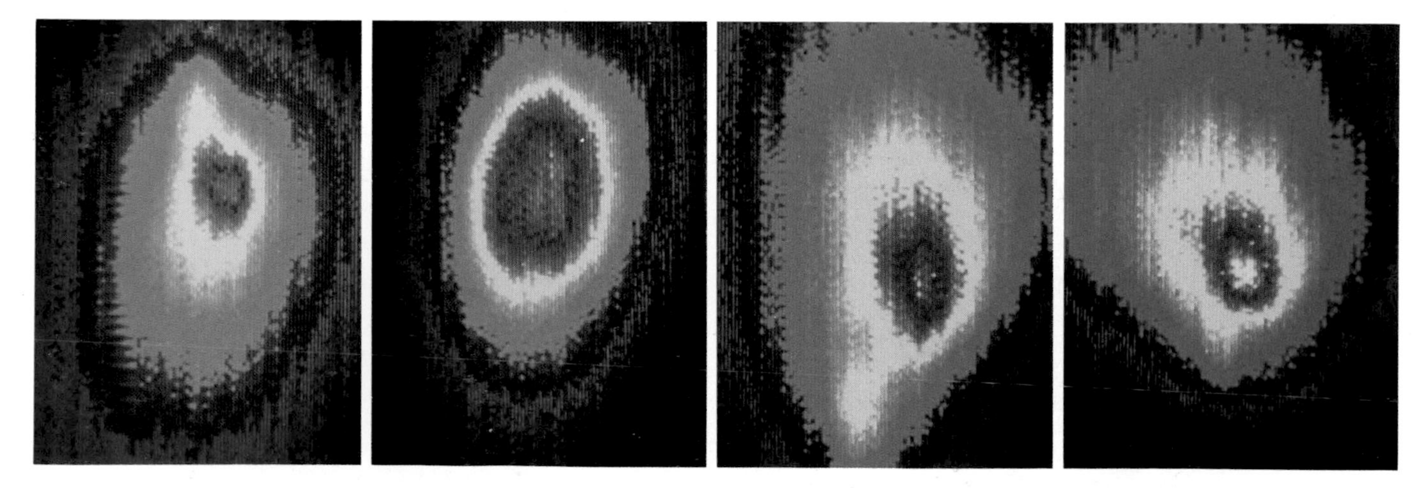

Figure 5. **The historic first images of Halley's nucleus and its surroundings from Vega 1. Different colors represent different brightness levels, with the brightest regions in the center. As the spacecraft sped past it viewed the irregularly shaped nucleus from several angles. The first image *(a)*, taken when Vega was 20,000 km from the nucleus, is 40 km across and shows a broad fan of dust and gas racing outward from the sunward (upper) side of the nucleus. The second *(b)*, from 11,000 km away, is 25 km across; the third *(c)*, acquired only seconds before closest approach from 8,900 km away, is 20 km across. In *d* the spacecraft was looking back at the nucleus and 25,000 km away; the image is 50 km across, and the Sun is now toward the bottom. The nucleus is indistinct because of light scattered by the cocoon of dust surrounding it.**

nucleus, and that we must be seeing the innermost region of the coma.

At 10:17, with the image resolution down to nearly 1 km, the audience was divided. "Maybe we see the nucleus here," said Galeev, indicating the red spot at the center. As if to question that idea, one listener noted that there was apparently no sharp boundary to the image. Finally, at 10:29, the screens displayed the most detailed images, which could distinguish features as small as 200 m. A voice from the audience asked, "What is the size of the red region?" "Three to 4 km," answered Galeev. His words triggered a burst of applause: this had long been considered just the right size for a cometary nucleus. By acclamation, it seemed, Vega had seen the heart of Halley's Comet.

Less than a minute thereafter, as Vega 1 sped away from the nucleus, the invited guests stood up and applauded once more, congratulating the participating scientists, engineers, and in particular IKI's director, Roald Sagdeev. Later came an intensive scientific press conference. Reporters pressed to know whether Vega 1 had indeed seen the nucleus. As Whipple put it, there was no longer any doubt that a nucleus lies at the center of the coma. But it was surrounded by dust, as if in a cocoon, and so we had not seen its outline clearly. Still, Whipple was all smiles, for Vega 1's images had proved his theory correct.

Three days later, on March 9th, Vega 2 gave a repeat performance. The closest approach was again at 7:20 Universal time, and the flyby's distance (8,030 km), speed, and approach direction were all similar to the first encounter. But the images obtained by the cameras on Vega 2 (Figure 6) looked surprisingly different. The central region around the nucleus was clearly elongated; some images even showed two central bright spots. Were there two nuclei, as some scientists in the audience wondered?

Over the next few days, as the pictures' analysis progressed, the answer became clear. Vega 2 had viewed the long side of an elongated nucleus facing the Sun and actively shedding gas and dust. What appeared to be a double nucleus on the images was in fact two prominent dust jets pointed almost directly at the camera. To determine the actual shape of the nucleus, the scientists combined the images from the twin spacecraft and found that it is indeed a single object, albeit an irregular one. The shape of Halley's nucleus has been compared, depending on individual preference, to an avocado, potato, or peanut. Mathematically inclined scientists call it "an ellipsoid with some deviations" measuring approximately 16 by 8 by 8 km, larger than had been generally expected.

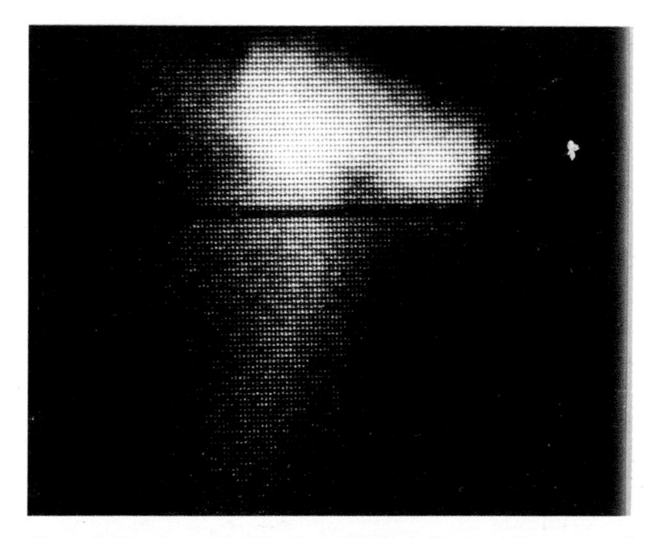

Figure 6. **Vega 2 recorded this view of Halley's Comet at the moment of its closest approach, when the two were 8,030 km apart. The nucleus appears peanut-shaped with overall dimensions of 16 by 8 km. A prominent dust jet appears to extend downward in projection, but actually it is spewing toward the Sun (and the camera).**

It was inevitable that the Vegas' solar-cell panels, which had to remain extended and exposed to sunlight at all times during the encounters, would be damaged by dust impacts. Afterward, engineers determined that the power produced by the panels had decreased by almost half. A few days after the encounters, the Soviets stopped tracking both Vegas, and the two automated explorers – their missions accomplished – began an endless, silent drift around the Sun.

"THE COMET IS BREATHING"

Between the two Vega encounters, on March 8th, Japan's Suisei spacecraft came its closest to Halley's Comet. (Suisei is the Japanese word for comet.) A spin-stabilized vehicle with a diameter of 1.4 m and a mass of 139 kg, Suisei carries an ultraviolet imager and an instrument to monitor the solar-wind plasma and cometary ions. As a test mission, the Japanese also launched a nearly identical craft called Sakigake ("Pioneer"), which carried a plasma-wave probe, a solar-wind monitor, and a magnetometer. Developed and built by ISAS, Sakigake and Suisei were the first Japanese spacecraft to reach interplanetary space. Their launches from Kagoshima Space Center on the island of Kyushu took place in 1985 on January 7th and August 18th, respectively.

Suisei's primary mission was to observe the huge corona, a cloud of hydrogen atoms that extended some 10 million km from the comet's nucleus. When excited by sunlight, the hydrogen atoms emit ultraviolet radiation at the Lyman-alpha wavelength of 1216 angstroms, which cannot be observed from the ground because it is mostly absorbed by the Earth's atmosphere.

In the months before the encounters Japanese scientists received up to six views per day from the ultraviolet camera aboard Suisei. These showed that the brightness of the hydrogen corona varied with an average period of 52.9 hours. ISAS scientists said it was as if the comet were "breathing." The cyclic brightening suggested that some parts of its surface were active and others quiet. Notably, the cycle length was very close to the rotation period for the nucleus that astronomers Zdenek Sekanina and Stephen Larson had derived a year earlier after reanalyzing digitally processed photographs of Halley's Comet taken in 1910.

For the few days prior to the encounter Suisei was actually inside the hydrogen corona, so at around midday on March 8th ultraviolet imagery ended and the plasma instrument began an intensive, 16-hour run of observations. (Suisei did not have the capacity to transmit images and plasma data simultaneously; the two instruments also required the spacecraft to spin at different rates.) Unfortunately, the plasma device was not operating when Suisei first crossed the cometary bow shock, which probably occurred on the 8th at about 11:00 Universal time. But it relayed good data during the outbound crossing between 14:43 and 14:49. Its record shows that the solar wind was deflected and decelerated from

400 km per second in interplanetary space to a minimum of 54 km per second when the spacecraft was closest to the nucleus.

Suisei passed 151,000 km from the nucleus at 13:06. At this distance, chosen to be much farther away than the Vegas and Giotto, the spacecraft was not expected to suffer collisions with dust particles during the flyby. Thus it came as quite a surprise when Suisei was hit by two grains, just before and after its closest approach, that were massive enough to alter noticeably the spacecraft's orientation and spin period. Nevertheless, Suisei survived the encounter undamaged and without any loss of data.

The second Japanese spacecraft, Sakigake, passed 6.99 million km from Halley at 4:18 Universal time on March 11th. Even at that large distance Sakigake's three instruments observed signs of cometary activity: long-period plasma waves induced by "pick-up" ions derived from water molecules. (Pick-up ions travel out to large distances from the nucleus as neutral atoms or molecules before they are ionized and "picked up" by the solar-wind plasma. They can be distinguished from solar-wind ions by their higher energy.)

A year after the encounters, in January and April 1987, respectively, Japanese ground controllers steered Sakigake and Suisei onto slightly different paths to return to the Earth. After flying past here in January 1992, June 1993 and October 1994, Sakigake will head off to a meeting with Comet Honda-Mrkos-Pajdusakova on February 4, 1996, and then reach Comet Giacobini-Zinner on November 29, 1998. Suisei will swing past Earth in August 1992 en route to encounters with comets Temple-Tuttle on February 28, 1998, and Giacobini-Zinner on November 24, 1998 (again beating its sibling Sakigake there by a few days).

Figure 7. **"Adoration of the Magi," one scene in a fresco cycle painted by the Florentine master Giotto di Bondone, for whom ESA's Halley spacecraft was named. The cycle decorates the interior of the Scrovegni Chapel in Padua, Italy and was probably begun in 1303. Halley's Comet appeared in 1301 and may have served as a model for Giotto's "Star of Bethlehem."**

NIGHT OF THE COMET: THE GIOTTO ENCOUNTER

The Soviet and Japanese spacecraft taught us much about Halley's Comet in a remarkably brief time. Giotto was to extend that knowledge even further by passing closer to the nucleus than any of its predecessors. With its flyby only a few days away, the attention of the scientific world shifted from Moscow and Tokyo to the European Space Operations Center (ESOC) in Darmstadt, a small town 30 km south of Frankfurt in the Federal Republic of Germany. As with the Japanese missions, Giotto was the first interplanetary mission for ESA. Giotto took its name from the Italian painter Giotto di Bondone, who in 1304 depicted Comet Halley as the "Star of Bethlehem" in one of his works (Figure 7).

Giotto left Earth on July 2, 1985, aboard an Ariane 1 rocket launched from Kourou, French Guiana, in South America. Measuring 1.85 m across and weighing 960 kg at launch, the spacecraft cruised through interplanetary space for 8 months before its head-on meeting with the famous comet. Tucked aboard were 10 scientific instruments: a CCD camera, three mass spectrometers to analyze the elemental and isotopic composition of cometary dust and gas, sensors to record the impact of dust particles, a photopolarimeter, and a suite of plasma instruments. The camera's rotation, together with the spacecraft's spin (once every 4 seconds), allowed viewing in all directions (Figure 8). As with the Vegas, Giotto's camera was designed to follow the brightest spot in its field of view, presumably the comet's nucleus.

Two days before the encounter, Roger Bonnet (ESA's director of scientific programs) faced a most important decision: how close to the nucleus to send Giotto. The camera experimenters wanted to go past on the sunward side, ideally 1,000 km away but no nearer than 500 km; if any closer, the camera would not be able to rotate fast enough during the flyby to track the rapid relative motion of the nucleus. A second group of experimenters wanted a pass as close as possible to the nucleus – even if it meant the spacecraft would not survive. They were mainly interested in detecting rare, short-lived "parent" molecules thought to exist only in the immediate vicinity of the nucleus. In fact, they advocated that Giotto be targeted right at the nucleus. A third group, those involved in the dust and most plasma experiments, wanted a close-in flyby that still maintained a high survival probability. This would give them a good chance to get data on the outbound pass as well.

As a compromise we agreed on a target point no closer than 500 km to the comet's sunward side. But an allowance for targeting uncertainty (40 km) was necessary, so we arrived at a final aim point 540 km from the nucleus. The actual flyby distance, derived from post-encounter analysis of camera rotation angles, was 596 km.

Such precise targeting would not have been possible without help from the two Vega spacecraft, which served as "pathfinders" for Giotto. Ground-based astrometric observations could establish the position of the comet's nucleus only to within about 400 km – rather large, considering Giotto's desired flyby distance. But we had known for a long time that the two Vegas would encounter Halley a few days earlier than Giotto's arrival, and that their cameras would be able to locate the nucleus with high precision. At the 1984 meeting of the IACG, the Soviet delegation agreed to provide the crucial navigation data shortly after the Vega 2 encounter, in time to aid us in aiming Giotto. This cooperative effort became known as the "pathfinder concept" (Figure 9).

One problem remained, however. The Soviets could only determine the Vegas' positions with respect to Earth within roughly 400 km. Only NASA was able to track the Vega spacecraft with sufficient accuracy, by using the widely separated radio dishes of its Deep Space Network (DSN) and precise very-long-baseline interferometry. Taken together, the Vega data and NASA's tracking effort reduced Giotto's targeting uncertainty to just 40 km (Figure 10). And so it was that early on March 12th, using this pooling of data, ground controllers commanded Giotto's last trajectory adjustment. One final maneuver the following day aligned the spacecraft's spin axis exactly along its direction of motion, so that during the encounter all dust particles would strike the spacecraft squarely on its protective dust shield. With these preparations complete, Giotto headed for its trip deep into Halley's coma. The moment of closest approach was predicted to occur on the 14th at 1:03 a.m. local time.

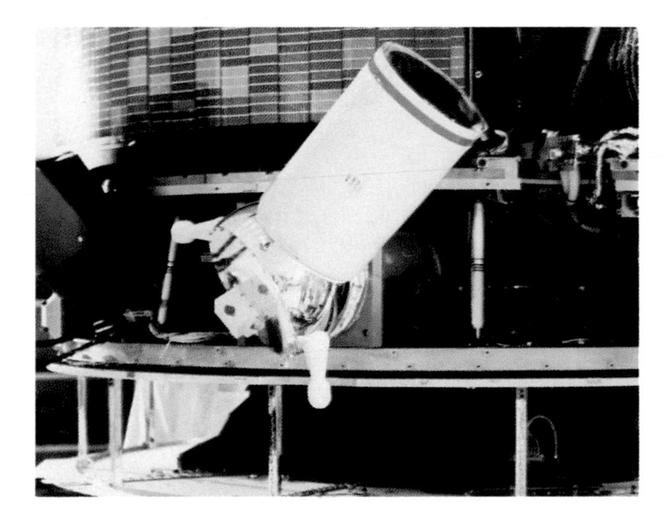

Figure 8a. **As did the two Vegas, Giotto carried a charge-coupled device (CCD) camera to photograph Halley's Comet. The camera optics are inside the spacecraft, fully protected from high-speed dust particles. Its white outer baffle, seen here, is about 30 cm long, and a 45° deflecting mirror is behind the oblique silvery part. The camera can be rotated through 180°, which, together with the spacecraft spin, allows viewing in all directions.**

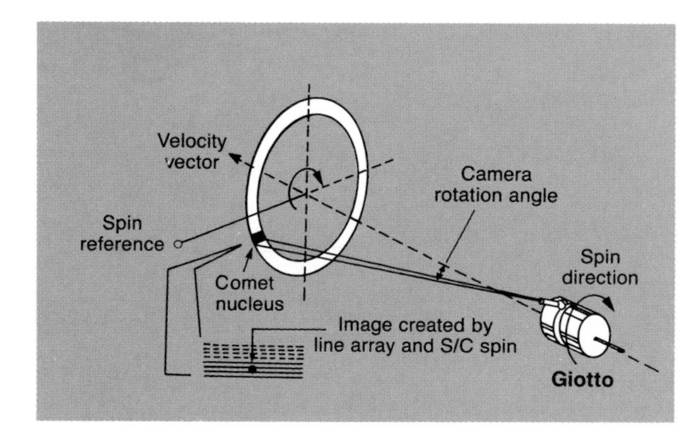

Figure 8b. **Giotto's on-board camera did not record its images instantaneously, but rather used spacecraft rotation to scan the nucleus and build up a picture line by line. Thus, the spacecraft produced a new image during each spin (every 4 seconds). The camera's direction of view depended on the angle of its rotation; during the flyby this angle steadily increased as the camera tracked the nucleus, and the exposure times became as short as 0.3 millisecond.**

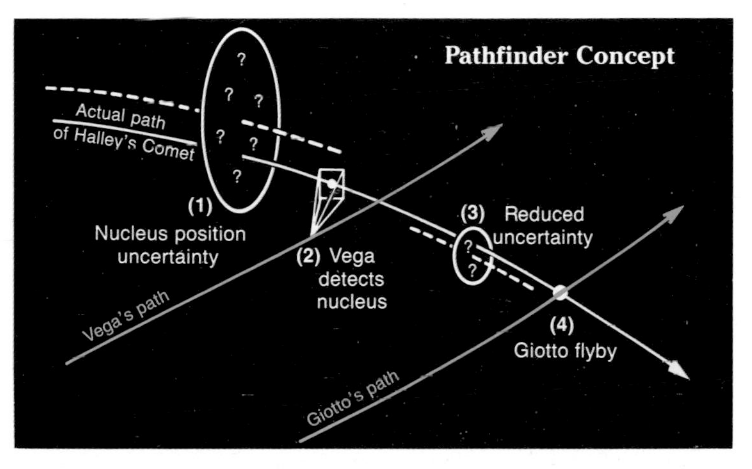

Figure 9. **The "pathfinder" concept was an international effort to target Giotto as precisely as possible.**

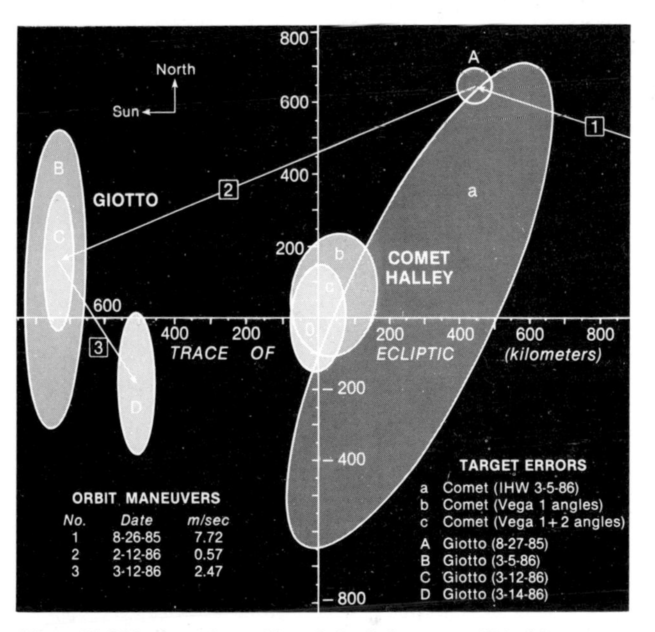

Figure 10. **This chart shows Giotto's final placement *(D)* with respect to the nucleus, which itself had been located with improved accuracy thanks to data from the Vega spacecraft and tracking support from NASA's Deep Space Network. As projected onto a plane perpendicular to Giotto's trajectory, each ellipse defines an area in which the object should be found with 97-percent certainty. Once positions *c* and *C* had been determined, ground controllers at the European Space Agency made one last move *(3)* to get Giotto closer and slightly south of the comet.**

To withstand the 68-km-per-second dust storm, Giotto carried a two-layer shield: a sheet of aluminum 1 mm thick and, 23 cm away, a second layer consisting of sheets of the plastic Kevlar sandwiched together. In theory, a dust particle striking the aluminum would pass through but become completely vaporized in the process. The vapor cloud would then expand before striking the rear sheet, where its energy would be absorbed over a large area. ESA's engineers calculated that the 50-kg shield should withstand the impact of dust particles up to at least 0.1 g in mass.

During the encounter scientific data had to be transmitted rapidly, at 40,000 bits per second, over a distance of 1 AU. This required using X-band (8.4-gigahertz) radio transmissions and a large receiving antenna, preferably in the Southern Hemisphere where the comet would be highest in the sky. Since ESA's ground station at Perth, Australia, has a diameter of only 15 m, we employed the 64-m station at Parkes, near Sydney, as our main receiver. The three 64-m antennas of NASA's Deep Space Network provided round-the clock coverage for a few days before and after the encounter, and provided backup during the critical flyby itself. Data from the spacecraft were forwarded from Parkes directly to Darmstadt, where scientists waited to begin their analyses.

On the chance that Giotto's solar-cell arrays might be damaged during the flyby, batteries were included both as an added precaution and to supply electricity during the peak periods when all spacecraft systems were operating. To save battery power, and to keep the spacecraft from becoming too warm, Giotto's experiments remained off for as long as possible. The plasma instruments were switched on first, immediately after the last trajectory-correction maneuver. These devices soon sensed the comet's presence, in the form of hydrogen pick-up ions, on the afternoon of March 12th, with Giotto still 7.8 million km from Halley's nucleus. The hydrogen ions slowly increased in number over the next 24 hours, as did the intensity of associated wave disturbances in the solar wind.

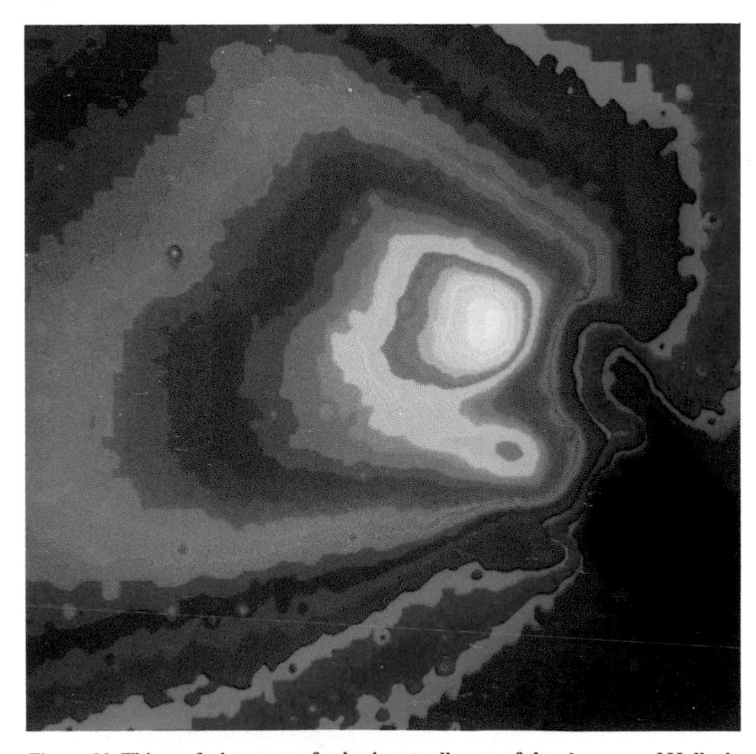

Figure 11. **This confusing maze of color is actually one of the close-ups of Halley's nucleus transmitted by Giotto shortly before it raced past the comet. The nucleus extends from the white spot in the center to the dark spot in the lower right corner (the frame measures 30 km on a side, and the Sun is roughly toward upper left). Displayed in this format, as they were the night of the encounter, Giotto's images were nearly unintelligible to project scientists and the public alike.**

By the evening of March 13th, ESOC literally buzzed with activity. The 350 Giotto experimenters had gathered in the Science Center; among them once again was Whipple, who spent most of the night analyzing the incoming images. Also present was Jan Hendrik Oort, well known among the scientists for his insightful deduction four decades ago that a huge swarm of comets (now called the Oort cloud) envelops the entire solar system. The cafeteria bulged with many of the 1,400 ESA officials and other invited visitors. In the makeshift press center 600 news representatives prepared to bring the "Night of the Comet" to a waiting world, linked to the operations center by 160 television monitors and 20 km of special cabling. ESA's public-relations personnel estimated that $1\frac{1}{2}$ billion people would see the live television coverage of the flyby.

At 7:30 p.m. Giotto crossed Halley's bow shock. Over the next few hours, with the spacecraft still well outside the coma, the rest of its instruments came to life. With all 10 switched on and working well, the level of excitement rose. The first images to arrive at Earth already had resolution 20 to 30 times better than photographs made with ground-based telescopes. But because the CCD exposures were only a few milliseconds long, we saw little detail in these first images.

The first dust particles hit Giotto when it was still 290,000 km from the nucleus – significantly farther out than expected. One question always on our collective mind was whether Giotto would survive the encounter. The spacecraft could be lost in two ways: (1) a "big" dust grain, at least 1 g in mass, could destroy it outright; or (2) a lesser particle, striking the shield off-center, could twist the spacecraft and expose its unprotected surfaces to the hypervelocity barrage of dust particles. Based on our expectations of the coma environment, we predicted Giotto's probability of survival to be about 90 percent.

Beginning at 11:00 p.m. new images appeared on the monitors every 4 seconds, showing ever more detail. We knew the nucleus had to be somewhere in the luminous center, but as with the Vega images its outline was not clearly visible. However, our immediate problem had nothing to do with the comet. As the Soviets had done, our camera team displayed the images in false color, which looked spectacular (Figure 11) but were impossible to interpret. We scientists found ourselves just as confused as the reporters already clamoring for explanations. For example, the British Broadcasting Corporation was televising the event live, using experts in a London studio to provide interpretation of the incoming data. But the incomprehensible images ruined everything. As the *Daily Telegraph* later reported, "The television pictures were colored splotches accompanied by a noise [an audio representation of the dust impacts] like frying bacon. The real row began later in 10 Downing Street, where the Prime Minister was watching the program. She was angry at what she saw and called Giotto 'a waste of money.' "

Giotto was approaching the nucleus at an incredible 246,000 km per hour, and things were happening extremely fast. With not much more than 1 minute until closest approach, the heavily bombarded Giotto crossed the comet's ionopause 4,700 km from the nucleus. A report from the magnetometer team quickly followed: the interplanetary magnetic field had suddenly dropped to zero. Giotto was so deep inside the coma that the solar wind could penetrate no further. Instead, the detectors registered only a stream of neutral molecules and cold ions flowing outward from the nucleus at 1 km per second.

Meanwhile, dust was sandblasting Giotto with ever greater ferocity and began to take its toll. At 33 seconds before closest approach we received the first report of damage to an instrument. At 21 seconds two more instruments stopped working. Three seconds later another failed. Then, 9 seconds before closest approach, images from Giotto stopped arriving – the camera had failed. The last image to be transmitted completely had been taken 1,930 km from the nucleus.

But the spacecraft itself pressed inexorably on, still in good general condition and with half the instruments still operating. The most exciting data of all streamed in until, 7 seconds before closest approach, our displays suddenly went blank. I stared at the monitors in puzzled disbelief. Was there a problem with the computers at ESOC, or the transmission lines from the tracking station, or perhaps with the Parkes receiving station itself? Or had our worst fears been realized – had Giotto been destroyed?

The NASA tracking antenna in Australia was not receiving any data either, so something had to be seriously wrong with the spacecraft. I had been convinced that the spacecraft would survive its encounter; thus its sudden silence was a great disappointment. Most television networks, assuming that Giotto was lost, quickly ended their live coverage.

Most of us did not notice when, 20 seconds after the transmissions stopped, data returned for a second, then disappeared again for a few seconds, then came back for a second, and so on. Clearly, the spacecraft was still working. We had calculated that if a 0.1-g dust particle were to strike the rim of the spacecraft, it would knock Giotto off its prior alignment by 1° and cause a nutation or wobble about this new axis with a period of 3.2 seconds. As Giotto oscillated, the highly focused radio beam from its antenna would briefly pass across the Earth (Figure 12), and only during those moments would we receive any telemetry.

Having at last recognized the meaning of this intermittent transmission pattern, we relaxed and waited. Giotto carried two tubes, 60 cm long, filled with a thick liquid. In time they would dampen the oscillation, and our steady stream of data would return. Only then could we assess the full extent of Giotto's damage (Figure 13). Half an hour later, with communications fully reestablished, we found that the spacecraft remained functional but had a slight residual wobble, which indicated that it was no longer properly balanced. (Months later, we would determine that the camera had lost its outer baffle.) About half of the experiments were still working and returning good data, though Giotto was by then too far from the nucleus to return meaningful images. Nevertheless, we did try to look back at our target, but the camera gears seized before completing the 180° slew to the rear.

Giotto's plunge through the dusty coma had slowed its

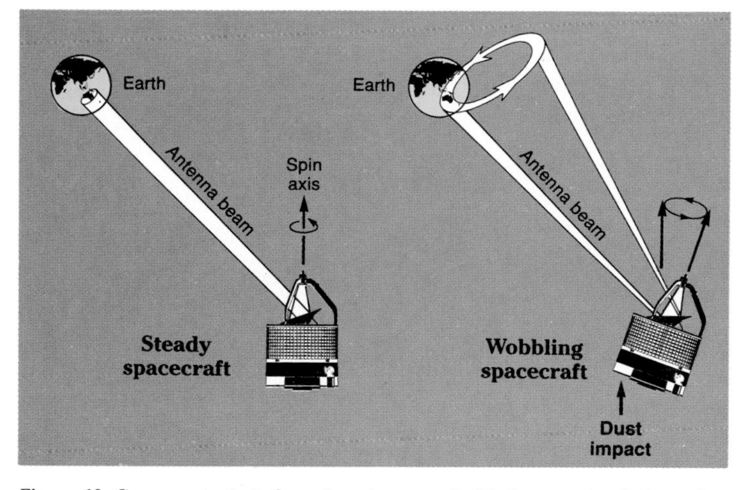

Figure 12. **Seven seconds before closest approach Giotto was struck by a dust particle that set the spinning spacecraft wobbling like a top, with a period of 3.24 seconds and an amplitude of 1° (exaggerated here for clarity). Consequently, the spacecraft's antenna beam, which was only 1° wide, no longer pointed continually toward Earth, and stable contact was lost for half an hour until on-board nutation dampers steadied the probe.**

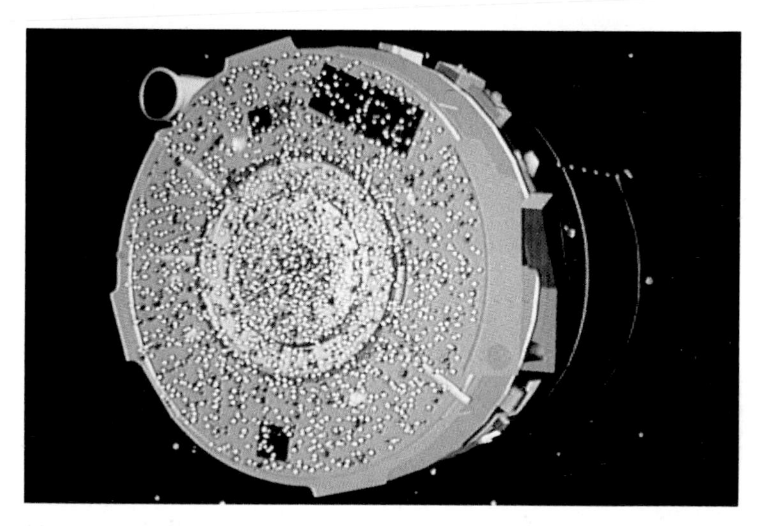

Figure 13. **After its passage through the comet's dusty coma, Giotto's protective shield undoubtedly had the battered appearance of this computer-generated picture. The white circle at center is the cover for the nozzle of the burned-out rocket used to inject Giotto into interplanetary space; black patches are dust impact sensors. The camera baffle appears at upper left, though in reality much of it was ripped away during the intense pummeling.**

Figure 14. **A jubilant H. Uwe Keller, leader of the Giotto camera team, points to features in an image of Halley's Comet in the early morning of March 14, 1986, shortly after the historic flyby. David Dale, Giotto project manager, holds the microphone for Keller as plasma investigator Alan Johnstone looks on.**

Figure 15. **Black-and-white images produced by the Giotto camera team early on the morning of March 14th show the outline of the nucleus much more clearly than color-coded ones like that in Figure 11. In this image, taken from 17,980 km away just seconds before the one in Figure 11, several surface features are visible.**

Figure 16. **One year after the encounter, members of Giotto's camera team produced this enhanced composite of two separate images taken at distances of 25,000 and 10,000 km. The dark nucleus is clearly visible against the brighter coma. Among the surface features visible are a hill on the comet's nightside and several pits along the terminator.**

velocity measurably, by 23 cm per second. It crossed the comet's ionopause outbound at 0:04 Universal time and the bow shock between 2:30 and 3:00. Instruments continued to detect the comet's pick-up ions and plasma waves out to a distance of 5 million km. The last experiment was switched off 26 hours after the closest approach and at a point 6.5 million km from the nucleus. As Bonnet noted at a summary press conference, Giotto had accomplished its mission "beyond all expectations" (Figure 14).

One week after the Halley flyby, mission controllers retargeted Giotto to return to Earth's vicinity, then put it in hibernation. ESA's plan is to reactivate the spacecraft in February 1990. If the spacecraft and at least some key instruments are still functional, ESA will probably extend the Giotto mission. After flying past Earth on July 2, 1990, precisely 5 years after its launch, Giotto could be redirected to encounter Comet Grigg-Skjellerup on July 10, 1992.

As a final note, it became clear very soon after the flyby that black-and-white displays of the Giotto images (Figure 15) were much more meaningful than the false-color creations that had bewildered us in the hours before. For example, the outline of the nucleus seemed much better defined; it was elongated in shape, about 15 km long and 8 wide. Surprisingly, Halley's nucleus was very dark, essentially black, and almost overwhelmed by the bright dust jets seen streaming toward the Sun. On closer inspection (Figure 16), individual surface features, such as a crater, could be identified. For me, it was the final proof that we had indeed come face-to-face with the heart of Halley's Comet.

17
Comets

John C. Brandt

THE APPEARANCE of a bright comet in the sky often triggers great interest among scientists and the public alike. Recently, this interest was heightened by the appearance and extensive study of the most famous of all these cosmic visitors, Halley's Comet. Astronomers, space physicists, geologists, and even biologists study them. To the eye, comets' most distinguishing features are their tails, which routinely stretch across space for several tens of millions of kilometers – and occasionally for 1 AU (about 150 million km) or more. Photographs frequently capture this spectacular sight (Figure 1).

Comets can affect the Earth in more direct ways, too. For example, their debris is responsible for most meteor showers. Until recently, a small comet was thought to have caused the Tunguska, Siberia, explosion in 1908. But now it seems likely that such a body would have disintegrated high in the Earth's atmosphere, long before it could have produced the fireball and other effects observed. Nevertheless, cometary bodies must certainly enter our atmosphere on occasion, and the largest chunks presumably strike the Earth from time to time. Swarms of small ones could explain some enigmatic phenomena observed in space, and in aeons past they may have been the source of some of the Earth's volatiles. Comets also represent key pieces in our efforts to understand the origin and early evolution of the solar system. Partially for this reason, the scientific study of comets is increasing, after decades of relative obscurity.

Public interest in comets is a mixture of curiosity, suspicion, and awe, and the spectacle of a bright one often evokes strong reaction. For example, in April 1986, when Comet Halley was brightest, the Griffith Planetarium in Los Angeles logged more than 1,000 telephone calls each day on its public-information line. The exceptional interest in this celebrated object occurs for many reasons. Observations of Halley's Comet go at least as far back as Chinese records of its appearance in 240 BC, and it has been observed at each perihelion passage for more than two millennia.

Until the time of Edmond Halley, however, comets were tacitly assumed to pass through the inner solar system only sporadically, and no serious attention was given to the possibility of their periodic return. Halley used Isaac Newton's then-new theories of gravitation and planetary motion to compute the orbits of several comets. He noted that the orbits of those observed in 1531, 1607, and 1682 were quite similar and assumed that the sightings probably referred to the same object at successive apparitions. On this basis, Halley boldly announced that it would return again in 1758–1759. Its return occurred as he predicted, and the comet was later named in his honor. The physical study of Halley's Comet was begun by F. W. Bessel at the 1835 apparition. In 1910, it was observed by E. E. Barnard, among others, and there was a worldwide effort to organize the observations made during its passage.

The orbit of Halley's Comet has an average period of 76 years, with a perihelion of 0.59 AU (inside Venus' orbit) and an aphelion of 35 AU (beyond Neptune's). This object displays the full range of cometary phenomena, including a long, spectacular tail (Figure 2). We owe the comet's repeated naked-eye visibility over many apparitions to fortunate circumstances. First, the orbit is favorably placed with the perihelion between the Sun and the Earth's orbit. Second, the comet is large, and the resultant activity is unusually spectacular. These two factors have combined through the centuries to make Halley's Comet readily visible and memorable. In the popular view, it has been associated with numerous historical events. Among its many historical representations are the depiction in the Nürnberg Chronicles (the apparition of 684 AD), the Bayeux Tapestry (1066 AD; see Figure 3), and the naturalistic rendering in one of Giotto's Arena Chapel frescoes in Padua (1301). Ironically, the 1986 apparition was poor for naked-eye viewing because the comet passed perihelion (February 9, 1986) while almost exactly on the side of the Sun opposite Earth.

Thus, Halley's Comet is unique. It possesses an unquestioned place in human history and is, at the same time, a large, dependable object with exciting scientific possibilities. This is the only comet that exhibits the entire range of classic, observable phenomena *and* occupies a predictable orbit.

Not surprisingly, therefore, efforts to launch a spacecraft toward a comet inevitably focused on Halley. Six spacecraft were sent to comets in 1985 and 1986, and Chapter 16 is devoted to them entirely. But a few words of elaboration are in order. The total instrument complement consisted of approximately 50 experiments. The scope of their investigations was vast, from detecting plasma waves far from the nucleus to obtaining images of the nucleus itself. This imaging capability was available on both Vegas and on Giotto.

About six months before the Halley flybys, the International Cometary Explorer (ICE) spacecraft passed through Giacobini-Zinner's tail and thus provided a valuable complement to the extensive sunward data obtained by the Halley "armada." This object was previously recognized as the source of the Giacobinid (or Draconid) meteor shower. But it is now famous as the first to be probed by a spacecraft.

The scientific results from the space measurements are discussed, where appropriate, in the sections below. However, our large advance in understanding does not rest solely on these missions. Other, existing spacecraft such as the International Ultraviolet Explorer (IUE) and the Pioneer Venus orbiter (PVO) made important contributions. Extensive ground-based observations were organized by the "International Halley Watch" (IHW). Its eight networks of observers utilized a variety of techniques to study the comet and its meteor streams. The IHW data will be preserved in an extensive archive of data banks and publications.

OBSERVATIONS AND MEASUREMENTS

The principal parts of active comets are (in decreasing size) the *tails, hydrogen cloud, coma,* and *nucleus*, which are illustrated schematically in Figure 4 and discussed individually below. All evidence available at present indicates that the central body and the ultimate source of all cometary phenomena is a lump of snow and dust with a typical dimension of roughly 1 km. The ability of this "dirty iceball" to interact with solar radiation and the solar wind to produce features up to 1 AU long is remarkable indeed.

Tails. Photographs of comets usually show two distinct kinds of tail (Figure 5): one containing dust and the other plasma (ions and electrons). The dust tail appears yellow because the light reaching us from it is reflected sunlight. The plasma tail looks blue because radiation emitted by fluorescing ions of carbon monoxide (CO^+) within the tail peaks at about 4200 angstroms. Dust and plasma tails can be found alone or together in a given comet.

Usually observed as sweeping arcs, dust tails typically have a homogeneous appearance and lengths ranging from 1 million km to perhaps 10 times that. The component particles are usually about 1 micron across (the size of smoke particles). Their composition is discussed below in the section on the coma. Sunward-pointing features, the so-called *antitails* seen in comets Arend-Roland in 1957, Kohoutek in 1973, Halley in 1986 (Figure 6), and others, are not directed at the Sun at all. They merely result from our seeing a dust tail projected ahead (sunward) of the Earth-comet line.

Plasma tails are usually straight, contain a great deal of

Figure 1a. **Halley's Comet on March 8, 1986. This beautiful photograph shows the distinctive colors of its dust and gas tails. Here the tails appear undisturbed, but compare this view with how the comet looked two days later (Figure 8a).**

Figure 1b. **The famous comet on April 14, 1986. Its plasma tail points straight up, while the dust tail forms a fan directed up and to the left. At bottom right is the giant globular cluster Omega Centauri. William Liller obtained these photographs on Easter Island as part of the International Halley Watch. Each frame is about 4½° wide.**

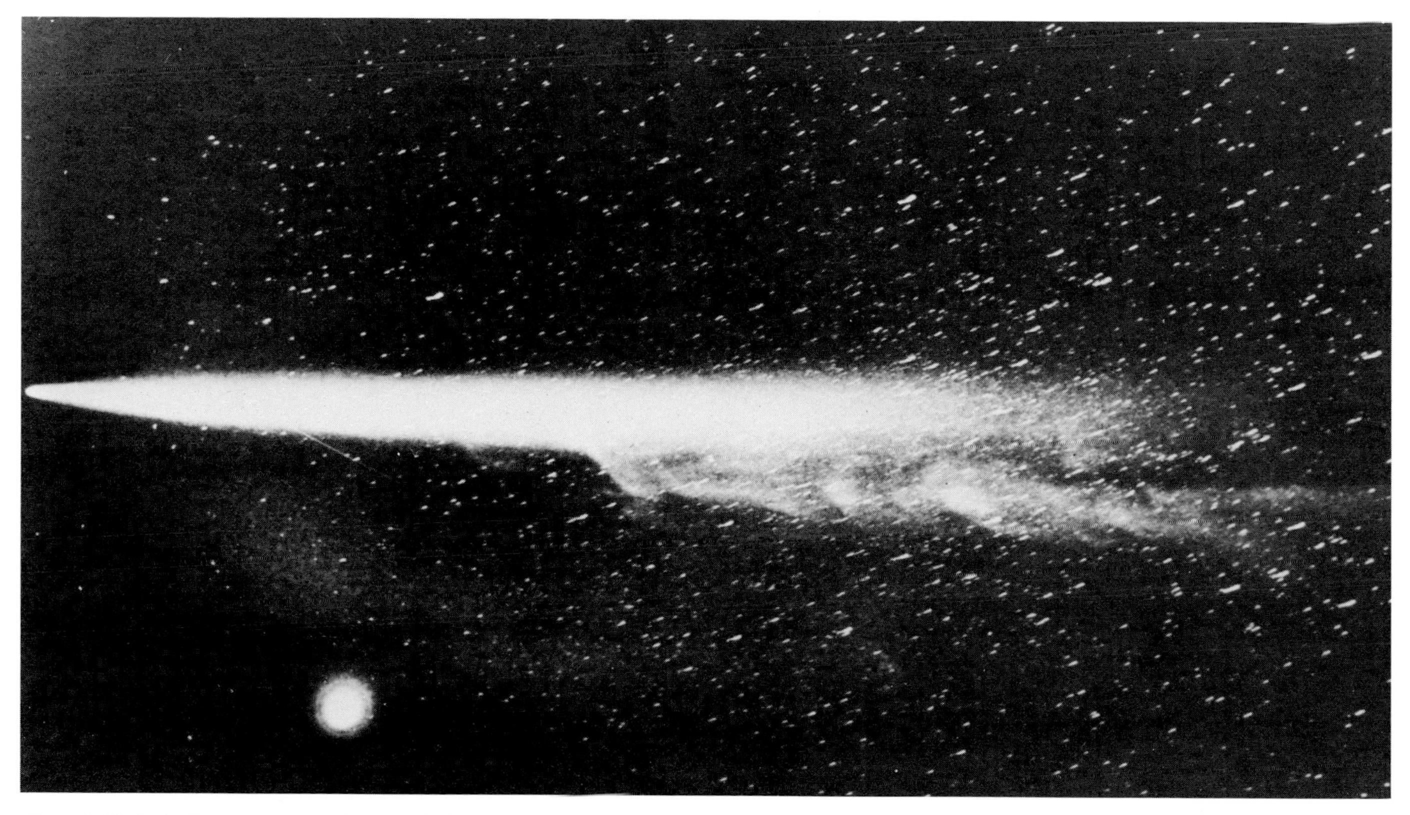

Figure 2. **Halley's Comet, as recorded at Lowell Observatory on May 13, 1910, 24 days after perihelion. This photograph captured the comet's 45°-long tail, as well as the planet Venus (seen at lower left).**

Figure 3. **An apparition of Halley's Comet is depicted in the celebrated Bayeux Tapestry, which commemorates the Norman Conquest in 1066. The comet made a close and impressive approach to Earth that year, about the time that William the Conqueror invaded England from Normandy (France). Its appearance was considered an evil omen for King Harold of England; in fact, Harold was killed later that year during the Battle of Hastings. The Latin inscription *Isti Mirant Stella* translates "They marvel at the star."**

H, OH, O, S, S_2, H_2O, H_2CO, $(H_2CO)_n$	CO^+, CO_2^+
C, C_2, C_3, CH, CN, CO, CS, N_2, NH_3	H_2O^+, OH^+, H_3O^+
NH, NH_2, HCN, CH_3CN	H^+, CN^+, N_2^+
Na, Fe, K, Ca, V, Cr, Mn, Co, Ni, Cu, Si, Mg, Al, Ti	C^+, Ca^+

***Table 1.* A surprising variety of atoms and molecules (left) and ions (right) have been found in the comas and tails of comets.**

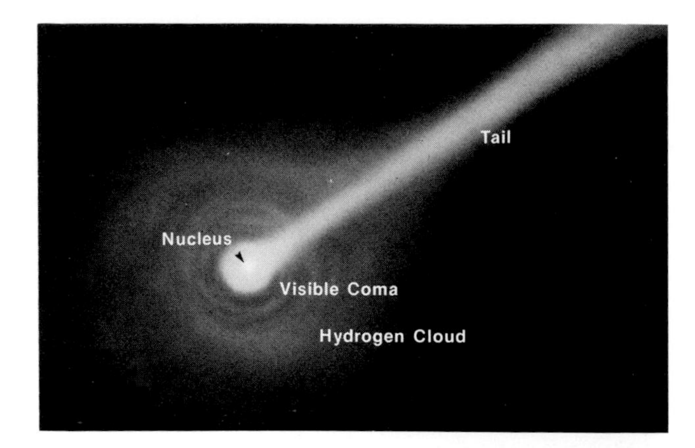

***Figure 4.* A schematic representation of the basic parts of a comet.**

fine structure, and attain lengths roughly 10 times that of their dusty siblings – up to 100 million km. The plasma races outward almost directly away from the Sun, lagging the true antisolar direction by a few degrees in the sense opposite the comet's motion. Locally the plasma becomes concentrated into thin bundles called *rays* or *streamers*. Such ubiquitous details provide convincing evidence that a magnetic field threads the tail's entire length. Consisting of a dense, cold mixture of electrons and molecular ions, the plasma-tail streamers seem to be rooted in a limited zone on the Sun-facing side of the nucleus. Their turning and lengthening (Figure 7) provide a good hint as to the origin of the magnetic field.

The firsthand measurements provided by spacecraft have removed any lingering doubt concerning the importance of the magnetic field in cometary phenomena. The field's orientations are what we would expect if solar-wind field lines were draped over the comet's ionosphere. The visible tail corresponds to the region where fields of opposite polarity come together, and along this boundary a *current sheet* forms. Spacecraft measurements also indicate the existence of a *bow shock* within the plasma on the sunward side of both comets. Finally, the disturbance that comets cause in the solar wind extends to large distances, as evidenced by the detection of plasma waves and energetic particles.

Plasma tails routinely become detached from the head, during a phenomenon called a *disconnection event*, or DE, that will be explained more fully later. During a DE, the old tail drifts away and a new one forms. Approximately 30 of these episodes were recorded during Comet Halley's most recent appearance, 19 of which were easily seen; in fact, one obvious DE occurred during the comet's encounter with the Halley armada (Figure 8).

Hydrogen cloud. In 1970, observations made above the atmosphere at the Lyman-alpha wavelength of 1216 angstroms (deep in the ultraviolet) indicated that comets Tago-Sato-Kosaka and Bennett were surrounded by huge clouds of hydrogen. Similar clouds have accompanied several other comets and span many million kilometers, making them substantially larger than the Sun (Figure 9). Astronomers have used these observations to estimate the production rate of hydrogen for several bright comets approaching the Sun. By the time they cross the Earth's orbit, these objects were producing more than 10^{29} hydrogen atoms per second! This escaping material cannot originate directly from the icy nucleus because the cloud's observed outflow speed is roughly 8 km per second, about 10 times faster than predicted for material simply sublimating (evaporating) from the nucleus' surface. Instead, most of this hydrogen

***Figure 5.* Comet West (1975n), as photographed on March 9, 1976, by amateur astronomer John Laborde. A blue plasma tail (right) and pale yellow or "whitish" dust tail (left) are apparent.**

probably comes from the dissociation of the hydroxyl radical, OH, by sunlight.

Coma. This spherical envelope of gas and dust surrounds the nucleus, extending from 100,000 to 1 million km from it and flowing away at an average speed of 0.5 to 1.0 km per second. It is the outflow of coma gas that drags dust particles away from the nucleus. Comas usually do not appear until comets come to within about 3 AU of the Sun.

Gases within the coma, as determined by spectroscopy

from Earth and by mass spectrographs aboard spacecraft, are principally neutral molecules. For example, Figures 10 and 11 show the spectra of two comets obtained in the ultraviolet from 1200 to 3400 angstroms and at visual wavelengths. The visual spectrum (3000 to 5600 angstroms) of Comet Bradfield reproduced in Figure 12 exhibits many emission lines frequently observed in comets from Earth. As may be apparent from these examples, many aspects of comet spectra generally all look alike.

Neutral molecules and atoms detected in comas to date are listed in Table 1, along with the ionized molecules found near the nucleus and in plasma tails. Note the presence of relatively complex compounds like HCN, CH_3CN, and $(H_2CO)_n$ or polymerized formaldehyde. Heavy metals begin to appear in the coma as a comet nears the Sun.

During its recent perihelion passage, the inner coma of Comet Halley consisted of the following compounds, listed by number of molecules: water (H_2O), about 80 percent; carbon monoxide (CO), about 10 percent (determined from a rocket experiment); carbon dioxide (CO_2), about 3½ percent; polymerized formaldehyde ($(H_2CO)n$), a few percent; and trace amounts of other substances. The discovery of polymerized formaldehyde is especially exciting. Chains of formaldehyde molecules,

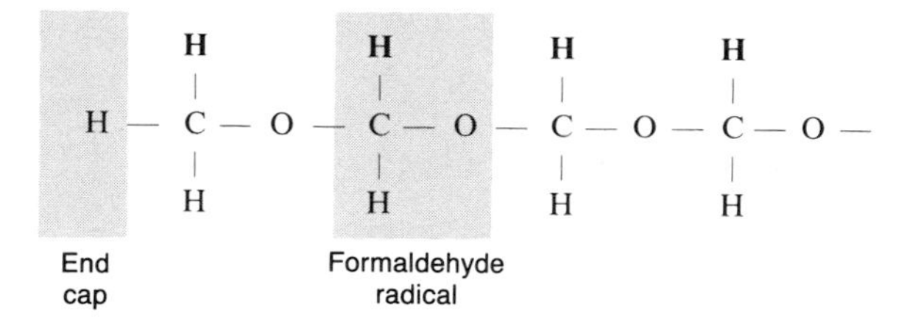

when broken and terminated with suitable "end caps" such as hydrogen, accurately reproduce the series of mass peaks recorded by mass spectrometers while flying through the

Figure 6. **On February 22, 1986, two weeks after passing perihelion, Comet Halley was a spectacular sight in the skies of the Southern Hemisphere. At that time its dust tail was a broad fan about 3° long with distinct sections, including one that seemed to point back toward the Sun (aimed toward lower left). K. S. Russell obtained this stunning photograph with the 1.2-m Schmidt telescope near Coonabarabran, Australia.**

comet. Although this identification is straightforward, some caution is warranted because other polymers or combinations may also provide a good match to the data. Thus, it is best to think of polymerized formaldehyde, a highly likely candidate, as the surrogate for a class of complex organic compounds until any doubt is resolved.

The Halley probes also measured the composition of the dust particles, which seem to fall into three general categories. The first type consist primarily of the elements H, C, N, and O; these are termed "CHON" particles. The second type has a silicate composition, that is, a mineralogy similar to the rocks in the crusts of the Earth, Moon, Mars, and most meteorites. The third type, and the most common, consists of mixtures of the other two groups. These particles are compositionally similar to primitive meteorites called carbonaceous chondrites, except they are enriched in the CHON elements. They could resemble the so-called "Brownlee particles" that are collected by research aircraft flying in the Earth's upper atmosphere (Figure 13).

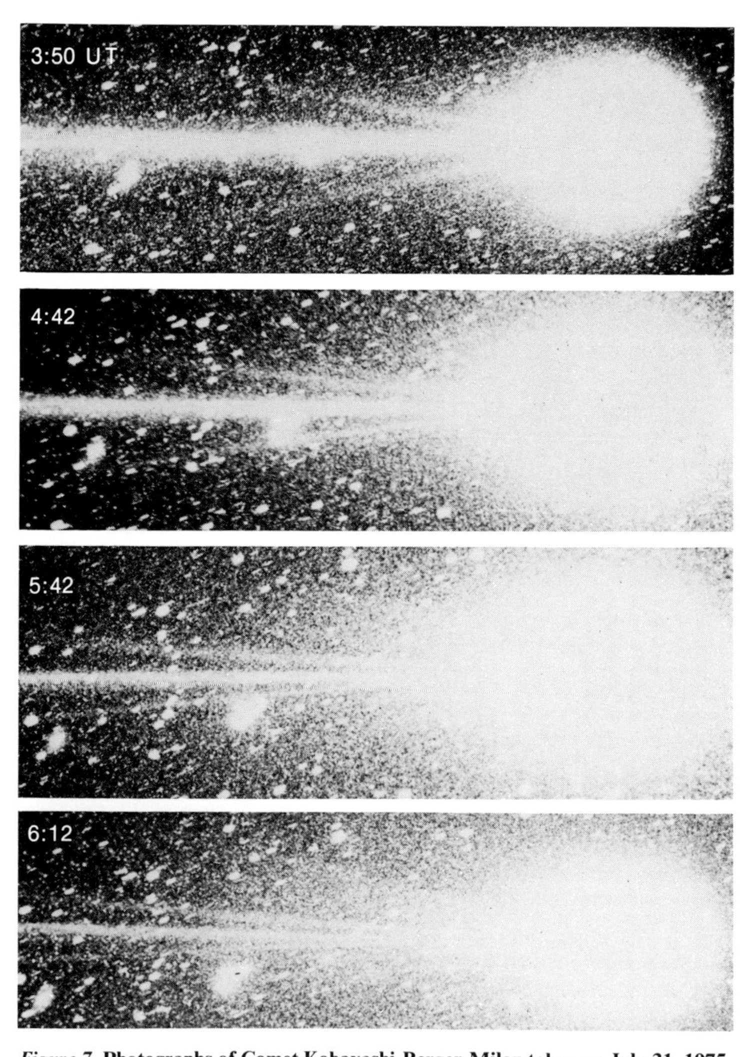

Figure 7. **Photographs of Comet Kobayashi-Berger-Milon taken on July 31, 1975, show the plasma tail's capture of magnetic field lines from the solar wind. The two most prominent streamers on either side of the tail's axis can be seen to lengthen and turn toward the axis in this sequence. (A comet is named for the individual who discovers it; in this case the object was found by three independent observers. Comets also receive sequential designations based on the year of their discovery, such as 1975h for this one.)**

Nucleus. Until the advent of space missions, we had no photograph of a cometary nucleus – only strong circumstantial evidence had implied the existence of a central source of all cometary gas and dust. Spacecraft have now supplied us with a large volume of data on exactly two comets, but only one's nucleus was examined at close range. The nucleus of Halley's Comet is not what most researchers expected (Figure 14). It is larger, darker, and more irregularly shaped than our previous, idealized concept (something roughly spherical with a diameter of about 5 km and an albedo as high as 60 percent). Compare these preconceptions with the observations described below.

In spacecraft images the irregular shape is clearly visible and has been said to resemble a peanut (the Vega model) or a potato (the Giotto model). The nucleus is roughly 16 km long and 8 km across at its widest point. It is not smooth (Figure 15) but shows features that can be likened to hills, craters, and so forth. The surface is very black – darker than coal or black velvet, with an albedo close to 3 percent. This means that the nucleus of Halley's Comet is one of the darkest bodies in the solar system.

Infrared radiation emanating from the surface indicates a temperature of approximately 330° K. This value is close to what we would expect for a slowly rotating blackbody situated 0.8 AU from the Sun (as the comet was during the spacecraft encounters). Such temperatures are appropriate for a dark, dusty crust, but a surface of sublimating ice should be roughly 100° colder. Thus, we may reasonably infer that the sublimation of ices takes place below the surface.

The bright jets visible in the images originated from a limited number of locations on the nucleus (perhaps one-tenth of the total area) and were confined to the sunward side. The jets appeared bright, presumably from sunlight reflecting off the entrained dust. These particles are dragged off the nucleus by the expanding, freshly sublimated gas. Once free of the surface, the gas quickly expands laterally to fill the gaps between the jets, but the dust coasts outward and retains the jets' original configuration.

The jets' rapid turn-on (at sunrise) and turn-off (at sunset) implies that a thin crust lies over the jets and a thick crust exists elsewhere. Evolution of the "pits" from which the jets originate could easily produce Brownlee-type particles; edges of the dust crust carried away by the gas flow are good candidates.

One major area of controversy exists concerning Halley's nucleus, namely, its rotation period. A value of approximately 2.2 days (53 hours) was widely assumed to be correct throughout most of the apparition, because it was supported by several lines of evidence. Recurring dust jets observed in 1910 are compatible with this period. Early in the 1986 apparition, the Suisei spacecraft noted variations in the comet's brightness at the Lyman-alpha wavelength that were interpreted as pulsations or "breathing" from the nucleus with this period. A 2.2-day rotation was also indicated by a variety of photometric investigations and by the imagery of the nucleus itself obtained by Giotto and the Vegas.

Consequently, the discovery of substantial evidence for a 7.4-day period, relatively late in the apparition, came as a real surprise. It was based on observations made by ground-based telescopes and by the IUE spacecraft, combined with the fact that astronomers had encountered difficulty in locating what should have been a rather obvious axis of

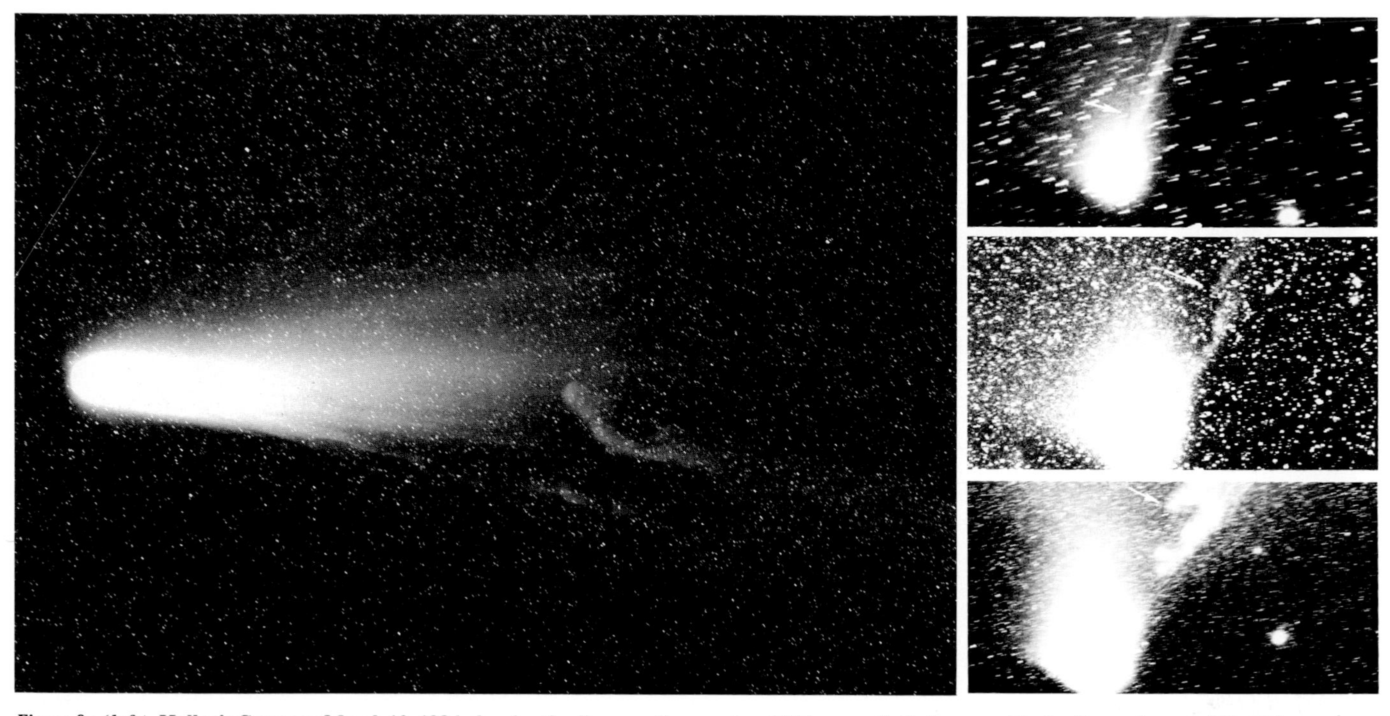

Figure 8a (left). **Halley's Comet on March 10, 1986, showing the disconnection event, or DE (arrowed), that occurred during "armada week." For this exposure, astronomers used the 1-m Schmidt telescope of the European Southern Observatory in Chile.** ***Figure 8b (right).*** **This sequence captures the initiation of a disconnection event on April 11, 1986. The first photograph was obtained by Patrick Malloy and Marlene Spector on Tahiti; the second one some 13 hours later by observers on Reunion Island; and the last one 6 hours later still by Freeman Miller using the Curtis Schmidt telescope in Chile. These photographs were also part of activities of the International Halley Watch.**

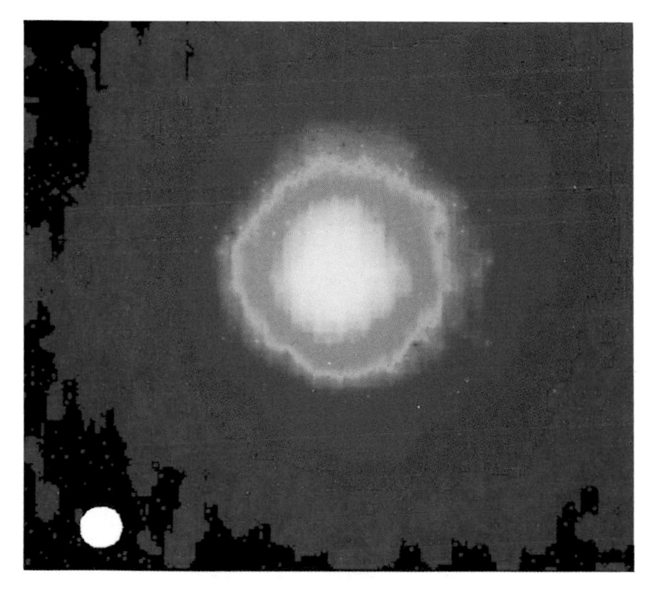

Figure 9. **Comet Halley's neutral-hydrogen cloud as recorded at the wavelength of 1216 angstroms by the ultraviolet spectrometer on the Pioneer Venus orbiter. Colors represent different brightness levels. The filled white circle at lower left is the Sun's disk as it would appear at the comet's distance.**

rotation. In fact, those attempting to model the rotation of the nucleus have not been able to agree on a general solution. But this should not be altogether surprising. The nucleus is highly asymmetrical (Figure 16) – it could be rotating about all three axes simultaneously, so its motion and the resulting light curve produced as it reflects sunlight could be very complex. Images obtained by the Vegas and Giotto, while extremely valuable, nonetheless represent only three snapshots in time. A further complication is that bright sources (the jets) turn on and off irregularly.

While no model is generally accepted, the long dimension probably rotates about an axis roughly perpendicular to it with a period of 2.2 days. The 7.4-day period could result from a rolling motion around the long dimension or a nodding motion. Other physical processes such as the precession of the spin axis may also be important. Astronomers plan to monitor the light curve carefully as Halley's Comet recedes from the Sun – a time, we hope, when jets of material are not adding to the confusion by switching off and on.

PHYSICAL MODELS

Any comprehensive and acceptable theory of comets must explain not only their major features (Figure 17) but also the ways in which they change with heliocentric distance. This theory should explain existing observations in a simple way and, if necessary, point out critical observations or measurements needed for its validation. While we believe that our general understanding is in good shape, experience shows that additional surprises could easily occur.

The basis of current theory is Fred L. Whipple's "icy conglomerate" model of the nucleus, which he proposed in 1950 and was later extended by Armand Delsemme (Figure 18). As a comet approaches the inner solar system, sunlight falls on the surface of the nucleus and heats it. When the comet is still far from the Sun, all the radiant energy goes into

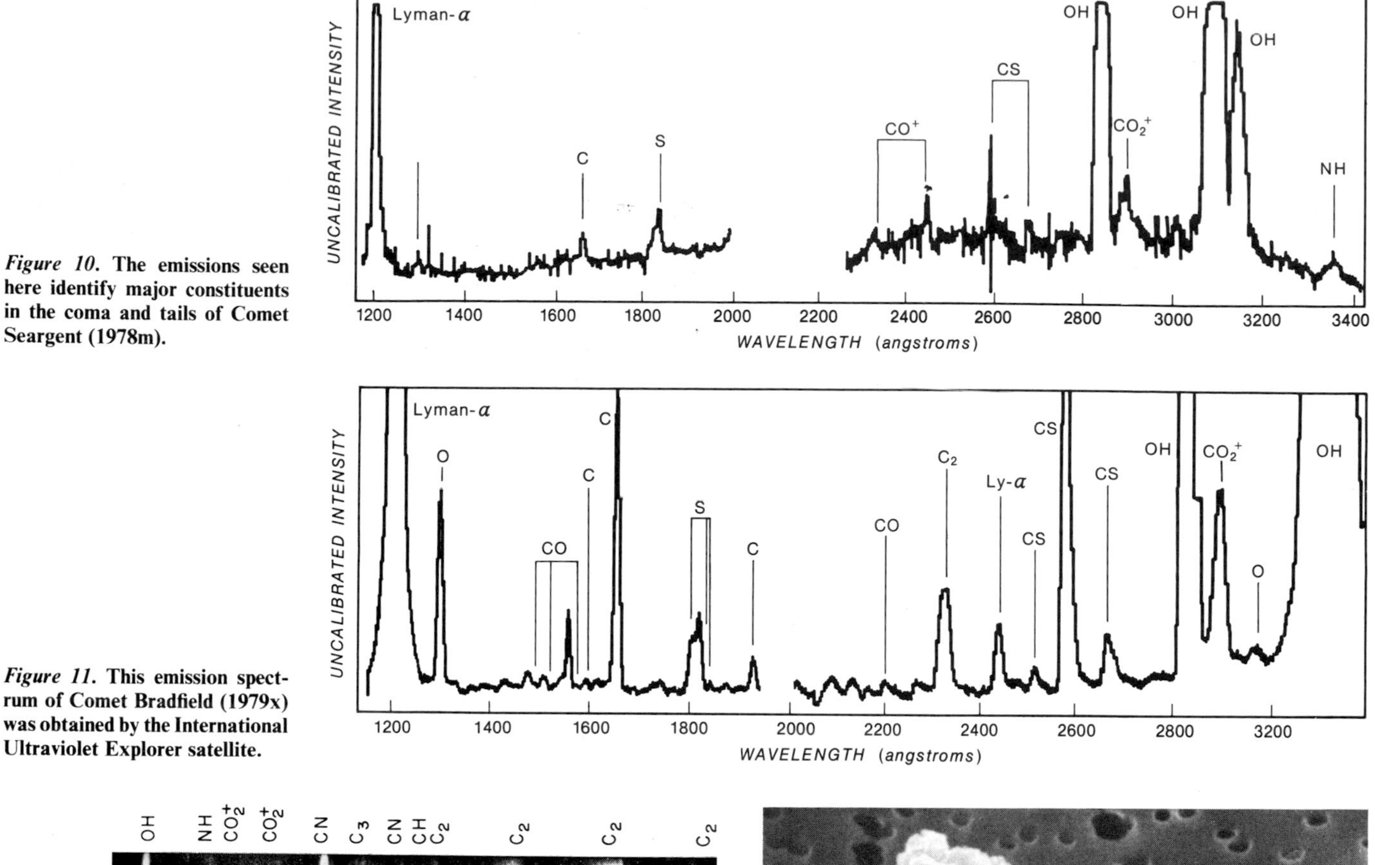

Figure 10. **The emissions seen here identify major constituents in the coma and tails of Comet Seargent (1978m).**

Figure 11. **This emission spectrum of Comet Bradfield (1979x) was obtained by the International Ultraviolet Explorer satellite.**

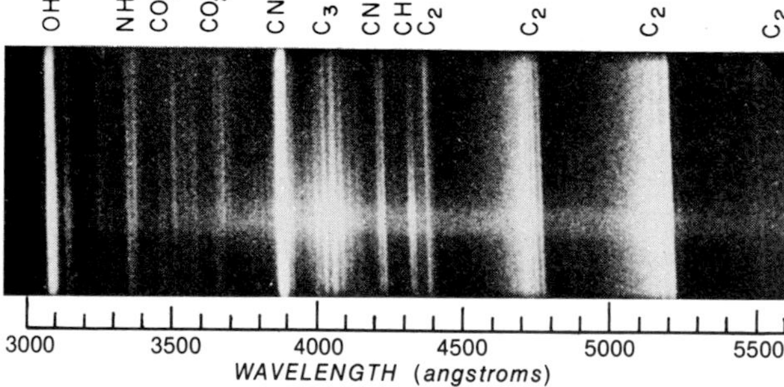

Figure 12. **A visual spectrum from Comet Bradfield (1979x) exhibits most of the major emission lines seen in comets. The bright horizontal band corresponds to the comet's nucleus, where the emissions are generally much stronger. The top edge of the spectrum marks a point in the tail 43,500 km from the nucleus.**

heating the nucleus. But the physical situation changes as the comet gets closer, because eventually its surface layers become warm enough to trigger the sublimation of ices. Then almost all solar radiation goes into maintaining that conversion process. As the ices vaporize, a dusty crust forms that insulates the deeper layers and regulates the sublimation process (now occurring a few centimeters below the surface). Irregularities in the materials cause sublimation to occur faster in some areas, a situation that can produce jets and ultimately the irregular shape and surface features of the nucleus.

Historically, the masses of comets have been estimated very simply by assuming a size and a density (usually 1 g/cm³). However, some indirect determinations of the mass of Halley's nucleus, when taken with its larger-than-expected dimensions, imply an overall density of about 0.25 g/cm³. As

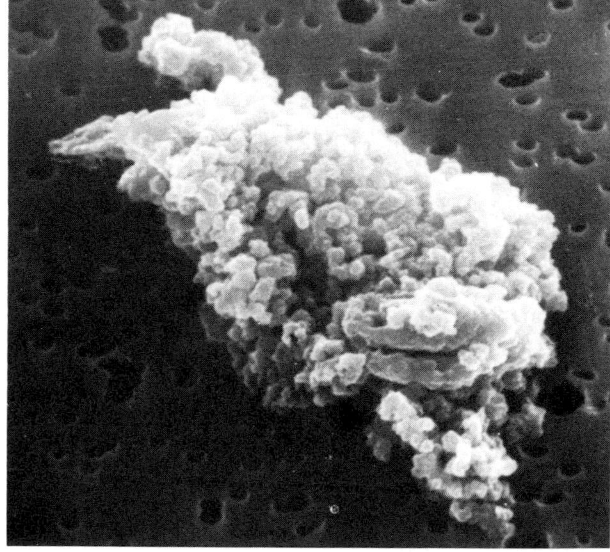

Figure 13. **This tiny aggregate of cosmic dust, termed a Brownlee particle, is actually a speck of interplanetary debris swept up by the Earth. It was collected by a special research airplane flying through the stratosphere at an altitude of about 18 km. The particle's tiny spheres may have once had ice between them, and the entire assembly – only a few microns long – may have been ejected from a comet's nucleus.**

such, the nucleus would be a porous, perhaps fragile structure. However, this low density is based on somewhat uncertain data and should be accepted only with some reservation.

The fact that comas appear when comets are near 3 AU is most *consistent* with water ice being the principal constituent of the nucleus, at least in its outer layers. Spectroscopic

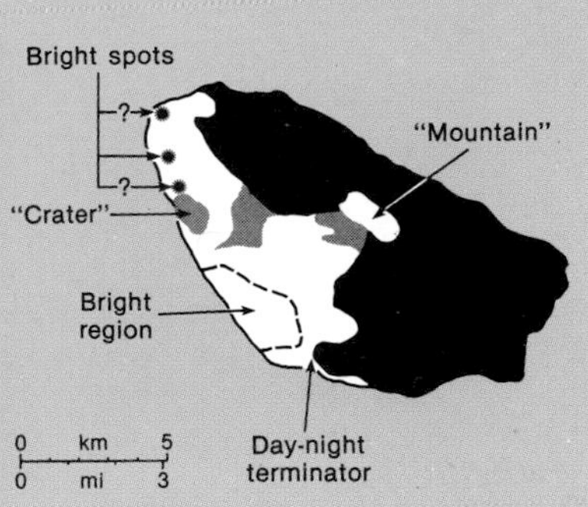

Figure 14. **The nucleus of Halley's Comet, as seen in a composite of 60 images from the Giotto spacecraft. Resolution varies from 800 m at lower right to 80 m at the base of the jet at upper left. The Sun is at the left, and material in the bright jets streams sunward. Features labeled in the accompanying key chart are not necessarily "craters" or "mountains" but are simply called such names for convenience.**

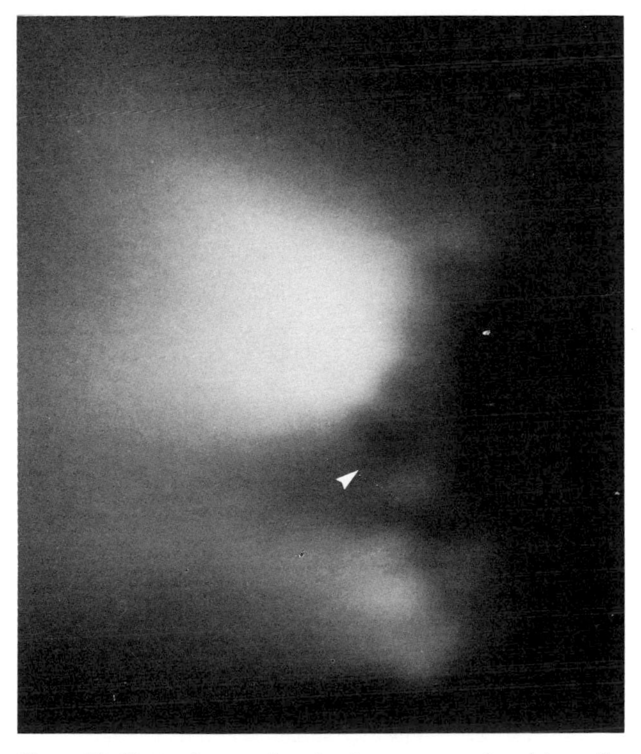

Figure 15. **Giotto close-up view showing an area referred to as the "crater," though its actual origin is unknown.**

observations and direct measurements have confirmed the predominance of "water," meaning both H_2O and water-derived molecules like OH, H_3O^+, H_2O^+, and OH^+. Many emissions from molecules such as CN and C_2 are also observed in spectra when the coma first appears. This makes sense if the ice occurs as a what is termed a clathrate hydrate, in which minor constituents are trapped in cavities within the water-ice crystal lattice. Thus, the sublimation of the ice also controls the release and escape of all substances from the nucleus.

Some of these minor constituents can be important in observations. CN and C_2, noted above, are often the first emissions observed as a comet approaches the Sun. Plasma tails are bright in the blue spectral bands of CO^+ (whose source may be CO in the clathrate lattice or the by-product of a more complex molecule). Polymerized formaldehyde may explain why the surface of Halley's Comet is so extremely dark. Formaldehyde (H_2CO) is a likely original constituent, and a coating of polymerized formaldehyde on the surface (either produced long ago or more recently by exposure to ultraviolet light) should be very dark. It could also be that the nucleus' surface, while intrinsically dark, is made to appear more so by the multiple reflections (and absorption) of light in a porous crust.

If this overall view is correct, then the outer crust of a comet near the orbit of Earth should have a temperature of about 300° K (as was observed for Halley), and its sublimating ice should be about 215° K. Curiously, there are areas on our planet with ice fields at this temperature. One example is Plateau Station in Antarctica (latitude 78° S, longitude 41° E, altitude 3,624 m, ice thickness 3,165 m). Such areas may provide the opportunity for testing instruments designed to probe inside the nucleus on some future mission to a comet.

The dominance of water ice in cometary nuclei may not apply to comets approaching the inner solar system for the first time or for the overall composition of comets. First, let us consider the new comets. The water-ice lattice can store other atoms, as mentioned, but no more than 17 percent of the number of molecules forming the lattice itself. If other substances (like carbon dioxide) exceed this value, their sublimation becomes controlled by their own thermodynamic properties rather than those of water. Since CO_2 and most other plausible minor constituents sublimate at lower temperatures than water does, their presence in fresh or new comets could cause a "turn on" quite far from the Sun and thus account for the abnormally bright comets observed at heliocentric distances of more than 3 AU.

Second, consider the overall composition of comets. Clearly, water is an important constituent, but is it always near the 80 percent value measured for Halley? Probably not. Comets that pass close to the Sun show a much lower ratio of H_2O:CO_2 than Halley's high value of 80:3½. Qualitatively, this is easy to explain. When the outer dust crust of a comet is

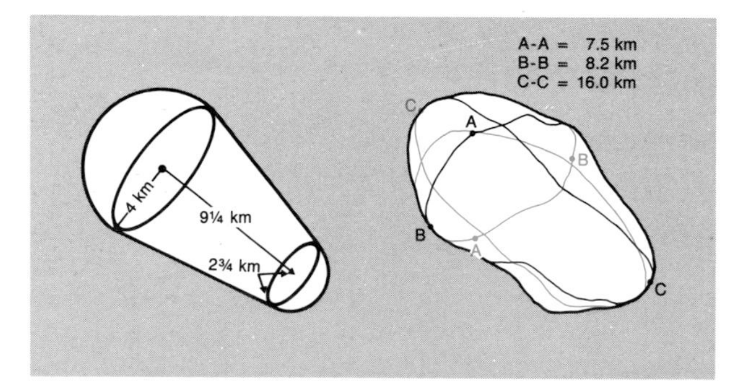

Figure 16. **What is the actual shape of the nucleus of Halley's Comet? Is it *(a)*, a cone with hemispherical ends and the dimensions shown? This simple figure has symmetry around the long axis. Or is it the more realistic shape in *(b)*, with all its geometrical axes unequal?**

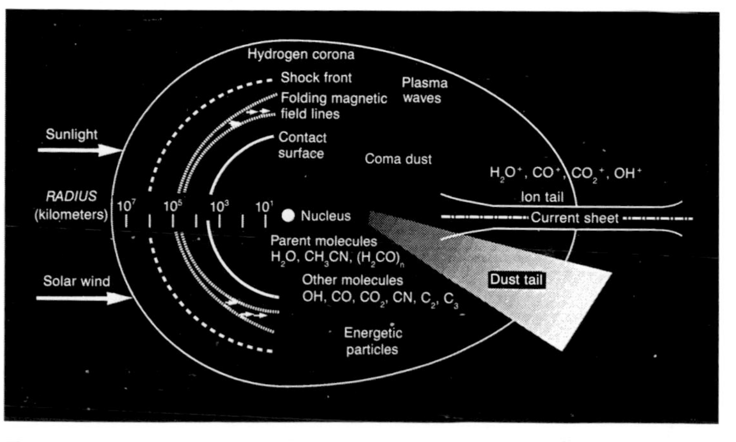

Figure 17. **This schematic drawing shows the principal features of a typical comet. The plot uses a logarithmic scale, so that both the nucleus (about 10 km across) and the huge hydrogen cloud (millions of times larger) can be included.**

heated, a wave of thermal energy travels inward, successively raising the temperature of deeper and deeper layers. For comets passing close to the Sun, the higher temperatures penetrate farther inward than normal and cause some of the deep-lying pristine ices to sublimate. Thus, we should seriously consider the possibility that the composition measurements made at Halley, a comet that has passed through the inner solar system many times, may not be representative of the pristine ices in comets.

SOLAR-WIND INTERACTIONS

The neutral molecules continuously produced by the sublimation of ices flow away from the nucleus in a manner physically similar to the flow of the solar wind away from the Sun. They drag some dust particles away with them, and this dusty gas forms the coma. The relatively simple compounds observed in cometary spectra may not be the ones initially released (so-called "parent" molecules), because the gases are dense enough near the nucleus that chemical reactions may occur among them. Gas flowing away from the nucleus ultimately interacts with the solar wind, the importance of which was demonstrated by Ludwig Biermann in the early 1950s (see Chapter 3). Actually, the existence of a solar wind was then unknown; it was inferred by Biermann from observations of plasma tails. Then, in 1957, Hannes Alfvén discovered that the magnetic field carried along by the solar wind plays a vital role in cometary interactions. Sunlight ionizes some of the molecules in the tail, and these ions become trapped on the magnetic field lines. This causes the field, now burdened with more mass, to decelerate in the vicinity of the comet. The field lines with the trapped plasma wrap around the nucleus like a folding umbrella (recall Figure 7), forming the current sheet and the plasma tail. In this scenario, the plasma tail is normally connected to the region near the nucleus by this "captured" magnetic field. These phenomena can be photographed because trapped molecular ions like CO^+ serve as visual tracers of the field lines.

Disconnection events (DE's) occur when the plasma tail's attachment to the region near the nucleus is disrupted. This can happen three general ways. First, a drop in ion production could weaken the ionosphere and allow the field lines to slip around the nuclear region. Second, a great increase in solar-wind pressure could force the entire ionosphere (and the field lines embedded within it) off the nuclear region, or it may simply compress the ionosphere until slippage is possible. Finally, magnetic field lines of opposite polarity could touch and become reconnected, either on the sunward side of the comet or in the tail across the current sheet (Figure 19). The segment of the plasma tail lying beyond the reconnection point would thus be severed and drift away. Results from the Halley apparition seem to indicate that DE's occur when the polarity of the solar-wind magnetic field changes (at what is called a sector boundary), a situation that apparently disrupts the connection to the near-nuclear region by reconnection on the sunward side. The dramatic DE that occurred during the week of the spacecraft encounters (see Figure 8) is entirely consistent with this mechanism. However, the situation is complex, and more analysis will be required to settle the issue.

The shock fronts around comets Halley and Giacobini-Zinner arise because the comets are obstacles in the solar wind. The solar-wind speed is higher than any wave speed in the plasma, such as the speed of sound. Thus, the bow shock lowers the wind's speed and allows it to flow smoothly around the comet.

Solid particles jetted free of the nucleus have varying fates. The smallest ones, only about 1 micron across, are blown in the antisolar direction by the Sun's radiation pressure and form the dust tail. Somewhat larger particles, not as strongly affected by radiation pressure, continue to orbit the Sun. The sunlight they reflect is seen by us as a faint glow in the night sky called the *zodiacal light.*

The largest solid particles would probably be produced when a comet has spent a long time near the Sun. This occurs if a comet travels in a short-period orbit (for example, Comet Encke has a period of only 3.3 years) or even if it makes many perihelion passages in a long-period orbit. For the latter case, a comet loses roughly 1 percent of its mass on each round, though for the largest comets, this value may be only 0.1 percent. For a typical comet with a diameter of 2 km, a passage through perihelion causes the loss of its outer layers to a depth of about 3 m. Sooner or later, all the ices sublimate, leaving behind a mass of fluffy fragments. If the comet initially had a rocky core, the end result becomes a member of an extinct-comet class of asteroids. In fact, dynamicists estimate that roughly one-third to one-half of all asteroids that cross or approach Earth's orbit are extinct comets (see Chapter 21).

Gravitational perturbation disperses the fragments along the comet's orbit, and some of these remnants enter the Earth's upper atmosphere, producing meteor showers. Most of the debris responsible for meteors is believed to be light and fluffy (as in Figure 13). Fragments liberated from the postulated core would be denser, but no meteorite in our possession has characteristics that appear "cometlike."

A summary of the physical processes believed to be important in comets is given in Figure 20. The basic model of the nucleus originated by Whipple and our understanding of the solar-wind interaction developed by Biermann and Alfvén have been severely tested by the space missions. As our knowledge has expanded, these concepts have remained sound.

THE ORIGIN OF COMETS

Here theory attempts to answer the question, "Where do these icy bodies originate?" Much of the information on this subject comes from the study of their orbits, an area of cometary science not really addressed by the space missions. Detailed orbital information is available for over 750 individual comets. Of this total, approximately 150 have periods of less than 200 years and are classified as "short-period" comets. These have mostly direct (prograde) orbits with inclinations to the plane of the ecliptic of 30° or less. Most short-period comets have aphelia (where they are farthest from the Sun) near the orbit of Jupiter and probably assumed their present orbits after numerous gravitational interactions with that planet. Short-period comets were thought to derive from among the population of long-period comets, those with periods exceeding 200 years but this view is probably incorrect. Currently, the inner edge of the "inner cloud" (discussed below) is considered the most likely source.

The 600 long-period comets are relatively unaffected by gravitational interactions with the planets, and their orbital planes are oriented approximately at random with respect to the plane of the Earth's orbit. Thus, there are essentially as many retrograde-moving comets as direct ones. In addition, very careful examination of the orbits they occupied before entering the inner solar system discloses that none came in

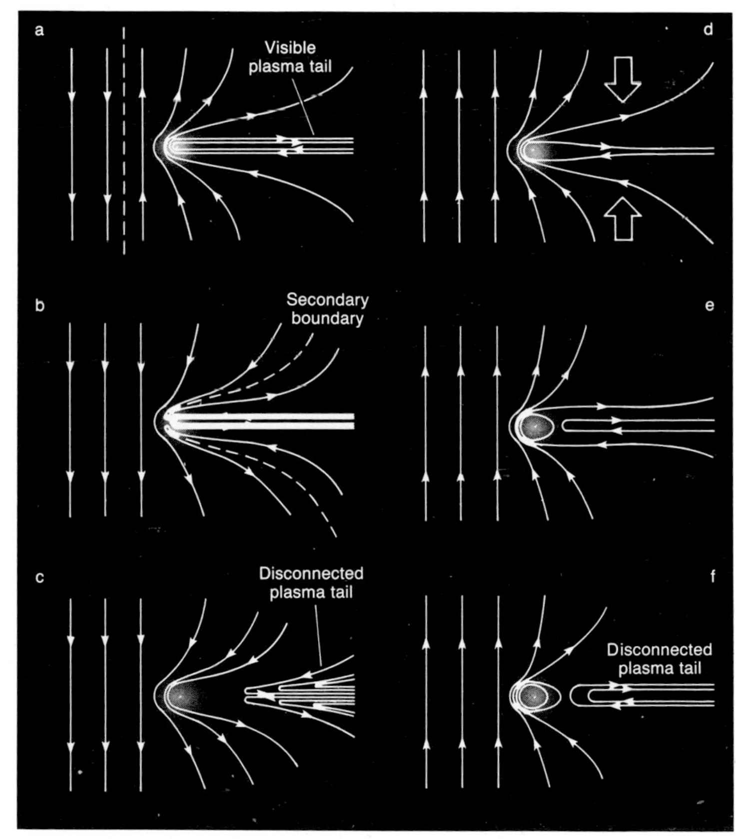

Figure 19. **Shown here are two of several possible theories that explain disconnection events in comets' plasma tails. In 1978, Malcolm Niedner and the author proposed that they result from sector boundaries (reversals) in the solar wind's magnetic field. The field lines compress and wrap around the comet *(a)* as they encounter the ionized coma. At the point of reversal (arrows denote field direction), the field lines "pinch off" ahead of the nucleus and slide around it *(b)*, carrying the tail away *(c)*. While studying Halley's Comet in 1986, a team of Japanese scientists suggested that disconnections can occur without the comet crossing a magnetic-field reversal, a concept that had been proposed in the past. As with the Niedner-Brandt theory, field lines compress and wrap around the inner coma *(d)*. Turbulence forces the lines to pinch together *behind* the comet *(e)*, where they carry the tail away *(f)*.**

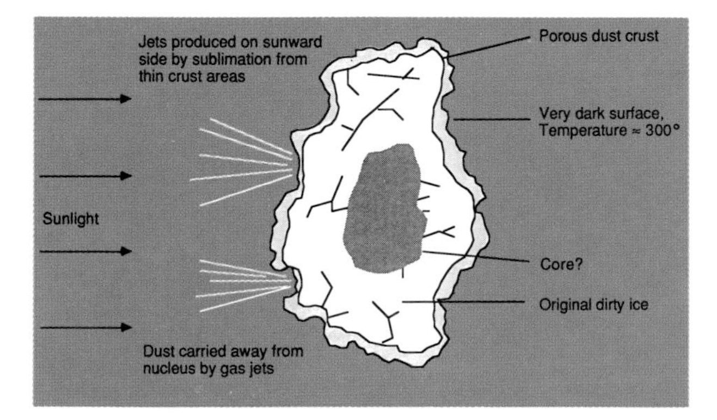

Figure 18. **Fred Whipple's "icy conglomerate" model, as extended by Armand Delsemme, and spacecraft observations of Halley's nucleus are the basis for this rendering of a cometary nucleus.**

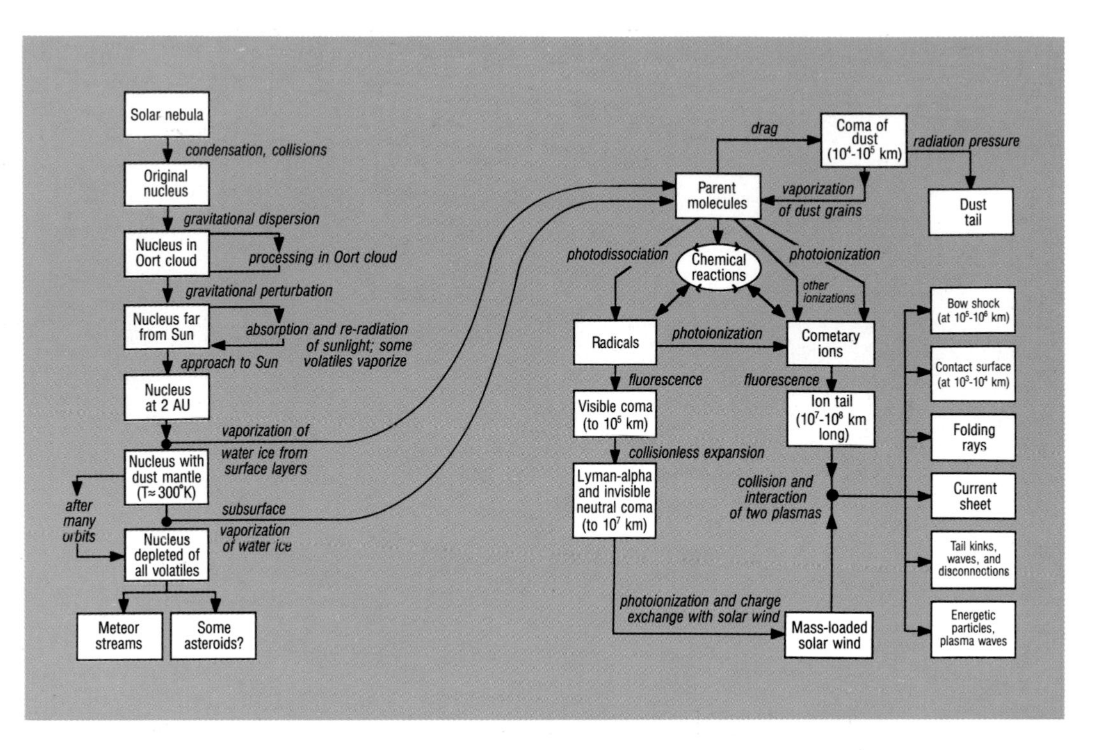

Figure 20. **Features and processes involved in a comet's interaction with sunlight and the solar wind are shown schematically. A comet making a single pass through the inner solar system should evolve to a "nucleus with dust mantle" (box at left center) and exhibit the phenomena in the boxes at right. If captured into a short-period orbit, the comet ultimately evolves to the end states at bottom of the left column. However, many comets never reach these end states, because whenever one chances to pass near a massive planet, gravitational perturbation may eject it from the solar system altogether.**

with hyperbolic trajectories – in other words, none of them were initially in interstellar space. These facts strongly imply that comets are gravitationally bound to the Sun like other members of the solar system and that they likely formed early in its evolution.

In 1950, the Dutch astronomer Jan Oort made a careful study of the statistics of comet orbits. These objects all seemed to originate very far (but not *infinitely* far) from the Sun. He dramatically concluded that the overwhelming majority of comets reside in an essentially spherical cloud around the Sun with a radius of perhaps 20,000 to 100,000 AU. (By comparison, the nearest stars to the solar system, the Alpha Centauri system, are some 275,000 AU distant.) Oort's proposed comet cloud has never actually been seen, and any guesses about its physical characteristics are bound to be uncertain (Figure 21). But the orbital statistics are very convincing, and the cloud's existence seems inescapable.

According to dynamicist Paul Weissman, the Oort cloud contains an estimated 1 *trillion* comets or more, with a total mass some 25 times that of the Earth. Gravitational disturbances produced by passing stars and interstellar gas clouds have several effects on the cloud's members. Most important for us, the disturbances sometimes rob some comets of their orbital energy, causing them to "fall" inward toward the Sun. Only upon reaching the neighborhood of the Earth do they produce their multifarious phenomena and become observable. In addition, such disturbances tidally limit the size of the cloud and tend to randomize the orbits.

Determining the origin of comets is an active area in cometary research. At present, the consensus is that they condensed from the solar nebula at approximately the same time that the Sun and planets did. Comets seem to be a natural by-product of the physical processes responsible for the creation of the solar system. Of course, many details of the process are lacking. The difficulty lies in assigning the location of the initial formation zone. They probably did not form in the Oort cloud itself, even though this is the current source of new comets for the inner solar system. Instead, theorists now tend to view the Oort cloud as a steady-state reservoir that loses comets by gravitational disturbances and gains them from a massive "inner cloud" – with up to 10 times more comets and mass – located between the orbit of Neptune and the traditional inner boundary of the cloud. Suggestions that periodic comet showers could result from a faint solar companion, "Nemesis," orbiting in the Oort cloud, or from a tenth planet in the inner cloud, are intriguing but have not gained general acceptance (see Chapter 21).

The traditional view is that objects in the Oort cloud have existed in a deep freeze and undergone little or no change for billions of years. However, recent calculations indicate that collisions among the cloud's members, irradiation by cosmic rays, and heating by passing stars may have caused significant alterations. Thus, comets may be ancient relics, but they are not entirely pristine.

Some theories ascribe additional roles for comets in the evolution of the solar system. They may have been an important source of the atmospheres of the terrestrial planets (see Chapter 8), and, further, supplied the original organic molecules necessary for the development of life on Earth. These viewpoints have gained support from the Halley apparition. The comet's ratio of deuterium to hydrogen (D:H), as measured by the Giotto mass spectrometer, is the same as terrestrial ocean water (within the errors of measurement). If comets did indeed supply a significant fraction of the Earth's volatiles, then by logical extension they could have similarly have been the source of complex organic species like polymerized formaldehyde. These ideas, though unproven, serve to illustrate the breadth of the scientific interest in comets – and some of the excitement, too.

FUTURE MISSIONS TO COMETS

We have learned a great deal about comets in a relatively brief time. So naturally one might question why the intensive study of these objects should continue and, in particular,

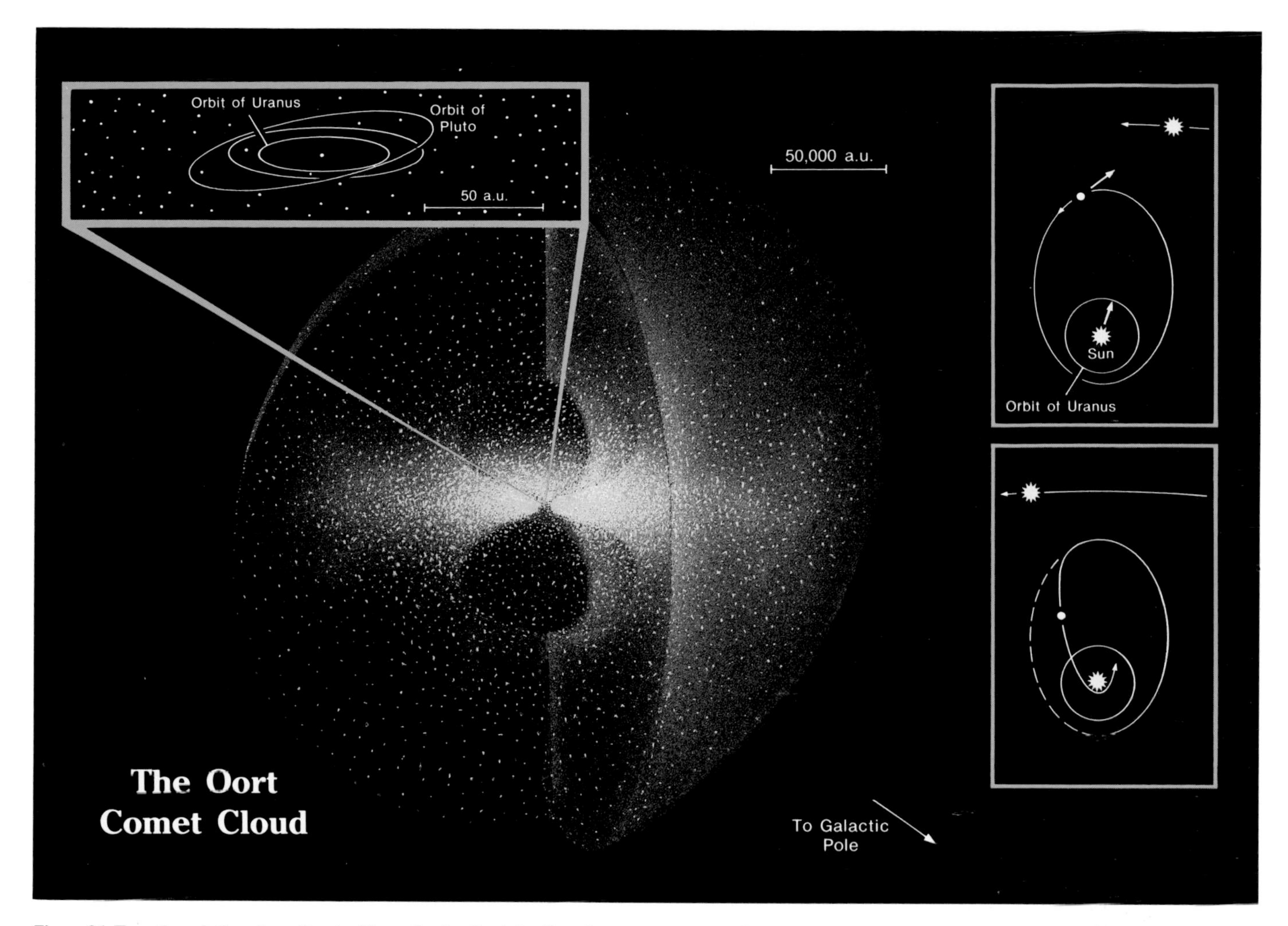

Figure 21. **Even though there is no direct evidence for the Oort cloud's existence, astronomers have pieced together enough theory and circumstantial observations to speculate on its probable structure. The cloud's innermost members (inset at upper left) may actually encroach on the orbits of Uranus, Neptune, and Pluto. However, most of what may be a trillion comets reside some 10,000 to 20,000 AU from the Sun, loosely concentrated near the ecliptic plane. Comets in the outer cloud have become scattered into random orbits due to passing stars and other external influences. At its outermost extent, perhaps 200,000 AU away, the cloud may have an ellisoidal shape induced by the gravitational tug of mass lying in the galactic plane. When a star comes close enough (insets at right), comets in the outer cloud can lose orbital energy and "fall" into the inner solar system. Only then do they become visible to us.**

what the rationale might be for future comet missions. To provide an answer, I wish to emphasize one idea: comets are individuals. The probability that all of them are like Halley is low, so we must everlastingly resist the temptation to build this preconception into our thinking. No matter how complete the theory described in the previous sections might seem, we have so much yet to learn.

Missions to comets fall broadly into three classes: flybys, rendezvous missions, and sample-return missions. Until late 1979, it was believed possible to rendezvous with (not just fly by) Halley's Comet during its recent apparition. But this plan required the rapid development by NASA of a new solar-electric ("ion drive") propulsion system, approval for which was not granted. Alternate plans were developed, but in the end NASA mounted no dedicated mission to Halley's Comet. Fortunately, its exploration *was* carried out by the flyby spacecraft of the European Space Agency (ESA), Soviet Union, and Japan.

The rendezvous phase of cometary exploration could be undertaken by the proposed Comet Rendezvous and Asteroid Flyby (CRAF) mission now under consideration by NASA. This spacecraft would rendezvous with and remain near a comet (to be chosen based upon the launch date) for an extended period of time while the comet passes through its inactive and active phases. An instrumented penetrometer may be fired into the nucleus itself.

Even if the CRAF mission succeeds, we still will have probed only three comets, and only two with respect to the nucleus. The highly individualistic nature of comets warns us that trying to reach general conclusions from such a small sample may be hazardous. To expand our experience, other missions could visit multiple targets or obtain actual samples of cometary material for study back on Earth. NASA, ESA, and Japan are considering such ventures, which would involve slow flybys and collection of a coma sample. Ultimately, of course, we would like to send a spacecraft to land on a comet's nucleus and return samples of the nuclear material to Earth for laboratory analysis. Such a mission, called Rosetta, is being contemplated by ESA for the 21st century.

The last few years have provided dramatic advances in the study of comets. But the progress needs to be sustained by the

Figure 22. **"The Comet is coming!" exclaims a German postcard issued at the time of Comet Halley's appearance in 1910. Considerable public alarm was generated when it was learned that the Earth would pass through the comet's tail. Casual references by some astronomers to "cyanogen in the tail" produced nothing short of panic in some areas. Entrepreneurs took advantage of the hysteria, promoting gas masks and "comet pills."**

Figure 23. **Despite decades of scientific enlightenment, the appearance of Halley's Comet in the 1980s still prompted speculation on its religious and psychic significance – as this humorous magazine cover from February 1986 attests.**

continued collection and archiving of ground-based observations and by sending space missions to additional comets. We can hope that these efforts will continue the halcyon days of comet studies.

Finally, the study of comets is *fun*, as typified by Figures 22 and 23. When Comet Halley returns in 2061, we can expect it to be greeted not only by a greatly advanced understanding of comets, but also by humorous commentary reflecting the times.

18

Asteroids

Clark R. Chapman

THE ASTEROIDS are a multitude of "minor planets" orbiting the Sun at distances ranging from inside the Earth's orbit to beyond Saturn's. They predominate in a large torus (the *main belt*) that has a volume exceeding the sphere of interplanetary space inside the orbit of Mars. While individually small, asteroids are ubiquitous throughout the inner solar system. By the late 1980s roughly 4,000 had been discovered, numbered, and named, a total that grows annually by some 150 to 200. And the existence of the enigmatic object 2060 Chiron, orbiting beyond Saturn, hints that additional asteroid belts may be present in the outer solar system.

Until the past two decades, these bodies were dismissed as uninteresting planetary "dregs" or as "the vermin of the skies." But we now realize they hold important clues concerning the nature of the planetary system and its earliest history. And public curiosity about asteroids, sparked by a hypothesis linking them to the dinosaurs' demise (see Chapter 21) has paralleled a developing appreciation among scientists of their relevance to the evolution of life on Earth and to the dawning age of human exploration and utilization of the planets.

ORIGIN OF ASTEROIDS

Apparently, the asteroids and their precursors were planetesimals just like those growing elsewhere in the solar nebula during planetary accretion. Before they could form a planet, however, these bodies were gravitationally perturbed into tilted, elongated orbits. So instead of slowly accumulating into a single whole, asteroids began to strike each other at speeds of kilometers per second (Figure 1), often resulting in catastrophic fragmentation and disruption rather than coalescence. This process of collisional destruction continues at a much diminished rate today.

What kinds of gravitational perturbations affected planetesimals in the asteroidal zone more so than those in other planetary zones? We cannot be sure, but two plausible scenarios have been advanced. Several scientists have hypothesized that one or more large planetesimals in Jupiter's zone were gravitationally "scattered" by close approaches to that partly formed giant into eccentric orbits that penetrated the asteroidal region. They would have come close to most of the asteroids before once again encountering Jupiter and being ejected from the solar system. Close passes to asteroids may have stirred up their velocities and altered their orbits. Most early asteroids may have been destroyed by either collisions with such Jupiter-scattered planetesimals or collisions with each other.

An alternate idea holds that distant gravitational forces from Jupiter itself stirred up and depleted the asteroids, during the period when the Sun was divesting the primordial solar system of its nebular gases. We see today numerous spaces (the *Kirkwood gaps*) and other lacunae in distributions of asteroidal orbital elements that are due to commensurabilities and resonances with Jupiter (Figure 2). Resonant effects occur at fixed locations today but might have swept through the asteroidal region while the solar system was losing mass early in its history. Such resonant interactions with Jupiter might have pumped up asteroidal velocities. Either way, it seems likely that massive Jupiter was responsible for the absence of an asteroidal planet.

REMNANT PLANETESIMALS

The reservoir of asteroids persists today, having endured these destructive and dissipative forces for billions of years. Small "leakages" do occur, however, creating stray bodies

Figure 1. **Two medium-size asteroids, a stony-iron S type and a dark, carbon-rich C type, destroy each other in a head-on collision hundreds of millions of kilometers from the Sun. Some of the fragments from this cataclysm will reach the Earth as meteorites.**

such as those whose orbits cross and come near the Earth's and the meteorite fragments that fall onto the planets and their moons. Presumably, there were a great many small bodies left wandering throughout the solar system after the planets formed. But those in orbits that made close approaches to planets risked hitting them (the final dribble of accretion) or ejection from the solar system. Only in orbits far from any planet or strong gravitational resonances could remnant bodies have remained until now. The asteroid belt is one such place. Another exists within the orbit of Mercury – if material ever condensed and accreted there in the first place. A few other bodies, such as the Trojans, which orbit ahead of and behind Jupiter in its orbit, are protected in special resonant orbits. Still others (for example, Chiron) may exist in the vast volumes of space between the outer planets.

One final population of remnant planetesimals is the comets, which were ejected from their place of origin (presumably in or beyond the outer solar system) by close planetary encounters. Those that did not quite escape the solar system have been preserved for aeons in the deep-freeze of outer space. Chance perturbations by passing stars and subsequent encounters with the outer planets bring a few comets into the inner solar system. Here their ices sublimate, become ionized, and produce the flashy comas and tails that are their hallmarks (see Chapter 17). But after a cosmically short time, a comet's volatiles are depleted, and it dies. If it contains a solid core, that body is then termed an "asteroid."

As discussed more fully in Chapter 20, we can only speculate about whether asteroids and comets might originally have been the same type of planetesimal. Most asteroids remain in nearly their original orbits, where any surface ices would long since have sublimated away. Prior to entering the inner solar system, "new" comets are better preserved than asteroids, but all trace of their origin has been lost due to their orbital wanderings. Most scientists believe that comets and their remnant cores were formed mainly in the outer solar system, while most asteroids were formed in the asteroid belt, with a few asteroids implanted into the belt from other places in the inner solar system. It is therefore rather ironic that these two groups of remnant planetesimals, with very different histories, are roughly equally represented among those bodies that venture near the Earth.

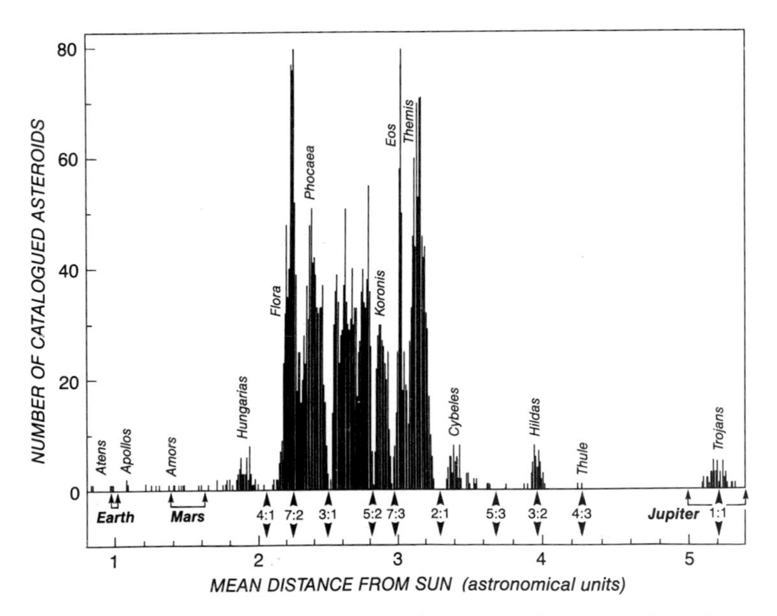

Figure 2. **The population of catalogued (numbered) asteroids is shown for various distances from the Sun. Our census of asteroids is very strongly biased by observational effects: small, bright bodies orbiting near the inner edge of the asteroid belt are favored, while very dark asteroids farther out are underrepresented. Also indicated are locations where an asteroid's orbital period is in a simple ratio with Jupiter's and is subject to a gravitational "resonance" with the giant planet. Inside roughly 3.5 AU, these resonances are the locations of gaps in the asteroids' distribution, called *Kirkwood gaps*. One very apparent gap is located at about 2.5 AU, where an asteroid orbits the Sun three times for every revolution Jupiter makes. Beyond 3.5 AU, for reasons not well understood, resonances correspond not to gaps but to isolated *groups* of asteroids, at ratio values of 3:2, 4:3, and 1:1.**

An object whose orbit approaches or crosses the orbit of the Earth is called, respectively, an *Amor* or *Apollo* asteroid. About 80 Amor-Apollos have been discovered so far, including a few so-called *Atens*, which have semimajor axes of less than 1 AU. Eugene Shoemaker estimates there may be a total of about 1,000 Amor-Apollos that are about 1 km or greater in size and capable of crossing the Earth's orbit, plus another 80 or so Atens.

Although we have understood for a long time how comets die and remain in Earth-crossing orbits, it has been a mystery how to get much material from the main belt into Amor- or Apollo-type orbits. Complicated processes involving resonances, collisions, and perturbations by Mars or Jupiter were investigated, but these fell short of providing the observed flux of small asteroids and meteorites. Recently, theorist Jack Wisdom discovered that the mathematics of chaos, applied to asteroids orbiting near the 3:1 Kirkwood gap (see Figure 2), readily explains how such asteroids may be extricated. He has shown that their orbits undergo dramatic changes at unpredictable times, inducing sudden increases in their eccentricities that permit them to cross the orbit of Mars, and then that of the Earth. Some scientists now believe that most of the commonest meteorites (the ordinary chondrites) are derived from so-called S-type asteroids near the 3:1 Kirkwood gap. But, as we will see, others doubt that any of those asteroids are made of the minerals in the same proportions as ordinary chondrites.

Comets and asteroids still impact the Earth and other planets. Earlier populations of remnant planetesimals from distant places presumably were the last to coat the surfaces of the nearly completely formed planets. They may have contributed preferentially to planetary crusts and hydrospheres. Certainly, their impacts formed craters, which remain the dominant topographic features on all but the most geologically active terrestrial bodies (see Chapter 21). There have been conjectures about whether volatile-rich comets and asteroids might have helped life get started on Earth. But, without doubt, life has been *affected* by these bodies. Mass extinctions (possibly including the dinosaurs') 66 million years ago apparently resulted from impact of an Earth-crossing asteroid: some clays at the boundary of the Cretaceous and Tertiary epochs have been enriched in certain trace elements, especially the rare metal iridium, that could have come only from an extraterrestrial source.

It is uncertain what role asteroids may play in our future. There is a remote but real probability of an impact-triggered disaster in the foreseeable future. A small asteroid's unexpected collision with Earth could be mistaken for a nuclear attack, with frightening consequences. More likely, however, asteroids will serve as humankind's stepping-stones to the planets. Some of the small, Earth-approaching asteroids found to date are easily accessible with present-day rockets. From them, future astronauts may mine materials necessary for living and working in space. Materials expected to be abundant within some asteroids include water, organic compounds, and metals, and it will be far cheaper to mine these from nearby asteroids than to hoist them up from Earth.

It is widely supposed that since asteroids are so small, they have been spared the kinds of processes that have virtually destroyed records of primordial history on the Earth and Moon – the extreme heat, pressure, chemical alteration, and crustal motions that a large planet experiences in response to its vast reservoirs of primordial and radioactive heat. Are the asteroids primitive planetesimals, arrested in their growth and preserved intact for $4\frac{1}{2}$ aeons? Or have their clues about planetary origins been disturbed as well? In short, what are asteroids like and what can we infer about their evolution?

PHYSICAL CHARACTERISTICS

We turn first to the astronomical evidence, gleaned from study of the time-variable behavior and spectral characteristics of reflected and emitted radiation coming from their distant, starlike images. Asteroid sizes are now well established (Figure 3). The largest, 1 Ceres (discovered in 1801), is about 930 km across and constitutes more than a quarter of the mass of all asteroids combined. The next largest, 2 Pallas and 4 Vesta, are each a bit over 500 km in diameter. Still smaller ones are increasingly numerous, grading down to countless kilometer-size asteroids and smaller boulders too small to detect (unless they happen to pass very close to the Earth).

Asteroid shapes and configurations are a matter of active controversy. It had long been supposed that they ranged from roughly spherical to elongated and irregular, bespeaking a fragmental origin. This was inferred from variations in asteroids' brightness as they spin. Rotational periods often fall near roughly 9 hours, with extremes from under 3 hours to many days. Recent evidence suggests that a few asteroids – especially some of those previously believed to be the most elongated bodies – may in fact be double or multiple bodies (Figure 4). In some instances, asteroidal satellites were "detected" by the occultation of a star as an asteroid passed between Earth and the star; unfortunately, such observations are either poorly confirmed or unconfirmed. Several asteroidal light curves are explained better in terms of eclipsing or "contact" binaries rather than by elongated single bodies. Undeniable evidence for contact binary asteroids may exist in recent radar data obtained for several asteroids, including 1627 Ivar (Figure 5) and 216 Kleopatra. A few asteroids with very long rotation periods, like 288 Glauke, which takes 48 days, may have had their spins slowed by the tidal drag of as-yet-unseen satellites. Direct optical searches, using coronagraphic (masking) techniques, have hunted for satellites around selected asteroids. These have been unsuccessful to date, but more searches are planned. Hence, it remains to be determined what fraction of asteroids are binary or multiple systems.

COMPOSITIONS

Sixty years ago at Lick Observatory, Nicholas Bobrovnikoff made an important discovery about asteroids: they differ in color. Yet the full import of his discovery was not appreciated until the 1970s, when asteroid spectrophotometry blossomed. Spurred by new detector technology, astronomers have recorded spectra of sunlight reflected from asteroids from the ultraviolet out to the mid-infrared, where heat radiated from the warm surface begins to overwhelm the corresponding reflected component of sunlight. Not only do the visible colors of asteroids differ, but many of their spectra exhibit absorption bands that are due to different minerals and hydrated compounds; some examples appear in Figure 6. The diversity of colors and spectra imply different surface compositions. If we could relate these derived mineralogies to the different types of meteorites being studied by cosmochemists (see Chapter 19), then we could tie the meteorites' implications about primordial environments and events to bodies in particular parts of the solar system.

There are two chief types of meteorites. Those whose nonvolatile chemical elements occur in roughly "cosmic" abundances are inferred to be relatively unaltered condensates from the primordial solar nebula. Those highly enriched or depleted in certain elements imply they were created from material greatly modified by processes of "planetary" evolution. One such process, called differentiation, was the melting and physical segregation that occurred early in the Earth's history; metals sank to form a core, and lighter materials floated and cooled to form the crust (Figure 7). Most primordial types of meteorites are called chondrites, while most of the geochemically altered ones occur in the iron, stony-iron, and achondrite classes.

An important conclusion about asteroids is that both the primordial and altered mineralogies are represented in the main-belt population. At least one asteroid, 4 Vesta, has a surface mineralogy similar to the so-called basaltic achondrites – meteorites similar in physical properties and chemistry to the basaltic lava flows common on the surfaces of the Earth and Moon. Many other asteroids are extremely black and show infrared absorption bands indicative of the water- and carbon-rich mineralogy of the most primitive meteorites – the carbonaceous chondrites. Therefore, some asteroids apparently melted and became geochemically differentiated, just like the larger terrestrial planets, while in others the initial chemistry has been preserved more-or-less intact. It remains a profound mystery how some asteroids could have been so altered while others of similar size in nearby orbits could have escaped modification.

The simple fact that some small asteroids melted after they formed establishes important constraints on thermal conditions and processes operating in the early solar system. It had been supposed that the high surface-area-to-volume ratio of small bodies would have allowed any internally or externally generated heat to radiate away quickly, keeping them relatively cool. But evidently some bodies became much warmer than can be ascribed to this traditional view. More exotic thermal sources must be considered, including intense pulses of heat due to the decay of short-lived radionuclides. Aluminum-26, for example, may have been synthesized in a nearby supernova explosion (which itself may have triggered the onset of the solar system's formation). But in order for the heat from its decay (into a stable form of magnesium) to melt an asteroid, aluminum-26 had to be incorporated into an accreting body within only a few million years of its synthesis. In a way, then, the telescopic detection of "basalt"

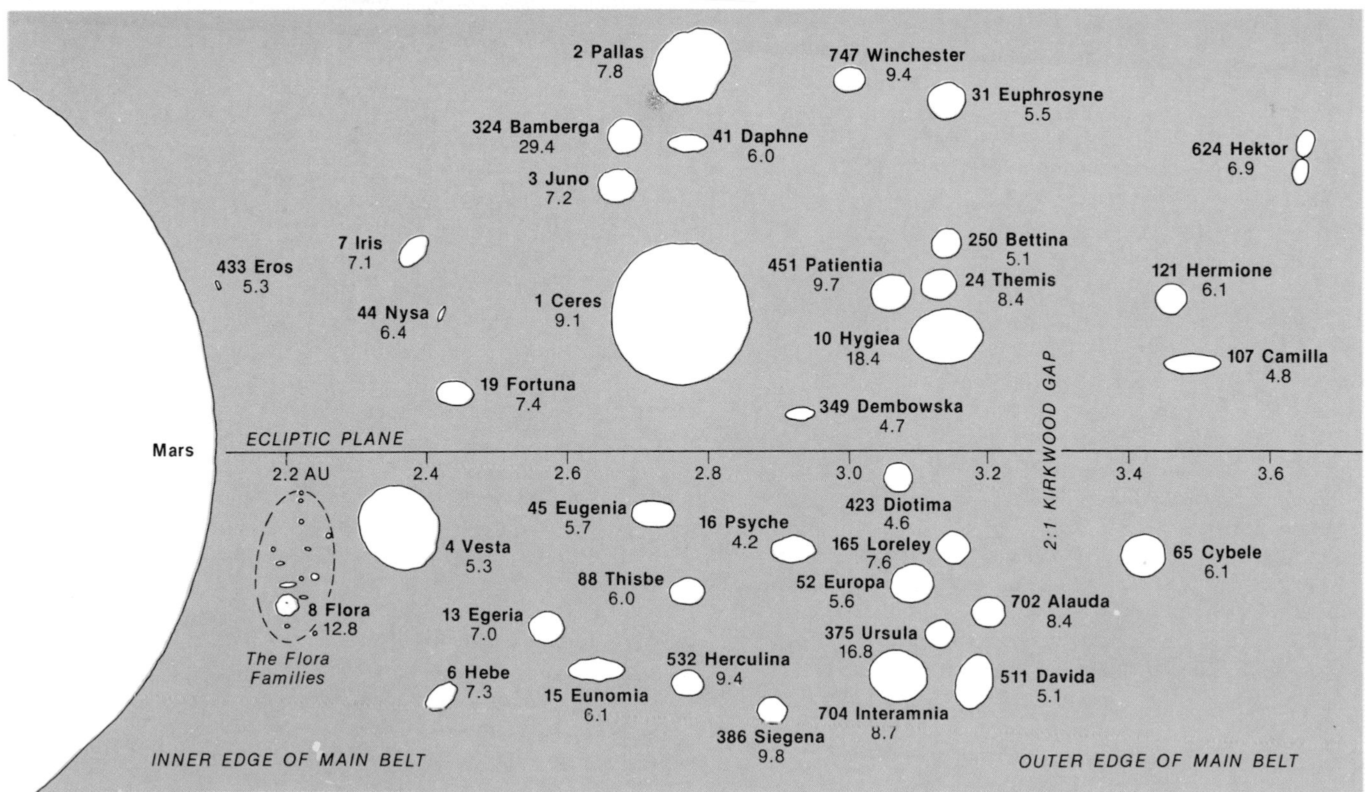

Figure 3. **This representation of the physical properties of interesting asteroids includes most of the asteroids larger than about 200 km in diameter. They are portrayed in their correct relative sizes and shapes (the limb of Mars is shown for comparison); colors and albedos are also indicated. The bodies are positioned at their correct relative distances from the Sun. Asteroids located near the top or bottom of the diagram occupy relatively eccentric or inclined orbits (or both), while those shown near the ecliptic plane move in relatively circular, noninclined orbits. Rotation periods, in hours, are given in the lower panel. Among the special smaller asteroids indicated are members of the Flora families larger than 15 km in diameter, but this illustration would be hopelessly cluttered if all asteroids of comparable size were shown – an estimated 1,150 asteroids in the main belt alone have diameters larger than 30 km, yet only five Flora family asteroids attain that size. Note the contact-binary Trojan asteroid 624 Hektor.**

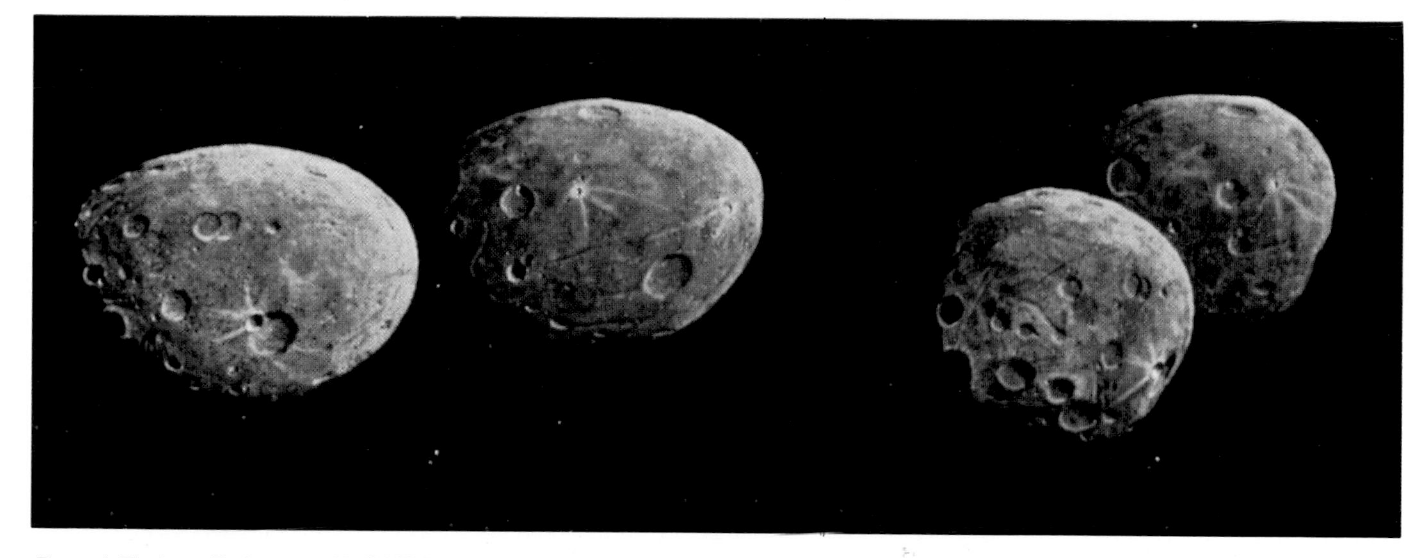

Figure 4. **The large Trojan asteroid 624 Hektor is a very elongated body that may, in fact, be a tidally deformed contact binary. Two views of a possible configuration for Hektor are shown; 216 Kleopatra may have a similar appearance.**

absorption bands in Vesta's spectrum may be thought of as helping to establish the time scale for the formation of the solar system.

Such ground-based determinations of asteroids' colors and albedos, augmented by data from the Infrared Astronomical Satellite, have been used to group the asteroids into taxonomic classes. Members of each class probably have similar, though not identical, surface mineralogy (Table 1).

More than three-fourths of the asteroids are extremely dark (with typical geometric albedos of 3.5 percent) and, so far as we can tell, have mineralogies analogous to carbonaceous chondrites. There are significant spectral variations among these so-called *C-type* asteroids. About two-thirds of them appear hydrated, probably indicating that they were heated to the point that water soaked their minerals, as is commonly noted in carbonaceous meteorites. Two classes of low-albedo asteroids show reddish colors; such *P-* and *D-type* spectra are common for bodies near and beyond the outer edge of the main belt, including many Trojans and some of Jupiter's small outer satellites (see Chapter 20). These colors may be due to "ultraprimitive" organic compounds, perhaps like the non-icy components of comets.

Roughly one-sixth of the asteroids have moderate albedos (typically 16 percent) and reddish colors. The spectra of such *S-type* asteroids imply assemblages of iron- and magnesium-bearing silicates (pyroxene and olivine) mixed with pure metallic nickel-iron. Unfortunately, these spectral characteristics are shared by two radically different types of metal-bearing stony meteorites: (1) the ordinary chondrites, relatively unaltered primitive objects believed to have formed closer to the Sun than the carbonaceous chondrites; and (2) the stony-iron meteorites, which are enriched in metal and other compounds due to extensive melting and geochemical fractionation within their precursor (or "parent") body. The most straightforward interpretation of S-type reflectance spectra, especially for a handful of well-observed objects like 8 Flora, suggests that their metal content is higher than that in ordinary chondrites, that their olivine content is also a bit

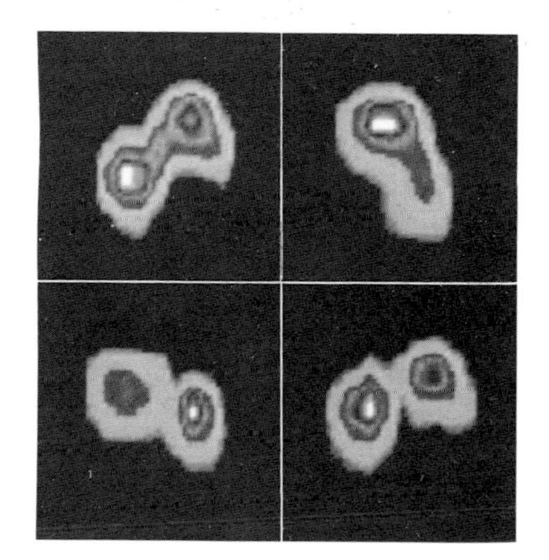

Figure 5. **These are the best "pictures" ever taken of what may be a small, double asteroid designated 1989 PB. They are actually maps of the echo delays and frequency shifts introduced when radar pulses were bounced off the asteroid. The halves could be just touching and rotating around each other every 4 hours, or 1989 PB may simply be one object with a strongly dumbbell shape. These data were obtained at Arecibo Observatory in Puerto Rico, when the asteroid passed within a few million kilometers of Earth in August 1989.**

higher, and that other mineralogical inconsistencies exist. On the other hand, the degree to which "space weathering" might be responsible for these relatively minor spectral differences is unclear – as mentioned earlier, there are other good reasons to suspect that many S-types are ordinary chondrites after all.

A third type of asteroid spectrum is also ambiguous. So-called *M-type* asteroids have moderate albedos. Their spectra exhibit the signature of metallic nickel-iron, with no hint of silicate absorption bands. These may well be wholly metallic asteroids, perhaps the remnant cores of differentiated precursor bodies stripped of their rocky mantles and crusts by collisions. But one type of stony meteorite, the enstatite chondrites, mimics the spectral signature of nickel-iron. These primitive meteorites formed in a highly reducing (oxygen-poor) environment, perhaps close to the Sun or deep within a protoplanet. They consist of grains of nickel-iron embedded in a clear matrix of the magnesium-rich silicate mineral enstatite. Since enstatite is colorless and lacks absorption bands, an enstatite-chondritic asteroid would have a spectrum just like that of a nickel-iron asteroid. However, radar soundings of asteroids seem to be resolving this ambiguity. The radar data have already shown that the

Type	*Albedo*	*Reflectance spectrum*	*Meteorite analog(s)*
C	0.03–0.07	Fairly flat longward of 0.4μ; UV and sometimes 3μ absorption bands	Carbonaceous chondrites (CM)
B	0.04–0.08	C-like, but slightly brighter and more neutral in color	Carbonaceous chondrites (?)
F	0.03–0.06	Flat (neutral color) with no ultraviolet absorption	?
G	0.05–0.09	C-like, but brighter and with very strong ultraviolet absorption	Carbonaceous chondrites (?)
P	0.02–0.06	Linear and slightly reddish (like M, but with very low albedo)	None
D	0.02–0.05	Redder than P's, especially longward of 0.6μ; very low albedo	None (kerogens?)
T	0.04–0.11	Reddish, esp. at shorter wavelengths; intermediate between D and S	?
S	0.10–0.22	Reddish short of 0.7μ; weak to moderate absorptions near 1μ and 2μ	Stony-irons or ord. chondrites
M	0.10–0.18	Linear and slightly reddish (like P, but with moderate albedo)	Irons; enstatite chondrites
E	0.25–0.60	Linear, flat or slightly reddish	Aubrites
(A)	0.13–0.40	Strong absorptions in UV and near 1.1μ due to olivine	Chassignites (Brachina)
(Q)	moderate	Like S, but with stronger absorptions	Ord. chondrites
(R)	mod. high	Like S, but with stronger absorptions (particularly due to olivine)	LL chondrites?
(V)	mod. high	Like S, but with stronger absorptions (particularly due to pyroxene)	Basaltic achondrites

***Table 1.* Astronomers find that asteroids exhibit a number of characteristics that can be used to subdivide them into taxonomic classes. The most important of these are the shape and slope of their reflectance spectra (the listed albedos are typical but do not define the classes). Designations in parentheses correspond to very rare types that are represented by only one or at most a few asteroids. The symbol μ stands for microns.**

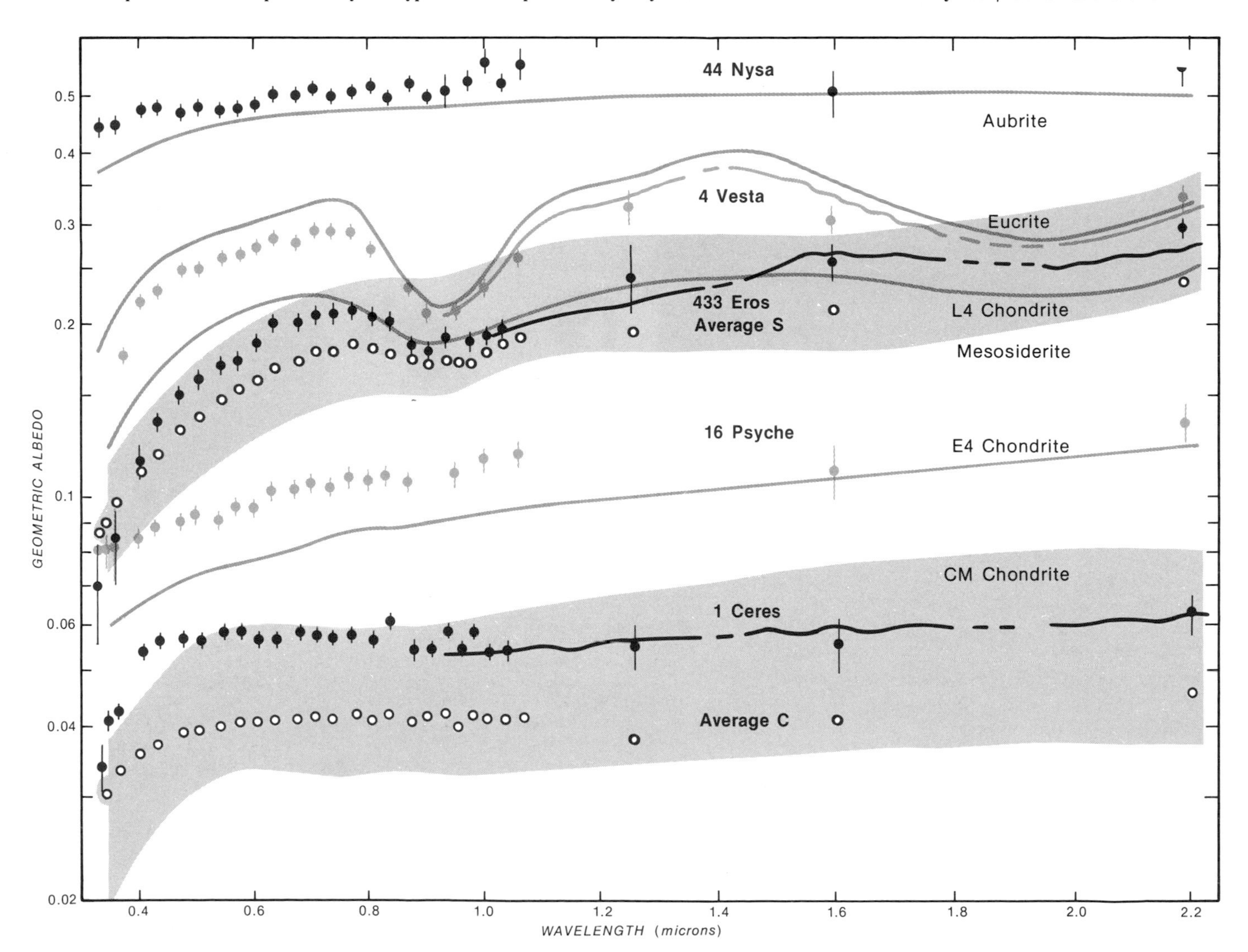

***Figure 6.* Reflectance spectra of asteroids and meteorites are compared for visible through infrared wavelengths. Asteroid data, shown in shades of red, consist of points with error bars (from filter spectrophotometry), lines (from Fourier spectroscopy), and open circles (average values for the S- and C-type asteroids). Laboratory measurements of meteorite powders are shown in black; two classes occupy the ranges of values indicated by gray bands. Investigators deduce surface mineralogy primarily from the shapes of these curves rather than from the objects' precise albedos: S-type asteroids appear to have spectra intermediate between the mesosiderites and L4-type ordinary chondrites; 433 Eros is more like an ordinary chondrite than the typical S-type asteroid is. Evidently, the diverse mineral assemblages found in our meteorite collections are also represented in the asteroid belt.**

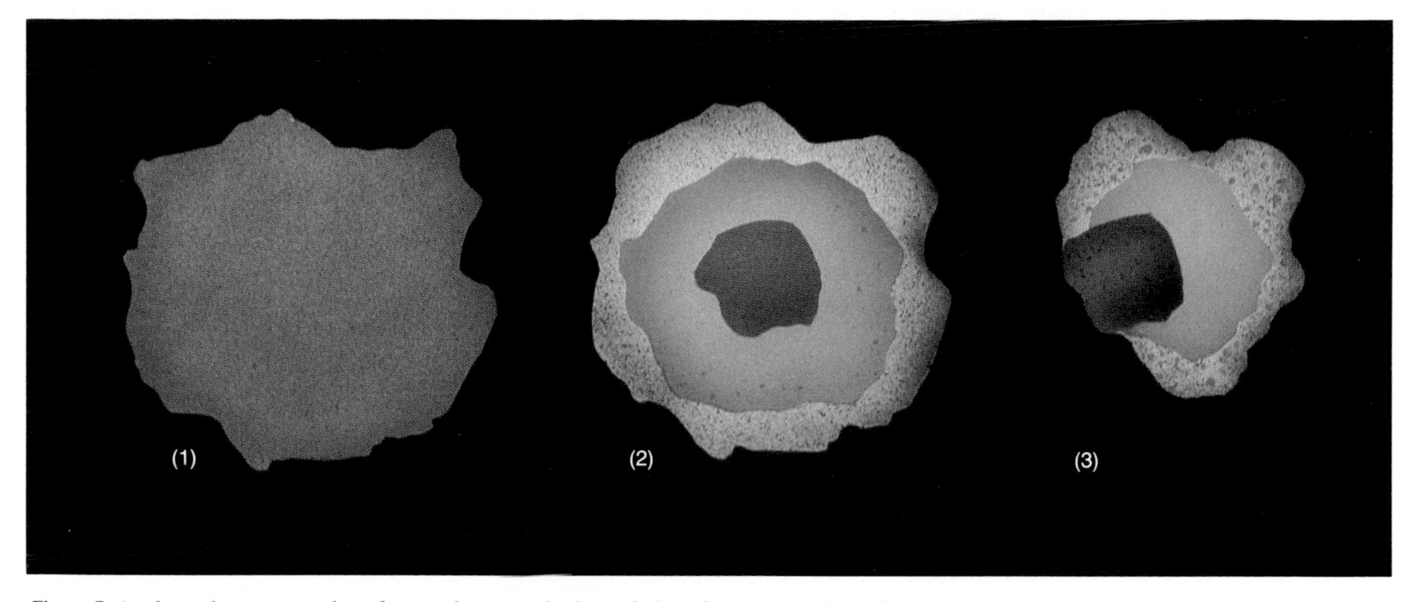

Figure 7. **A schematic representation of successive stages in the evolution of an asteroid that is heated early in its history. The original body of primitive composition *(1)* is heated to the point that constituent iron separates and sinks to its center, forming a core *(2)*. Partially melted rock from the mantle floats upward through cracks in the crust, erupting onto the surface as basaltic lava flows. As heat radiates away, the body cools, the iron solidifies, olivine accumulates in the deep interior, and crustal magmas solidify. Repeated collisions fragment the mantle and crustal rocks into a "megaregolith" and ultimately eject the rocks, exposing the iron core (and any embedded rocks) to space *(3)*. Most asteroids were not heated much beyond the first stage; 4 Vesta reached stage *2*, but was not fragmented thereafter. Some M- or S-type asteroids may be the parent bodies *(3)* for iron and stony-iron meteorites.**

M-type object 16 Psyche plus two small Earth-approaching M-types are unusually metal-rich.

In addition to the common types described above, observers have identified a number of rarer low-albedo asteroid classes. Also, several percent of asteroids are oddballs of one sort or another, like the olivine-rich body 349 Dembowska and basalt-covered 4 Vesta.

Only within the last two decades have researchers realized that the compositional distinctions among asteroids show a remarkable distribution with distance from the Sun (Figure 8). While examples of most types span a large range of solar distances, there is a clear progression from high-albedo E's to S's in the inner belt, then to M's and C's in the mid and outer belt; P's occur mostly near and beyond the belt's outer edge, and D's dominate among the Trojans. It is generally thought that this gradation with heliocentric distance reflects primordial properties of the solar nebula, with high-temperature minerals preferentially condensing near the inner planets and cometlike assemblages of ices and organics toward the outer solar system. But, conceivably, the variation could reflect subsequent evolutionary processes.

The preceding sections describe nearly everything known about what asteroids are like and what they are made of. Of course, we can analyze theoretically the processes that have been affecting asteroids during the life of the solar system and speculate on the implied geologic morphologies of asteroids. However, until a spacecraft visits an asteroid, no one can know what surprises await us. Since such a multi-asteroid rendezvous mission or even a simple, single flyby (Figure 9) will not happen until 1991 at the earliest, let me proceed with such speculative theory.

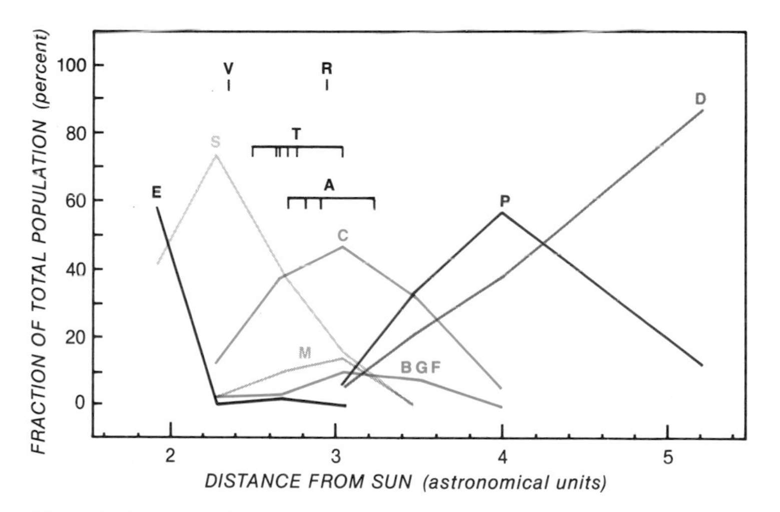

Figure 8. **A census of asteroids of different compositions shows that they are systematically distributed with respect to distance from the Sun. Raw data from a survey made in eight colors (and from other observational programs) have been corrected for the survey biases against dark asteroids in order to derive the approximate fraction of asteroids above a certain size (determined by observational limitations and varying with distance) within each class. The letter designations, which refer to different spectral types, are summarized in Table 1.**

COLLISIONAL EVOLUTION

Planets have radioactively driven internal "heat engines" that keep some of them actively evolving aeons after they formed. Yet it is difficult to imagine how this mechanism can be relevant to the comparatively tiny asteroids. In all probability, asteroids are cold, dead, airless bodies whose destinies are shaped solely by external forces. Their orbits may occasionally be perturbed by gravitational interactions with distant planets. Their surfaces are certainly bombarded by micrometeorites, solar-wind particles, and cosmic rays. And while their surface temperatures rise and fall in rapid day-night cycles as they spin, the feeble light of the distant Sun never warms them much above 200° K. By far the most important effects on asteroids, however, are due to those rare events when they encounter one another while hurtling through space.

Despite the tiny sizes of asteroids compared with the immensity of the torus of space through which they revolve about the Sun, there are enough asteroids traveling sufficiently fast that major collisions are inevitable during the life of the solar system for all but a lucky few asteroids. The typical collision velocity is about 5 km per second. Laboratory simulations suggest that for the largest impacts a typical asteroid might experience, the total energy is much more than sufficient to fracture and fragment an object having the material strength of rocks. The only bodies that might be expected to survive such collisions more-or-less unscathed are (1) those with a cohesive strength exceeding that of iron and (2) the very largest asteroids, whose rocks are strengthened under gravitational compression.

In "super-catastrophic" collisions, most of the fragments achieve the escape velocity of the target body and are lost to space. But unless an asteroid is quite small, lesser collisions can probably fragment it without destroying it. After such marginal collisions, the target object's gravity may keep most fragments bound in orbit around the center of mass, and they will probably coalesce back into a single body again (Figures 10,11). Or if a rapidly spinning asteroid were hit off-center, the combined angular momentum might be too great for a single body to reform, and a binary or multiple system would result.

A super-catastrophic collision provides sufficient energy and momentum to disperse the fragments into independent, but still similar, heliocentric orbits. These would rarely or never meet again, and the asteroid population would have gained a family of smaller asteroids at the expense of the larger target body (the projectile, by comparison, is most likely negligibly small). Several such families of asteroids in similar orbits were discovered by the Japanese astronomer Kiyotsugu Hirayama about 70 years ago and are now called *Hirayama families*. Many smaller asteroids have been discovered since, and more than 100 families have been tabulated. Some dust bands discovered by the Infrared Astronomical Satellite appear to be associated with major families (Figures 12,13).

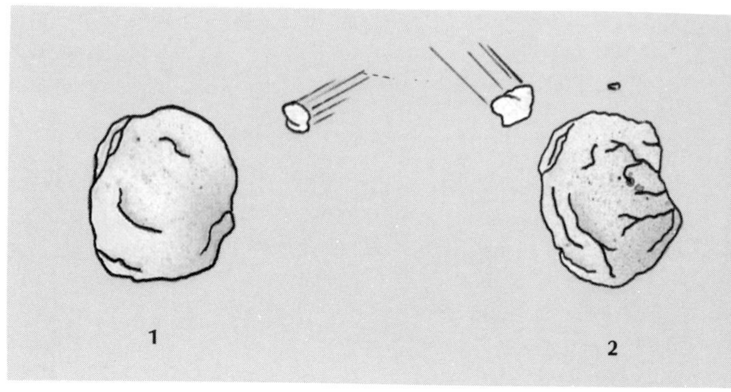

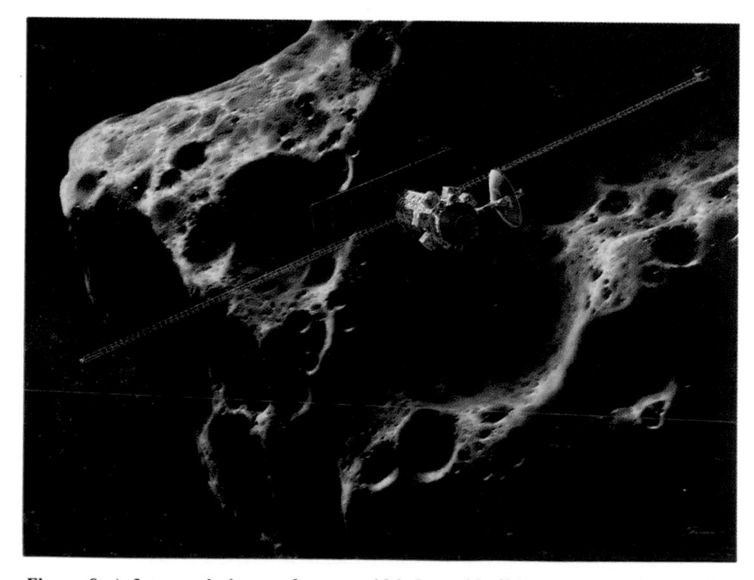

Figure 9. **A future mission to the asteroid belt could allow a spacecraft to visit a number of objects over several years. Multiple asteroids or those thought to be rich in metals would rate high as candidates for such a mission. "Flybys of opportunity," using spacecraft en route to the outer solar system, are also possible; two such encounters are planned in the early 1990s during NASA's Galileo mission to Jupiter.**

Study of the spectra of members of a single Hirayama family could help us learn about the interior composition of its now-shattered precursor asteroid, provided that observed families are, in fact, the by-products of collision rather than chance clusterings of asteroids in similar orbits due to some subtle dynamical process. Some families, including the populous Themis, Eos, and Koronis families, have members with similar spectra, which implies that their precursors were homogeneous throughout. But others exhibit a variety of spectral types, which are rarely easy to "put back together" into a single precursor body that makes physical and geochemical sense. This raises questions about (1) the reality of the smaller families as collisional by-products, (2) the mineralogical interpretations for some of the spectral types, or (3) the geophysical and cosmochemical models for asteroids and meteorite parent bodies.

Our theoretical models now allow us to estimate confidently the frequencies and energies with which asteroids of different sizes collide with each other. But we do not know how the energy and momentum are partitioned into fracturing the material, heating it up, ejecting fragments into space, and so on. The problem is that we lack evidence from laboratory experiments or terrestrial explosions even remotely approaching the magnitude of asteroidal collisions. So theory must be combined with risky extrapolations from laboratory-scale experiments to infer the outcomes of asteroidal collisions. This tenuous marriage has yielded a number of interesting – if uncertain – conclusions about the collisional evolution of asteroids. The correct story almost certainly lies somewhere between the following two scenarios.

It may be that asteroids have been broken up into generations of successively smaller fragments, gradually grinding themselves down to meteoroids and eventually to dust that is swept out of the asteroid belt, out of the solar system, or into the Sun. In this case, which assumes efficient conversion of impact energy into ejecta velocities, the present-day asteroids might represent a small remnant of a much larger earlier population. The larger asteroids still existing would be considered the lucky few that have by chance escaped destruction.

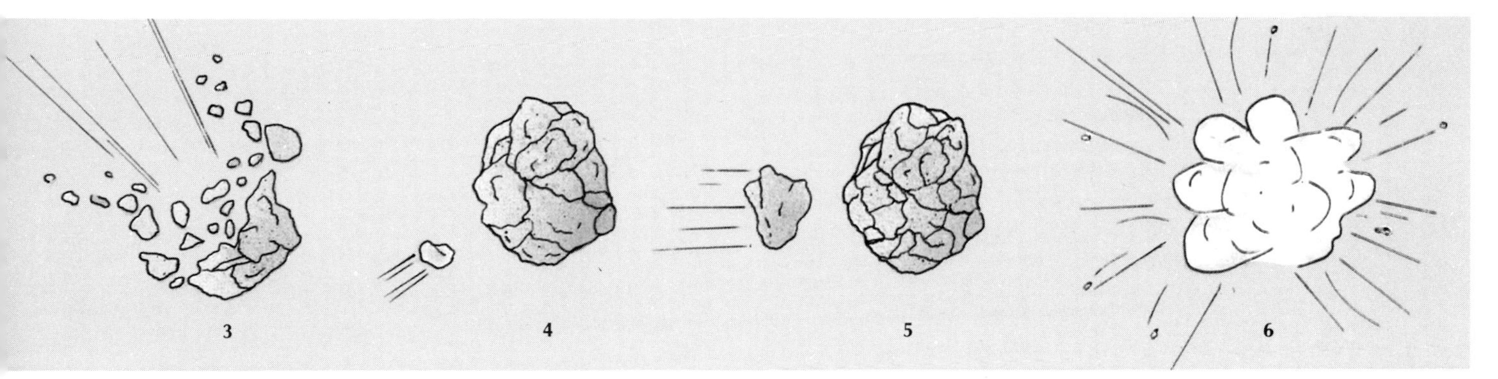

Figure 10. **Stages in the fragmentation history of a moderately large asteroid. Originally composed of strong rock *(1)*, the asteroid is cratered *(1,2)*, then catastrophically fragmented by a more energetic impact *(3)*. Most of the ejecta fail to reach escape velocity, and the body reassembles *(4)*. Later impacts further fragment the body, converting it to a gravitationally bound pile of boulders. Finally *(5,6)*, a sufficiently gigantic collision occurs to completely disrupt and destroy the asteroid; its remnants then become scattered through space.**

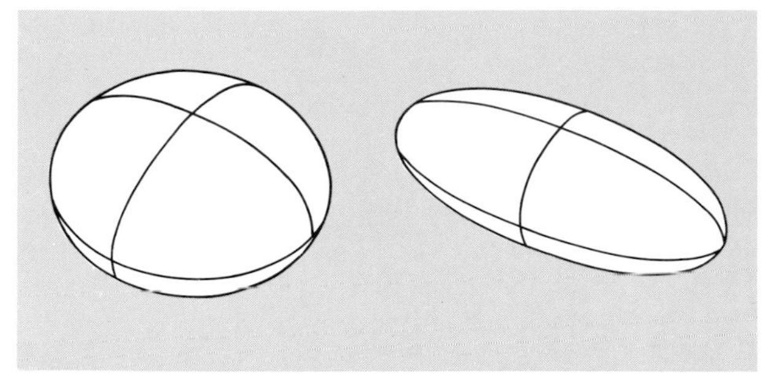

Figure 11. **Asteroids may be structurally weakened by collisional fragmentation. Stuart Weidenschilling and his colleagues think such bodies may form conglomerations that are quasi-stable when pieces reassemble following particularly large impacts. If they are spinning rapidly, they might adopt the flattened, oblate shape of a Maclaurin spheroid (left). If collisions add more angular momentum to the system, the equilibrium shape is a Jacobi ellipsoid (right). Still more angular momentum would make the conglomerate body unstable, perhaps resulting in a contact binary (like 624 Hektor; see Figure 4), a more separated binary, or a multiple configuration.**

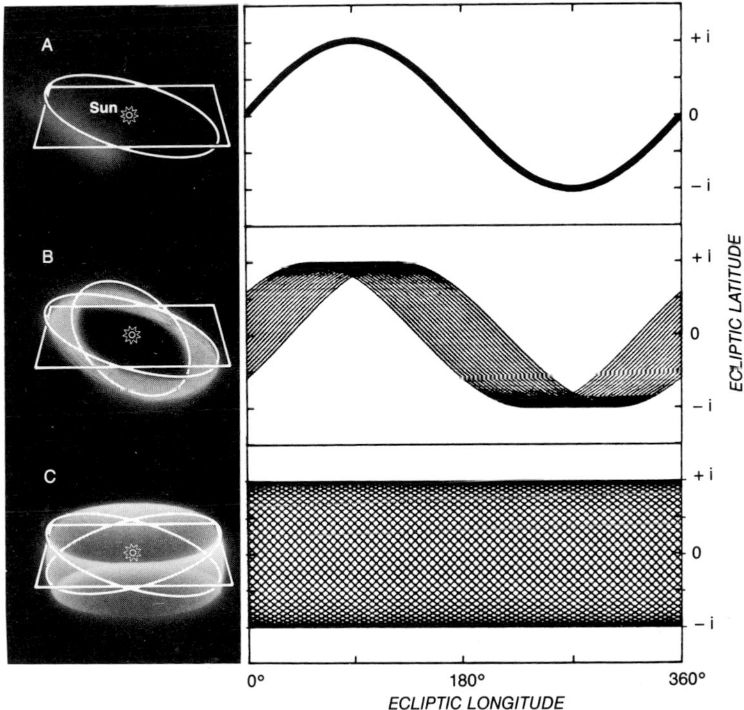

Figure 12. **Mark Sykes and Richard Greenberg propose that faintly visible bands of dust can be created in the sky after two good-size asteroids have collided. Small differences in the orbits of the resulting fragments will spread the material into a ring within 100 to 1,000 years *(A)*. Over time, gravitational perturbations induced by Jupiter should spread the particles' orbits around the ecliptic *(B)*. Within a million years of the initial collision, particles have been smeared into a large torus or doughnut *(C)*. Because pieces of the dusty debris are seldom actually near the ecliptic plane, they give the appearance of two separate bands on either side of the ecliptic.**

The second alternative (if ejecta velocities are generally low), is that the number of asteroids larger than 50 km in diameter hasn't changed much over the aeons. But individual asteroids have been damaged repeatedly, as depicted schematically in Figure 10. When such asteroids are smashed to bits by repeated collisions, their fragments often reaccumulate into a gravitationally bound collection of rubble. Some of these asteroids will be spun up by repeated collisions and may become binary or multiple systems following an energetic off-center impact.

Some scientists now prefer the latter scenario, since it seems to explain aspects of the distribution of asteroid sizes we observe today. But it means that asteroidal materials may have been badly shocked and rearranged during repeated impacts, making it more difficult to read the evidence about primordial events. Also, how could one very large asteroidal precursor – the parent body of the large, metallic asteroid 16 Psyche – have been stripped to its core, while the similar-size object 4 Vesta has apparently preserved its thin basaltic crust intact? Although the collisional environment should have been the same for both, Psyche's exposed core implies torrential bombardment rates, whereas Vesta's crust has endured little damage over the presumably 4.5 billion years since it solidified. There are other difficulties with collisional evolution that we do not yet understand. For example, if many asteroids were melted and geochemically differentiated, much of their now-shattered interiors should have been pure olivine – yet only a handful of pure olivine asteroids (the A's) are known. Where did it all go? If the

Figure 13. **Tenuous traces of asteroidal dust, here color-coded blue, were discovered in 1983 by the Infrared Astronomical Satellite. The broad bands straddle the ecliptic plane and correspond to doughnut-shaped clouds of dust particles associated with the Koronis and Themis asteroid "families" in the outer asteroid belt (see Figure 12). According to theorists, the dust bands are transient features that result either from the violent collision of two asteroids or from the gradual erosion among a family's members. The red (cold) emission at upper right is from distant interstellar clouds, and the thin line (arrowed) below the dust band is debris lying along the orbit of the periodic comet Tempel 2. North is up and east to the left.**

answer is "the olivine in the S-types," then where are the parent bodies for the less olivine-rich ordinary chondrites?

The asteroids continue to collide with each other today. Smaller impacts crater and crack their surfaces, which are gradually covered over again by regional or even global ejecta blankets from the larger cratering events. Infrequent large impacts destroy smaller asteroids or reassemble the configurations of larger ones. Even rarer super-catastrophic collisions yield new Hirayama families; if some of these fragments are sprayed into resonant orbits, they can be quickly perturbed into elliptical paths that cross the orbits of other planets. Such small asteroids and meteoroids lead a transitory existence of only about 10 million years before they strike a planet or become gravitationally yanked into a radically different trajectory.

This collisional and dynamical evolution is just the tail end of the accretionary processes that gave rise to the terrestrial planets earlier in the solar system's history. As some of these long-protected remnant planetesimal fragments finally enter our atmosphere to crash down as meteorites, they produce a spectacular flash in the nighttime sky before coming to rest on the Earth's surface. Ultimately a few of them are found and dissected in laboratories for clues about the earliest history of the solar system. Very rarely, whole kilometer-size asteroids strike the Earth itself, as one evidently did 66 million years ago. It has been estimated that the chance of a civilization-destroying impact occurring during our lifetime may be as high as one in a few thousand. That sobering statistic makes the study of asteroids of more than just theoretical interest.

19

Meteorites

John A. Wood

SOMETIME AFTER MIDNIGHT on February 8, 1969, the editor of *El Heraldo* in Chihuahua City, Mexico, was called from his office by an excited night watchman, who wanted him to witness an alarming spectacle outside. The night sky was illuminated by a brilliant pulsating blue-white light that moved across the southern horizon, leaving a glowing trail behind it. The object lit the sky so brightly that it was seen in parts of the United States, hundreds of kilometers north of Chihuahua. Soon the editor's phone began to ring. Agitated voices rushed to tell him what they had seen. Some were convinced the end of the world was at hand.

Even greater was the fright of residents 240 km to the south, directly beneath the dazzling phenomenon – which was an exceptionally bright fireball. Moving northeast, the brilliant flare broke into two major pieces. Each of these shortly exploded into a fireworks display of diverging lights. All this time a chain of tremendous detonations (sonic booms) assailed the ears of onlookers. Thousands of stones and pebbles dropped from the dark winter sky.

The first of these was found the next day, dug slightly into the ground, only a few steps from a house in the small village of Pueblito de Allende. Regional newspapers told of its discovery. These reports, together with accounts of the spectacular fireball, brought meteorite hunters to the area: scientists from Mexican and American museums and universities and from the NASA Lunar Receiving Laboratory in Houston (which was gearing up to process the first samples from the Moon in that same year), mineral dealers, and amateur collectors. An untold number of specimens (such as those seen in Figure 1) were found, roughly 2 tons in total. Many stones went undiscovered, of course; probably no more than half of the material that fell was recovered.

The discovery sites of meteorite specimens were distributed in an elliptical *strewnfield* approximately 50 km long and 10 km wide. Large specimens, least decelerated by the atmosphere, were found at the strewnfield's northeast end; the largest of all (110 kg) defined its northeast tip. Smaller and more easily decelerated specimens fell farther to the southwest. This is the classic fall pattern for members of a meteorite shower.

THE ALLENDE METEORITE

All stones from this particular shower are named "Allende" for the locality of first discovery. Other meteorites are similarly named for their recovery sites. Allende has come to be the most intensively studied of all meteorites. This is partly because it is available in such copious quantities, and partly because in 1969 a number of research groups were preparing to study the cherished Apollo rocks, and they welcomed a new extraterrestrial sample on which to practice. But Allende is also an exceptionally interesting meteorite, and a number of very important discoveries have been made about it.

The individual specimens of Allende are coated with thin black layers of slaggy material that melted during atmospheric deceleration. Minerals immediately beneath these *fusion crusts* show no sign of thermal damage, however, because ablation peels surface material away from a meteorite as fast as it becomes hot during flight through the atmosphere.

When broken open as in Figure 2, an Allende stone is found to consist of dense, reasonably hard, dark gray material devoid of cavities or visible porosity. Its substance is

Figure 1. **Specimens of the Allende meteorite found shortly after it fell in February 1969. Schools were let out and children mobilized into search parties that combed the area for meteorites. Manuel Gomez, a local engineer, points to the impression in the ground made by the stone behind it.**

perceptibly heavier (specific gravity: 3.67 g/cm^3) than that of an ordinary terrestrial rock of equivalent size. Close examination reveals that the stone is a conglomeration of small objects, mostly about a millimeter in size, ranging from spherical to highly irregular in shape and from white to dark gray in color (Figure 3). These small stony inclusions are embedded in a matrix of very-fine-grained, dark gray, earthy-looking matter.

An obvious first thing to do with an unknown extraterrestrial sample like this is submit it to chemical analysis. When this was done with Allende, the meteorite exhibited the pattern of chemical-element abundances shown in Figure 4. It is interesting to plot Allende's composition against the spectroscopically derived composition of the Sun's atmosphere: the two are very similar. It is as if a mass of material had been ripped out of the Sun and allowed to cool and condense. Only a few elements depart far from the 45° line of equal concentrations. Hydrogen, carbon, nitrogen, oxygen, and the noble gases are so volatile, or form compounds so volatile, that they are incapable of condensing in the inner solar system. Consequently, we would expect them to be underrepresented in condensed solar material. Lithium, on the other hand, is destroyed by thermonuclear reactions in hot stellar interiors; thus, after 4.5 billion years of attrition in the Sun, this element has been depleted in the solar atmosphere.

Another interesting measurement to make on an exotic stone is to determine its radiometric age, the time that has elapsed since it crystallized and cooled to the point where the isotopic daughters of radioactive decay in it could begin to accumulate in the crystals where their parent nuclides had resided. The amount of argon-40 found in the meteorite corresponds to that which would be generated by 4.57 billion years of decay of the potassium-40 in it. This is, to within the error of the measurement, the accepted age of the solar system.

The great ages and unfractionated chemical compositions of *chondritic* meteorites (of which Allende is an example) are convincing evidence that they are specimens of primitive planetary material, preserved in the same state they assumed when the planets first began to accrete. By contrast, rocks from geologically active planets like the Earth are highly fractionated and have younger ages (less than 500 million years for most Earth rocks).

ALLENDE COMPONENTS AND EVENTS LONG AGO

Most of the excitement Allende has generated stems from the small objects or inclusions, mentioned earlier, with which it is studded. On closer study, each of these is found to be an integral mineral system that must have formed more or less independently of the meteoritic material now surrounding it (Figure 5). It seems clear that the inclusions were dispersed in space when they formed, and subsequently they accreted, together with fine mineral dust, into planetary objects of some type. The dust, packed together, comprises the earthy matrix of the present meteorite.

Many of the inclusions consist of minerals that are rather uncommon on Earth, and that had not previously been recognized as abundant meteoritic phases: melilite, perovskite, hibonite, fassaite, spinel, and others. What these minerals have in common are high concentrations of calcium, aluminum, and titanium relative to the mean abundances of these elements in bulk Allende material. And the significance of Ca, Al, and Ti is that they are the most refractory of the major elements in meteoritic or planetary matter. That is, if meteoritic material were heated to progressively higher temperatures in a gas of solar composition, these would be the last major elements to vaporize; if the hot vapor were cooled, they would condense first.

This intrigued meteoriticists in 1969 because it agreed with some preconceptions held at that time about the formation of planetary matter as a by-product of the Sun's origin. In 1962 theorist Alastair G. W. Cameron developed a model of solar-system formation, with roots extending back to Immanuel

Figure 2. **A large stone from the Allende shower with a 2-cm scale bar. The near end of the stone has been broken off, exposing unaltered meteoritic material. Other surfaces, rounded by ablation during entry into the atmosphere, are coated with a fusion crust and Mexican clay.**

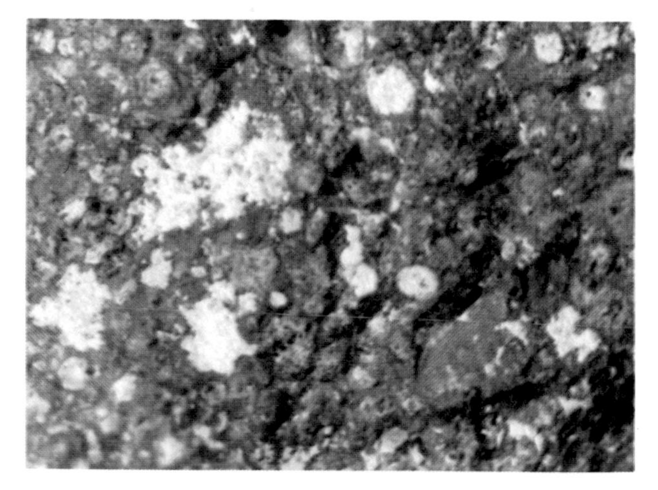

Figure 3. **A closer view of the broken surface on the Allende specimen, showing inclusions and chondrules (one or a few millimeters across) embedded in a dark matrix.**

Kant and Pierre Simon Laplace, wherein the gravitational collapse of a cloud of interstellar gas and dust gave rise both to the proto-Sun and to a rotating disk or nebula (Figure 6), also of solar composition. In Cameron's model compression of the nebular gas as it fell together heated it temporarily, so that in the region of the terrestrial planets temperatures were high enough to completely vaporize the interstellar dust incorporated in the nebula. As the hot nebula cooled by radiation, the first things expected to condense from it would be grains rich in the least volatile elements, namely calcium, aluminum, and titanium – just what was found in the Allende calcium- and aluminum-rich inclusions *(CAI's)*.

The correspondence between theory and observation is actually much more striking than that just indicated. The tool of chemical thermodynamics can be used to predict which minerals would condense under equilibrium conditions, and in what order, in a hypothetical Cameron nebula that cooled from a totally vaporized state. Lawrence Grossman worked this scheme out (while a graduate student) in 1972, and a simplified depiction of the condensation sequence of minerals appears in Figure 7. The minerals at the top of the diagram (highest temperatures) closely approximate the minerals actually found in the Allende CAI's.

In addition to CAI's, Allende contains another class of inclusions. These consist largely of the minerals olivine and pyroxene, which appear lower in the condensation sequence than the CAI minerals. Their principal elements are magnesium, iron, and silicon; their compositions much more nearly reflect elemental abundances in Allende overall (and in the solar atmosphere) than CAI compositions do. Named *chondrules*, they are actually much more abundant in Allende than CAI's, consistent with the higher concentrations of Mg, Fe, and Si than Ca, Al, and Ti in cosmic matter. (Chondrules are conspicuous in chondritic meteorites generally, not just Allende.) The textures of chondrules show that most if not all of them have been melted; they tend to be more regular in shape than CAI's, and some are nearly spherical. An English gentleman scientist, Henry C. Sorby, concluded as early as 1877 that the chondrules in chondrites once must have been dispersed molten droplets.

Could chondrules represent matter that condensed from the cooling solar nebula at lower temperatures than the CAI's, as the "olivine, pyroxene of intermediate iron content" box between 800° and 1,300° K in Figure 7 suggests? Unfortunately this simple concept is inconsistent with the clear textural evidence that chondrules formed as molten droplets. The minerals in chondrules would not have been molten at temperatures as low as 800° to 1,300° K. In the neighborhood of 1,800°, where they *would* have been molten, Grossman's calculations predict the melt could not condense. Consequently, it is thought that chondrules were formed not by condensation, but by the remelting of preexisting condensed matter of appropriate composition, and that the droplets were not molten for long enough to permit them to achieve equilibrium by evaporating away into the nebula. Thus, two substantially different modes of origin seem needed to account for the CAI's and chondrules.

Since the fall of Allende and the publication of Grossman's work in 1972, much has changed in our understanding of chondritic meteorites and the solar nebula (see Chapters 20,23), though it is not clear that we have come much closer to a correct picture of their roles in the origin of the solar system. Cameron and other astrophysicists no longer believe interstellar matter fell together all at once to form the nebula, generating high temperatures by adiabatic compression as it did so. Instead the collapse was a protracted process: gas and dust continued to fall onto the nebula for many thousands of years. During this time the disk evolved in a complex way, as the viscous coupling of fast-revolving gas near the proto-Sun

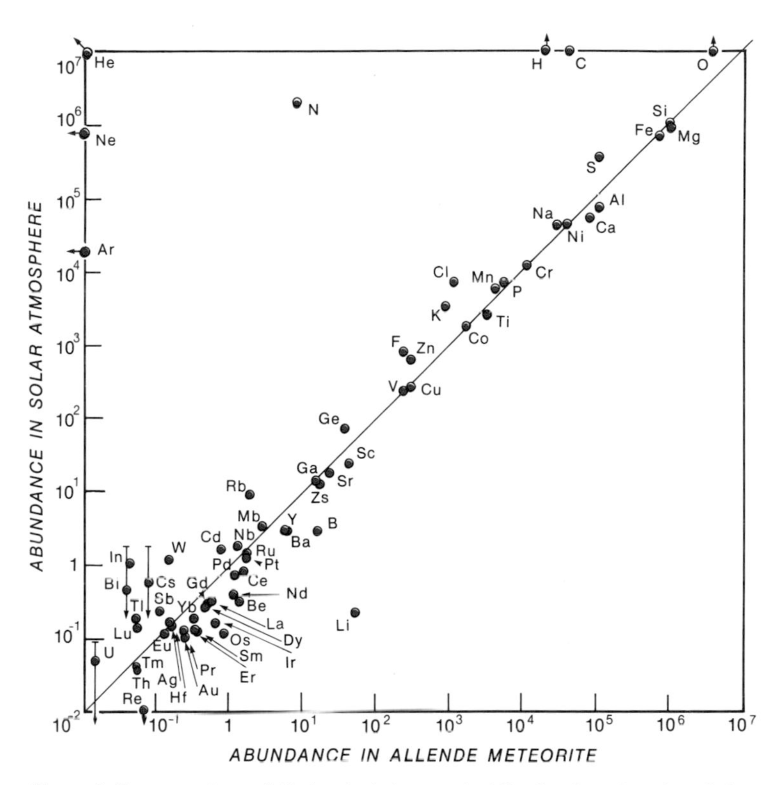

Figure 4. **Concentrations of 69 chemical elements in Allende, plotted against their concentrations in the solar atmosphere (using logarithmic axes). In both cases, abundances are relative to 1 million silicon atoms.**

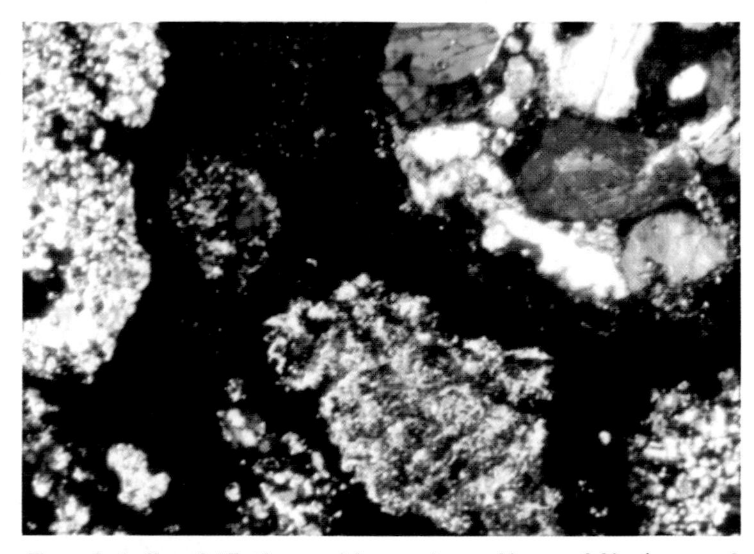

Figure 5. **A slice of Allende material, ground to a thinness of 30 microns and illuminated from behind. Light-colored (relatively transparent) objects are chondrules. Most CAI's (some are indicated) appear darker and finer-grained. Matrix material, which fills in between the chondrules and CAI's, contains opaque minerals that make it appear black. The field of view is 1.5 mm wide. These are not the minerals' true colors but are caused by introducing polarizing filters in the optical train.**

Figure 6. **An impression of the solar system in the process of formation. The infant Sun is believed to have been surrounded by a revolving disk of gas and dust (the solar nebula). Dust in the nebula was transformed by a complex series of processes into planets; the chondrules and CAI's in chondrites are thought to represent intermediate products of these processes.**

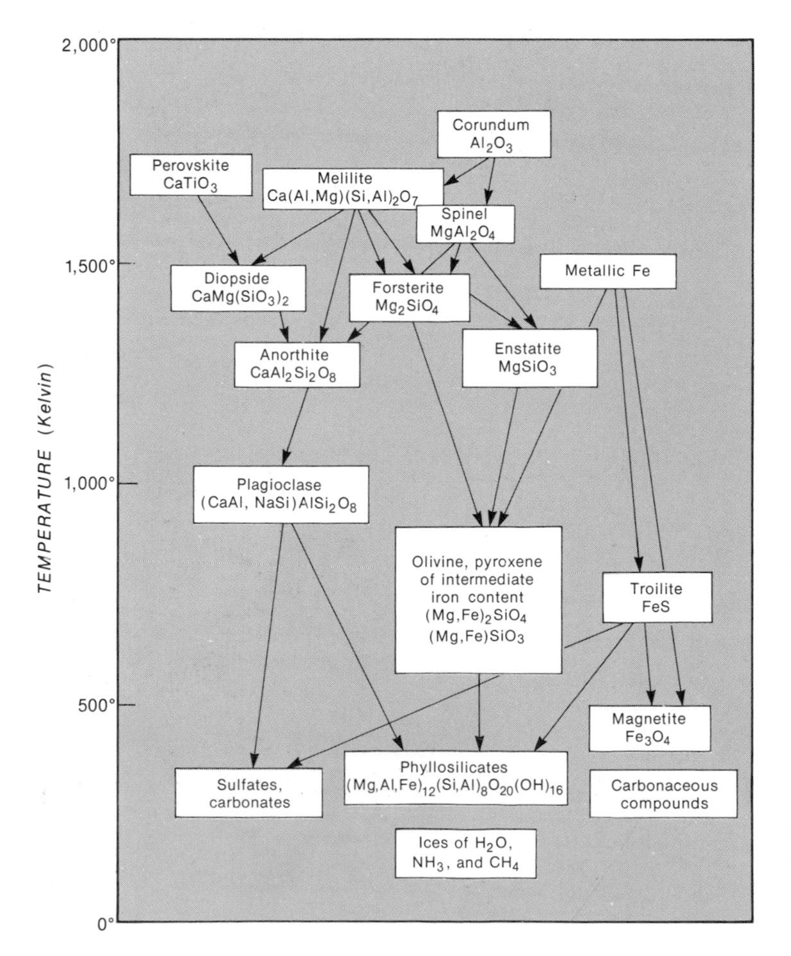

Figure 7. **A simplified depiction of the sequence in which minerals would condense from a cooling gas of solar composition. Arrows signify that continuing reaction with cooling residual gases would transform minerals from the upper boxes into those minerals beneath them. Rapid cooling would prevent complete transformations, accounting for the preservation of the spinel, melilite, perovskite, and so on in Allende CAI's.**

with slower-revolving gas farther out in the disk tended to decelerate the motion of the former and accelerate the latter. The collapsing nebula is no longer thought to have been heated by the compression of its gases. Rather, astrophysicists now attribute nebular heating to friction in the viscous couplings just mentioned, or possibly to compression of the gas as density waves passed through it.

Several properties of chondritic meteorites have emerged that now seem inconsistent with the 1972 picture of a hot nebula that precipitated CAI's and other meteoritic components as it cooled. In laboratory simulations, it has been found that silicate droplets must be cooled at a rate between 100° and 2,000° K per hour to acquire the properties of crystals found in meteoritic chondrules. This rate may seem rather gradual, but in the context of the solar nebula it is very rapid. The scale of the solar nebula was so vast that it could not possibly have lost its heat this fast, nor could droplets embedded within it. Instead, the chondrules must have been heated and melted somehow in small pockets of the nebula that were able to dissipate their heat rapidly to cooler surroundings.

The CAI's in Allende and other chondrites of similar subtype have proven to be a mine of enigmatic discoveries. Some of the chemical elements in CAI's contain unusual proportions of their isotopes, different from the standard isotopic allotments found in the Earth, or in lunar rocks, or in bulk samples of the meteorites themselves. Isotopic anomalies have been observed in the oxygen, magnesium, silicon, calcium, titanium, strontium, barium, samarium, and neodymium of CAI's. Chondrules also contain oxygen with unusual isotopic ratios, but they have not been extensively investigated for other anomalous elements.

These isotopic peculiarities cannot be attributed to radioactive decay or to any other process known to occur in the solar system. One is driven to conclude that they are relics of isotopic differences that existed among the interstellar grains that fell together at the outset to form the solar system. These grains originated in many different presolar astrophysical environments (such as stellar atmospheres and supernovae), and the material condensing in each environment would have had a different history of nucleosynthesis and thus a different isotopic makeup. The isotopic compositions of elements we observe in the Earth and Moon are a grand averaging of the compositions of the incalculable number of dust particles that joined the solar nebula. Apparently, however, not all of the infalling grains got mixed and homogenized: the isotopically anomalous CAI's must have incorporated above-average proportions of dust from one type of astrophysical source or another. This observation contradicts the simple concept that CAI's condensed from cooling vapors. If the presolar dust grains that joined the nebula had been totally vaporized, their isotopic compositions *would* have been homogenized; isotopic anomalies like those observed in CAI's could not have been preserved.

Most of the packed-together dust grains (typically only a few microns across) that constitute the matrices of chondrites probably are of nebular origin: condensates from the hot zones where chondrules were melted, and finely broken debris from the collisions in space between chondrules and CAI's. But isotope anomalies have been found here as well. When dissolved in strong acids, the matrix material leaves behind a minuscule (less than 1 percent) residue of particles in which the isotopes of silicon, carbon, nitrogen, neon, and xenon depart markedly from their average solar-system proportions. In particular, the xenon is enriched in the isotopes ^{128}Xe, ^{130}Xe, and ^{132}Xe, nuclides known to be created in the interiors of red-giant stars. Evidently these grains condensed from cooling gas that was emitted as stellar winds by ancient red giants, resided for a time in the interstellar medium, then joined the gas and dust that collapsed to form the solar system.

The grains consist at least in part of diamond (Figure 8) and silicon carbide (Carborundum). We think of diamond as a mineral that forms at high pressures. The interstellar diamond may have been formed this way too, from graphite or amorphous carbon grains that collided at high velocity after being accelerated by shock waves in the interstellar medium. However, laboratory experiments have shown that it is also possible for diamond to form metastably in hot, low-pressure gases.

METEORITES GENERALLY

Meteorites fall to Earth all the time. Most are so small that they are consumed by ablation in the atmosphere, or if one survives, it is such a diminutive pebble as to have no chance of being found on the ground. (A rare exception is the Ras Tanura chondrite, weighing only 6 g, which dropped on the end of a petroleum-loading dock in Saudi Arabia in 1961.) According to one recent estimate, 560 meteorites weighing 100 g or more drop onto every million square kilometers of Earth's surface each year. But only about a half dozen or so observed falls are recovered annually, and another dozen or two meteorites are found that fell unobserved.

Natural history museums have been collecting meteorites since the early 19th century, when their extraterrestrial origin was first appreciated. By now roughly 3,000 discrete meteorites have been catalogued. (A meteorite shower, like Allende, counts as one meteorite in this reckoning.) A recently discovered bountiful source of "finds" is Antarctica (Figure 9). Meteorites fall in the Antarctic snow, and over the years glacial flow carries them toward the margins of the continent, where in certain critical areas they tend to accumulate on the wind-scoured surface in remarkable numbers. Meteorites also fall on the Moon, of course. Lunar soil samples contain a component of smashed-up meteorite material amounting to as much as 4 percent by weight.

There are many varieties of meteorites (Figure 10). The traditional breakdown is into stones, irons, and stony-irons. A more meaningful division, however, would be into *undifferentiated meteorites* (chondrites) and *differentiated*

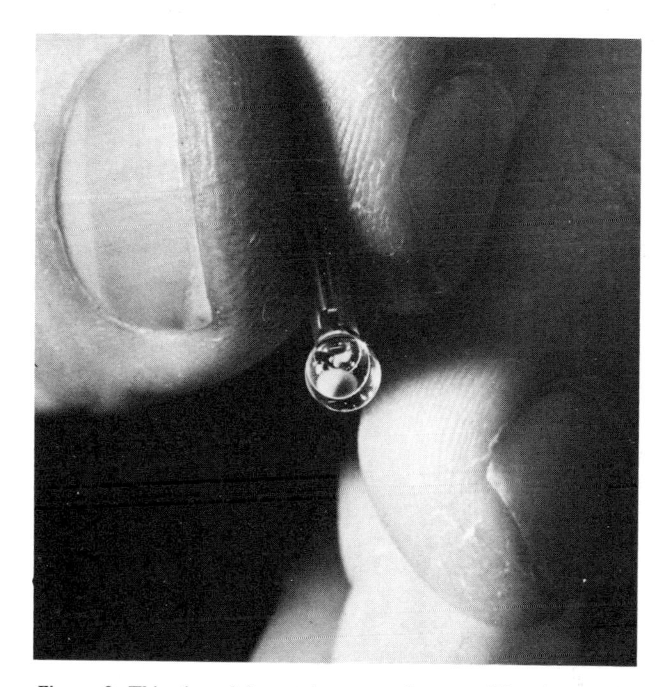

Figure 8. **This tiny vial contains more than a trillion interstellar diamonds (the small lump of white). Roy Lewis' research team extracted them from a primitive, carbon-rich meteorite, but the minute crystals apparently formed in the vicinity of another star. If so, they are older than the solar system itself.**

Figure 9. **A Japanese-American research team collects meteorites in Antarctica during the summer of 1978–1979. Here the specimen is a tiny fragment of carbonaceous chondrite. Kazuyuki Suraishi photographs the meteorite before moving it; Ursula Marvin holds a scale and numbering device that will appear in the photograph for identification.**

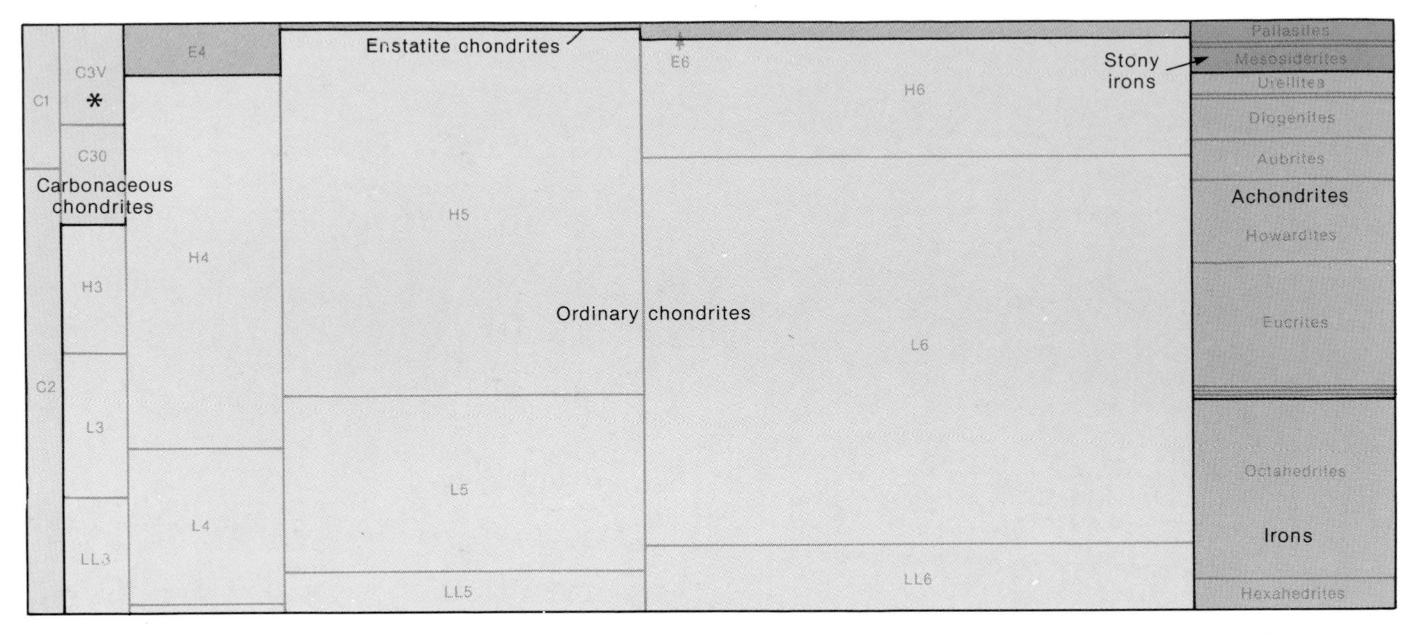

Figure 10. **The six major categories of meteorites, broken down into subtypes. The area of each box is proportional to the abundance of that meteorite type among all recovered falls. (An asterisk marks the position of Allende in this scheme.) Ordinary chondrites contain major elements in solar proportions (like Allende), but have been thermally metamorphosed to varying degrees. The degree of such metamorphism increases to the right.**

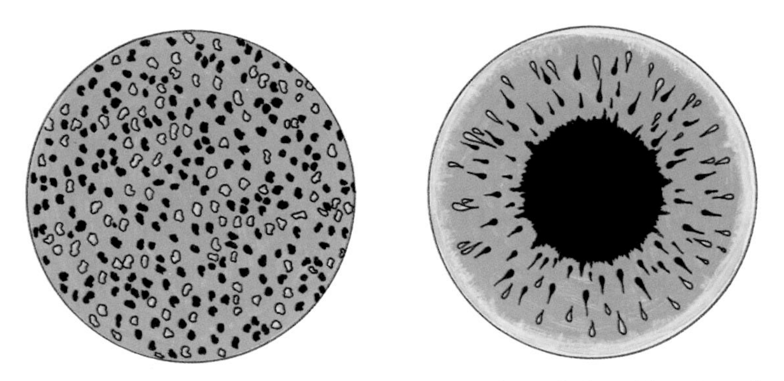

Figure 11. **Two schematic cross-sections suggest the evolution of interiors of meteorites' "parent" planets. These accrete (left) as chondritic mixtures of metal and sulfide grains (black), minerals of relatively low melting temperature (yellow), and abundant high-melting Mg and Fe silicates (green). Internal heating of the planets causes them to melt partially (right). Molten metal and sulfide drain to the centers, forming cores (the source of iron meteorites?). Silicates that melt at low temperatures produce a low-density magma, which tends to erupt to planetary surfaces (the source of Ca-rich achondrites?).**

meteorites. The most abundant chemical elements are present in all chondrites, as they are in Allende, in the same proportions as they are in the solar atmosphere. The significance of this is that the chondrites must be samples of primitive planetary material that have never been melted, since molten rock has an irresistible tendency to differentiate (separate) into layers or volumes of more specialized composition (see Figure 11). The subtypes of chondrites vary in their content of minor and trace elements, particularly elements that are considered "volatile." This has been interpreted to mean they are samples of material that accreted at various temperatures in the nebula, and some materials accreted at temperatures sufficiently high that the most volatile elements were left in the gas and did not join the accreting particles.

Differentiated meteorites apparently *are* the products of melting and separation of more primitive planetary material. Meteorites in this major category include irons, stony-irons, and some stones (the achondrites). Three such samples are seen in Figures 12, 13, and 14.

THE PLANETARY HISTORY OF METEORITES

Where do meteorites come from? That is, where do they reside between the nebular and Antarctic stages of their careers? The great majority of them are fragments of asteroids. Trajectories of several meteorites in the atmosphere have been photographed with sufficient precision to allow the meteorites' orbits in space to be calculated (Figure 15), and these closely resemble the orbits of Apollo (Earth-crossing) asteroids. At one time Apollo asteroids were apparently main-belt asteroids situated between Jupiter and Mars, with orbital periods that happened to be integral fractions of the period of Jupiter's orbit. Jupiter's gravitational influence, applied resonantly to the motion of such asteroids, is capable of reshaping the asteroid's orbit so it dives into the inner solar system and crosses the orbit of the Earth.

The relationship between meteorites and asteroids has been convincingly demonstrated. The spectra of sunlight reflected from asteroid surfaces are very similar to laboratory-derived reflection spectra of pulverized meteorite specimens (Figure 16). In both cases the characteristic shape of the spectrum is caused by the selective absorption of light by the minerals present in the reflecting material.

There is abundant evidence that meteorites have not always wandered through interplanetary space in the form of rocks as small as they are now, but were once parts of planetary bodies. Presumably impacts among the bodies broke them up and released fragments, a few of which have fallen to Earth.

First, there is the relationship between meteorites and asteroids, many of which are hundreds of kilometers in

Figure 12. **This fragment of the Nakhla, Egypt, achondritic meteorite (an igneous rock) consists entirely of silicate minerals; no metal alloys are present. Nakhla is one of the eight known SNC meteorites, which are thought to have originated on Mars.**

dimension and so qualify as small planets. Many or perhaps all of the smaller asteroids may be fragments of once-larger objects that collided with one another.

Second, cosmic rays have the effect of generating certain isotopes (such as ^{3}He, ^{20}Ne, and ^{38}Ar) in meteorites while they are at large in space as small rocks – but not when they are buried more than a few meters deep inside larger objects, where they are largely shielded from cosmic radiation. Measurement of the abundances of such isotopes in meteorite specimens shows that they were shielded from cosmic radiation during most of the life of the solar system, typically until a few tens of millions of years ago.

Third, the great majority of meteorites apparently endured high temperatures for relatively long periods of time. Some have been melted and chemically differentiated, as already noted. A great many more (most of the chondrites) were not melted but rather *metamorphosed,* that is, held at temperatures of approximately 1,300° K for long enough (many years) to make their silicate textures coarsen and become more granular, and to cause the compositions of their silicate minerals to become very uniform (by the slow diffusion of elements from one mineral grain to another). We do not have certain knowledge of the source of this heat, but planetary interiors are better able to retain heat, once generated, than are small dispersed rocks.

The decay of the radioactive elements uranium and thorium, and of potassium's radioactive isotope ^{40}K, is an important source of heat in large planets. But asteroids are not large enough to retain heat generated as slowly as these isotopes decay. The heat is conducted to their surfaces and radiates away into space as fast as it is created. The heating of asteroids could be understood if during accretion they incorporated a radioactive nuclide with a much shorter half-life than the isotopes of U, Th, and K. Such a nuclide would have delivered its heat in a pulse short enough to have heated small planets substantially, then decayed to undetectable levels by today. Evidence has been found in the CAI's of Allende and other chondrites that they once contained aluminum-26, a nuclide that decays very rapidly (its half-life is only 720,000 years). If all the aluminum in accreting planetesimals contained proportionately as much ^{26}Al as the aluminum of these Allende CAI's did, its decay would have released enough energy to melt the centers of objects as small as 20 km in diameter.

Figure 13. **A polished slice of the Pavlodar, Siberia, meteorite. It is a pallasite – a stony-iron meteorite that consists of a close-packed accumulation of rounded crystals of olivine, with metal alloys filling the spaces between them. The metal is similar in composition and structure to that in iron meteorites.**

Figure 14. **A slab of the Edmonton, Kentucky, iron meteorite has been sawed, ground, polished, and chemically etched to reveal the *Widmanstätten structure*, which consists of intersecting plates of kamacite alloy with taenite in the interstices. The irregular inclusion toward the left is an iron-nickel phosphide mineral.**

There is a way of gauging the rate at which meteoritic material cooled from the high temperatures it experienced. Most meteorites contain metallic nickel and iron in varying (and in some cases dominant) amounts. These metals form two alloys, kamacite and taenite (Figure 14), the proportions and compositions of which vary with temperature if the alloys are allowed to react with one another to equilibrium. The degree to which they *do* continue reacting and equilibrating in a crystallizing meteoritic system depends on how fast the system cools. If it cools rapidly, alloy compositions appropriate to high temperatures are frozen in. If it cools more slowly, the alloys take up lower-temperature configurations before the diffusive motions of Ni and Fe atoms are frozen in. There is substantial uncertainty in the method, but the cooling rates it points to (typically about 10° to 20° K per million years) are those that would be experienced 30 to 50 km deep in a small body after its heat source was "turned off" and its cooling restrained only by the insulating effect of the rock above it.

Granted that the meteorites are fragments of planets, is it possible that they and the asteroids all came from various parts of a single large "parent," perhaps a now-missing fifth terrestrial planet that once orbited between Mars and Jupiter but was somehow demolished? This concept is astonishingly popular and durable everywhere except among meteorite and asteroid researchers. There are a number of difficulties. First, the depths of burial indicated by the metal-alloy cooling rates just discussed are not all that great. Second, rock in the interior of largish planets is subjected to high pressure; this produces certain characteristic minerals such as spinel and garnet, which are not found in the meteorites that appear to have equilibrated in planetary interiors. Third, important chemical differences found among the various meteorite subclasses make it awkward to require that they once coexisted in a single body. Rather, meteorites are much easier to understand if they came from one or two *dozen* parent bodies, each of which was an isolated chemical system.

ASTEROID REGOLITHS

Many stony meteorites, if broken or sawed so an interior section can be examined, are found to consist not of integral rocky material but of a collection of angular fragments, tightly welded together. The structure referred to is coarser and different than the conglomeration of inclusions and matrix seen in meteorites like Allende, described earlier. The fragments can be centimeters in size, and if the meteorite in question is a chondrite, each of these will itself contain numerous chondrules and inclusions. It appears that the chondritic system went through discrete episodes of (1) assembly of chondrules, inclusions, and matrix; (2) lithification (welding together) of this conglomeration into a hard mass; (3) shattering of the chondritic rock into fragments; (4) lithification of the fragmental debris into a new hard mass; and (5) another, later shattering event that broke free a piece of the rock and left it in an orbit that eventually intersected Earth's.

Such aggregations of rock fragments (Figure 17) are called *breccias*, an Italian word. It is not hard to imagine how the breaking-up occurred. The asteroid belt is a crowded place, and collisions take place with calculable frequency. At least for the present distribution of asteroid orbits, the mean collision velocity is about 5 km per second, more than adequate to smash hard rock.

Our instincts tell us that the debris from such collisions would scatter into space in all directions and thus be unable to reassemble and lithify into a breccia. After all, asteroids are so small that their feeble gravity fields would have little chance of holding onto the flying debris fragments from a high-energy impact. Recent consideration of the asteroid-impact process has shown that this is not entirely true, however, as explained in Chapter 18. Impacts produce fragments with a wide spectrum of velocities, ranging from very high (comparable to the impact velocity) down to almost zero. Although most of the debris may indeed be lost from the asteroid, some fraction of the fragments will not achieve escape velocity. These fall back to the surface and join a layer of residual, unconsolidated debris. A layer of this sort, which is called the *regolith*, blankets the Moon to a depth of at least a few meters.

The process that tends to lithify loose asteroidal regolith fragments into a breccia like the Cumberland Falls meteorite is the same type of event that broke them up in the first place: an impact. Beneath the point of impact, particles and fragments are momentarily jammed against one another at tremendous pressures. The stresses may cause some melting along fragment surfaces, and this liquid acts as a glue when it resolidifies. Breccias that are not "well-glued" may exist in space, but they would be too weak to survive passage through

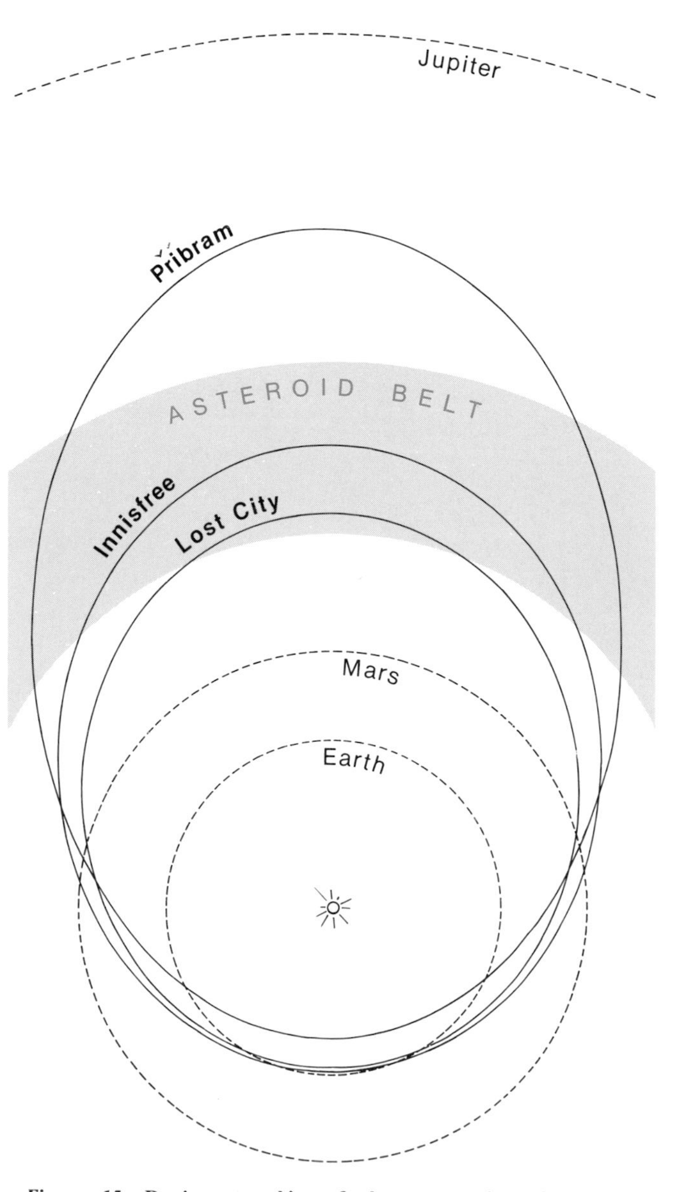

Figure 15. **Pre-impact orbits of three meteorites that were photographed during atmospheric entry.**

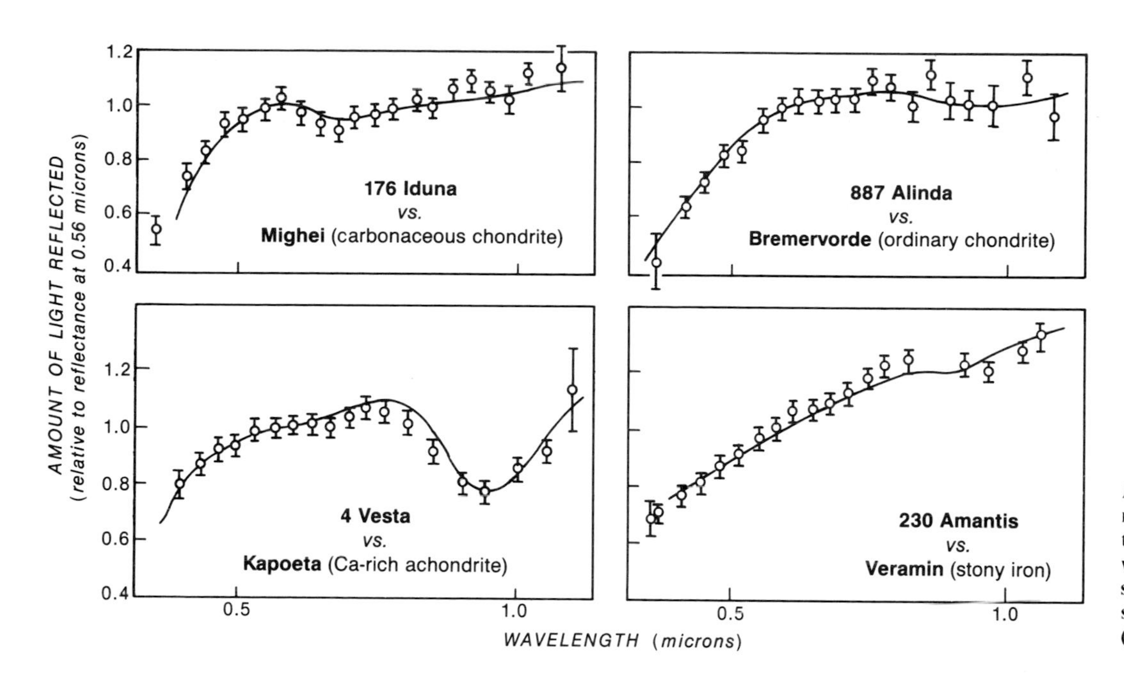

Figure 16. **A comparison of the reflection spectra of four asteroids (points with error bars) with the laboratory-determined spectra of meteorites representing four major subtypes (continuous curves).**

the Earth's atmosphere and so are not represented in our meteorite collections.

Breccias from both lunar and asteroidal regoliths have similar properties: the surfaces of fragments contain particular nuclides implanted in them by ancient and modern solar winds, and tiny glass-lined pits attributable to micrometeorite impacts. Many meteoritic breccias consist of fragments all of the same type of material, but in some cases a rich variety of meteorite types are present, occasionally types completely outside our previous experience. Evidently a complex history of impacts, with mingling of debris from both projectiles and target, gave rise to these.

Figure 17. **This fragment of the Cumberland Falls, Kentucky, meteorite is a breccia: a welded mass of angular fragments that aggregated long ago and far away. The internal components are derived from diverse types of meteorites: achondrites (white) and enstatite chondrites (black).**

METEORITES FROM THE MOON AND MARS

One small subset of achondritic (igneous) meteorites has a curious property: its members have much younger radiometric ages than other meteorites, only about 1.3 billion versus 4.5 billion years. This subset involves only a handful of achondrites and is too small to be labeled in the orange achondrite field of Figure 10. It consists of the minor subtypes called shergottites, nakhlites, and the sole chassignite known to have fallen; collectively they are referred to as *SNC* ("snick") *meteorites*.

As noted earlier, ages this young are common among terrestrial rocks, because the Earth's dynamic geologic activity is constantly creating new rocks. But, lacking an enduring source of internal heat, the tiny asteroids would have quickly cooled to an inert state. So both the asteroids and the meteoritic fragments they spawned should continue to register very old radiometric ages. The great majority of meteorites *are* old, dating back almost to the beginning of the solar system. Since small asteroids could not have continued to make new igneous rock as recently as 1.3 billion years ago it seems almost necessary that the SNC meteorites came from some planet larger than an asteroid. In fact it must have been larger even than the Moon, since none of the Apollo rocks are younger than 3 billion years old (about that time the Moon cooled to the point where new igneous rocks were no longer being created near the surface; see Chapter 4).

What fairly large, recently dynamic planet could they have come from? There is really only one candidate. In 1979 several research groups cautiously suggested that the SNC meteorites might be pieces of Mars that had been blasted off the planet by one or more major cratering events.

The idea was greeted with profound skepticism. The key objection was that it had been "proven" that an impact cannot knock an intact fragment off a body even as large as the Moon, let alone Mars. For an impact to accelerate a debris fragment almost instantaneously to the escape velocity of the Moon (2.4 km per second), the great stress involved would have pulverized, melted, or otherwise destroyed the fragment. This was why rocks similar to the Apollo lunar samples had never been observed to fall to Earth as meteorites, which should occur if impacts were chipping fragments off the Moon. No less an authority than this writer asserted in the 1981 edition of *The New Solar System* that it could not happen.

All this changed dramatically in 1982, when a peculiar-looking "meteorite" (Figure 18) collected on the Antarctic ice cap the year before turned out to be a piece of the Moon. Comparison with the Apollo samples made the identification unequivocal. In the years since, five more lunar samples have been found in Antarctica. All are superficially similar regolith breccias from the lunar highlands. Somehow cratering events *did* eject them, intact, from the Moon.

Figure 18. **This 31-g chunk of breccia, only 3 cm across, was found in January 1982 during a meteorite-hunting expedition in the Allan Hills region of Antarctica. The white fragments are abundant in a calcium-rich silicate called anorthosite, which (among other characteristics) proved that the small stone formed on the Moon and was later blasted off its surface.**

Figure 20. **A mosaic of Viking photographs showing the 220-km crater Lyot on Mars, at latitude 50° N, longitude 331° W. This crater – and the impact that formed it – are believed by some to be the most promising source of the SNC meteorites. However, based on the abundance of later, small craters on Lyot's ejecta blanket, others have argued that Lyot is slightly too old to be consistent with the meteorites' 1.3-billion-year ages.**

Figure 19. **In December 1979 geologists found this 8-kg meteorite, a shergottite, in Antarctica's Elephant Moraine region. Later analysis at NASA's Johnson Space Center showed that the rock had been infused with a mixture of gases strikingly like those found in the Martian atmosphere. It appears that Mars was the source of this and a handful of other known meteorites.**

These finds largely invalidated the argument against ejecting debris from a large body, though the problem remains formidable in the case of Mars. There the escape velocity is greater (5 km per second), and escaping fragments would be retarded by the planet's atmosphere. Even so, interest in the concept of meteorites from Mars increased, and additional observations have reinforced the concept. For example, pockets of shock-melted glass in one SNC meteorite (Figure 19) were found to contain noble gases and nitrogen in the same proportions, and with the same isotopic compositions, as had been found in the atmosphere of Mars by instruments in the Viking spacecraft as they descended to the planet's surface in 1976. By the late 1980s scientific opinion on the issue had swung around: SNC meteorites probably *do* come from Mars.

It is still not clear how a cratering event can accelerate intact fragments to the necessary escape velocities. But we now realize that debris spalled from particularly shallow depths during cratering can attain high velocities, yet experience relatively little destructive shock pressure. Several properties of SNC meteorites constrain the nature and location of a hypothetical source crater on Mars. It must be in one of the youngest Martian terrain units, to be consistent with the 1.3-billion-year SNC age. It must also be large enough to have ejected sufficient material into space to account for the rate at which SNC meteorites are now falling to Earth. The required crater diameter has been estimated to be 200 km, give or take a factor of two, and about 14 craters on Mars are possible candidates. The largest and most promising of these is Lyot (Figure 20), 227 km across.

Meteorites are one of our best sources of information about the cosmos. Free of charge, we have been presented with actual samples of asteroids, the Moon, and (very probably) Mars. We can examine them in exquisite detail in our laboratories, reading in their microstructures something about how the solar system came together and the planets formed. With enough cleverness, we can even catch glimpses of galactic history before there was a solar system. They still hold secrets that are beyond our present imagining.

20
Small Bodies and Their Origins

William K. Hartmann

UNTIL ABOUT a century ago, the solar system seemed like an orderly place, with just a few well-defined classes of inhabitants. The planets were the eight large objects going around the Sun. By 1840, 13 satellites had been discovered that were all intermediate in size and moving in direct orbits around their respective planets (traveling counterclockwise as seen from "above," or north of, the solar system's plane). The solar system of the 1800s was also populated by asteroids, which seemed to be an entirely different class of objects; all of them were small and neatly grouped in the planetary niche between Mars and Jupiter that had been predicted by Titius and Bode. A fourth class of objects, the comets, were again obviously different with their fuzzy comas and enormous diffuse tails.

SMALL BODIES: CONFUSION AMONG TYPES

But some problematic objects soon appeared. In 1846, Neptune's large satellite, Triton, was discovered in a retrograde (clockwise) orbit. Triton seemed perhaps only a bizarre exception, since the next six satellites found were again all moving in direct orbits. This group, however, included asteroid-size objects like Phobos and Deimos (discovered circling Mars in 1877) and Amalthea, Jupiter's fifth satellite (discovered in 1892). The tidy state of affairs began to break down even more dramatically soon thereafter, for 1898 brought the discovery of Phoebe, the small, retrograde, outermost satellite of Saturn. By 1900 the list of known satellites had grown to 21; most of these were large and moving prograde, but two had retrograde motion and two others were far below 100 km in diameter. Planetary astronomers began to recognize that the solar system was quite different from what they had thought. By 1960, nine more satellites had been discovered, of which four moved retrograde and six were less than 100 km across. Thus, satellites became an unholy mixture of planet-size, asteroid-size, prograde, and retrograde bodies.

To make matters worse, asteroids began popping up in unexpected parts of the solar system. In 1898, a German astronomer discovered asteroid 433 Eros, whose perihelion lies inside the orbit of Mars. Asteroid 624 Hektor, spotted in 1907, turned out to be moving in the same orbit as Jupiter but 60° ahead of the planet. Numerous other asteroids were later found sharing Jupiter's orbit, either 60° ahead of or 60° behind the planet. These Trojan asteroids (named after characters in Homer's Trojan war epics) occupy gravitationally stable niches termed Lagrangian points. In 1932, asteroid 1862 Apollo was found passing inside the Earth's orbit. Others of this type came to be called Apollo asteroids. By then it had become clear that asteroidal bodies resided not just in their "belt" but elsewhere in interplanetary space as well.

The neat physical distinction between asteroids and comets also began to break down when telescopic observers realized that as comets recede from the Sun, they eventually lose their tails and become indistinguishable from asteroids. Furthermore, dynamical studies during the 1950s showed that comets could be scattered by Jupiter's gravity into orbits like those of many asteroids, including the Apollos.

Such findings introduced confusion: How do we know an asteroid is really an asteroid and not an inactive comet? Astronomers argued about the differences between the two groups. Asteroids, they decided, inhabit the inner solar system (out to the orbit of Jupiter), while comets drop in for visits from farther away. Meanwhile, by the 1970s spectroscopic observations indicated that all "true" asteroids have surfaces made out of stony materials, whereas comets emit gases and thus must have a substantial component of icy material either on their surfaces or in their interiors. Nonetheless, we could not rule out the possibility that some asteroids with stony-looking surfaces contained ices, or that comets might contain rocky material. For example, the surface soils of many large, black asteroids contain minerals with chemically bound water, and some may contain ice crystals as well. Also, observers have noted similarities in the colors of cometary dust and certain reddish-black asteroids of the outer solar system. Such data are blurring the distinction between comets and remote asteroids; both may contain icy and stony material (Figure 1).

Orbital distinctions between these objects became even more blurred in 1977, when Charles Kowal discovered the enigmatic object 2060 Chiron moving in a cometlike orbit between Saturn and Uranus. Chiron is unlike any known comet – with a diameter of about 250 km, it is 15 times larger than the nucleus of Halley's Comet. Many comets interact with giant planets as they dip into the inner solar system from the Oort cloud. Chiron fits this pattern, as it occasionally comes close to Saturn. Such interactions will cause its orbit to evolve, and perhaps one day it will dip into the inner solar system.

Talking of Chiron in the same breath as comets illustrates our problem. When discovered, Chiron was catalogued as an "asteroid," and that was that; no other scientific certification was necessary. Yet we can't assume it is a mass of silicate rock (as are most main-belt asteroids). Instead, Chiron's colors are similar to those of other blackish asteroids and comet nuclei of the outer solar system. Moreover, after brightening unexpectedly in 1988, Chiron was found to have a distinct coma the following year (see Chapter 15). Apparently, it is a giant comet awaiting a chance to venture nearer to the Sun and blaze spectacularly into view.

In addition to asteroids and comets, astronomers began to distinguish a third class of small solar-system bodies: the outer satellites of Jupiter and Saturn. Gerard P. Kuiper, who pioneered modern astronomical studies of the solar system, pointed out as early as the 1950s that these objects seem distinct from the large, prograde inner satellites. Saturn's Phoebe and Jupiter's sixth through thirteenth moons are relatively small, have especially inclined and eccentric orbits, and in many cases move with retrograde motion. Kuiper suggested that these moons are captured interplanetary wanderers. Perhaps they originated among the interplanetary assortment of comets and asteroids.

LOOKING FOR PATTERNS

Scientists are taught to classify things as a first step toward understanding them, but this process can introduce misconceptions. In the past astronomers looked at the solar system in a way I call "the nine-planets gestalt." In this view of things the planets were supremely important; all other bodies were looked upon as less important and less interesting. This hierarchical ordering – and particularly the disinterest accorded small objects – probably provided the basis for the misleading distinctions among comets, asteroids, and certain moons.

Figure 1. **Astronomers once believed asteroids, comets, and small planetary satellites to be totally separate classes of objects. But as symbolized here, the distinctions among them have become blurred, and many are interrelated in ways that only now are being recognized and understood. For example, some comets may be nothing more than icy asteroids and some asteroids nothing more than burned-out comets.**

Now we are experiencing the breakdown of "the nine-planets gestalt." The Sun's family is a complex system of worlds, 25 of which exceed 1,000 km in diameter. Today we know that some satellites are larger and more geologically active than some so-called planets. And, rather than distinct categories, we are finding more of a continuum among the smaller bodies (Figure 2).

Admittedly, our new view of the solar system may seem no better organized; at first glance, it looks hopelessly confused because the physical distinctions between comets, asteroids, and small captured moons in particular seem unclear. They *are* all small bodies, after all, even if one type tends to be icy and the other rocky. Yet such confusion is common to any new stage of scientific exploration, and from it we are beginning to make some sense of the patterns observed in the solar system (Figure 3).

To help clarify the situation, let us go back to the beginning. Studies of unaltered, aeons-old meteorites show that the primordial solar system was crowded with small bodies ranging from hundreds of kilometers across down to 1 km or less, collectively called *planetesimals.* Any attempt to understand the present solar system – especially its small bodies – must be based on a determination of the compositions and destinies of these original planetesimals. As discussed more in Chapter 23, different minerals condensed from nebular gas and dust, depending on their distance from the infant Sun. In the inner solar system, the gas never got cold, and planetesimals accumulated almost exclusively from "high-temperature" minerals such as nickel-iron flecks and silicate rock grains. In turn, Earth and the other terrestrial planets aggregated from these smaller objects.

In the outer asteroid belt the temperature was lower, allowing in addition two basic types of material to condense at distances more than about $2\frac{1}{2}$ AU from the Sun. The first was ice, mostly ordinary frozen water; ices are very bright and whitish in color. The second was very black material rich in elemental carbon and carbonaceous compounds; it looks like soot and reflects very little (only 2 to 10 percent) of the sunlight that hits it. Also, it occurs in various shades from "neutral" (flat black) to brownish-black, subtle color differences that are probably due to organic compounds.

Our only samples of this second type of material may be the crumbly, black meteorites known as carbonaceous chondrites. Many of them show evidence that liquid water once trickled through parts of their interiors, thus affirming that they originated in mixtures containing water or ice. This is a striking but little recognized mineralogical association that may have application throughout the outer solar system. In particular, it offers a new and quite different insight on the Victorian categorization of "comets" versus "asteroids." For example, as recently as the 1970s most astronomers visualized comets (more or less correctly) as dirty icebergs and therefore expected (incorrectly) that comets would have bright white surfaces like icebergs, quite different from the "rocky" asteroids. But during the 1980s spacecraft images and other data finally confirmed a controversial suspicion: comets are black. In fact, *all* of the interplanetary bodies we've observed beyond the asteroid belt (all Trojan asteroids, Hilda asteroids, comet nuclei, and probably Chiron) and *all* small satellites believed to be captured (Jupiter's outer moons, Phoebe, Phobos, and Deimos) have blackish surfaces thought to be rich in carbonaceous minerals.

In contrast, objects in the inner half of the asteroid belt, as well as the terrestrial planets, contain more ordinary rocky material and have higher reflectivities (10 to 20 percent). The remaining asteroids, whose orbits dip inside the orbits of Mars and Earth, can have either bright or dark surfaces. They are an interesting grab bag consisting of objects ejected from various parts of the asteroid belt (generally by gravitational interactions with Jupiter) and probably a few burned-out comet nuclei as well. Thus, they represent both the high- and low-reflectivity classes of small objects.

The explanation of these gross patterns is simply that familiar rocky materials – that is, silicate minerals – dominated everything that formed near the Sun, while the black material dominated the colors of all planetesimals beyond $2\frac{1}{2}$ AU. Hence the familiar gray, tan, and rust-red rocks of Mercury, Venus, Earth, Moon, Mars, and the closest asteroids appear only in the inner solar system. These minerals also condensed farther out, but they were swamped by the black sooty material that condensed at lower temperatures. Ices were also added to the mix at larger solar distances. Indeed, in the outer solar system, ices may constitute more than half of many bodies, but the black carbonaceous soot is so effective a coloring agent that even a little of it can make a "dirty ice" mixture look very black. So any object in the outer solar system still showing a relatively fresh unaltered surface (the Trojan asteroids, comet nuclei, and so forth) has a black appearance. And thus the color match found among comets and very distant asteroids becomes understandable. Indeed, comets may be objects similar to distant black asteroids – carbonaceous planetesimals that merely had different orbital histories. Comet nuclei may simply be the ones that passed close enough to a giant planet to get kicked into the inner solar system.

HISTORY OF THE SOLAR SYSTEM'S PLANETESIMALS

With this background, we can interpret better the histories and distributions of different types of asteroids, comets, and satellites. Asteroids in the main belt are leftover rocky planetesimals of the inner-to-middle solar system, most of which formed between Mars and Jupiter and remained trapped there. But they are the minority – most planetesimals were fated to be added to the growing planets or to get thrown out of the solar system, due to the gravitational scattering by the planets. Jupiter was especially efficient at perturbing objects from the asteroid belt either inward or outward.

In the outer solar system, many *icy* planetesimals had close encounters with giant planets and were thrown not quite free of the Sun's gravitational grip but rather into long, elliptical orbits. Occasionally they fall back into the inner system and are seen as comets.

Both the asteroidal and cometary objects may have their orbits further altered by additional close encounters with various planets. A comet can be thrown into an asteroid-like orbit and an asteroid into a comet-like orbit. This explains why active comets, "burned-out" comets, and rocky asteroids can all end up with similar paths around the Sun (Figure 4). Remarkably, certain objects in known cometary orbits may be temporarily captured by Jupiter and become its outer satellites for periods ranging from months to decades or longer. This finding comes from computer projections of the motions of such bodies backward or forward in time, and it may begin to explain the captured satellites – the bodies with irregular orbits on the outskirts of the Jovian and Saturnian systems.

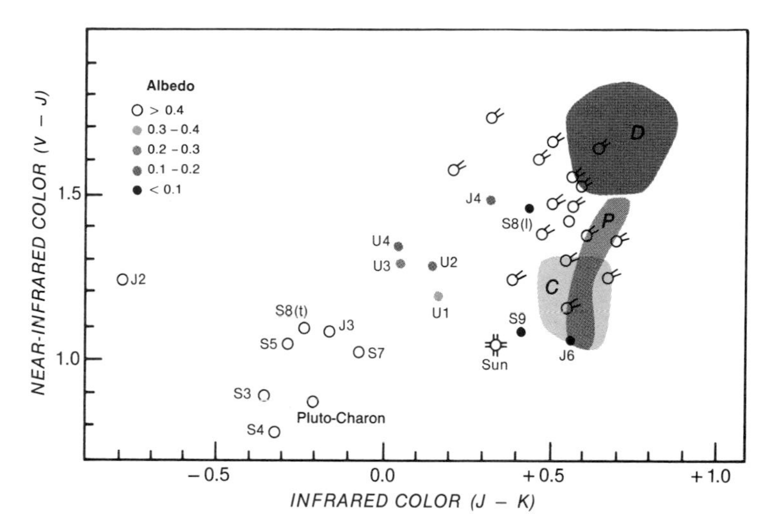

Figure 2. **Small bodies in the outer solar system can be characterized in terms of three parameters: near-infrared color (visual brightness minus the brightness at the near-infrared *J* wavelength of 1.25 microns); color further in the infrared (the difference in brightness between *J* and *K*, at 2.2 microns); and albedo (surface reflectivity) in visible light. In this diagram different types of surface materials are clearly distinguished: bright ices cluster at lower left and black carbon-rich soils at upper right. Therefore, where an object plots relative to these extremes may indicate the ice-soil ratio in its surface materials. Objects intermediate between the bright icy bodies and black asteroids include Jupiter's satellite J4 (Callisto) and the Uranian satellites U1-U4. Active and inactive comets cluster near the black asteroids, evidence that led to the successful prediction that the nucleus of Halley's Comet would be very dark. The most distant asteroids fall into one of three spectral classes: C's dominate the outer edge of the asteroid belt, D's are common farther out among the Trojans, and P's appear to be a transitional class.**

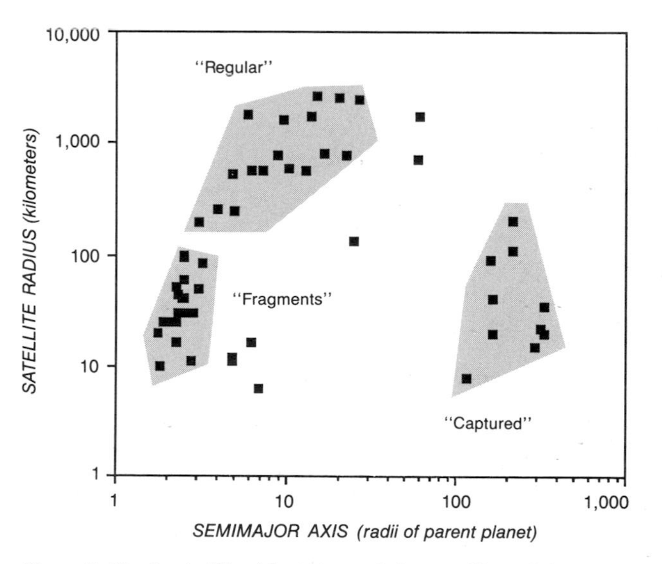

Figure 3. **Charles A. Wood has observed that satellites of the outer solar system form relatively compact clusters when grouped according to size and distance from their parent planet. "*Regular*" moons tend to be large and occupy prograde orbits. "*Fragments*" are very small inner satellites, often found near or even within a planet's ring system, that may be the collisional remnants of larger predecessors. "*Captured*" objects have distant, unusual orbits; perhaps they were once interplanetary wanderers that became captured after venturing too close to the planet they now orbit.**

Belt asteroids were once viewed as a swarm of objects hopelessly mixed by gravitational scattering as well as by collisions among asteroids that fragmented many and produced strange surfaces: cross-sectional exposures of the fragmented bodies' interiors. Today we see that, in spite of these processes, an underlying systematic compositional trend has survived.

Imagine a voyage outward through the asteroid belt. As we cross its inner edge, about 70 percent of the asteroids we encounter have stony (silicate) surfaces – essentially the same type of rocky material that originally formed in the inner solar system. By the time we reach the middle belt at about $2\frac{1}{2}$ AU, the fraction of black "C-type" asteroids (whose spectra resemble those of carbonaceous chondrite meteorites) has increased to about two-thirds. As we pass into the outer belt, the fraction of C-type objects is still increasing and probably exceeds 85 percent. Meanwhile, the percentage of ordinary stony-type asteroids, the ones that resemble most meteorites that fall on Earth, has dropped to nearly zero.

As we move to greater distances, there is a slight reddening of the color of the asteroids encountered, though they remain dark. In the 1970s, the Dutch-American astronomer Johan Degewij discovered this effect and referred to the reddest of the black asteroids as a new "RD" (reddish-dark) class. For historical reasons, this was shortened to class "D" and an additional intermediate transition category "P" was introduced. The C, P, and D designations, which range in color from neutral black to reddish black, apply to nearly every asteroid beyond the middle belt. Reaching Jupiter's orbit, we find that virtually all the Trojan asteroids are classes P and D, fitting the idea that more red (organic?) minerals condensed at these distances. (see Chapter 18).

The seemingly captured satellites present one of the most interesting mysteries – or clues – among the small bodies. As mentioned, they are all quite dark, which fits with the idea that they originated in the outer solar system asteroids, then became scattered by the giant planets into other regions. But the paradox is this: if Jupiter's outer moons are captured asteroids, we predict that the ones most easily snared would be local (the Trojans). This is so because these asteroids share

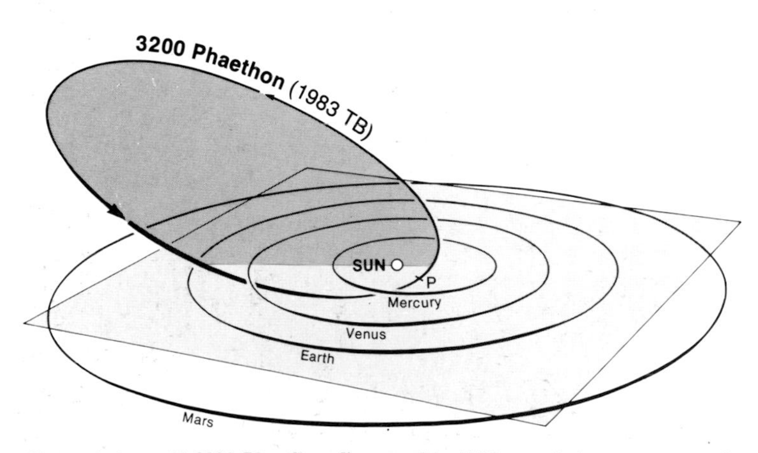

Figure 4. **Asteroid 3200 Phaethon, discovered in 1983, travels in a very eccentric orbit and passes closer to the Sun than any known minor planet. The object could conceivably skirt the Earth well inside the Moon's orbit. Is Phaethon a dead comet? Most observational evidence suggests that it is not. But the object's orbit is virtually the same as that traveled by the Geminid meteors, and some meteor showers closely follow the orbits of known comets. The Geminids might be dust knocked off Phaethon by a meteoritic impact.**

Jupiter's orbit, so they would approach the planet more slowly than any other type of object. This makes them easier to capture. And since we know the Trojans are spectral classes P and D, we would assume the outer black moons, on close study, to be P and D objects also. The surprise is that the four or five captured Jovian moons bright enough to have been studied well are all C's! So are the seemingly captured moons Phobos, Deimos, and Phoebe! Yet C-type objects would seem to have formed only in the outer part of the main belt. The mystery, therefore, is how these bodies ended up as satellites of Mars, Jupiter, and Saturn.

Recently, I proposed an explanation for these conflicting observations. As Jupiter and the other giants formed, they possessed extended atmospheres that slowed and captured P and D asteroids in their vicinity. But the gas drag was too great, causing all the captured objects to spiral too far inward and crash into their host planets. No P- or D-type moons survived. Only at the very end of planetary formation, when Jupiter reached its final size and the nebula around it dissipated, could the planet capture moons that would survive. However, exactly at this "moment" (an interval of perhaps a few tens of millions of years) Jupiter also became most efficient at perturbing asteroids from the belt. The planet plucked them from regions where C's are common – from the so-called Kirkwood gaps within the main belt and from the zone between the belt's outer edge (about $3\frac{1}{2}$ AU from the Sun) and the Hilda asteroid family (3.9 AU). Thus, at the very close of planetary formation, Jupiter may have scattered a host of black, C-type asteroids around the solar system, and this may explain why Mars, Jupiter, and Saturn – the planets closest to the belt – captured C-class moons. This scenario is very sketchy; more dynamical work is needed to understand how the Kirkwood gaps were cleared, how Trojan asteroids were captured in Lagrangian clouds, and how interplanetary bodies' orbits evolve.

Jupiter's outer moons, which orbit at the limit of the planet's gravitational sphere of influence, make a strange constellation of two distinct groups. Leda, Himalia, Lysithea, and Elara all have prograde orbits about 11.5 million km from Jupiter, with inclinations between 26° and 29°. On the other hand, Ananke, Carme, Pasiphae, and Sinope move in retrograde orbits about 22 million km from Jupiter, with inclinations between 16° and 33°. Perhaps only two asteroids were captured, one at 11.5 million and the other at 22 million km; then each underwent a collision that broke it into fragments. Alternatively, the capture mechanism may be highly selective, allowing objects to end up in only one size of prograde orbit and another size of retrograde orbit. The exact history of Jupiter's outer moons remains intriguing.

Dark Phoebe (Figure 5) lies on the outskirts of Saturn's gravitational grip and lends strong support for the capture hypothesis. It travels in a retrograde orbit some 26 million km across and inclined 30° to Saturn's equator. Other major Saturnian moons are larger, more icy, and have prograde orbits with inclinations not exceeding 15°. One intriguing observation is that Phoebe and Chiron have very similar sizes and surface colors. Phoebe was probably captured $4\frac{1}{2}$ billion years ago at the end of Saturn's formation (when it had a massive, extended atmosphere). Chiron does come close to Saturn but is unlikely to be captured by it. Even so, Chiron may offer a present-day example of the kind of interplanetary wanderer that gave rise to Phoebe.

Mars' two small satellites, Phobos and Deimos (Figures 6,7), are also dark bodies with C-type surface materials. With diameters of roughly 22 and 12 km, respectively, they look much like asteroids from the pool of blackish carbonaceous (and icy?) objects that formed in the outer part of the belt. This supports an earlier conjecture by several scientists that Mars captured Phobos and Deimos when it was surrounded by a massive primitive atmosphere.

SURFACE EVOLUTION OF SATELLITES: WHY ARE SOME SURFACES ICY?

Outer-solar-system bodies are truly alien worlds to us rock-lovers. Both theory and observation suggest strongly that from Jupiter's realm outward worlds formed with a large percentage of ices in their interiors. Voyager pictures have made some of these worlds familiar to us, and indeed frozen water covers the major satellites of Jupiter and Saturn. In addition, recent spectroscopic observations of the frigid, outermost worlds Triton and Pluto indicate that frozen methane (CH_4) dominates in that distant region. But why is it that all the big moons look bright and icy, yet none of the interplanetary bodies do? If moons of Jupiter, Saturn, and Uranus condensed from the same mixture of soot and ice as the most distant asteroids, why aren't they also black?

Interestingly, the solar system's only good exposures of fresh, clean ice (besides the poles of Earth and Mars) are on the larger satellites, Saturn's ring particles, and the Pluto-Charon pair. The common denominator in this pattern may be a higher degree of geologic processing of the original dirty ice material. For example, if a world has been heated moderately (by tidal interactions or the decay of its radioactive minerals, for example), the ice component would melt first, and the "lavas" erupting onto the surface would indeed be water or other melted volatiles. Even if a watery eruption carried sooty dirt with it, that dirt would sink, and the flow would freeze with a clean, white icy surface. Then,

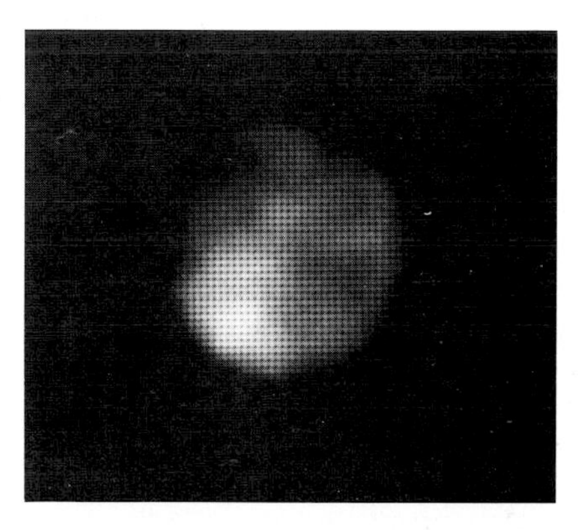

Figure 5. **Voyager 2 took this image of Saturn's dark, outermost satellite Phoebe in 1981 from a distance of 2.2 million km. It reveals the moon's spherical form and mottled surface, but other details are indistinct. Phoebe's surface reflects no more than 6 percent of the sunlight that strikes it.**

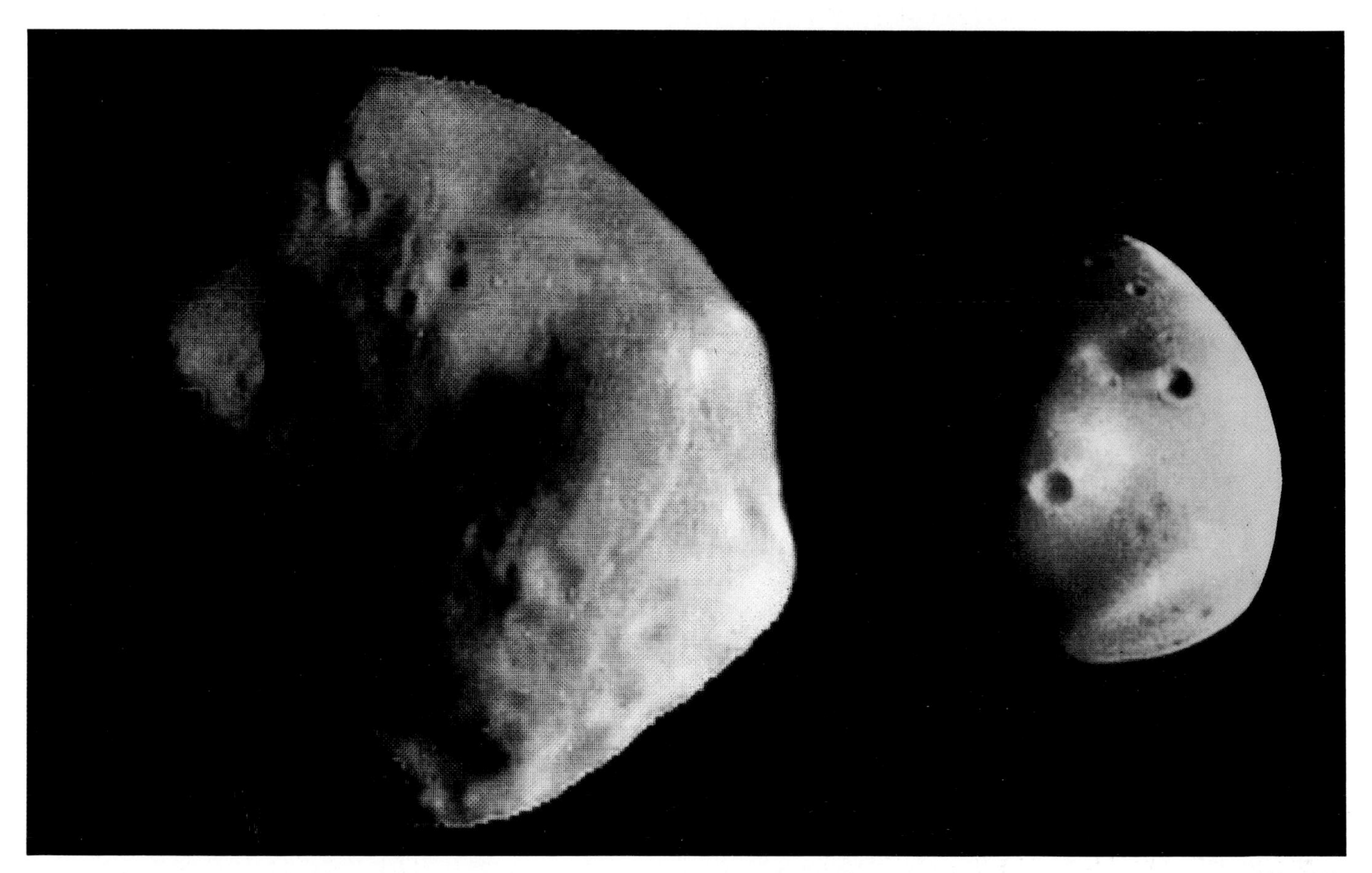

Figure 6. **The two Martian moons Phobos (left) and Deimos are reproduced at about their correct relative sizes in these Viking 1 orbiter images obtained in 1977. Rugged Phobos is shown from the south, with the crater Hall (about 6 km across) near the terminator. Deimos, seen south up, exhibits dark craters up to 1.3 km in diameter. These bodies reflect only a few percent of the sunlight striking them and are much darker than many other asteroids in the inner asteroid belt beyond Mars. Phobos and Deimos have spectra resembling those of certain "C-type" asteroids common to the outer asteroid belt and to Jupiter's Trojan asteroids; such material is believed to have a composition similar to carbonaceous chondrite meteorites. This suggests that perhaps Phobos and Deimos were at one time farther from the Sun, but Jupiter perturbed them into orbits that allowed their capture by Mars – a theory that is inviting but remains unproven.**

over time, contamination from carbonaceous meteorites would darken it. This idea seems to be borne out by the fact that the oldest, most heavily cratered parts of Callisto, Ganymede, and Europa are the darkest, while the young and uncratered parts appear bright and icy.

Let us take another imaginary trip, this time from the outskirts of Jupiter's large moon system in toward the planet. After we pass the black, captured moons, the outermost "native" satellite encountered is Callisto. Its density tells us that it is roughly half ice and half rocky material. Furthest of the four Galilean satellites from Jupiter, Callisto is least likely to have been geologically altered by heating effects from proto-Jupiter itself, or the tidal effects that heat inner moons like Io. And, sure enough, its surface is the most cratered and the darkest of the large moons (its reflectivity is 19 percent). Ganymede, the next moon inward, is clearly not as primitive as Callisto. Some areas are dark and cratered, but other areas are more sparsely cratered, brighter, and exhibit long fissures. Apparently wherever the old crust was broken, lighter, fresher ice resurfaced the cracked areas.

We next encounter Europa, which also supports the pattern. Europa's very bright surface reflects 69 percent of the light striking it and consists of almost pure water ice. It is virtually uncratered, meaning that the surface is relatively young and fresh. Heating has apparently melted much of the interior, causing eruptions of water that resurfaced the whole satellite with bright ice. But what was the source of the heat? All the Galilean moons may have experienced some heating due to radioactive decay in their interiors. However, the degree of eruptive activity has increased with each step toward the planet. This suggests that Jupiter itself may have played a role, either by radiating a large amount of heat as it formed or by inducing strong tidal flexure within the satellites nearest it. Striking confirmation of the latter idea comes from the amazing inner Galilean satellite Io and its continuous, tidally driven volcanoes. This volcanic activity, in turn, explains the absence of ice on Io: so much of the material has been heated and erupted that all water has long since vanished from the satellite's outer layers. Io's surface is dominated by neither ice nor rock, but by the sulfurous lavas that continue to flow from its interior.

Moving inward from Io, we encounter Amalthea (Figure 8). It too has an anomalous surface – reddish, dark, and cratered, but unlike C- or D-type asteroids or satellites. Joseph Veverka and his co-workers have suggested that Amalthea is simply coated with sulfur compounds blown off of Io and toward Jupiter. The tiny neighboring satellites Metis, Adrastea, and Thebe, discovered during the Voyager flybys, may also have anomalous surfaces for the same reason. Metis and Adrastea orbit near the edge of Jupiter's ring, a relationship that sheds further light on the evolution of satellites in general. The two moonlets are continually sandblasted by micrometeorites attracted toward Jupiter (which are most concentrated and fastest moving close to the planet). Under the influence of certain dynamical forces, particles blown off the satellites spiral in toward the planet. Apparently, these minuscule dust grains are the main constituents of the ring, which is described more fully in Chapter 12. Rings and small moons have close associations at Saturn and Uranus as well (Figures 9,10), though we cannot yet say if these satellites are sources of ring particles.

On Saturn's moons, the cratered surfaces do not seem to have darkened as much as those in Jupiter's system. Perhaps the proportion of ices was even higher so far from the Sun, with too little dust available to darken these satellites much. However, there are trends that support the ideas discussed above. The reflectivity of Saturn's moons generally increases as the planet is neared. On Enceladus, one of the interior moons, some areas seem to have been resurfaced with sparsely cratered, fresh ice (see Chapter 15); its surface is one of the most reflective known in the solar system. And Saturn's outermost satellite, Phoebe, is also the darkest.

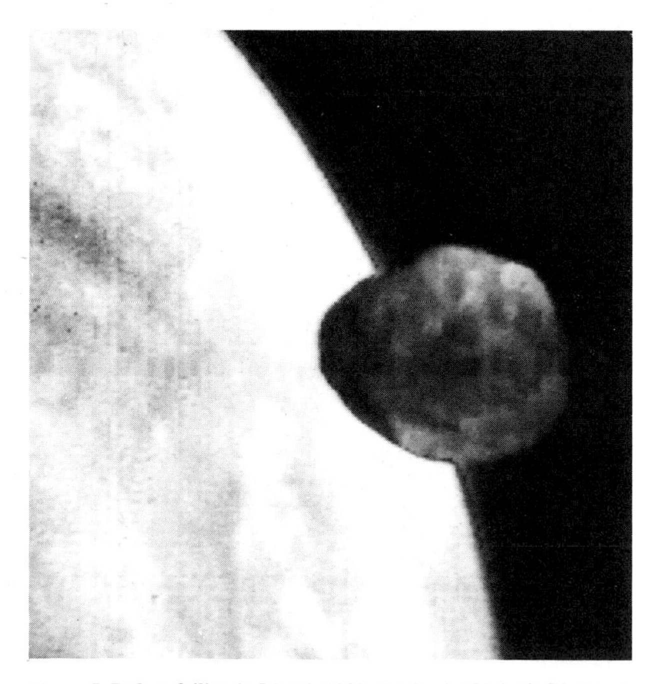

Figure 7. **Before failing in March 1989, the Soviet Union's Phobos 2 spacecraft obtained a few dozen photographs and other data on Mars and its satellite Phobos. In this view, taken from 500 km away, Phobos is silhouetted against the planet's limb. The satellite's surface shows details down to about 500 m across.**

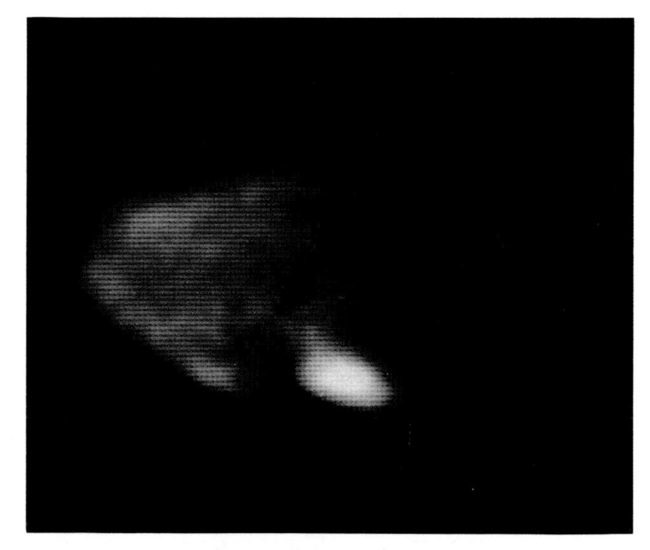

Figure 8. **Jupiter's satellite Amalthea, as viewed by Voyager 1 in 1979. Note the satellite's red color, believed to be the result of a coating of sulfur particles ejected from nearby Io. Amalthea's elongated shape, with the longest axis always pointed toward Jupiter, provided varying views for the passing spacecraft.**

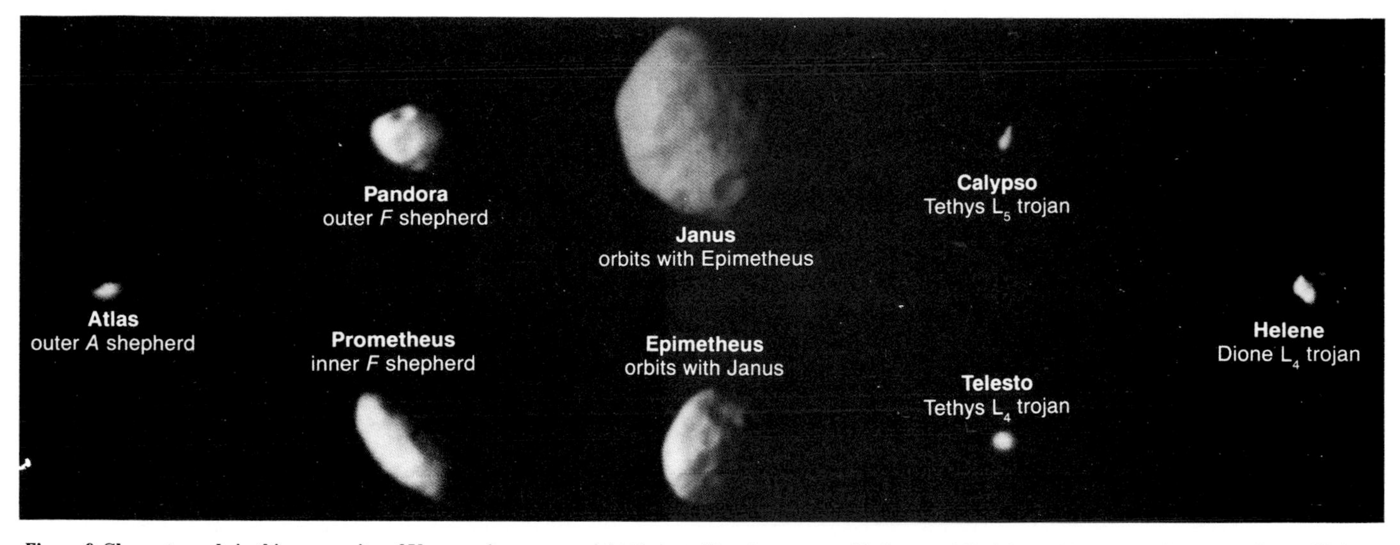

Figure 9. **Shown to scale in this composite of Voyager images are eight tiny satellites known to orbit Saturn. All of them either "share" orbits or "guard" rings.**

The solar system's most dramatic example of the contrast between black dust and white ice is found in the Saturnian satellite system. Meteorites that strike small, retrograde-moving Phoebe knock black material off its surface, and once free of the moon's weak gravitational field, the dust spirals inward under various dynamical influences (Figure 11). It approaches prograde-moving Iapetus from the retrograde direction, striking it head-on at high speed. Iapetus' original surface was presumably bright and icy, but over time the sooty Phoebe dust has apparently coated the leading hemisphere. This has made the leading side of Iapetus extremely black – yet its trailing side remains white (see Chapter 15 for more details). We cannot yet say to what extent Phoebe's dust has contaminated the surface of the next inward moon, potato-shaped Hyperion (Figure 12).

It is not clear whether the black-white systematic trend among small bodies extends out to the systems of Uranus and Neptune. For one thing, new types of ice (like methane) condensed in these cold regions and may affect the balance between dark and bright materials. But the Uranian satellites do confirm our general rule, for they have brighter surfaces (reflectivities of 20 to 30 percent) than do interplanetary bodies. As in the Jovian and Saturnian systems, the innermost moons show signs of geologic activity; Miranda is heavily fissured and has young, sparsely cratered regions. Curiously, the middle member of the five major Uranian moons, Umbriel, is the darkest (Figure 13).

How does the "double planet" of Pluto and its moon Charon fit into the systematic pattern among small, distant bodies? So little is known about Pluto that one hesitates to speculate. But it does have a bright surface, and some might argue that Pluto violates the notion that small "interplanetary" bodies in the outer solar system are all black. Then are our general ideas wrong? Not necessarily. Charon is a giant when measured in proportion to the size of its planet, and its orbit is not very large. Thus, tidal forces between the two bodies may have generated enough heat in the past to cause some melting and eruptions, which covered their presumably black initial surfaces with extensive flows that quickly froze into bright ice. A fascinating clue about further

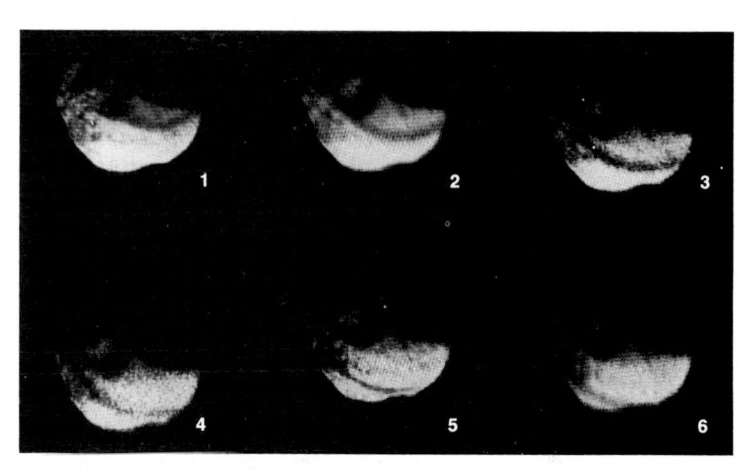

Figure 10. **As Voyager 1 passed Saturn in 1980, it chanced to photograph the small inner satellite Epimetheus just as the shadow of Saturn's F ring crossed it. These images were taken 3.2 minutes apart. Epimetheus and its companion moon Janus occupy virtually the same orbit and may be two halves of what was once a single object.**

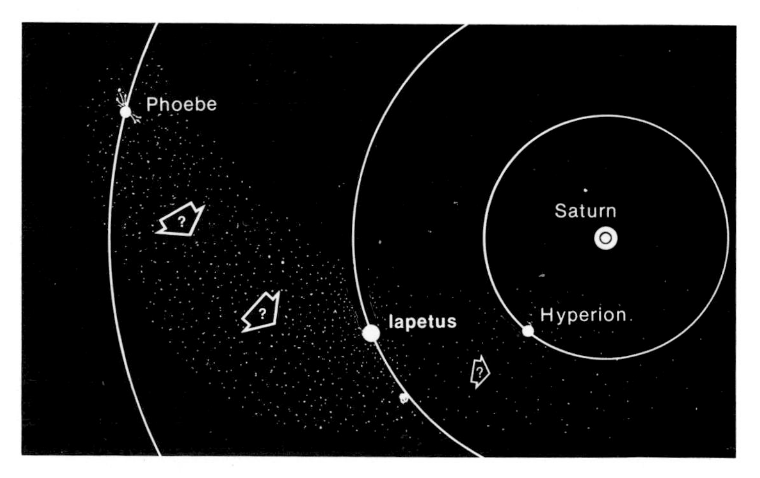

Figure 11. **Black dust, thrown from Phoebe's surface during meteoritic collisions, slowly spirals toward Saturn over time and may be the source of dark material on the leading hemisphere of Iapetus. A fraction of this debris should get past Iapetus and continue inward toward Hyperion. But apparently it is not enough to create a similar coating on Hyperion, which has a bright surface.**

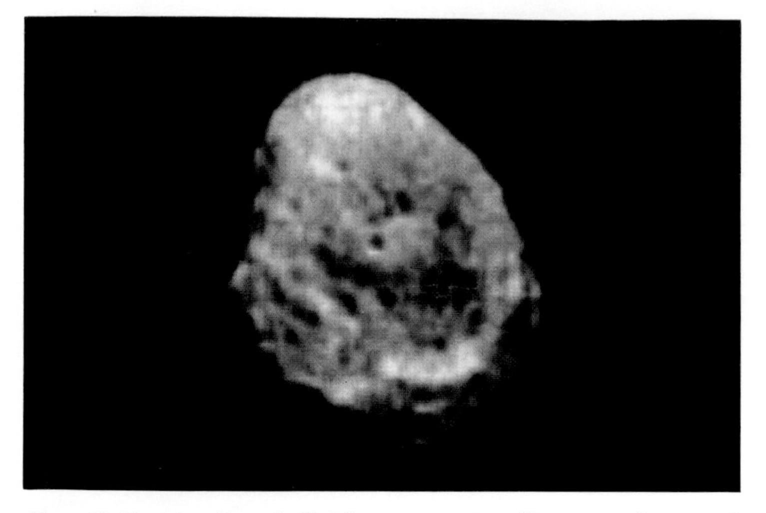

Figure 12. **Hyperion, Saturn's third-from-outermost satellite, as seen face-on and side-on by Voyager 2. A major collision on this moon may have blown away part of its surface, leaving it far from spherical. Hyperion's potato shape (350 by 240 by 200 km) and eccentric orbit make it subject to gravitational forces from Saturn that have set it tumbling out of control. In 1983, theorists Jack Wisdom, Stanton Peale, and Francois Mignard deduced that the spin of Hyperion is chaotic – its rotation period is not constant, but varies from one orbit to the next. Their suspicions were confirmed in 1988, after observer James Klavetter monitored Hyperion's changing brightness on a near-continuous basis for $3\frac{1}{2}$ months.**

evolution comes from the recent discovery that much if not all of Pluto has bright methane ice on its surface, while Charon's ice is darker and primarily frozen water. Presumably, their initial ice compositions must have been the same. According to observations and interpretations by Robert Marcialis and his co-workers, many of the methane molecules leaving the surface of tiny Charon have enough velocity to escape the moon's slight gravitational field. Some of the methane must eventually become trapped as frost deposits on Pluto, and some of it remains in the planet's thin atmosphere. Over time, this migration has preferentially built up methane on Pluto, leaving water ice and darker soil behind on its satellite. This explains why Charon is darker than Pluto.

All of this goes to confirm our most basic tenet: the small bodies of the solar system are much more diverse and individual than astronomers had suspected a few years ago. Again, these differences result from the degree of internal heating, the consequent geologic activity on the surface, and the roles of various surface materials with distinct compositions, densities, and mechanical properties.

SUMMARY

We have looked at the small bodies of the solar system in a new way. For a moment, we put aside the old observational categories of asteroids, comets, moons, and so on, which were originally based on telescopic appearance and present-day orbits. Instead, we looked upon these bodies as planetesimals with different initial compositions dependent mainly on their original location in the solar system. Some formed in orbit around planets and were moons from the outset, possibly altered early by heat sources associated with nearby planets. Some formed in the depths of interplanetary space and were perturbed into various final locations, somewhat diffusing the original orderly arrangement of compositionally distinct groups.

The planetesimals that formed in the inner asteroid belt were primarily rocky objects, like ordinary meteorites. Some odd surface types among them may be associated with major collisions that split them and revealed interior materials such as metallic cores. The ones that formed beyond $2\frac{1}{2}$ AU incorporated two important additional components: very dark, carbon-rich "soot" and bright ices. Trojan asteroids, situated 5 AU from the Sun, are good examples. The dark material probably dominated the colors of all these bodies when they first formed. Thus, all outer-solar-system planetesimals were initially blackish. Those that chanced to pass through the inner solar system, where the ice component can sublime and produce tails of gas and dust, came to be known as comets – the others are catalogued as asteroids, a diverse group with varied orbits and compositions.

The proportion of ice varied with location, and from the outermost asteroid belt to Uranus the icy component was mostly frozen water. If these planetesimals were not disturbed by heating, their surfaces were black and grew still darker as ice was vaporized during impacts and escaped from surface soils, leaving a dark regolith of material akin to carbonaceous chondrites. This applies to most outer-solar-system bodies smaller than a few hundred kilometers across. If heating was important, the ice melted into a watery lava that may have erupted, producing bright ice surfaces instead of black ones on many of the outer solar system's larger, inner moons. A number of small satellites may have formed as C-type planetesimals in the asteroid-to-Jupiter region, only to be perturbed elsewhere by gravitational scattering and then captured by planets.

In short, the processes of planetary evolution have conspired to present us with a rich variety of mini-worlds.

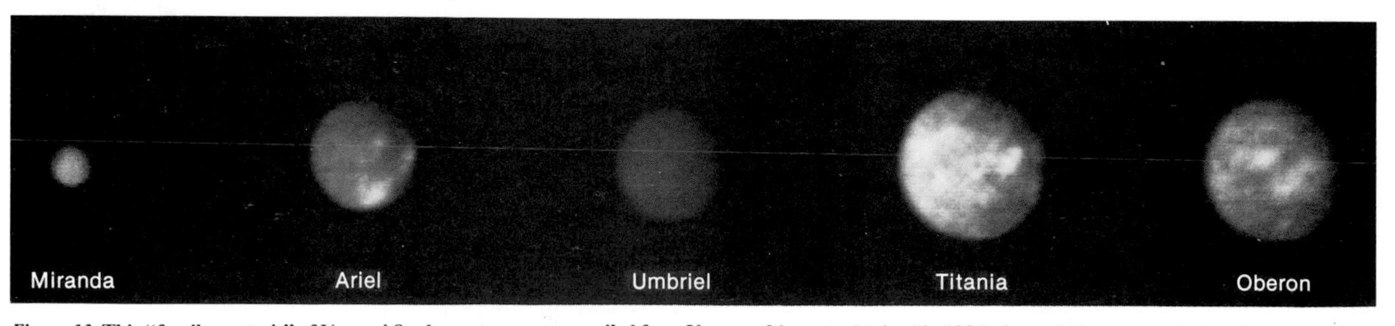

Figure 13. **This "family portrait" of Uranus' five largest moons, compiled from Voyager 2 images obtained in 1986, shows their correct relative sizes and surface reflectivities. Astronomers do not understand why Umbriel is so much darker than the others; it may have had less geologic activity and thus fewer eruptions of bright ices.**

21

The Collision of Solid Bodies

Eugene M. Shoemaker and Carolyn S. Shoemaker

ONE OF THE MOST STRIKING DISCOVERIES from three decades of space exploration is that most of the solid surfaces of planets and satellites – from Mercury to the satellites of Uranus – are heavily cratered. Oddly, as late as 1960, the significance of impact in the sculpturing of the Moon's surface was a subject of controversy. But today there is no longer any doubt. Indeed, impact seems to have been the most fundamental process on the terrestrial planets – a group of bodies whose very formation probably depended on the collision and accretion of smaller objects. Impact craters are by far the dominant landforms on most of the rocky and icy bodies yet surveyed in the solar system.

Our examination of the impact history of the planets and satellites will begin with the present situation and work backward in time. We can look through telescopes into interplanetary space and estimate the populations of "stray bodies," as Ernst J. Öpik once called them, that cross a planet's orbit. These estimates, in turn, yield the rates at which small bodies collide with planets and satellites. The rates calculated for Earth are consistent with the count of recent impact craters on its surface, which provides some assurance that our estimating methods are correct. We then infer the cratering history of other terrestrial planets by extrapolating the present distribution of stray bodies and using the Earth-Moon cratering record as a benchmark.

PLANET-CROSSING BODIES TODAY

Of the small bodies that have been found crossing the orbits of the planets, objects of asteroidal appearance are most common in the vicinity of the Earth and Mars, while comet nuclei predominate at and beyond the orbit of Jupiter. Traditionally, we designate a small body a "comet" if it exhibits an envelope, or coma, of gas and dust. However, a number of coma-lacking "asteroids" that cross the orbits of giant planets are very probably inactive comets.

Roughly 200 of the numbered asteroids, about 5 percent of the ones with well-determined orbits, are planet-crossing. One of these, 944 Hidalgo, crosses the orbits of both Jupiter and Saturn. It travels on a path similar to the orbits of certain short-period comets, which characteristically are ejected from the solar system after close passes by Jupiter on time scales of less than a million years. About a half dozen other Jupiter crossers have been discovered, and these also cross Mars' orbit. The ones whose colors have been measured, including Hidalgo, are similar in color as well as in orbital characteristics to weakly active comets. Another "asteroid," 2060 Chiron, crosses the orbits of Saturn and Uranus. It is very probably a large comet nucleus that remains too far from the Sun to exhibit any detectable coma, though its brightness fluctuates in a manner suggestive of possible cometary activity.

All other known planet-crossing asteroids make close approaches to one or more of the terrestrial planets. About half of these objects do not overlap any planet's orbit at present, but over time their orbits and Mars' orbit evolve in ways that lead to slight overlap and possible collisions. Most of these "shallow Mars crossers" can be distinguished from other main-belt asteroids only by elaborate theoretical investigation of their motion. Combining these studies and the results of photographic surveys, we calculate that there are about 20,000 Mars crossers brighter than absolute visual magnitude 18. (Absolute magnitude is how bright something would appear if placed 1 AU from both the observer and the Sun.) Each of these objects has about a 50-percent chance of ultimately hitting a planet or being thrown entirely out of the solar system. Most are probably fragments of bodies stranded in very shallow Mars-crossing orbits after the planets formed.

All but four of the remaining numbered, planet-crossing asteroids currently overlap the orbit of Mars and also venture inside or come close to the orbit of the Earth. Objects that currently come within 1.3 AU of the Sun but do not pass inside the orbit of the Earth have been named *Amor* asteroids; those with orbits overlapping (but larger than) Earth's are the *Apollo* asteroids. In addition, seven *Aten* asteroids have been found whose orbits have semimajor axes smaller than 1 AU (Figure 1). Most of the Aten orbits lie entirely inside the orbit of Mars, and they all overlap Earth's. Small and intrinsically faint, most known members of these three asteroid classes are observable only when relatively close to the Earth. The majority are less than 2 km in diameter. The closest known passage of any such object to Earth occurred in March 1989, when the previously unknown asteroid 1989 FC passed a scant 690,000 km away – less than twice the Moon's distance!

About 80 numbered and unnumbered Earth-crossing

asteroids have been found; these include about half of the known Amors as well as all the Apollos and Atens. Based on their discovery rate during systematic photographic surveys, the estimated total of Earth-crossing asteroids brighter than absolute visual magnitude 18 is about 1,100 (Table 1). Thus, the Earth resides within an asteroid swarm, though one about a thousand times smaller than the collection of main-belt asteroids of comparable size. Still, the production of craters on Earth by asteroid collisions occurs at one-fourth the rate as on the large main-belt asteroids Ceres and Vesta. If the orbits of the known Earth crossers are representative of the entire population's, about five objects brighter than absolute magnitude 18 strike the Earth every million years, on the average. A comparable number collide with Venus, and some collide with Mars, Mercury, or the Moon. The typical dynamic lifetime of these maverick bodies is several tens of millions of years, though some probably last much longer; over time, about a third are ejected from the solar system.

Clearly, a large majority of the Earth-crossing asteroids cannot have survived in their present orbits over most of geologic time. Yet the cratering records of the Earth and Moon suggest that the population has been in approximate equilibrium over the last 3 billion years. Apparently these objects arrive from other regions of interplanetary space at about the same rate at which they are swept up or ejected. A fraction were once shallow Mars crossers that were deflected inward during close encounters with Mars. Most have come chiefly from the main asteroid belt, where catastrophic collisions among asteroids create large numbers of fragments. Some of these then become subject to strong resonant perturbations, such as when their orbital periods are simple fractions of the orbital period of Jupiter (see Chapter 18). Over time, perturbations and encounters with Mars combine to redirect some of these toward Earth's vicinity. In addition, some Earth-crossing asteroids may be extinct comets that have been driven into very short-period orbits by close encounters with Jupiter and by the rocketlike thrust provided by their own escaping gases. Comet Encke, for example, has been propelled into an orbit like those of some Earth-crossing asteroids.

Active comets also play an important role in impact cratering on the Earth and the other terrestrial planets. Each year, on average, astronomers find about three comets with nearly parabolic orbits that pass within 1 AU of the Sun. A comet's nucleus becomes nearly stellar ("asteroidal") in appearance as observers watch it recede far from the Sun. Many lines of evidence show that the nucleus is a rotating solid body, and that the coma is derived from the nucleus by the sublimation of ices and ejection of dust (see Chapter 17). In fact, the collision of comets on the Galilean satellites of Jupiter has yielded fresh ray craters that look remarkably like ray craters on the Moon.

Combining the discovery rate of bright comets with the brightness of dormant comet nuclei observed far from the Sun, we estimate that about 30 comets with solid nuclei brighter than absolute blue magnitude 18 pass inside the orbit of the Earth each year. These appear to be coated with dark material left behind when the ices sublimate. A typical dark comet nucleus (blue geometric albedo of 3 percent) of magnitude 18 has a diameter of 2.5 km. Given the present flux, comet nuclei of this size and larger should collide with the Earth about once every 10 million years, on average. On the same basis, we expect comet nuclei at least 10 km in diameter to hit the Earth about once every 100 million years and, about once every 200 years, to pass within 4.7 million km of the Earth. (This is the miss distance of Comet IRAS-Araki-Alcock, which in 1983 passed closer to Earth than any observed comet since Comet Lexell in 1770. Radar observations showed it to have a nucleus with a mean diameter of 9.3 km.)

Today small comet nuclei pass Earth's vicinity about 10 times less often than do asteroids of the same size. But the mean impact velocity with Earth is 3.3 times higher for comets than for asteroids, so the kinetic energy per unit of mass is an order of magnitude higher. In other words, if comet nuclei and asteroids had equal densities, they would be producing impact craters now at about the same rate. Comets' densities are very poorly known, but most estimates are in the range of 0.5 to 1.2 g/cm^3 – a fraction of the probable densities of stony asteroids. Consequently, our best guess is that comets currently produce about one-fourth of the impact craters larger than 20 km in diameter (Table 2). However, since Earth-crossing asteroids are rarely more than 3 km across, most craters larger than 60 km probably result from cometary impact.

A complication in deducing the contribution of comets to the long-term crater production rate arises from past

Asteroid type	*Number discovered*	*Percent discovered*	*Estimated population*
Atens	4	(5)	80 ± 50
Apollos	36	5	700 ± 300
Amors (Earth-crossing)	15	(5)	300 ± 150
All Earth crossers	55	5	1,080 ± 500

Table 1. **By the end of 1988, astronomers had discovered several dozen Earth-crossing asteroids brighter than absolute visual magnitude 18. The diameters of objects at this limiting magnitude generally range from 0.9 to 1.7 km. *Percent discovered* values are based on the rate of discovery in systematic sky surveys with the 46-cm Schmidt telescope at Palomar Observatory, California; values in parentheses have been assumed.**

Figure 1. **The Earth-crossing asteroid 3554 Amun, a member of the Aten class, was discovered by the authors in 1986. It is 2.0 km across and composed of metal. Moving along an orbit slightly smaller but more eccentric than the Earth's, the asteroid left a trailed image 33 arc-seconds long in this 8-minute exposure taken with the 46-cm Schmidt telescope at Palomar Mountain in California. Amun probably will collide with Earth in the next 100 million years. If it hits a continent, it will make a crater at least 30 km across.**

variations in the comet flux. During intervals of tens to hundreds of millions of years, stars passing near the solar neighborhood have probably produced comet "showers" of varying intensity. Even the background flux of comets probably changes by as much as a factor of two over intervals of 10 million years. We cannot determine from comet observations alone how near the present flux is to the average. But surveys of large ray craters on the Moon suggest that comets are now coming in at about twice the mean background rate, and that the present flux roughly equals the average (including showers) over the last billion years. As we shall see, there is even a hint that we are now experiencing the tail of a weak comet shower that reached a peak about a million years ago.

IMPACT RECORD OF THE EARTH

If a projectile is large enough, it can survive passage through the Earth's atmosphere more or less intact and strike the ground or the ocean at high velocity. The threshold size for survival depends on the material strength and density of the body and on its velocity at the time of encounter; for a stony body, this size appears to be about 150 m, close to the diameter of the smallest Earth-crossing asteroid for which we have an accurate orbit. Most smaller objects are sheared apart and then slowed dramatically by aerodynamic drag (see Chapter 19). A recent example is the meteoroid that produced the great Tunguska fireball over western Siberia on June 30, 1908. This object, probably about 50 m across, failed to produce a crater but did generate a tremendous, destructive shock wave in the atmosphere. On the other hand, fragments from disrupted, relatively small iron meteoroids sometimes reach the ground with a substantial fraction of their cosmic velocity and produce swarms of small impact craters. Half a dozen such swarms are known, including a cluster of craters produced by a fall observed in the Sikhote-Alin region of the eastern U.S.S.R. in 1947. In other cases, a single large iron fragment or a very compact cluster survives and strikes the ground with sufficient energy to produce a single crater.

When a solid body strikes the ground at high speed (Figure 2), shock waves propagate into the target rocks and into the impacting body. At collision speeds of tens of kilometers per second, the initial pressure on the material engulfed by the expanding shock waves is millions of times the Earth's normal atmospheric pressure. This can squeeze even dense rock into one-third of its usual volume. Stress so overwhelms the target material that the rock initially flows almost like a fluid. A rarefaction (decompression) wave follows the advancing shock front into the compressed rock, allowing the material to move sideways. As more and more of the target rock becomes engulfed by the shock wave, which expands more or less radially away from the point of impact, the flow of target material behind the shock front is diverted out along the wall of a rapidly expanding cavity created by the rarefaction wave. The impacting body, now melted or vaporized, moves outward with this divergent flow and lines the cavity walls. Decompressed material sprays out of the cavity as an expanding conical sheet, and rocky material continues to flow outward until stresses in the shock wave drop below the strength of the target rocks. In the case of small craters, the ground motion is arrested at this time.

An iron asteroid 40 to 50 m in diameter survived atmospheric entry 50,000 years ago to form Meteor Crater, Arizona (known officially as Barringer Crater). The assortment of relatively unshocked meteorites found in its vicinity suggests that numerous fragments were detached and decelerated during atmospheric passage of the principal body, which may have broken apart just before impact. Meteor Crater is 1.2 km in diameter and 200 m deep (Figure 3). The surrounding raised rim consists partly of rocky fragments ejected from the crater and partly of bedrock that has been lifted up and shoved radially outward. The rim has a gently rolling, hummocky character that reflects the irregular distribution of large lumps in the ejected debris. Roughly half of the volume of the crater stems from material thrown out and half from the outward displacement of rocks in the crater walls. Ejected debris lies stacked on the crater rim in inverse stratigraphic order; that is, the original sequence of layers in the target bedrock is preserved but upside down.

Beneath the crater floor is a lens-shaped body of *breccia*, rock that has been smashed by the shock wave. Abundant

	Minimum crater diameter (km)						
Impacting object	*10*	*20*	*30*	*50*	*60*	*100*	*150*
Asteroids	820	180	73	10	4.5	0.3	0
Comets	(270)	60	24	8	5.3	1.7	1
All objects	(1,090)	240	97	18	10	2	1

Table 2. **The estimated production of impact craters on Earth over the last 100 million years, at or larger than the indicated diameters. Values are uncertain to about a factor of two. Comet nuclei tend to break up as they pass through the atmosphere, so the rate at which they produce craters smaller than 20 km may have been suppressed. Earth's continents should bear roughly one-third of the listed crater estimates.**

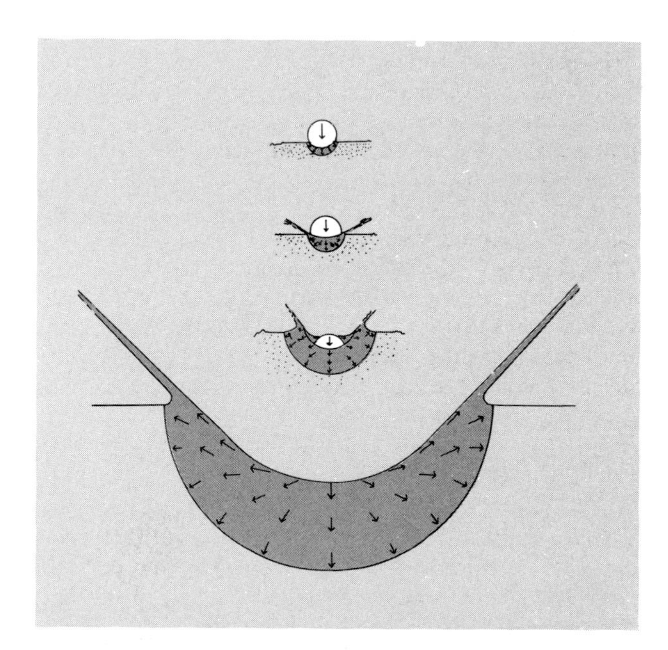

Figure 2. **A schematic representation of the stages of an impact event like the one that formed Meteor Crater, discussed in the text. A compression wave spreads outward from the point of impact; arrows indicate the flow of material. A wave of rarefaction (decompression) moving behind the shock front allows material mobilized in the event to be ejected in a conical sheet. Engulfed by the shock wave, the colliding meteoroid melts and partly vaporizes. Simulations indicate that for all but very low angles of impact, less than about 20°, the crater produced is circular.**

Figure 3. **Meteor Crater, near Flagstaff, Arizona, is one of the youngest impact craters found on the Earth. The crater formed about 50,000 years ago when an iron mass (or perhaps several) struck flat-lying sedimentary rocks at more than 11 km per second. Between 15 and 20 megatons of kinetic energy were released during the impact, which left a bowl-shaped crater 1.2 km in diameter and 200 m deep, surrounded by an extensive blanket of ejecta. The crater now has a somewhat squarish outline, lake sediments blanket the floor, and erosion has removed 15 to 25 percent of the ejecta.**

glass, produced by shock melting of the rocks, occurs near the base of the breccia. It contains microscopic spheres of meteoritic iron – the metamorphosed remains of the impacting body. The shock initially excavated a cavity to a depth of nearly 400 m. But the breccia, which was at first smeared along the cavity walls, collapsed toward the center to produce a much shallower final crater. Relatively fine-grained ejecta, arrested in its flight through the atmosphere, then showered down and left a layer of mixed debris about 10 m thick on the crater floor. Subsequently, the rim and upper walls of the crater were eroded, and lake-bed sediments 30 m thick now lie in the center.

The altered rocks found at Meteor Crater provide important clues for the identification of impact craters and structures elsewhere on Earth. In addition to shock-melted glass and distinctive macroscopic and microscopic deformation of unmelted rocks, two high-pressure forms of crystalline silica were discovered there. These shock-formed minerals, coesite and stishovite, occur at many other impact localities around the world.

Crater-hunters identify other impact sites on Earth primarily by their general structure and by evidence of shock metamorphism in the rocks. The craters themselves are ephemeral; they tend to fill in or erode away very quickly. If erosion has not encroached too deeply, some of the breccia may be preserved. Impact structures larger than about 30 km in diameter commonly have a fairly thick layer of congealed shocked-melted material, collected into a pool on the crater floor (Figure 4). As shown by Edward Anders and his colleagues, the contamination of this melt by material from the impacting body can be recognized by looking for trace elements such as the noble metals (including platinum, iridium, and gold) that are relatively abundant in meteorites but greatly depleted in the crust of the Earth.

Except for small, very young craters associated with iron or stony-iron meteorites, the impact record of the Earth generally consists of craters or eroded circular structures about 3 km in diameter and larger. A 3-km crater would be formed by a stony body roughly 150 m across – 20 times smaller – if it arrives at a velocity typical of Earth-crossing asteroids (about 17.5 km per second).

In most known impact structures, the rock walls slumped inward soon after excavation of the initial cavity. This has occurred in all craters larger than 3 km in diameter that formed in soft sedimentary rocks, and in most craters larger than about 4 km across in strong crystalline rocks. As the slumping material converged inward it produced a pronounced central hill or peak in most cases; a few large craters exhibit a ring-shaped inner ridge or more complex central structure. Evidence of uplifted rocks at the center and of subsidence and inward flow from the sides are important clues for the recognition of the deeply eroded impact structures.

Approximately 100 impact structures on Earth between 3

and 150 km in diameter have been recognized by the two basic criteria of structural form and shock-induced metamorphism. Another 30 exhibit suggestive but inconclusive evidence for impact origin. Earth's known impact craters are generally younger than 500 million years, but some of the largest are more ancient. (The geologic record is generally more complete for recent times than for older ones.)

Most of the recognized impact structures occur on the continental shields, structurally stable areas where the Earth's oldest rocks are preserved and exposed. Geologists have studied the shields in North America, Europe, and Australia most thoroughly, and that is where they have identified the majority of well-documented impact structures. An analysis by Richard Grieve suggests that on the American and European shields, an average of five craters larger than 20 km in diameter have been produced per million square kilometers per billion years. Where the geologic record is most complete in North America, the production of craters down to 10 km diameter has been about four times as great, in agreement with the size distribution of young craters observed on the Moon. These estimates pertain to the last several hundred million years of Earth's history.

Of course, the number of craters seen on the Earth should match our estimates based on the number of objects destined to collide with it. For Earth-crossing asteroids, 3.5 craters greater than 20 km in diameter should be produced per million square kilometers per billion years. If we include the contribution expected from collisions with comets, the total predicted cratering rate is in excellent agreement with the geologic record. The correspondence is good at smaller sizes, too. On average, asteroid impact should create about three craters at least 10 km across on Earth's land areas every million years, and geologists have indeed found two 10-km impact craters no older than 1.1 million years. One is in the U.S.S.R., and the other is occupied by Lake Bosumtwi, the sacred lake of Ghana's Ashanti tribe. In addition, one or perhaps two great *strewnfields* of impact glass tektites have been formed in the last million years. This glass was probably thrown from a crater or craters much larger than 10 km, but the craters themselves have not yet been recognized.

Figure 4. **The Manicouagan impact structure in Quebec, Canada, as seen from a Landsat satellite. About 210 million years old and 70 km in diameter, this structure has a central peak of shock-metamorphosed rock, surrounded by a thick layer of frozen impact melt that pooled on the original crater floor. The ring-shaped reservoir fills a valley carved by glaciers out of soft sedimentary rocks that had slumped inward along the original walls of the crater.**

IMPACT RECORD OF THE MOON

Unlike the Earth, where erosion, deformation, and renewal of the crust tend to obscure the effects of impact, the surface of the Moon preserves a pristine record of bombardment by solid bodies that extends back several billion years. Because of the absence of an atmosphere, even microscopic particles strike the Moon at high speed, producing countless tiny craters on exposed rocks. Larger particles pound and churn the surface into a layer of ground-up rocky debris (called the regolith) that blankets nearly all of the Moon. This layer averages about 3 m thick where it covers lava flows some 3 billion years old. Most of the Moon's surface has been darkened by accumulation in the regolith of black, impact-produced glass.

Asteroids and comet nuclei striking the Moon have produced conspicuous craters long familiar to telescopic observers (Figure 5). The youngest of these are surrounded by bright deposits of freshly excavated rock that extend radially outward in discontinuous bright streaks called *rays*. One of the youngest and perhaps the most spectacular of the Moon's large ray craters is Tycho, 85 km in diameter (Figures 6,7). A continuous deposit of ejecta surrounds Tycho beyond its rim crest for another 85 km on average. The rays form a great asymmetric splash whose fingers can be traced to distances of up to about one-fourth of the lunar circumference.

Within the rays and the outer part of the continuous ejecta deposit are abundant small *secondary craters*, formed by chunks of rock and clots of debris flung out of Tycho. Just outside the crater rim's crest, which rises 1 km or so above the average level of the surrounding terrain, ejected debris flowed down portions of the outer slopes in viscous, lobate masses. Some of the most fluid, shock-melted material (termed *impact melt*) filled small depressions with smooth, dark deposits resembling frozen ponds or lakes. The terraced crater walls arose when great slabs of rock slumped inward during collapse of the initial cavity. Shock-melted rock pooled on the terraces and created frozen rivulets several kilometers long down to the main crater floor, itself filled with a deep layer of once-molten material. A prominent central peak, formed by the inward and upward flow of material as the crater walls slumped, rises about 2 km from this now-solid "lake bed."

All of the events that led to Tycho's present appearance took place rather quickly. Some, like the formation of impact melt and the excavation of the initial cavity, happened within a few minutes. Slumping of the crater walls and formation of the central peak may have taken an hour or so. The rain of ejecta, formation of secondary craters, and emplacement of impact melt probably occurred within a few hours after the impact. The times involved for some of these events vary at other impact sites, taking proportionally longer as the crater size increases.

Other large ray craters on the Moon closely resemble Tycho. However, craters smaller than about 20 km in diameter generally have smooth walls devoid of terraces. Evidently, up to this size the bedrock walls are strong enough to prevent collapse in the weak lunar gravity (which at the surface is about one-sixth that at the surface of the Earth).

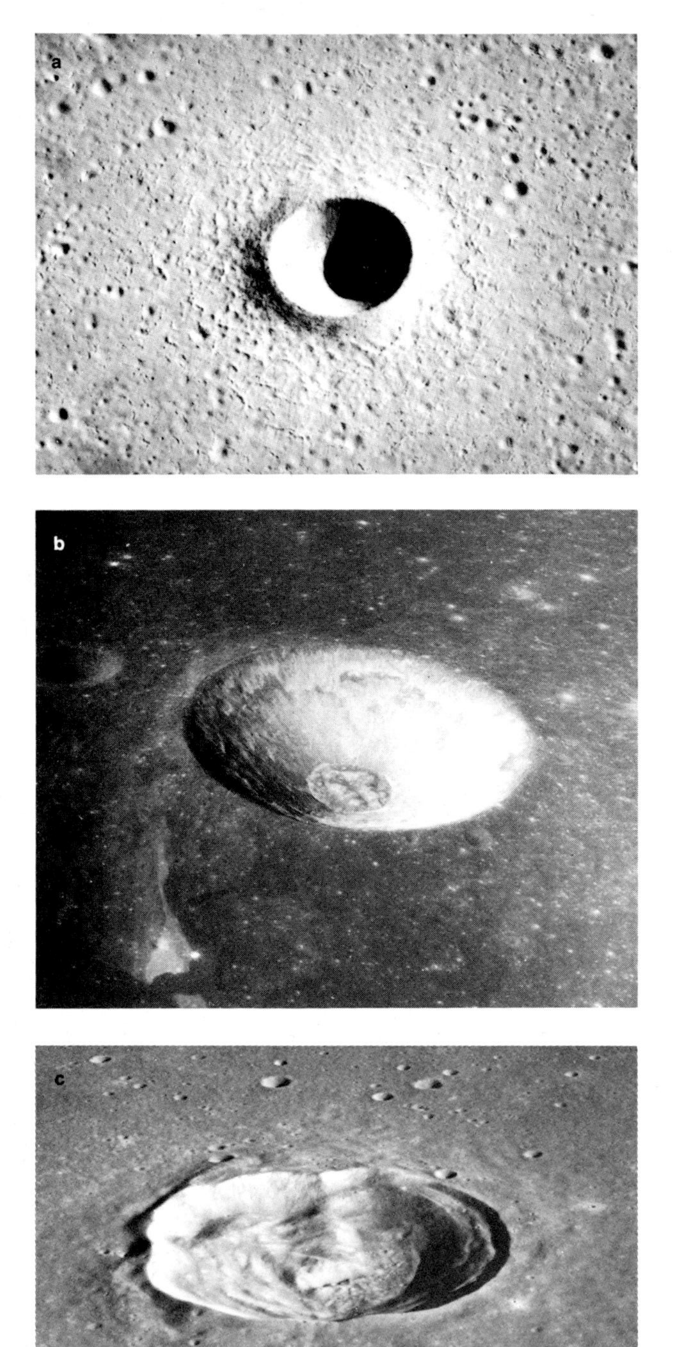

Figure 5. **Large lunar craters exhibit more complex characteristics than their smaller siblings. Linné *(a)*, photographed from Apollo 15, is 2.5 km across and has a smooth bowl shape. As seen by Apollo 10, Taruntius H (8.5 km across) also has a smooth profile *(b)*, but its floor has been filled and leveled by rubble from the rim. After the impact that formed the crater Arago *(c)*, the rim walls apparently slumped, creating terraces and enlarging the diameter to 26 km (Apollo 10 photograph). Note that the primitive central peak merges with the crater's walls. Compare these craters with the much larger Tycho (Figure 7).**

Where terraces are missing, central peaks are absent as well. Craters 7 to 20 km across generally have fairly smooth, level floors, which probably consist of impact breccia that settled from the walls of the initial cavity, as at Meteor Crater. Pools of frozen impact melt are not easily recognizable in or around these smaller craters. Below diameters of about 7 km, lunar ray craters have a simple bowl shape.

Certain features are characteristic of all ray craters. The exterior slopes of each rim are marked by a pattern of small hills or hummocks (again resembling Meteor Crater) and by lower-lying ridges that are roughly radial in orientation. Abundant small secondary craters invariably occur near the outer limit of the ejecta blanket and in the rays. As small meteoroids continued to pound the lunar surface, the rim deposits and floors of old craters gradually darkened and the rays disappeared. About two-thirds of the large craters found on the Moon's great lava plains (the maria) lack rays, but hummocky rim deposits and surrounding swarms of secondary craters confirm their impact origin.

Our inventory of lunar samples comes from only a small number of landing sites. But thanks to isotopic dating techniques, the ages of these samples can be determined very accurately (see Chapter 4). Meanwhile, geologic units at the landing sites can be correlated in relative age with similar units in other areas using orbital photography and the relationships of superposition of the mapped units. Age correlations can also be made by matching different units' areal densities of small craters (the older a given surface, the more superposed small craters it will have of any specific size).

Using these methods, Don E. Wilhelms has mapped all craters at least 10 km across that formed on the lunar nearside in the last 3.3 billion years, after many of the Moon's maria had appeared. He concludes that since then an average of 10 craters in this size range have been produced per million square kilometers per billion years; the corresponding rate for craters at least 20 km in diameter is 2.4. It is possible to extrapolate these rates to the Earth, but one must take into account the differences in gravity here and on the Moon, which affects sizes of the initial cavities and how readily their walls slump (thus enlarging the crater diameters). Additionally, Earth's stronger gravitation draws in the bodies in its vicinity better and increases their collision speed. When all these corrections have been made, Wilhelms' lunar values yield rates on Earth of 12.5 and 2.3 craters per million square kilometers per billion years (for craters at least 10 km and 20 km across, respectively). These are about half what we find either from the present flux of asteroids and comets or from the Earth's geologic record of impact over the last several hundred million years.

Is this discrepancy real, or is it due to uncertainties either in our calculations or in the statistics of Earth's cratering record? Quite possibly, the difference *is* real – collisions on the Earth may well have occurred twice as often during the last several hundred million years as during the previous 3 aeons. We believe the long-term flux of asteroids derived from the main belt has remained steady to within about 10 percent, so variations in their arrival rate are probably not the cause. Rather, we must look to the comets to explain a doubling of the cratering rate in recent geologic time. Perhaps a few relatively strong comet showers caused the average number of comets passing near the Earth to increase, or maybe the "background" rate of Earth-crossing comets has risen. Possible sources of near-Earth comets will be discussed in detail later.

In tracing the geologic record of the Moon, we find that collisions occurred very frequently prior to the eruption of the mare lavas but dropped off rapidly between 3.9 and 3.3 billion years ago (Figure 8). For example, the crater density found on a great sheet of ejecta surrounding the Imbrium basin, formed 3.85 billion years ago, is six times greater than on lava plains that erupted only 550 million years later. The abrupt decrease in the intervening impact rate can be mimicked mathematically by combining one cratering rate that decayed exponentially (a 50-percent decrease every 100 million years) with another that remained steady with time. According to this model, the rate of formation of craters dropped off by a factor of 35 between 3.85 and 3.3 billion years ago. An even steeper decline apparently occurred between the formation of the Nectaris basin, estimated at 3.92 billion years ago, and the time of the Imbrium basin. Evidently, 11 of the Moon's largest impact basins (from 300 to 1,060 km across, or roughly one-fourth of its total) formed in this one 70-million-year interval; moreover, about two times more smaller craters appeared in that brief span than in the billions of years that followed it.

From these observations we can infer the existence of two general populations of Earth-crossing bodies. One group became depleted with time as they collided with the planets or

Figure 6. **The ray pattern of the crater Tycho, 85 km across and situated in the Moon's southern hemisphere, is strikingly distinct.**

Figure 7. **Lunar Orbiter 5 photographed Tycho under oblique lighting that emphasized its topographic detail, which is much more complex than that of Arago (Figure 5). Clearly visible are hummocky debris on the rim, terraced walls, a prominent central peak, and the frozen pool of impact melt on the crater floor.**

were ejected from the solar system. The second group corresponds to the Earth-crossing objects we observe today – a population that is apparently renewed from other regions of the solar system at approximately the same rate that it is lost. Without this renewal, the "steady" population would quickly dwindle away.

The first, rapidly decaying population may have originated in at least three ways: (1) the impacting bodies were a remnant of the principal batch of small *planetesimals* from which most of the Earth accreted; (2) they were injected into Earth-crossing orbits from the asteroid belt, perhaps following major collisions there; (3) a large planetesimal of Uranus or Neptune was perturbed into an Earth-crossing orbit and became tidally disrupted (fragmented) during a close approach to one of the terrestrial planets. The first two sources might suffice to account for the population's 100-million-year half-life. But only the sudden injection of objects on collision-prone orbits, as in the third case, would satisfactorily account for the apparently abrupt drop in the cratering rate just after 3.92 billion years ago.

The episode of rapid cratering near 3.9 billion years ago has been referred to as the *late heavy bombardment* (LHB). One view holds that the heaviest pummeling occurred as the Moon formed about 4.5 billion years ago and that the bombardment and cratering simply declined in a generally steady way for about $1\frac{1}{4}$ billion years thereafter. From this standpoint, the LHB simply represents the final vestiges of planetary accretion. All evidence of the Moon's earlier, more intense bombardment has been lost or obscured by emplacement of the final large basins and their regional deposits of ejecta.

But the very fast decay of the cratering rate between 3.92 and 3.85 billion years ago argues instead that a discrete pulse occurred, as first suggested by Fouad Tera, Dimitri Papanastassiou, and Gerald Wasserburg. In our view, it seems most likely that multiple pulses were superimposed on a general decline in cratering rate and that the LHB was the last of these pulses. The best explanation for pulses is the breakup of large objects on Earth-crossing orbits, either by collision or tidal disruption. As we show later, a likely source for the last large broken object is the region of the outermost planets.

MERCURY AND VENUS

The production of craters on Mercury by comets and asteroids appears to be roughly comparable with the rate here on Earth (Table 3), though the proportions due to various classes of impactors differ. Fewer than one-fifth of the known asteroids traversing Earth's orbit pass inside Mercury's as well. But the confluence of their orbits occurs in the relatively small volume of space between Mercury and the Sun, and this increases their likelihood of impact. Moreover, about 10 percent of the asteroids that can hit Mercury should occupy orbits that lie entirely inside the Earth's. Finally, an object nearing Mercury travels faster, because of the Sun's gravitational acceleration; when it strikes the planet, the impact energy is greater and a proportionately larger crater is formed (Mercury's relatively weak surface gravity compared with Earth is also a factor). Taken together, these effects partly offset the markedly fewer asteroids that pass in the vicinity of Mercury and yield a cratering rate per unit area of surface about half that on the Earth.

Roughly one-third of the comets passing inside the Earth's orbit also penetrate inside Mercury's average aphelion (the point in its orbit farthest from the Sun. The same orbital congestion that increases the chance of asteroidal impact also holds for comets, resulting in a cratering rate from cometary objects about 30 percent higher for Mercury than for the Earth.

Since the same basic classes of objects strike the Moon, Earth, and Mercury, some idea of Mercury's cratering history can be deduced from the impact records of the other two. For example, one would expect that, if the flux of crater-forming objects near the Earth changed, at least in recent eras, a corresponding change should have occurred for Mercury. We calculate that the present production of craters greater than 20 km diameter on this innermost planet is about two-thirds of that on the Moon, a proportion that may have remained nearly the same for several billion years. It was thus no surprise when Mariner 10 revealed in 1974 that Mercury has a heavily cratered surface very similar in appearance to the Moon's. Fresh craters on Mercury look remarkably like their counterparts on the lunar maria: the youngest ones have rays, but older ones do not. The most distinctive difference exists in the patterns of secondary craters; these lie closer to the rims of their primaries on Mercury than on the Moon, because Mercury's higher surface gravity confines crater ejecta to shorter ballistic ranges.

Mercury's surface, like the Moon's, bears a cratering record that probably extends far back in time. Geologist Paul

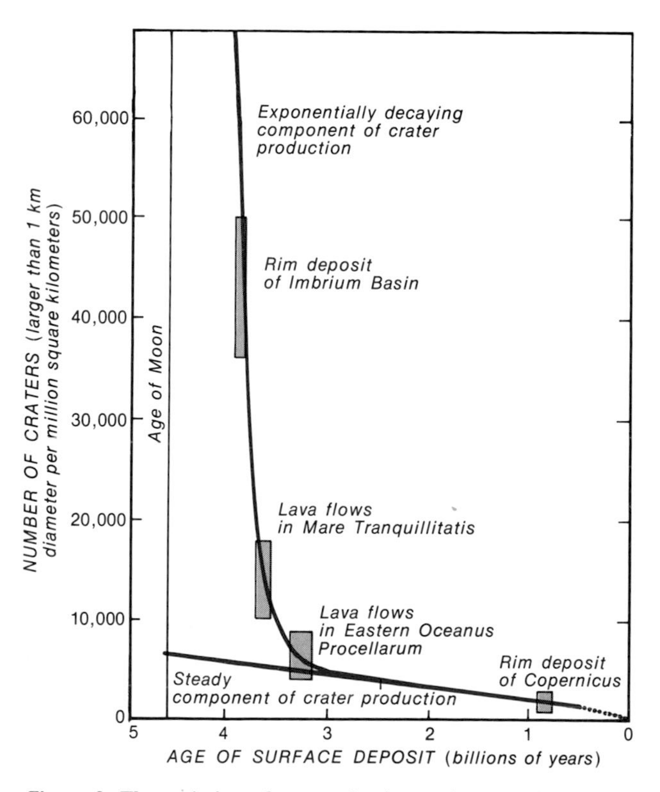

***Figure 8.* The variation of crater density on lunar surfaces with different ages. Widths of the small rectangles (which correspond to Apollo landing sites) indicate the uncertainty or possible range in age of each dated surface, and their heights indicate the statistical uncertainty in crater density. The high cratering rate during the late heavy bombardment dropped rapidly between 3.9 and 3.3 billion years ago, giving way to a slower, steady rate of crater production.**

Impacting object	*Mercury*	*Venus*	*Earth*	*Moon*	*Mars*
"Venus-crossing asteroids"	0.4	0.4			
Earth-crossing asteroids	1.5	3.1	3.5	3.5	1.2
Mars-crossing asteroids					≈6
Comets	1.6	(1.3)	1.2	1.5	0.6
All objects	3.5	3.5	4.7	5.0	≈8

Table 3. **Estimates of the current cratering rates on the inner planets caused by collisions with asteroids and comets. Each entry is the average number of craters larger than 20 km in diameter produced on a surface area of 1 million square kilometers every billion years. These rates are uncertain by a factor of about two. *"Venus-crossing" asteroids* have not yet been observed but should theoretically exist; they cross the orbit of Venus (and in some cases Mercury's) but not the orbit of Earth. *Mars-crossing asteroids* do not cross the orbit of any other terrestrial planet. The entry in parentheses refers to the theoretical crater production due to comets if Venus had no atmosphere.**

Spudis finds that craters 20 km across are about six times more abundant on the smooth plains of Mercury than on the 3.3-billion-year-old lunar maria. This suggests that the smooth Mercurian plains were in place about 3.9 billion years ago, somewhat earlier than when the Imbrium basin formed on the Moon. In terms of packing density and size distribution, Mercury's highland craters match those of the lunar highlands fairly closely (Figure 9), which has led Robert Strom to conclude that the same population of bodies created most of the craters in both regions. If the orbital distribution of the impacting bodies resembled what we now observe among the Earth-crossing asteroids, the cratering rates on Mercury may have been about half the rate on the Moon during the late heavy bombardment. The enormous Caloris basin, which is probably the last giant impact structure created on Mercury, exhibits about three-fourths the density of craters as the Nectaris basin on the Moon and probably predates it slightly.

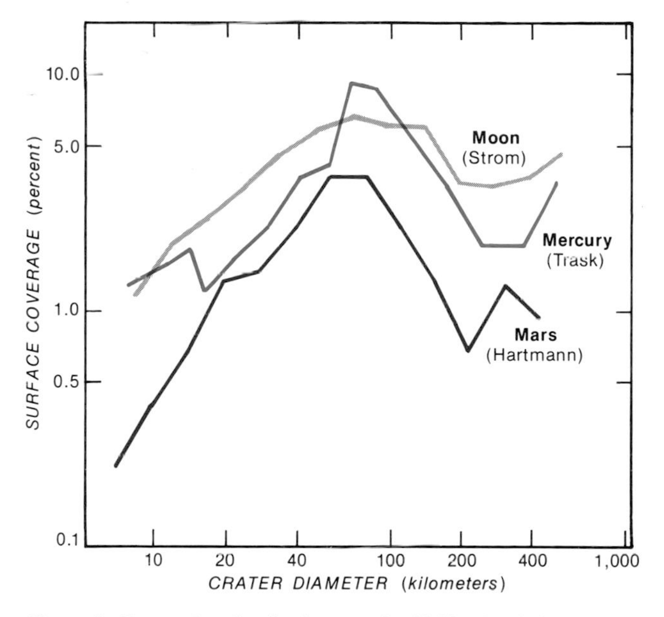

Figure 9. **Crater size distributions on the highlands of the Moon, Mercury, and Mars (authors whose data have been used are named by each curve). The ordinate shows the percentage area of the surface covered by craters of a given diameter. The coverage on all three bodies peaks for craters with diameters of 40 to 100 km, and the distributions' similar shapes suggest that the same population of impacting projectiles produced most of the craters on the highlands of all three planets.**

The impact record of cloud-veiled Venus remained a mystery until very recently. During the 1970s, low-resolution images from Earth-based radar telescopes displayed many circular features on the planet that were thought to be large craters and impact basins. From these observations, geologists concluded that the density of large craters was somewhat similar to that of the heavily cratered regions of the Moon and Mercury. It was a puzzling result. Venus is similar to the Earth in size, mass, and presumably its heat budget, and the planet's surface is hundreds of degrees hotter than Earth's. Consequently, if Venus' geology were truly Earthlike, all that heat should have made its lithosphere (the upper surface layer) relatively thin and susceptible to deformation during the convective stirrings of the interior. How, then, could ancient craters be preserved?

The mystery was resolved by the Soviet spacecraft Veneras 15 and 16, the results of which are covered more fully in Chapter 7. The Veneras' radar images of the planet's northern hemisphere reveal about 150 features, ranging from 20 to 140 km across, that resemble impact craters (Figure 10). The areal density of these features turns out to be much lower than the crater density on the lunar highlands – but remarkably similar to the density of impact structures found on the American and European continental shields of Earth. The Venera orbiters also observed huge, enigmatic ring structures called ovoids or coronas. These do not resemble impact craters but may instead be a consequence of internal convection and volcanism (which, conceivably, might have been triggered by impacts).

Almost precisely half of the discovered Earth-crossing asteroids are Venus-crossing as well. These objects are more likely to collide with Venus than the Earth crossers are with Earth. Also, impact speeds are greater at Venus, and its surface gravity is slightly lower. Finally, about 5 percent of the asteroids striking Venus must belong to an as-yet undiscovered class of objects that remain entirely inside the orbit of Earth. When these factors have been accounted for, the calculated cratering rate by asteroidal impact is nearly the same on Venus and on Earth (Table 3). Similar calculations show that the cratering rate by cometary impact would be about 14 percent higher on Venus than here – *if* that planet did not have its dense atmosphere.

In fact, the degree to which Venus' atmosphere "shields"

its surface is the single largest uncertainty in estimating the planet's cratering history. Geophysicist Jay Melosh has calculated that the breakup of stony asteroids penetrating the atmosphere of Venus would inhibit the creation of craters much smaller than 20 km across (as seems to be the case from the Venera images); similarly, only the largest comet nuclei would survive, producing craters no smaller than about 100 km. Above the 20-km threshold, the size distribution of Venus' putative impact craters resembles that of young craters on the Moon. Since comets, rather than asteroids, probably create most of the young craters on the Earth and Moon larger than 60 km in diameter, we suspect that the largest craters on Venus have been produced by comet hits.

Based on the craters observed in the Venera radar images, Soviet investigators Boris A. Ivanov, Aleksandr T. Bazilevskiy, and their colleagues estimate that Venus' northern hemisphere has a mean age of about 1 billion years, give or take half of that value. However, Gerald Schaber, Eugene Shoemaker, and Richard Kozak suspect that the surface is actually no more than 400 million years old. Their estimate was extrapolated from Earth's record of recent asteroid-produced craters at least 20 km across; moreover, it assumes that Venus' atmosphere has cut the production rate of 20-km craters by up to 50 percent and that only the largest comet nuclei ever reach the surface. Perhaps not coincidentally, the mean age of Earth's continental and seafloor crust is 450 million years. Should some fraction of the craters on Venus ultimately prove to be of volcanic origin, the average surface age could be younger still.

IMPACTS ON MARS

At present, Mars is being bombarded mostly by a group of asteroids that do not cross the Earth's orbit. This means that the cratering history of Mars does not relate to Earth-Moon system's as closely as those of Mercury and Venus do. The present formation rate of 20-km craters on the red planet probably is somewhat higher than on Earth. Comets contribute about 10 percent of the impacting bodies, Earth-crossing asteroids another 15 percent, and Mars-crossing asteroids the remainder (Table 3).

The Martian landscape abounds with impact craters. However, many have been significantly modified by erosion, sediment deposition, or volcanism; those that have not resemble, in most respects, craters on the Moon and Mercury. One rather remarkable difference is that the ejecta around most large, fresh Martian craters appears to have hugged the ground as it flowed radially outward, rather than being deposited by ballistic ejection (Figure 11). This fluidlike behavior may indicate that the target material contained ice, which melted or vaporized when struck, or perhaps atmospheric gases became trapped beneath the outward-moving ejecta blanket.

The geology of Mars is much more diverse than the geology of the Moon or Mercury (see Chapter 5). Many of the Martian plains are only sparsely cratered, while others are cratered about as heavily as the lunar maria (Figure 12). This diversity implies that the Martian surface varies greatly in age from place to place, from relatively recent geologic time to several billion years. If the flux of Mars-crossing asteroids has been steady with time, some of the most ancient plains should be about 3 billion years old. However, this age is similar to the typical lifetimes of very shallow Mars-crossing asteroids. Through collisions with Mars and ejection from the solar system, the number of Mars crossers has dwindled steadily. So Mars' cratering rate was probably higher in the past, and some of the more ancient plains may actually be no more than about 1 or 2 billion years old.

The craters in the heavily pummeled Martian highlands, found almost entirely in the southern hemisphere, resemble in both abundance and size distribution the highlands of the Moon and Mercury. As with Mercury, the lunarlike size

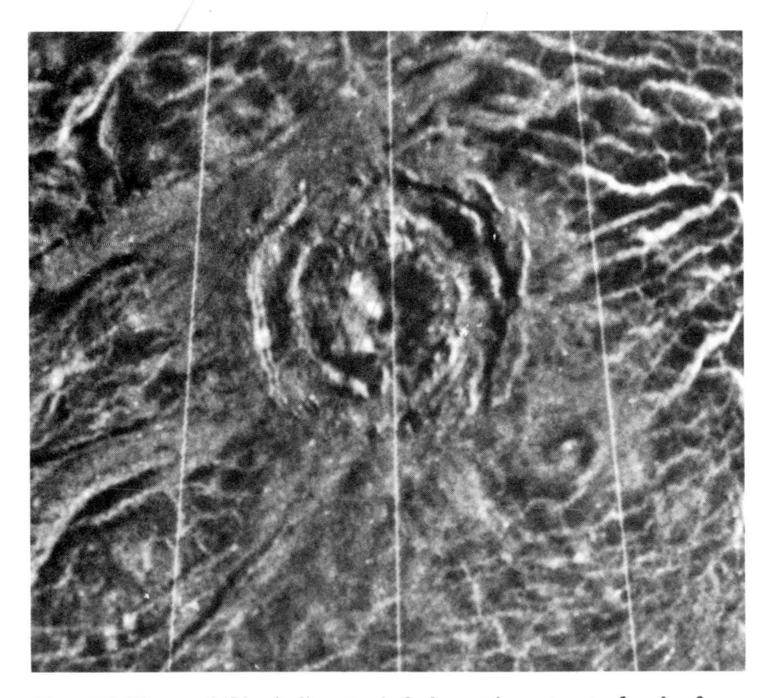

Figure 10. **Klenova, 142 km in diameter, is the largest impact crater found so far on Venus. It is shown in a radar image obtained by the Soviet Union's Venera 15 and 16 spacecraft. The multiple-ring structure is typical of all giant impact features on the terrestrial planets.**

Figure 11. **Part of the rim deposit of the Martian crater Arandas (28 km in diameter) as seen from a Viking orbiter. The rim deposit apparently flowed as a fluid or fluidized system, forming several superposed lobes.**

distribution suggests that one population of impacting bodies was responsible for both the Moon's late heavy bombardment and most of the highland craters on Mars. If these bodies had orbits like those of modern-day Earth-crossing asteroids, their cratering rate on Mars would have been only 40 percent of their rate on the Moon during the late heavy bombardment. On the other hand, if enough of them originated in the outer solar system, Mars and the Moon could have been subjected to about equally intense battering at that time. Either way, the Martian highlands bear a record of impact scars that may span nearly 4 billion years.

OUTER-PLANET SATELLITES

The Voyager missions have extended our firsthand knowledge of collisions from the orbit of Mars to the orbit of Uranus and beyond. In these ever-colder regions of the solar system the record preserved for our examination is written in ice. It is the history of icy planetesimals and comet nuclei, and of swarms of icy shards slamming at high speed into the frozen crusts of the outer-planet satellites. In recent times, the bombardment has been dominated almost entirely by comets traversing this region of the solar system.

Some important members of the cast of impacting bodies are the so-called Jupiter family of short-period comets. These objects were likely long-period comets that became "captured" by Jupiter after a succession of close encounters with it and perhaps other major planets. If a comet revolves around the Sun in the same direction the planets do, its orbital period shortens if it chances to pass close to and ahead of Jupiter; encounters behind Jupiter lengthen the period. An orbital period of less than 20 years (typical of the Jupiter family) arises only after an unlikely series of passes ahead of Jupiter or an even less likely single pass at extremely close range. Understandably, this happens to only a very few long-period comets, and the entire process takes, on average, about 200,000 years.

Usually, once trapped by Jupiter, a comet evolves rapidly. Ices near the surface of its nucleus are lost during repeated close approaches to the Sun. In a few hundred to a few thousand revolutions, the surface becomes coated with a residue of rocky and organic material that insulates the comet's interior during its passages through perihelion. Hence the comet nucleus no longer provides a source for the distinctive coma – it "dies," becoming an inert body that resembles an asteroid. The asteroids 944 Hidalgo, 3552 (1983 SA), and 1984 BC are examples of extinct comets. Only a few of these objects have been discovered because they are difficult to find once they become inactive, and they usually remain relatively far from Earth. Since most of these objects stop producing a coma long before they collide with a planet or are ejected by Jupiter's strong gravitation, the Jupiter family contains perhaps 10 times as many extinct comets as active ones.

Table 4 lists the estimated cratering rates due to the collision of active and extinct comet nuclei with the Galilean satellites. The greatest uncertainties here lie in determining the fluxes of extinct and long-period (active) comets in the neighborhood of Jupiter. Recent finds of Jupiter-crossing asteroids and distant long-period comets have convinced us that these bodies are three times more numerous than previously believed. Despite the uncertainties, the estimated *relative* rates of cratering among the different satellites should be fairly accurate.

The paths of comets near Jupiter, particularly the short-period ones, are strongly focused by the planet's gravitational field. As objects "fall" toward the planet, their

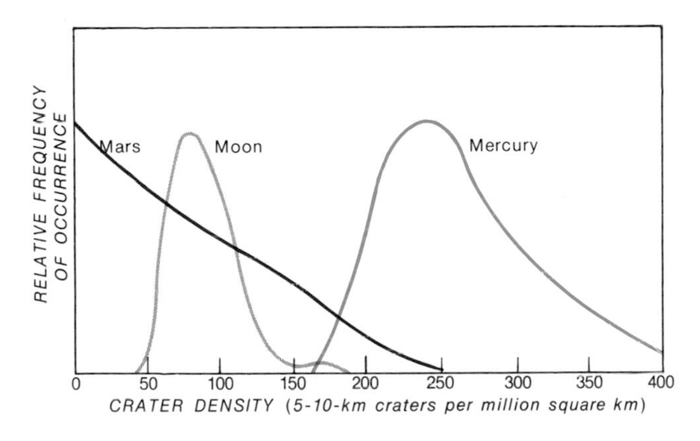

Figure 12. **Compare the densities of craters 5 to 10 km in diameter on the plains of three terrestrial planets. The plains on Mercury generally have the highest crater densities, while Mars' broad range of crater densities indicates that its plains have diverse surface ages.**

	Satellites of Jupiter				*Satellites of Saturn*				*Satellites of Uranus*				
Impacting object	*Io*	*Europa*	*Ganymede*	*Callisto*	*Mimas*	*Tethys*	*Dione*	*Rhea*	*Miranda*	*Ariel*	*Umbriel*	*Titania*	*Oberon*
Long-period comets	3.9	4.6	3.2	2.5	3.3	2.0	1.2	0.8	0.7	0.4	0.4	0.3	0.3
Jupiter-family comets	3.5	2.8	1.3	0.7	0.2	0.1	0.05	0.03					
Extinct comets	30	25	12.0	5.5									
Saturn-family comets					≈10	≈4	≈2	≈1.3					
Uranus-family comets									≈23	≈9.2	≈5.6	≈2.3	≈1.6
All objects	37	32	16.5	8.7	≈14	≈6	≈3	≈2	≈24	≈10	≈6	≈3	≈2

Table 4. **Estimates of the current cratering rates on the satellites of Jupiter, Saturn, and Uranus by cometary impact. Each entry is the average number of craters larger than 20 km in diameter produced on a surface area of 1 million square kilometers every billion years. Most craters on the satellites of Saturn and Uranus mark the impacts of comets captured into short-period orbits by these planets; these objects' populations and orbits are chiefly theoretical derivations.**

velocities also increase. Consequently, the present cratering rate on Io, the innermost Galilean satellite, is about four times that on Callisto, the outermost one. In turn, Callisto's rate (at least for craters larger than 10 km across) is presently about twice that on the Moon. However, no impact craters at all have been found on Io and very few on nearby Europa, due to the obliterating effects of volcanism and other surface processes (see Chapter 13).

Voyagers 1 and 2 revealed that the icy crusts of Ganymede and Callisto, the two largest Galilean satellites, are densely cratered. The satellites' ray craters look enough like their lunar counterparts to show that they are of impact origin. As on the Moon, these satellites developed regoliths that have gradually darkened with time, and around older craters the rays have disappeared. Rays persist longer on Callisto than on Ganymede, though the retention time varies over the surface of each satellite. An even more interesting effect is the slow disappearance of the craters themselves by plastic flow of the ice in which they were created. Many large craters have vanished completely, leaving only vague, discolored patches on the surface, called *palimpsests*, and telltale swarms of secondary craters. Most of the remaining craters have been flattened by flow of the icy crusts. We estimate that the ray craters on Callisto and Ganymede were created no more than about 100 million to 1 billion years ago.

Callisto has more craters than can be readily accounted for by extrapolating the recent average rate of cometary impacts back over geologic time; the same is true of dark, heavily cratered areas on Ganymede. Apparently, a period of heavy bombardment took place in this region of the solar system at some time in the past. It probably did not occur within the last 3.3 billion years, because many of these Jupiter-crossing bodies would have crossed the Earth's orbit and left their signature in the cratering record of the Moon. However, the excess collisions could easily have taken place earlier, perhaps roughly coincident with the late heavy bombardment of the Moon. Plastic flow in the icy crusts has prevented us from learning whether the original size distribution of ancient craters on Ganymede and Callisto resembled that of the lunar highlands. But if the breakup of a large body orbiting among the terrestrial planets was responsible for the LHB pulse on the Moon, its fragmentation should not have significantly affected the Galilean satellites.

Callisto also has several enormous sets of nested concentric ridges. At their centers are palimpsests, which probably mark the sites of former impact basins. Remnants of another multi-ring system are found on Ganymede. If the past variations of impact rates on Ganymede and Callisto more or less mimic those observed in the Earth-Moon system, then the great multi-ring impact sites may be about 3.6 to 3.8 billion years old – somewhat similar to the ages of the last large lunar impact basins formed during the LHB.

At Saturn, just as at Jupiter, collision-bound objects are now almost entirely comets. Long-period comets may account for about 25 to 35 percent of recent craters on the inner satellites of Saturn, and Jupiter-family comets less than 1 percent (Table 4). The remaining craters are produced by a family of objects captured by Saturn from the long-period population. Most of these remain too far from the Sun to generate observable tails or comas, so their existence is inferred almost entirely from theoretical arguments. The large "asteroid" 2060 Chiron is the only apparent member discovered so far, but the total number with nuclei larger than 2.5 km across may be about 10,000.

Images acquired during the Voyager encounters with Saturn disclosed another set of icy satellites with heavily cratered surfaces. Once again, only a fraction of the observed craters can be accounted for by extending the present collision rate of long-period and Saturn-family comets back into geologic time. One or more episodes of heavy bombardment are indicated.

In contrast with the icy satellites of Jupiter, there is little evidence that plastic flow of ice has erased any except the largest craters on most of Saturn's very cold satellites. The giant crater Odysseus, preserved on the leading hemisphere of Tethys, has clearly been flattened, though the crater is still easily recognized. Parts of heavily cratered Rhea resemble the lunar highlands, yet large impact basins seem to be absent and probably have been lost or nearly erased by plastic flow. On Enceladus, remnants of a heavily cratered surface are preserved. Here, relatively *small* craters have been flattened by viscous flow, perhaps indicating an episode of internal heating that was associated with complete resurfacing of part of the satellite.

Mimas and parts of Tethys and Dione have crater populations quite distinct from those found in the inner solar system. Craters 5 to 20 km across pepper these surfaces, but they are practically unaccompanied by craters larger than about 40 km. The implication is that most collisions on these surfaces involved fragments derived from the satellites themselves or from other small satellites sharing the same orbits. Countless icy shards of this type would be created during high-energy impacts on Mimas. One such collision is marked by the crater Herschel, which is about one-third the satellite's diameter. The scene probably was repeated many times during heavy bombardment. Indeed, Mimas, Enceladus, Tethys, and Dione probably were all "destroyed" by large impacts, only to reassemble themselves from rings of debris left in orbit around Saturn. Evidently, some of these fragments were stranded near Lagrangian points along the orbits of the reassembled satellites and are preserved today as tiny "Trojan" satellites sharing the orbits of Tethys and Dione.

If a satellite lies too close to Saturn, the individual fragments produced during its catastrophic breakup will not reform as a single object. Instead, tidal forces will constrain the debris to remain in a ring. This almost certainly is how the well-known rings of Saturn came to exist. A former satellite 60 percent of the size of Tethys (about 600 km in diameter) took a large hit and disintegrated. The resulting fragments, in turn, were shattered by impacts again and again until almost the entire mass was ground to the sizes of the present ring particles (see Chapter 12). A few remaining coarser fragments survived to form tiny satellites that have been discovered near the outer edge of the ring system.

Beyond Saturn, our ability to estimate small-body populations and the rates at which they strike distant satellites becomes progressively weaker. Nevertheless, we can make a stab at the problem by extrapolating from our knowledge of comets in the inner solar system. At the orbit of Uranus, the flux of long-period comets is roughly one-fourth that at the orbit of Jupiter. On the other hand, we estimate that Uranus has captured a current population of about a million comet nuclei larger than 2.5 km in diameter into moderately short-period orbits. So cratering on the Uranian satellites presently is controlled by the planet's family of captured comets (Table 4). As in Jovian and Saturnian systems, the cratering rate increases steeply toward the innermost satellites, owing to gravitational focusing. Oberon is now struck only about half as often as the Moon is, while Miranda is hit at about five times the Moon's rate. Collisions happen even more often on the small satellites discovered by Voyager 2. Puck, the largest of these, probably is a fragment of a larger body broken apart by impact about 3½ billion years ago, and the innermost moons may be collisional fragments spawned within the last 500 million years.

Voyager 2 revealed that abundant impacts have scarred the surfaces of all the large Uranian satellites, which in most cases cannot be explained by extrapolating the present cratering rate back over solar-system history. Hence these cratered surfaces also seem to preserve a partial record of an episode of heavy bombardment that probably occurred during the first billion or so years of dispersal of a swarm of Uranus and Neptune planetesimals. As with some of Saturn's inner satellites, it appears as though Titania and Ariel were pummeled by fragmental debris in orbit around Uranus. Probably all of the planet's satellites, except possibly Oberon, were catastrophically disrupted by collisions with planetesimals. The rings of debris so produced then reassembled into the satellites we see today. From the record of impacts preserved on Oberon, we calculate that Miranda endured a bombardment 12 times more intense and that, remarkably, it was shattered and reassembled about five times! As at Saturn, collisional debris produced very close to Uranus has remained dispersed as rings. The collisions that formed most of the observed rings probably occurred within the last few hundred million years.

THE OORT CLOUD OF COMETS

In 1950, the Dutch astronomer Jan Oort showed by carefully analyzing the most accurate orbits that the source of long-period comets is a gigantic, diffuse cloud of comets surrounding the Sun at an average distance of about 40,000 AU. The cloud's outer limit extends about halfway to the nearest known star, so its members are only weakly bound by gravity to the Sun. A combination of galactic tidal forces and passing stars steadily shuffles the orbits of the comets in the cloud, redirecting some of them into the inner solar system, where they can be observed. To account for the three or so long-period comets discovered annually near the Earth, Oort estimated that there must be about 100 billion comets in the cloud. He also suggested that the comets originated as planetesimals of Jupiter, thrown into the cloud long ago by a succession of perturbations with the young, massive planet.

But more detailed study indicates that Jupiter's perturbations were actually *too* strong, slinging most of the planetesimals in its vicinity completely out of the solar system. Although some Jupiter planetesimals may reside in the Oort cloud, most of the objects there probably originated near Uranus and Neptune. Emplacement of these planetesimals in the Oort cloud occurred over a protracted period of early solar-system history. Their orbital eccentricities were gradually amplified by close encounters with Uranus and Neptune. After a while, many Uranus planetesimals were perturbed into Saturn-crossing orbits, ultimately falling under the controlling influence of Saturn and then Jupiter. Similarly, Neptune planetesimals were occasionally perturbed into Uranus-crossing orbits, and some of these, in turn, passed to the control of Saturn and Jupiter. All together, about half of the Uranus-Neptune planetesimals assumed this evolutionary track and were ultimately expelled from the solar system.

Most of the remainder were perturbed into huge, long-period orbits. Encounters with passing stars then dispersed part of this swarm into the diffuse halo around the Sun that we recognize as the Oort cloud (see Chapter 17). The neighborhood of the planets gradually emptied out, as roughly half of the comets were ejected or found their way into the Oort cloud between 4.55 and 4.4 billion years ago and all but a few percent of the rest in the following billion years (Figure 13). We estimate that about 20 percent of the Uranus-Neptune planetesimals initially entered the Oort cloud and that roughly 10 percent remain there today. (In a sense, the long-period comets allow us to sample the Oort cloud directly and define its boundaries.)

Only about 5 to 10 percent of the planetesimals actually accumulated to form Uranus and Neptune. This implies that the original mass of the planetesimal swarm was about 10 to 20 times the two planets' combined mass, which would represent by far the solar system's largest amount of condensed solid material (equal to one or two times the present mass of Jupiter).

Although they were but a small fraction of the total population, a great many planetesimals nevertheless collided with the planets and the satellites. Enough objects probably crossed the orbits of Jupiter, Saturn, and Uranus about 4 billion years ago to account for the heavy bombardment recorded on the satellites of these planets. Jupiter must have

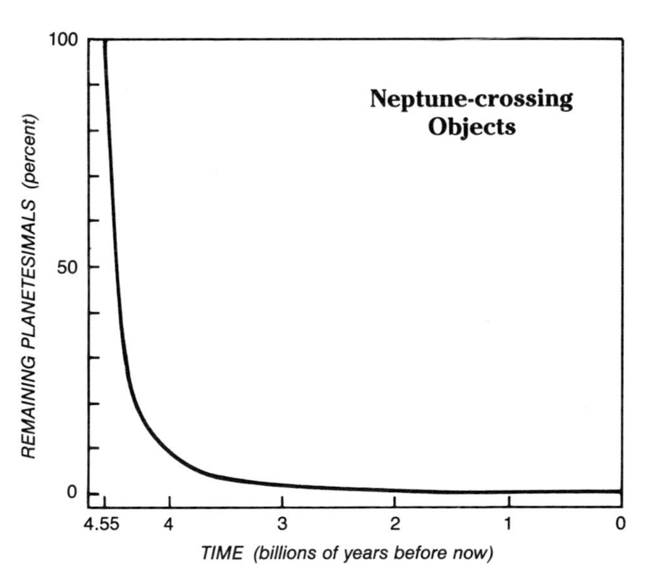

Figure 13. **Depletion of planetesimals in the region of Uranus and Neptune, as calculated by Eugene Shoemaker and Ruth Wolfe. By far the largest amount of solid material in the infant solar system condensed near these planets. Even 4 billion years ago, long after the planets formed, up to 25 Earth masses of cometlike bodies probably remained on Neptune-crossing orbits and about 0.01 Earth mass on Jupiter-crossing orbits. Only a very small fraction of this material collided with the outer-planet satellites and the terrestrial planets, but even so it was a period of torrential bombardment whose consequences can still be seen in these bodies' battered surfaces.**

captured many of these passing planetesimals into orbits like those of the current short-period comets. Earth-crossing comets almost certainly evolved from this group, including some that assumed very short-period orbits. The breakup of one such body, somewhat larger than Chiron, may have produced the pulse in cratering we recognize as the late heavy bombardment.

The volatile-rich, cometary material that collided with the Earth when the Uranus-Neptune planetesimals dispersed may have equaled about 10 times the present mass of our oceans. Therefore, these objects may well have supplied the Earth with almost its entire inventory of water. Other volatile substances probably also reached our planet in this way, and a similar batch was delivered to Venus. Of course, these incoming objects must also have struck the Moon, Mercury, and Mars – so why has no water been found in the rocks brought back from the Moon? The answer seems to be that the planetesimals came in at such high velocities that their volatile substances simply didn't stick to the Moon or Mercury, but instead were driven off into space as high-temperature gases. Also, the Moon's gravity is so weak that any volatiles remaining on the Moon would ultimately have been swept away by the solar wind.

Very likely, the traces of complex organic molecules found today in certain carbonaceous meteorites (see Chapter 19) were also incorporated into the Uranus and Neptune planetesimals – and probably reside now in comet nuclei as well. The larger bodies struck Earth at speeds fast enough to destroy most of these compounds. But an abundance of organic material probably ended up here anyway, borne by the small fragments in meteor streams and sporadic meteoroids. Such particles were decelerated more gently when they struck Earth's atmosphere, and their organic component may thus have been the principal flavoring of the oceanic broth from which life arose. In other words, planetesimals that were deflected sunward from the vicinity of Uranus and Neptune may have supplied the Earth with not only water but also a fairly rich mixture of complex molecules to fuel the process of biological evolution (see Chapter 22).

RECENT COLLISIONS OF LARGE BODIES WITH THE EARTH

It seems probable that, over the past several hundred million years, several comet nuclei at least 10 km across have collided with the Earth. Projectiles of this size produce craters more than 150 km in diameter when they hit, or many large craters if they break up, and eject enormous amounts of material into the atmosphere. The odds are that about two-thirds of such impacts occurred in the ocean, creating giant tsunamis ("tidal waves") and driving huge volumes of water vapor into the upper atmosphere. About 40 years ago Ralph Baldwin pointed out that large impact events may have triggered mass extinctions of living species recognized from the paleontological record.

It remained, however, for Luis and Walter Alvarez and their colleagues Frank Asaro and Helen Michel to establish the link between a major impact and a global biological catastrophe that occurred 66 million years ago. They found that the boundary between the rocks of the Cretaceous and Tertiary geologic periods at several localities around the world is marked by a thin layer of clay. This boundary marker was anomalously enriched with certain elements – particularly iridium – that are rare in terrestrial rocks but relatively common in meteorites. The percentages of these elements corresponded to what could have been supplied by a stony asteroid 6 to 10 km across or possibly a somewhat larger comet. Fossils found below and above the boundary show that more than half the species of living organisms disappeared at about the time the claystone was laid down. All this evidence led the scientists to conclude that the claystone represented dust thrown up from a great impact and distributed worldwide by atmospheric circulation, and that while airborne the dust blocked enough sunlight to arrest photosynthesis for a period of months or even years, interrupt the food chain, and cause many species to simply starve to death.

The evidence for a major impact at the end of the Cretaceous has been greatly strengthened since the Alvarezes first announced their remarkable discoveries in 1980. Iridium anomalies have been found in the boundary claystone at more than 80 localities on the continents and in the ocean basins. The relative proportions of other metals, including platinum, osmium, and gold, have been shown to be close to the relative proportions in primitive stony meteorites. Of special significance was the discovery by geologists Bruce Bohor and Glenn Izett of shocked mineral grains and rock fragments precisely at the Cretaceous-Tertiary boundary. The shocked grains occur worldwide but are most abundant, by far, on North America – was this continent, therefore, the site of the Alvarezes' impact?

The telltale but heavily eroded crater has perhaps been found. Geochronologist Michael Kunk and his colleagues determined that the 35-km-diameter Manson impact structure of Iowa, the largest known in the United States, has an age that coincides with the Cretaceous-Tertiary boundary to within the 1 million year probable error of measurement. Further, the shock-deformed minerals and rock types observed at the boundary are like those excavated by the Manson impact. The size of the Manson structure, however, appears to be too small to be consistent with the size of the projectile thought to be necessary to explain the noble-metal excesses. It may turn out that the projectile broke up just prior to encountering Earth and produced a number of impact craters (Figure 14). Isotopic studies of the boundary clay, moreover, suggest that at least one large crater was formed somewhere in the ocean floor. There is evidence for a tsunami at the Cretaceous-Tertiary boundary around the Gulf of Mexico, so perhaps a large impact occurred at that time either in the Gulf or possibly in the early Atlantic Ocean.

Numerous possible global climatic consequences of large impacts have been recognized in the past 10 years (see also Chapter 8). For example, a large impact in the ocean would throw enough water into the stratosphere to wash out any accompanying dust quickly and initiate a period of greenhouse warming. The evaporation of more ocean water and release of carbon dioxide would ensue, possibly raising temperatures in the lower atmosphere and the uppermost oceans by more than 10° K. Such a pulse of heat could explain why certain species were exterminated at the end of the Cretaceous, while others living in protected environments (deep in the ocean or at high latitudes) survived.

Enormous quantities of atmospheric nitrogen would have

been burned by the impact fireball, creating nitrogen oxides and strong nitric acid rain that greatly increased the acidity of soils, lakes, and shallow ocean waters. Carbon soot found in the boundary clay suggests that the continents were engulfed in widespread forest fires. Vegetation killed by a combination of darkness and cold – or subsequent heat, acid rain, or possibly other causes – may have provided an abnormal supply of tinder that fueled this conflagration. (Paleontological evidence shows that flowering plants largely disappeared, at least locally, for several thousand years.)

At the same time, conditions in the surface layer of the ocean also were inimical to many forms of life. Temperatures fluctuated widely and biological productivity plummeted, probably in response to dramatically changing climatic conditions and acid rain. Kenneth Hsü has referred to this situation as the "Strangelove ocean." Various species of marine organisms gradually died off over perhaps several tens of thousands of years. The altered marine ecology, with various organisms disappearing forever, may have affected the survivors as significantly as did the physical changes in their environment.

Another episode of mass extinction that is related in time to large impacts occurred late in the Eocene Epoch, about 35 million years ago. Paleontologist Gerta Keller and her colleagues have found tiny glass spherules of impact melt in layered deep-sea sediments that suggest at least three large impacts occurred over an interval of about 0.5 to 1.0 million years. A strong iridium anomaly is associated with the middle spherule layer. The mass extinction of the late Eocene occurred in a series of discrete steps that evidently coincide with climatic crises, and some of these in turn are associated in time with the major impacts. Besides the layers with impact glass, geologists are aware of a number of impact craters that formed about 35 million years ago. Apparently the Earth was subjected at that time to a pulse of bombardment that is most readily explained as a mild comet shower.

There are hints of other possible comet showers from the geologic record of impacts and extinctions. The largest peak in the distribution of ages determined for known large impact structures is found for craters younger than 5 million years. This apparent surge in the cratering rate may be due, in part, to the fact that it is easier to recognize and date young craters than it is to find and date old eroded ones. However, two known large craters and several strewnfields of impact glass were all formed between about 0.7 and 1.1 million years ago. One of the latter, the Australasian tektite field, is among the largest glass strewnfields known. It may have been produced by two large impacts separated in time by about 100,000 years. The close spacing in time of these impact events suggests that there may have been a genuine short-lived increase in the flux of Earth-crossing bodies between 0.7 and 1.1 million years ago. Such a brief surge, if it is real rather than a statistical fluke of our observations, is indicative of a weak comet shower.

Quite different and much less direct evidence for a possible comet shower is provided by a mass extinction that occurred about 91 million years ago near the boundary of the Cenomanian and Turonian stages of Earth history. As shown by paleontologists William Elder and Earle Kauffman, this extinction took place in a series of discrete steps over a time interval of the order of a million years. The extinction is similar in this respect to the one that occurred in the late Eocene, and a noble-metal anomaly has been discovered at the stratigraphic horizon of one of the Cenomanian-Turonian extinction steps. It is too soon to say for sure, but the anomaly's presence suggests that an impact may have occurred at the time of at least one extinction step, and it is possible that most of the steps were triggered by large impacts. If so, the close spacing of the steps in time may be indicative of surge in bombardment due to a comet shower.

Noble-metal enrichments have now been discovered in the stratigraphic section that are correlated with each of four global mass extinctions of the last 100 million years. These took place about 11, 35, 66, and 91 million years ago. In the case of two extinctions, the ones at 35 million years (late Eocene) and at 66 million years (Cretaceous-Tertiary

Figure 14. **One day, 66 million years ago, a large comet nucleus (or possibly an asteroid) about 10 km across approached the Earth. The comet may have broken up and produced a number of impact craters – at least one, it seems, on North America. One consequence was the eradication of numerous life forms (probably including the dinosaurs). Don Davis has rendered that catastrophic moment, with the flora and fauna shown representative of the era.**

boundary), there is definite independent evidence of large impacts in the form of impact glass or shocked minerals and rock fragments. The odds are good that at least three and possibly all four mass extinctions are related either to impacts of large comets or asteroids or to surges in the impact rate caused by comet showers.

A lively debate has arisen as to whether mass extinctions and impact events are periodically distributed in time. From their analysis of mass extinctions in the last 250 million years, paleontologists David Raup and John Sepkoski suggested in 1984 that such events recur regularly, about once per 26 million years. Their conclusion sparked a flurry of papers in which various astrophysical mechanisms that might produce periodic comet showers were explored. Two studies, by Marc Davis, Piet Hut, and Richard Muller, and by Daniel Whitmire and Albert Jackson, suggested that the Sun has a faint undiscovered companion star that revolves on a highly eccentric orbit with a period of 26 million years. The companion star, dubbed Nemesis by the first-mentioned team, plunges deep within the Oort cloud and dispatches a shower of comets on Earth-crossing orbits each time it passes near perihelion.

Whitmire and colleague John Matese then proposed an alternative hypothesis: a fairly massive undiscovered planet that revolves beyond the orbit of Pluto periodically perturbs an unseen disk of comets in its neighborhood. The planet has already cleared a gap in the comet disk, but the orbit of the planet is inclined to the plane of the disk; according to Whitmire and Matese, comet showers are generated when the perihelion and aphelion of the planet's orbit lie close to this plane and thus near the edges of the gap. This occurs every 26 million years as a result of precession of the long axis of the planet's orbit.

A third theory was advanced by Michael Rampino and Richard Stothers. They point out that the Sun oscillates up and down across the central plane of the Milky Way, a cycle that requires about twice the purported periodicity of Earth's mass extinctions. Each time the Sun passes near the galactic plane, they reason, the probability of an encounter with clouds of massive molecular gas lying near the plane is increased. These encounters would perturb the Oort cloud and send in a shower of comets.

Much discussion followed the publication of these intriguing hypotheses. One problem common to all of them is establishing accurate ages for the extinction events, which are poorly known for those that occurred more than 100 million years ago. Consequently, the apparent periodicity found by Raup and Sepkoski may not be statistically significant. Similarly, an apparent periodicity found in Earth's impact record, which seems to peak at roughly 30-million-year intervals, may not be significant either.

Moreover, there are serious difficulties with each of the proposed astronomical "clocks." A Nemesis-like object would tend to become unbound from the Sun due to galactic tides and encounters with molecular clouds and other stars – the odds are no better than 1 in 1,000 that a companion star exists with an orbit large enough to have the required 26- or 30-million-year period. Whitmire and Matese's distant undiscovered planet would not form a particularly sharp-edged gap in a comet disk and thus its comet showers would not be well-defined. The hypothesis of Rampino and Stothers is weakened by the fact that the present height of oscillation of the Sun above and below the galactic plane is too small with respect to the vertical distribution of molecular clouds or stars. The periodic comet showers expected from this effect should be too weak to be noticed either in the present statistics of crater ages or in the record of mass extinctions of the last 250 million years.

We conclude that excellent evidence exists for a correlation between large impacts and some specific mass extinctions, and that as many as four or five weak-to-mild comet showers may have occurred in the last 100 million years. (By our reckoning, a "shower" corresponds to a surge in the comet flux at least three times above average.) The number of possible mild comet showers suggested by geologic evidence is about that expected from the random passage of stars through the Sun's close neighborhood. For the period in which geologic ages can be determined accurately, the times between apparent peaks in the cratering rate, or between the corresponding mass extinctions, tend to be moderately uniform but no more so than might be expected by chance.

The essential point is that evidence accumulated during the 1980s suggests that the collisions of large objects with the Earth have played a major role in the destruction (and evolution) of life here. When an impact triggers the loss of species and even whole families of organisms, ecological or environmental space opens up for new ones. Various species of mammals, for example, multiplied rapidly after the Cretaceous-Tertiary extinction. It can be argued that the presence of the human race on Earth may be due, in no small part, to chains of events initiated by large impacts about 66 and 35 million years ago.

22

Life in the New Solar System?

Gerald A. Soffen

WITHIN the solar system, only the planet Earth is known to have indigenous life. But for centuries we have pondered whether other planets are inhabited as well, all the while trying to understand the processes of biogenesis. Only within the past few decades have biologists had the means to perform relevant laboratory experiments and, more recently, to venture across the solar system to begin the search for the inhabitants of other worlds.

In some sense, we are not certain what to search *for,* since our understanding of the nature and definition of "life" is still controversial. There is no universal agreement, but most biologists would recognize terrestrial life by two characteristics: its ability to reproduce, and its ability to evolve through natural selection. Researchers have investigated many possible life-initiating chemistries, but they are struck by the one important fact that terrestrial life is made up of the most abundant elements of the cosmos: hydrogen, oxygen, carbon, and nitrogen.

Most scientists who study the origin of the life and its chemical processes believe that all terrestrial beings are the result of a single sequence of events. After Earth cooled, chemical evolution, aided by an abundance of solar energy and simple materials, led to the formation of an organic broth (Figure 1). These components, concentrated in a watery solution with an inorganic chemical milieu, combined to form very complex macro-molecules.

Beyond that step the picture is very incomplete, and we really do not understand the story. Somehow, the first self-replicating "biological entity" emerged. It began to reproduce and evolve through natural selection. Certain advantageous mistakes during reproduction produced inexact copies of the infant organism that proved to be better survivors in the hostile world around them. Initially, to make more of itself the first organism had only to use the chemical materials and sources of energy surrounding it. But alas, the second law of thermodynamics began to catch up; the surrounding milieu was being comsumed and changed, and the general state of energy was decreasing. To cope with environmental crises and assure its survival, the new creature and its progeny had to make some very difficult discoveries: a food source, efficient internal chemical machinery, and rapid responses to the changing conditions.

Terrestrial life succeeded. At least one form survived all the disasters, eventually spawning the 500,000 species observed today. But why only one beginning, one origin? Was there a successful second genesis on Earth? We do not believe so. The evidence is strong that all known contemporary life descended from that one occurrence. All terrestrial life has the same biochemistry, uses the same genetic code, and the same unique set of organic *stereoisomers.* (Asymmetric organic molecules come in pairs, something akin to the mirror images seen in Figure 2; living organisms use only one

Figure 1. **Did life on Earth begin deep in its oceans? This "black smoker," a mineral-laden hot spring lying 2.5 km below the surface of the Pacific Ocean, may be similar to the primordial oases that provided the energy and nutrients necessary for primitive life forms. Dudley B. Foster took this historic first-ever photograph of a black smoker in April 1979, after literally bumping into it with the submersible craft Alvin. The geyser's toppled chimney rests behind the smoke column, and some of Alvin's equipment is visible in the foreground.**

Name	Formula	Year of discovery
Two-atom molecules		
Methyladine	CH	1937
Cyanogen	CN	1940
Methyladyne	CH^+	1941
Hydroxyl	OH	1963
Carbon monoxide	CO	1970
Molecular hydrogen	H_2	1970
Carbon monosulfide	CS	1971
Silicon monoxide	SiO	1971
Sulfur monoxide	SO	1973
Nitrogen sulfide	NS	1975
Silicon monosulfide	SiS	1975
Diatomic carbon	C_2	1977
Nitric oxide	NO	1978
Hydrogen chloride	HCl	1985
Three-atom molecules		
Water	H_2O	1968
Formyl ion	HCO	1970
Hydrogen cyanide	HCN	1970
Hydrogen isocyanide	HNC	1971
Carbonyl sulfide	OCS	1971
Hydrogen sulfide	H_2S	1972
Ethynyl radical	C_2H	1974
Diazenylium	N_2H^+	1974
Formyl radical	HCO	1975
Sulfur dioxide	SO_2	1975
Thioformylium	HCS^+	1980
Silcyclopropyne	SiC_2	1984

Name	Formula	Year of discovery
Four-atom molecules		
Ammonia	NH_3	1968
Formaldehyde	H_2CO	1969
Isocyanic acid	$HNCO$	1971
Thioformaldehyde	H_2CS	1971
Acetylene	C_2H_2	1976
Cyanoethynyl radical	C_3N	1976
Isothiocyanic acid	$HNCS$	1979
Protonated carbon dioxide	$HOCO$	1980
Protonated hydrogen cyanide	$HCNH^+$	1984
Proponyl radical	C_3H	1984
Tricarbon monoxide	C_3O	1984
Protonated hydrogen cyanide	$HCNH^{2+}$	1985
Five-atom molecules		
Formic acid	$HCOOH$	1970
Cyanoacetylene	HC_3N	1970
Methanimine	CH_2NH	1972
Cyanamide	NH_2CN	1975
Ketene	CH_2CO	1976
Butadiynyl radical	C_4H	1978
Silane	SiH_4	1984
Cyclopropenylidene radical	C_3H_2	1985

Name	Formula	Year of discovery
Six-atom molecules		
Methyl (wood) alcohol	CH_3OH	1970
Methyl cyanide	CH_3CN	1971
Formamide	NH_2CHO	1971
Methyl mercaptan	CH_3SH	1979
Seven-atom molecules		
Methylacetylene	CH_3C_2H	1971
Acetaldehyde	CH_3CHO	1971
Methylamine	CH_3NH_2	1974
Vinyl cyanide	CH_2CHCN	1975
Cyanodiacetylene	HC_5N	1976
Eight-atom molecules		
Methyl formate	$HCOOCH_3$	1975
Methyl cyanoacetylene	CH_3C_3N	1983
Nine-atom molecules		
Ethyl alcohol	CH_3CH_2OH	1974
Dimethyl ether	$(CH_3)_2O$	1974
Ethyl cyanide	C_2H_5CN	1977
Cyanotriacetylene	HC_7N	1977
Methyl diacetylene	CH_3C_4H	1984
Eleven-atom molecules		
Cyanoctatetrayne	HC_9N	1977
Thirteen-atom molecules		
Cyanodecapentyne	$HC_{11}N$	1981

Table 1. **By the mid-1980s, astronomers had detected the signatures of more than 60 organic compounds in interstellar space. Note the dominance of carbon, hydrogen, nitrogen, and oxygen – these four atoms make up what researcher Gerrit L. Verschuur terms "the simple alphabet of life."**

of these, the so-called "left-handed" set, and reject the other.) The best explanation for this is that once the first successful living entity gained a foothold on Earth, the process of evolution was so effective and rapid that no later organism had a chance to succeed. One form rapidly became dominant, and its success inhibited the others. There is a good deal of conjecture here, but given our state of knowledge, this is where we stand.

So our solar system evolved this one great event, *life on the Earth,* and now we question, "Did biogenesis happen elsewhere, as a separate event from our own beginning?" Those who ponder the extent of life in the universe can draw upon only this singular known event. Therefore, the discovery of even a simple bacterium of indisputably different origin would shock the intellectual world.

There exists a great deal of circumstantial evidence that our search for life elsewhere may not go unrewarded. Dozens of simple organic molecules have been discovered in interstellar space by sensitive spectroscopic techniques (Table 1), with more compounds added to the list each year. Carbonaceous chondrites, a particular class of meteorites, contain several percent organic material including numerous amino acids. In the laboratory the precursor organic chemicals of living things are synthesized with relative ease.

THE LURE OF MARS

Based upon our knowledge of life on the Earth, on the cosmic abundance of the elements, and on our understanding of the fundamentals of organic chemistry, by the late 1960s most biochemists had drawn the conclusion that Mars is by far the most plausible planet (besides Earth) that might be inhabited. A number of biologists fully anticipated finding the molecular precursors of life – or, with long odds, perhaps some primitive organisms – on Mars. This conclusion led to an ambitious American program to search for life there. In 1976 two automated laboratories, the Viking landers, were sent to the red planet to perform this investigation.

The Viking missions did not find terrestrial type of life at either of the two landing sites. The evidence further suggests that probably all of Mars is lifeless, and that Earth retains its claim as the only known life-bearing planet in the solar system. But science demands a more rigorous proof, and providing it beyond doubt is exceedingly difficult; some scientists, for example, still believe the search for *fossil* life would prove worthwhile. Thus, the question of life on Mars remains enigmatic. We still do not know the complete answer!

The Viking landers fell silent years ago, but the lure of Martian biology continues to stimulate the work of many scientists. To some, the question of indigenous biota on Mars is an extension of the historically fascinating but outdated ideas posed by Giovanni Schiaparelli, Alfred Wallace, and Percival Lowell. To others, Martian biogenesis is a link toward understanding the origin of life on the Earth. To more philosophical scientists, cosmobiology sheds light on the question of our aloneness in the universe. Rather than asking "Is there life on Mars?" some feel a more important question is "If there is life on Mars, is it of different origin than terrestrial life?" It is a topic so profound that Viking scientist Norman Horowitz once reflected, "The discovery of life on

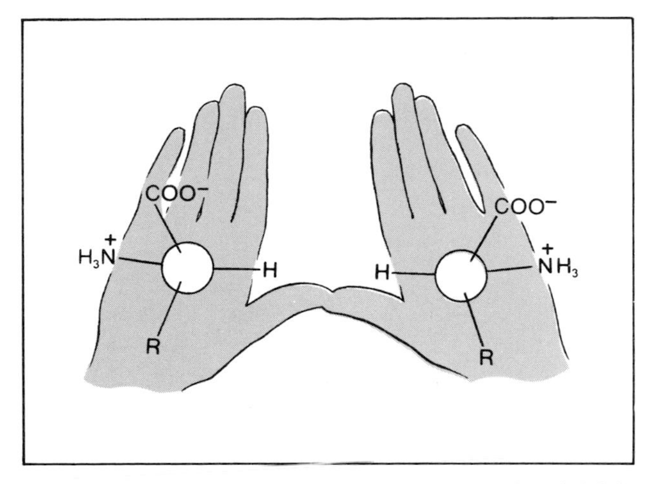

Figure 2. **A schematic illustration demonstrates the meaning of "left-" and "right-handed" amino acids. The large circles represent carbon atoms, and R denotes carbon-hydrogen chains. In 1979, biologists S. C. Bondy and M. E. Harrington pointed out that certain left-handed organic compounds on Earth were selectively absorbed by bentonite clay over their right-handed (and biologically uncommon) counterparts. Such preferential binding may have influenced Earth's early stereochemistry.**

Mars would be hailed as one of the most significant discoveries of the 20th century."

Project Viking's eight years of goal-setting, hardware specification, computer programming, data collection, and analysis by scores of scientists and hundreds of engineers resulted in an almost perfect set of experiments, based on the limited assumptions. Many of the Viking instruments could have detected macroscopic or microbial terrestrial-type life – had there been any to detect. For example, the orbiters' cameras could have seen cities or the lights of civilization (Figure 3). The infrared mappers could have found an unusual heat source. The water-vapor sensor could have detected watering holes or moisture from some great metabolic source. Mass spectrometers activated during atmospheric entry could have identified gases that were considerably outside the limits of chemical equilibrium (as oxygen is on Earth). Seismometers could have detected a nearby elephant.

Viking's life-seeking investigation, its main objective, utilized three instrumental approaches on each lander (Figure 4): a pair of cameras, a pyrolytic gas chromatograph and mass spectrometer (GC/MS), and a trio of biology experiments designed to look directly for metabolic activity.

Lander cameras. Two cameras were used to take pictures from the base of the lander to the horizon, in a complete azimuthal circle, at all times of the day and a few at night for two complete Martian years. Conscientiously examined for any signs of life, the pictures contain countless interesting forms, several subtle changes in the terrain, and some very suggestive colors – but nothing to suggest that life was present at either of the landing sites. One unusual marking, known as the "B" on the rock seen in Figure 5, provided a delightful interruption during the mission's difficult schedule, but no scientist seriously considered it anything more than an odd coincidence of geometric form and lighting.

The pyrolytic GC/MS. One of the most important results obtained from the Viking landers came from this instrument, which determined the nature of organic chemicals on the surface of Mars. Both landers obtained several samples of surface and subsurface material that were heated (pyrolyzed) in steps to 773° K; gases driven off from the samples were then analyzed with a combined gas chromatograph and mass spectrometer to detect indigenous organic compounds. The instrument could have identified concentrations as low as one part per billion (and down to tenths of this for some compounds). But no organic molecules were detected by either of the spacecraft at either landing site – a remarkable result in light of our expectation that organic matter derived

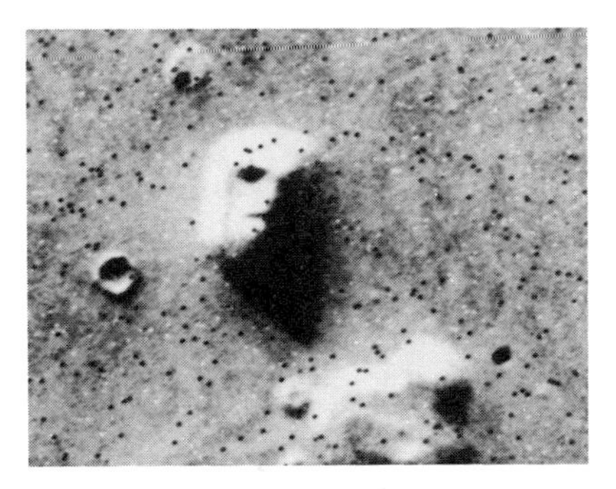

Figure 3. **The "Face of Mars" or "Cydonia Face," which was photographed by the Viking 1 orbiter on July 25, 1976. Shadows on one of these mesalike landforms, about 1.5 km across, have given it the appearance of a human head. Twelve years later, at the time of the launch of the Soviet Union's two Phobos spacecraft, interest in this feature was rekindled by the American news media.**

Figure 4. **A model of the Viking lander is commanded to test its 3-m-long sampling arm, seen extended to the left. The scoop on the end delivered handful-sized samples of soil to three experiment hoppers.**

Figure 5. **One of the rocks near the Viking 1 lander appears to have B etched onto its face. This is an illusion caused by shadowing of surface grains on the rock, or possibly it is a dark stain caused by weathering.**

from falling meteorites should have been scattered all over the surface by Mars' frequent dust storms.

The biology experiments. The Viking landers contained three different biology experiments (Figure 6), each based on a different set of premises. This was done to broaden the search as much as possible. Because the lander did not have wheels, only samples collected by an extendible scoop from an area of about 12 m^2 could be analyzed (Figures 7,8).

The pyrolytic release experiment (PR) specifically sought Martian life forms able to assimilate and reduce carbon dioxide or carbon monoxide. This search was tried several ways: in both dark and light (as in the photosynthesis of terrestrial life), under dry conditions, and under humid conditions. The strength of this experiment lies in the very few assumptions it makes about Martian life; in fact, we assumed only that CO_2 or CO was needed as a carbon source. The experiment used versions of these two gases that had been labeled with radioactive carbon-14, a relatively easy isotope to trace. To prevent obtaining a false positive result, some samples were first sterilized by heating them; this provided a control to distinguish metabolic uptake from chemical reactions that might simulate a biological result. On its first attempt, the PR experiment appeared to give a weakly positive result. But to the consternation of the Viking scientists, the effect could not be repeated *on Mars*, though subsequent laboratory experiments have demonstrated possible scenarios that could explain that first unusual result. We believe it was probably related to unusual iron compounds in the samples, but we still do not fully understand the anomalously positive reading.

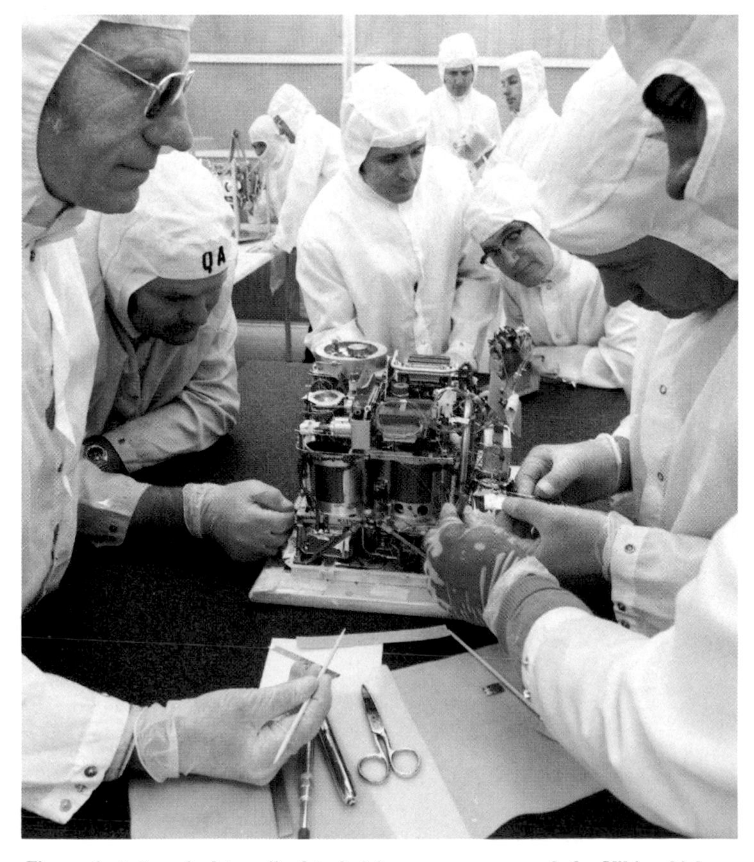

Figure 6. **A "surgical team" of technicians swarms around the Viking biology instrument. This compact package occupied less than one cubic foot of space, yet it contained thousands of individual components.**

A second device was called the labeled-release experiment (LR). It also used radioactive ^{14}C as a sensitive tracer, but here the carbon was in an organic nutrient broth that was "fed" to the surface samples. The biologists assumed that if Martian microbes were present, they would eat the food containing the organics and breathe out $^{14}CO_2$ or ^{14}CO, which could be measured. Again, we got a strange result. Part of the organic mixture was rapidly converted to CO_2 and CO, but then the conversion stopped. (This is unlike terrestrial microbes that continue to feed until the supply runs out, or until the toxic products of their own metabolism inhibit them.) The sample was given a second injection of the organic nutrient, then a third, but to no avail! The resulting reaction, shown in Figure 9, in no way resembled the way terrestrial life handles organic material. The LR also employed heat sterilization to distinguish chemical reactions from biology. Heat seemed to destroy whatever was converting the organic nutrient. Results of this experiment by itself could be interpreted either by a biological or chemical explanation, though the Martian biota (if they exist) would have to consume and expel organic material in a manner very different from terrestrial life. The answer was found in the results of the third biology experiment.

In the third direct test for biology, the gas-exchange experiment (GE), a very rich organic nutrient broth (which we called "chicken soup") was combined with Martian soil, and the incubation chamber's headspace monitored for gaseous metabolic changes. In order not to shock the would-be Martian microbes, which inhabit a very dry world, the soil samples were first only humidified with water vapor prior to the inoculation. Here, the results were most remarkable! When the sample was humidified, it released an extraordinary burst of oxygen. This was not anticipated, and no terrestrial soils have ever done this. But when the actual nutrient broth was added, very little happened except that a small amount of CO_2 was evolved – no more explosive oxygen and no continuing exchange of gases due to biological metabolism. The GE data suggest that an unusual oxidizing substance exists on the Martian surface. A great deal of speculation (and some laboratory work) supports the idea that this oxidizing agent could be some kind of iron peroxide or superoxide.

An important point is that such an interpretation would explain why the organic compounds used in the labeled-release experiment were changed, and why its results resembled a strange biological reaction. One (or maybe two) of the organic compounds in the nutrient is rather sensitive to these kinds of oxidizing agents. Most of the biologists involved believe that during the LR tests the oxidizing substances were consumed, so that subsequent addition of nutrient showed no effect.

After combining these results, the biology team made the statement that *the Viking results do not permit any final conclusion about the presence of life on Mars*. However, more recently a National Academy of Sciences' panel published a report that takes a less circumspect view. It concluded that the Viking results have lowered the possibility of life on Mars, and that further exploration of the question must await samples of Mars returned to Earth laboratories.

Viking's biology experiments stimulated a brilliant line of

research in chemical evolution to understand the origin of life on the Earth. It now appears that the organic world of carbon chemistry was very tightly coupled to the inorganic world of aqueous and gaseous chemistry. Reactions involving salts, ions, metals, and especially the early formation of clays were part of the events that led to the first large polymers of organic molecules, then to information-bearing molecules, and finally to the self-replication that is considered the hallmark of living organisms.

I have devoted 20 years to thinking about this question. In my role as the Viking project scientist during the primary mission, I began with an optimistic view of the chances for life on Mars. I now believe that it is very unlikely. But one doubt lingers: we have not explored the planet's polar regions. I have always believed that in the search for life we must go where the water is. The permanent polar caps of Mars are frozen water and would act as a splendid "cold trap" where organic molecules would condense. Moreover, the oxidizing agents might be absent – due to the presence of water.

But before a polar mission can occur, some intermediate steps will be completed. The first of these began in July 1988, when the Soviet Union launched two ambitious automated spacecraft toward Mars. Soviet scientists planned to have these craft rendezvous with the planets' small moons Phobos and Deimos. Unfortunately, however, contact with the first spacecraft, Phobos 1, was lost several weeks after launch and with Phobos 2 after it began orbiting the red planet. Had they operated as designed, these robotic explorers would have come as close as 50 m to each moon's surface. Cameras would have looked on as a small laser vaporized tiny patches of soil, creating gaseous wisps to be analyzed to determine the atomic masses of elements in the surface minerals. Each Phobos craft would have also fired a lander-penetrator into the surface from close range, and one carried a remarkable probe literally able to hop around to different locations.

The U.S. has its own mission planned, called the Mars Orbiter, which will circle the planet for at least two years. It is to create a detailed global map of the distribution of elements and minerals on the Martian surface, a necessary step before sites can be chosen for sample-return missions in the late 1990s. The spacecraft will also photograph portions of the landscape below with exquisite detail, as well as study the planet's gravity field, internal density distribution, atmosphere, ionosphere, and magnetic field.

Figure 7. **At one point, Viking 2's sampling arm pushed aside a rock to obtain soil not exposed to large doses of ultraviolet radiation. It was thought that the protected soil would provide a safe habitat for Martian organisms.**

Figure 8. **A partial view of the landscape surrounding the Viking 1 lander includes a series of trenches excavated by the surface sampler. The tall white mast is topped with a suite of miniaturized weather instruments.**

SEARCHING ELSEWHERE

An amazing discovery in the past few decades is that extremely complex organic materials can be formed by exposing mixtures of simple gases such as methane, ammonia, water, carbon monoxide, and carbon dioxide to various sources of energy. The energy sources can be as varied as a spark discharge (mimicking lightning), ultraviolet radiation (sunlight), or to high temperature and pressure (volcanism). The organic compounds so generated number in the many hundreds – many of the precursors of living organisms. If organic substance are formed so easily, why are they not more common on the other planets? And why haven't some of them been organized into living organisms other than here on Earth.

First, we *do* anticipate finding organic substances elsewhere in the solar system, but living organisms are exceedingly complex, much more than a mere bagful of mixed chemicals. An analogy might be like having a few musical instruments and wishing for a symphony. Life is a system of highly organized chemical processes that has the ability to replicate itself and evolve through mutation and natural selection. Biologists are careful to state that until we

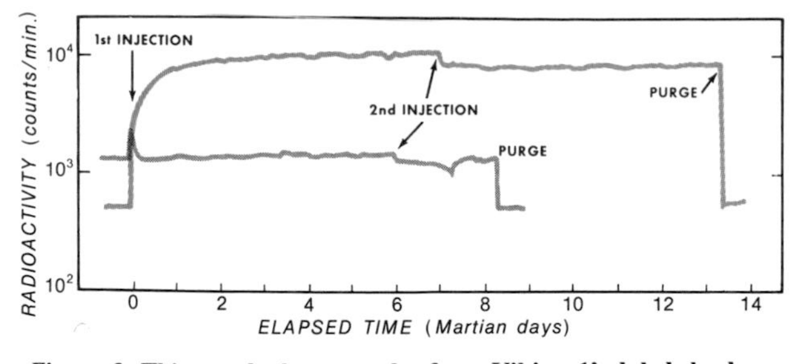

Figure 9. **This graph shows results from Viking 1's labeled-release experiment, for both active (orange) and sterilized (gray) samples of Martian soil. The curves record the accumulation of radioactive carbon monoxide and carbon dioxide after the injection of a nutrient broth labeled with carbon-14.**

actually find living organisms or their remains, we still do not know whether any other forms of life exist in the solar system.

Mars appears to lack organic material at present because ultraviolet sunlight is intense enough to form metallic superoxides on the surface. But, in earlier times, when water coursed across Mars' surface, the formation of stable organic material would have been quite plausible. Some of it may have been trapped in the frozen polar regions; if so, we will discover it someday.

What about living systems or organic materials on the other planets, or their satellites? We have visited the Moon and concluded that there is no life there. Mercury has essentially no atmosphere, is extremely hot on the side facing the Sun, and extremely cold on its night side. Some fanciful scientists have speculated that the twilight zone between hot and cold could be an interesting region to search for organic debris, but no one speculates that there may be life on Mercury.

Venus has a huge atmosphere, 90 times denser than Earth's, that consists mostly of carbon dioxide gas and clouds of sulfuric acid droplets. Because heat cannot easily escape, the mean temperature is well above 800° K. Given the strongly acidic environment, the existence of organic compounds seems impossible, since they would all break down. There is no significant water on Venus and probably no living organisms (at least as we know them).

In striking contrast with Earth, the giant planets Jupiter, Saturn, Uranus, and Neptune are composed mostly of gases and ice. These planets are so large that the envelopes of hydrogen and helium they acquired aeons ago remain bound to them by gravity. Cosmologists believe that their elemental composition is rather similar to the Sun. The giant planets themselves are not believed to be inhabited by living organisms. Nevertheless, scientists who study primitive organic chemistry are anxious to learn about the possible reactions taking place in these planets' atmospheres.

Of particular interest are the comets, asteroids, and the *moons* of the giant planets. Comets are dominated by the ices of condensed gases (like H_2O, CO, and CO_2) and other volatile materials (hydrogen cyanide and methyl cyanide). But based on the recent analyses made at Halley's Comet by spacecraft (see Chapter 17), more complex organic molecules are anticipated. Comets are the most unaltered bodies of the solar system and offer an opportunity to examine some of the pristine solid materials from which the planets accreted. One popular hypothesis is that organic materials from comets contributed to the "primordial soup" that gave rise to life here in the inner solar system. So biochemists are enthusiastic about a mission that would return a sample of a comet to their laboratories for analysis. We could then answer many questions about the materials and conditions that existed in the period of planetary history that preceded biology.

Carbonaceous chondrites, meteorites that have reached Earth from the asteroid belt, have been found to contain a wealth of organic material – as much as several percent of their mass. One of these, the Murchison meteorite, fell in Australia in 1969 and was soon recovered and analyzed. In its interior geochemists found 16 amino acids, the basic building blocks of proteins. Analysis of these chemicals showed that they did not come from living organisms but result instead from natural reactions that are probably widespread in the universe. Meteorites falling on the early Earth, therefore, must have also contributed to the "primordial soup."

Numerous moons of the giant planets have attracted the attention of biologists and biochemists. For example, Europa, a satellite of Jupiter, has peculiar surface markings which suggest the possibility that frozen or liquid water lies just below them. Titan, Saturn's largest satellite, is of enormous interest to the scientists who are planning future interplanetary missions. It is enveloped by a dense atmosphere of nitrogen, methane, and traces of ethane, acetylene, ethylene, and cyanide. The surface of Titan is expected to have oceans of liquid hydrocarbons, and its atmosphere contains what appear to be aerosol hazes made up of organic polymers (see Chapter 14).

Perhaps Titan's cold, murky landscape holds clues to the pathways that cause organic chemical processes to trigger the actual formation of living organisms. The great series of events involved in this conversion is the experimental and intellectual challenge of the next century, and it will guide our search outward into the solar system.

What about life *beyond* the solar system? We now have the technology to listen for signals coming from other stellar systems. Assuming that our Sun is an average star and that planetary systems are numerous, we can imagine that many planets of the cosmos have evolved and nurtured living organisms. Some, like Earth, may have developed technological civilizations. If so, we anticipate that they communicate with each other by radio signals. A program called SETI (the *S*earch for *E*xtra*T*errestrial *I*ntelligence) is being pursued to detect the signals of these communicating civilizations. While recognizing that this kind of eavesdropping is fraught with uncertainty, SETI's proponents maintain that the payoff is so large that a modest effort is warranted, and most scientists agree.

Our search for life in the solar system (and beyond) continues . . . .

23
Putting it all Together

John S. Lewis

THE GREATEST CHALLENGE faced by students of the solar system is that of understanding the extraordinary diversity of objects it includes. It is not easy to see the similarities and relationships between bodies as different as Vesta, Earth, Io, Saturn, Halley's Comet, and the Allende meteorite. And yet, in the most fundamental sense, they are all siblings in a single family, all sharing a common inheritance.

BEFORE THE PLANETS

On the strength of astronomers' observations of active star-forming regions, meteoriticists' analyses of thousands of extraterrestrial rocks, and theorists' complex computer simulations of how stars form, we can piece together some general idea of the events that preceded the formation of these bodies. They, along with every other member of the solar system, were shaped by physical and chemical processes in a large, dense cloud of gas and dust that surrounded the forming Sun. This condensation in turn probably followed the collapse of a vastly larger interstellar cloud (Figure 1), which typically has thousands of times the mass of the entire solar system.

Theorists believe the cloud that spawned the Sun and its planets was a heterogeneous, chaotic combination of matter. Hydrogen dominated this primordial mixture, as it does the universe in general, but other elements were present as well. In particular, the cloud contained debris that had been injected into the interstellar medium both by numerous stellar explosions and by less spectacular ejections of gas by stars. Through the process called *nucleosynthesis,* heavier elements are constructed from lighter ones in stars' interiors, where temperatures reach billions of degrees. For example, these stellar furnaces fuse hydrogen atoms into helium (the most common reaction), helium into carbon and oxygen, and so forth. Consequently, when a star sheds matter or explodes, heavy elements escape their confinement and spew into the interstellar medium.

The solar system's giant parent cloud was also turbulent and gravitationally unstable; over time it broke up repeatedly, collapsing around local concentrations of matter to produce thousands of stellar and planetary systems. Most of the fragments probably gave rise to double, triple, and more complex star systems. In our portion of the galaxy, only 5 to perhaps 20 percent of all the stars seen today are *not* members of multiple-star systems. Yet single stars (plus very close double stars and very widely separated multiple stars) are the ones most likely to be accompanied by "solar systems," because planets that formed with them could have orbits that remain stable over billions of years.

Sometimes the cloud produced large fragments, or *prestellar nebulae,* that possessed little angular momentum. These collapsed quickly and evolved into stable hydrogen-burning stars in only about 100,000 years – a remarkably brief time relative to the age of the universe. Today we see this process being repeated elsewhere in the galaxy. A few of the stars thus formed are huge; some are at least 20 times more massive and 40,000 times more luminous than the Sun. Such stars often run through their entire hydrogen-fusion lifetimes in less than 1 million years, dying in violent explosions. (In

Figure 1. **Excited by the fierce radiation from hot stars like those in the closely spaced quartet at center (the Trapezium), dense clouds of gas and dust glow with the bright green emissions of oxygen in the heart of the Orion nebula. Here, as elsewhere in countless galaxies, we are witnessing the collapse of an interstellar cloud and the formation of new stars. This photograph was taken in 1961, using the 3-m reflector of Lick Observatory.**

contrast, stars comparable to the Sun last some 10,000 times longer, or about 10 billion years, and expire uneventfully.) The heat and pressure generated during a supernova's demise create heavier elements that had not existed before either in the cloud or the star's interior; in fact, nucleosynthesis during supernovas appears to be responsible for the bulk of all elements heavier than iron. Among the new elemental creations are a number of short-lived radioactive isotopes such as aluminum-26 and cobalt-56.

The point is that enormously massive stars are born, live their entire lifetimes, and die spectacularly in cataclysmic explosions – while other fragments of the same interstellar cloud are still collapsing. The shock waves from supernovas race outward, plowing through the surrounding interstellar medium at a few percent of the speed of light (Figures 2,3). They quickly dissipate the more tenuous outer portions of the cloud complex, while at the same time violently compressing the denser inner region. Such compression encourages and may even cause the collapse of individual prestellar nebulae, hence acting as a trigger for star formation.

Vaporized debris from the explosion rides the shock wave outward, cooling rapidly and condensing into tiny smoke-like grains of dust. First to condense are *refractory* minerals (those that are solid at high temperatures), which incorporate many of the short-lived isotopes. These radioactive grains can then strike prestellar nebulae in the supernova's vicinity. The hot, low-density shock front becomes unstable upon entering a cool, dense medium and breaks up into narrow fingers that can penetrate deeply into the nebular gas. As a result, the distinctive supernova grains do not mix uniformly with the nebula but instead end up in very inhomogeneous pockets of gas and dust. If the nebulae are dense enough, the mineral grains decelerate, become captured, and find their way into larger aggregates of material as the nebula cools and solidifies.

We do not know for certain whether a supernova triggered the formation of the solar system, but strong circumstantial evidence does exist. The key lies in an excess of the isotope magnesium-26 found in the Allende meteorite, which fell to Earth in 1969. The only viable source for this excess is aluminum-26, which is created during supernovas. Since the half-life of aluminum-26 is a brief 700,000 years, it must have become incorporated in Allende soon after being spawned by a supernova – and the supernova itself must have been quite near the pre-solar nebula. This observation, combined with the known effects of shock-wave propagation through a dense medium, have made a strong case for a cause-effect relationship.

As it collapses, a prestellar nebula eventually begins to rotate (due to its inherent angular momentum) about a well-defined spin axis. Angular momentum acts to oppose the nebula's collapse toward the spin axis, but it does not hinder the concentration of matter onto its "equatorial plane." The escape of infrared (heat) radiation from the nebular disk leads to further shrinkage, because as gas cools its pressure drops. This cooling, especially in the early stages, is partially offset by the continuing flow of captured gas into the disk. But as cooling progresses and infall ceases, the disk must ultimately collapse and flatten.

Temperatures in such a disk range from well over 1,000° K near the center to no more than about 50° K near its periphery (Figure 4). The corresponding pressures range from about 0.1 bar to a mere one-millionth of that (1 bar is the atmospheric pressure at sea level on Earth). Under these conditions, matter in the innermost portion of the disk is vaporized completely, permitting the elements contained locally in dust and gas to mix thoroughly. Thus, the nebula's composition becomes well homogenized near its center. Farther out, however, not all mineral grains inherited from the interstellar medium vaporize. Some chemical or isotopic inhomogeneities survive wherever the temperature falls short of outright vaporization. In general, the pre-solar characteristics most easily preserved are those carried by highly refractory mineral grains, or by any mineral grain that never comes close enough to the heart of the nebula to be heated strongly.

This, then, is our best-guess scenario. And while it will certainly seem oversimplified in years to come, we can still use it to attempt explanations for the general features of solar-system bodies. The goal of such a *cosmogony* is to generate soundly based physical and chemical models of the evolution of the nebular disk, the composition and the accretion of solid objects within the disk, and the internal evolution of these bodies to their current states. However, working the problem backward from the present has proven extremely difficult (and is perhaps impossible). Planetary accretion, violent disruptive events, and the capture of some bodies into resonant, repetitive motions has clouded our window into the past because these processes have caused a wide variety of early logical possibilities to converge into the system we now observe.

THE SOLAR SYSTEM TODAY

The preceding chapters provide a many-faceted discussion of the present state of our solar system, drawing conclusions

***Figure 2.* On February 24, 1987, the brightest supernova in 383 years blazed into view and within a few weeks became a prominent naked-eye star visible from the Southern Hemisphere. The exploding star, located roughly 160,000 light-years away in the Large Magellanic Cloud, appears against a background of nebulosity in this photograph made with the 3.9-m Anglo-Australian Telescope. Some theorists believe our solar system began when a supernova's shock wave passed through a cloud of interstellar matter, inducing the cloud to collapse and, ultimately, to spawn the Sun and its planets.**

from data on the Earth, meteorites, other planets, the Sun itself, and astronomical observations of distant objects. The quality of this information varies, however. For example, we have enormously more data on Earth's structure, composition, and history than for any other body. Even so, we are limited in our knowledge of the Earth by two important factors. First, because of the difficulty of studying our planet's deep interior, we rely heavily on inferences from seismic data. Perhaps 99 percent of our geological and geochemical data on Earth pertain to the most accessible 1 percent of the mass of the planet. Second, we often lack the planetary data needed to determine whether a particular terrestrial feature is a very general and even universal property of planets – or a rare, idiosyncratic feature unique to Earth.

These difficulties are alleviated somewhat by comparison of Earth data with meteorites. Most of the roughly 3,000 known meteorites, members of the class called *chondrites*, are primitive objects; they have not undergone melting and geochemical differentiation. Historically, much of our understanding and discussion of the composition of Earth's mantle and core has been based on the study of the compositions of chondritic meteorite classes. The diversity and heterogeneity of meteorites make this a complex but rewarding task (see Chapter 19).

It is difficult to relate the trends in meteorite compositions directly to the planets because of our ignorance of where the various meteorite classes formed. But careful chemical and mineralogical studies of each class reveal a great deal about physical conditions at *some* location in our collapsing nebular disk at a well-defined and very ancient time – about 4.6 billion years ago. Until the last few years meteoriticists commonly assumed that very abundant, volatile-poor *ordinary chondrites* came from the asteroid belt and that rare, volatile-rich *carbonaceous chondrites* came from comets. We are now skeptical of this stereotyping, since photometric and spectroscopic data show not only that the asteroid belt is dominated by carbonaceous material, but also that ordinary chondritic material is extremely rare and in fact may be wholly absent from the belt (see Chapters 18,20). This means that the main meteoritic contribution on Earth is probably due to a small, local population of Earth-crossing asteroids with diameters of 1 to 20 km that are not representative samples of the belt. These bodies are in relatively unstable orbits that doom most of them to collisions with Earth or another terrestrial planet within 10 to 100 million years. The

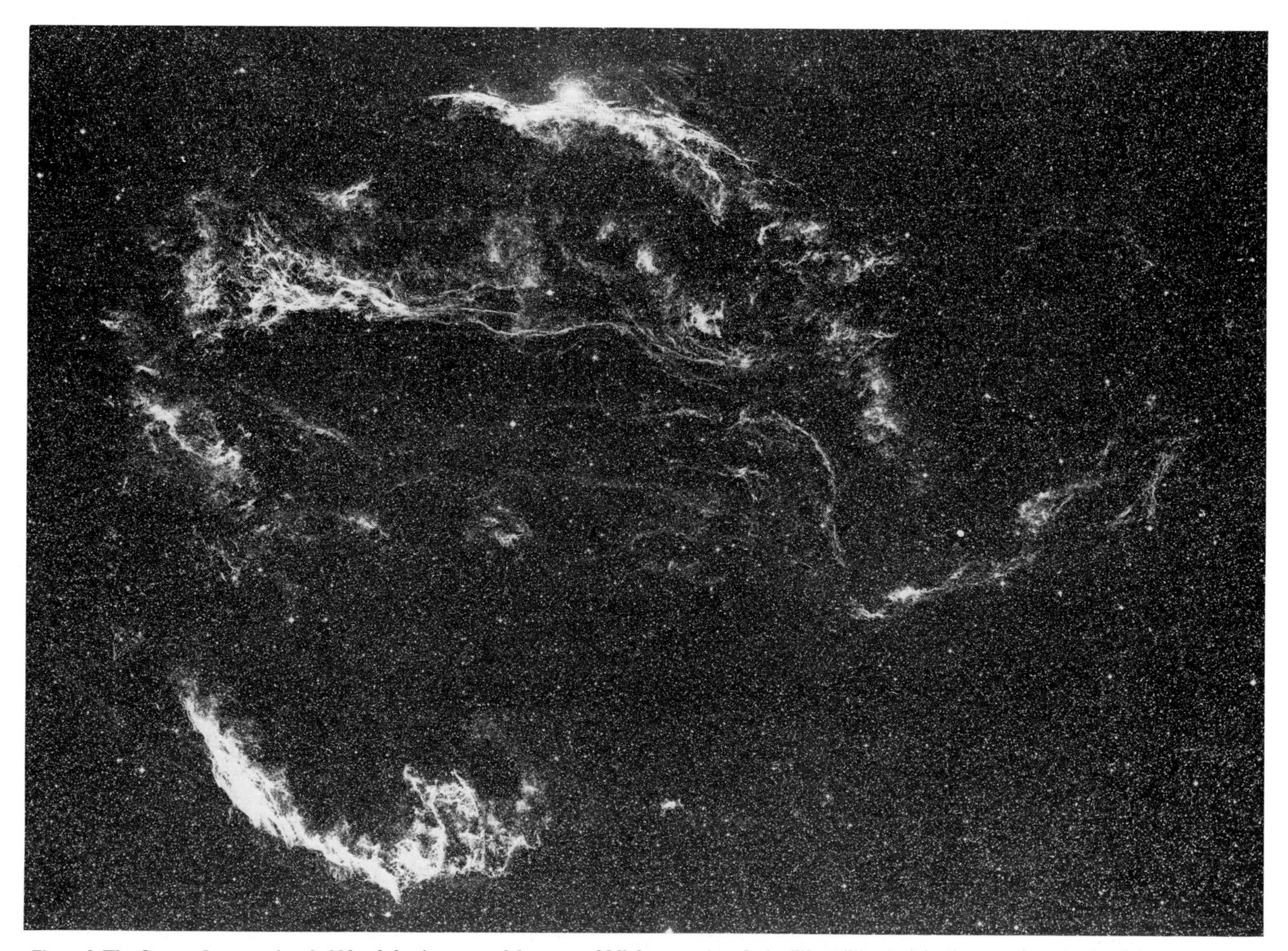

Figure 3. **The Cygnus Loop, a giant bubble of glowing gas and dust some 98 light-years (nearly 1 million billion km) in diameter, is probably all that remains of a supernova explosion that occurred about 50,000 years ago. The delicate filaments seen in red light consist mostly of interstellar matter swept up and compressed by the expanding blast.**

Earth-crossing population must therefore constantly recruit new members from among the belt asteroids and the short-period comets.

Lunar data are derived mainly from our analyses of a large suite of samples returned by the Apollo and Luna landers (see Chapter 4). Such data show the Moon to be a differentiated and extremely volatile-poor body with strong chemical similarity to chondritic meteorites. Unfortunately, our sampling of the Moon is limited to nine sites, all on the nearside and mostly in basalt-flooded mare basins, which cover only a small fraction (16 percent) of our satellite's total surface area. Information on the Moon's deep interior is still limited because of the rarity of natural seismic events, which serve as probes of the structure. The very existence of a lunar core is uncertain, yet one with 20 percent of the Moon's diameter is theoretically possible. A further problem concerns the genetic relationship of the Moon to other solar-system bodies. While the three major pre-Apollo theories of lunar origin (fission from Earth, simultaneous accretion, and capture) remain tenable, in the last few years catastrophic (and extremely messy) scenarios, such as giant collisions that mix fission and capture, have gained increased acceptance.

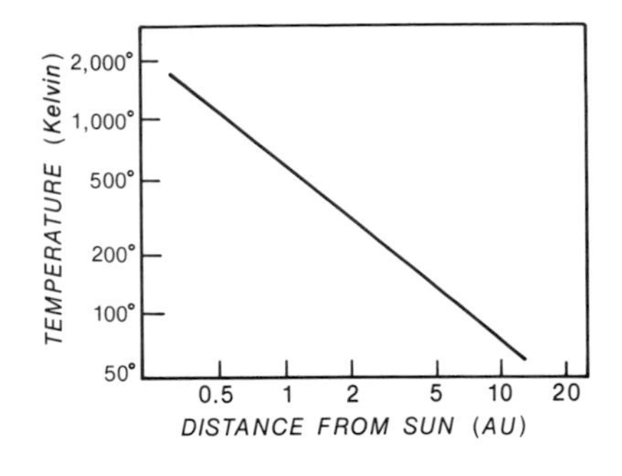

Figure 4. **As material in an accumulating nebula collapses inward toward its gravitational center, it begins to transform from a shapeless mass into a spinning, flattened disk. The compression generates heat, and the small graph demonstrates how temperatures probably varied throughout the early pre-solar nebula.**

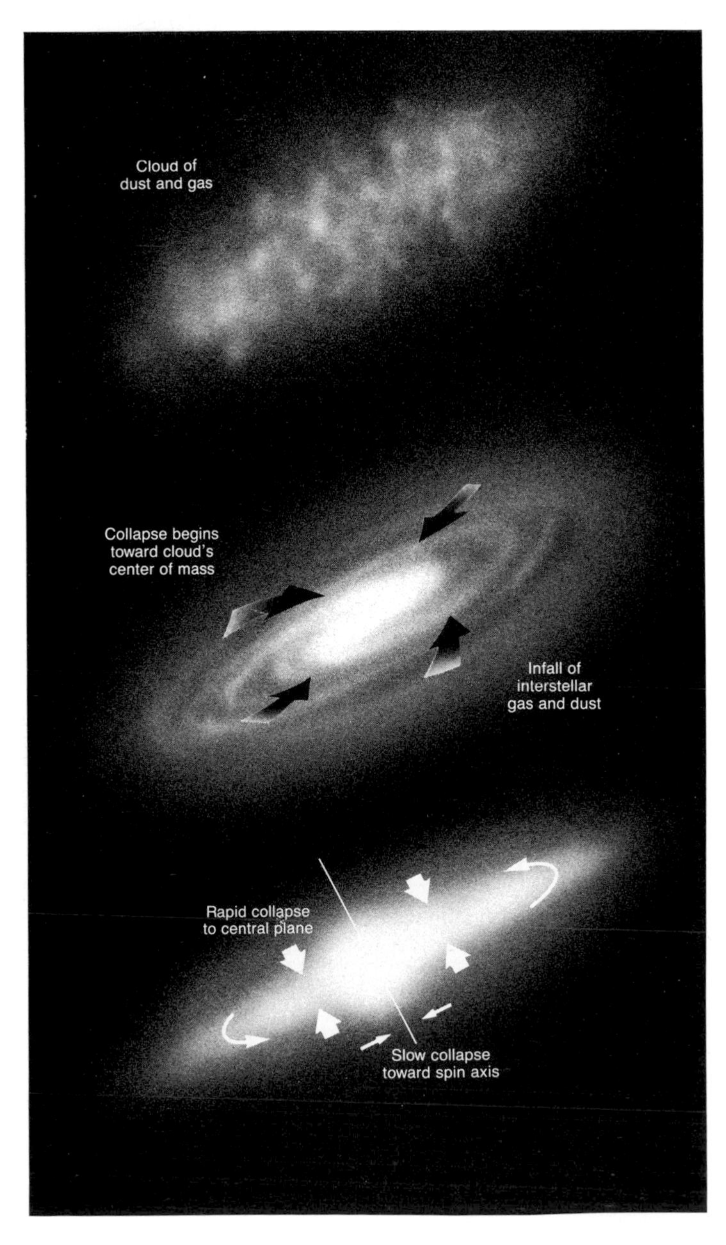

Instrumented Soviet and American probes have analyzed the atmospheres of both Venus and Mars a number of times. The large enrichment of deuterium ("heavy" hydrogen) around Venus suggests to some that the planet possessed oceans in the distant past, and to others the ongoing capture, through collision, of water-rich comets (see Chapter 8). In contrast, Mars has a wealth of surface detail implying strongly that it is endowed with great volumes of water, which is now frozen but long ago flowed freely across its surface. After searching for (but failing to find) life on Mars, two Viking spacecraft studied the local meteorological conditions for a full Martian year. Radar mapping of Venus through its clouds has disclosed a topography strikingly unlike ours, with no clear evidence of global tectonism. Spacecraft have provided limited glimpses of the abundances of radioactive and rock-forming elements on the surfaces of Venus and Mars, but overall the dearth of geochemical and historical data hinders our interpretive efforts. In fact, knowledge of the chemical, physical, and thermal states of their interiors borders on calculated guesses, with no seismic data on Venus and virtually none on Mars.

Even less is known about Mercury, which superficially resembles the Moon but has a far higher density. We know almost nothing of its crustal composition beyond a marginal hint of ferrous oxide (FeO) detected spectroscopically. A very large accumulation of metal in the interior is clearly indicated by Mercury's high bulk density, yet once again the very existence of a discrete core is still a matter of conjecture (Mariner 10 discovered a small magnetic field that *may* require a core dynamo).

During the 1970s, our interplanetary wanderings also reached outward to the Jovian planets. Voyager and Pioneer spacecraft have provided a wealth of information about the atmospheres, magnetospheres, and satellites of Jupiter and Saturn. The Voyagers' exciting observations of Jupiter's Galilean satellites and of Saturn's Titan-dominated family of moons open new chapters in our study of the solar system. The astonishing spectacle of massive eruptions and sulfurous lava flows on Io demonstrates behavior encountered on no other world. Likewise, revelations of tectonic activity in the rock-hard crusts of icy satellites challenge our theories and broaden our concepts of satellite formation. Yet, despite all this, we are no closer to understanding the gas giants and their satellites than we were in the study of Venus and Mars 15 years ago. Nothing is known directly about the interior structures of these bodies, and we have only inferences to guide us in understanding their thermal states and chemical compositions. For example, what explains the striking colors in Jupiter's atmosphere? The heat emanating from both it and Saturn? Their sporadic radio emissions? Perhaps atmospheric probes like that included on the Galileo mission will resolve some of these puzzles.

Thanks to Voyager 2, we have obtained modest improvements in our knowledge of the internal heat source and cloud-top atmospheric structure of Uranus and a broader perspective on its ring system. Also acquired was an enormous body of data on the Uranian moons, as exemplified by the discovery of the extremely complex and diverse surface structure of the satellite Miranda. The Neptune system was Voyager 2's final target; awaiting explanation are the planet's odd satellites and, in particular, how massive Triton ended up in its striking retrograde orbit.

Unfortunately, another century may pass before any spacecraft reaches Pluto and its satellite Charon – a "double planet" in an eccentric, Neptune-crossing orbit. Only a few years ago, little more than crude density and compositional estimates existed for Pluto. Recently, however, Earth passed through the plane of that system, resulting in a run of mutual eclipses of Pluto and Charon. From the analysis of these events have come greatly improved estimates of the sizes, densities, and albedos of both bodies. More recently, astronomers watched the planet cross in front a faint star and in the process discovered that Pluto possesses a thin atmosphere.

Comets remain the object of remote observation from Earth, largely by infrared and ultraviolet spectroscopy. Their evaporation, photolysis, and interactions with the solar wind, while very interesting, are complex and very poorly understood. The most exciting advances in recent years have come from the five Soviet, European, and Japanese spacecraft that flew by (or through) Halley's Comet in 1986. Three of them photographed its irregular, extremely dark nucleus, and organic polymers were found among the gases in its coma.

Data from beyond the solar system have also begun to throw light on the early evolution of planetary systems. For example, astronomers have mounted fruitful observational attacks on very young, energetic stars. They have also recorded what may be another solar system in the making surrounding the star Beta Pictoris (Figure 5). Its disk seems to represent an early stage in planetary accretion, in which rock rubble and asteroids are abundant, but full-size rocky planets have not yet had time to accrete. Also, the immensely successful Infrared Astronomical Satellite (IRAS) has revealed an important class of bright infrared sources that can be unambiguously identified as extremely flattened dust disks in orbit about young stars.

GENERALIZATIONS AND GENESIS

As varied as objects in the solar system are, they do appear to fit rather well into a general compositional scheme. The silicate surfaces of inner solar-system bodies and the water-ice surfaces of many of the satellites surrounding Jupiter and Saturn give way to methane farther from the Sun. Similarly, the oxidized atmospheres of Venus, Earth, Mars, and Io give way to the reduced gases (such as hydrogen, methane, and ammonia) of the Jovian planets and the large, most distant satellites. The densest planet, Mercury, gives way to less dense Venus and Earth, then to the even less dense Moon, Mars, and Io. Next come ice-rich bodies, which populate the outer solar system, and finally the Jovian planets, which are least dense of all.

Harrison Brown suggested over 30 years ago that the main materials of the solar system could be conveniently grouped as gases (like H_2, He, and Ne), ices (H_2O, NH_3, CH_4), and rock (Fe, FeS, $(Fe,Mg)_2SiO_4$, and so on). When seeking trends in the distributions of these compounds, we quickly come to realize that the compositions of objects in the solar system reflect the differences in the vapor pressures of the rocky, icy, and gaseous components. Far from the infant, still-condensing Sun, nebular temperatures were low enough for the condensation of ices.

Armed with such fundamental guidelines, we can now examine ideas that try to explain the rich range of data described in this book. While reviewing these theories, keep in mind that all of them probably contain large elements of truth, but that no single model should reasonably be expected to explain all we presently know. These genetic models have been proposed in the spirit that intentionally simplified descriptions with few variables can readily generate specific and testable predictions.

Beta Pictoris, like other stars seen during the IRAS mission, suggests that flattened dust disks commonly accompany young stars. Virtually all models for the origin of the solar system begin with such a disk forming from the primitive solar nebula. But theorists have suggested two radically different ways of accreting the disk's gas and dust into larger bodies, both of which may have been very important in different parts of the solar system.

The first model (Figure 6) works best far from the disturbing tidal pull of the Sun. Dynamic instabilities in the nebular disk split off rings of material from its outer edge. These rings then subdivide into several large, gravitationally bound, gaseous protoplanets, which eventually accumulate into planets with about the same elemental composition as we now find in the Sun. This concept seems particularly appropriate for the Jovian planets, which mimic the Sun's composition very closely. Also, as they came together these giant worlds may have been immersed in dense nebular clouds of their own, which led to the formation of massive

Figure 5. **The young star Beta Pictoris lies hidden under a coronagraphic mask in the center of this image, but a highly flattened disk of matter extends about 60 arc-seconds (about 150 billion km) to either side and appears edge-on from our perspective. This remarkable red-light image was taken in 1985 with the 2.5-m du Pont reflector in Chile. The bright and dark concentric bands are processing artifacts. While no planet has yet been discovered around any star but the Sun, Beta Pictoris is one of a growing number of stars known to have disks of matter orbiting around them.**

Figure 6. **One model of the solar system's formation involves the detachment of unstable rings of material from the contracting nebula's outer edge. These eventually coalesce into the planets and other bodies.**

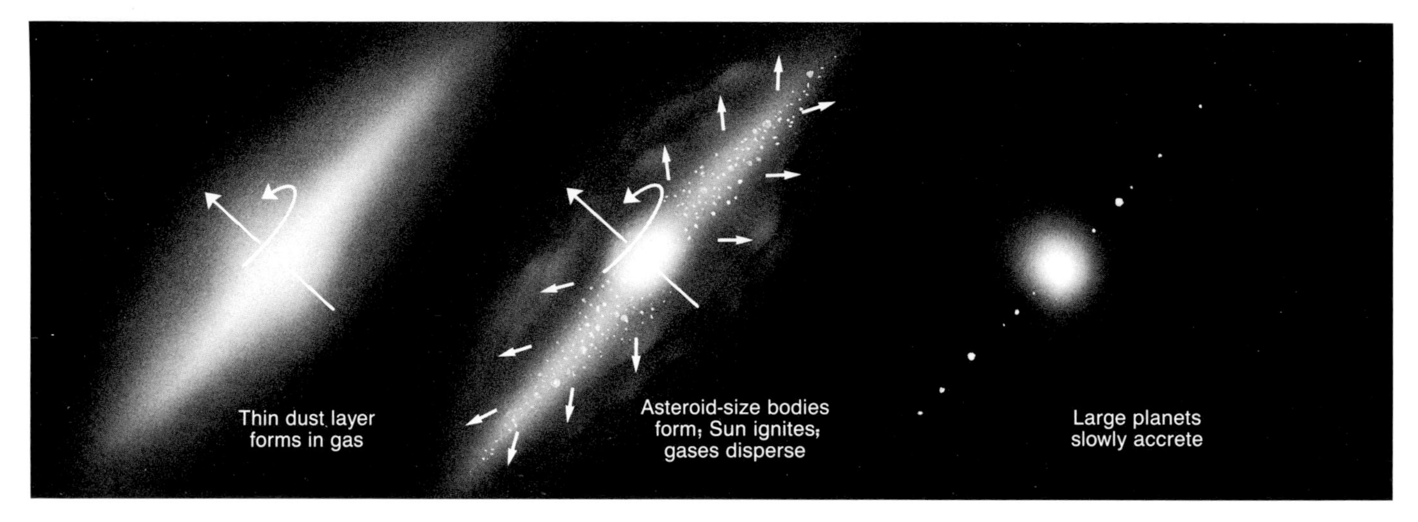

Figure 7. **Another formation scenario envisions the accretion of asteroid-sized bodies, from which the planets ultimately accumulate. Residual material in the nebula dissipates quickly once the embryonic Sun ignites and begins radiating energy; planets then accrete in the absence of gas.**

satellites and the capture (through drag) of many smaller ones.

In the second method (Figure 7), dust settles rapidly onto the central plane of the nebula, forming grains by some combination of adhesive forces (electrostatic, magnetostatic, and so on) and gravitational instability. This leads to a buildup of asteroid-sized bodies, which in a few million years coalesce further into a swarm of roughly lunar-sized objects in nearly circular orbits. By this time, the Sun has ignited and blown away the uncondensed nebular gases. "Stirring" of the orbits of the small solid bodies by Jupiter's gravity and by their mutual perturbations induces collisions, and they slowly accrete to form the terrestrial planets. The entire process takes about 100 million years to form Earth, a little less for Mercury and Venus, and a little longer for Mars. By comparison, the nebular disk itself lasts only 100,000 years.

In this dust-accretion model, the material available for each planet's formation is dominated by local dust, which remains in virtual chemical equilibrium with gases in the nebula around it. This equilibrium is very sensitive to temperature: brief exposure to intense heat permits the dust-gas mixture to react more completely than even long periods of exposure to temperatures several hundred degrees lower. We assume, therefore, that the resulting dust compositions reflect the maximum temperature achieved in each locality during the nebula's evolution (Figure 8).

What actually happens is certainly more complex. Very small grains may continue to react with the surrounding gas as it cools, long after the formation of larger grains rich in refractory elements. This would preferentially infuse the small grains with volatiles like water and compounds incorporating carbon and the halogen elements. The volatiles reside in minerals such as hydroxyl silicates, sulfides, halides, and reactive forms of iron oxide. This general explanation works well for primitive chondritic meteorites, which show a higher concentration of all volatile elements in their fine-grain ("matrix") material than in their larger crystals of high-temperature compounds. Meanwhile, turbulence can preferentially carry these small grains far from the nebula's central plane into sparsely populated regions, where radiative cooling is more effective and the temperature lower. Or it can mix compositionally distinct grains formed at various distances from the Sun, so that both volatile-rich and refractory-rich dust populations may be found in any location (Figure 9).

Volatile-rich minerals are but a small fraction of the total supply of planetary "building blocks" close to the Sun. A terrestrial planet's entire volatile inventory may be concentrated in as little as 1 percent of its mass – a quantity too small to influence the bulk density of the body detectably. Instead, large solid objects that came together, say, anywhere inside the inner edge of the asteroid belt, would have compositions dominated by material that had condensed locally. This means that the chemistry and major-element mineralogy of each planet should reflect, rather closely, the equilibrium composition of solid particles formed nearby in the solar nebula, at the highest temperature experienced by the dust-gas mixture. Likewise, volatile elements in the same planet should bear evidence of processes that occurred at lower temperatures.

So far in this scenario, we have mentioned only the consequences of chemical equilibration between dust and gas. There are very strong theoretical reasons to believe that, at the temperatures prevalent throughout much of the nebula, certain important chemical reactions *cannot* reach equilibrium over its 100,000-year lifetime. For example, close to the Sun, CO is the most stable compound of carbon and N_2 the most stable of nitrogen. But in regions cooler than 680° K, methane (CH_4) is more stable than CO; and below 330° K, ammonia (NH_3) is more stable than N_2. So as the temperature dropped, carbon and nitrogen should have converted to these more reduced forms. However, at such low temperatures the reactions involved are extremely sluggish, such that no more than a few percent methane or ammonia evolves even over the entire lifetime of the nebula. Intermediate products of the reduction reactions, species that ordinarily would be unstable, accumulate instead. These compounds include solid polymorphs of carbon and a rich variety of organic matter. Edward Anders has proposed this process as the source of organic matter in the carbonaceous chondrite meteorites.

Low-temperature chondritic meteorites do contain these species, but the *most* volatile-rich meteorites (the C1 and C2 carbonaceous chondrites) also incorporate much larger amounts of magnetite, carbonates, sulfates, hydroxyl silicates, and elemental sulfur. Yet no satisfactory scenario for making any of these species in the nebula is known! Moreover, strong evidence argues that these meteorites formed from the infusion of liquid water, carbon dioxide, oxidizing agents, and other volatiles into some other more conventional type of chondritic material. For example, the oxidation of certain reduced forms of sulfur (primarily sulfides) has apparently occurred in the presence of liquid water near its freezing point. Such processes are conceivable on or below the surface of a parent body, but they are improbable in a cold, tenuous nebula. Clearly, if these C1- and C2-type materials can be made only in already-accreted solid bodies, we must question whether they were available when the inner planets came together.

ASSEMBLING THE PLANETS

Ideas about chemical behavior within the nebula and laboratory data on meteorite composition are used in different ways by those who model the formation of the terrestrial planets. While theorists may agree that solids with different formation conditions accreted to form each planet, they still debate the finer details of their individual models. At one extreme, large amounts of volatile-laden C1 material, some 10 to 20 percent of each planet's mass, could have accumulated together in a homogeneous mixture with refractory (volatile-poor) material. A massive escape of the volatiles followed, leaving only the amounts now found on the terrestrial planets. This concept, the work of A. E. Ringwood, is an adaptation of his earlier theory in which each planet began as 100-percent C1 material. Taking a different approach, Anders suggests that the distribution of

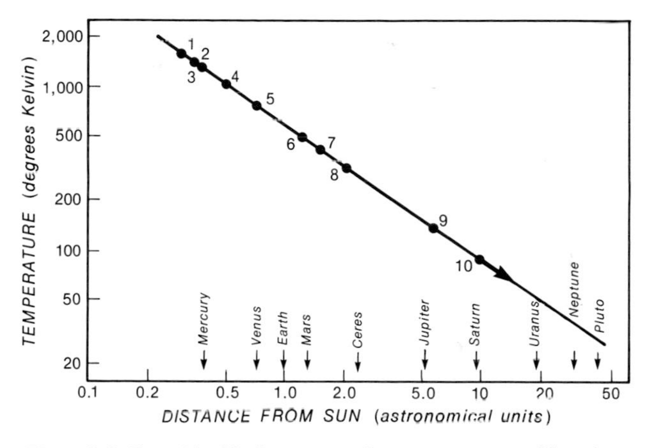

Figure 8. **Indicated in this diagram are the temperatures and locations at which major planetary constituents would be expected to condense from the primordial solar nebula (at a more evolved and cooler stage than in Figure 4): *1*, refractory minerals like the oxides of calcium, aluminum, and titanium, and rare metals like tungsten and osmium; *2*, common metals like iron, nickel, cobalt, and their alloys; *3*, magnesium-rich silicates; *4*, alkali feldspars (silicates abundant in sodium and potassium); *5*, iron sulfide; *6*, the lowest temperature at which unoxidized iron metal can exist; *7*, hydrated minerals rich in calcium; *8*, hydrated minerals rich in iron and magnesium; *9*, water ice; and *10*, other ices.**

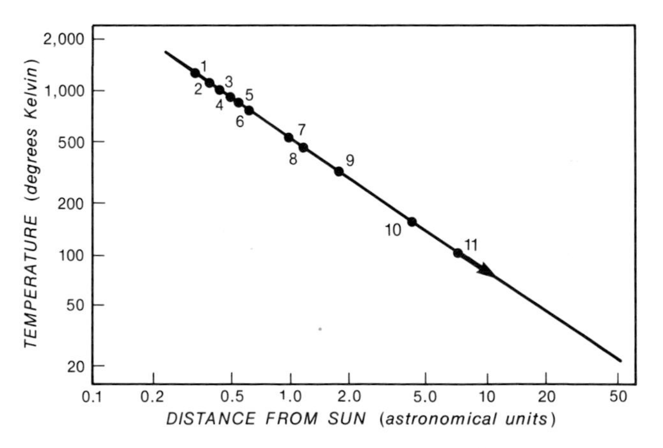

Figure 9. **A second diagram identifies the condensation temperatures of those minerals expected to be carriers of volatile elements: *1*, uranium and thorium (the source of helium through radioactive decay); *2*, iron phosphide; *3*, potassium compounds (the source of argon-40 through the decay of potassium-40); *4*, sodalite (rich in chlorine); *5*, fluorapatite (rich in fluorine); *6*, iron sulfide; *7*, the lowest temperature at which unoxidized iron metal can exist; *8*, hydrated silicates abundant in calcium; *9*, hydrated silicates abundant in iron and magnesium; *10*, water ice and simple ammonium and carbonate salts; and *11*, other ices.**

material outward from the Sun allowed each planet to accrete in layers that differ in composition outward from its center. Karl Turekian favors a scenario in which high-temperature condensates accumulate first, followed by successive layers of ever more volatile-rich material that accretes as it condenses in the solar nebula. In Turekian's view, the final "veneering" of the Earth by water-rich, highly oxidized C1 material brings in most of the volatiles. Alternatively, Robert Pepin proposes a late veneering that consisted of very dry, chemically reduced E chondrites.

My co-workers and I find no convincing reason to make C1 chondrites (which contain 18 to 22 percent water) an important component of the forming Earth. Instead, as it accreted the Earth may have captured solid material with a "tail" that extended out beyond its orbit. Most of its volatiles would then come from C3 chondritic material (containing no more than 3 percent water) near the orbit of Mars. Other accretion studies show that this method of collection should work, but the composition of these hypothetical captured solids is open to debate. As a point in its favor, this model correctly predicted the asteroid belt's domination by matter similar to carbonaceous chondrites – a conclusion later confirmed by astronomical observations.

Furthermore, since the planets almost certainly took much longer to come together than the nebula's 100,000-year existence, it seems unlikely that accretion could have taken place in step with the cooling nebula. The idea of strong compositional layering would then be ruled out. We would still expect the terrestrial planets to have collected volatile-enriched material near the end of their accretion, but in a different way. Small bodies near the fringes of the early solar system, composed largely of icy substances, in time would have been perturbed inward by gravitational tugs from the nearly completed planets. Migrating into the vicinity of the terrestrial planets, they would soon be swept up, supplying the Earth and its neighbors with a late surge of volatiles to supplement the unavoidable trace of C3-like material that had been present from the beginning.

The change in planetary composition with increasing distance from the Sun was mimicked to some extent in the pre-planetary nebula from which the Jovian planets and their satellites condensed. For example, Jupiter's own nebula was apparently too hot near its center for ices to condense, so it became dominated by rocky material that contained only a small proportion of water (in hydroxyl silicates). Water reacts with (oxidizes) metallic iron, but given little enough iron in the initial assembly some water would have been left over. If so, heating from radioactive decay soon drove water vapor and carbon dioxide from the interior, which then easily escaped from the body's gravity field. Once all hydrogen compounds were lost, sulfur and sulfur dioxide remained as the principle volatiles. In this way, a body with a roughly Mars-like initial composition could evolve to resemble Io if it is small enough – but not *so* small that the sulfur gases can also escape.

LOOKING TO THE FUTURE

The study of the origin and evolution of the solar system is still in its infancy. As our theoretical models of the origin and evolution of planets become more complex, our sketches of (qualitatively) plausible histories seem generally consistent with the growing body of observational data. While this is encouraging, these models are far from the last word – new observations may either buttress or undermine them at any moment. As we have seen, the realms of our present ignorance about the solar system are vast indeed.

Any satisfactory description of the early solar system will almost certainly draw upon several of the competing theories we have discussed. New elements will surely be introduced by both experiment and theory. Perhaps in your lifetime we shall know in detail why Mercury has such a high metal content, whether Venus ever had oceans, where the Moon came from, whether there was once a benign climate on Mars, what materials color the clouds of Jupiter, why planetary ring systems are so diverse, whether massive deposits of organic matter exist on Titan, why Uranus' spin axis is tipped at right angles to the axis of its orbit, what Pluto is made of, where comets were born, and which asteroids spawn meteorites. For now, all these questions are debated. Even where clear answers are known, not all can be sensibly integrated into the overall picture of solar-system origin and evolution. Because such a vast amount must yet be learned, the exploration of the solar system will continue irrespective of the level of effort of any one nation – and the intellectual and material rewards of that endeavor will fall to those who have the vision to continue to explore.

Planet, Satellite, and Small-body Characteristics

Characteristics of the inner planets

	MERCURY	VENUS	EARTH	MOON	MARS
Reciprocal mass[1]	6,023,600	408,524	328,900	27,069,000	3,098,710
Mass[2] (Earth = 1)	0.0553	0.8149	1.0000	0.01230	0.1074
Mass[2] (g)	3.303×10^{26}	4.870×10^{27}	5.976×10^{27}	7.349×10^{25}	6.421×10^{26}
Equatorial radius (Earth = 1)	0.382	0.949	1.000	0.272	0.532
Equatorial radius (km)	2,439	6,051	6,378	1,738	3,393
Ellipticity[3]	0.0	0.0	0.0034	0.002	0.0052
Mean density (g/cm^3)	5.43	5.25	5.52	3.34	3.95
Equatorial surface gravity (m/s^2)	2.78	8.60	9.78	1.62	3.72
Equatorial escape velocity (km/s)	4.3	10.4	11.2	2.4	5.0
Sidereal rotation period	58.65 days	243.01 days	23.9345 hours	27.322 days	24.6229 hours
Inclination of equator to orbit	(2°)[4]	177°.3[5]	23°.45	6°.68	25°.19

Characteristics of the outer planets

	JUPITER	SATURN	URANUS	NEPTUNE	PLUTO
Reciprocal mass	1,047.355	3,498.5	22,869	19,424	135,300,000
Mass (Earth = 1)	317.938	95.181	14.531	17.135	0.0022
Mass (g)	1.900×10^{30}	5.688×10^{29}	8.684×10^{28}	1.024×10^{29}	1.29×10^{25}
Equatorial radius[6] (Earth = 1)	11.209	9.449	4.007	3.883	0.180
Equatorial radius[6] (km)	71,492	60,268	25,559	24,764	1,150
Ellipticity	0.0649	0.0980	0.0229	0.017	(0.0)
Mean density (g/cm^3)	1.33	0.69	1.29	1.64	2.03
Equatorial surface gravity (m/s^2)	22.88	9.05	7.77	11.0	0.4
Equatorial escape velocity (km/s)	59.6	35.5	21.3	23.3	1.1
Sidereal rotation period at equator	9.841 hours[7]	10.233 hours[8]	17.9 hours[9]	19.2 hours[10]	6.3872 days
Inclination of equator to orbit	3°.12	26°.73	97°.86[5]	29°.6	122°.46[5]

Characteristics of planetary orbits

	Mean distance from Sun		*Sidereal period*		*Synodic period*	*Mean orbital velocity*	*Orbital eccentricity*	*Inclination to the ecliptic*
	(AU)	*(10^6 km)*	*(years)*	*(days)*	*(days)*	*(km/s)*		*(degrees)*
MERCURY	0.3871	57.91	0.24085	87.969	115.88	47.89	0.2056	7.004
VENUS	0.7233	108.20	0.61521	224.701	583.92	35.03	0.0068	3.394
EARTH	1.0000	149.60	1.00004	365.256	—	29.79	0.0167	0.000
MARS	1.5237	227.94	1.88089	686.980	779.94	24.13	0.0934	1.850
JUPITER	5.2028	778.33	11.8623	4,332.71	398.88	13.06	0.0483	1.308
SATURN	9.5388	1,426.98	29.458	10,759.5	378.09	9.64	0.0560	2.488
URANUS	19.1914	2,870.99	84.01	30,685	369.66	6.81	0.0461	0.774
NEPTUNE	30.0611	4,497.07	164.79	60,190	367.49	5.43	0.0097	1.774
PLUTO	39.5294	5,913.52	248.54	90,800	366.73	4.74	0.2482	17.148

Satellite of Earth

Name	Discoverer	Year of discovery	Magnitude (V_O)[11]	Mean distance from Earth (km)	Sidereal period (days)	Orbital inclination (degrees)	Orbital eccentricity	Radius (km)	Mass (g)	Mean density (g/cm^3)
Moon	?	?	–12.7	384,400	27.322	18.3–28.6	0.05	1,738	7.35×10^{25}	3.34

Satellites of Mars

Name	Discoverer	Year of discovery	Magnitude (V_O)	Mean distance from Mars (km)	Sidereal period (days)	Orbital inclination (degrees)	Orbital eccentricity	Radius (km)	Mass (g)	Mean density (g/cm^3)
Phobos	A. Hall	1877	11.3	9,380	0.319	1.0	0.01	14 × 10	1.08×10^{19}	2.0
Deimos	A. Hall	1877	12.4	23,460	1.263	0.9–2.7	0.00	8 × 6	1.8×10^{18}	1.7

Satellites of Jupiter

Name	Discoverer(s)	Year of discovery	Magnitude (V_O)	Mean distance from Jupiter (km)	Sidereal period (days)	Orbital inclination (degrees)	Orbital eccentricity	Radius (km)	Mass (g)	Mean density (g/cm^3)
Metis	S. Synnott	1979	17.5	127,960	0.295	(0)	0.00	(20)	?	?
Adrastea	D. Jewitt, E. Danielson	1979	18.7	128.980	0.298	(0)	(0)	12 × 8	?	?
Amalthea	E. Barnard	1892	14.1	181,300	0.498	0.4	0.00	135 × 75	?	?
Thebe	S. Synnott	1979	16.0	221,900	0.675	(0.8)	0.01	(50)	?	?
Io	S. Marius, Galileo	1610	5.0	421,600	1.769	0.04	0.00	1,815	8.94×10^{25}	3.57
Europa	S. Marius, Galileo	1610	5.3	670,900	3.551	0.47	0.01	1,569	4.80×10^{25}	2.97
Ganymede	S. Marius, Galileo	1610	4.6	1,070,000	7.155	0.19	0.00	2,631	1.48×10^{26}	1.94
Callisto	S. Marius, Galileo	1610	5.6	1,883,000	16.689	0.28	0.01	2,400	1.08×10^{26}	1.86
Leda	C. Kowal	1974	20.2	11,094,000	238.72	27	0.15	(8)	?	?
Himalia	C. Perrine	1904	15.0	11,480,000	250.57	28	0.16	(90)	?	?
Lysithea	S. Nicholson	1938	18.2	11,720,000	259.22	29	0.11	(20)	?	?
Elara	C. Perrine	1905	16.6	11,737,000	259.65	28	0.21	(40)	?	?
Ananke	S. Nicholson	1951	18.9	21,200,000	631	147	0.17	(15)	?	?
Carme	S. Nicholson	1938	17.9	22,600,000	692	163	0.21	(22)	?	?
Pasiphae	P. Melotte	1908	16.9	23,500,000	735	147	0.38	(35)	?	?
Sinope	S. Nicholson	1914	18.0	23,700,000	758	153	0.28	(20)	?	?

Satellites of Saturn

Name	Discoverer(s)	Year of discovery	Magnitude (V_O)	Mean distance from Saturn (km)	Sidereal period (days)	Orbital inclination (degrees)	Orbital eccentricity	Radius (km)	Mass (g)	Mean density (g/cm^3)
Atlas	R. Terrile	1980	18.0	137,640	0.602	(0)	(0)	20 × 15	?	?
Prometheus	S. Collins and others	1980	15.8	139,350	0.613	(0)	0.00	70 × 40	?	?
Pandora	S. Collins and others	1980	16.5	141,700	0.629	(0)	0.00	55 × 35	?	?
Epimetheus	R. Walker	1966	15.7	151,422	0.694	0.34	0.01	70 × 50	?	?
Janus	A. Dollfus	1966	14.5	151,472	0.695	0.14	0.01	110 × 80	?	?
Mimas	W. Herschel	1789	12.9	185,520	0.942	1.53	0.02	195	3.8×10^{22}	1.17
Enceladus	W. Herschel	1789	11.7	238,020	1.370	0.02	0.00	250	8.4×10^{22}	1.24
Tethys	G. Cassini	1684	10.2	294,660	1.888	1.09	0.00	525	7.55×10^{23}	1.26
Telesto	B. Smith and others	1980	18.7	294,660	1.888	(0)	(0)	(12)	?	?
Calypso	B. Smith and others	1980	19.0	294,660	1.888	(0)	(0)	15 × 10	?	?
Dione	G. Cassini	1684	10.4	377,400	2.737	0.02	0.00	560	1.05×10^{24}	1.44
Helene	P. Laques, J. Lecacheux	1980	18.4	377,400	2.737	0.2	0.01	18 × 15	?	?
Rhea	G. Cassini	1672	9.7	527,040	4.518	0.35	0.00	765	2.49×10^{24}	1.33
Titan	C. Huygens	1655	8.3	1,221,850	15.945	0.33	0.03	2,575	1.35×10^{26}	1.88
Hyperion	W. Bond	1848	14.2	1,481,000	21.277	0.43	0.10	175 × 100	?	?
Iapetus	G. Cassini	1671	10.2-11.9	3,561,300	79.331	14.72	0.03	720	1.88×10^{24}	1.21
Phoebe	W. Pickering	1898	16.5	12,952,000	550.48	175.3	0.16	110	?	?

Satellites of Uranus

Name	Discoverer	Year of discovery	Magnitude (V_O)	Mean distance from Uranus (km)	Sidereal period (days)	Orbital inclination (degrees)	Orbital eccentricity	Radius (km)	Mass (g)	Mean density (g/cm³)
Cordelia	Voyager 2	1986	24	49,750	0.335	(0.14)	(0)	(15)	?	?
Ophelia	Voyager 2	1986	24	53,760	0.376	(0.09)	(0.01)	(15)	?	?
Bianca	Voyager 2	1986	23	59,160	0.435	(0.16)	(0)	(20)	?	?
Cressida	Voyager 2	1986	22	61,770	0.464	(0.04)	(0)	(35)	?	?
Desdemona	Voyager 2	1986	22	62,660	0.474	(0.16)	(0)	(30)	?	?
Juliet	Voyager 2	1986	22	64,360	0.493	(0.06)	(0)	(40)	?	?
Portia	Voyager 2	1986	21	66,100	0.513	(0.09)	(0)	(55)	?	?
Rosalind	Voyager 2	1986	22	69,930	0.558	(0.28)	(0)	(30)	?	?
Belinda	Voyager 2	1986	22	75,260	0.624	(0.03)	(0)	(35)	?	?
Puck	Voyager 2	1985	20	86,010	0.762	(0.31)	(0)	75	?	?
Miranda	G. Kuiper	1948	16.5	129,780	1.414	3.40	0.00	235	6.89×10^{22}	1.35
Ariel	W. Lassell	1851	14.4	191,240	2.520	0.00	0.00	580	1.26×10^{24}	1.66
Umbriel	W. Lassell	1851	15.3	265,970	4.144	0.00	0.00	585	1.33×10^{24}	1.51
Titania	W. Herschel	1787	14.0	435,840	8.706	0.00	0.00	790	3.48×10^{24}	1.68
Oberon	W. Herschel	1787	14.2	582,600	13.463	0.00	0.00	760	3.03×10^{24}	1.58

Satellites of Neptune

Name	Discoverer	Year of discovery	Magnitude (V_O)	Mean distance from Neptune (km)	Sidereal period (days)	Orbital inclination (degrees)	Orbital eccentricity	Radius (km)	Mass (g)	Mean density (g/cm³)
1989 N 6	Voyager 2	1989	25	48,000	0.296	(0)	(0)	(25)	?	?
1989 N 5	Voyager 2	1989	24	50,000	0.312	(4.5)	(0)	(40)	?	?
1989 N 3	Voyager 2	1989	23	52,500	0.333	(0)	(0)	(90)	?	?
1989 N 4	Voyager 2	1989	23	62,000	0.429	(0)	(0)	(75)	?	?
1989 N 2[12]	Voyager 2	1989	21	73,600	0.554	(0)	(0)	(95)	?	?
1989 N 1	Voyager 2	1989	20	117,600	1.121	(0)	(0)	(200)	?	?
Triton	W. Lassell	1846	13.6	354,800	5.877	157	0.00	1,350	2.14×10^{25}	2.07
Nereid	G. Kuiper	1949	18.7	5,513,400	360.16	29	0.75	(170)	?	?

Satellite of Pluto

Name	Discoverer	Year of discovery	Magnitude (V_O)	Mean distance from Pluto (km)	Sidereal period (days)	Orbital inclination (degrees)	Orbital eccentricity	Radius (km)	Mass (g)	Mean density (g/cm³)
Charon	J. Christy	1978	16.8	19,640	6.387	98.8	0.00	595	(1.77×10^{24})	(2.0)

Rings of Jupiter

Name	Distance from Jupiter's center (R_J)	(km)	Radial width (km)	Thickness (km)	Optical depth	Total mass (g)	Albedo
"Halo"	(1.40)-1.72	(100,000)-122,800	22,800	(20,000)	6×10^{-6}	?	0.05
"Main"	1.72-1.81	122,800-129,200	6,400	<30	10^{-6}	(10^{16})	0.05
"Gossamer"	1.81-(3)	129,200-(214,200)	850,000	?	10^{-7}	?	0.05

NOTES AND EXPLANATIONS

[1]The mass of the Sun divided by the mass of the planet (including its atmosphere and satellites). [2]Satellite masses not included. [3]The ellipticity is $(R_e-R_p)/R_e$, where R_e and R_p are the planet's equatorial and polar radii, respectively. [4]Values in parentheses are uncertain by more than 10 percent. [5]By IAU convention, each planet's north pole is the one lying north of the ecliptic plane; as such, Venus, Uranus, and Pluto are considered to have retrograde rotation. [6]Since the outer planets have no solid surfaces, these are the radii at the 1-bar pressure level in their atmospheres. [7]Jupiter's internal (System III) rotation period is 9.925 hours. [8]Saturn's internal rotation period is 10.675 hours. [9]Uranus' internal rotation period is 17.240 hours. [10]Neptune's internal rotation period is 16.11 hours. [11]V_o is an object's magnitude in visible light at opposition. [12]Probably detected by H. Reitsema and others during an occultation in 1981. [13]Although the discovery of this gap is often ascribed to J. Encke, J. Keeler was likely the first to observe it. [14]Objects whose semimajor axes are less than 1 are termed Aten asteroids.

Rings of Saturn

Name	Distance from Saturn's center (R_S)	(km)	Radial width (km)	Thickness (km)	Optical depth	Total mass (g)	Albedo
D	1.11-1.24	67,000-74,500	7,500	?	(0.01)	?	?
C	1.24-1.52	74,500-92,000	17,500	?	0.08-0.15	1.1×10^{21}	0.25
Maxwell gap	1.45	87,500	270				
B	1.52-1.95	92,000-117,500	25,500	(0.1-1)	1.21-1.76	2.8×10^{22}	0.65
Cassini division	1.95-2.02	117,500-122,200	4,700	?	0.12	5.7×10^{20}	0.30
A	2.02-2.27	122,200-136,800	14,600	(0.1-1)	0.70	6.2×10^{21}	0.60
Encke gap[13]	2.214	133,570	325				
Keeler gap	2.263	136,530	35				
F	2.324	140,210	30-500	?	0.01-1	?	?
G	2.75-2.88	165,800-173,800	8,000	100-1,000	10^{-4}-10^{-5}	$6\text{-}23 \times 10^9$	?
E	(3-8)	(180,000-480,000)	(300,000)	(1,000)	10^{-6}-10^{-7}	?	?

Rings of Uranus

Name	Distance from Uranus' center (R_U)	(km)	Radial width (km)	Thickness (km)	Optical depth	Total mass (g)	Albedo
1986U2R	(1.49)	(38,000)	(2,500)	(0.1)	<0.001	?	(0.03)
6	1.597	41,840	1-3	(0.1)	0.2-0.3	?	(0.03)
5	1.612	42,230	2-3	(0.1)	0.5-0.6	?	(0.03)
4	1.625	42,580	2-3	(0.1)	0.3	?	(0.03)
Alpha (α)	1.707	44,720	7-12	(0.1)	0.3-0.4	?	(0.03)
Beta (β)	1.743	45,670	7-12	(0.1)	0.2	?	(0.03)
Eta (η)	1.801	47,190	0-2	(0.1)	0.1-0.4	?	(0.03)
Gamma (γ)	1.818	47,630	1-4	(0.1)	1.3-2.3	?	(0.03)
Delta (δ)	1.843	48,290	3-9	(0.1)	0.3-0.4	?	(0.03)
1986U1R	1.909	50,020	1-2	(0.1)	0.1	?	(0.03)
Epsilon (ϵ)	1.952	51,140	20-100	<0.15	0.5-2.1	?	(0.03)

Rings of Neptune

Name	Distance from Neptune's center (R_N)	(km)	Radial width (km)	Thickness (km)	Optical depth	Total mass (g)	Albedo
1989N3R	1.69	41,900	(15)	?	10^{-4}	?	(low)
1989N2R	2.15	53,200	(15)	?	0.01-0.02	?	(low)
1989N4R	2.15-2.4	53,200-59,100	5,800	?	10^{-4}	?	(low)
1989N1R	2.53	62,930	<50	?	0.01-0.1	?	(low)

Selected comets

Name	Year of discovery	Perihelion distance (AU)	Orbital eccen-tricity	Orbital inclination (degrees)	Orbital period (years)	Next perihelion	Associated meteor shower(s)
Arend-Roland	1956	0.316	1.000	119.95	–	–	
Bennett	1970	0.538	0.996	90.04	–	–	
Biela	1772	0.861	0.756	12.55	6.62	(lost)	Andromedids
Brorsen-Metcalf	1847	0.479	0.972	19.33	70.6	2060	
d'Arrest	1851	1.291	0.625	19.43	6.38	1995	
Donati	1858	0.578	0.996	116.96	–	–	
Encke	1786	0.341	0.846	11.93	3.31	1990	Taurids
Giacobini-Zinner	1900	1.028	0.708	31.88	6.59	1992	Draconids (October)
Halley	?	0.587	0.967	162.24	76.0	2061	η Aquarids, Orionids
Ikeya-Seki	1965	0.008	1.000	141.86	880	2845	
IRAS-Araki-Alcock	1983	0.991	0.990	73.25	–	–	
Kohoutek	1973	0.142	1.000	14.31	–	–	
Lexell	1770	0.674	0.786	1.56	5.60	(lost)	
Morehouse	1908	0.945	1.001	140.18	–	–	
Mrkos	1957	0.355	0.999	93.94	–	–	
Schwassmann-Wachmann 1	1908	5.448	0.105	9.75	15.0	2004	Perseids
Swift-Tuttle	1862	0.963	0.960	113.56	120	(lost?)	
Tempel 2	1873	1.381	0.545	12.44	5.29	1994	
West	1976	0.197	1.000	43.07	–	–	

The largest asteroids

Number	Name	Year of discovery	Discoverer	Radius (km)	Rotation period (hours)	Albedo	Mean distance from Sun (AU)	Mean distance from Sun (10^6 km)	Orbital period (years)	Orbital eccen-tricity	Orbital inclination (degrees)
1	Ceres	1801	G. Piazzi	457	9.08	0.10	2.767	413.9	4.61	0.097	10.61
2	Pallas	1802	H. Olbers	261	7.81	0.14	2.771	414.5	4.61	0.180	34.81
4	Vesta	1807	H. Olbers	250	5.34	0.38	2.362	353.4	3.63	0.097	7.14
10	Hygiea	1849	A. De Gasparis	215	(18.4)	0.08	3.144	470.3	5.59	0.136	3.84
511	Davida	1903	R. Dugan	168	5.13	0.05	3.178	475.4	5.67	0.171	15.94
704	Interamnia	1910	V. Cerulli	167	8.73	0.06	3.062	458.1	5.36	0.081	17.30
52	Europa	1858	H. Goldschmidt	156	5.63	0.06	3.097	463.3	5.46	0.119	7.44
15	Eunomia	1851	A. De Gasparis	136	6.08	0.19	2.644	395.5	4.30	0.143	11.76
87	Sylvia	1866	N. Pogson	136	5.18	0.04	3.486	521.5	6.52	0.051	10.87
16	Psyche	1852	A. De Gasparis	132	4.20	0.10	2.922	437.1	5.00	0.100	3.09
31	Euphrosyne	1854	J. Ferguson	(124)	5.53	(0.07)	3.156	472.1	5.58	0.099	26.34
65	Cybele	1861	E. Tempel	123	6.07	0.06	3.429	513.0	6.37	0.129	3.55
3	Juno	1804	K. Harding	122	7.21	0.22	2.670	399.4	4.36	0.218	13.00
324	Bamberga	1892	J. Palisa	121	29.43	0.06	2.683	401.4	4.41	0.285	11.14
107	Camilla	1868	N. Pogson	118	4.84	(0.06)	3.488	521.8	6.50	0.084	9.93
532	Herculina	1904	M. Wolf	116	9.41	0.16	2.772	414.7	4.61	0.184	16.36
451	Patientia	1899	A. Charlois	115	9.73	0.07	3.063	458.2	5.36	0.059	15.24
48	Doris	1857	H. Goldschmidt	113	11.89	(0.06)	3.112	465.5	5.48	0.064	6.54
29	Amphitrite	1854	A. Marth	120	5.39	0.16	2.554	382.1	4.08	0.066	6.10
423	Diotima	1896	A. Charlois	119	4.62	0.04	3.068	459.0	5.37	0.052	11.25
121	Hermione	1872	J. Watson	109	(6.1)	0.04	3.451	516.3	6.37	0.089	7.56
13	Egeria	1850	A. De Gasparis	107	7.04	0.10	2.576	385.4	4.14	0.121	16.52
45	Eugenia	1857	H. Goldschmidt	107	5.70	0.05	2.721	407.1	4.49	0.115	6.60
94	Aurora	1867	J. Watson	106	7.22	0.04	3.158	472.4	5.63	0.068	8.01
7	Iris	1847	J. Hind	102	7.14	0.21	2.386	356.9	3.68	0.210	5.51
702	Alauda	1910	J. Helffrich	101	(8.36)	0.06	3.194	477.8	5.71	0.041	20.57
372	Palma	1893	A. Charlois	98	(6.58)	0.05	3.146	470.1	5.57	0.156	23.85
128	Nemesis	1872	J. Watson	97	39	0.04	2.750	411.4	4.56	0.088	6.25
6	Hebe	1847	K. Hencke	96	7.27	0.25	2.425	362.8	3.78	0.146	14.79
154	Bertha	1875	Prosper Henry	96	?	0.07	3.184	476.3	5.68	0.094	21.11
76	Freia	1862	H. d'Arrest	95	9.98	0.03	3.390	507.1	6.30	0.186	2.12
130	Elektra	1873	C. Peters	95	5.22	(0.09)	3.119	466.6	5.49	0.218	22.88
22	Kalliope	1852	J. Hind	94	4.15	0.12	2.910	435.3	4.97	0.109	13.70
259	Aletheia	1886	C. Peters	92	?	0.04	3.139	469.6	5.60	0.140	10.77
41	Daphne	1856	H. Goldschmidt	91	5.99	(0.07)	2.765	413.6	4.59	0.279	15.78
2060	Chiron	1977	C. Kowal	(90)	?	?	13.716	2,051.9	50.80	0.382	6.93
747	Winchester	1913	J. Metcalf	89	9.40	0.05	2.998	448.5	5.18	0.245	18.18
120	Lachesis	1872	A. Borrelly	89	(>20)	0.04	3.118	466.4	5.47	0.088	6.96
790	Pretoria	1912	H. Wood	88	10.37	0.03	3.406	509.5	6.29	0.169	20.55
566	Stereoskopia	1905	P. Gotz	88	?	0.03	3.387	506.7	6.27	0.065	4.92
911	Agamemnon	1919	K. Reinmuth	88	(7)	0.04	5.201	778.1	11.90	0.068	21.83
153	Hilda	1875	J. Palisa	87	?	0.06	3.969	593.8	7.92	0.142	7.84
194	Prokne	1879	C. Peters	87	(15.67)	0.05	2.616	391.3	4.23	0.166	18.53
96	Aegle	1868	J. Coggia	87	?	0.04	3.051	456.4	5.33	0.164	16.00
59	Elpis	1860	J. Chacornac	87	13.69	0.05	2.713	405.9	4.47	0.094	8.64
386	Siegena	1894	M. Wolf	87	9.76	0.06	2.896	433.2	4.93	0.169	20.27

Aten asteroids[14]

Number	Name or designation	Year of discovery	Discoverer(s)	Radius (km)	Distance from Sun: minimum (AU)	mean (AU)	maximum (AU)	Orbital period (years)	Orbital eccentricity	Orbital inclination (degrees)
2340	Hathor	1976	C. Kowal	0.1	0.464	0.844	1.224	0.775	0.450	5.86
2100	Ra-Shalom	1978	E. Helin	1.7	0.469	0.832	1.195	0.759	0.436	15.76
3753		1986	D. Waldron	?	0.484	0.998	1.511	0.997	0.515	19.81
	1954 XA	1954	G. Abell	0.3	0.509	0.777	1.046	0.685	0.345	3.93
3362	Khufu	1984	R. Dunbar, M. Barucci	0.5	0.526	0.990	1.453	0.985	0.469	9.92
3554	Amun	1986	C. and E. Shoemaker	?	0.701	0.974	1.247	0.961	0.280	23.36
2062	Aten	1976	E. Helin	0.5	0.790	0.966	1.143	0.950	0.183	18.94

Earth-crossing asteroids

Number	Name or designation	Year of discovery	Discoverer(s)	Radius (km)	Distance from Sun minimum (AU)	Distance from Sun mean (AU)	Distance from Sun maximum (AU)	Orbital period (years)	Orbital eccentricity	Orbital inclination (degrees)
3200	Phaethon	1983	IRAS (satellite)	3	0.140	1.271	2.403	1.433	0.890	22.03
1566	Icarus	1949	W. Baade	0.7	0.187	1.078	1.969	1.119	0.827	22.91
2212	Hephaistos	1978	L. Chernykh	4.4	0.357	2.165	3.972	3.185	0.835	11.88
	1974 MA	1974	C. Kowal	(3)	0.423	1.775	3.128	2.365	0.762	37.78
	5025 P-L	1960	Van Houten, T. Gehrels	?	0.439	4.201	7.961	8.609	0.895	6.20
2101	Adonis	1936	E. Delporte	(0.5)	0.442	1.875	3.307	2.567	0.764	1.36
3838		1986	A. Maury	?	0.449	1.505	2.560	1.846	0.701	29.31
	1982 TA	1982	E. Helin	(2)	0.523	2.299	4.076	3.487	0.772	12.19
	1984 KB	1984	C. and E. Shoemaker	2	0.523	2.216	3.910	3.299	0.764	4.85
1864	Daedalus	1971	T. Gehrels	1.7	0.563	1.461	2.359	1.766	0.615	22.16
1865	Cerberus	1971	L. Kohoutek	(1)	0.576	1.080	1.585	1.123	0.467	16.09
4034		1986	E. Helin	?	0.589	1.060	1.530	1.091	0.444	11.17
	1987 KF	1987	C. and E. Shoemaker	?	0.590	1.836	3.082	2.488	0.679	11.87
	1987 SY	1987	C. and E. Shoemaker	?	0.596	1.442	2.288	1.731	0.587	5.52
	1987 OA	1987	J. Mueller	?	0.606	1.497	2.388	1.831	0.595	9.03
	Hermes	1937	K. Reinmuth	(0.5)	0.617	1.639	2.662	2.099	0.624	6.22
1981	Midas	1973	C. Kowal	(1.0)	0.622	1.776	2.930	2.367	0.650	39.84
2201	Oljato	1947	H. Giclas	(1.5)	0.626	2.173	3.721	3.204	0.712	2.52
3360		1981	E. Helin, S. Dunbar	1	0.628	2.458	4.289	3.855	0.744	22.02
	1988 EG	1988	J. Alu	?	0.635	1.269	1.903	1.430	0.499	3.49
	1983 TF2	1983	A. Maury, C. Pollas	0.5	0.643	2.439	4.235	3.809	0.736	14.70
1862	Apollo	1932	K. Reinmuth	0.7	0.647	1.471	2.295	1.784	0.560	6.35
	1979 XB	1979	K. Russell	0.3	0.649	2.262	3.876	3.403	0.713	24.87
	1989 FC	1989	H. Holt, N. Thomas	0.2	0.657	1.023	1.388	1.034	0.357	4.91
2063	Bacchus	1977	C. Kowal	(0.5)	0.701	1.078	1.454	1.119	0.349	9.42
	1959 LM	1959	C. Hoffmeister	1.5	0.718	1.980	3.242	2.787	0.637	6.77
	1987 SB	1987	E. Elst, V. Ivanova	?	0.748	2.204	3.661	3.272	0.661	3.04
	1988 XB	1988	Y. Oshima	?	0.761	1.466	2.172	1.776	0.481	3.12
	1983 LC	1983	E. Helin	(0.5)	0.765	2.632	4.498	4.269	0.709	1.52
1685	Toro	1948	C. Wirtanen	2.4	0.771	1.367	1.963	1.599	0.436	9.37
	1989 FB	1989	H. Holt, N. Thomas	?	0.781	1.042	1.303	1.064	0.251	14.16
	1988 VP4	1988	C. Shoemaker	?	0.786	2.263	3.740	3.405	0.651	11.66
2135	Aristaeus	1977	S. J. Bus	0.5	0.795	1.600	2.405	2.024	0.503	23.04
	1988 TA	1988	J. Mueller, J. Phinney	?	0.803	1.541	2.279	1.913	0.479	2.54
	1983 VA	1983	IRAS	1.5	0.805	2.611	4.416	4.218	0.692	16.24
3361	Orpheus	1982	C. Torres	0.3	0.819	1.209	1.600	1.330	0.323	2.69
	6743 P-L	1960	Van Houten, T. Gehrels	0.5	0.821	1.620	2.418	2.061	0.493	7.29
2329	Orthos	1976	H.-E. Schuster	1.5	0.821	2.405	3.988	3.729	0.659	24.39
1620	Geographos	1951	A. Wilson, R. Minkowski	1.0	0.827	1.245	1.662	1.389	0.335	13.32
	1950 DA	1950	C. Wirtanen	1.5	0.838	1.683	2.528	2.184	0.502	12.15
1866	Sisyphus	1972	P. Wild	5	0.872	1.894	2.915	2.606	0.539	41.15
	1987 QA	1987	J. Mueller	?	0.875	1.647	2.419	2.114	0.469	40.71
	1989 AZ	1989	C. Shoemaker	?	0.875	1.648	2.422	2.116	0.469	11.78
	1973 NA	1973	E. Helin	3	0.878	2.427	3.976	3.781	0.638	68.00
	1978 CA	1978	H.-E. Schuster	1.0	0.883	1.125	1.366	1.193	0.215	26.12
1863	Antinous	1948	C. Wirtanen	1.5	0.889	2.260	3.630	3.397	0.606	18.42
	1986 JK	1986	C. and E. Shoemaker	?	0.897	2.802	4.706	4.690	0.680	2.14
	1989 AC	1989	J.-L. Heudier and others	?	0.900	2.509	4.117	3.974	0.641	0.47
2102	Tantalus	1975	C. Kowal	1.5	0.905	1.290	1.675	1.465	0.298	64.01
3103		1982	M. Lovas	2	0.908	1.407	1.905	1.668	0.355	20.94
	1989 JA	1989	E. Helin	?	0.912	1.772	2.630	2.358	0.485	15.26
	6344 P-L	1960	Van Houten, T. Gehrels	0.1	0.940	2.576	4.211	4.134	0.635	4.62
	1982 DB	1982	E. Helin, E. Shoemaker	0.5	0.953	1.489	2.026	1.818	0.360	1.42
3752		1985	E. Helin, M. Barucci	?	0.986	1.414	1.841	1.681	0.303	55.55
	1989 DA	1989	J. Phinney	?	0.987	2.166	3.346	3.189	0.545	6.45
4015		1979	E. Helin	1.6	0.996	2.641	4.285	4.291	0.623	2.79
3671	Dionysius	1984	E. and C. Shoemaker	1	1.010	2.198	3.385	3.258	0.540	13.61
3757		1982	E. Helin	0.2	1.016	1.836	2.656	2.488	0.446	3.87

Glossary

accretion The accumulation of gas and dust into larger bodies, such as stars, planets, and moons.

achondrite A stony meteorite that lacks chondrules; most achondrites appear to be the products of igneous differentiation.

adiabatic Occurring without the gain or loss of heat.

advection Horizontal motion of atmospheric gases, which provides a mechanism for heat transfer.

airglow An emission of light caused by the excitation of neutral molecules in the upper atmosphere.

albedo The reflectivity of an object, expressed as the ratio of light it reflects to the light incident upon it and ranging from 1.0 (a white, perfectly reflecting surface) to 0.0 (an absolutely black surface that absorbs all incident radiation).

albedo feature A light or dark marking on the surface of an object that may or may not be associated with topographic or geologic features.

Amor A class of asteroids having perihelion distances between 1.017 and 1.3 AU.

anorthosite A granular igneous rock composed almost wholly of anorthite, a calcium-rich variety of plagioclase feldspar.

anticyclone A system of winds rotating around a region of relatively high atmospheric pressure. On a planet with prograde rotation (like Earth) anticyclones rotate clockwise in the northern hemisphere and counterclockwise in the southern hemisphere.

antitail The part of a comet's dust tail that, due to an observer's viewing angle, appears to emanate from the sunward side of the nucleus.

Apollo A class of asteroids having semimajor axes greater than 1.0 AU and perihelion distances less than 1.017 AU.

asthenosphere A zone of plastic and possibly partially molten rock located below the lithosphere.

astronomical unit (AU) 149,597,870 km, which is very nearly the average Earth-Sun distance.

Aten A class of asteroids having semimajor axes less than 1.0 AU and aphelion distances greater than 0.983 AU.

aurora A glow in a planet's ionosphere produced by the interaction of the planet's magnetosphere with charged particles from the Sun.

backscatter The redirection of radiation or particles at angles less than 90° from the initial direction of travel.

bar A unit of pressure equal to 10^5 newtons per m^2, or 0.987 atmosphere.

baroclinic An atmospheric condition in which surfaces of equal pressure do not coincide with surfaces of equal density.

barotropic An atmospheric condition in which surfaces of equal pressure coincide with surfaces of equal density.

basalt A dark, fine-grained, mafic igneous rock commonly produced in volcanic eruptions.

blackbody An idealized body that absorbs all radiation incident on it, then reemits the energy solely as a function of its temperature.

bow shock The boundary around an object's magnetosphere across which the solar wind's velocity drops and its plasma becomes heated, compressed, and deflected.

breccia Rock composed of broken rock fragments welded together by the extreme heat and pressure of a meteoroid impact.

CAI A *c*alcium- and *a*luminum-rich *i*nclusion in a meteorite.

caldera A large crater formed by the explosion, collapse, or modification of a preexisting volcanic vent.

central peak The mountain or group of mountains at the center of a large impact crater.

chaotic terrain Jumbled depressions and isolated hills found on Mars, possibly produced by collapse after the removal of subsurface water or ice.

chassignite A class of meteorite thought to have originated on Mars.

chondrite A class of stony meteorites that usually contain chondrules. *Carbonaceous chondrites* also include carbon compounds, while those of the *Type I* or *C1* variety contain no chondrules.

chondrules Small spherical grains, usually composed of iron-, aluminum-, or magnesium-rich silicates, found in abundance in primitive stony meteorites.

chromophores Groups of atoms or molecules responsible for pigmentation.

chromosphere The region of the solar atmosphere between the photosphere and corona.

clathrate A compound in which one molecular component is trapped in cavities in the crystalline lattice of the "host" component.

coesite A very dense form of silica produced by the extreme heat and pressure of a meteoroid impact.

coma The spherical envelope of gas and dust surrounding the nucleus of an active comet.

commensurability A situation in which two orbits have periods whose ratio is a simple fraction.

continental drift The gradual motion of the Earth's continents due to plate tectonism.

convection Circulation within a fluid that is driven by strong temperature gradients.

co-orbital satellites Two satellites that share almost the same orbit.

corona (1) The Sun's hot, highly ionized, and luminous outermost atmosphere. (2) A circular formation of ridges enclosing a central area of jumbled relief (as occurs on Venus).

co-rotation The rotation of plasma in a planetary magnetosphere at the same angular rate as the planet's magnetic field.

cosmic rays High-energy atomic nuclei (mostly protons) that enter the solar system from interstellar space.

cosmogony The study of the origin and evolution of stars and planets.

crater density A measure of impact-crater crowding on a planetary surface, usually given as the number of craters of a given size per unit area.

crater ray A deposit of ejected material extending radially beyond a crater's rim.

crust The outer solid layer of a planet, composed of relatively low-density materials.

cyclone A storm or system of winds rotating around a region of relatively low atmospheric pressure. On a planet with prograde rotation (like Earth) cyclones rotate counterclockwise in the northern hemisphere and clockwise in the southern hemisphere.

dendritic Having a branching, treelike pattern, like that of a river system.

deuterium The isotope of hydrogen containing one proton and one neutron.

differentiation The processes by which a planet (or, alternatively, a volume of magma) develops layers or zones of different chemical and mineralogical composition.

dipole The simplest form of magnetic field, in which lines of force run between one north pole and one south pole, as in a bar magnet.

eccentricity A value that defines the shape of an ellipse: the ratio of the distance between the foci to the major axis.

eclogite A dense igneous rock, composed essentially of garnet and pyroxene, thought to form at depth in the Earth's mantle.

ejecta blanket The material surrounding an impact crater that was thrown out during its formation.

eucrite A class of basaltic meteorites believed to have originated on the asteroid Vesta.

exosphere The outermost level of an atmosphere, in which atoms and molecules collide so infrequently that they can escape directly into space.

fault A fracture in the crust at which one side has moved relative to the other.

feldspar One of the most common rock-forming silicate minerals, rich in aluminum, with varying amounts of sodium, calcium, or potassium.

forwardscatter The redirection of radiation or particles at angles greater than 90° from the initial direction of travel.

geoid The equipotential surface of the Earth's gravitational field that coincides with mean sea level.

graben A linear valley bounded by high-angle faults.

granulation The mottling of the Sun's photosphere caused by cells of convecting gas packed "shoulder to shoulder."

greenhouse effect The trapping of infrared radiation by gases in a planet's atmosphere, which raises the surface temperature.

heliopause The edge of the heliosphere, where the pressure of the solar wind equals that of the interstellar medium.

helioseismology A technique used to study the Sun's interior by analyzing oscillations at the Sun's surface.

heliosheath The transition region from the heliosphere to the interstellar medium.

heliosphere The region in space permeated by the Sun's gases, magnetic field, and the solar wind.

Hirayama family Any group of asteroids with similar orbital elements, implying a common collisional origin in the past.

homopause The level in an atmosphere above which gases cease being uniformly mixed and become layered; often called the *turbopause*.

hydrostatic equilibrium The state of balance in a fluid between gravity and pressure gradients.

hydroxyl The chemical radical OH, consisting of an oxygen and hydrogen atom bound together.

igneous rock Any rock produced by cooling from a molten state.

impact basin An extremely large impact crater, often characterized by concentric rings of mountains.

interstellar grains Small particles thought to have been among the raw materials that formed the solar system.

ionopause The boundary in a comet's coma marking the transition from neutral gases to ions.

ionosphere The region of a planet's upper atmosphere in which many atoms are ionized.

iridium A rare-earth element that is scarce in Earth's crust but relatively abundant in meteorites.

iron meteorite A meteorite composed primarily of metallic iron and nickel, thought to represent material from the core of a differentiated parent body.

isostasy A state of equilibrium in which the buoyancy of crustal rocks sitting atop the mantle counteracts the pull of gravity.

kamacite An alloy of iron and a small amount of nickel, found in some meteorites.

kerogen Fossilized, insoluble organic material found in rocks.

Kirkwood gap One of several gaps in the asteroid belt where an object's orbital period is a simple fraction of that of Jupiter.

KREEP Lunar material rich in certain radioactive elements: *K* (potassium), *REE* (rare-earth elements), and *P* (phosphorus).

Lagrangian points The five equilibrium points in the restricted three-body problem; two Lagrangian points (L_4 and L_5), located at the vertices of equilateral triangles formed by the two primaries, are stable; the other three are unstable and lie on the line connecting the primary and secondary bodies.

Laplacian plane The plane in which a satellite's or ring's orbit lies because of perturbations by planetary oblateness, the Sun, and nearby satellites.

libration A small oscillation, or apparent rocking, exhibited by a synchronously rotating satellite.

lithify To change into rock.

lithophile Denoting an element that tends to concentrate in oxygen-containing compounds, particularly silicates.

lithosphere The rigid, outermost layer of the Earth.

Lorentz force The force experienced by a charged particle moving in a magnetic field.

luminosity A measure of the total radiation emitted by an object.

mafic Denoting rocks and minerals rich in magnesium and iron.

magma Molten rock deep within a planet; the term also refers to other molten materials (like water) involved in planetary volcanism.

magnetohydrodynamics The study of the motion of electrically conductive fluids in magnetic fields.

magnetopause The outer boundary of a planetary magnetosphere.

magnetosheath The region between a planetary bow shock and magnetopause, within which solar-wind plasma flows around the magnetosphere.

magnetosphere The region of space in which the planet's magnetic field dominates that of the solar wind.

magnetotail The portion of a planetary magnetosphere pulled downstream by the solar wind.

mantle The part of a planet between its crust and core, composed of relatively dense materials.

mare Any of several dark and relatively smooth regions on the Moon (and Mars) that are composed of basalt flows.

mascon A subsurface *mass concentration* that causes large-scale gravity anomalies on the Moon.

Maunder minimum The period of irregular and relatively low solar activity that occurred between about 1640 and 1710.

megaregolith The very deep regolith thought to characterize asteroids.

mesosphere An atmospheric region between the stratosphere and the thermosphere characterized by decreasing temperature with height.

metamorphic rock A rock produced by the alteration of preexisting rock by heat, pressure, shearing, or the chemical action of contained fluids.

micrometeorite A meteoritic particle less than a millimeter in diameter.

mixing ratio The fraction of any one component in a mixture of gases, expressed as the ratio of the masses of the gas and the entire mixture's.

Mohorovičić discontinuity The seismically distinct boundary between the Earth's crust and upper mantle; sometimes called the *Moho.*

nakhlite A class of meteorite thought to have originated on Mars.

neutrino A neutral elementary particle thought to have little, if any, rest mass.

nitrile Any of a group of organic molecules containing the cyanide radical CN.

noble gases The inert gases helium, neon, argon, krypton, xenon, and radon.

nucleosynthesis The process by which chemical elements are produced from light elements; it occurs in stellar interiors and during supernovas.

nucleus The core of a comet, typically a few kilometers in diameter, consisting of a mixture of ices and solid silicate and carbonaceous grains.

nuclide A particular species of atom, characterized by the number of protons and neutrons in its nucleus.

oblateness The rotational distortion of an otherwise spherical planet, defined as the difference between the equatorial and polar radii, divided by the equatorial radius.

obliquity The angle between an object's rotation axis and the pole of its orbit.

octupole A complex magnetic field with four north poles and four south poles.

olivine A metal-rich silicate mineral common in mafic rocks.

Oort cloud The region extending more than 100,000 AU from the Sun, where perhaps a trillion cometary nuclei are thought to reside.

opacity The ability of a medium to block light or other radiation.

optical depth A measure of the absorption of radiation passing through a gaseous medium; the medium is termed *transparent* when the value is zero, *optically thin* when less than 1, and *optically thick* when greater than 1.

organic Denoting a compound containing carbon, though the term is not necessarily associated with life.

outgassing The release of gas from a planet's interior, often in association with volcanic activity.

P wave A seismic disturbance whose oscillations are in the same direction as its direction of travel (a compression wave).

paleomagnetism Ancient magnetic fields recorded in rocks.

palimpsest A roughly circular spot on icy satellites, thought to identify a former impact crater.

Pangea The hypothetical supercontinent that contained all of Earth's land masses about 225 million years ago.

phase angle The angle between the Sun, an object, and an observer.

photodissociation The breakdown of molecules due to the absorption of light, especially ultraviolet light.

photosphere The visible surface of the Sun.

plagioclase A variety of the mineral feldspar and the most common rock-forming mineral.

planetesimals Primordial bodies of intermediate size that accreted into planets or asteroids.

planetology The study of the physical and chemical properties of planets.

plasma A highly ionized gas, consisting of almost equal numbers of free electrons and positive ions.

plasma tail The narrow, ionized component of a comet's tail, which extends directly away from the Sun.

plate tectonism An interpretation of features and processes in the Earth's crust based on the movement of segments of the lithosphere.

polarization A property of radiation in which its electromagnetic vibrations are not randomly oriented.

polymict breccia Breccia that originated from two or more unrelated rocks.

polymorph Any of two or more chemical species that possess the same chemical formula but different crystal structure (for example, diamond and graphite).

Poynting-Robertson effect The loss of orbital energy experienced by a small orbiting particle (between a micron and a centimeter in size) when it absorbs sunlight on one hemisphere, then reradiates the energy in all directions.

Precambrian The earliest era in geologic time, encompassing all Earth history before the beginning of the Cambrian era 570 million years ago.

precession A slow, periodic, conical motion of the rotation axis of a spinning body.

primary crater A crater produced by a meteoroid impact.

primordial Denoting the time at or near the formation of the solar system, about 4.6 billion years ago.

prograde Orbital motion (or rotation) in a counterclockwise direction as viewed from the north pole of the ecliptic (or of the rotating object).

protoplanet An accumulation of material in the solar nebula that represented the embryonic precursor to a planet.

protosun The accumulation of material in the solar nebula that later became the Sun.

pyroxene A rock-forming silicate mineral characteristic of mafic igneous rocks and high-temperature metamorphic rocks.

quadrupole A magnetic field with two north poles and two south poles.

rad A measure of exposure to ionizing radiation, equal to 100 ergs of energy per gram of absorbing material.

radiative diffusion The process by which energy produced in a star's core is carried outward by a series of emissions and reabsorptions.

radiogenic heat Heat produced by the fission of potassium, uranium, thorium, and other radioactive isotopes.

radionuclide A radioactive isotope of an element.

rarefaction A decrease in density and pressure in a medium that occurs after the passing of a compression wave.

refractory Denoting an element or compound that vaporizes at high temperatures, such as uranium, calcium, and aluminum.

regolith The layer of dust and fragmented rocky debris, produced by meteoritic impact, that forms the uppermost surface on planets, satellites, and asteroids.

remote sensing Any technique for investigating an object from a distance.

resonance A state in which one orbiting object is subject to periodic gravitational perturbations by another.

retrograde Orbital motion (or rotation) in a clockwise direction as viewed from the north pole of the ecliptic (or of the rotating object).

rift A fracture in the crust caused by the slow separation of tectonic plates.

rille A trenchlike valley; rilles on the Moon are up to several hundred km long and 1 to 2 km wide.

Roche limit The critical distance inside which an idealized satellite with no tensile strength would shatter due to tidal forces exerted by its parent planet.

S wave A seismic disturbance whose oscillations are perpendicular to its direction of travel.

saltation The transport of a particle by a series of leaps that occurs when a wind or current is too weak to keep it suspended.

scale height The altitude over which atmospheric density decreases by the factor *e* (2.7183).

scarp A cliff produced by faulting or erosion.

secondary crater An impact crater produced by ejecta from a primary impact.

seismic tomography A technique used to produce three-dimensional maps of the Earth's interior by analyzing seismic waves.

semimajor axis One-half of the longest dimension of an ellipse.

shepherd satellite A satellite that constrains the extent of a nearby planetary ring through gravitational influence.

shergottite A class of meteorite thought to have originated on Mars.

siderophile Denoting an element soluble in molten iron and often occurring in the native state (like cobalt, nickel, or gold).

shield volcano A broad-based volcano built up through the repeated eruptions of highly fluid basalt lavas.

silicate A rock or mineral whose crystalline structure is dominated by bonded silicon and oxygen atoms.

SNC meteorites Three small classes of basaltic meteorites (*s*hergottites, *n*akhlites, and *c*hassignites) thought to have been ejected from Mars' surface during an impact.

solar flare A sudden, violent release of magnetic energy in or near the Sun's photosphere that often sends great amounts of radiation and highly accelerated charged particles into interplanetary space.

solar wind The high-speed outflow of energetic charged particles and entrained magnetic field lines from the solar corona.

spalling The chipping or fracturing of rock caused, for example, by shock waves.

spectroscopy The study of the light emitted from a body (its spectrum).

spherule A round particle of glass or metal thrown into the air as a droplet of molten material during a meteoritic impact or volcanic eruption.

spokes Dark, radial streaks in Saturn's ring system.

sputtering A process of chemical alteration caused by atomic particles striking a surface at high speed.

stereoisomers Two compounds of the same composition that differ in the orientation of their molecular structures, in some cases as mirror images of each other.

stishovite An extremely dense form of silica produced by the extreme heat and pressure of a meteoroid impact.

stratigraphy Aspects of the geologic history of a region deduced from the study of its strata (layers of rock).

stratosphere The layer of the Earth's atmosphere, above the troposphere and below the ionosphere, in which the temperature gradually increases with altitude.

subduction The process by which sections of a planet's lithosphere are forced down into the mantle.

sunspot A relatively cool, dark, and highly magnetic area on the solar photosphere.

supergranulation A pattern of very large-scale convection cells in the solar chromosphere.

superrotation Atmospheric rotation that is faster than that of the planet itself.

synchronous orbit An orbit whose period equals the rotation period of the primary body it encircles.

synchrotron radiation Electromagnetic radiation emitted when a charged particle spirals around magnetic field lines.

tectonic Associated with the forces acting in a planet's crust.

tektites Small, rounded pieces of silicate glass, thought to have become airborne during terrestrial impact events.

tidal heating The frictional heating of a satellite's interior caused by repeated flexure induced by the gravitational field of its parent planet.

thermosphere The region of an atmosphere in which the temperature increases with altitude due to ionospheric heating.

Titius-Bode law A numerical series derived in the 18th century that approximates the distances of the planets from the Sun.

Trojan asteroids Asteroids located near the two stable Lagrangian points of Jupiter's orbit (60° preceding and following the planet).

tropopause The boundary between the troposphere and the stratosphere in Earth's atmosphere.

troposphere The lowest level of the Earth's atmosphere, where most weather takes place; also, the convection-dominated region of other planetary atmospheres.

volatiles Elements or compounds with low melting temperatures, such as potassium, sodium, water, and ammonia.

WIMP *W*eakly *i*nteracting *m*assive *p*article, a hypothetical subatomic particle able to travel freely through matter.

Zeeman effect The doubling, tripling, or broadening of lines in the spectrum of light passing through a magnetic field.

zodiacal light A faint glow caused by sunlight scattering off interplanetary dust near the plane of the ecliptic.

Author biographies and suggested further reading

Introduction

CARL SAGAN is the director of the Laboratory for Planetary Studies and David Duncan Professor of Astronomy and Space Sciences at Cornell University. He was on the science teams for the Mariner 9 and Viking missions and is on the Voyager imaging team. Sagan has served as chairman of the American Astronomical Society's Division of Planetary Sciences, and for 12 years he was Editor-in-chief of the journal *Icarus*. He is now president of the Planetary Society (65 N. Catalina Ave., Pasadena, California 91106), a nonprofit organization devoted to promoting planetary exploration. In addition to 400 published scientific and popular articles, Sagan has authored, coauthored, or edited more than a dozen books, including *The Dragons of Eden*, for which he won a Pulitzer prize in 1978.

Frazier, K., *Solar System*, part of the "Planet Earth" series (Time-Life, 1985)

Gore, R., "The Planets: Between Fire and Ice" (*National Geographic, 167*, 4-51, 1985)

Morrison, D., and T. Owen, *The Planetary System* (Addison-Wesley, 1988)

Whipple, F. L., *Orbiting the Sun* (Harvard Univ. Press, 1981)

Wood, J. A., *The Solar System* (Prentice-Hall, 1979)

Chapter 1: The Golden Age of Solar System Exploration

NOEL W. HINNERS retired from NASA in 1989 after serving as the agency's Associate Deputy Administrator. He earned a Ph.D. in geochemistry and geology at Princeton University in 1963. Hinners then served as the head of the lunar-exploration department at Bellcomm, Inc., where he evaluated scientific tasks for the first manned lunar landing. He was also involved in the selection of Apollo landing sites and in developing the Apollo Lunar Surface Experiment Packages. In 1974, Hinners was named NASA's Associate Administrator for Space Science, in which position he oversaw lunar and planetary programs, as well as those in physics, astronomy, and life sciences. Hinners served as director of the Smithsonian Institution's National Air and Space Museum from 1979 to 1982, and thereafter as director of NASA's Goddard Space Flight Center. Hinners was the founding editor of the journal *Geophysical Research Letters*.

Beyond Earth's Boundaries: Human Exploration of the Solar System in the 21st Century (U. S. National Aeronautics and Space Administration, 1988)

Logsdon, J. M., "An Apollo Perspective" (*Aeronautics and Astronautics, 17*, December 1979)

Newell, H. E., *Beyond the Atmosphere* (U. S. National Aeronautics and Space Administration, SP-4211, 1981)

Pioneering the Space Frontier: The Report of the National Commission on Space (Bantam, 1986)

Planetary Exploration Through Year 2000: A Core Program (U. S. Government Printing Office, 1983)

Ride, S. K., *Leadership and America's Future in Space* (U. S. National Aeronautics and Space Administration, 1987)

Chapter 2: The Sun

ROBERT W. NOYES is a professor of astronomy at Harvard University and an astrophysicist at the Smithsonian Astrophysical Observatory. He received his Ph.D. in physics from the California Institute of Technology in 1963, where he wrote his thesis on the discovery of five-minute oscillations on the Sun. Since then his research interests have stressed solar physics and the study of Sun-like phenomena on other stars. He was part of a team using telescopes at Mount Wilson Observatory to study magnetic activity on nearby stars. Noyes has authored many research articles on solar and stellar physics, as well as a popular book on the Sun.

Eddy, J. A., *A New Sun: The Solar Results from Skylab* (U. S. National Aeronautics and Space Administration, SP-402, 1979)

Giovanelli, R., *Secrets of the Sun* (Columbia Univ. Press, 1984)

Harvey, J. W., J. R. Kennedy, and J. W. Leibacher, "GONG: To See Inside Our Sun" (*Sky & Telescope, 74*, 470-476, November 1987)

Leibacher, J. W., *et al.*, "Helioseismology" (*Scientific American, 253*, 48-57, September 1985)

Noyes, R. W., *The Sun, Our Star* (Harvard Univ. Press, 1982)

Washburn, Mark, *In the Light of the Sun* (Harcourt Brace Jovanovich, 1981)

Wentzel, D. G., *The Restless Sun* (Smithsonian Institution, 1989)

Willson, R. C., H. Hudson, and M. Woodard, "The Inconstant Solar Constant" (*Sky & Telescope, 67*, 501-503, June 1984)

Chapter 3: Magnetospheres, Cosmic Rays, and the Interplanetary Medium

JAMES A. VAN ALLEN is one of the pioneers in space physics, having initiated scientific work with high-altitude rockets in late 1945. In the years since he has been a principal investigator on 24 Earth-orbiting, interplanetary, and planetary spacecraft, including Pioneers 10 and 11. In 1958, he and students George H. Ludwig, Carl E. McIlwain, and Ernest C. Ray discovered the radiation belts of the Earth with their instrument on the first successful American satellite, Explorer 1. Van Allen's research since then has emphasized planetary magnetospheric physics, most recently that of Saturn. He has also done extensive work on solar X-rays, the propagation and acceleration of energetic solar electrons and protons in interplanetary space, and the intensity of primary cosmic rays to great heliocentric distances. He was a professor of physics and head of the Department of Physics and Astronomy at the University of Iowa from 1951 until his retirement in 1985.

Brandt, J. C., *Introduction to the Solar Wind* (W. H. Freeman, 1970)
Fimmel, R. O., J. Van Allen, and E. Burgess, *Pioneer: First to Jupiter, Saturn, and Beyond* (U. S. National Aeronautics and Space Administration, SP-446, 1980)
Friedlander, M., *Cosmic Rays* (Harvard Univ. Press, 1989)
Lanzerotti, L. J., and C. Uberoi, "Earth's Magnetic Environment" (*Sky & Telescope, 76*, 360-362, October 1988) and "The Planets' Magnetic Environments" (*Sky & Telescope, 77*, 149-152, February 1989)
Russell, C. T., "Planetary Magnetism" (*Geomagnetism, 2*, 457-523, 1987)
Van Allen, J. A., *Origins of Magnetospheric Physics* (Smithsonian Institution, 1983)

Chapter 4: The Moon

PAUL D. SPUDIS is a geologist for the U.S. Geological Survey in Flagstaff, Arizona. He received his Ph.D. in geology from Arizona State University in 1982. Prior to that, as a member of the site-selection team for the Viking 2 lander, Spudis made geologic maps of Mars and evaluated possible hazards. Using data from planetary spacecraft, he has mapped geologic features on the Moon, Mercury, Venus, Mars, and Io. He is currently a member of the Lunar and Planetary Geology and Geophysics Review Panel, which evaluates proposals to NASA's Planetary Geology Program. His current research includes analysis of lunar rock samples and studies leading to the establishment of a permanent base on the Moon.

Cortwright, E. M., ed., *Apollo Expeditions to the Moon* (U. S. National Aeronautics and Space Administration, SP-350, 1975)
Hartmann, W. K., *et al.*, eds., *Origin of the Moon* (Lunar and Planetary Institute, 1986)
Lewis, R. S., *The Voyages of Apollo: The Exploration of the Moon* (Quadrangle, 1974)
Lunar Geoscience Working Group, *Status and Future of Lunar Geoscience* (U. S. National Aeronautics and Space Administration, SP-484, 1986)
Masursky, H., *et al.*, eds., *Apollo Over the Moon: A View From Orbit* (U. S. National Aeronautics and Space Administration, SP-362, 1978)
Mendell, W., ed., *Lunar Bases and Space Activities of the 21st Century* (Lunar and Planetary Institute, 1986)
Mutch, T. A., *The Geology of the Moon: A Stratigraphic Approach* (Princeton Univ. Press, 1972)
Rubin, A. E., "Whence Came the Moon?" (*Sky & Telescope, 68*, 389-393, November 1984)
Taylor, S. R., *Planetary Science: A Lunar Perspective* (Lunar and Planetary Institute, 1982)
Wilhelms, D. E., *The Geologic History of the Moon* (U. S. Geological Survey Professional Paper 1348, 1987)

Chapter 5: Mars

MICHAEL H. CARR lived in England before coming to the United States to obtain advanced geology degrees from Yale University (an M.S. in 1957 and a Ph.D. in 1960). He has been on the scientific staff of the U. S. Geological Survey since 1962 and served as chief of its Branch of Astrogeologic Studies in Flagstaff, Arizona, from 1974 to 1978. Carr's early research involved the Moon and meteoritic debris, but more recently he has modeled the various roles volcanism, water, and ice have played throughout Martian history and how volcanic processes occur on Jupiter's satellite Io. Carr has been very active in NASA's interplanetary explorations, having served on the scientific teams for the Lunar Orbiter, Mariner 9, Voyager, Galileo, and Mars Observer missions. He was also the leader of the Viking orbiter imaging team and oversaw the analysis of the 55,000 images acquired by those spacecraft.

Arvidson, R. E., *et al.*, "The Surface of Mars" (*Scientific American, 238*, 76-89, March 1978)
Carr, M. H., *The Surface of Mars* (Yale Univ. Press, 1981)
Cordell, B. M., "Mars, Earth, and Ice" (*Sky & Telescope, 72*, 17-22, July 1986)
Mutch, T. A., *et al.*, *The Geology of Mars* (Princeton Univ. Press, 1976)
Schultz, P. H., "Polar Wandering on Mars" (*Scientific American, 253*, 94-102, December 1985)
Scientific Results of the Viking Project (*Jour. Geophys. Res., 82*, 3959-4681, 1977)
Squyres, S. W., "Water on Mars" (*Icarus, 79*, 229-288, 1989)
Viking Orbiter Views of Mars (U. S. National Aeronautics and Space Administration, SP-441, 1980)

Chapter 6: The Earth

DON L. ANDERSON is director of the Seismological Laboratory and a professor of geophysics at the California Institute of Technology, where he received his Ph.D. in geophysics and mathematics in 1962. His interests lie in the origin, evolution, structure, and composition of the Earth and other planets. Anderson served with Chevron Oil Co., the Air Force Cambridge Research Laboratory, and the Arctic Institute of North America from 1955 to 1958. He was elected to the National Academy of Sciences in 1982, and has received the Emil Wiechert Medal of the German Geophysical Society, the Arthur L. Day Medal of the Geological Society of America and the Gold Medal of the Royal Astronomical Society. Anderson began serving a 3-year term as president of the American Geophysical Union in 1988.

Anderson, D. L., "The Earth as a Planet: Paradigms and Paradoxes" (*Science, 223*, 347-535, 1984)
Anderson, D. L, and A. Dziewonski, "Seismic Tomography" (*Scientific American, 251*, 60-68, October 1984)
Continents Adrift and Continents Aground, readings from *Scientific American* (W. H. Freeman, 1976)
Ozima, M., *The Earth, Its Birth and Growth* (Cambridge, 1981)
Special issue: "The Dynamic Earth" (*Scientific American, 249*, 46-78 and 114-189, September 1983)
Stacey F., *Physics of the Earth*, 2nd ed. (Wiley, 1977)
Wahr, J., "The Earth's Inconstant Rotation" (*Sky & Telescope, 71*, 545-549, June 1986)

Chapter 7: Surfaces of the Terrestrial Planets

JAMES W. HEAD, III, is a professor of geological sciences at Brown University, where he received his Ph.D. in 1969. From 1968 to 1972 he participated in selecting the lunar landing sites for the Apollo program while serving at Bell-comm, Inc. During this time Head also helped plan and evaluate the package of experiments to be deployed on the Moon. Head's research centers on the study of the processes that form and modify planets' surfaces and crusts, and how these processes vary with time and interact to produce the historical record preserved on the planet. His extensive involvement in unmanned planetary exploration includes cooperative research with scientists at the Vernadskiy Institute of the Soviet Academy of Sciences. Head serves on the science teams for NASA's Magellan, Mars Observer, and Galileo missions. He is also an editor of the journal *Earth, Moon, and Planets*.

Bazilevskiy, A. T., "The Planet Next Door" (*Sky & Telescope, 77,* 360-368, April 1989)
Chapman, C. R., *Planets of Rock and Ice* (Scribner's, 1982)
Greeley, R., *Planetary Landscapes* (Unwin and Allen, 1985)
Hartmann, W. K., *Moons and Planets* (Wadsworth, 1983)
Head, J. W., and L. S. Crumpler, "Evidence for Divergent Plate-Boundary Characteristics and Crustal Spreading on Venus (*Science, 238,* 1380-1385, 1987)
Head, J. W., and S. C. Solomon, "Tectonic Evolution of the Terrestrial Planets" (*Science, 213,* 62-76, 1981)
Murray, B. C., *et al., Earthlike Planets* (W. H. Freeman, 1981)
The Near Planets, part of the "Voyage Through the Universe" series (Time-Life, 1989)
Strom, R. G., *Mercury: The Elusive Planet* (Smithsonian Institution, 1987)

Chapter 8: Atmospheres of the Terrestrial Planets

JAMES B. POLLACK is a senior scientist at NASA's Ames Research Center in Moffett Field, California. He received a Ph.D. in astronomy from Harvard University in 1965. From 1965 to 1968 Pollack was at the Smithsonian Astrophysical Observatory in Cambridge, Massachusetts, and from 1968 to 1970 he was at Cornell University's Center for Radiophysics and Space Research. While at Cornell he was on the Mariner 9 imaging team, in charge of photographing the Martian moons Phobos and Deimos. Pollack studied the atmospheres of Mars and Venus while participating in the Viking and Pioneer Venus missions. He is on the Voyager imaging team and is also an interdisciplinary scientist for both Galileo and Mars Observer. Pollack's research centers on planetary atmospheres and on understanding climatic change on Earth and other planets. He is an associate editor of *Icarus* and *Journal of Geophysical Research.*

Atreya, S. K., J. B. Pollack, and M. S. Matthews, eds., *Origin and Evolution of Planetary and Satellite Atmospheres* (Univ. of Arizona Press, 1989)
Chapman, C., "The Vapors of Venus and Other Gassy Envelopes" (*Mercury, 12,* 130-141, 1983)
Fimmel, R. O., L. Colin, and E. Burgess, *Pioneer Venus* (U. S. National Aeronautics and Space Administration, SP-46, 1983)
Gillett, S. L., "The Rise and Fall of the Early Reducing Atmosphere" (*Astronomy, 13,* 66-71, July 1985)
Goody, R. H., and J. C. G. Walker, *Atmospheres* (Prentice-Hall, 1972)
Harberle, R. M., "The Climate of Mars" (*Scientific American, 254,* 54-62, May 1986)
Kasting, J. F., O. B. Toon, and J. B. Pollack, "How Climate Evolved on the Terrestrial Planets" (*Scientific American, 258,* 90-97, February 1988)

Chapter 9: The Voyager Encounters

BRADFORD A. SMITH is the team leader of the Voyager imaging-science experiment and a professor of planetary sciences at the University of Arizona in Tucson. In 1958 he joined the faculty of New Mexico State University, where he developed a program of planetary photography. In addition to receiving his Ph.D. in astronomy there, he later became the university's director of planetary research. Smith served on the imaging teams for the Mariner 6, 7, 9, and Viking missions to Mars and for the Soviet Vega mission. He is an Interdisciplinary Scientist on the Soviet Phobos mission and a member of the Wide-Field and Planetary Camera team for the Hubble Space Telescope.

Eberhart, J., "Jupiter and Family" (*Science News, 115,* 165-173, 1979)
Gore, R., "Voyager Views Jupiter's Dazzling Realm" (*National Geographic, 157,* 2-29, 1980)
Littmann, M., "The Triumphant Grand Tour of Voyager 2" (*Astronomy, 16,* 34-40, December 1988)
McLaughlin, W. I., "Voyager's Decade of Wonder" (*Sky & Telescope, 78,* 16-20, July 1989)
Morrison, D., *Voyages to Saturn* (U. S. National Aeronautics and Space Administration, SP-451, 1982)
Morrison, D., and J. Samz, *Voyage to Jupiter* (U. S. National Aeronautics and Space Administration, SP-439, 1980)
Special issue: Voyager 2 at Saturn, (*Science, 215,* 499-594, 1982)
Special issue: Voyager 2 at Uranus, (*Science, 233,* 39-109, 1986)
Washburn, M., *Distant Encounters* (Harcourt Brace Jovanovich, 1982)

Chapter 10: Outer-Planet Interiors

WILLIAM B. HUBBARD received a Ph.D. in astronomy from the University of California at Berkeley in 1967, where his dissertation addressed stellar structure and evolution. In 1973, after postdoctoral work at the California Institute of Technology and an assistant professorship at the University of Texas at Austin, Hubbard worked at the O. Yu. Schmidt Institute in Moscow, collaborating with Soviet scientists on determining the interior structure of the giant planets from gravitational-field measurements. Since 1973, Hubbard has been at the University of Arizona, where he is professor of planetary sciences in the department he headed from 1977 to 1981. His current work involves giant-planet interiors, brown dwarfs, and observations of occultations of stars by planets, satellites, rings, and asteroids.

Beatty, J. K., "A Place Called Uranus" (*Sky & Telescope, 71,* 333-337, April 1986)
Black, D. C., and Matthews, M., *Protostars and Planets II* (Univ. of Arizona Press, 1985), Parts VI and VII
Gehrels, T., and Matthews, M. S., *Saturn* (Univ. of Arizona Press, 1984)
Hubbard, W. B., *Planetary Interiors* (Van Nostrand Reinhold, 1984)
Kafatos, M., Harrington, R. S., and Maran, S. P., *Astrophysics of Brown Dwarfs* (Cambridge Univ. Press, 1985)
Stevenson, D. J., "Looking Ahead to Neptune" (*Sky & Telescope, 77,* 481-483, May 1989)

Chapter 11: Outer-Planet Atmospheres

ANDREW P. INGERSOLL's interest in oceans and atmospheres was nurtured at Harvard, where he was awarded his Ph.D. in 1966. That year Ingersoll joined the Division of Geological (and more recently, Planetary) Sciences at the California Institute of Technology, where he has been a Professor of Planetary Science since 1976. Ingersoll is still interested in how planetary atmospheres and oceans work – how they redistribute heat, why winds and currents flow as they do, and how their climates change. He has served on several spacecraft experiment teams, including those for the Pioneer Venus infrared radiometer, the Voyager imaging system, the Pioneer Jupiter and Pioneer Saturn radiometer experiments, and the Vega balloon experiment at Venus. Ingersoll is also involved in the Mars Observer and Galileo missions.

Alexander, A. F. O'D., *The Planet Saturn* (Dover, 1980)
Berry, R., "Voyager: Science at Saturn" (*Astronomy, 9,* 6-22, February 1981)
The Far Planets, part of the "Voyage Through the Universe" series (Time-Life, 1988)
Ingersoll, A. P., "Uranus" (*Scientific American, 256,* 38-45, January 1987)
Kerr, R. A., "Jovian Weather: Like Earth's or a Star's?" (*Science, 209,* 1219-1220, 1980)
Lindal, G. F., D. N. Sweetnam, and V. R. Eshleman, "The Atmosphere of Saturn: An Analysis of the Voyager Radio

Occultation Measurements" (*Astronomical Journal, 90,* 1136-1146, 1985)

Chapter 12: Planetary Rings

JOSEPH A. BURNS is Chairman of Theoretical and Applied Mechanics and Professor of Astronomy at Cornell University, where he received his Ph.D. in 1966. In addition, he held a postdoctoral fellowship at NASA's Goddard Space Flight Center in 1967-68, was a senior investigator during 1975-76 and 1982-83 at NASA's Ames Research Center, and was a Visiting Professor of Astronomy during 1982-83 at the University of California at Berkeley. Burns has also spent extended leaves in Moscow, Prague, and Paris. Presently he is on leave at the Lunar and Planetary Laboratory of the University of Arizona. His current research concerns planetary rings, the small bodies of the solar system (satellites, asteroids, and interplanetary debris), orbital evolution and tides, in addition to the rotational dynamics and strength of planets, satellites, and asteroids. He is Editor of the journal *Icarus*.

Brahic, A., and W. B. Hubbard, "The Baffling Ring Arcs of Neptune" (*Sky & Telescope, 77,* 606-609, June 1989)
Cuzzi, J. N., "Ringed Planets: Still Mysterious" (*Sky & Telescope, 68,* 511-516, December 1984, and *69,* 19-23, January 1985)
Cuzzi, J. N., and L. W. Esposito, "The Rings of Uranus" (*Scientific American, 257,* 52-66, July 1987)
Elliot, J., and R. Kerr, *Rings* (MIT Press, 1984)
Esposito, L. W., "The Changing Shape of Planetary Rings" (*Astronomy, 15,* 6-17, September 1987)
Greenberg, R., and A. Brahic, eds., *Planetary Rings,* (Univ. of Arizona Press, 1984)
Lissauer, J. J., "Shepherding Model for Neptune's Arc Ring" (*Nature, 318,* 544-545, 1985)
Showalter, M. R., *et al.*, "Jupiter's Ring System: New Results on Structure and Particle Properties" (*Icarus, 69,* 458-498, 1987)

Chapter 13: The Galilean Satellites

TORRENCE V. JOHNSON, a member of the Voyager imaging team, has long been interested in the Galilean satellites of Jupiter. He earned a Ph.D. in planetary science at the California Institute of Technology in 1970. Since then his research has centered on remote sensing of these bodies along with the Moon, the terrestrial planets, large planetary satellites, and asteroids. In the course of this research Johnson has used a variety of large telescopes including the 5-meter (200-inch) Hale reflector. He is a Senior Research Scientist at the Jet Propulsion Laboratory in Pasadena, California. In 1977 Johnson was named Project Scientist for the Galileo mission to Jupiter.

Beatty, J. K., "The Far-Out Worlds of Voyager 1" (*Sky & Telescope, 57,* 423-427 and 516-520, May-June 1979)
Johnson, T. V., *et al.*, "Io: Evidence for Silicate Volcanism in 1986" (*Science, 242,* 1280-1283, 1988)
Johnson, T., and L. A. Soderblom, "Io" (*Scientific American, 249,* 56-67, December 1983)
Morrison, D., "The Enigma Called Io" (*Sky & Telescope, 69,* 198-205, March 1985)
Morrison, D., ed., *Satellites of Jupiter* (Univ. of Arizona Press, 1982)
Peale, S. J., P. Cassen, and R. Reynolds, "Melting of Io by Tidal Dissipation" (*Science, 203,* 892-894, 1979)
Soderblom, L. A., "The Galilean Moons of Jupiter" (*Scientific American, 242,* 88-100, January 1980)
Special issue: Voyager 1 at Jupiter, (*Nature, 280,* 725-806, 1979)
Squyres, S. W., "Ganymede and Callisto" (*American Scientist, 71,* 56-64, January-February 1983)

Chapter 14: Titan

TOBIAS OWEN is a professor of astronomy at the State University of New York at Stony Brook. He was a member of the Viking science team and is presently on the Voyager imaging-science team and the mass-spectrometer teams for the Galileo and CRAF (Comet Rendezvous and Asteroid Flyby) missions. He is the leader of the U. S. scientists involved in the study of the Cassini mission. In collaboration with American and French astronomers, Owen has maintained a program of ground-based observations of planets, satellites, and comets at several U. S. and foreign observatories. This program has included a systematic investigation of deuterium in the outer solar system, the first detection of carbon monoxide on Titan, and the discovery of deuterium enrichment on Mars. In addition to 200 published scientific and popular articles, Owen has coauthored two books.

Hunten, D. M., *et al.*, "Titan" in *Saturn,* T. Gehrels and M. S. Matthews, eds. (Univ. of Arizona Press, 671-759, 1984)
Lindal, G. F., *et al.*, "The Atmosphere of Titan: An Analysis of the Voyager 1 Radio Occultation Measurments" (*Science, 53,* 348-363, 1983)
Morrison, D., T. Owen, and L. A. Soderblom, "The Satellites of Saturn" in *Satellites,* J. A. Burns and M. S. Matthews, eds. (Univ. of Arizona Press, 764-801, 1986)
Owen, T., "Titan" (*Scientific American, 246,* 98-109, February 1982)
Toon, O. B., *et al.*, "Methane Rain on Titan" (*Icarus, 75,* 255-284, 1988)
Tyler, G. L., *et al.*, "Radio Science Investigations of the Saturnian System with Voyager 1" (*Science, 212,* 201-206, 1981)

Chapter 15: Icy Worlds

DALE P. CRUIKSHANK spent the years following his undergraduate studies working for the late Gerard P. Kuiper, first at Yerkes Observatory, then at the University of Arizona, where he received a Ph.D. in 1968. A year of study at the Soviet Academy of Sciences followed. In 1970 Cruikshank joined the staff of the Institute for Astronomy at the University of Hawaii, where he spent nearly 18 years observing the planets' surfaces and atmospheres, their satellites, asteroids, and comets. In early 1988 he transferred to NASA's Ames Research Center, where he continues to study Io, Iapetus, Pluto, Triton, unusual asteroids, and comets. Cruikshank is a member of experiment teams on Voyager, CRAF (Comet Rendezvous and Asteroid Flyby), and the Hubble Space Telescope. He is the author or coauthor of some 150 articles and serves as an Associate Editor of *Icarus*.

DAVID MORRISON directs the Space Science Division of NASA's Ames Research Center. Prior to that he was a planetary astronomer at the University of Hawaii, where he was a Professor and the director of NASA's 3-meter Infrared Telescope Facility atop Mauna Kea. His research involves the asteroids and the satellites of the outer planets. Morrison is a member of the Voyager imaging-science team and serves as an Interdisciplinary Scientist on both the Galileo and CRAF missions. He was also the chair of the NASA Solar System Exploration Committee, and in 1981 he served at NASA's Washington headquarters as Acting Deputy Associate Administrator for Space Science. Morrison has authored or coauthored more than 100 technical papers, numerous popular articles, and eight books.

Beatty, J. K., "Pluto and Charon: The Dance Goes On" (*Sky & Telescope, 74,* 248-251, September 1987)
Beatty, J. K., and A. Killian, "Discovering Pluto's Atmosphere" (*Sky & Telescope, 76,* 624-627, December 1988)
Brown, R. H., and D. P. Cruikshank, "The Moons of Uranus, Neptune, and Pluto" (*Scientific American, 253,* 38-47, July 1985)
Cruikshank, D. P., and R. H. Brown, "Satellites of Uranus and

Neptune, and the Pluto-Charon System," in *Satellites*, J. Burns and M. S. Matthews, eds. (Univ. of Arizona Press, 836-873, 1986)
Hoyt, W. G., *Planets X and Pluto* (Univ. of Arizona Press, 1980)
Johnson, T. V., R. H. Brown, and L. A. Soderblom, "The Moons of Uranus" (*Scientific American, 256*, 48-60, April 1987)
Littmann, M., *Planets Beyond: Discovering the Outer Solar System* (Wiley, 1988)
Lunine, J. I., and D. J. Stevenson, "Physical State of Volatiles on the Surface of Triton" (*Nature, 317*, 238-240, 1985)
Tombaugh, C. W., and P. Moore, *Out of the Darkness: The Planet Pluto* (Stackpole, 1980)

Chapter 16: The Halley Encounters

RÜDEGER REINHARD is a scientist at the European Space Research and Technology Center (ESTEC) in Noordwijk, the Netherlands, which is the main scientific facility of the European Space Agency. Appointed Giotto Project Scientist in July 1980, Reinhard coordinated the development of the spacecraft's scientific objectives and numerous instruments. His research in the propagation of solar plasma in interplanetary space earned him a Ph.D. from the University of Kiel in the Federal Republic of Germany in 1976, after which he joined the ESTEC staff. Reinhard is also an investigator on the forthcoming Ulysses and WIND missions, during which he will investigate the distribution of low-energy plasma in space. He serves as the Executive Secretary of the Inter-Agency Consultative Group for Space Science.

Balsiger, H., H. Fechtig, and J. Geiss, "A Close Look at Halley's Comet" (*Scientific American, 259*, 96-103, September 1988)
Beatty, J. K., "An Inside Look at Halley's Comet" (*Sky & Telescope, 71*, 438-443, May 1986)
Berry, R., and R. Talcott, "What Have We Learned from Comet Halley?" (*Astronomy, 14*, 6-22, September 1986)
Reinhard, R., and B. Battrick, *Space Missions to Halley's Comet* (European Space Agency, SP-1066, 1986)
Special issue: Comet Halley, (*Nature, 321*, 259-366, 1986)
Special issue: Comet Halley, (*Sky & Telescope, 73*, 238-270, March 1987)

Chapter 17: Comets

JOHN C. BRANDT is Senior Research Associate, Laboratory for Atmospheric and Space Physics, and Professor (Attendant Rank), Department of Astrophysical, Planetary and Atmospheric Sciences, University of Colorado, Boulder. He was previously at NASA's Goddard Space Flight Center in Greenbelt, Maryland. Brandt has been involved in such space projects as the Solar Maximum Mission and the International Cometary Explorer mission. He is also principal investigator on the Goddard High Resolution Spectrograph to be flown on NASA's Hubble Space Telescope. Brandt received a Ph.D. in astronomy and astrophysics from the University of Chicago in 1960. Subsequently, he served at Mount Wilson and Palomar Observatories, Kitt Peak National Observatory, and taught astronomy and physics at several major universities. Brandt has authored or coauthored more than 300 publications on such topics as astrophysics, solar physics, planetary atmospheres, and comets.

Battrick., B., E. J. Rolfe, and R. Reinhard, eds., *20th ESLAB Symposium of the Exploration of Halley's Comet* (European Space Agency, SP-250, 1986)
Brandt, J. C., and R. D. Chapman, *Introduction to Comets* (Cambridge, 1981)
Brandt, J. C., and M. B. Niedner, Jr., "The Structure of Comet Tails" (*Scientific American, 254*, 48-56, January 1986)
Delsemme, A. H., "Whence Come Comets?" (*Sky & Telescope, 77*, 260-264, March 1989)
Grewing, M., F. Praderie, and R. Reinhard, eds., *Exploration of Halley's Comet* (Springer-Verlag, 1988)
Whipple, F. W., *The Mystery of Comets* (Smithsonian Institution, 1985)
Wilkening, L. L., ed., *Comets* (Univ. of Arizona Press, 1982)

Chapter 18: Asteroids

CLARK R. CHAPMAN is a research scientist at the Planetary Science Institute, a division of Science Applications International Corp. in Tucson, Arizona. He is the author of about 100 articles on terrestrial planets, asteroids, Jupiter, and other topics relating to the origin and evolution of planets. Chapman has written three popular books and is at work on two others. He has been a member of the National Academy of Sciences Space Science Board's planetary advisory committee, chaired the Division for Planetary Sciences of the American Astronomical Society, and is on the imaging team of the Galileo mission to Jupiter. Chapman received his Ph.D. from the Massachusetts Institute of Technology.

Binzel, R. P., ed., *Asteroids II* (Univ. of Arizona Press, 1989)
Chapman, C. R., "The Nature of Asteroids" (*Scientific American, 232*, 24-33, January 1975)
Cunningham, C. J., *Introduction to Asteroids* (Willmann-Bell, 1988)
Gehrels, T., ed., *Asteroids* (Univ. of Arizona Press, 1979)
Kowal, C. T., *Asteroids: Their Nature and Utilization* (Wiley, 1988)

Chapter 19: Meteorites

JOHN A. WOOD received a Ph.D. in geology at the Massachusetts Institute of Technology in 1958. While there he became interested in the geologic properties of meteorites, about which little was known at that time. Since then Wood has pursued this subject at the Smithsonian Astrophysical Observatory in Cambridge, Massachusetts, where he is a member of its research staff. He has also worked at the Enrico Fermi Institute, University of Chicago, on meteoritic research. Wood participated in the analysis of lunar samples brought back by the Apollo astronauts. Since 1973 Wood has taught solar system studies at Harvard University. He is the author of two books and is at work on a third.

Beatty, J. K., "Stardust on Earth" (*Sky & Telescope, 73*, 610, June 1987)
Dodd, R. T., *Thunderstones and Shooting Stars* (Harvard Univ. Press, 1986)
Kerridge, J. F., and M. S. Matthews, eds., *Meteorites and the Early Solar System* (Univ. of Arizona Press, 1988)
Lewis, R. S., and E. Anders, "Interstellar Matter in Meteorites" (*Scientific American, 249*, 66-77, August 1983)
McSween, H. Y., Jr., *Meteorites and their Parent Planets* (Cambridge Univ. Press, 1987)
Rubin, A. E., "Chondrites and the Early Solar System" (*Astronomy, 12*, 17-22, February 1984)
Weaver, K. F., "Meteorites: Invaders from Space" (*National Geographic, 170*, 390-418, 1986)
Wood, J. A., "Chrondritic Meteorites and the Solar Nebula" (*Ann. Rev. Earth Planet. Sci., 16*, 53-72, 1988)

Chapter 20: Small Bodies and Their Origins

WILLIAM K. HARTMANN is a senior scientist at the Planetary Science Institute, a division of Science Applications International Corp. in Tucson, Arizona. He was a project scientist for Mariner 9, which in 1971-72 mapped Mars from orbit for the first time. His research involves the origin and evolution of planetary surfaces. He has published dozens of technical and popular articles, and is the author of two textbooks. In addition to his research activities, Hartmann is an artist specializing in astronomical subjects; he has coauthored (with fellow artists Ron Miller and Pamela Lee) three popular illustrated books on astronomy and space exploration. Hartmann received advanced degrees in geology and astronomy at the University of Arizona, Tucson, in 1965.

Binzel, R. P., ed., *Asteroids II* (Univ. of Arizona Press, 1989)

Burns, J. A., and M. S. Matthews, eds., *Satellites* (Univ. of Arizona Press, 1986)

Cunningham, C. J., *Introduction to Asteroids* (Willmann-Bell, 1988)

Hartmann, W. K., *Moons and Planets* (Wadsworth, 1983)

Veverka, J., and J. A. Burns, "The Moons of Mars" (*Ann. Rev. Earth Planet. Sci., 8,* 527-558, 1980)

Chapter 21: The Collision of Solid Bodies

EUGENE M. SHOEMAKER is a geologist with the U.S. Geological Survey, in Flagstaff, Arizona. He conducted the first detailed geologic mapping of Meteor Crater in Arizona, which helped provide a foundation for cratering research on the Moon and planets. He developed methods of lunar geologic mapping and established the lunar geologic time scale. Shoemaker participated in the Ranger missions and was principal investigator for the television experiment on the Surveyor lunar landers. He also headed the geologic-investigations team for the first Apollo landings, and he is a member of the Voyager imaging team. Shoemaker received his Ph.D. from Princeton in 1960. His current research concerns impact process in the solar system and the effects of large-body impacts on the evolution of life.

CAROLYN S. SHOEMAKER is a planetary astronomer at the U.S. Geological Survey in Flagstaff, Arizona. In collaboration with her husband Eugene, she has been engaged since 1980 in a systematic survey of asteroids and comets. With the discovery of her 14th comet in 1988, she became the most successful comet hunter of the 20th century. She is also a leading discoverer of planet-crossing asteroids. Her rates of discovery of comets and asteroids per unit area of sky photographed provide the foundation for some of the estimates of collision and cratering rates cited in this chapter. Her discoveries frequently include Trojan asteroids, which share Jupiter's orbit. Shoemaker received her B.A. in 1949 and her M.A. in 1950 from Chico State College, California.

Alvarez, L. W., *et al.,* "Extraterrestrial Cause for the Cretaceous-Tertiary Extinction" (*Science, 208,* 1095-1108, 1980)

Chapman, C.R., and D. Morrison, *Cosmic Catastrophes* (Plenum, 1989).

Grieve, R. A. F., and M. R. Dence, "The Terrestrial Cratering Record II. The Crater Production Rate" (*Icarus, 38,* 230-242, 1979)

Melosh, H. J., *Impact Cratering* (Oxford Univ. Press, 1989)

Öpik, E. J., *Interplanetary Encounters: Close-Range Gravitational Interactions* (Elsevier, 1976)

Shoemaker, E. M., "Asteroid and Comet Bombardment of the Earth" (*Ann. Rev. Earth Planet. Sci., 11,* 461-494, 1983)

Wetherill, G. W., "Late Heavy Bombardment of the Moon and Terrestrial Planets" (*Proc. 6th Lunar Sci. Conf.*, Pergamon, 1539–1561, 1975).

Chapter 22: Life in the Solar System?

GERALD A. SOFFEN is an associate director in the Space and Earth Sciences Directorate at the NASA-Goddard Space Flight Center in Greenbelt, Maryland, and Senior Project Scientist for NASA's Earth Observing System. Before these appointments, he directed the Office of Life Sciences at NASA Headquarters, where he was responsible for all medical and biological research within that agency. Soffen served as the Project Scientist for the Viking mission; from the start of the project in 1969 to the successful landing of two spacecraft on Mars in 1976, he was responsible for all the spacecraft's scientific investigations and the activities of the science teams. Prior to Viking, Soffen managed biological instrument development at the Jet Propulsion Laboratory in Pasadena, California. He received a Ph.D. in biophysics from Princeton University in 1960. Soffen is the author of many papers on biomedical problems in space and the search for life on Mars.

Carroll, M., "Digging Deeper for Life on Mars" (*Astronomy, 16,* 6-15, April 1988)

Cooper, H. S. F., Jr., *The Search for Life on Mars* (Harper and Row, 1980)

Dickerson, R. E., "Chemical Evolution and the Origin of Life" (*Scientific American, 239,* 70-86, September 1978)

Goldsmith, D., and T. Owen, *The Search for Life in the Universe* (Benjamin-Cummings, 1980)

Horowitz, N. H., *To Utopia and Back* (W. H. Freeman, 1986)

Klein, H. P., "The Viking Mission and the Search for Life on Mars" (*Reviews of Geophysics and Space Physics, 17,* 1655, October 1979)

Papagiannis, M. D., "Bioastronomy: The Search for Extraterrestrial Life" (*Sky & Telescope, 67,* 508-511, June 1984)

Chapter 23: Putting It All Together

JOHN S. LEWIS is a professor of planetary sciences at the University of Arizona in Tucson. He has been deeply involved in exploring the outer planets and has advised NASA on a number of topics ranging from planetary probes to interstellar communication. He has also served on the Space Science Board of the National Academy of Science. Lewis received a Ph.D. in geochemistry and physical chemistry from the University of California at San Diego in 1968. He is an associate editor of *Icarus.* His research interests include applications of inorganic and physical chemistry to space science, including solar system cosmogony, cool stellar atmospheres, the origin of life, meteoritics, space resources, geochemistry, the mechanisms of mass-extinction events, and atmospheric thermodynamics.

Cameron, A. G. W., "The Origin and Evolution of the Solar System" (*Scientific American, 233,* 66-75, September 1975)

Dermott, S. F., ed., *The Origin of the Solar System* (Wiley, 1978)

Falk, S. W., and D. N. Schramm, "Did the Solar System Start with a Bang?" (*Sky & Telescope, 58,* 18-22, July 1979)

Lewis, J. S., and R. G. Prinn, *Planets and Their Atmospheres: Origin and Evolution* (Academic Press, 1984)

Smoluchowski, R., J. N. Bahcall, and M. S. Matthews, eds., *The Galaxy and The Solar System* (Univ. of Arizona Press, 1986)

Illustration credits

Introduction1: Jon Lomberg **Chapter 1** *1:* NASA-JPL *2:* Carl Sagan (Cornell) *3:* JPL *4: Sky & Telescope 5:* NASA-JPL *6:* David Hardy *7:* NASA *8:* Hoppy Price (JPL) **Chapter 2** *1: Sky & Telescope 2:* R. Noyes *3:* Brookhaven National Lab. *4.* Larry Webster, UCLA *5:* Harold Zirin (CIT) *6:* Alan Title (Lockheed Solar Obs.) *7:* NASA-JSC *8:* CIT *9:* John Eddy (UCAR) *10:* Richard Willson (JPL) *11:* Royal Greenwich Obs. *12:* William Livingston (NOAO) *13,14:* R. Noyes *15:* Steven Musman (Sacramento Peak Obs.), David Rust (Johns Hopkins) *16,17:* John Harvey (NOAO) *18:* Kenneth Libbrecht (CIT) *19:* Kazuo Shiota *20:* NASA, Amer. Sci. & Eng. *21:* Naval Res. Lab. *22:* R. Noyes *23:* Arthur Vaughn (JPL) *24:* R. Noyes **Chapter 3** *1:* J. Van Allen *2: Sky & Telescope 3:* J. Van Allen *4: Sky & Telescope 5:* J. Van Allen *6:* Louis Lanzerotti (AT&T Bell Labs.) *7,8:* J. Van Allen *9: Sky & Telescope 10:* Louis Frank (U. Iowa) *11:* NASA-JSC *12:* J. Van Allen *13:* Herbert Bridge (MIT) *14:* Glenn Berge (CIT) *15:* J. Van Allen *16: Sky & Telescope 17:* Norman Ness (U. Delaware) *18: Sky & Telescope 19:* J. Van Allen **Chapter 4** *1:* NASA-JSC *2:* NASA-JSC, P. Spudis *3:* Lick Obs. *4:* NASA-JSC *5:* NASA-JSC, P. Spudis *6:* NASA, Lunar and Planetary Inst. *7:* Bevan French (NASA) *8-9:* NASA-JSC, P. Spudis *10,11:* P. Spudis *12: Sky & Telescope 13:* William Hartmann *14:* Willy Benz, Alastair Cameron (Center for Astrophysics) *15:* P. Spudis *16:* Don Wilhelms **Chapter 5** *1:* Jean Lecacheux, Eric Thouvenot, and Christian Buil *2-5:* NASA-JPL *6:* NASA, Brown U. *7:* NASA, M. Carr *8:* Dale Schneeberger (JPL) *9:* NASA-JPL *10: Sky & Telescope 11:* NASA, M. Carr *12,13:* Alfred McEwen (USGS, Flagstaff) *14,15:* NASA-JPL *16:* NASA, M. Carr *17:* NASA-JPL *18: Journal of Geophys. Res. 19:* NASA-JPL *20:* Alfred McEwen (USGS, Flagstaff) *21,22:* NASA-JPL **Chapter 6** *1:* Thomas Hunt, *Astronomy 2:* D. Anderson *3: upper pair:* Henri-Claude Nataf, Ichiro Nakanishi, Don Anderson (CIT); *lower pair:* Adam Dziewonski, John Woodhouse (Harvard) *4:* USGS, Flagstaff *5:* D. Anderson *6:* Adam Dziewonski, John Woodhouse *7:* William Haxby (Columbia), NOAA *8: Scientific American 9:* D. Anderson *10:* Hua-Wei Zhou, Robert Clayton *11:* Kelly Beatty *12:* Christopher Scotese, Paleomap Project *13:* R. Rapp, *Journal of Geophys. Res. 14:* D. Anderson **Chapter 7** *1:* John McHone (Arizona State) *2,3:* NASA-JPL *4:* Bruce Hall (Goodyear Aerospace) *5:* J. Head *6-8:* NASA-JPL *9:* Willy Benz, Alastair Cameron (Center for Astrophysics) *10:* USGS, Flagstaff *11:* Vernadskiy Inst., U.S.S.R. Academy of Sciences *12:* Peter Ford (MIT) *13,14:* Vernadskiy Inst., U.S.S.R. Academy of Sciences *15:* J. Head, Larry Crumpler (Brown) *16:* J. Head *17:* Sean Solomon (MIT), J. Head *18:* J. Head, Sean Solomon **Chapter 8** *1:* NASA-JPL; Don Davis *2:* Mark Schoeberl (NASA-GSFC), A. Krueger *3: Sky & Telescope 4,5:* J. Pollack *6:* NASA-ARC *7:* J. Pollack *8:* James Kasting (Penn. State) *9:* J. Pollack *10:* J. Pollack *11:* James Kasting (Penn. State) *12:* NASA-JPL *13:* James Walker *14,15:* J. Pollack *16:* NASA-JPL **Chapter 9** *1:* Stephen Larson (U. Ariz.) *2-5:* NASA-JPL *6:* NASA-JPL, *Sky & Telescope 7-9:* NASA-JPL *10:* Stephen Larson (U. Ariz.) *11:* NASA-ARC *12:* NASA-JPL *13: Sky & Telescope 14:* B. Smith *15,16:* NASA-JPL *17:* B. Smith *18-24:* NASA-JPL **Chapter 10** *1: Sky & Telescope 2-6:* W. Hubbard *7: Sky & Telescope 8-11:* W. Hubbard *12:* Jonathan Lunine (U. Ariz.) *13:* W. Hubbard **Chapter 11** *1-3:* NASA-JPL *4:* Heidi Hammel (JPL) *5,6:* NASA-JPL *7-9:* A. Ingersoll *10:* Richard Terrile (JPL), NASA *11:* NASA-JPL *12:* A. Ingersoll, David Godfrey (NOAO) *13:* NASA-JPL *14:* A. Ingersoll *15:* Gareth Williams (NOAA) *16:* NASA-JPL *17:* NASA, A. Ingersoll *18:* A. Ingersoll *19:* Joel Sommeria, Steven Myers, and Harry Swinney (U. Texas) *20:* A. Ingersoll **Chapter 12** *1:* J. Burns, NASA *2:* J. Burns *3: Sky & Telescope 4: top:* Jeffrey Cuzzi (NASA-ARC); *bottom:* Frank Shu (UC Berkeley) *5:* NASA-JPL *6:* JPL *7: Sky & Telescope 8:* NASA-JPL *9:* Mark Showalter (Stanford), J. Burns *10:* James Elliot (MIT) *11,13:* NASA-JPL *14: Sky & Telescope 15:* Jack Lissauer (SUNY, Stony Brook) *16,17:* NASA-JPL *18:* Lonne Lane (JPL), Len Tyler (Stanford) *19:* NASA-JPL *20:* Mark Showalter (Stanford), J. Burns *21-25:* NASA-JPL **Chapter 13** *1: Sky & Telescope 2:* T. Johnson *3:* Don Davis *4:* JPL *5:* Roger Clark (USGS, Denver) *6:* NASA-JPL *7: Scientific American 8-14:* NASA-JPL *15:* Laurence Soderblom (USGS, Flagstaff) *16:* NASA-JPL *17-20:* Alfred McEwen (USGS, Flagstaff) *21:* NASA-JPL *22:* Alfred McEwen (USGS, Flagstaff) *23:* NASA-JPL *24,25:* Alfred McEwen (USGS, Flagstaff) *26:* Nicholas Schneider, Robert Strom (U. Ariz.) *27:* Alfred McEwen (USGS, Flagstaff) *28:* T. Johnson *29:* Charles Yoder, *Nature 30:* T. Johnson *31:* Bruce Goldberg (JPL) *32:* John Trauger (JPL) **Chapter 14** *1:* Paul Doherty *2:* Robert Danehy, T. Owen *3:* James Pollack (NASA-ARC) *4:* NASA-JPL *5:* Jonathan Lunine (U. Ariz.) *6:* Don Davis **Chapter 15** *1: Sky & Telescope 2:* Karen Meech (U. Hawaii), Michael Belton (Kitt Peak National Obs.) *3-8:* NASA-JPL *9:* William Sinton (U. Hawaii) *10,11:* NASA-JPL *12:* Laurence Soderblom (USGS, Flagstaff) *13:* NASA-JPL *14,15:* U. Ariz. *16:* James Christy (U.S. Naval Obs.) *17-18: Sky & Telescope 19:* David Tholen (U. Hawaii) *20:* Marc Buie (Space Telescope Sci. Inst.) *21:* Mark Sykes (U. Ariz.) *22:* Allan Meyer (NASA-ARC) **Chapter 16** *1:* R. Reinhard *2: Sky & Telescope 3:* European Space Agency *4-6:* Space Res. Inst., U.S.S.R. Academy of Sciences *7:* R. Reinhard *8,9:* European Space Agency *10:* Frank Jordan (JPL) *11:* Harold Reitsema (Ball Aerospace), Uwe Keller, and the Halley Multicolor Camera team *12:* R. Reinhard *13:* Science Photo Library *14:* MPG Photographs, Munich *15,16:* Halley Multicolor Camera team **Chapter 17** *1:* William Liller, International Halley Watch *2:* Lowell Obs. *3: Sky & Telescope 4:* J. Brandt *5:* John Laborde *6:* Ken Russell (Royal Obs. Edinburgh) *7:* J. Brandt *8: left:* European Southern Obs.; *right:* International Halley Watch *9:* Ian Stewart (U. Colorado) *10:* M. Jackson *11:* Paul Feldman (Johns Hopkins) *12:* Stephen Larson (U. Ariz.) *13:* NASA-JSC *14:* Halley Multicolor Camera team *15:* Halley Multicolor Camera team *16:* Space Res. Inst., U.S.S.R. Academy Sciences *17,18:* J. Brandt *19: Sky & Telescope 20:* J. Brandt *21: Sky & Telescope 22:* J. Brandt *23:* Kim Passey, Worldwide Church of God **Chapter 18** *1:* William Hartmann *2:* Richard Binzel (MIT) *3:* Andrew Chaikin, Clark Chapman *4:* William Hartmann *5:* Steven Ostro (JPL) *6,7:* C. Chapman *8:* Jeffrey Bell (U. Hawaii) *9:* Paul Hudson, NASA *10,11:* C. Chapman *12,13:* Mark Sykes (U. Ariz.) **Chapter 19** *1:* Roy Clarke, Jr. *2-5:* J. Wood *6:* William Bond, National Geographic Society *7:* J. Wood *8:* Keith Swinden (U. Chicago) *9:* William Cassidy (U. Pittsburgh) *10-15:* J. Wood *16:* Clark Chapman (Planetary Sci. Inst.) *17:* J. Wood *18,19:* NASA-JSC *20:* Peter Schultz (Brown) **Chapter 20** *1:* W. Hartmann *2:* W. Hartmann, Dale Cruikshank (NASA-ARC) *3:* Charles Wood (NASA-JSC) *4: Sky & Telescope 5:* NASA-JPL *6:* NASA, *Sky & Telescope 7:* Space Res. Inst., U.S.S.R. Academy of Sciences *8-10:* NASA-JPL *11: Sky & Telescope 12,13:* NASA-JPL **Chapter 21** *1:* E. and C. Shoemaker *2:* Donald Gault *3:* David Roddy, Karl Zeller (USGS, Flagstaff) *4:* NASA *5:* Brown Univ., Lunar and Planetary Inst. *6:* Lick Obs. *7:* NASA-LaRC *8:* E. Shoemaker *9:* Robert Strom (U. Ariz.) *10:* Vernadskiy Inst., U.S.S.R. Academy of Sciences *11:* NASA-JPL *12:* Laurence Soderblom (USGS, Flagstaff) *13:* E. Shoemaker *14:* Don Davis **Chapter 22** *1:* Woods Hole Oceanographic Inst. *2:* Charles Pellegrino, Jesse Stoff *3-5:* NASA-JPL *6:* TRW Space Systems *7,8:* NASA-JPL *9:* Gilbert Levin (Biospherics, Inc.) **Chapter 23** *1:* George Herbig (U. Hawaii), Lick Obs. *2:* David Malin (Anglo-Australian Obs.) *3:* Hale Observatories *4:* J. Lewis *5:* Bradford Smith (U. Ariz.), Richard Terrile (JPL) *6-9:* J. Lewis

Planetary Maps

The following 16 pages contain pictorial maps of most of the solid bodies in the solar system visited by spacecraft. Usually, the portrayal is a Mercator projection; as a result, the cited scales decrease away from the equator and polar features appear disproportionately large. Maps of the Galilean and Saturnian satellites are preliminary versions that will be upgraded with time.

In general, these illustrations were prepared using the cartographic and airbrush techniques of the U. S. Geological Survey in Flagstaff, Arizona. Exceptions are (1) the Earth map, prepared by Bruce C. Heezen and Marie Tharp (available from the latter at 1 Washington Avenue, South Nyack, N.Y. 10960), and (2) the two lunar maps, prepared by the Defense Mapping Agency in St. Louis, Mo. The U.S.G.S. maps can be obtained through the agency's distribution centers at the Denver Federal Center, Denver, Colo. 80225; and at 1200 S. Eads St., Arlington, Va. 22202.

Mercury

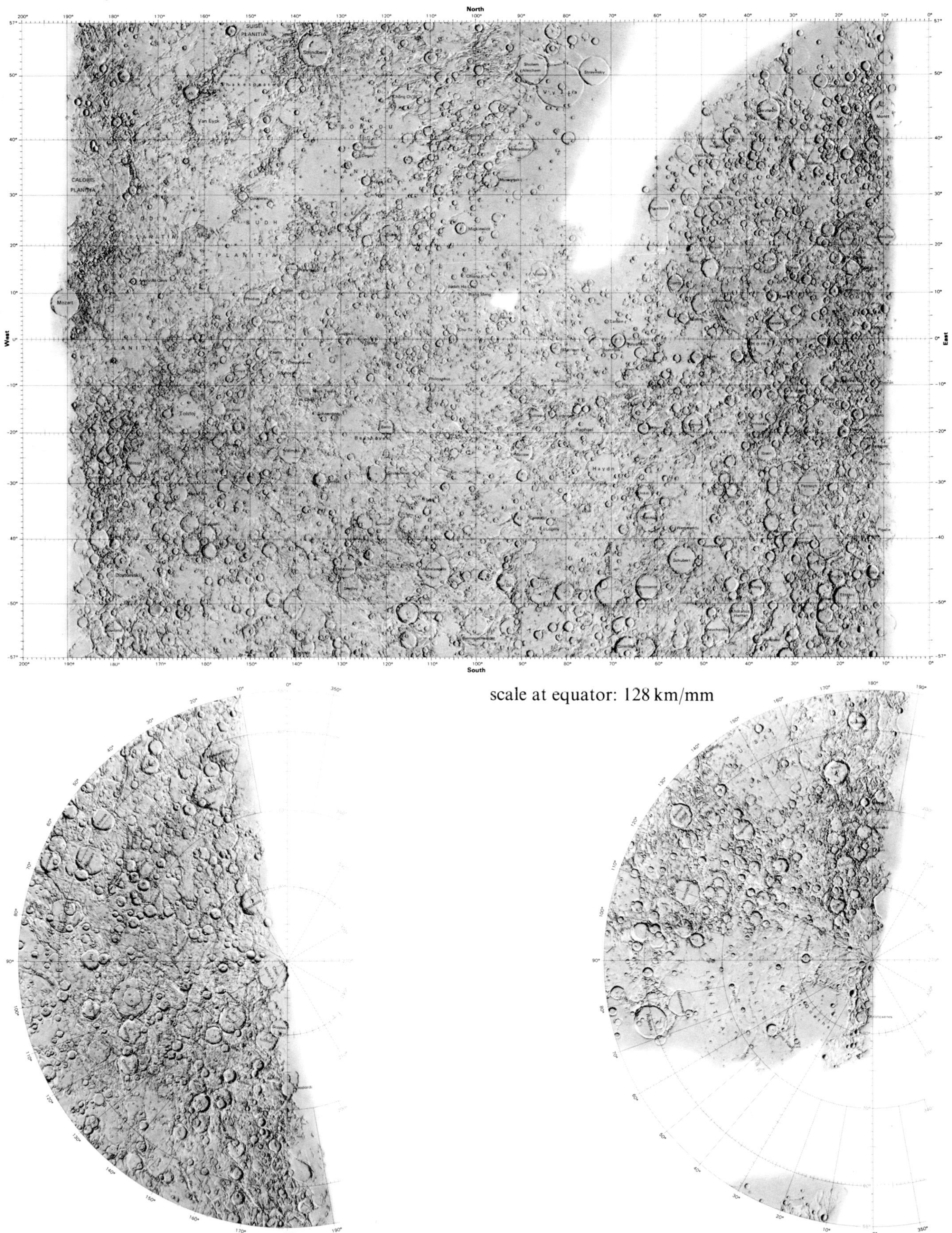

scale at equator: 128 km/mm

SOUTH POLAR REGION

NORTH POLAR REGION

Earth

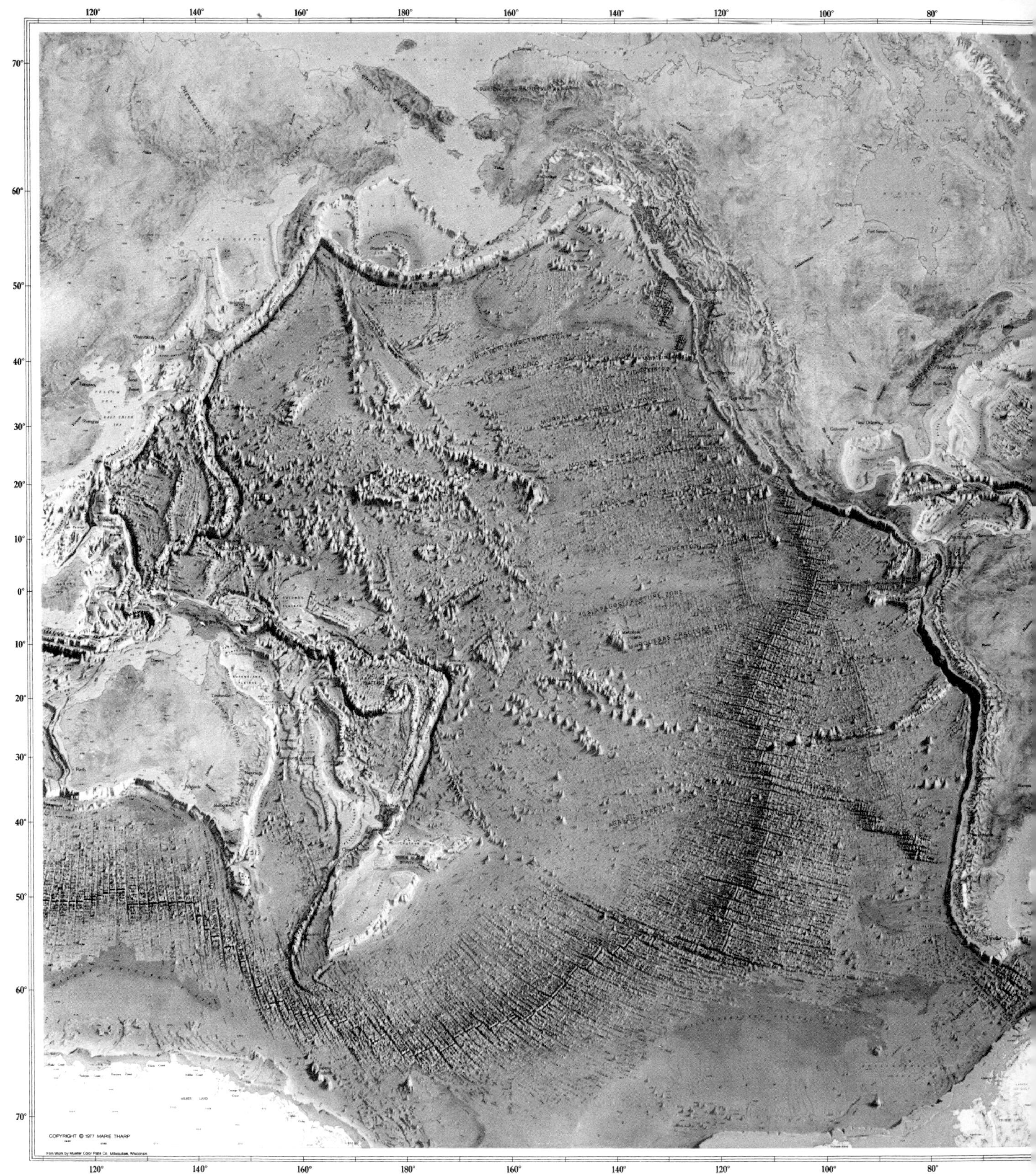

scale at equator: 115 km/mm

WORLD OCEAN FLOOR
BY BRUCE C. HEEZEN AND MARIE THARP
Based on Research and Exploration Initiated and Supported by the
UNITED STATES NAVY
OFFICE OF NAVAL RESEARCH
1977
URAL MOUNTAINS
CARPATHIAN MTS
CAUCASUS MTS
40°
20°
0°
20°
40°
60°
80°
100°
120°
70°
60°
50°
40°
30°
20°
10°
0°
10°
20°
30°
40°
50°
60°
70°

Moon (near side)

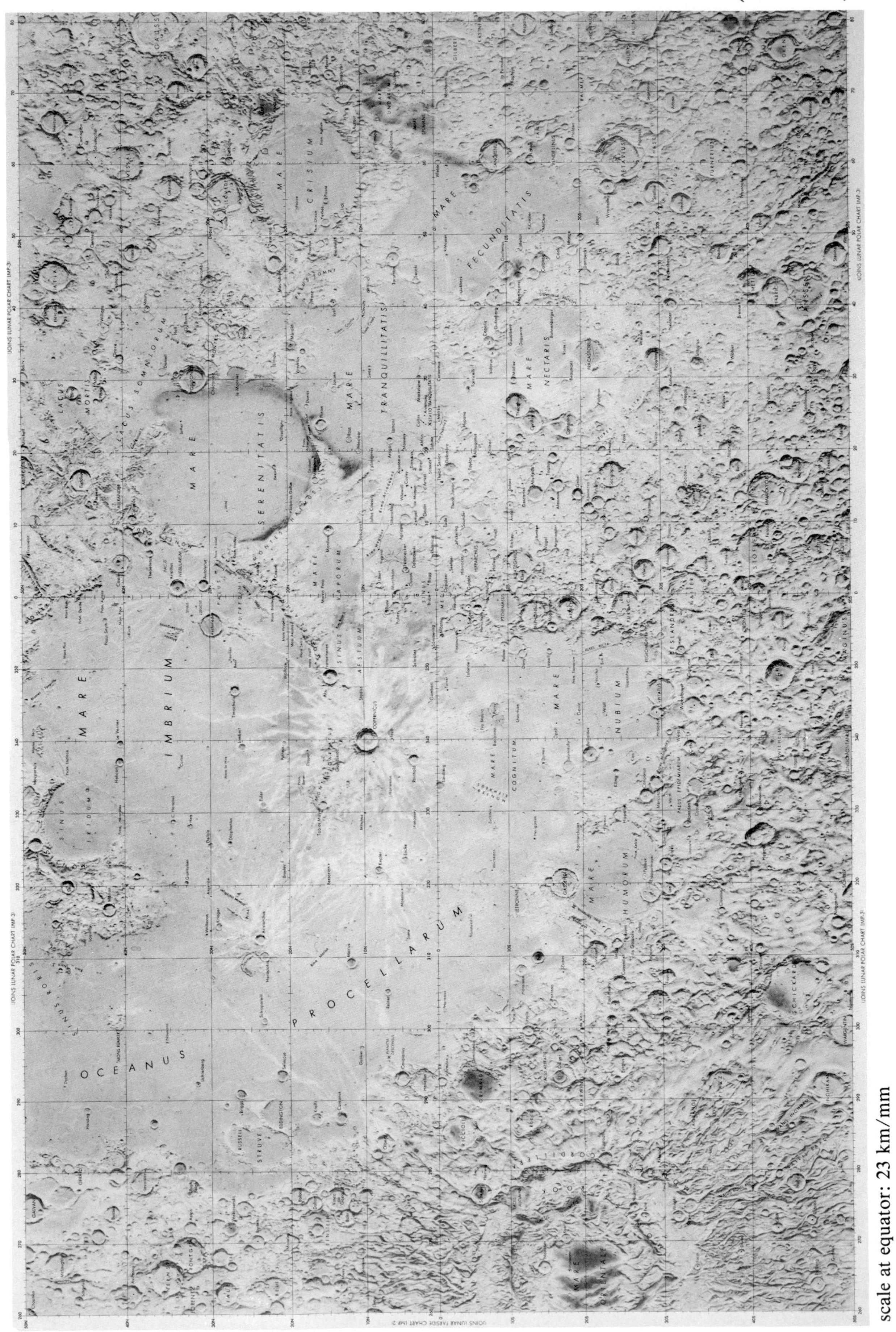

scale at equator: 23 km/mm

Moon (far side)

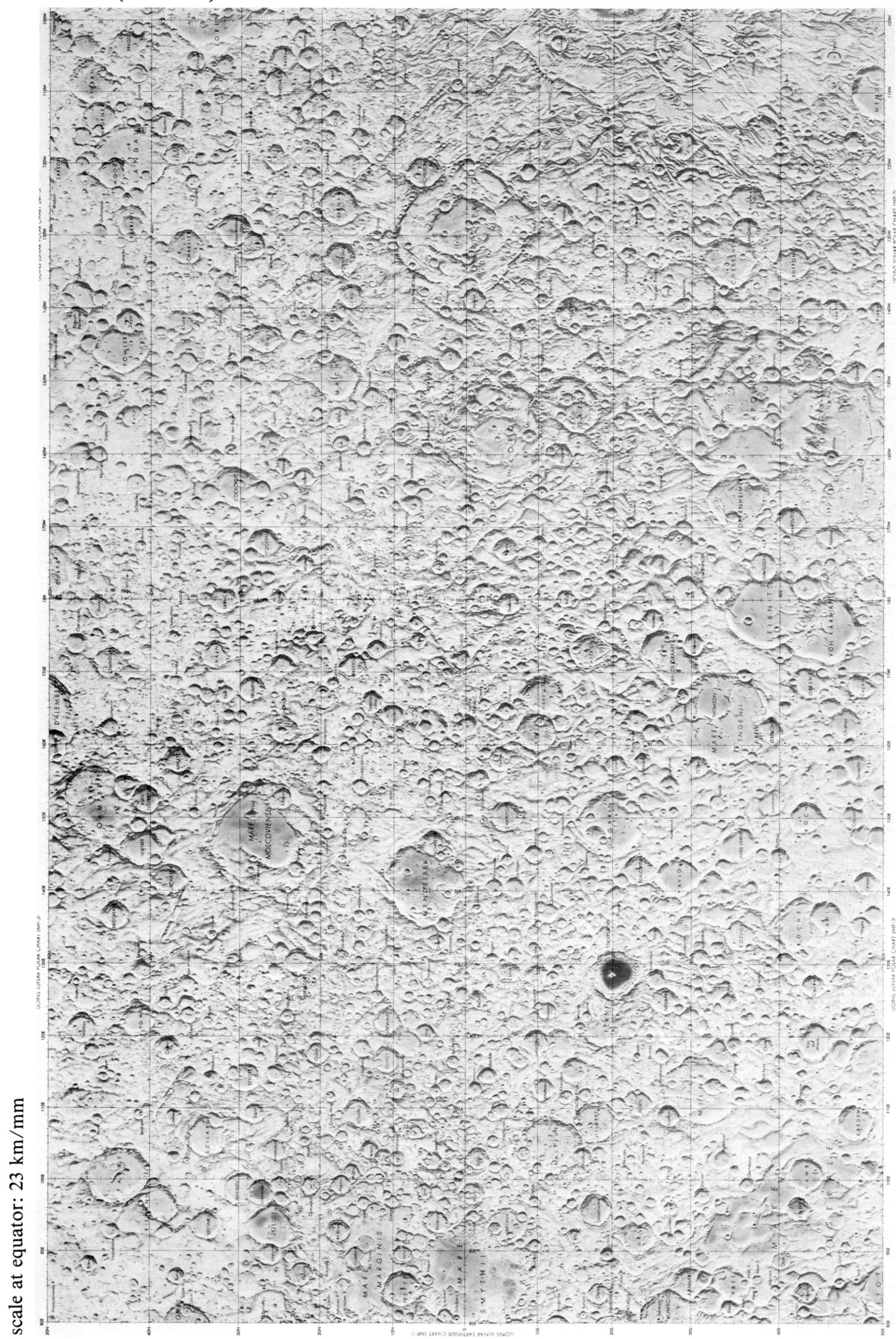

scale at equator: 23 km/mm

Io

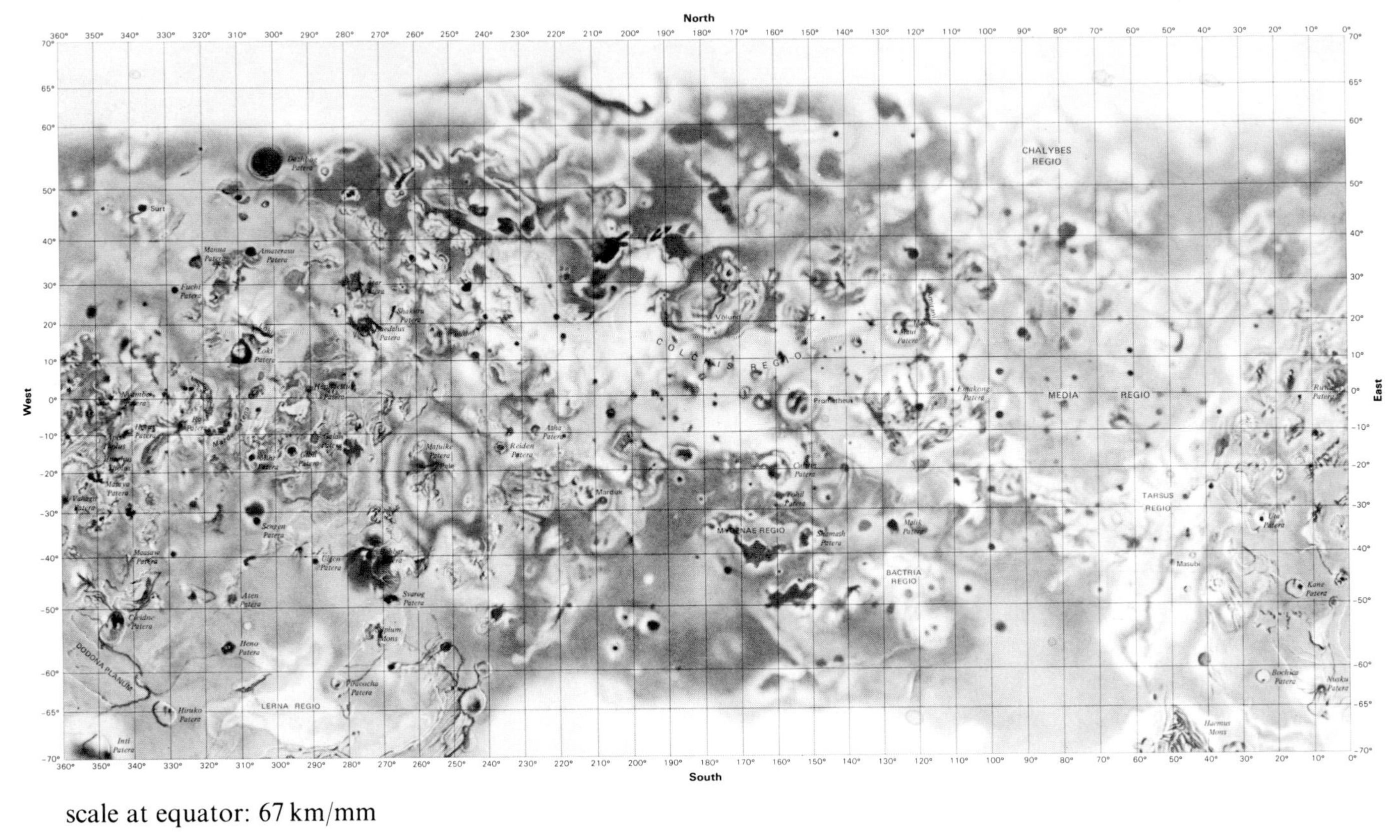

scale at equator: 67 km/mm

Europa

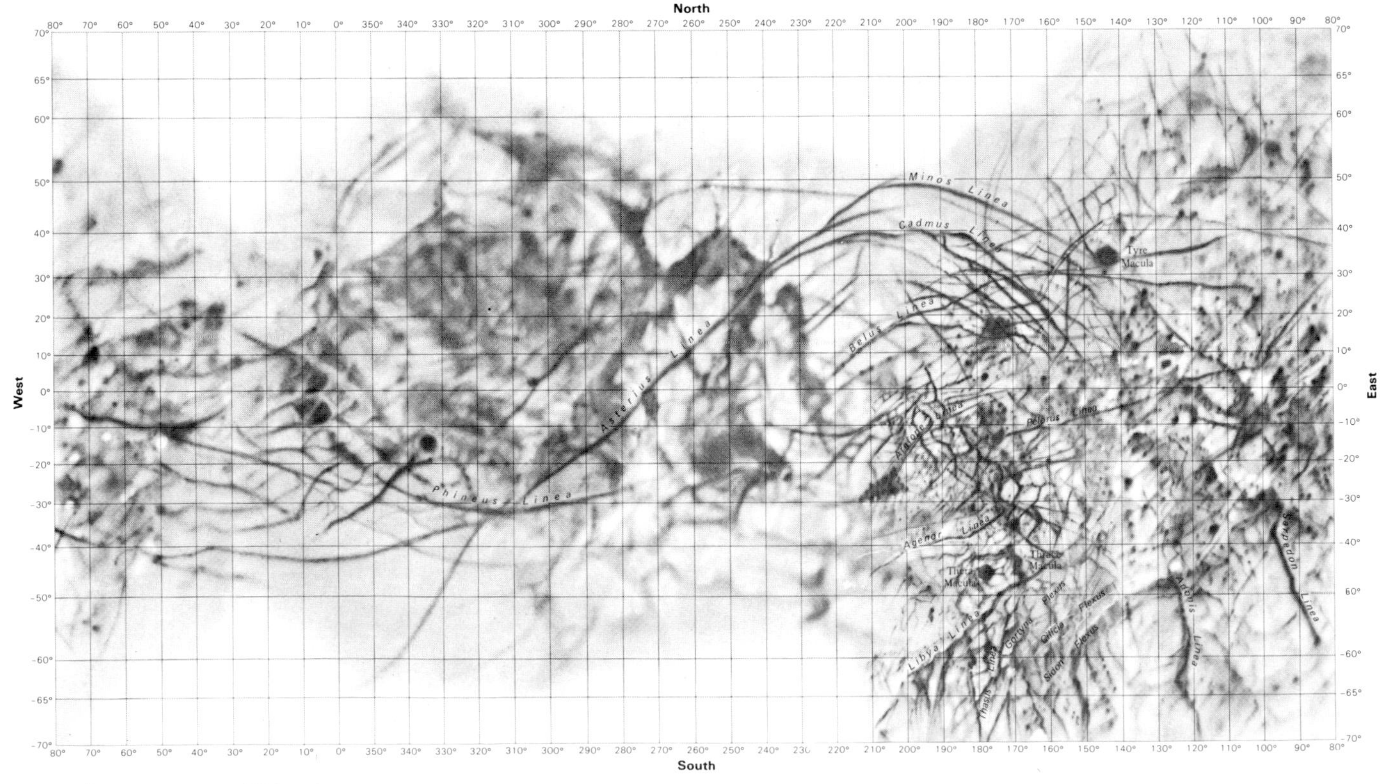

scale at equator: 58 km/mm

Ganymede

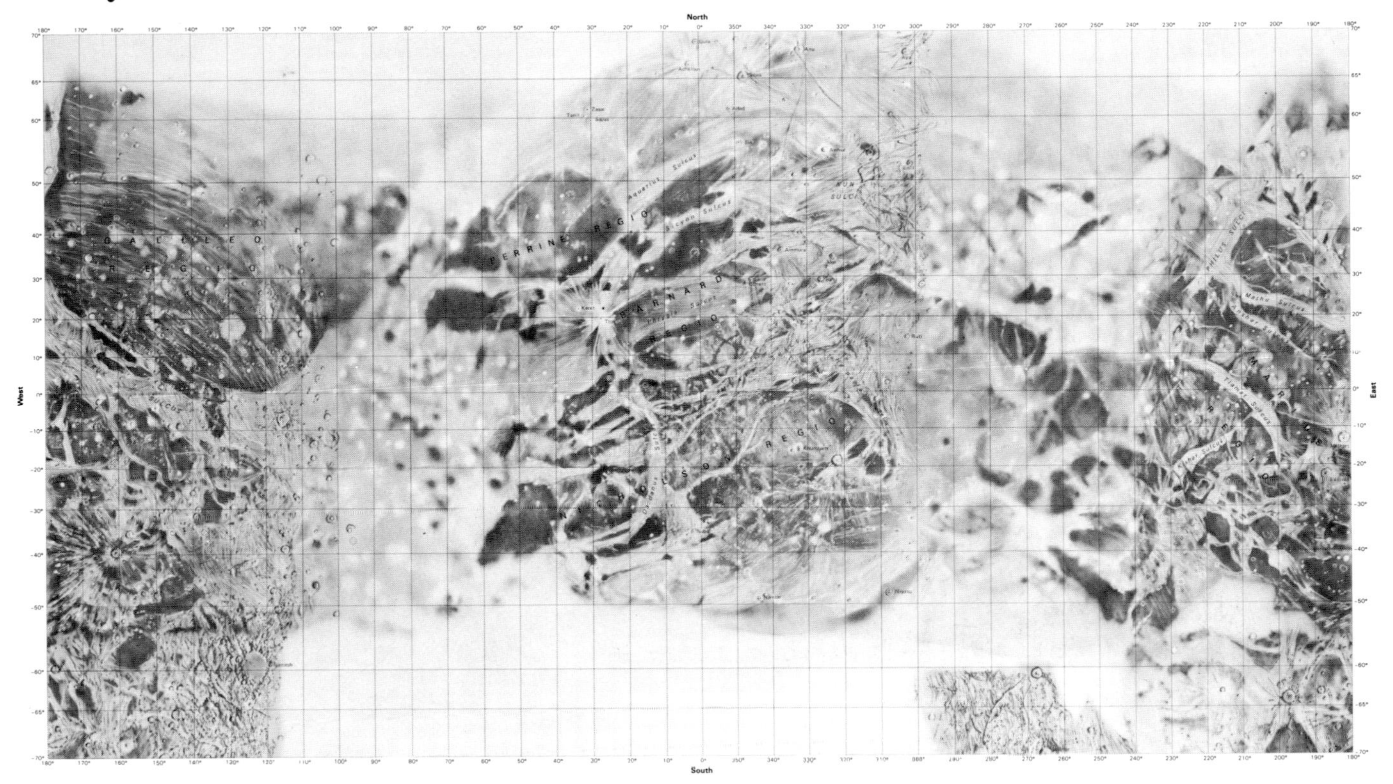

scale at equator: 96 km/mm

Callisto

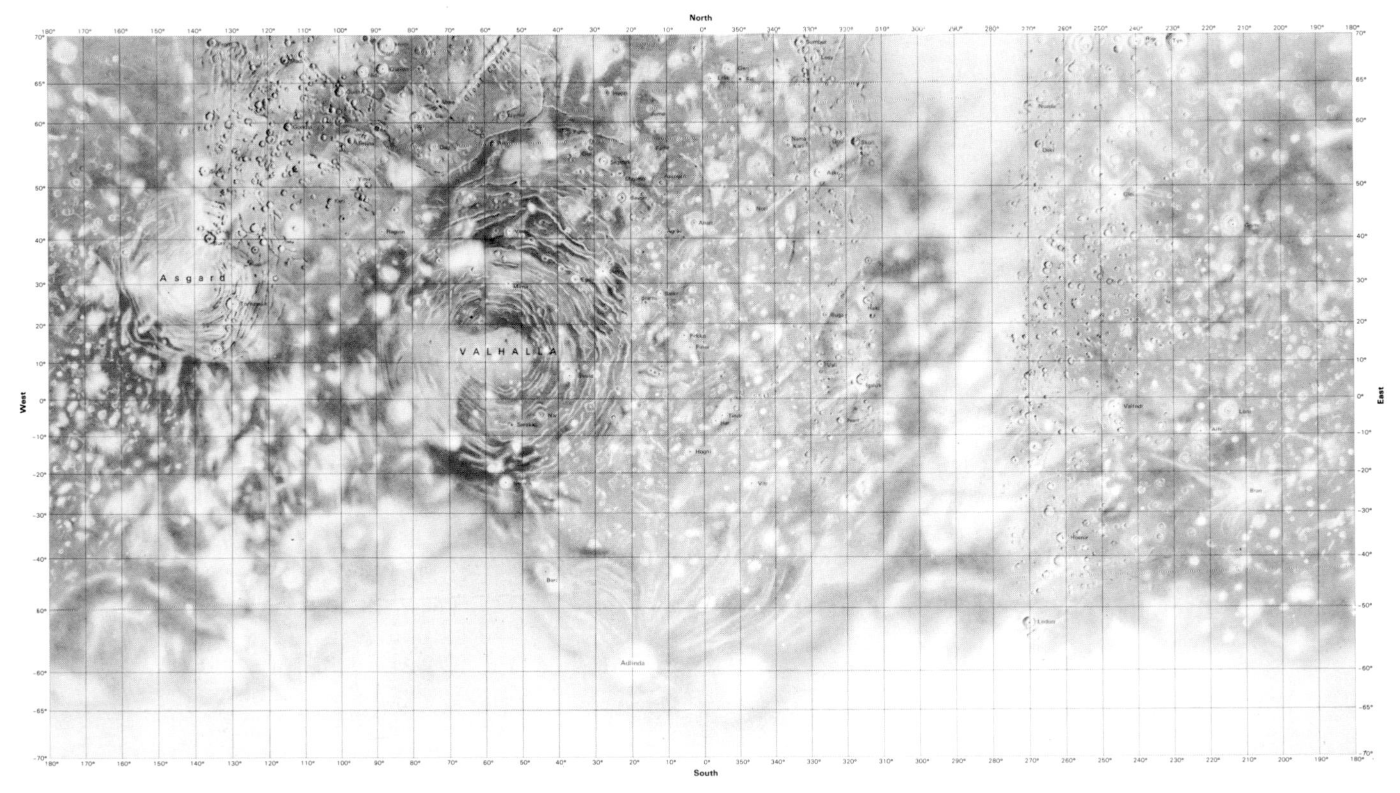

scale at equator: 88 km/mm

Mimas

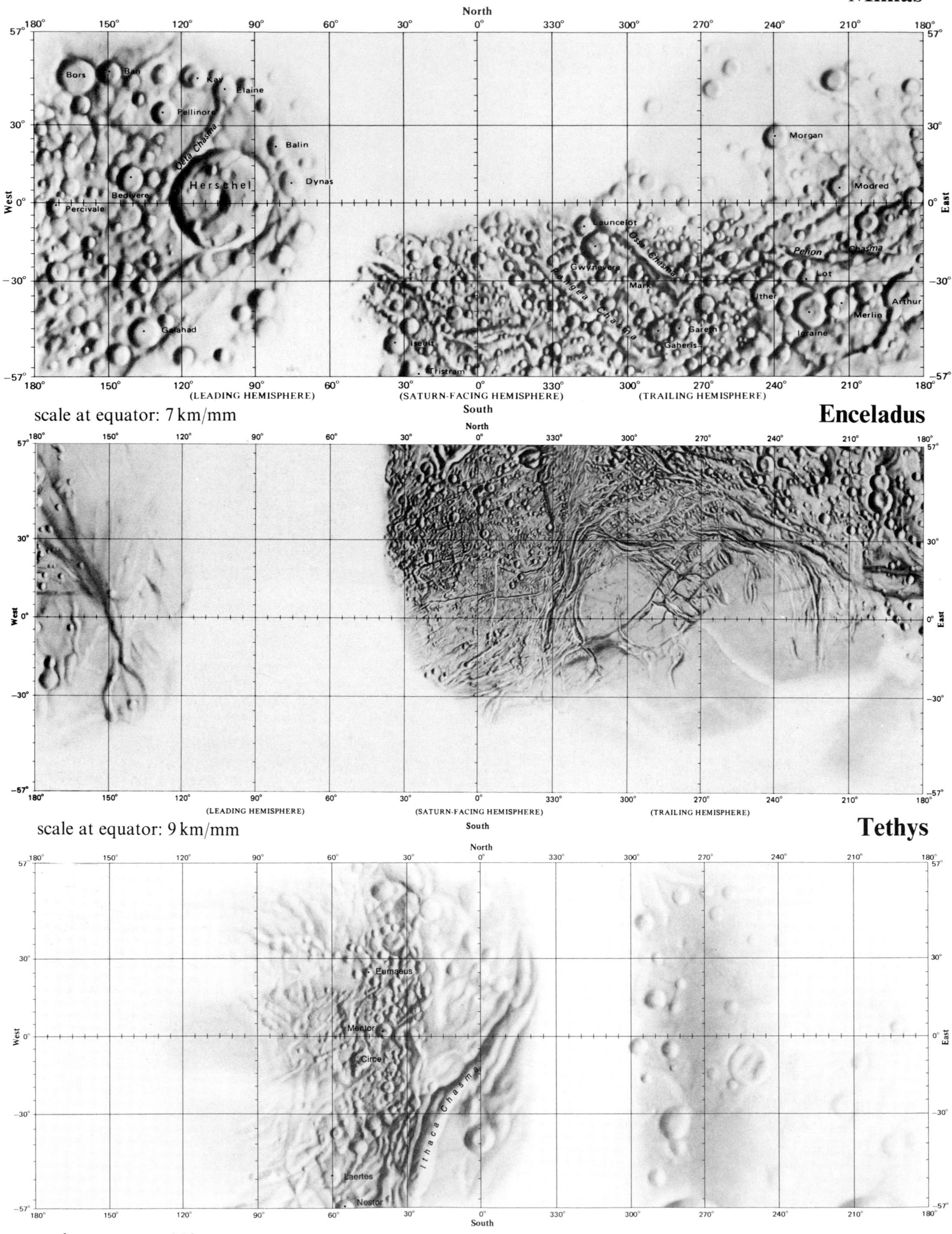
North
180° 150° 120° 90° 60° 30° 0° 330° 300° 270° 240° 210° 180°
57° 30° 0° −30° −57°
West
East
Bors
Ban
Kay
Elaine
Pellinore
Oeta Chasma
Balin
Herschel
Dynas
Bedivere
Percivale
Galahad
Iseult
Tristram
Launcelot
Ossa Chasma
Gwynevere
Pangea Chasma
Mark
Gareth
Gaheris
Uther
Igraine
Merlin
Arthur
Lot
Pelion Chasma
Morgan
Modred
(LEADING HEMISPHERE)
(SATURN-FACING HEMISPHERE)
(TRAILING HEMISPHERE)
South
scale at equator: 7 km/mm
Enceladus
North
180° 150° 120° 90° 60° 30° 0° 330° 300° 270° 240° 210° 180°
57° 30° 0° −30° −57°
West
East
(LEADING HEMISPHERE)
(SATURN-FACING HEMISPHERE)
(TRAILING HEMISPHERE)
South
scale at equator: 9 km/mm
Tethys
North
180° 150° 120° 90° 60° 30° 0° 330° 300° 270° 240° 210° 180°
57° 30° 0° −30° −57°
West
East
Eumaeus
Mentor
Circe
Ithaca Chasma
Laertes
Nestor
South
scale at equator: 15 km/mm

Dione

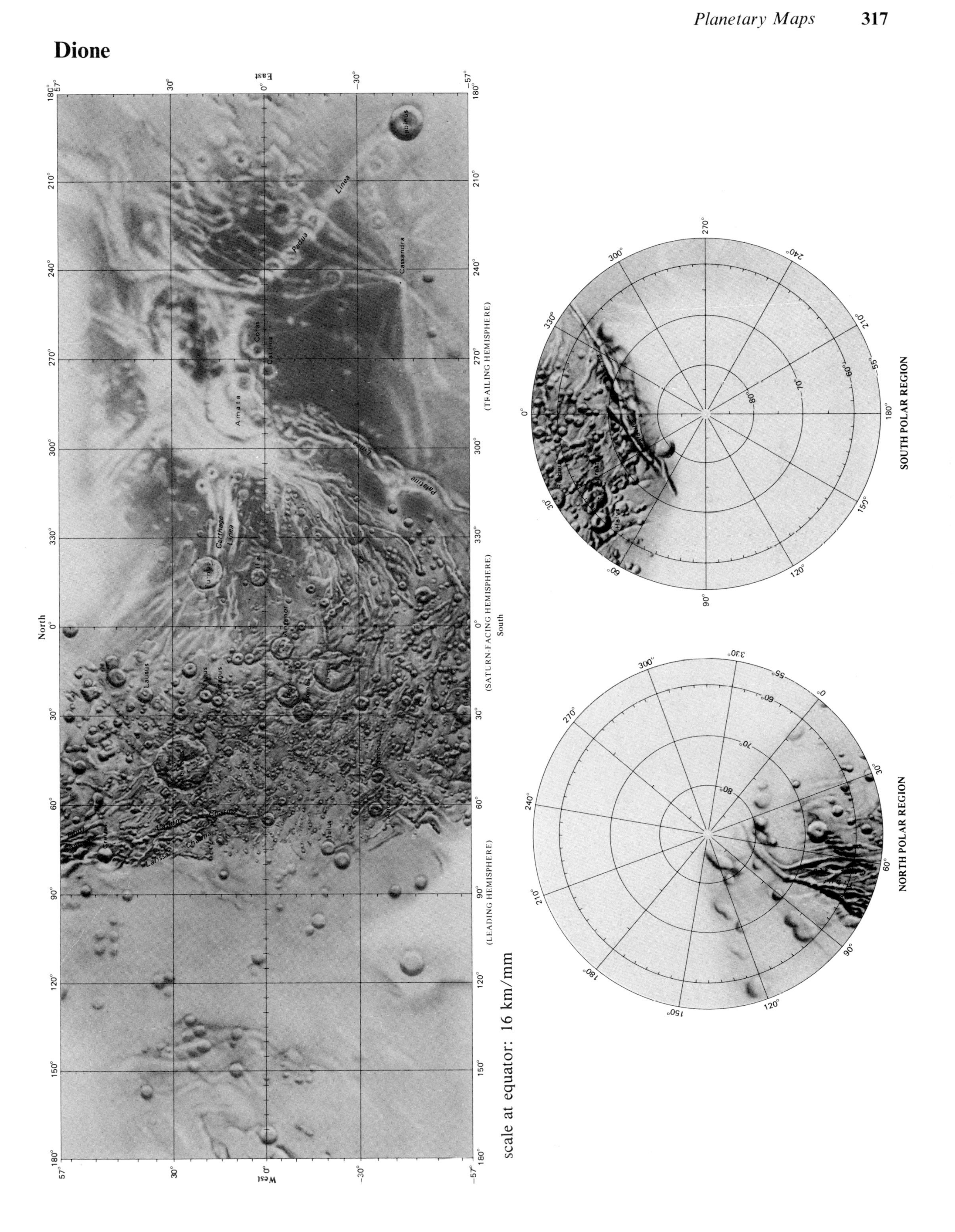
North
South
East
West
(TRAILING HEMISPHERE)
(SATURN-FACING HEMISPHERE)
(LEADING HEMISPHERE)
Padua
Linea
Cassandra
Coras
Catillus
Amata
Palatine
Carthago
Linea
Antenor
Dido
Lausus
Latium
Chasma
scale at equator: 16 km/mm
SOUTH POLAR REGION
NORTH POLAR REGION

Rhea

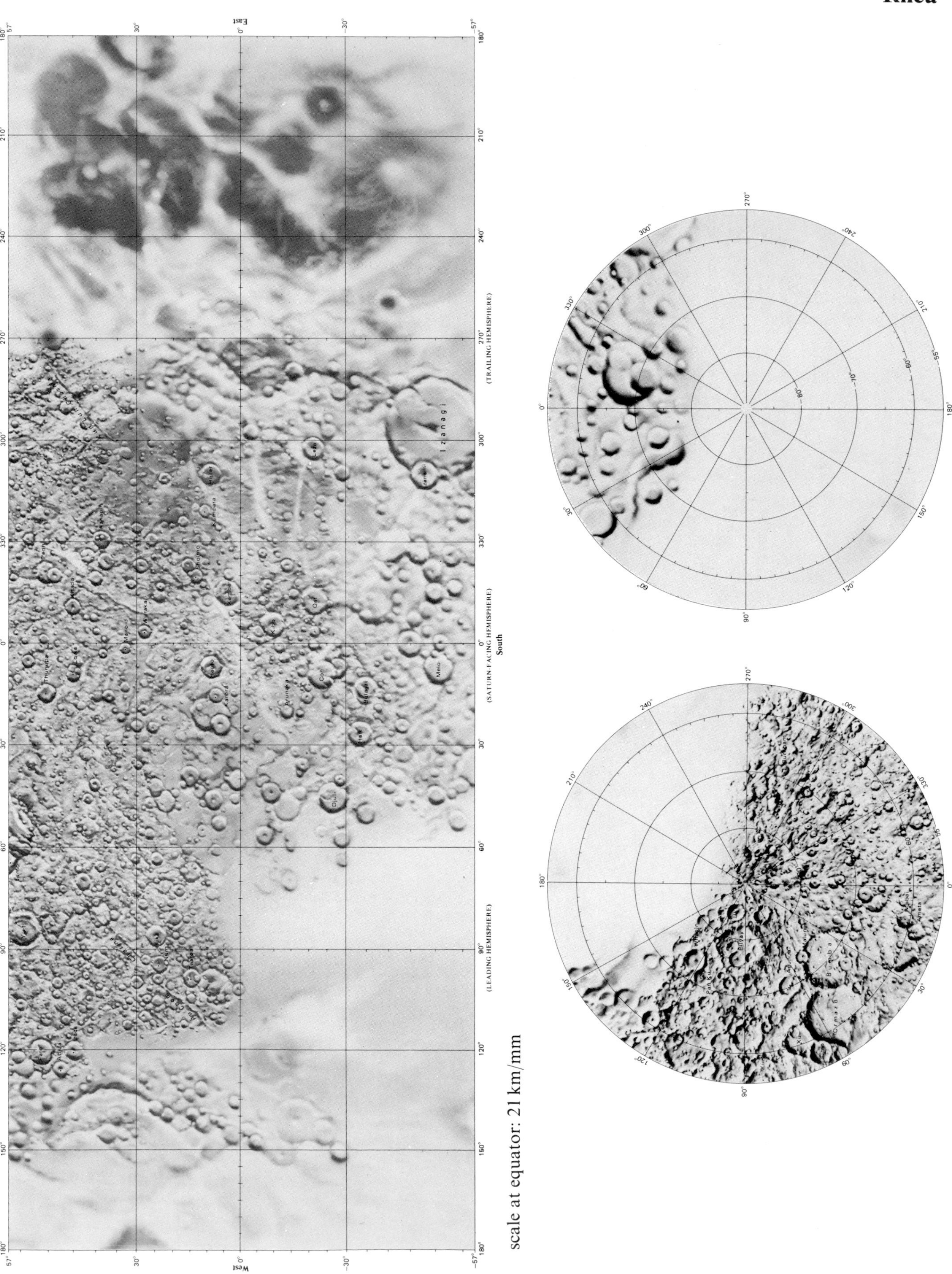
SOUTH POLAR REGION
NORTH POLAR REGION
North
South
East
West
(TRAILING HEMISPHERE)
(SATURN-FACING HEMISPHERE)
(LEADING HEMISPHERE)
Izanagi
scale at equator: 21 km/mm

Iapetus

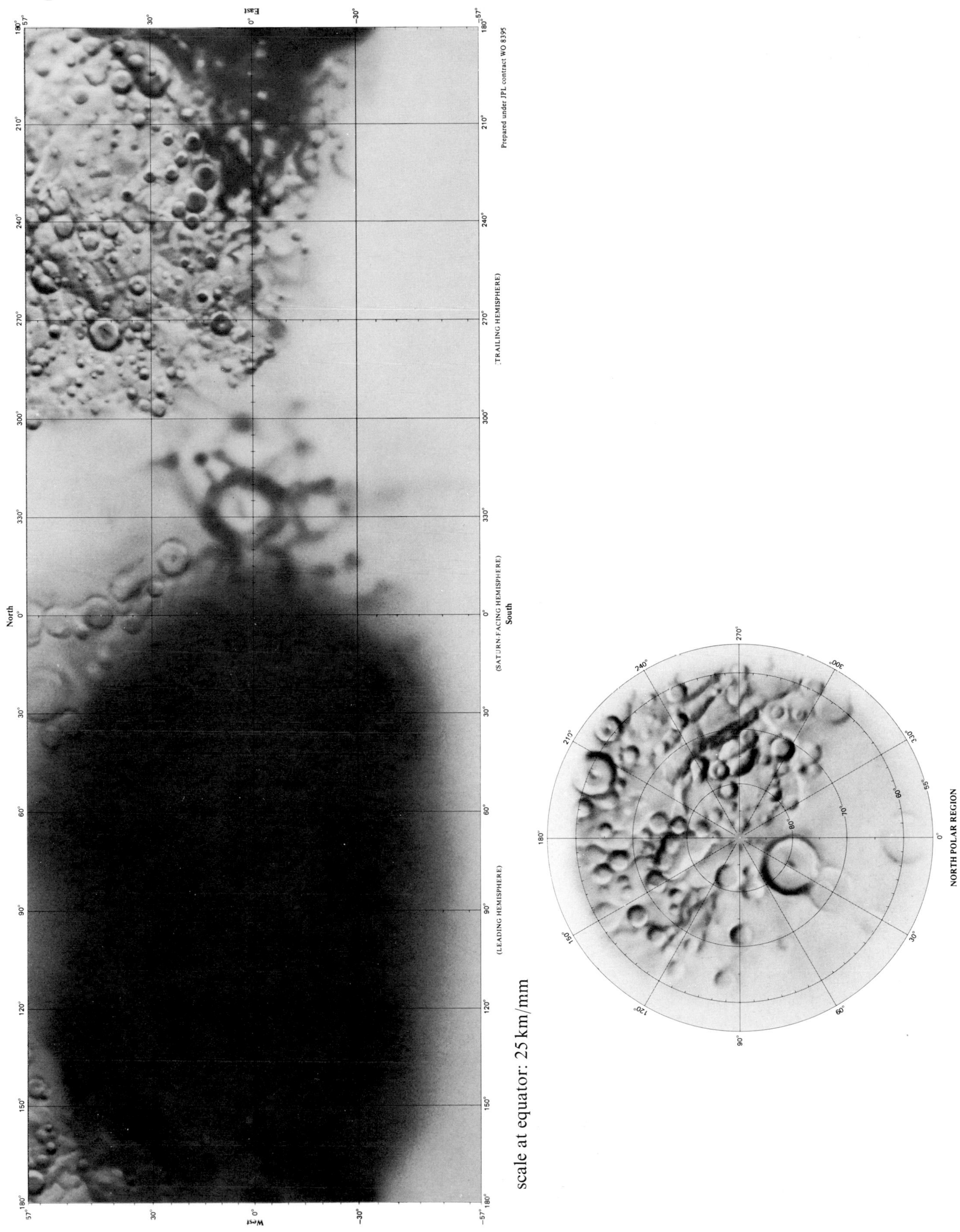
Prepared under JPL contract WO 8395
(TRAILING HEMISPHERE)
(SATURN-FACING HEMISPHERE)
(LEADING HEMISPHERE)
North
South
East
West
scale at equator: 25 km/mm
NORTH POLAR REGION

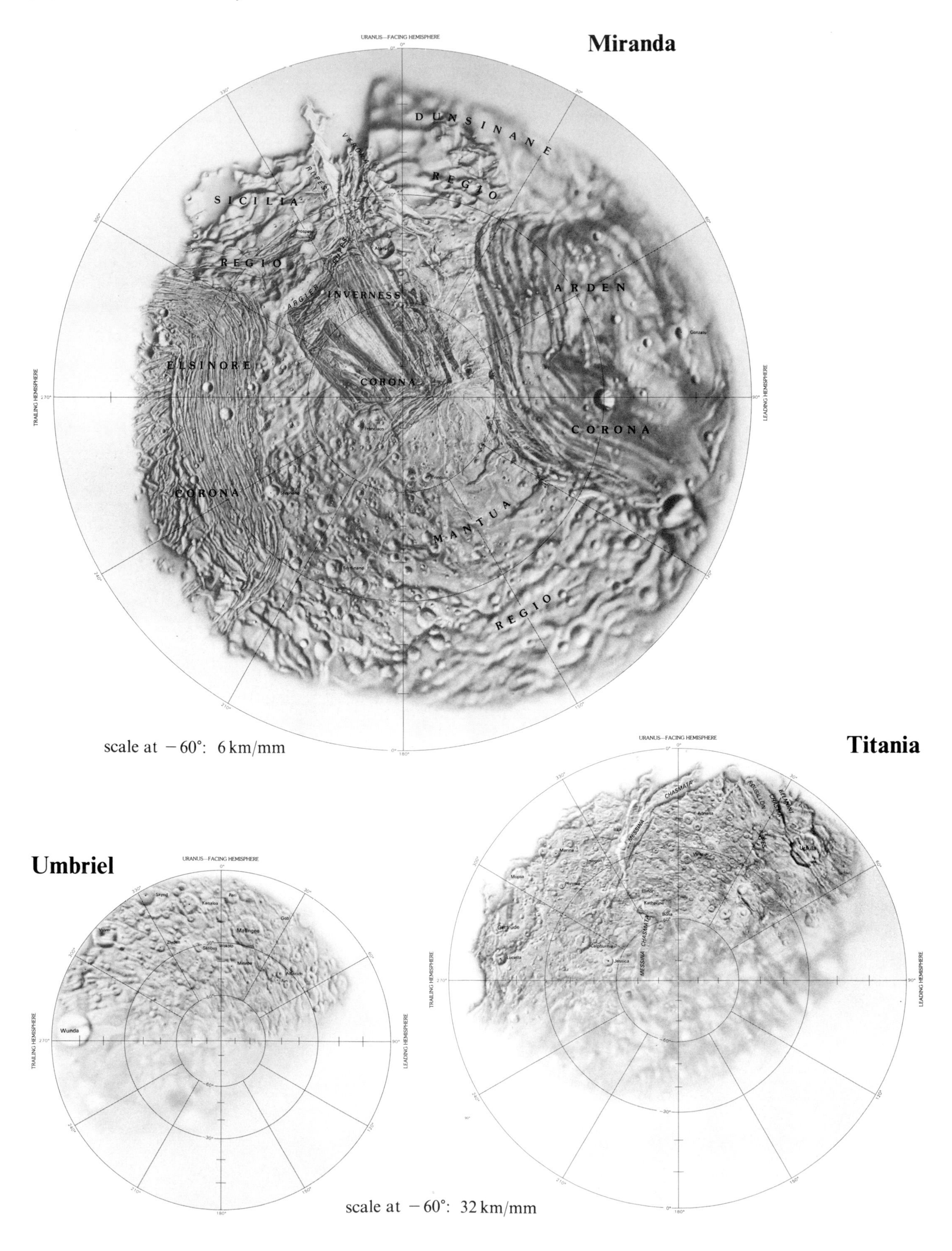
Miranda
URANUS—FACING HEMISPHERE
DUNSINANE REGIO
VERONA RUPES
SICILIA REGIO
ARGIER RUPES
INVERNESS CORONA
ELSINORE CORONA
ARDEN CORONA
MANTUA REGIO
TRAILING HEMISPHERE
LEADING HEMISPHERE
scale at −60°: 6 km/mm
Titania
URANUS—FACING HEMISPHERE
MESSINA CHASMATA
Gertrude
TRAILING HEMISPHERE
LEADING HEMISPHERE
Umbriel
URANUS—FACING HEMISPHERE
Wunda
TRAILING HEMISPHERE
LEADING HEMISPHERE
scale at −60°: 32 km/mm

Ariel

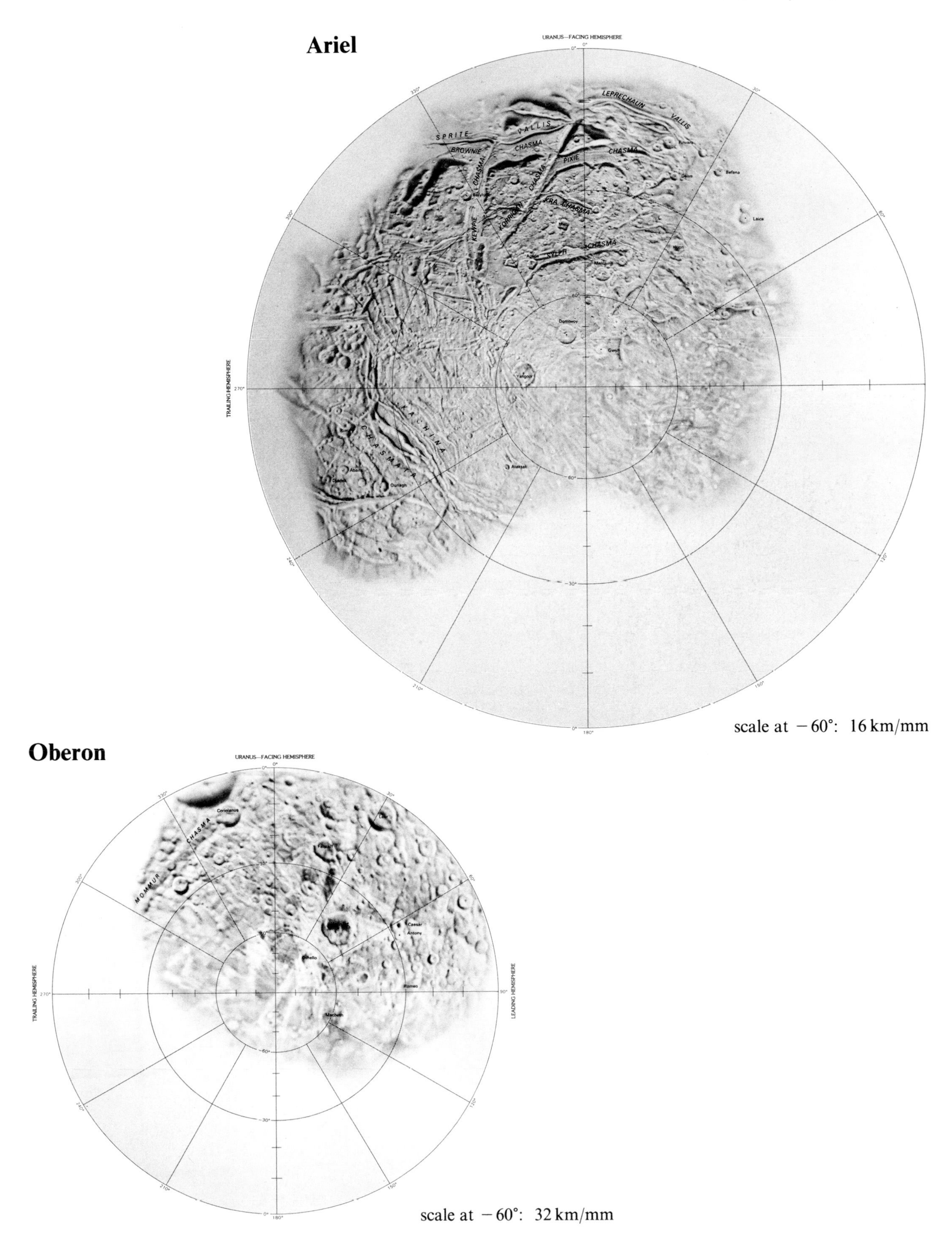

scale at $-60°$: 16 km/mm

Oberon

scale at $-60°$: 32 km/mm

Index